Pharmaceuticals

Volume 4

J. L. McGuire (Editor)

Pharmaceuticals

J. L. McGuire (Editor)

Volume 1 Introduction
         Cardiovascular Drugs

Volume 2 Neuropharmaceuticals
         Gastrointestinal Drugs
         Respiratory Tract

Volume 3 Antiinfectives
         Endocrine and Metabolic Drugs

Volume 4 Miscellaneous Drugs
         Related Technology
         Indexes

Pharmaceuticals

Classes, Therapeutic Agents, Areas of Application

J. L. McGuire (Editor)

Volume 4

Miscellaneous Drugs
Related Technology
Indexes

WILEY-VCH

Weinheim · New York · Chichester · Brisbane · Singapore · Toronto

Dr. J. L. McGuire (Editor)
Johnson & Johnson
Science & Technology/Business Development
One Johnson & Johnson Plaza
New Brunswick, NJ 08933
USA

> This book was carefully produced. Nevertheless, editor, authors and publisher do not warrant the information contained therein to be free of errors. Readers are advised to keep in mind that statements, data, illustrations, procedural details or other items may inadvertently be inaccurate.

Cover illustration: courtesy of BASF Aktiengesellschaft, Ludwigshafen, Germany.

Library of Congress Card No.: Applied for.

British Library Cataloguing-in-Publication Data: A catalogue record for this book is available from the British Library.

Die Deutsche Bibliothek – CIP Cataloguing-in-Publication-Data:
A catalogue record for this publication is available from Die Deutsche Bibliothek.

ISBN 3-527-29874-6

© WILEY-VCH Verlag GmbH, D-69469 Weinheim (Federal Republic of Germany), 2000

Printed on acid-free paper.

All rights reserved (including those of translation in other languages). No part of this book may be reproduced in any form – by photoprinting, microfilm, or any other means – nor transmitted or translated into a machine language without written permission from the publishers. Registered names, trademarks, etc. used in this book, even when not specifically marked as such, are not to be considered unprotected by law.

Composition: Rombach GmbH, D-79115 Freiburg
Printing: Strauss Offsetdruck GmbH, D-69509 Mörlenbach
Bookbinding: Wilhelm Osswald & Co., D-67433 Neustadt (Weinstraße)

Contents

Miscellaneous Drugs ... 1669

1 Anti-inflammatory– Antirheumatic Drugs ... 1671

1. Introduction ... 1671
2. Glucocorticoids ... 1675
3. Nonsteroidal Anti-inflammatory Drugs and Selective COX-2 Inhibitors ... 1676
4. Antirheumatic Agents ... 1702
5. Inhibitors of Tumor Necrosis Factor ... 1707
6. Natural Products ... 1708
7. References ... 1709

2 Drugs Used in Dermatology ... 1713

1. Introduction ... 1713
2. Functions and Structure of the Skin ... 1714
3. Topical Therapy ... 1715
4. Drugs Used in Dermatology by Class ... 1718
5. References ... 1733

3 Gout Remedies ... 1737

1. Introduction ... 1737
2. Gout and Hyperuricemia ... 1739
3. Treatment ... 1741
4. References ... 1749

4 Immunotherapy and Vaccines ... 1753

1. Introduction ... 1755
2. Bacterial Vaccines ... 1764
3. Viral Vaccines ... 1787
4. Vaccines against Parasites ... 1805
5. Immunotherapy ... 1821
6. Immunotherapeutic Uses of Monoclonal Antibodies ... 1839
7. References ... 1844

5 Cancer Chemotherapy ... 1865

1. Introduction ... 1866
2. Antimetabolites ... 1868
3. Alkylating Agents ... 1885
4. Anthracyclines ... 1897
5. Intercalating Anthracenes and Analogs ... 1902
6. Antitumor Antibiotics Other than Anthracyclines ... 1904
7. Antitubulin Agents ... 1909
8. Heavy-Metal Complexes ... 1918
9. Hormonally Active Anticancer Drugs/ Antihormones ... 1922
10. Signal Transduction Inhibitors ... 1937
11. Economic Aspects ... 1939
12. References ... 1940

6 Interferons … 1957

1. Introduction … 1958
2. Classification and Origin … 1959
3. Biological Activity … 1960
4. Mechanism of Action … 1963
5. Physicochemical Properties … 1964
6. Purification … 1968
7. Quality Specifications … 1972
8. Clinical Studies … 1975
9. References … 1978

7 Monoclonal Antibodies … 1981

1. Introduction … 1981
2. Derivation of Murine Monoclonal Antibodies by Somatic Cell Fusion … 1983
3. Human Monoclonal Antibodies … 1987
4. Production … 1989
5. Purification … 1991
6. Monoclonal Antibody Patents … 1993
7. Uses … 1994
8. References … 1997

8 Ophthalmological Agents … 2001

1. Introduction … 2002
2. Anesthetics … 2002
3. Antimicrobial Agents … 2004
4. Anti-inflammatory Agents … 2015
5. Antiglaucomatous Agents … 2019
6. Mydriatics and Cycloplegics … 2026
7. Vasoactive (Adrenergic) Agents … 2029
8. Diagnostic Agents … 2030
9. Dry Eye Medications … 2032
10. Miscellaneous Ophthalmic and Contact Lens Preparations … 2035
11. References … 2040

9 Prostaglandins … 2045

1. History … 2045
2. Structure, Nomenclature, and Physical Properties … 2046
3. Occurrence and Biosynthesis … 2050
4. Pharmacological Effects and Uses … 2055
5. Syntheses … 2058
6. References … 2071

Related Technology … 2075

10 Pharmaceutical Dosage Forms … 2077

1. Dosage Forms as Drug Delivery Systems … 2078
2. Routes of Administration … 2081
3. Types of Dosage Forms … 2085
4. Pharmaceutical Excipients … 2101
5. Quality Assurance … 2105
6. Pharmaceutical Containers … 2106
7. Development of the Ideal Drug Delivery System … 2109
8. References … 2122

11 Drug Testing 2125

1. Preclinical Testing............. 2126
2. Clinical Trials.................. 2134
3. Quality Control of Pharmaceutical Products 2140
4. References 2155

12 Veterinary Drugs 2157

1. Antibiotics and Antibacterial Drugs . 2158
2. Antimycotics 2171
3. Coccidiostats 2172
4. Parasiticides.................. 2173
5. Anti-inflammatory Drugs 2175
6. Respiratory Stimulants 2179
7. Anesthetics 2180
8. Narcotic Agonists and Antagonists.. 2182
9. Antihistamines................ 2184
10. References 2185

Index 2187

1. Author Index................. 2187
2. CAS Registry Number Indes 2191
3. Subject Index................. 2211

VII

Miscellaneous Drugs

Several other groups of drugs, no less important than the major categories already reviewed but not as easily grouped into the traditional drug categories, are reviewed in this section. Examples include some major classes of drugs such as analgesics and vaccines, smaller classes of drugs such as gout remedies or opthamological agents and specific agents such as interferons and prostaglandins.

Anti-inflammatory – Antirheumatic Drugs

BURKHARD HINZ, Department of Experimental and Clinical Pharmacology and Toxicology, Friedrich Alexander University Erlangen-Nürnberg, Erlangen, Germany

CONRAD P. DORN, JR., Merck Sharp & Dohme Research Laboratories, Rahway, New Jersey 07065, United States

TSUNG YING SHEN, Merck Sharp & Dohme Research Laboratories, Rahway, New Jersey 07065, United States

KAY BRUNE, Department of Experimental and Clinical Pharmacology and Toxicology, Friedrich Alexander University Erlangen-Nürnberg, Erlangen, Germany

1.	Introduction.	1671	3.7.	Nonacidic Drugs	1699
2.	Glucocorticoids	1675	3.8.	Selective COX-2 Inhibitors . .	1700
3.	Nonsteroidal Anti-inflammatory Drugs and Selective COX-2 Inhibitors . .	1676	4.	Antirheumatic Agents	1702
			4.1.	Gold.	1702
			4.2.	Antimalarials	1704
3.1.	Salicylates	1677	4.3.	Sulfhydryl Compounds	1705
3.2.	Arylacetic Acids	1679	4.4.	Immunoregulants	1707
3.3.	Arylacetic Acid Prodrugs . . .	1686	5.	Inhibitors of Tumor Necrosis Factor.	1707
3.4.	Arylpropionic Acids	1687
3.5.	Acidic Enolic Compounds . .	1694	6.	Natural Products	1708
3.6.	Anthranilates	1697	7.	References	1709

1. Introduction

The *inflammatory response to injury* is a normal host defense mechanism that serves to isolate and remove the damage. Early events in this process include release of mediators, such as histamine and serotonin that dilate the blood vessels and increase their permeability. Subsequently, fluid leaking into the surrounding tissue space causes inflammation (i.e., swelling, pain, redness, and heat). Included in this fluid exudate are proteins, such as fibrinogen, which, when converted to fibrin, helps to seal off the affected area. Also present are polymorphonuclear neutrophils (PMNs) and leukocytes that phagocytize the infectious or toxic agent.

The inflammatory process is triggered by several interrelated cascade systems in the body. These include:

1) The complement system
2) The plasmakinin system
3) The clotting system
4) The arachidonic acid cascade

The *complement system* represents a cascade of approximately 25 serum proteins. The complement system mediates lysis of antibody-coated targets (bacteria, viruses, cells), recruits inflammatory cells to the sites of inflammation, and increases the efficiency of phagocytosis through opsonization.

Activation and interaction of the *clotting* and *plasmakinin systems* of the body also contribute to the inflammatory process (→ Blood). Activation of the Hageman factor (clotting factor XII), in addition to its effects on the coagulation system of the host, in turn activates the circulating protein, prekallikrein, to its enzymatically active form, kallikrein. The subsequent cleavage of another plasma substrate, kinonogen, by kallikrein liberates bradykinin, a linear nonapeptide. This peptide is a potent vasodilator that also has the ability to increase blood vessel permeability. In addition, bradykinin acts as a mediator of pain (elicits a pain response).

The *arachidonic acid cascade* has been the most actively studied of all the physiologic components contributing to inflammation. Both cyclooxygenase and lipoxygenase pathway of arachidonic acid metabolism have been investigated for their effects on inflammation. Arachidonic acid is generated from cellular membrane phospholipids by the action of the enzyme phospholipase A_2. In the cyclooxygenase pathway, it is converted first to a short-lived endoperoxide and then to prostaglandins (→ Prostaglandins). In mammalian cells two cyclooxygenase (COX) enzymes exist which are encoded by different genes, but share a 60% identity in amino acid sequence. COX-1 is constitutively expressed as a "housekeeping" enzyme in most tissues and mediates physiological responses such as regulation of renal and vascular homeostasis and cytoprotection of the stomach. In comparison, COX-2 is primarily considered as an inducible immediate-early gene product whose synthesis can be up-regulated by various proinflammatory agents, including endotoxin, cytokines, and mitogenes. COX-2 is the major isoform expressed by inflammatory cells and has, accordingly, been shown to release the high levels of prostanoids present under pathological conditions, such as acute and chronic inflammation [1]. The role of prostaglandins in inflammation is profound. Those of the E type increase blood vessel permeability and have the ability to sensitize various tissues (e.g., blood vessels, pain receptors) to the effects of other mediators, such as bradykinin.

The alternate pathway of arachidonic acid metabolism, the 5-lipoxygenase pathway, leads to the generation first of a hydroperoxide derivative, 5 HPETE, that stimulates histamine release from basophiles. Subsequent conversions lead to compounds known as leukotrienes that synergize with prostaglandins to cause increased blood vessel permeability and pain. One such leukotriene, LTB_4, is a potent chemotactic agent and a hyperalgesic agent (pain sensitizer) that shows additive effects with prostaglandins in pain mediation.

Much of the tissue destruction occurring during the inflammatory cycle can be attributed to substances released from the activated phagocytes that have been recruited to the site. These substances include free radicals and radical precursors, such as hydroxyl radicals and superoxide anion. Free radicals are nonspecific in their action and can cause destruction of membrane components, degradation of connective tissue, and depolymerization of hyaluronic acid leading to collagen damage. They may potentiate the action of proteolytic enzymes by oxidative destruction of naturally occurring inhibitors, such as α-1-proteinase inhibitor. In addition they feed the inflammatory cycle by further stimulation of leukocyte and macrophage functions.

Proteolytic enzymes, such as elastase and collagenase, are elaborated by PMNs at the inflammatory site either by active secretion or by death of the cell. ElastaseElastase is a neutral protease capable of degrading elastin and other connective tissue components, such as collagen, proteoglycan, and basement membrane. It may also interact with the complement and plasmakinin systems to liberate peptide mediators of inflammation, such as kinins, from soluble precursors. Collagenase catalyzes a specific cleavage of collagen and its precursor tropocollagen.

Furthermore, several *cytokines* (→ Blood) appear to play an essential role in the inflammatory process. Cytokines mediate effects induced by infectious agents, regulate lymphocyte growth, stimulate growth of precursor cell populations, and mediate inflammatory reactions. Interleukin (IL)-1 and tumor necrosis factor (TNF) possess very similar physiological actions in that both cytokines co-stimulate the process of antigen-induced activation of T cells, cause fever, and act on hepatocytes to induce the synthesis of plasma proteins, termed acute phase reactants. Under conditions of rheumatoid disease, IL-1 and TNF induce bone resorption by osteoclasts and increase proliferation of fibroblasts and synovial cells. Accordingly, increased levels of IL-1 and TNF have been detected in joint fluids from patients with rheumatoid arthritis. IL-6, which is produced by macrophages, monocytes, endothelial cells, and fibroblasts in response to IL-1 and TNF, stimulates the production of fibrinogen by hepatocytes and acts as a B cell growth factor. Other cytokines and growth factors contributing to the manifestation of the inflammatory response are IL-2, IL-8, and granulocyte-macrophage colony-stimulating factor (GM-CSF). Interestingly, some cytokines and growth factors possess anti-inflammatory activity (e.g., IL-10 and transforming factor (TGF)-β_1).

Despite the complexity and apparent propensity for endless self-perpetuation, most bouts of acute inflammation are self-limiting and result in eventual resolution and healing. In some cases, however, persistent stimulation or immunologic derangement occurs, leading to a chronic and degenerative inflammation. Anti-inflammatory drugs are commonly used for these situations.

The most prevalent clinical inflammatory disorders are a variety of arthritic conditions, such as rheumatoid arthritis, osteoarthritis, ankylosing spondylitis, gout, and systemic lupus erythematosus, which together affect about 7% of the population worldwide.

Rheumatoid arthritis is a chronic inflammatory disease occurring in about 2% of the population, with more women than men being affected. Most cases are relatively mild, and spontaneous remissions of the disease often occur. A minority of patients, however, develop severe and crippling deformities. Rheumatoid arthritis often affects peripheral joints, such as the knee, the wrist, and the fingers' proximal joints. The ankle, elbow, and hip may become involved but much less frequently. In rheumatoid arthritis, the membrane that lines the joint becomes inflamed and swollen, eventually covering and invading the cartilagenous surfaces of the joint. Progressive destruction of articular and periarticular structures follows, leading to loss of joint function.

The etiology of rheumatoid arthritis is largely unknown. One of the main characteristics of rheumatoid arthritis is an accumulation of T cells, plasma cells, macrophages, and fibroblast-like cells in inflamed joints. T lymphocytes specific for an unknown autoantigen have been suggested to play a crucial role in the induction and maintenance of synovial inflammation. However, the T cell centered hypothesis has been challenged by recent data showing that T cell targeted therapies were rather disappointing, particularly the use of anti-CD4 monoclonal antibodies.

Osteoarthritis, or degenerative joint disease, probably occurs as the result of chronic trauma or biomechanical stress. This disease also shows a predilection for certain joints, namely, the hip, the knee, and the fingers' distal joints. The elbow, shoulder, and ankle are rarely affected. Stress damage or simple "wear and tear" of the joint leads to chondrocyte (type of cartilage cell) damage and to the release of proteolytic enzymes that destroy the articular cartilage. The body's attempt to repair this damage causes formation of cartilagenous lumps that harden and become the osteophytes (bone knobs) characteristic of this disease.

Ankylosing spondylitis is a chronic and progressive inflammation that usually begins at the entheses (sites where muscle and tendon are joined to bone) of the sacroiliac joint and slowly spreads up the spinal column. Bony outgrowths gradually fuse one vertebral joint to the next, leading to restricted movement and eventual immobility of the spine. This disease occurs primarily in men, particularly those whose genetic make-up results in the presence of the HL-A W27 blood group antigen. Such individuals are approximately 600 times more likely to contract ankylosing spondylitis than those not carrying this antigen.

Gout is an arthritis of peripheral joints resulting from the deposition of sodium urate crystals in and about the joints and tendons. Ingestion of these crystals by PMNs results in the release of lysosomal enzymes that induce inflammation. Urate crystals can also activate the Hageman factor and the complement cascade. This condition is initially an acute problem, but repeated attacks lead to chronicity and deformation of the affected joints.

Systemic lupus erythematosus is an autoimmune disease of unknown origin resulting in a generalized inflammation of connective tissue throughout the body. It affects the joints, lungs, skin, heart, and kidneys and is primarily a disease of young women.

The anti-inflammatory drugs commonly used to treat the above conditions are generally divided into three classes:

1) Corticosteroids (→ Hormones)
2) Nonsteroidal anti-inflammatory–analgesic drugs (NSAIDs) and Selective COX-2 Inhibitors
3) Disease-modifying antirheumatic agents

2. Glucocorticoids

The glucocorticoids are oxygenated pregnenes bearing an α-ketol side chain at position 17. They include the natural adrenal cortical hormones, cortisol and cortisone, as well as a host of synthetic analogs. The potent and dramatic anti-inflammatory effect of cortisone was first demonstrated in rheumatoid arthritic patients in 1948 and thus began a new era of antiarthritic therapy. In addition corticosteroids also are used to treat dermatitis and often preserve sight when used for ocular inflammation.

Molecular-biological studies performed in the 1990s have provided new insights into the mode of action of glucocorticoids in controlling inflammation. Glucocorticoids bind to a cytoplasmic receptor within the target cells thus inducing a rapid nuclear translocation of the activated glucocorticoid receptor-steroid complex into the nucleus. After binding to DNA at consensus sites (i.e., glucocorticoid response elements) in the promoter region of glucocorticoid-responsive genes glucocorticoids can either increase or decrease transcription of numerous genes involved in inflammatory reaction. Furthermore, glucocorticoid receptors directly interact with transcription factors activated under chronic inflammatory conditions. Accordingly, glucocorticoid receptor complexes inhibit the activated form of nuclear factor κB (NF-κB) via a direct protein-protein interaction thereby repressing gene transcription. As NF-κB induces many inflammatory genes encoding cytokines, chemokines, adhesion molcules, and the inflammatory enzymes inducible nitric oxide synthase (iNOS) and COX-2, the inactivation of activated NF-κB in the nucleus represents an important effect of glucocorticoids in repressing gene transcription of various proinflammatory mediators [2], [3].

The use of glucocorticoids is accompanied by serious *adverse effects*. Changes are seen in the metabolism of carbohydrate (increased gluconeogenesis and hyperglycemia), of protein (increased catabolism and negative nitrogen balance), and of fat (increased synthesis and deposition). Disturbances in mineral balance include sodium retention, potassium depletion, and loss of bone calcium. Effects on the CNS range from euphoria to mania. Finally, suppression of endocrine function, particularly of the pituitary and hypothalamus, occurs.

Glucocorticoids used for systemic or local anti-inflammatory activity include hydrocortisone, cortisone acetate, prednisolone, dexamethasone, betamethasone, and triamcinolone. Derivatization provides injectable compounds having a rapid onset of action. For this purpose, esters of the 21-hydroxy group with phosphoric or succinic acid are used as their sodium salts. Depot forms of glucocorticoid drugs can be obtained by preparing slowly hydrolyzable esters of the 21-hydroxy group with hindered acids, such as pivalic or *tert*-butylacetic acid.

Glucocorticoids used extensively for topical administration include hydrocortisone acetate, betamethasone acetate, and triamcinolone acetonide.

For structure, properties, and preparation → Hormones.

3. Nonsteroidal Anti-inflammatory Drugs and Selective COX-2 Inhibitors

Nonsteroidal anti-inflammatory drugs (NSAIDs) inhibit the activity of COX which catalyzes the synthesis of prostanoids by converting arachidonic acid and molecular oxygen into prostaglandin H_2, the common substrate for specific prostaglandin synthases [4]. Most of the NSAIDs inhibit both COX-1 and COX-2 at therapeutic doses, although they vary in their relative potencies against the two isozymes [5]. Whereas the mechanism-based side effects of NSAIDs (e.g., gastrointestinal toxicity, platelet dysfunctions) are due to a suppression of COX-1-derived prostanoids, compelling evidence suggests that inhibition of prostanoids produced by COX-2 can be ascribed to the anti-inflammatory, analgesic, and antipyretic effects of NSAIDs. Accordingly, treatment with NSAIDs with a higher affinity to the COX-1 isozyme (e.g., piroxicam, sulindac) results in increased gastrointestinal adverse reactions. Specific COX-2 inhibitors are expected to exert anti-inflammatory and analgesic effects without causing gastric ulcerogenic effects or platelet dysfunction. *Per definitionem*, a substance may be regarded as a specific COX-2 inhibitor if it causes no clinically meaningful COX-1 inhibition at maximal therapeutic doses. Such compounds usually reveal a more than 100-fold difference in the concentration that inhibits recombinant COX-2 versus COX-1 in respective biochemical in vitro assays [68]. X-ray crystallographic studies of the three-dimensional structures of COX-1 and COX-2 have provided more insight into how COX-2 specificity is achieved. Within the hydrophobic channel of the COX enzyme, a single amino acid difference in position 523 (isoleucin in COX-1, valin in COX-2) has been detected, that is critical for the COX-2 selectivity of several drugs. Accordingly, the smaller valin molecule in COX-2 gives access to a "side pocket", which has been proposed to be the binding site of COX-2 selective substances. Consequently, the total NSAID-binding site is about 25 % larger in COX-2 than in COX-1 [7]. Thus, the increased NSAID-binding pocket of the COX-2 isozyme can bind bulky inhibitors more readily than the COX-1 isoform. Celecoxib (SC58635) and rofecoxib (MK-966) are novel selective COX-2 inhibitors of the diarylheterocyclic family (see Section 3.8). The 4-methylsulfonylphenyl and 4-sulfonamoylphenyl groups of these compounds have been shown to interact with specific residues within the "side pocket" of the COX-2 isozyme. Both compounds have been shown to be effective analgesics in dental pain models, and effective anti-inflammatory and analgesic substances in patients with rheumatoid arthritis and osteoarthritis [8]. Celecoxib and rofecoxib have been associated with fewer adverse gastrointestinal effects than traditional NSAIDs. The first selective COX-2 inhibitor, celecoxib (SC58635) was approved in December 1998 by the U.S. Food and Drug Administration for rheumatoid arthritis and osteoarthritis. Rofecoxib became available in 1999 for the indications osteoarthritis and pain. In recent studies, COX-2 has been shown to be also expressed under basal conditions in organs such as the ovary, uterus, brain, spinal cord, kidney, cartilage, and bone, suggesting that this isozyme may play a more complex physiological role than formerly expected. Accordingly, recent findings suggest that COX-2 may be implicated in physiological processes such as ovulation [9] and delivery [10], [11]. Moreover, COX-2 is induced in tissue on the edges of ulcers, and in animals studies selective COX-2 inhibitors have been shown to retard ulcer healing [12]. Thus, in patients with NSAID-associated ulcers, it will be obligatory to show whether effective ulcer healing occurs in those patients that switched to treatment with selective COX-2 inhibitors. Controlled clinical trials will gain insights into other possible side effects of selective COX-2 inhibitors in humans. On the other hand, the involvement of the COX-2 isozyme in pathological states suggests that selective COX-2 inhibitors may have further indications in conditions such as colonic polyposis, colorectal cancer, and Alzheimer's disease.

3.1. Salicylates

The therapeutic use of salicylates, in the form of extracts of willow bark, for an antipyretic effect dates back to early civilization. Elucidation of the chemical structure of the active ingredient, salicin, led to the use of sodium salicylate for the treatment of both rheumatic fever and gout in the 1870s. This was followed by the introduction of the less irritating O-acetyl derivative of salicylic acid, aspirin, in 1899. Acetylsalicylic acid served for many years as the cornerstone in the therapy of rheumatoid disease, being the safest, least expensive yet effective agent available. This efficacy is now attributed to the ability of aspirin to irreversibly inhibit arachidonic acid cyclooxygenase by transacetylation of this enzyme. However, the theory that suppression of prostaglandin biosynthesis accounts for the pharmacological actions and the side effects of NSAIDs has been questioned by comparing the actions of salicylate and aspirin [13]. Salicylate, being used as an NSAID for more than one century now, does not, unlike aspirin, inhibit COX-1 and COX-2 activity in vitro [5], [14]. However, despite lacking inhibitory activity against purified COX-1 or COX-2, salicylate exerts comparable anti-inflammatory properties as aspirin [15], and has considerable inhibitory effects on prostaglandin synthesis in intact cells [5], [16], [17] as well as in vivo [18] and ex vivo [19] at sites of inflammation. Several suggestions have been made to describe how salicylates exert their anti-inflammatory and side effects. One proposal is that salicylates act by inhibition of the transcription factor NF-κB [20]. However, relatively high concentrations of sodium salicylate (i.e., higher than that obtained after therapeutic doses) were required to provide significant effects. In 1997 is was demonstrated that sodium salicylate is an effective inhibitor of COX-2 activity at concentrations far below those required to inhibit NF-κB activation and it has been suggested that aspirin and sodium salicylate exerts their pharmacological effects via a weak competitive inhibition with arachidonic acid in the active site of the enzyme, rather than via inhibition of NF-κB activation. Furthermore, in IL-1β-stimulated endothelial cells sodium salicylate was shown to inhibit COX-2 expression at pharmacological concentrations [16]. However, these results could not be reproduced in other studies [17], [21], [22], [23] despite the fact that much higher salicylate concentrations were investigated.

Years of chemical effort to modify the salicylic acid molecule in order to find a safer and more effective derivative have yielded only a few therapeutically useful salicylates. Newer salicylic acid derivatives such as trilisate, benorylate, and chlorthenoxazine are available for use primarily as acute analgesic–antipyretic agents. Diflunisal with a biphenyl structure is a salicylate analog possessing greater potency, better tolerance, and longer duration of action in anti-inflammatory and analgesic studies.

For the structure and syntheses of acetylsalicylic acid, benorylate, and diflunisal, → Analgesics and Antipyretics,

Acetylsalicylic Acid [50-78-2], 2-(acetyloxy)benzoic acid, aspirin, $C_9H_8O_4$, M_r 180.15, mp 135 °C.
Trade Names: Acetylin (Bristol Myers Squibb, Germany), Acimetten (Pharmonta, Austria), Adprin (Pfeiffer, USA), Alka Seltzer (Bayer, Austria, Switzerland, Czech Republic), Angettes (Bristol-Myers, UK), Asaped (Sanofi-Winthrop, USA), Aspirin (Bayer, Germany), Aspro (Roche Nicholas, Germany), Colfarit (Bayer Pharma Deutschland, Germany), Zorprin (Knoll, USA).

Benorylate [5003-48-5], 4-acetamidophenyl 2-acetoxybenzoate, $C_{17}H_{15}NO_5$, M_r 313.32, mp 175–176 °C.
Trade Names: Benoral (Sterling-Winthrop, UK), Benortan (Winthrop, The Netherlands), Duvium (Inpharzam, Belgium, Switzerland), Longalgic (Evans Medical, France), Salipran (Evans Medical, France), Spierifex (Winthrop, The Netherlands).

Benorylate is the acetaminophen ester of acetylsalicylic acid, designed to provide a less irritating form of aspirin. Gastrointestinal blood loss in patients taking benorylate is less than that during therapy with soluble aspirin. Benorylate has been found to be as effective as aspirin for the treatment of chronic rheumatic disorders.

Diflunisal [22494-42-4], 2′,4′-difluoro-4-hydroxy-(1,1′-biphenyl)-3-carboxylic acid, $C_{13}H_8F_2O_3$, M_r 250.20, mp 210–211 °C.
Trade Names: Biartac (Merck Sharp & Dohme, Belgium), Diflonid (Dumex, Denmark, Sweden), Diflusal (Merck Sharp & Dohme, Belgium), Dolobid (Frosst, Canada; Logos, South Africa; Merck, USA; Morson, UK), Dolobis (M.S.D.-Chibret, France), Fluniget (Merck Sharp & Dohme, Austria).

Diflunisal is a salicylate with greater anti-inflammatory and analgesic potency and with a long plasma half-life that allows for twice-a-day dosage. Without an *O*-acetyl group it is a reversible inhibitor of cyclooxygenase. It is comparable in efficacy to aspirin in the treatment of osteoarthritis and rheumatoid arthritis but is much better tolerated.

3.2. Arylacetic Acids

Indomethacin [*53-86-1*], 1-(4-chlorobenzoyl)-5-methoxy-2-methyl-1*H*-indole-3-acetic acid, $C_{19}H_{16}ClNO_4$, M_r 357.81, *mp* 155 °C or 162 °C depending on crystal form.

Synthesis: acylation of sodium 2-(4-methoxyphenyl)hydrazine-1-sulfonate with 4-chlorobenzoyl chloride followed by heating yields 1-(4- chlorobenzoyl)-1-(4-methoxyphenyl)hydrazine. Condensation with levulinic acid in a Fischer indole synthesis affords indomethacin [24].

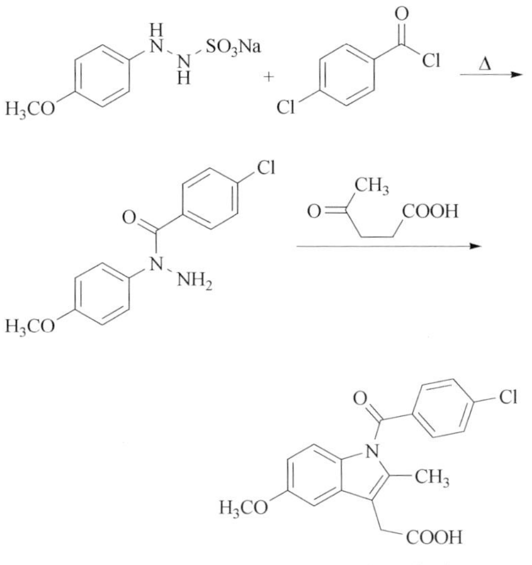

Indomethacin

The introduction of indomethacin in 1963 represented a milestone in the therapy of rheumatic diseases. For 20 years this drug has served as the clinical standard against which new NSAIDs are evaluated. It has also been a valuable biochemical–pharmacological tool in the study of the inflammatory process. It is a potent inhibitor of prostaglandin biosynthesis.

Clinically it is indicated for the treatment of rheumatoid arthritis, ankylosing spondylitis, osteoarthritis, and gout. Its potency as an anti-inflammatory, analgesic, antipyretic agent is counterbalanced by accompanying adverse effects, the most common of which are gastrointestinal irritation and headache.

Trade Names: Amuno (MSD Sharp & Dohme, Germany), Arthrexin (Lennon Generics, South Africa), Confortid (Dumex, Denmark, Sweden, Finland), Doctucid (Coctum, Greece), Dynamectin (Dynamed, South Africa), Flexidin (Mundipharma, Austria), Inacid (Merck Sharp & Dohme, Spain), Indobene (Merckle, Austria, CIS), Indocin (Merck, USA), Reumadolor (Bros, Greece), Zoflam (Norpharma, Denmark).

Attempts to develop a safer indomethacin have led to the introduction of the following four derivatives.

Acemetacin [53164-05-9], {[1-(4-chlorobenzoyl)-5-methoxy-2-methylindol-3-yl]acetoxy}acetic acid, $C_{21}H_{18}ClNO_6$, M_r 415.60, *mp* 149.5 – 150.5 °C

Synthesis: alkylation of indomethacin with benzyl bromoacetate in $K_2CO_3/N,N$-dimethylformamide gives the corresponding benzyl glycolate ester, which is hydrogenated over 10 % palladium on charcoal in acetic acid to yield acemetacin [25].

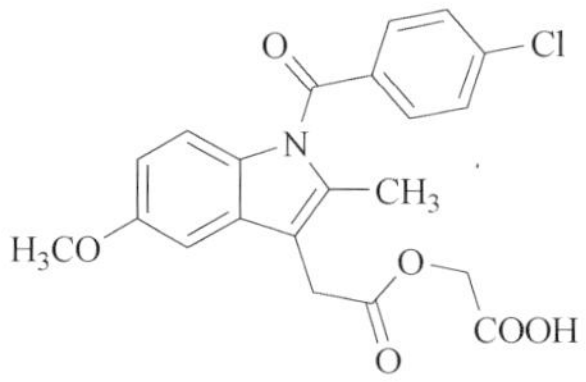

Acemetacin

Acemetacin is used in the treatment of pain and restricted mobility resulting from chronic articular rheumatism, degenerative articular disease, gout, and inflammation of muscle, joints, and tendons. Acemetacin causes less gastrointestinal blood loss than indomethacin. Its anti-inflammatory activity results from liberation of the parent compound, indomethacin.

Trade Names: Acemetacin (Heumann, Stada, Germany), Acemix (Bioprogress, Italy), Altren (Rhône-Poulenc Rorer, Belgium), Emflex (Merck, UK), Rantudil (Bayer Pharma Deutschland, Germany).

Glucametacin [52443-21-7], 2-({[1-(4-chlorobenzoyl)-5-methoxy-2-methyl-1*H*-indol-3-yl]acetyl}amino)-2-deoxy-D-glucose, $C_{25}H_{27}ClN_2O_8$, M_r 518.96, *mp* 218 °C decomp.

Synthesis: glucametacin is prepared by acylation of D-glucosamine with indomethacin acid chloride (prepared from indomethacin and thionyl chloride) in the presence of sodium hydroxide [26].

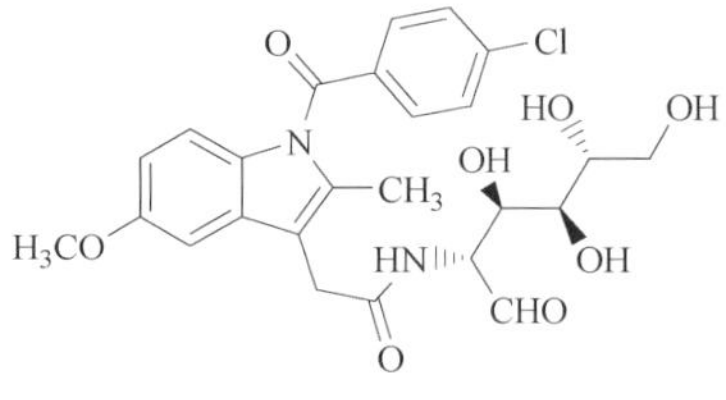

Glucametacin

Glucametacin is used to treat inflammatory and degenerative arthropathy. The compound is well tolerated with significantly less gastrointestinal distress than indomethacin. Glucametacin does not appear to be metabolized to indomethacin.
Trade Names: Euminex (Asta Medica, Spain), Teoremin (Labofarma/Degussa, Brazil).

Oxametacine [*27035-30-9*], 1-(4-chlorobenzoyl)-*N*-hydroxy-5-methoxy-2-methyl-1*H*-indole-3-acetamide, $C_{19}H_{17}ClN_2O_4$, M_r 372.81, *mp* 181–182 °C

Synthesis: oxametacine is prepared by the acylation of hydroxylamine with indomethacin acid chloride [27].

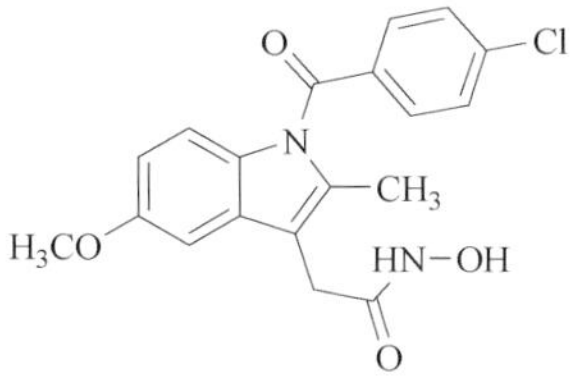

Oxametacin

Oxametacine at a dose of 100 mg is reported to be as effective as 50 mg of indomethacin in reducing pain and inflammation with a lower incidence of adverse effects.
Trade Names: Flogar (UCB, Belgium; ABC, Portugal).

Proglumetacin maleate [*59209-40-4*], *N*-{2-[1-(4-chlorobenzoyl)–5-methoxy-2-methyl-3-indolylacetoxy]ethyl}-*N*′-[3-(*N*-benzoyl-*N*′,*N*′-di-*n*-propyl-DL-isoglutaminyl)oxypropyl]piperazine (±)-dimaleate, $C_{46}H_{58}ClN_5O_8 \cdot 2\ C_4H_4O_4$, M_r 1075.98, *mp* 146–148 °C.

Synthesis: alkylation of the anticholinergic agent proglumide with *N*-(3-chloropropyl)-*N*′-(2-hydroxyethyl)piperazine using sodium methoxide in dimethyl sulfoxide gives *N*-[3-(*N*-benzoyl-*N*′,*N*′-di-*n*-propyl-DL-isoglutaminyl)oxypropyl]-*N*′-(2-hydroxyethyl)piperazine, which is then condensed with indomethacin using dicyclohexylcarbodiimide. The resulting ester is treated with maleic acid to give proglumetacin maleate [28].

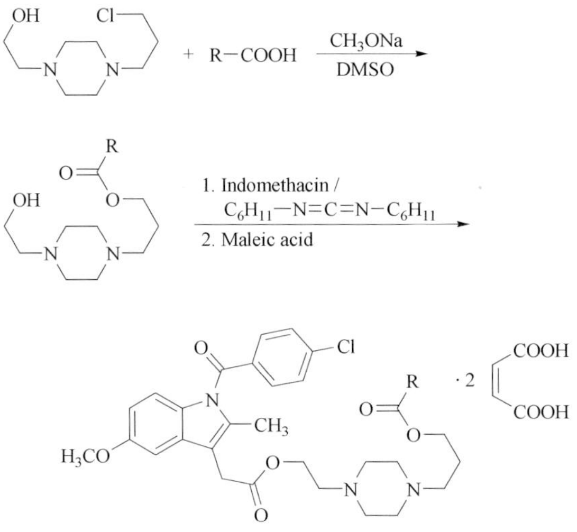

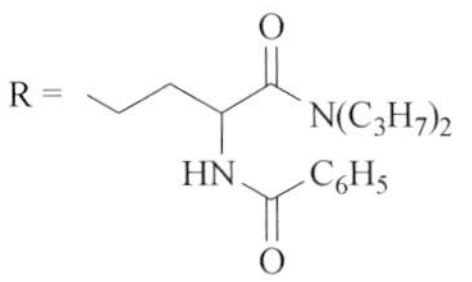

Proglumetacin at a dose of 450 mg/d appears to be as effective as 150 mg/d of indomethacin in the treatment of a wide variety of rheumatic conditions, and is faster acting with significantly lower incidence and severity of adverse effects. It is metabolized to several derivatives, including indomethacin, in humans.

Trade Names: Afloxan (Rotta Research, Italy), Protaxil (Rottapharm, Spain), Protaxon (Opfermann, Germany), Proxil (Rottapharm, Italy), Tolindol (La Meuse, Belgium).

Sulindac [*38194-50-2*], (*Z*)-5-fluoro-2-methyl-1-{[4-(methylsulfinyl)phenyl]-methylene}-1*H*-indene-3-acetic acid, $C_{20}H_{17}FO_3S$, M_r 356.42, *mp* 182–185 °C (decomp.).

Synthesis: Friedel–Crafts reaction of fluorobenzene and α-bromoisobutyryl bromide gives 5-fluoro-2-methylindan-1-one, which is treated with 4-methylthiobenzylmagnesium chloride to yield 5-fluoro-2-methyl-1-(4-methylthiobenzyl)indene. Condensation with glyoxylic acid in the presence of *N*-benzyltrimethylammonium hydroxide (Triton B) gives 3-carboxymethylene-5-fluoro-2-methyl-1-(4-methylthiobenzyl)indene, which is isomerized in acid to 5-fluoro-2-methyl-1-(4-methylthiobenzylidene)indene-3-acetic acid. Oxidation with hydrogen peroxide affords sulindac [29], [30].

Sulindac is an indene isostere of indomethacin and is a reversible prodrug. Its anti-inflammatory activity depends on metabolic conversion to the corresponding sulfide. The long serum half-life of the sulfide metabolite allows it to be given only once or twice a day. Sulindac is much better tolerated than indomethacin in all five of its indications: rheumatoid arthritis, osteoarthritis, gout, ankylosing spondylitis, and acute painful shoulder. There is a low incidence of adverse effects, consisting chiefly of gastrointestinal disturbances.

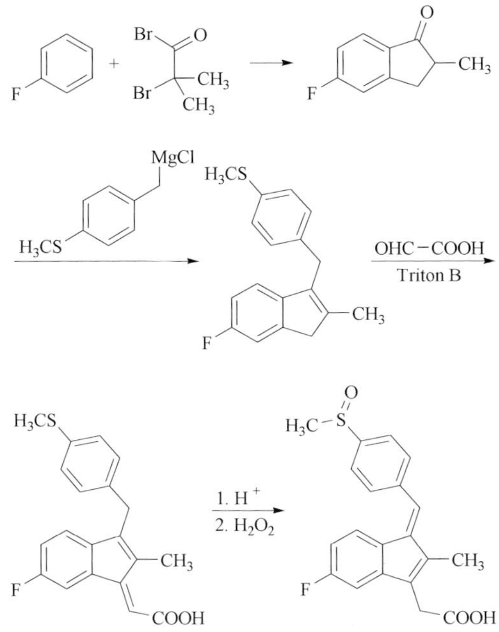

Trade Names: Aflodac (Biotekfarma, Italy), Algocetil (Francia Farm., Italy), Clinoril (Frosst, The Netherlands; Merck Sharp & Dohme, Austria, Belgium, Denmark, UK, Slovacian republic); Dorindac (Chibret, Portugal), Zirofalen (Farmalen, Greece).

Tolmetin [*26171-23-3*] 1-methyl-5-(4-methylbenzoyl)-1*H*-pyrrole-2-acetic acid, $C_{15}H_{15}NO_3$, M_r 257.30, mp 155 – 157 °C.

Synthesis: Friedel – Crafts acylation (aluminum chloride – carbon disulfide) of ethyl 1-methyl-1*H*-pyrrole-2-acetate with 4-toluyl chloride gives ethyl 1-methyl-5-(4-methylbenzoyl)-1*H*-pyrrole-2-acetate, which on saponification yields tolmetin [31].

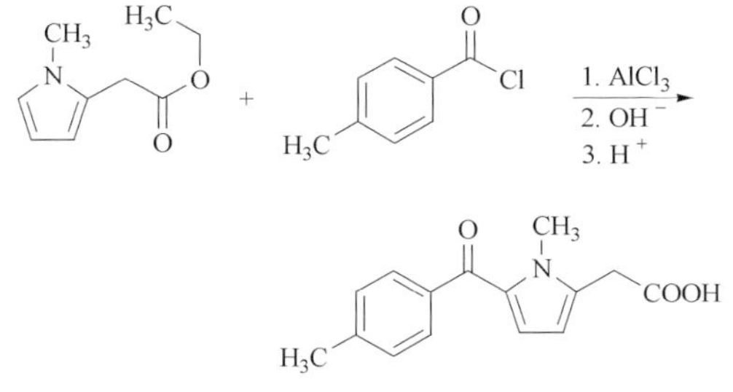

Tolmetin is an anti-inflammatory, analgesic – antipyretic agent indicated for acute

and long-term treatment of rheumatoid arthritis. Its activity is comparable to aspirin and indomethacin, but it causes fewer and milder gastrointestinal disturbances than the former and fewer CNS disturbances than the latter. Other adverse effects include dizziness, tinnitus, edema, and rash. It is administered as its sodium salt dihydrate.
Trade Names: Artrocaptin (Estedi, Spain), Tolectin (Cilag, Belgium; Janssen-Cilag, Austria; McNeil, USA).

Metiazinic acid [*13993-65-2*], 10-methylphenothiazine-2-acetic acid, $C_{15}H_{13}NO_2S$, M_r 271.34, mp 146 °C.
Synthesis: Metiazinic acid is prepared via the Willgerodt reaction (sulfur and morpholine) from 2-acetyl-10-methylphenothiazine [32].

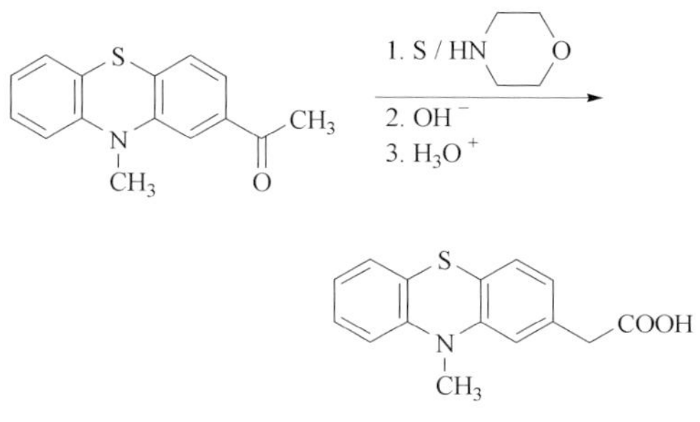

Metiazinic acid

Metiazinic acid has been beneficial to arthritis patients when given at a daily dose of 0.75 – 1.5 g/d.
Trade Names: Roimal (Kantoishi, Japan), Soripal (Torii, Japan).

Alclofenac [*22131-79-9*], 3-chloro-4-(2-propenyloxy)benzeneacetic acid, $C_{11}H_{11}ClO_3$, M_r 226.66, mp 92 – 93 °C.
Synthesis: alkylation of ethyl 3-chloro-4-hydroxyphenylacetate with allyl bromide in acetone containing potassium carbonate followed by saponification of the ester group gives the product [33].

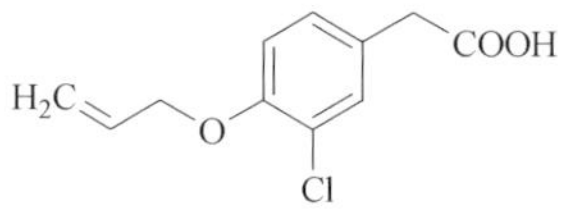

Alclofenac

Alclofenac when used at 3 g/d in the treatment of rheumatoid arthritis is as effective as 75 mg/d of indomethacin with fewer adverse effects. In the treatment of osteoarthritis 1500 mg/d is as effective and is better tolerated than 300 mg/d of phenylbutazone. Alclofenac has been withdrawn from the market in the United Kingdom because of skin rash and vasculitis (blood vessel inflammation).
Trade Names: Mervan (Continental, Belgium).

Diclofenac sodium [*15307-79-6*], 2-[(2,6-dichlorophenyl)amino]benzeneacetic acid monosodium salt, $C_{14}H_{10}Cl_2NNaO_2$, M_r 282.68, *mp* 283–285 °C.

Synthesis: acylation of *N*-phenyl-2,6-dichloroaniline with chloroacetyl chloride gives the corresponding chloroacetanilide, which is fused with aluminum chloride to give 1-(2,6-dichlorophenyl)-2-indolinone. Hydrolysis of the indolinone with dilute aqueous-alcoholic sodium hydroxide affords the desired sodium salt directly [34].

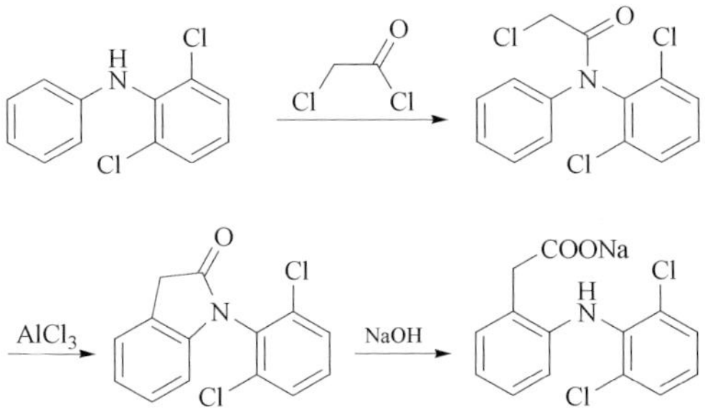

Diclofenac sodium

The structure of diclofenac is a hybrid of a fenamate and an arylacetic acid. The compound is used as its sodium salt for the symptomatic relief of rheumatoid arthritis and osteoarthritis, including degenerative joint disease of the hip. The recommended dose is 75–150 mg/d which is clinically equivalent to 3.6 g/d of aspirin. Gastrointestinal problems (ulceration and bleeding) and adverse CNS reactions (dizziness and headache) are the most commonly encountered adverse effects (for review see [35]).
Trade Names: Arthrex (BASF GENERICS, Germany), Benfofen (Sanofi Winthrop, Germany), Diclac (Hexal, Germany), Diclobene (Merckle, CIS, Austria), Diclophlogont (Azupharma, Germany), Diclosifar (Siphar, Switzerland), Dolgit (Dolorgiet, Germany), Rewodina (Asta Medica AWD, Germany), Voltaren (Novartis Pharma, Germany), Arthrotec (Searle, UK, Italy).

Fentiazac [*18046-21-4*], 4-(4-chlorophenyl)- 2-phenyl-5-thiazoleacetic acid, $C_{17}H_{12}ClNO_2S$, M_r 329.81, *mp* 161–162 °C.

Synthesis: refluxing a mixture of 3-bromo- 3-(4-chlorobenzoyl)propionic acid and thiobenzamide in ethanol gives ethyl 2-phenyl-4-(4-chlorophenyl)thiazole-5-acetate, which is then saponified to yield fentiazac [36].

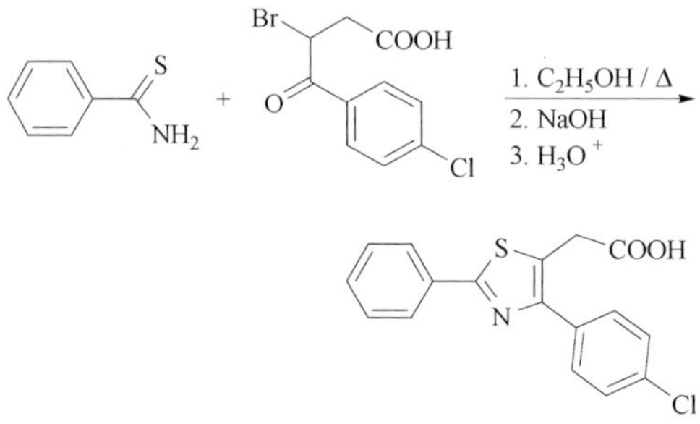

Fentiazac is reported to have anti-inflammatory, analgesic, and antipyretic activity. It has been given once or twice a day at levels between 100 and 200 mg/dose in the treatment of postoperative pain, including that following orthopedic surgery. The most common adverse effect is gastrointestinal intolerance, including epigastric pain, nausea, and vomiting. Effects on the CNS, such as headache and dizziness, also have been reported.

Trade Names: Atilan (Zambon, Brazil), Donorest (Fontoura-Wyeth, Brazil), Flogene (Polifarma, Italy), Norvedan (Boehringer Mannheim, Austria; LPB, Italy).

3.3. Arylacetic Acid Prodrugs

Metabolic precursors to arylacetic acids have been introduced into rheumatoid therapy in the form of 4-aryl-4-oxobutyric acids. ω-Phenyl fatty acids undergo normal two-carbon degradation of the side chain to give either benzoic or phenylacetic acids depending on the number of carbon atoms in the original molecule. Therefore, ω-phenyl-γ-oxoacids were presumed to behave similarly; e.g., 4-phenyl-4-oxobutyric and 6-phenyl-4-oxo-hexanoic acid are degraded in vivo to phenylacetic acid.

The oxoacids are not very potent inhibitors of cyclooxygenase and their in vivo activity is probably dependent on metabolic conversion to the corresponding phenylacetic acid.

Fenbufen [*36330-85-5*], γ-oxo(1,1′biphenyl)-4-butanoic acid, $C_{16}H_{14}O_3$, M_r 254.29, mp 185–187 °C.

Synthesis: fenbufen is prepared by the Friedel-Crafts (aluminum chloride – nitrobenzene) acylation of biphenyl with succinic anhydride [37].

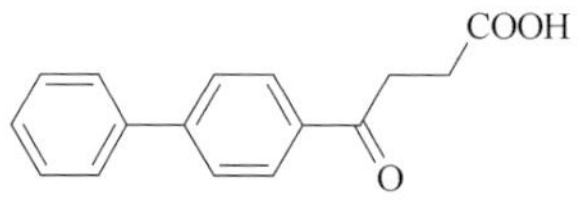

Fenbufen has been found to be an effective, well-tolerated drug for the treatment of rheumatoid arthritis, osteoarthritis, and ankylosing spondylitis. The compound is metabolized in humans first to 4-hydroxy-4-biphenylbutyric acid (t_{max} 2.5 h) then to 4-biphenyl acetic acid (t_{max} 7.5 h). Both metabolites are more active than fenbufen itself and circulate for several hours ($t_{1/2}$ 10 h). This slow conversion of fenbufen to active metabolites having relatively long plasma half-lives allows for once a day dosing with this agent.
Trade Names: Clincopal (Lederle, Spain), Lederfen (Lederle, UK), Napanol (Lederle, Japan).

3.4. Arylpropionic Acids

Ibuprofen [*15687-27-1*], α-methyl-4-(2-methylpropyl)benzeneacetic acid, $C_{13}H_{18}O_2$, M_r 206.27, mp 75–77 °C.

Synthesis: treatment of ethyl 4-isobutylphenylacetate and diethyl carbonate with sodium ethoxide gives diethyl 4-isobutylphenylmalonate, which is methylated using methyl iodide and sodium ethoxide. Saponification followed by decarboxylation of the resulting malonic acid derivative affords ibuprofen [38].

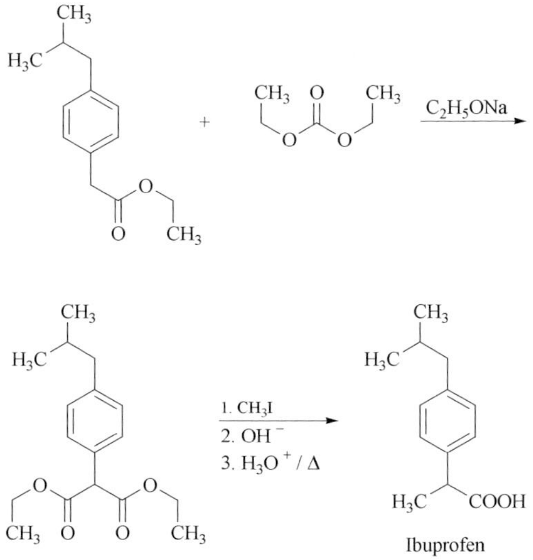

Ibuprofen is indicated for both acute and long-term management of rheumatoid arthritis, osteoarthritis, and psoriatic arthritis. Anti-inflammatory efficacy depends on relatively high dosage, 2.4 g/d. The drug is well tolerated, the chief adverse effect being a low incidence of gastrointestinal disturbance. The drug is rapidly absorbed and eliminated with a serum half-life of about 2 h. Ibuprofen does not interfere with the protein binding of warfarin and can be used safely in patients undergoing anticoagulant therapy. It is available as a nonprescription drug.

Trade Names: Advil (Whitehall, USA; Whitehall-Robins, Canada); Aktren (Bayer Selbstmed, Germany), Algifor (Vifor, Switzerland), Ardinex (Boots, Finland), Brufen (Boots, Portugal; Knoll, Belgium), Dismenol (Merz Consumer Division, Germany), Dolgit (Dolorgiet, Germany), Dolo-Puren (Isis Puren, Germany), Dolormin (Woelm, Germany), Duafen (Pharmed, Austria), Haltran (Upjohn, USA), Ibufen (Amino, Switzerland), Ibunet (Nettopharma, Denmark), Rufen (Knoll, USA).

Naproxen [*22204-53-1*], (+)-6-methoxy-αmethyl-2-naphthaleneacetic acid, $C_{14}H_{14}O_3$, M_r 230.26, *mp* 155.3 °C, $[\alpha]_D$ + 65.5° (*c* 1.0, $CHCl_3$).

Synthesis: Friedel-Crafts acylation (aluminum chloride - nitrobenzene) of β-naphthol methyl ether affords 2-acetyl-6-methoxynaphthalene, which, when treated with either dimethylsulfonium or dimethylsulfoxonium methylide, gives 2-(6-methoxynaphthalen-2-yl)propylene oxide. Treatment of the latter with boron trifluoride etherate in tetrahydrofuran gives 2-(6-methoxynaphthalen-2-yl)propionaldehyde, which is oxidized using Jones reagent (4 M chromic acid) to yield the racemic 2-(6-methoxynaphthalen-2-yl)propionic acid. Resolution and isolation of the dextrorotatory enantiomer is accomplished via its cinchonidine salt [39].

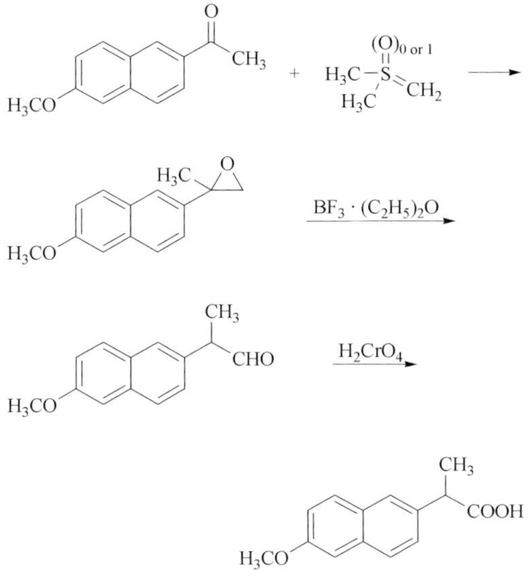

Naproxen is used in the treatment of rheumatoid arthritis, osteoarthritis, gout, and ankylosing spondylitis at a daily dose of 500 – 700 mg. Absorption from the gastrointestinal tract is virtually complete, and extensive binding (99%) to serum albumin provides for a relatively long plasma half-life (13 h) and twice-a-day dosing. Naproxen is reasonably well-tolerated with gastrointestinal disturbances being the principal adverse

effect. Its more rapidly absorbed sodium salt is marketed as an analgesic providing earlier pain relief.
Trade Names: Anaprox (Syntex, Canada, USA), Apranax (Roche, France), Bonyl (Ercopharm, Denmark), Miranax (Syntex, Finland), Novo-Naprox (Novorpharm, Canada), Proxen (Hoffmann La Roche, Germany).

Ketoprofen [*22071-15-4*], 3-benzoyl-α-methylbenzeneacetic acid, $C_{16}H_{14}O_3$, M_r 254.29, *mp* 94 °C.

Synthesis: bromination (bromine–ultraviolet light) of 3-methylbenzophenone and nucleophilic substitution of the resulting bromomethyl compound with sodium cyanide gives 3-benzoylphenylacetonitrile. Treatment with diethyl carbonate and sodium ethoxide followed by alkylation of the resulting 2-cyano ester gives ethyl 2-(3-benzoylphenyl)-2-cyanopropionate. Acid hydrolysis and decarboxylation yields ketoprofen [40].

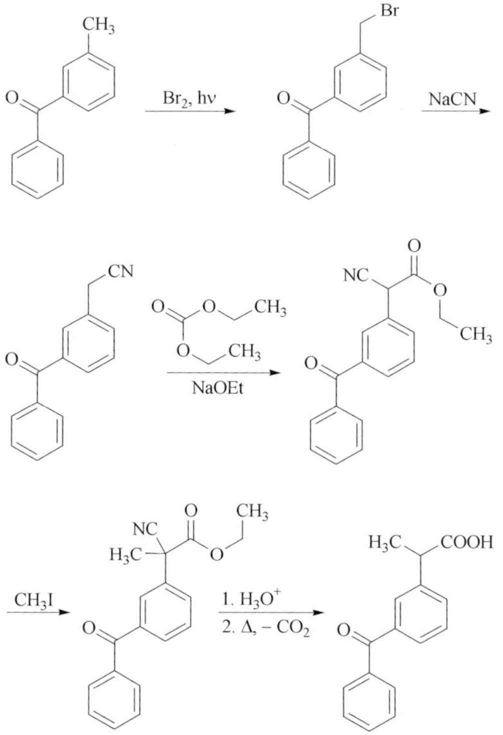

Ketoprofen is indicated in the treatment of rheumatoid arthritis, ankylosing spondylitis, and osteoarthritis at a daily dose of 150–300 mg. At 150 mg/d its clinical efficacy is comparable to a similar dose of indomethacin, but fewer and less severe adverse effects are observed.

Trade Names: Alrheumat (Bayer, United Kingdom), Alrheumun (Bayer Pharma Deutschland, Germany), Gabrilen (Kreussler, Germany), Orudis (Rhône-Poulenc Rorer, Canada, Denmark; Wyeth-Ayerst, USA).

Fenoprofen [*31879-05-7*], α-methyl-3-phenoxybenzeneacetic acid, $C_{15}H_{14}O_3$, M_r 242.28, *bp* 168 – 171 °C (0.015 kPa).

Synthesis: sodium borohydride reduction of 3-phenoxyacetophenone followed by bromination of the resulting alcohol with PBr_3 gives α-methyl-3-phenoxybenzyl bromide. Reaction of this bromide with sodium cyanide in dimethyl sulfoxide gives the corresponding nitrile, which is hydrolyzed using sodium hydroxide. Acidification affords fenoprofen [41].

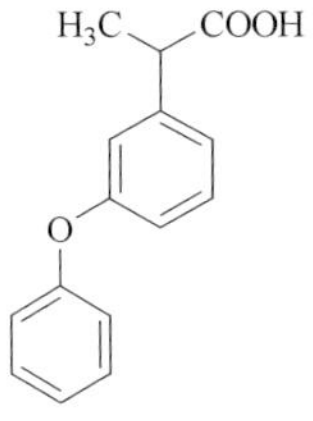

Fenoprofen

Fenoprofen is used as its calcium salt dihydrate in the treatment of rheumatoid arthritis and osteoarthritis at a daily dose of 1.2 – 3.0 g. The drug is rapidly absorbed and excreted with a plasma half-life of about 3 h despite being extensively bound (99 %) to plasma protein. Fenoprofen is well tolerated, with dyspepsia being the chief adverse effect.

Trade Names: Fenopron (Lilly, South Africa, United Kingdom), Fepron (Lilly, Italy), Nalfon (Dista, USA, Austria, Canada), Nalgesic (Lilly, France), Progesic (Lilly, United Kingdom).

Flurbiprofen [*5104-49-4*], 2-fluoro-α-methyl(1,1′-biphenyl)-4-acetic acid, $C_{15}H_{13}FO_2$, M_r 244.27, *mp* 110 – 111 °C.

Synthesis: ethyl 2-fluoro-4-biphenylacetate, prepared from 4-acetyl-2-fluorobiphenyl via the Willgerodt reaction and subsequent esterification, is treated first with diethyl carbonate and sodium ethoxide, then with dimethyl sulfate to give diethyl 2-(2-fluoro-4-biphenylyl)-2-methyl malonate. Saponification followed by acidification affords the malonic acid, which is decarboxylated at 180 – 200 °C to yield flurbiprofen [42].

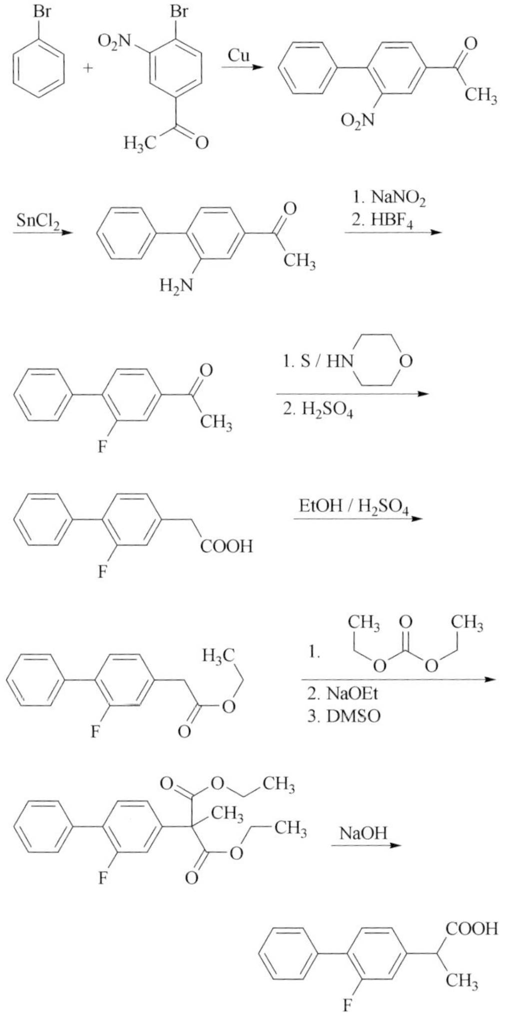

Flurbiprofen has been used at 150–300 mg/d in the treatment of rheumatoid arthritis, osteoarthritis, and ankylosing spondylitis. Gastrointestinal disturbances, particularly dyspepsia, are the principal adverse effects.

Trade Names: Ansaid (Pharmacia & Upjohn, CIS; Upjohn, Canada, Czech Republic, Poland, USA), Cebutid (Knoll, France), Froben (Kanoldt, Germany; Ebewe, Austria; Boots, Spain, The Netherlands, Portugal), Tulip (Upjohn, Spain).

Clidanac [*34148-01-1*], 6-chloro-5-cyclohexyl-2,3-dihydro-1*H*-indene-1-carboxylic acid, $C_{16}H_{19}ClO_2$, M_r 278.78, mp 150.5–152.5 °C.

Synthesis: formylation of cyclohexylbenzene using dichloromethyl methyl ether and titanium tetrachloride gives 4-cyclohexylbenzaldehyde, which is condensed with diethyl malonate to yield diethyl 4-cyclohexylbenzylidenemalonate. Treatment of the latter with potassium cyanide in ethanol at 70 °C gives ethyl 3-cyano-3-(4-cyclohexylphenyl)propionate, which is converted to 4-cyclohexylphenylsuccinic anhydride following acid hydrolysis and treatment with refluxing acetic anhydride. An intramolecular Friedel–Crafts acylation (aluminum chloride–methylene chloride) affords 5-cyclohexyl-3-oxo-1-indanecarboxylic acid, which is catalytically reduced using 10 % palladium on charcoal in the presence of perchloric acid to give 5-cyclohexyl-1-indanecarboxylic acid. Chlorination using N-chlorosuccinimide in N,N-dimethylformamide (DMF) gives clidenac [43].

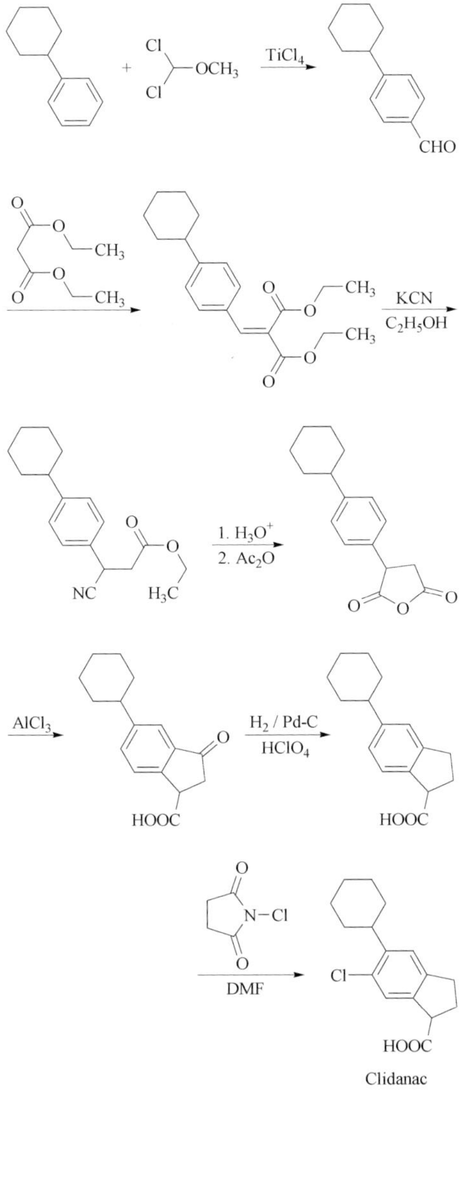

Clidenac has been recently introduced for the relief of pain and inflammation related to arthrosis deformans, periarthritis humeroscapularis, cervical syndrome, and lumbago.
Trade Names: Britai (Bristol Banyu, Japan), Indanal (Takeda, Japan).

Pirprofen [*31793-07-4*], 3-chloro-4-(2,5-dihydro-1*H*-pyrrol-1-yl)-α-methylbenzeneacetic acid, $C_{13}H_{14}ClNO_2$, M_r 251.71, *mp* 98–100 °C.

Synthesis: Treatment of the sodium salt of diethyl methylmalonate with 2,4-dichloronitrobenzene yields diethyl (3-chloro-4-nitrophenyl)methylmalonate. Saponification, decarboxylation, and subsequent reesterification followed by catalytic reduction gives ethyl 4-amino-3-chloro-α-methylbenzeneacetate hydrochloride. Treatment of the latter with 1,4-dichloro-2-butene in the presence of sodium carbonate followed by saponification affords pirprofen [44].

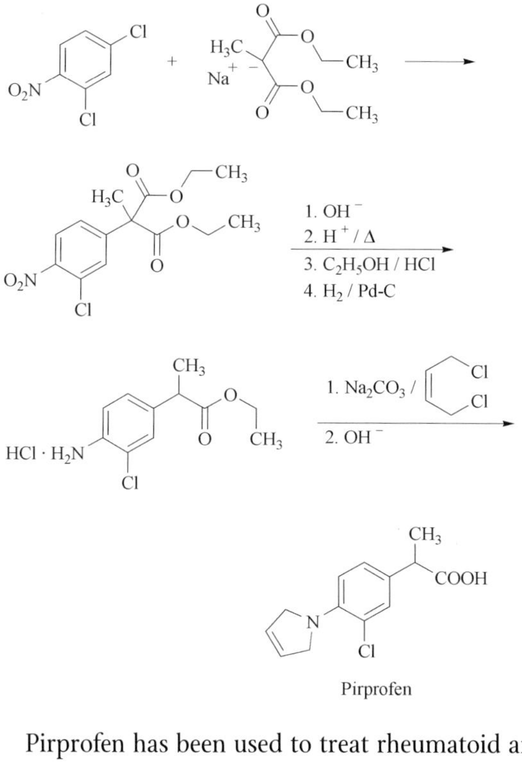

Pirprofen has been used to treat rheumatoid arthritis, osteoarthritis, and ankylosing spondylitis. An optimal dosing regimen of 200 mg three times a day has been developed for maximal activity with minimal adverse effects. Pirprofen also is effective in relieving pain from malignant disease and oral surgery.
Trade Name: Rengasil (Ciba, Greece), Seflenyl (Geigy, Argentina).

3.5. Acidic Enolic Compounds

Phenylbutazone [*50-33-9*], 4-*n*-butyl-1,2-diphenyl-3,5-pyrazolidinedione, $C_{19}H_{20}N_2O_2$, M_r 308.37, *mp* 105 °C.

Synthesis: Heating diethyl-*n*-butylmalonate and 1,2-diphenylhydrazine in the presence of sodium ethoxide at 150 °C followed by acidification affords phenylbutazone in good yield. Alternatively, *n*-butylmalonyl chloride in pyridine and ether at 0 °C is treated with 1,2-diphenylhydrazine to give the desired product [45].

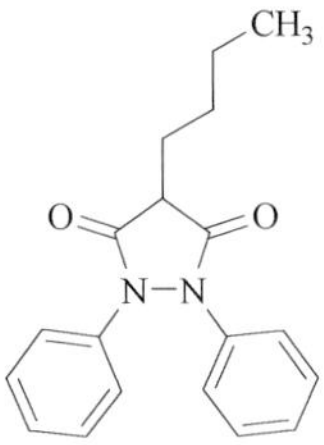

Phenylbutazone

Phenylbutazone, one of the earliest NSAIDs introduced, is now indicated for the symptomatic relief of rheumatoid arthritis, osteoarthritis, psoriatic arthritis, ankylosing spondylitis, gout, and acute superficial thrombophlebitis.

The gastrointestinal and bone marrow toxicity observed in its early use have been greatly reduced by lower dosage (300 mg/d). Nevertheless, it is used primarily where other drugs have failed and then only for short-term therapy. The drug has a long serum half-life of about 100 h. It is a moderately active cyclooxygenase inhibitor and it suppresses both spontaneous and chemotactic motility of neutrophils. In addition to the serious gastrointestinal and hematological adverse effects, sodium and water retention, rash, vertigo, and dermatitis are observed.

Trade Names: Ambene (Merckle, Germany), Butazolidin (Novartis Pharma, Germany), Butazone (Major, USA), Butrex (Noristan, South Africa), Ticinil (Boehringer Ingelheim, Italy).

Oxyphenbutazone [*129-20-4*], 4-*n*-butyl-1-(4-hydroxyphenyl)-2-phenyl-3,5-pyrazolidinedione, $C_{19}H_{20}N_2O_3$, M_r 324.37, *mp* 124–125 °C.

Synthesis: heating 1-(4-benzyloxyphenyl)-2-phenylhydrazine with diethyl-*n*-butylmalonate in the presence of sodium ethoxide at 140 °C followed by acidification gives 1-(4-benzyloxyphenyl)-4-*n*-butyl-2-phenyl-pyrazolidine-3,5-dione, which is unblocked by catalytic reduction (Raney nickel) to yield oxyphenbutazone [46].

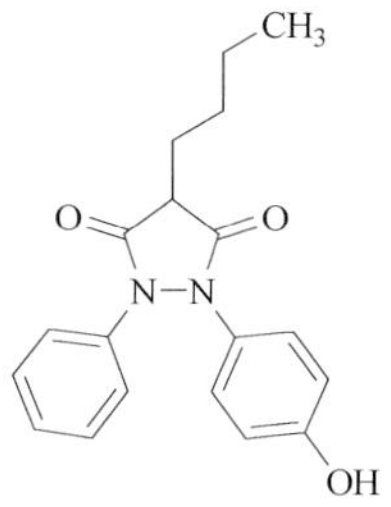

Oxyphenbutazone

Oxyphenbutazone is one of the metabolites formed in the liver following administration of phenylbutazone. Given orally it causes fewer gastrointestinal adverse effects than phenylbutazone and is used at 300–400 mg/d for the same indications as the parent drug.

Trade Names: Diflamil (Belmac, Spain), Phlogont (Azupharma, Germany), Tabazone (Major, USA).

Apazone, azapropazone [*13539-59-8*], 5-(dimethylamino)-9-methyl-2-propyl-1*H*-pyrazolo[1,2*a*][1,2,4]benzotriazine-1,3-(2*H*)-dione, $C_{16}H_{20}N_4O_2$, M_r 300.37, *mp* 228 °C.

Synthesis: A mixture of 1,2-dihydro-3-dimethylamino-7-methylbenzotriazine, diethyl propylmalonate, and sodium ethoxide in xylene is heated over a period of time to 150 °C, cooled, and acidified to give the desired product [47].

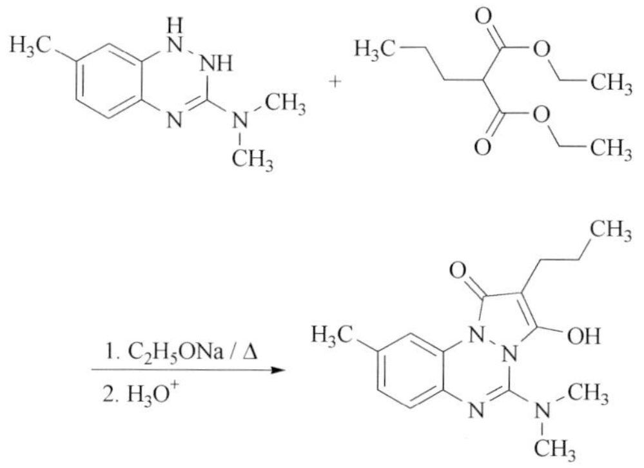

Apazone

Apazone is a complex pyrazolidinedione used at 600–1200 mg/d to achieve an anti-inflammatory effect comparable to phenylbutazone. It is tightly bound to plasma protein and may displace anticoagulants and hypoglycemic agents. Its plasma half-life is 8–12 h. It is well tolerated with only a low incidence of gastric disturbances. As with all pyrazolone derivatives, the potential for bone marrow suppression exists and careful monitoring of the patient's hematologic state is recommended.

Trade Names: Cinnamin (Chemiphar, Japan), Prolixan (Siegfried, Greece), Rheumox (Wyeth, UK), Xani (Farmakos, Yugoslavia).

Piroxicam [36322-90-4], 4-hydroxy-2-methyl-*N*-2-pyridinyl-2*H*-1,2-benzothiazine-3-carboxamide 1,1-dioxide, $C_{15}H_{13}N_3O_4S$, M_r 331.35, *mp* 198–200 °C.

Synthesis: An improved procedure using 2- methoxyethyl 2-chloroacetate in place of methyl 2-chloroacetate for the alkylation of sodium saccharin has been described [48]. The resulting 2-methoxyethyl saccharin-2-acetate is treated with sodium 2-methoxyethoxide in dimethyl sulfoxide, then acidified to give 2-methoxyethyl 4-hydroxy-2*H*-1,2-benzothiazine-3-carboxylate 1,1-dioxide, which is *N*-alkylated with methyl iodide in acetone–aqueous sodium hydroxide. The resulting 2-methoxyethyl 4-hydroxy-2-methyl-2*H*-1,2-benzothiazine-3-carboxylate 1,1-dioxide is heated with 2-aminopyridine in xylene to give piroxicam [49].

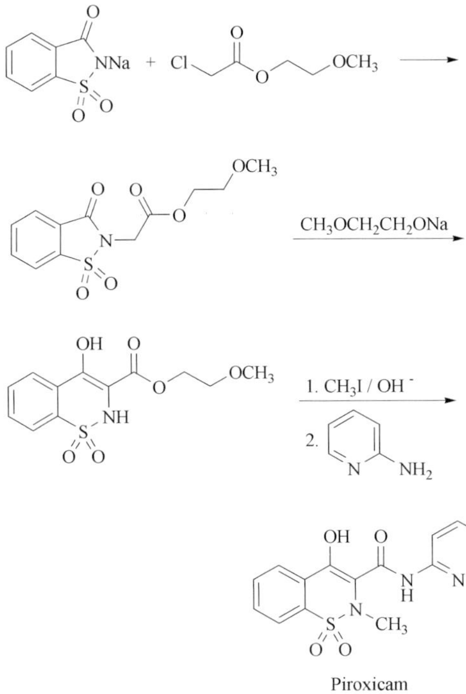

Piroxicam

Piroxicam, the first of a new class of NSAIDs known as "oxicams," was introduced in 1979 for the treatment of rheumatoid arthritis, osteoarthritis, ankylosing spondylitis, gout, and acute musculoskeletal disorders [48]. This compound was the result of a long and intensive effort to develop a potent NSAID having a novel structure and a plasma half-life sufficiently long to maintain continuous therapeutic blood levels on a once-a-day dosing regimen. Piroxicam is a potent competitive inhibitor of cyclooxygenase and lowers blood levels of prostaglandins in humans. Its ability to inhibit the infiltration of polymorphonuclear leukocytes into inflamed joints has been demonstrated in animals. Its potency and long plasma half-life of 40 h in humans allow for once-a-day dosing at 20 mg. A high incidence of gastric irritation results from this agent.

Trade Names: Brexidol (Pharmacia & Upjohn, Germany), Brexin (Pharmacia & Upjohn, Austria), Doblexan (Organon, Spain), Durapirox (Durachemie, Germany), Felden (Mack/Pfizer, Germany), Piro-Phlogont (Azupharma, Germany), Ruvamed (Coup, Greece).

3.6. Anthranilates

Mefenamic acid [*61-68-7*], 2-[(2,3-dimethylphenyl)amino]benzoic acid, $C_{15}H_{15}NO_2$, M_r 241.28, mp 230–231 °C.

Synthesis: Mefenamic acid is prepared via the Jourdan–Ullmann–Goldberg synthesis utilizing either anthranilic acid and 3-bromo-1,2-dimethylbenzene or 2,3-dimethylaniline and an *o*-halobenzoic acid in the presence of a copper catalyst and a proton acceptor [50].

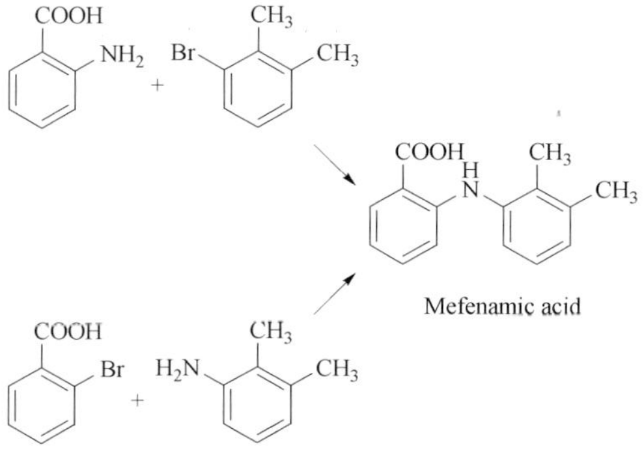

Mefenamic acid has mild anti-inflammatory properties and is used primarily as a short-term analgesic. Gastrointestinal disturbances, including possibly allergic diarrhea and potential renal toxicity, limit its use.

Trade Names: Lysalgo (SIT, Italy), Opustan (Opus Pharm, UK), Parkemed (Parke Davis, Germany), Ponstan (Werner-Lambert, Switzerland), Ponstel (Parke Davis, USA), Pontal (Sankyo, Japan).

Flufenamic acid [*530-78-9*], 2-{[3-(trifluoromethyl)phenyl]amino}benzoic acid, $C_{14}H_{10}F_3NO_2$, M_r 281.24, mp 125 °C.

Synthesis: refluxing a mixture of 2-iodobenzoic acid and 3-trifluoromethylaniline in aqueous potassium carbonate containing copper-bronze as the catalyst affords, after acidification, the desired flufenamic acid [51].

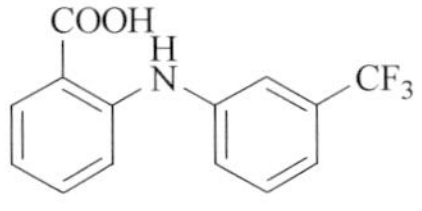

Flufenamic acid

Flufenamic acid is used at 600–800 mg/d to provide a beneficial therapeutic effect in chronic polyarthritis. The adverse effects most often encountered are gastrointestinal disturbances.
Trade Names: Ansatin (Ono, Japan), Arlef (Sankyo, Japan), Felunamin (Hokuriko, Japan), Romafen (Biofarma, Turkey).

Meclofenamic Acid [644-62-2], 2-[(2,6-dichloro-3-methylphenyl)amino]benzoic acid, $C_{14}H_{11}Cl_2NO_2$, M_r 296.15, *mp* 257–259 °C.
Synthesis: Heating a mixture of potassium 2-bromobenzoate, 2,6-dichloro-3-methylaniline, morpholine, diglyme (diethylene glycol dimethyl ether), and cupric bromide at 145–155 °C and subsequent acidification afford meclofenamic acid [52].

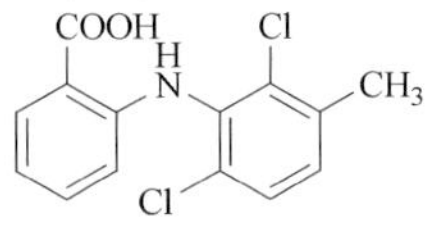

Meclofenamic acid

Meclofenamic acid is available as its sodium salt hydrate. This drug is indicated for the relief of acute and chronic rheumatoid arthritis and osteoarthritis. Gastrointestinal distress, including diarrhea, nausea, vomiting, and abdominal pain, is the most commonly reported adverse effect.
Trade Names: Meclodol (Parke Davis, Italy), Meclomen (Parke Davis, Portugal, USA), Stadium (Menarini, Belgium).

Niflumic acid [4394-00-7], 2-{[3-(trifluoromethyl)phenyl]amino}-3-pyridinecarboxylicacid, $C_{13}H_9F_3N_2O_2$, M_r 282.23, *mp* 204 °C.
Synthesis: Heating a mixture of 2-chloronicotinic acid, 3-trifluoromethylaniline, and a catalytic amount of potassium iodide at 140 °C affords the desired niflumic acid [53].

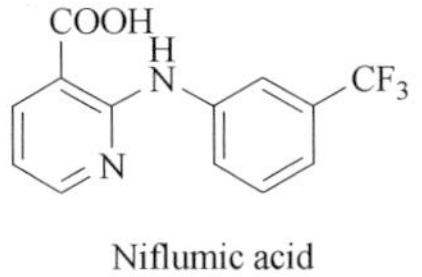

Niflumic acid

Niflumic acid has been used effectively to treat rheumatoid arthritis, psoriatic arthritis, and hypertrophic osteoarthritis of the hip and knee. Gastric complications are the chief adverse effects of this drug.

Trade Names: Actol (Mayrhofer, Austria; Upsamedica, Spain), Donalgin (Gedeon Richter, Hungary), Livornex (Genepharm, Greece), Niflam (Upsamedica, Italy), Nifluril (UPSA, France; Upsamedica, Belgium, Switzerland, Portugal).

3.7. Nonacidic Drugs

Proquazone [*22760-18-5*], 7-methyl-1-(1-methylethyl)-4-phenyl-2(1*H*)-quinazolinone, $C_{18}H_{18}N_2O$, M_r 278.35, *mp* 137–138 °C.

Synthesis: Proquazone is prepared by the alkylation of 2-amino-4-methylbenzophenone with 2-iodopropane in the presence of sodium carbonate to give 4-methyl-2-[(1-methylethyl)amino]benzophenone, which is condensed with urethane in the presence of zinc chloride at 190 °C [54].

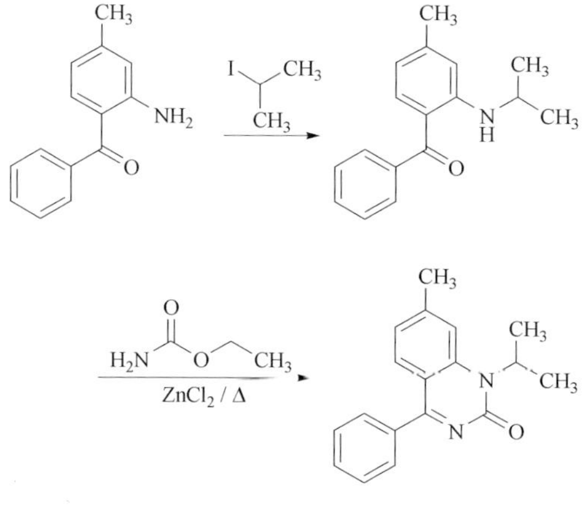

Proquazone

Proquazone has been used successfully to treat rheumatoid polyarthritis. At 900 mg/d it is comparable in effect to 150 mg/d of indomethacin. Gastrointestinal disturbances are the principal adverse effects observed with its use.

Trade Names: Biarison (Swiss-Pharma, Egypt; Biochemie, Austria; ICN Alkaloida, Hungary; Sandoz, Finland, Turkey).

Ditazole [*18471-20-0*], 2,2′-[(4,5-diphenyl-2- oxazolyl)imino]diethanol, $C_{19}H_{20}N_2O_3$, M_r 324.38, *mp* 96–98 °C (monohydrate).

Synthesis: Ditazole is prepared by heating 2-chloro-4,5-diphenyloxazole and diethanolamine in ethanol [55].

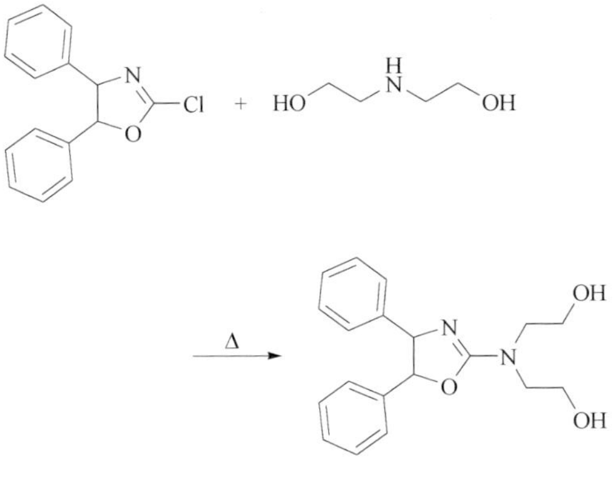

Ditazole is an anti-inflammatory analgesic agent that is of interest primarily for its ability to inhibit platelet aggregation. For example, it normalizes the enhanced platelet activity in cerebrovascular patients suffering transient ischemic attacks.
Trade Name: Ageroplas (Lepori, Spain; Serono, Italy), Fendazol (Lepori, Portugal).

3.8. Selective COX-2 Inhibitors

Celecoxib [*169590-42-5*], 4-[5-(4-methylphenyl)-3-(trifluoromethylpyrazol-1-yl] benzenesulfonamide, $C_{17}H_{14}F_3N_3O_2S$, M_r 381.37, *mp* 157–159 °C.

Synthesis: Celecoxib is prepared by condensation of 4-methylacetophenone with ethyl trifluoroacetate to give 4,4,4-trifluoro-1-(4-methylphenyl)butane-1,3-dione, which is cyclized with 4-hydrazinophenylsulfonamide [56]:

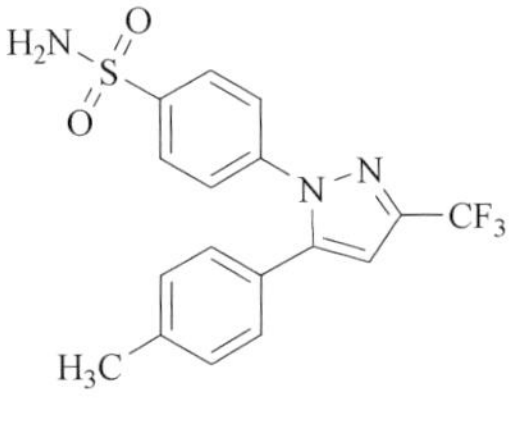

Celecoxib (SC-58635) is a selective inhibitor of the COX-2 isozyme. Based on human recombinant enzyme assays celecoxib was shown to be 375-fold more selective for COX-2 (IC_{50} for COX-2: 4×10^{-8} M vs. IC_{50} for COX-1: 1.5×10^{-5} M). Celecoxib displays a plasma half-life of 10–12 h. Celecoxib is indicated for relief of the signs and symptoms of osteoarthritis (recommended oral dose is 200 mg per day administered as a single dose or as 100 mg twice per day) and rheumatoid arthritis in adults (recommended oral dose is 100 to 200 mg twice per day). In comparison to traditional NSAIDs, celecoxib

causes fewer gastrointestinal side effects. Celecoxib has been associated with adverse effects such as headache, change in bowel habits, abdominal discomfort, and dizziness.
Trade Name: Celebrex (G. D. Searle & Co., USA).

Rofecoxib [*162011-90-7*], 4-[4-(methylsulfonyl)phenyl]-3-phenylfuran-2(5*H*)-one, $C_{17}H_{14}O_4S$, M_r 314.36, *mp* 208 °C.

Synthesis: Rofecoxib can be obtained by different synthetic routes [57], e.g., by condensation of phenylacetic acid with ethyl bromoacetate to ethyl 2-phenylacetoxyacetate, which is then cyclized to a hydroxyfuranone. Subsequently, the hydroxyfuranone reacts with trifluoromethanesulfonic (triflic) anhydride to the corresponding triflate which reacts with LiBr to yield a bromofuranone. The bromofuranone is condensed with 4-(methylsulfanyl)phenylboronic acid to give 4-[4-(methylsulfanyl)phenyl]-3-phenylfuran-2(5*H*)-one which is finally oxidized to rofecoxib.

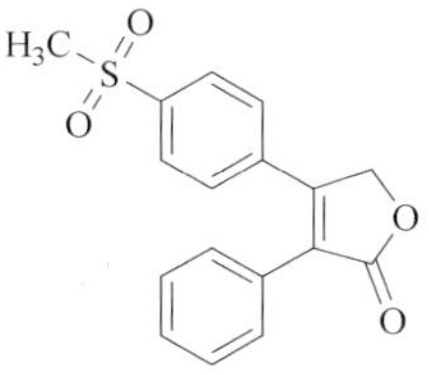

Rofecoxib

The selective COX-2 inhibitor rofecoxib is a methylsulphonylphenyl derivative. In comparison to celecoxib, rofecoxib is slightly more potent and has a longer half life of ca. 17 h. Based on human recombinant enzyme assays rofecoxib was shown to be > 800-fold more selective for COX-2 (IC_{50} for COX-2: 1.8×10^{-8} M vs. IC_{50} for COX-1: 1.5×10^{-5} M). Rofecoxib is indicated for relief of the signs and symptoms of osteoarthritis (recommended starting dose is 12.5 mg once daily, maximum recommended dose is 25 mg/d), for the management of acute pain in adults and for the treatment of primary dysmenorrhea (recommended initial doses are 50 mg once daily, use of rofecoxib for more than 5 d in management of pain has not been studied). Rofecoxib causes a significantly lower incidence of upper-gastrointestinal adverse effects (perforations, ulcers, and bleeding) than conventional NSAIDs. Most common adverse events associated with rofecoxib are diarrhoea, headache, nausea, and upper respiratory tract infection.
Trade Name: Vioxx (Merck & Co., USA).

4. Antirheumatic Agents

This class of agents encompasses a wide variety of structural and pharmacologic types, including gold, antimalarials, sulfhydryl compounds, and immunoregulants (both immunosuppressants and immunostimulants). These drugs are characterized in many instances by a slow onset of activity and, in contrast to the anti-inflammatory agents discussed heretofore, do have a beneficial effect on the underlying disease process. The acronyms SAARDs (slow acting antirheumatic drugs) and DMARDs (disease modifying antirheumatic drugs) have been applied to these substances.

4.1. Gold

Until the advent of the orally administered auranofin, gold preparations used in the treatment of rheumatoid arthritis were administered parenterally. Ionic gold in its monovalent state is preferred. Gold(I) has a strong affinity for sulfur relative to other elements and the antirheumatic preparations used contain gold bound to an organo-sulfur carrier. Indeed, the ability of gold to bind to physiologic sulfhydryl groups is thought to be an important factor in its bioactivity. Nevertheless, the actual mechanism of action remains to be determined.

The following activities of gold related to inflammation have been observed:

1) protection of collagen against collagenase
2) inhibition of lysosomal enzymes
3) prevention of denaturation of macro-globulins and formation of antigen–antibody complexes
4) uncoupling of oxidative phosphorylation
5) inhibition of leukocyte chemotaxis

Chrysotherapy (treatment with gold salts) is successful in approximately 80% of patients. Profound modification and remission of the disease, including retardation of the abnormal growth of synovial tissue over the joint surface, are observed. Unfortunately, a high incidence of adverse effects (25–50%) counterbalances the beneficial effects. Cutaneous hypersensitivity and lesions of mucous membranes are most common. Blood disorders including thrombocytopenia, leukopenia, agranulocytosis, and aplastic anemia occur. Renal effects, as evidenced by proteinuria, are observed.

Aurothioglucose [*12192-57-3*], (1-thio-D-glucopyranosato)gold, $C_6H_{11}AuO_5S$, M_r 392.22.

Synthesis: Gold thioglucose is prepared by adding a solution of gold bromide to an aqueous solution of thioglucose that contains sulfur dioxide. After heating, the product is precipitated by the addition of ethanol [58].

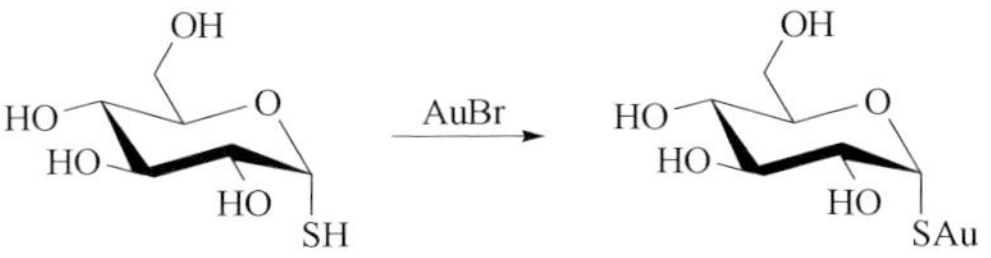

[Structure: Aurothioglucose synthesis from thioglucose + AuBr]

Aurothioglucose

Aurothioglucose is an antirheumatic used to treat active progressing rheumatoid arthritis and nondisseminated lupus erythematosus. The drug is administered at weekly intervals by intramuscular injection (10 mg, 25 mg, then 50 mg) until 800 mg to 1 g has been given. If improvement takes place, the drug is then administered at levels that balance the urinary excretion of gold. During this maintenance therapy the interval between injections is lengthened to 3–4 weeks.

Trade Names: Aureotan (BYK Gulden, Germany), Solganal (Schering, USA, Yugoslavia).

Sodium aurothiomalate [*12244-57-4*], mercaptobutanedioic acid monogold(I) sodium salt, $C_4H_4AuNaO_4S/C_4H_3AuNa_2O_4S$.

Synthesis: A solution of thiomalic acid and 3 equivalents of sodium hydroxide are mixed with an aqueous suspension of gold(I) iodide. The product, a mixture of the mono- and disodium salts, is precipitated by the addition of ethanol [59].

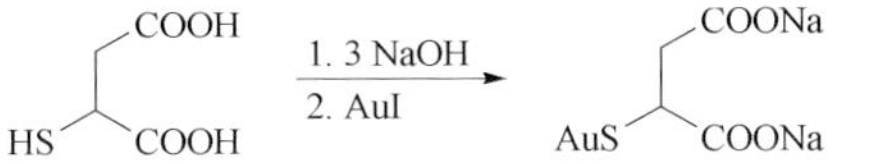

[Structure: Sodium aurothiomalate synthesis]

Sodium aurothiomalate

Sodium aurothiomalate is administered in the same manner and for the same indications as aurothioglucose.

Trade Names: Aurolate (Pasadena, USA), Myocrisin (Rhône-Poulenc Rorer, Denmark, Sweden, Finland), Taured0n (BYK Gulden, Germany).

Auranofin [*34031-32-8*], (2,3,4,6-tetra-*O*-acetyl-1-thio-*β*-D-glucopyranosato-*S*)(triethylphosphine)gold, $C_{20}H_{34}AuO_9PS$, M_r 678.49, *mp* 110–111 °C.

Synthesis: Ethanolic thiodiglycol is treated first with aqueous gold(I) acid chloride trihydrate, then with ethanolic triethylphosphine to give triethylphosphine gold(I) chloride, which is added to an aqueous solution of *S*-(2,3,4,6-tetra-*O*-acetylglucopyranosyl)pseudothiourea hydrobromide and potassium carbonate to give the desired auranofin [60].

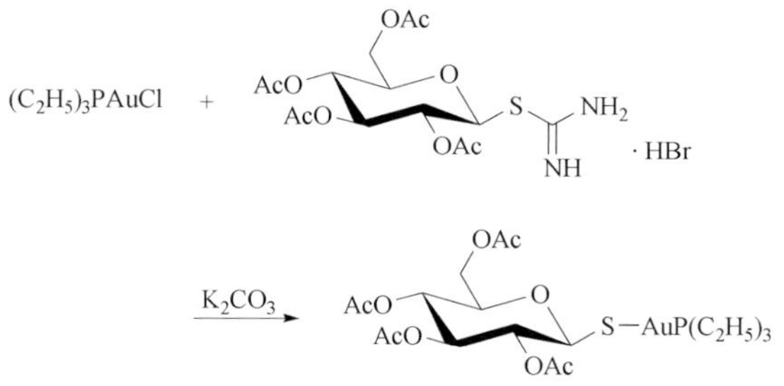

$(C_2H_5)_3PAuCl$ + [tetraacetyl thioglucose pseudourea] · HBr

$\xrightarrow{K_2CO_3}$ Auranofin [tetraacetyl-1-thio-β-D-glucopyranosato-S-(triethylphosphine)gold]

Auranofin was the first orally active gold preparation available for treatment of rheumatoid arthritis. Although it is incompletely absorbed, administration of 3 mg twice a day has resulted in profound improvement in patients, approaching the efficacy of parenteral gold, after a period of 8–12 weeks. The principal adverse effects with auranofin are diarrhea and loose stools. Mucocutaneous effects, such as rash and pruritis, also were seen, as were abnormalities in taste; but in general, it seems less toxic than parenteral gold.

Trade Names: Aktil (Lek, Yugoslavia), Ridaura (Smith Kline Beecham, USA), Ridauran (Robapharm, France).

4.2. Antimalarials

Several antimalarial agents — chloroquine, hydroxychloroquine, and dapsone — possess clinically useful anti-inflammatory properties. Hydroxychloroquine is preferred over chloroquine for the treatment of rheumatoid, but not other types of arthritis. When used clinically at a low dose of 4–6 mg/kg, safety, moderate effectiveness, and a low patient dropout rate can be achieved. Retinopathy can be minimized by careful ophthalmologic monitoring. The clinical effect is delayed and good-to-moderate suppression of the disease is observed after 6–9 months of therapy. Hydroxychloroquine also is effective in treating systemic lupus erythematosus.

Dapsone is clinically effective in treating rheumatoid arthritis. Decreases in both erythrocyte sedimentation rate and plasma levels of *C*-reactive protein have been observed. Unfortunately, its propensity to induce blood disorders and its carcinogenic potential preclude its widespread use.

The mechanism of the antirheumatic action of antimalarials is still under investigation. Inhibition of various cellular functions of lymphocytes and macrophages has been observed.

For structure, properties, and preparation → Synthetic Chemotherapeutic Agents.

4.3. Sulfhydryl Compounds

D-Penicillamine [52-67-5], 3-mercapto-D-valine, $C_5H_{11}NO_2S$, M_r 149.21, *mp* 198.5 °C.
Synthesis: D-Penicillamine can be synthesized in a multistep process that begins with heating isobutyraldehyde, pyridine, sulfur, and ammonia in benzene to form 5,5-dimethyl-2-isopropyl-Δ^3-thiazoline. Treatment with hydrogen cyanide gives 4-cyano-5,5-dimethyl-2-isopropylthiazolidine, which on acid hydrolysis gives D,L-penicillamine hydrochloride [61]. Resolution is accomplished by conversion of the racemate to D,L-3-formyl-2,2,5,5-tetramethylthiazolidine-4-carboxylic acid by treatment first with acetone, then with acetic formic anhydride. The enantiomers are separated in the usual manner, using, for example, L-lysine [62] or D-(−)-threo-1-(4-nitrophenyl)-2-aminopropane-1,3-diol [63]. Acidification liberates D-3-formyl-2,2,5,5-tetramethylthiazolidine-4-carboxylic acid, which is hydrolyzed with hydrochloric acid to yield D-penicillamine hydrochloride. Neutralization with ethanolic triethylamine affords D-penicillamine.

Improvements have also been made in the preparation of D-penicillamine from penicillins by conversion to the penicilloic acid, which when treated with concentrated hydrochloric acid and methanolic mercuric chloride gives D,L-penicillamine as its mercuric thiolate. Treatment with hydrogen sulfide in hydrochloric acid gives D,L-penicillamine hydrochloride [64], which is resolved as described previously.

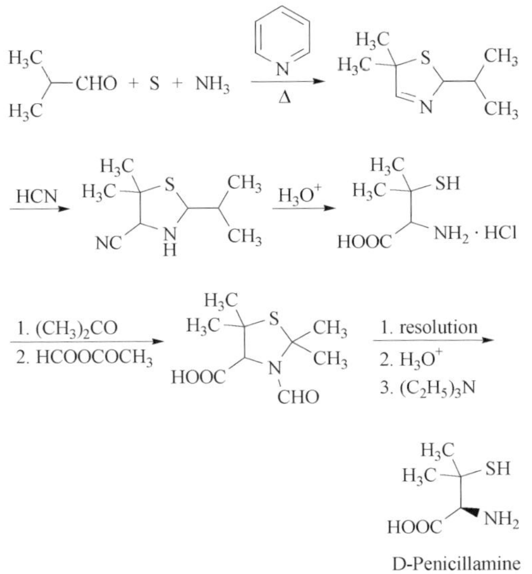

D-Penicillamine is as effective as gold in bringing about remission in severe rheumatoid arthritis. As with gold it is administered in gradually increasing amounts over a long period of time (4 – 6 months) before beneficial effects are seen. Its mechanism of action is unknown. Speculations concerning the basis for its effectiveness include: copper chelation and mobilization as well as effects on collagen biosynthesis, lympho-

cyte and monocyte function, antibody response, and cellular immunity. The profound therapeutic benefits of D-penicillamine are offset by a high incidence of adverse reactions. Skin rashes similar to those seen during chrysotherapy constitute the most common adverse effect. Dyspepsia and hypogusia also are encountered. The most serious problem is bone marrow suppression, evidenced by thrombocytopenia and agranulocytosis. Renal toxicity with resultant albuminuria also is observed frequently.
Trade Names: Artamin (Sanabo, Austria), Atamir (Sandoz, Denmark), Dimetylcystein (Lilly, Denmark), Metalcaptase (Knoll, South Africa), Reumacillin (Leiras, Finland), Trisorcin (Merckle, Germany), Trolovol (Asta Medica, Germany).

Pyritinol, pyrithioxin [*1098-97-1*], 3,3′-(dithiodimethylene)bis(5-hydroxy-6-methyl-4-pyridinemethanol), $C_{16}H_{20}N_2O_4S_2$, M_r 368.48, mp 218–220 °C.
Synthesis: Treatment of pyridoxine with hydrobromic acid gives 4,5-bis(bromomethyl)-3-hydroxy-2-methylpyridinium bromide, which when treated in the cold with potassium ethyl xanthate gives ethyl 4-hydroxymethyl-3-hydroxy-2-methyl-3-pyridyl methylxanthate. Hydrolysis and oxidation are carried out in aqueous-alcoholic ammonia to give the disulfide pyritinol [65].

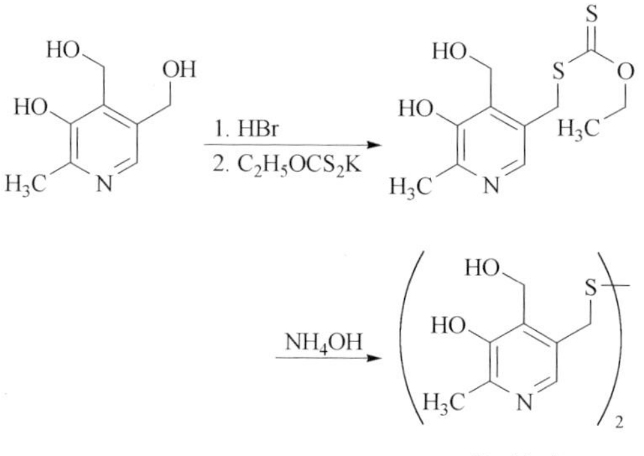

Pyritinol, the disulfide of pyridoxine-5-thiol, is marketed as a cerebral stimulant in Europe. Both the thiol and the disulfide have been shown to have D-penicillamine-like activity in the treatment of rheumatoid arthritis. Adverse effects are similar to those seen with D-penicillamine. Occasionally, pyritinol seems effective and better tolerated in some individuals who fail to respond to D-penicillamine.
Trade Names: Ansefal (Anka, Turkey), Bonifen (Merck, Portugal), Enbol (Chugai, Japan), Encefabol (Merck, Brazil, Greece, Czech Republic), Encephabol (Merck, Austria, CIS, South Africa), Neuroxin (Yamanouchi, Japan).

4.4. Immunoregulants

The idea that autoimmune mechanisms might be involved in the pathogenesis of rheumatoid arthritis led to the use of immunosuppressive therapy as far back as the 1950s. Only recently, however, has its use in severe forms of the disease been studied carefully.

The folic acid analogue *methotrexate* is currently a very promising drug for the treatment of rheumatoid arthritis. At doses used for the treatment of rheumatic diseases its principal mode of action probably relates to an inhibition of aminoimidazolecarboxamide ribonucleotide transformylase and thymidylate synthase, plus enhanced adenosine release.

The alkylating agent *cyclophosphamide* is sometimes used in rheumatoid arthritis. A more frequent use is precluded by its considerable toxicity. Cyclophosphamide nonspecifically kills cells by reacting with DNA and RNA molecules.

The immunosuppressive drug *azathioprine* is FDA-approved for treatment of rheumatoid arthritis. Azathioprine is believed to interfere with cell division by inhibiting metabolism and synthesis of proteins and DNA.

Novel therapeutic approaches for treatment of autoimmune rheumatic diseases include *cyclosporin A* and *leflunomide*. Cyclosporine A is an immunosuppressive drug used to prevent organ rejection which has been shown to provide benefit in rheumatoid arthritis. Most common adverse effects of this drug are renal damage, hypertension, and increased body hair. Leflunomide is an isoxazol derivative that has no structural relationship to other known immunoregulatory drugs [66], [67]. In vivo leflunomide is rapidly metabolized to its active form, a malononitrilamine termed A771726. The substance may serve as an immunomodulatory agent in rheumatoid arthritis by preferentially causing cell arrest of autoimmune lymphocytes through inhibition of the enzyme dihydroorotate dehydrogenase, which catalyzes a critical step in the production of uridine monophosphate (rUMP). A decrease in rUMP leads to cell cycle arrest in lymphocytes. Furthermore, leflunomide inhibits tyrosine kinase activation in T cells after stimulation of the IL-2 receptor. Tyrosine kinase inhibition has been proposed to partially explain the immunosuppressive action of leflumide.

5. Inhibitors of Tumor Necrosis Factor

Recent clinical trials suggest that *monoclonal antibodies* directed against TNF provide a novel approach for the treatment of rheumatoid arthritis [68], [69]. The proinflammatory cytokine TNF has been implicated in the pathogenesis of rheumatoid arthritis. TNF concentrations are increased in the synovial fluid of persons with active rheumatoid arthritis, and increased TNF plasma levels have been shown to be associated with joint pain. TNF increases synoviocyte proliferation, and triggers the release of secondary mediators involved in the recruitment of inflammatory cells during neoangiogenesis and in the process of joint destruction. Infliximab is a chimeric TNF monoclonal antibody (CA2) which was the first monoclonal antibody to be investigated for the treatment of rheumatoid arthritis. Etanercept is a recombinant version of the soluble p75 TNF receptor linked to the Fc portion of human immunoglobulin G1. Etanercept competitively inhibits the binding of TNF to its cell surface receptors, thereby inhibiting the biological activity of TNF in a reversible manner.

6. Natural Products

Colchicine [*477-27-0*], (*S*)-*N*-(5,6,7,9-tetrahydro-1,2,3,10-tetramethoxy-9-oxobenzo[*a*]-heptalen-7-yl)acetamide, $C_{22}H_{25}NO_6$, M_r 399.43, *mp* 142–150 °C.

Production: Colchicine is obtained by extracting the seeds and/or corm of *Colchicum autumnale* Linné [70].

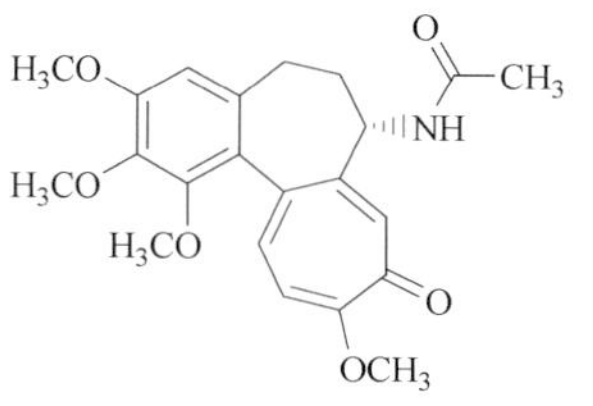

Colchicine

Colchicine is a specific drug for the treatment of gouty arthritis. It is especially useful in acute attacks but also is used prophylactically in conjunction with other drugs, such as allopurinol or phenylbutazone. The drug is believed to act by disrupting microtubular assembly in inflammatory cells, thereby impairing both their locomotion and their membrane function, including phagocytosis and the release of inflammatory mediators. The alkaloid is extremely toxic and must be used with care. Gastrointestinal adverse effects include nausea, vomiting, and diarrhea. Blood disorders, including agranulocytosis and aplastic anemia, sometimes are seen after prolonged administration of this drug.

Trade Names: Colchicina (Smith Kline, Brazil), Colchicine (Various, USA), Colchicum-Dispert (Solvay, Poland), Colabid (Major, USA).

Orgotein [*9016-01-7*], M_r 34000, is the name applied to a group of metalloproteins obtained from various tissue sources, such as hemolyzed plasma-free bovine erythrocytes [71] or beef liver [72]. They are copper–zinc/protein chelates with molecular masses of about 34 000. These compounds have potent superoxide dismutase activity. When administered directly into the joint at 4 mg/week for 6 weeks, marked improvement in the mobility of acute rheumatoid patients occurred. Synovial fluid levels of rheumatoid factor and prostaglandin E_2 were reduced.

Trade Names: Ontosein (Tedec-Meiji, Spain), Peroxinorm (Grünenthal, Austria), Serosod (Serono, Italy).

7. References

[1] H. R. Herschman, "Prostaglandin synthase 2," *Biochim. Biophys. Acta* **1299** (1996) 125–140.

[2] P. J. Barnes, I. Adcock, "Anti-inflammatory actions of steroids: molecular mechanisms," *Trends Pharmacol. Sci.* **14** (1993) 436–441.

[3] P. J. Barnes, M. Karin, "Nuclear factor-KB: a pivotal transcription factor in chronic inflammatory diseases," *N. Engl. J. Med.* **336** (1997) 1066–1071.

[4] W. L. Smith, L. J. Marnett, "Prostaglandin endoperoxide synthase: structure and catalysis," *Biochim. Biophys. Acta* **1083** (1991) 1–17.

[5] J. A. Mitchell et al., "Selectivity of nonsteroidal antiinflammatory drugs as inhibitors of constitutive and inducible cyclooxygenase," *Proc. Natl. Acad. Sci. USA* **90** (1994) 11693–11697.

[6] P. E. Lipsky, "Specific COX-2 inhibitors in arthritis, oncology, and beyond: where is the science headed?," *J. Rheumatol.* **26** (1999) Suppl. 56, 25–30.

[7] C. Luong et al., "Flexibility of the NSAID binding site in the structure of human cyclooxygenase-2," *Nat. Struct. Biol.* **3** (1996) 927–933.

[8] N. E. Lane, "Pain management in osteoarthritis: the role of COX-2 inhibitors," *J. Rheumatol.* **24** (1997) Suppl. 49, 20–24.

[9] J. Sirois, M. Dore, "The late induction of prostaglandin G/H synthase-2 in equine preovulatory follicles supports its role as a determinant of the ovulatory process," *Endocrinology* **138** (1997) 4427–4434.

[10] R. Sawdy et al., "Use of a cyclooxygenase type-2-selective nonsteroidal anti-inflammatory agent to prevent preterm delivery," *Lancet* **350** (1997) 265–266.

[11] H. Lim et al., "Multiple female reproductive failures in cyclooxygenase 2-deficient mice," *Cell* **91** (1997) 197–208.

[12] A. Schmassmann et al., "Effects of inhibition of prostaglandin endoperoxide synthase-2 in chronic gastro-intestinal ulcer models in rats," *Br. J. Pharmacol.* **123** (1998) 795–804.

[13] G. Weissmann, "Prostaglandins as modulators rather than mediators of inflammation," *J. Lipid Mediat.* **6** (1993) 275–286.

[14] J. R. Vane, "Inhibition of prostaglandin synthesis as a mechanism of action for aspirin-like drugs," *Nat. New Biol.* **231** (1971) 232–235.

[15] S. J. Preston et al., "Comparative analgesic and anti-inflammatory properties of sodium salicylate and acetylsalicylic acid (aspirin) in rheumatoid arthritis," *Br. J. Clin. Pharmacol.* **27** (1989) 607–611.

[16] K. K. Wu et al., "Aspirin inhibits interleukin 1-induced prostaglandin H synthase expression in cultured endothelial cells," *Proc. Natl. Acad. Sci. USA* **88** (1991) 2384–2387.

[17] J. A. Mitchell et al., "Sodium salicylate inhibits cyclo-oxygenase-2 activity independently of transcription factor (nuclear factor KB) activation: role of arachidonic acid," *Mol. Pharmacol.* **51** (1997) 907–912.

[18] B. J. Whittle et al., "Selective inhibition of prostaglandin production in inflammatory exudates and gastric mucosa," *Nature* **284** (1980) 271–273.

[19] G. A. Higgs, J. A. Salmon, B. Henderson, J. R. Vane, "Pharmacokinetics of aspirin and salicylate in relation to inhibition of arachidonate cyclooxygenase and antiinflammatory activity," *Proc. Natl. Acad. Sci. USA* **84** (1987) 1417–1420.

[20] E. Kopp, S. Ghosh, "Inhibition of NF-KB by sodium salicylate and aspirin," *Science* **265** (1994) 956–959.

[21] D. A. Kujubu, H. R. Herschman, "Dexamethasone inhibits mitogen induction of the TIS10 prostaglandin synthase/cyclooxygenase gene," *J. Biol. Chem.* **267** (1992) 7991–7994.

[22] M. G. O'Sullivan, E. M. Huggins Jr, C. E. McCall, "Lipopolysaccharide-induced expression of prostaglandin H synthase-2 in alveolar macrophages is inhibited by dexamethasone but not by aspirin," *Biochem. Biophys. Res. Commun.* **191** (1993) 1294–1300.

[23] M. Barrios-Rodiles, K. Keller, A. Belley, K. Chadee, "Nonsteroidal antiinflammatory drugs inhibit cyclooxygenase-2 enzyme activity but not mRNA expression in human macrophages," *Biochem. Biophys. Res. Commun.* **225** (1996) 896–900.

[24] Merck & Co., US 3527796, 1970 (M. Sletzinger, J. Chemerda, G. Gal).

[25] Tropenwerke Dinklage & Co., US 3910952, 1975 (K. Boltze, O. Brendler, H. Dell, H. Jacobis).

[26] SIR Laboratorie Chimico Biologici SpA, DE 2223051, 1973 (D. Antonia, F. Ganzina, M. Magi, E. Serino et al.).

[27] Instituto Biologico Chemiotherapico "ABC", US 3624103, 1971 (F. DeMartiis, E. Arrigoni-Martelli, T. Tamietto).

[28] Rotta Research Laboratorium SpA, US 3985878, 1976 (F. Makorec, P. Sevin, L. Rovati).

[29] Merck & Co., US 3870753, 1975 (R. J. Tull, R. F. Czaja, R. F. Shuman, S. H. Pines).

[30] Merck & Co., US 3994600, 1976 (R. J. Tull, R. F. Czaja, R. F. Shuman, S. H. Pines).

[31] McNeil Laboratories Inc., FR 1574570, 1969 (J. R. Carson).

[32] Rhône-Poulenc SA, US 34244748, 1969 (D. Farge, C. Jeanmart, M. N. Messer).

[33] Madan AG, BE 704368, 1968 (N. P. Buu-Hoi, C. Gillet, G. Lambelin).

[34] Geigy Chemical Corp., US 3558690, 1971 (A. Sallmann, R. Pfister).

[35] P. D. Fowler: "Diclofenac sodium," in E. C. Huskisson (ed.): *Anti-rheumatic Drugs*, Praeger Publishers, New York, 1983, p. 117.

[36] John Wyeth & Brother Ltd., US 3476766, 1969 (K. Brown).

[37] American Cyanamid Co., DE 2147111, 1972 (A. S. Tomcufcik, R. G. Child, A. E. Sloboda).

[38] Boots Pure Drug Co., Ltd.,US 3385886,1968 (J. S. Nicholson, S. S. Adams).

[39] Syntex Corp., US 3637767, 1972 (F. S. Alvarez).

[40] Rhône-Poulenc SA, US 3641127, 1972 (D. Farge, M. N. Messer, C. Moutonnier).

[41] Eli Lilly & Co., US 3600437, 1971 (W. S. Marshall).

[42] The Boots Co. Ltd., US 3755427, 1973 (S. S. Adams, J. Bernard, J. S. Nicholson, A. R. Blancafort).

[43] Bristol-Myers Co., US 3565943, 1971 (P. F. Juby, R. A. Partyka, T. W. Hudyma).

[44] Ciba Corp., US 3641040, 1972 (R. W. J. Carney, G. deStevens).

[45] J. R. Geigy AG, US 2562830, 1951 (H. Stenzl).

[46] J. R. Geigy AG, US 2745783, 1956 (F. Hafliger).

[47] Siegfried AG Zofingen, US 3482024, 1969 (L. Molnar, T. Wagner-Jauregfg, U. Jahn, G. Mixich).

[48] E. H. Wiseman, J. G. Lombardino in J. S. Bendra, D. Lednicer (ed.): *Chronicles of Drug Discovery*, vol. **1**, J. Wiley & Sons, New York 1982, Chapter 8.

[49] Pfizer Inc., US 4289879, 1981 (J. G. Lombardino).

[50] Parke, Davis & Co., BE 605302, 1961 (R. A. Scherner).

[51] J. H. Wilkinson, I. L. Finar, *J. Chem. Soc.* 1948 32–35.
[52] Parke, Davis & Co., US 3313848, 1967 (R. A. Scherrer, F. W. Short).
[53] Laboratories UPSA, US 3415834, 1968 (C. Hoffman, A. Faure).
[54] Sandoz, Inc., US 3925548, 1975 (H. Oh).
[55] Instituto Farmacologico Serono SpA, US 3557135, 1971 (E. Marchetti).
[56] *Drugs of the Future* **22** (1997) 711–714.
[57] *Drugs of the Future* **23** (1998) 1287–1296.
[58] P. Lebeau (ed.): *Traite de Pharmacie Chimique II,* 4th ed., vol. **2,** Masson et Cie, Paris 1956, p. 661.
[59] Societe des Usines Chimiques Rhône-Poulenc, US 1994213, 1935 (M. Delepine).
[60] SKF, US 3635945, 1972 (P. E. Nemeth, B. M. Sutton).
[61] Degussa Normals Roessler, US 4060548, 1977 (F. Ansinger, H. Offermanns, M. Ghyczy).
[62] Degussa Normals Roessler, US 3980665, 1976 (R. Fahnenstick, J. Heese, H. Offermanns).
[63] Degussa Normals Roessler, US 3980666, 1976 (P. Scherberick).
[64] Firma Heyl & Co. Chemisch-Pharmazeutische Fabrik, US 3894067, 1975 (M. Bock).
[65] E. Merck AG, US 3010966, 1961 (O. Zima, G. Schorre).
[66] R. I. Fox, "Mechanism of action of leflunomide in rheumatoid arthritis," *J. Rheumatol.* **25** (1998) Suppl. 53, 20–26.
[67] B. Rozman, "Clinical experience with leflunomide in rheumatoid arthritis," *J. Rheumatol.* **25** (1998) Suppl. 53, 27–32.
[68] L. W. Moreland, "Inhibitors of tumor necrosis factor for rheumatoid arthritis," *J. Rheumatol.* **26** (1999) Suppl. 57, 7–15.
[69] E. C. Keystone, "The role of tumor necrosis factor antagonism in clinical practice," *J. Rheumatol.* **26** (1999) Suppl. 57, 22–28.
[70] F. E. Hamerslag, *The Technology and Chemistry of Alkaloids,* D. Van Nostrand Co., Inc., New York 1950, pp. 75–78.
[71] Diagnostic Data Inc., US 3579495, 1971 (W. Huber).
[72] Diagnostic Data Inc., US 3624251, 1971 (W. Huber).

Drugs Used in Dermatology

JOHN L. MCGUIRE, Johnson & Johnson, New Brunswick, New Jersey 08933, United States

GERARD J. GENDIMENICO, Johnson & Johnson Consumer Products Worldwide, Skillman, New Jersey 08558, United States

GEERT CAUWENBERGH, Johnson & Johnson Consumer Products Worldwide, Skillman, New Jersey 08558, United States

JAMES A. MEZICK, Johnson & Johnson Consumer Products Worldwide, Skillman, New Jersey 08558, United States

1. Introduction	1713	
2. Functions and Structure of the Skin	1714	
3. Topical Therapy	1715	
3.1. Topical Dosage Forms	1715	
3.2. Percutaneous Absorption	1716	
3.3. Vehicles	1717	
4. Drugs Used in Dermatology by Class	1718	
4.1. Anti-Inflammatory Agents	1718	
4.2. Antiacne agents	1724	
4.3. Antipsoriatics	1727	
4.4. Antifungals	1730	
4.5. Antiaging	1732	
5. References	1733	

1. Introduction

The skin is the largest organ in the body. Often viewed as an inert covering, in reality, skin is a dynamic, interactive and multifunctional network of cells and tissues. A spectrum of diseases affects the skin, which is reflection of this functional complexity. Many of these diseases result from abnormal function of endogenous physiologic processes. Environmental stresses to which skin is exposed can also act as causative factors in skin pathology.

Drugs used in dermatology encompass an assortment of pharmacological classes including anti-infectives, anti-inflammatories, and hormonal agents. Some of these therapies were placed in use many years before understanding of basic skin physiology. Newer insights of skin function have allowed a more rational approach to dermatology drug discovery. This chapter will focus on the major drugs currently used in dermatology, both older and newer agents, all of which play an important role in treating skin disease.

2. Functions and Structure of the Skin

The skin is a three-compartment organ that serves several essential roles for the human body. As the major protective barrier between the body and the environment, the skin minimizes the potentially harmful effects of ultraviolet radiation, temperature extremes, microbial infection, and mechanical, chemical, or electrical trauma.

The outer protective layer of the skin, the epidermis, is a thin, stratified squamous epithelium composed primarily of keratinocytes, melanocytes, and Langerhans cells. The epidermis (ca. 0.07 – 0.1 mm thick) lacks blood vessels. Keratinocytes, the predominant epidermal cell type, synthesize the insoluble fibrous protein keratin, responsible for the protective qualities of the epidermis. These cells originate in the innermost or basal layer of the epidermis and undergo a multistage differentiation process, termed keratinization. The terminally differentiated cells form the stratum corneum, a flattened, dehydrated, compact outer layer (0.01 mm thick). It plays a critical role in the barrier functions of the skin and has significant water-retaining properties.

Melanocytes originate from the neural crest and produce melanin, a pigment that largely determines normal skin color. Together with keratin, melanin protects the underlying basal layer against ultraviolet light damage. Langerhans cells are dendritic cells responsible for uptake and transfer of antigens from the epidermal surface to T-cell lymphocytes, thereby initiating delayed hypersensitivity reactions.

The *dermis* forms the 2 – 3-mm-thick middle layer of the skin. It consists of dense connective tissue composed of collagenous and elastic fibers dispersed in an amorphous ground substance composed primarily of glycosaminoglycans.

Collagen accounts for the bulk of the dermis and lends strength and flexibility to the skin. Elastic fibers are primarily responsible for maintaining the skin's elasticity. Fibroblasts, the most abundant cell type in the dermis, produce connective tissue components and are required for normal tissue repair.

Mast cells are the other major cell type in the dermis; they produce histamine and other chemical mediators of inflammation. The dermis is rich in smaller blood vessels, lymphatic vessels, and nerves. Sebaceous glands, apocrine and eccrine sweat glands, and hair follicles originate in the dermis or subcutaneous tissue and extend upward to the skin surface.

3. Topical Therapy

Topical therapy generally does not result in appreciable levels of drug in the systemic circulation; however, some therapeutic agents, e.g., nitroglycerin [55-63-0] and scopolamine [51-34-3], have been successfully delivered transdermally.

3.1. Topical Dosage Forms

Topical therapy allows for selective treatment of skin disease while the possibility of systemic adverse effects is minimized. Although systemic absorption can be either a deliberate or unintentional consequence of topical application, the action of topically applied drugs is most often limited to the site of application. Local therapeutic action is enhanced by formulations that promote prolonged contact with the target area while reducing the likelihood of absorption. Semisolid preparations, such as ointments, pastes, creams, and gels, fulfill these criteria to varying degrees. Semisolids are used exclusively for topical drug delivery. They differ from other topical dosage forms such as hydroalcoholic solutions, powders, and suspensions, wherein the latter can also be used for other routes of administration.

Semisolid ointment bases are used primarily as vehicles to prepare pharmaceutical ointments, creams, and pastes. Ointment bases vary in viscosity and solubility and are classified as oleaginous, emulsifiable, emulsified, and water-soluble.

Ointments are produced by dissolving or suspending a drug in a semisolid base. Although the term "ointment" is sometimes used specifically for drug-containing oleaginous bases, emulsion bases (creams), water-soluble bases, and pastes are also considered ointments. *Pastes,* which contain large quantities of insoluble solids such as zinc oxide or talc, are stiffer than other types of ointments and are used often as protective barriers or absorbents.

Hydrocarbon-based ointments are typically hydrophobic, greasy, and occlusive, and inhibit the evaporation of water from the skin. In addition to this moisturizing effect, oleaginous ointments have emollient and protective qualities. When applied to intact skin, ointments form a thin film that is maintained at the application site for an extended period, providing prolonged delivery of the active ingredient.

Creams are semisolid emulsions that are more fluid and more easily spread than ointments. Emulsion bases include w/o emulsions, which absorb small quantities of water, and water-washable oil-in-water (o/w) emulsions.

Like oleaginous and emulsifiable bases, w/o emulsions are greasy and difficult to remove from the skin. They are effective emollients, but somewhat less occlusive than oleaginous bases.

Water-washable o/w emulsions are commonly referred to simply as "creams." Because they are nonocclusive, nongreasy, and easily washed from the skin, these creams have relatively high esthetic appeal and patient acceptance.

Typically, creams have (1) an oleaginous internal phase composed of petrolatum-based hydrocarbons and high molecular mass alcohols or fatty acids, (2) an aqueous external phase often containing preservatives, humectants, and buffers, and (3) one or more emulsifying agents.

Aqueous gels or jellies represent another class of water-soluble ointment bases in which a liquid internal phase is immobilized within a three-dimensional matrix. This system is formed by colloidal dispersions of small inorganic or large organic molecules in aqueous media. Gels can also be formed by semisolid w/o or o/w microemulsions, which may provide better solubilization of drug after application to the skin.

Although semisolids are an invaluable class of topical delivery system, liquid and solid dosage forms are also widely used for topical therapy. Liquid preparations applied to the skin include lotions and solutions. *Lotions* are generally aqueous preparations that spread easily on the skin. They may contain suspended solids or emulsified liquids. Topical *solutions* often contain aqueous vehicles with cosolvents, e.g., alcohol, acetone, and poly(ethylene glycol), as required for drug solubilization. Propylene glycol can also be used as a solvent, particularly in topical solutions of corticosteroids.

Topical agents are also formulated into *powders*. For example, antifungal agents, such as miconazole nitrate [22832-87-7], are widely used in powder form as an adjunct to treatment with antifungal solutions, creams, and ointments.

3.2. Percutaneous Absorption

Topically applied drugs often cannot reach disease sites in the epidermis or dermis in sufficiently high concentrations to exert a therapeutic effect [1], [2]. The stratum corneum is the major barrier to skin penetration. The interaction of the stratum corneum with the applied drug and vehicle largely determines the rate and extent of percutaneous absorption.

The two major processes in percutaneous absorption are *intraphase diffusion* and *interphase partitioning* [3]. They occur consecutively and repeatedly as the drug passively leaves the vehicle and permeates the various layers of the skin. The higher the diffusion and partition coefficients of a drug, the greater is the drug's ability to penetrate the skin. Drugs move from the vehicle to the viable epidermis primarily through the stratum corneum, but may also gain entry through pilosebaceous follicles or sweat ducts.

The rate of penetration is also directly proportional to the rate of vehicle release of drug, concentration of applied drug, surface area, duration of application, and degree of stratum corneum hydration. Percutaneous absorption is inversely proportional to stratum corneum thickness and, therefore, may vary, depending on anatomical site.

Transfer of drug from the vehicle to the skin surface, the initial step in percutaneous absorption, is determined by drug diffusion through the vehicle and partitioning into surface skin tissue (e.g., stratum corneum or sebum).

Solubility is a crucial factor in release of drug from the vehicle. First, dissolution of the drug in the vehicle is a prerequisite for its diffusion through the vehicle to the skin surface. Second, the relative solubility of the drug in the vehicle and skin surface is a major determinant of the partition coefficient between the two phases.

The vehicle–skin surface partition coefficient is an important factor in percutaneous absorption [4]. Increasing the lipid solubility of a drug increases its penetration into the stratum corneum and often enhances percutaneous absorption.

After reaching the skin surface, the drug must diffuse across the stratum corneum (or through the follicle) before moving relatively easily through underlying tissue. Drug diffusion through the stratum corneum is generally slow and represents the rate-limiting step in percutaneous absorption. Large lipophilic drugs, such as the corticosteroids, may be preferentially absorbed by the transfollicular route because of their relatively high diffusion rates in the sebum compared to that in the stratum corneum.

Drug transport is driven by a concentration gradient that is maintained by rapid removal of drug by the systemic circulation via dermal capillaries. The systemic circulation is often referred to as a "sink" for the drug.

3.3. Vehicles

A drug's formulation may have a significant impact on its topical potency, given the critical role of vehicle release of drug and the stratum corneum–vehicle partition coefficient in percutaneous absorption. Vehicles may enhance drug penetration by influencing solubility and release or by altering stratum corneum resistance [5].

Oleaginous vehicles, including petrolatum-based ointments and, to a lesser extent, water-in-oil emulsions, form an occlusive barrier on the skin, preventing the evaporation of sweat and insensible water loss and thereby promoting hydration of the stratum corneum. Hydration softens the stratum corneum, expanding the protein matrix by a "sponge" effect and increasing drug penetration.

The important role of drug solubility, and, therefore, choice of solvent, in determining drug release, penetrability, and bioavailability was demonstrated by numerous studies of topical corticosteroids. The solubility of corticosteroids is greatly enhanced by adding cosolvents, such as propylene glycol or dimethylsulfoxide (DMSO), to aqueous vehicles. For example, the release of fluocinonide (0.025 wt %) from propylene glycol–water gels is greatest from vehicles containing exactly enough propylene glycol to completely dissolve the drug [6]. Excess propylene glycol reduces the bioavailability of fluocinonide by decreasing the partition coefficient of the drug between the skin and the vehicle. Similarly, maximal penetration cannot be attained from vehicles containing insufficient solubilizer.

Modifications in the vehicle after application to the skin can affect drug bioavailability. For example, adding a volatile solvent such as isopropanol to the vehicle substantially increases the penetration of corticosteroids from nonvolatile solvents, such as propylene glycol or isopropyl myristate [7]. Rapid evaporation of isopropanol after application to the skin increases the drug concentration in the nonvolatile component.

In addition to their effect on solubility, solvents may also influence the barrier properties of the skin. Propylene glycol and aprotic solvents, such as DMSO, dimethylacetamide (DMA), and dimethylformamide (DMF), interact with the stratum corneum to increase the permeability of the skin to various compounds, including glucocorticosteroids and antibiotics.

Although the mechanism of their effect is not clearly defined, studies suggest that DMSO and related compounds induce reversible changes in the otherwise impenetrable protein matrix of the stratum corneum [8].

4. Drugs Used in Dermatology by Class

Each major class of drugs is categorized and discussed in the following sections. A brief summary of the pathophysiology of each disease is given to provide a rationale for the pharmacological basis of treatment.

4.1. Anti-Inflammatory Agents

Inflammation is a complex host response to cellular injury associated with many cutaneous disorders. Cutaneous inflammation is often mediated by immunologic mechanisms, including antigen-dependent mast cell release of histamine, leukotrienes, chemotactic factors, and other vasoactive and smooth muscle contracting substances. Inflammation can also be produced by prostaglandins and kinins [9].

Eczematous dermatitis, for example, represents a group of inflammatory skin disorders that develop in response to various exogenous or endogenous stimuli. The types of eczema include seborrheic, atopic, stasis, infectious eczematoid, and allergic or irritant contact dermatitis.

Eczema is often mediated by immune mechanisms. *Allergic contact dermatitis* is a delayed hypersensitivity reaction to an external substance (e.g., nickel, rubber, or poison ivy) in previously sensitized individuals. *Photoallergic contact dermatitis,* in which photoactivation of the compound or hapten (e.g., halogenated salicylanilides) is followed by conjugation with skin proteins and results in the formation of a complete antigen, is associated with a cell-mediated immune response similar to that in contact allergy. In contrast, *nonallergic contact dermatitis* (irritation) may be elicited immediately after initial exposure to primary chemical or mechanical irritants.

Atopic dermatitis is a common form of chronic eczema that often develops in genetically predisposed children in conjunction with hay fever or asthma. The underlying cause is not known, but it may be related to alteration in cellular and humoral immunity mediated by cyclic nucleotides or to autonomic nervous system function [10].

Immediate-type hypersensitivity reactions may also lead to inflammation. *Urticaria* or hives, characterized by pruritic, erythematous areas of localized edema known as wheals, may result from IgE-dependent mast cell degranulation on exposure to certain foods, drugs (e.g., penicillin), or inhaled allergens, or to physical agents, such as cold, sunlight, or pressure.

Although the pathophysiologic characteristics of eczematous dermatoses may vary, the *clinical features* are similar. These include erythema, blisters, crusting, and pruritus in acute conditions and epidermal thickening, accentuated skin lines, and dryness in chronic stages. Thus, *symptomatic treatment* of eczema is relatively uniform and often includes systemic antihistamines for pruritus and topical or systemic anti-inflammatory drugs. Adjunctive therapy may include topical antipruritics, occlusive emollients, and baths. Many of the same treatment modalities are employed in urticaria.

The following four classes of drugs are frequently used in the symptomatic treatment of inflammatory disorders and, as indicated, may also benefit various other cutaneous disorders.

Antihistamines. Histamine [51-45-6] is a substituted imidazole formed by enzymatic decarboxylation of the amino acid histidine. It is primarily synthesized and stored in mast cells and basophils. One of several chemical inflammatory mediators of immediate hypersensitivity reactions, histamine is actively secreted from IgE antibody-bound mast cells in response to a specific antigen. Histamine release is triggered by influx of calcium ions, which initiates a series of intracellular changes mediated by membrane and cytosolic signal transducing pathways [11], [12].

Histamine has a wide range of *physiologic effects* mediated through three distinct G protein-coupled receptors, H-1, H-2 and H-3 [13]. The H-1 and H-2 receptor subtypes are found in skin. Both of these receptors mediate the vasodilatory effects of histamine. H-1 receptors participate in histamine-induced increases in capillary permeability and edema. These responses are manifested in the skin as localized erythema and whealing (urticaria).

Histamine-induced stimulation of sensory nerve endings, predominantly an H-1 effect, completes the so-called "triple response" to histamine. This action causes axon-reflex vasodilation characterized by an erythematous flare. Nerve stimulation also accounts for histamine-induced pruritus and pain.

Widespread release of histamine in the body can cause extensive vasodilation and a marked fall in blood pressure. This response, combined with vascular fluid leakage, can lead to a condition resembling surgical or traumatic shock.

Extravascular smooth muscle contraction is H-1-mediated and leads to bronchoconstriction. In some cases, activating H-2 receptors will relax smooth muscle. In the heart, almost all of the effects of histamine on cardiac muscle, such as increases in heart rate and force of heart contraction, are through H-2 receptors. Gastric secretion is stimulated by histamine at H-2 receptors. The existence of these receptors was verified from development of specific H-2 receptor antagonists that were designed to attenuate gastric acid production [14].

Systemic H-1 antihistamines competitively and reversively bind to H-1 receptors. These belong to six chemical classes and all share in common a substituted tertiary amino group. The older H-1 compounds are effective histamine blockers but they also possess varying degrees of anticholinergic, antiemetic, and sedative properties. The most recent H-1 blockers have been designed so as not to penetrate the blood-brain barrier and thus cause little or no sedation. These newer compounds also lack anticholinergic activity.

The original H-1 antihistamines, such as tripelennamine [*91-81-6*], pyrilamine [*91-84-9*], diphenhydramine [*58-73-1*] and chlorpheniramine [*132-22-9*], have been widely used for symptomatic relief of histamine-induced pruritus and urticaria [15]. Antihistamines are generally not used topically because of their potential to induce contact or photocontact dermatitis. The antipruritic effects of oral H-1 antihistamines in eczematous dermatitis (e.g., atopic dermatitis) or nonallergic pruritic disorders may be primarily mediated by a sedative action. The nonsedating H-1 antihistamines, including loratadine [*79794-75-5*] and fexofenadine [*83799-24-0*], are useful for allergic rhinitis and chronic urticaria, but their effectiveness in dermatitic and pruritic conditions continues to be investigated.

Studies of H-2 blockers in the treatment of pruritus have produced mixed results. Hydroxyzine [*68-88-2*], a piperazine H-1 blocker, is a frequent choice for systemic treatment of pruritus and appears to be one of the most effective antihistamines for suppression of whealing. With the exception of cyproheptadine [*129-03-3*], H-1 antihistamines appear less effective in physical urticarias than in other forms of urticaria.

Cimetidine [*51481-61-9*] has been used together with H-1 antihistamines for treatment of chronic urticaria unresponsive to H-1 therapy. Evidence exists that such combinations are more effective than the use of H-1 blockers alone, although cimetidine may worsen symptoms by blocking feedback inhibition of histamine release. Likewise, diminished T-cell suppression by cimetidine may cause exacerbation of delayed hypersensitivity.

Cromolyn sodium [*15826-37-6*], a water-soluble bis(chromone) derivative, antagonizes histamine through stabilization of the mast cell membrane rather than receptor blockade. This agent may be used orally in mastocytosis, food allergies, and dermatitis herpetiformis. Topical formulations may be beneficial in atopic dermatitis, atopic keratoconjunctivitis, and various ulcerations.

Doxepin [*1668-19-5*], a tricyclic antidepressant, is a potent H-1 antagonist with mild H-2 blocking activity. Intradermally administered doxepin suppressed whealing in human skin, and topical applications were shown to be possibly effective for the control of pruritus [16]. In a double-blind study, oral doxepin was significantly more effective and less sedative than diphenhydramine in the treatment of chronic urticaria. When tested topically, doxepin was more effective than placebo in reducing pruritus in patients with atopic dermatitis [17]. Thus, compounds in this class may possibly prove useful for controlling allergic skin reactions.

Nonsteroidal Anti-Inflammatory Drugs. These drugs are widely used for treating internal inflammation and pain but may also have some utility for treating cutaneous inflammation. Drugs in this class inhibit the production of eicosanoids, inflammatory mediators derived primarily from arachidonic acid [18]. The major eicosanoid-producing enzymes are cyclooygenases (COX) and lipoxygenases (LOX). Two enzymes, COX-1 and COX-2, produce prostaglandins. Most traditional NSAIDs, such as acetylsalicylic acid [50-78-2] and indomethacin [53-86-1], are non-selective and inhibit both COX-1 and COX-2. Gastrointestinal and renal toxicity by NSAIDS is the result of inhibiting COX-1, an enzyme needed for normal physiological processes in these two organ systems. Because COX-2 is elevated at sites of inflammation, newer COX-2-selective inhibitors, such as celecoxib [169590-42-5] and rofecoxib [162011-90-7], appear to have fewer side effects than traditional NSAIDS.

Lipoxygenases produce several forms of lipid hydroxy acids, most notably hydroxyeicosatetraenoic acids (HETES) and leukotrienes. Inhibitors of 5-lipoxygenase, such as zilueton [111406-87-2], prevent the synthesis of leukotrienes. The leukotriene antagonists zafirlukast [107753-78-6] and montelukast [158966-92-8] block the action of leukotrienes at their receptors.

Studies in animals and humans show that prostaglandins and leukotrienes are produced in various skin inflammatory conditions, including ultraviolet radiation-induced erythema, contact dermatitis and psoriasis. Indomethacin has been shown to partially inhibit erythema induced by UV-B light (290 – 320 nm) when administered topically [19]-[21]. In contrast, topical corticosteroids were found to be inactive. Topical bufexamac [2438-72-4] has demonstrated some efficacy in steroid-responsive dermatoses other than psoriasis.

Glucocorticosteroids. Systemic and topical glucocorticosteroids are widely used to relieve inflammatory and pruritic manifestations of various acute and chronic skin diseases.

Glucocorticoids are 21-carbon pregnane derivatives that are synthesized in the adrenal cortex. Biologic activity is dependent on the presence of a double bond at the 4 – 5 position, two keto groups (positions 3 and 20), and three hydroxy groups (positions 11, 17, and 21).

Structural modification of the steroid nucleus greatly influences therapeutic potency as a result of changes in pharmacokinetic parameters, such as absorption, protein binding, and metabolism. For example, fluorination at the 9 position, addition of a 1,2-double bond, esterification of the 17- or 21-hydroxyl groups, and inclusion of an acetonide group at positions 16 and 17 greatly increase anti-inflammatory activity [22], [23]. Replacement of the 21-hydroxyl group with chlorine results in a highly effective class of corticosteroids; these synthetic steroids, which have two or more halogen groups, are the most potent of the synthetic glucocorticoids [24].

Other structural modifications of the steroid nucleus has led to development of moderately potent glucocorticoids with less propensity to cause adverse effects such as skin atrophy. These include mometasone furoate [*83919-23-7*], a 21-chloro compound with a 17-furoate moiety, and prednicarbate [*73771-04-7*], which is a non-halogenated, 17,21-diester. These agents either have low rates of transdermal absorption or are rapidly metabolized to compounds with minimal glucocorticoid activity.

Glucocorticoids inhibit inflammation by various mechanisms. Like other steroid hormones, they initiate their biological effects by binding to intracellular receptors. Once bound to the receptor, the glucocorticoid – receptor complex translocates to the nucleus. This complex then interacts alone or with other proteins to attenuate or enhance expression of numerous genes. Consequently, almost all aspects of the inflammation process are diminished by glucocorticoids. These include reduced leukocyte infiltration into tissues, reduced expression of cytokines, decreased expression of eicosanoid-generating enzymes (phospholipases, cyclooxygenases) and proteolytic enzymes. This in part accounts for the ability of glucocorticoids to suppress cell-mediated immune responses. In skin, glucocorticoids cause vasoconstriction and inhibit proliferation of keratinocytes.

Systemically administered glucocorticoids decrease the number of polymorphonuclear leukocytes (PMNS) at sites of tissue inflammation, as reflected by an increase in circulating PMNS. These changes are accompanied by decreases in circulating lymphocytes, monocytes, and eosinophils [25], [26].

Corticosteroids also exert their anti-inflammatory effect by inhibiting T-lymphocyte and B-lymphocyte function, and reduce the immune responsiveness and microbiocidal activity of macrophages. Although corticosteroids inhibit antibody synthesis and related B-cell functions, they have a more pronounced effect on T-cell-mediated, delayed hypersensitivity responses. In addition, corticosteroids depress antigen-induced proliferation of T-cells and cytokine release and also interfere with the action of several mediators secreted by T-cells.

Topical glucocorticosteroids are particularly beneficial in inflammatory skin disorders, such as atopic and contact dermatitis, eczema, seborrheic dermatitis, and psoriasis [27]. Daily applications are generally employed to limit the possibility of tachyphylaxis. Moreover, steroids apparently are slowly released from the stratum corneum for several days after a single application.

Choice of vehicle is determined by lesion type and drug solubility and is an important factor in epidermal penetration of glucocorticoids [28]. Hydration of the stratum corneum by ointment bases may enhance steroid penetration and may be useful in dry, scaling lesions. Occlusion of the skin with an impermeable plastic film has been shown to increase steroid penetration by approximately 10-fold. Glucocorticosteroids are highly soluble in gel vehicles and are equally effective in gel or ointment bases.

The anti-inflammatory potency of glucocorticosteroids closely parallels their capacity to inhibit the synthesis of essential structural cell proteins. Thus, skin atrophy may result from either topical or systemic corticosteroid administration, which is due to inhibition of connective tissue synthesis. In addition to epidermal thinning and stria formation, possible adverse effects of potent halogenated topical steroids include stellate pseudoscarring, purpura, telangiectasia, acneiform and rosacea-like eruptions, and perioral dermatitis.

Glucocorticosteroids are also administered by intralesional or intramuscular injection. Among the skin conditions responsive to intralesional steroids are acne vulgaris, lichen planus, psoriasis, localized eczema, keloids, and sarcoidosis. The intramuscular route is used extensively for various acute and chronic dermatoses, but may be associated with increased risk of steroid toxicity, especially alteration of the hypothalamic–pituitary–adrenal axis.

Similarly, systemic glucocorticosteroids are important in a broad range of acute and chronic disorders (e.g., allergic contact dermatitis, atopic dermatitis, lichen planus, and herpes zoster) [29]. Their use is limited by the risk of hypothalamic–pituitary–adrenal axis suppression, cushingoid symptoms, and a host of other potential metabolic, hematologic, and infectious complications.

Steroids may be administered in single daily doses or on alternate days to minimize the potential toxicity of long-term maintenance therapy. Although single-dose therapy may be an alternative to divided doses in many acute dermatoses, achieving adequate initial control of inflammation by using alternate-day regimens may be difficult. Likewise, high-potency topical steroids are generally used for initial therapy and are replaced by less potent agents when appropriate. Cessation of steroid therapy or conversion from daily to alternate-day therapy requires careful modulation of dosage to avoid rebound flare of disease or adrenal insufficiency.

Topical Antipruritics, Emollients, and Baths. Lotions containing antipruritics, such as menthol [*1490-04-6*], phenol [*108-95-2*], camphor [*76-22-2*], and astringents, such as zinc oxide [*1314-13-2*] and calamine [*8011-96-9*], are often used for symptomatic treatment of acute eczema and urticaria. Wet dressings containing aluminum acetate [*139-12-8*] solution or bath preparations containing colloidal solids, such as oatmeal or *sulfur* [*7704-34-9*], may also be beneficial.

Antipruritic tar baths and emollient oil baths are helpful for chronic lesions. Occlusive emollient creams or ointment bases, including anhydrous lanolin [*8006-54-0*] and petrolatum [*8009-03-8*], decrease evaporative loss of water from the skin surface and may be applied to relieve dryness. Urea [*57-13-6*] and lactic acid [*50-21-5*] are frequently incorporated into these emollient bases to enhance moisture retention and soften the stratum corneum.

Local anesthetics, including the ester derivative benzocaine [*94-09-7*], depress the conduction of nerve impulses and may provide relief of pain and itching. Their use in eczema is limited because they may cause contact sensitization.

4.2. Antiacne agents

Acne vulgaris is a common dermatological disorder characterized by noninflammatory comedones (open or closed) and inflammatory papules, pustules, and cysts.

Therapy may be directed at one or more of the following pathophysiologic processes associated with acne: increased sebum production, hyperkeratinization, *Propionibacterium acnes* (*P. acnes*) proliferation, and inflammation.

The initial *stimulus for acne development* appears to be an alteration in the normal pattern of follicular epithelial cell differentiation known as abnormal follicular keratinization [30]. The cause of this defect is unknown, but may relate to hormonal changes, increased sebum production, or bacterial colonization of the hair follicles. The altered keratinization process is characterized by production of a mass of tightly coherent keratinocytes and results in blockage of sebum flow, overgrowth of *P. acnes*, and follicular dilation to form microcomedones. Breakdown of the follicular wall initiates the inflammatory phase of acne, which is mediated by bacteria-derived enzymes and chemotactic factors [31].

Consequently, therapeutic agents that reverse the underlying keratinization defect are particularly effective in clearing microcomedones and preventing their formation. They also indirectly prevent inflammatory lesions. Antibacterial, anti-inflammatory, and sebostatic agents have a pronounced effect on inflammatory lesions, but do not correct the underlying abnormal follicular keratinization. Topical keratolytic or peeling agents, including sulfur, resorcinol [*108-46-3*], and salicylic acid [*69-72-7*], do not prevent lesion formation and, therefore, provide only temporary clearing.

Comedolytic Agents. The *retinoids*, including vitamin A and its derivatives, modify the altered epithelial replication process characteristic of acne, psoriasis, ichthyosis, and other keratinizing dermatoses [32].

In a study comparing the antikeratinizing effects of topical antiacne agents in rhino mouse skin, only the retinoids produced substantial reduction in the size of utriculi (keratinized follicles) [33]. Of the retinoids, all-*trans*-retinoic acid [*302-79-4*] (tretinoin) exhibited the greatest comedolytic potency. Tretinoin also gave greater reduction in comedones than motretinide [*56281-36-8*] in clinical studies [34]. Orally administered retinoids showed less comedolytic activity than topical retinoids in an earlier study using the rhino mouse model [35]. Oral tretinoin had more than twice the potency of oral 13-*cis*-retinoic acid [*4759-48-2*] (isotretinoin).

Tretinoin is a natural retinoid metabolite of vitamin A that binds to cellular retinoic acid binding proteins in the cytoplasm and receptors in the nucleus. The nuclear receptors are mechanistically considered the site of action of tretinoin. Tretinoin activates the three members of the retinoic acid (RAR) nuclear receptors (RARα, RARβ, RARγ) which act to modify gene expression, and subsequent protein synthesis, to account for its beneficial clinical effects [36].

Tretinoin is perhaps the most effective topical agent for treatment and prevention of comedonal and inflammatory acne. Tretinoin decreases horny cell cohesiveness and stimulates epidermal cell turnover, thereby removing existing microcomedones and preventing their reformation. Tretinoin may enhance resolution of inflammatory lesions by increasing blood flow. Moreover, its ability to reduce stratum corneum thickness and to potentiate the absorption of other topical compounds makes tretinoin a potentially valuable component in combination acne therapy.

Two new synthetic receptor-selective topical retinoids are now available to treat acne vulgaris. Adapalene [106685-40-9], a naphthoic acid derivative, and tazarotene [118292-40-3], an acetylenic retinoid, are selective for RARβ and RARγ [37], [38]. Tazarotene does not bind to the RARs but is metabolized by an esterase to the active form, tazarotenic acid, which binds to RARs and activates them.

Antibacterial Agents. The oxidizing agent benzoyl peroxide [94-36-0] is commonly used in the topical treatment of acne. Benzoyl peroxide possesses both comedolytic and antibacterial properties, but the latter is believed to be primarily responsible for its therapeutic efficacy. Release of free-radical oxygen by benzoyl peroxide decreases the levels of *P. acnes* and skin surface lipids, leading to reduction in inflammatory lesions. Benzoyl peroxide also reduces inflammatory lesions via stimulation of blood flow and may have a sebostatic effect. Acetone or water-based gel formulations may allow for enhanced penetration and efficacy of benzoyl peroxide. The combination of benzoyl peroxide with erythromycin [114-07-8], a macrolide antibiotic, is more effective than either ingredient alone in reducing inflammatory lesions.

Oral tetracyclines or erythromycin are the drugs of choice for treatment of moderate inflammatory acne, with trimethoprim [738-70-5] plus sulfamethoxazole [723-46-6] occasionally used in refractory patients. Topical antibiotics, including erythromycin, clindamycin [18323-44-9], tetracycline [60-54-8], and meclocycline [2013-58-3], are generally recognized as safe and effective alternatives to long-term systemic therapy [39], [40].

The mechanism of action of oral and topical antibiotics in acne is not fully understood. Previously, oral antibiotics were shown to suppress follicular bacterial colonization and production of free fatty acids. Later studies suggested that inhibition of bacterial metabolism (e.g., lipase production) may be the primary mode of action of topical antibiotics [41]. Topical antibiotics also possess anti-inflammatory properties mediated in part by inhibition of neutrophil chemotaxis [42].

Sulfur preparations containing *sulfacetamide* [144-80-9] are useful in the topical control of acne vulgaris. The sulfonamides act as competitive antagonists to *p*-aminobenzoic acid [150-13-0], an essential component to bacterial growth. *Azelaic acid* [123-99-9], a naturally occurring saturated dicarboxylic acid, is topically effective in treating mild to moderate inflammatory acne vulgaris. The exact mechanism is not known but it has antimicrobial activity against *P. acnes* and *S. epidermidis* with normalization of keratinization.

Systemic Retinoids. Oral isotretinoin provides excellent therapeutic results and long-term remissions in patients with severe cystic acne [43]. Its use is limited by a high incidence of mucocutaneous (xerosis, cheilitis) and systemic adverse effects (vertebral hyperostoses, hypertriglyceridemia) and by a potential for teratogenicity. Isotretinoin is teratogenic in humans. Women of child-bearing age must have a pretreatment pregnancy test and be ensured of adequate contraception for the duration of therapy.

The effects of isotretinoin in acne are related primarily to marked suppression in sebum production that results from a direct effect on the sebaceous gland [44]. The decrease in sebum production is accompanied by reduction in bacterial flora [45]. Less clearly characterized is the inhibitory effect of isotretinoin on follicular hyperkeratinization [46]. Isotretinoin may also exert an anti-inflammatory effect with inhibition of neutrophil function.

Hormones. Elevated levels of androgens are known to cause acne in some women. High levels of androgens lead to increased sebum production and follicular hyperkeratosis. Oral contraceptives can reduce androgen levels and are effective in treating acne in women. This therapy for acne targets sex hormone binding globulin (SHBG) and free testosterone [58-22-0]. A low dose oral contraceptive containing norgestimate [35189-28-7] and ethinyl estradiol [57-63-6] is a safe and effective treatment of acne with efficacy similar to benzoyl peroxide and topical or systemic antibiotics [47]. This combination increases SHBG and decreases free testosterone in healthy women. Norgestimate binds to progestin receptors and has low affinity for androgen receptors and SHBG. Ethinyl estradiol may have antiandrogenic effects by suppressing the secretion of gonadotropins and inhibiting the production of ovarian androgens.

Spironolactone is an antiandrogen that blocks the binding of androgen to the androgen receptor and is useful in treating recalcitrant acne in women. Combinations of synthetic estrogens with nonandrogenic progestogens or the synthetic antiandrogen cyproterone acetate [427-51-0] have been used successfully to treat female acne patients with acne conglobata.

Anti-Inflammatory Agents. Short-term therapy with oral corticosteroids (e.g., prednisone [53-03-2]) in conjunction with maintenance therapy with sulfones (e.g., dapsone [80-08-0]) is sometimes used in patients with severe acne conglobata. Dapsone has been shown to stabilize lysosomes and interfere with neutrophil function. Intralesional injections of triamcinolone acetonide [76-25-5] are frequently used for patients with severe scarring. Ibuprofen [15687-27-1] has also shown some efficacy in diminishing inflammatory acne.

4.3. Antipsoriatics

Psoriasis is a common, chronic skin disorder of variable and unpredictable course. It is characterized by accelerated epidermal cell turnover and inflammation to produce elevated, erythematous plaques covered with silver scale. The underlying causes of the disease are unknown, but various genetic, immunologic, and environmental factors influence its expression. A familial predisposition to psoriasis is evidence of one or several genes accounting for its expression. A dysregulation in the immune system, primarily with T-lymphocytes, is considered to be the fundamental mediator of the epidermal and inflammatory changes in the psoriatic lesion [48]. External stimuli, such as infectious disease, skin trauma, or psychological stress are known triggers of psoriatic plaque formation.

Topical therapy is most effective for short-term use in mild psoriasis and may be combined with ultraviolet radiation or systemic agents in extensive or resistant cases. Many of the therapeutic modalities for severe psoriasis (immune modulators, photochemotherapy, phototherapy, and cytotoxic therapy) directly suppress immune cell activity. Some of these agents also diminish epidermal hyperplasia by inhibiting DNA synthesis and mitosis. Development of therapeutic regimens that selectively normalize immune system and epidermal functions is a major focus of current research.

Topical Agents. Occlusive emollients, such as petrolatum and mineral oil [*8012-95-1*], are frequently used alone or in combination with keratolytic agents, such as salicylic acid [*69-72-7*] and urea [*57-13-6*], to facilitate hydration and removal of hyperkeratotic plaques.

The topical synthetic vitamin D analog, calcipotriene [*112965-21-6*] (calcipotriol) is frequently used for treating moderate plaque psoriasis. This agent acts to enhance keratinocyte differentiation and also exhibits immunomodulatory activity [49]. The major local side effect of calcipotriene is skin irritation; unlike corticosteroids, calcipotriene does not cause skin atrophy. Adverse systemic effects of topical calcipotriene are limited due to its rapid metabolic inactivation within 24 h after it is applied.

Potent topical corticosteroids (e.g., clobetasol propionate [*25122-46-7*], halobetasol propionate [*66852-54-8*]) are frequently administered intermittently in more severe cases of psoriasis because of their vasoconstrictive, antimitotic, and anti-inflammatory effects. Glucocorticoids also have significant inhibitory effects on immune cell activation. Intralesional steroids are also used in psoriasis. Long-term use of corticosteroids results in a loss of effectiveness and an increased risk of toxicity.

Coal tar preparations have long been used safely and effectively in psoriasis. Coal tar is a complex and variable mixture of at least 10 000 compounds; it contains many phenols, bases, and polyaromatic hydrocarbons, including several potent photosensitizers and carcinogens. Although its mechanism of action in psoriasis is not well-defined, coal tar inhibits DNA synthesis and may also have antipruritic and vasoconstrictive properties. Coal tar is commonly used in conjunction with ultraviolet radiation as described in the following "Phototherapy" section.

Anthralin [480-22-8] (dithranol), a synthetic derivative of the natural antipsoriatic agent chrysarobin [491-59-8], depresses DNA synthesis and epidermal replication by interfering with polyamine production. It may also exert anti-inflammatory and immunosuppressive effects. Like coal tar, anthralin may cause skin irritation and staining.

A synthetic retinoid, tazarotene, is topically effective for moderate psoriasis. Retinoic acid [302-79-4] has also been applied topically in psoriasis, as have several cytotoxic agents, including the sulfur and nitrogen mustards (e.g., mechlorethamine hydrochloride [55-86-7]).

Phototherapy. Ultraviolet radiation is well-known for its therapeutic value in psoriasis. Its biologic effects, thought to account for its efficacy in psoriasis, are wide-ranging; it can suppress the immune system and inhibit mitosis. These properties are primarily attributable to the UV-B wavebands (290 – 320 nm), although relatively high doses of UV-A (320 – 400 nm) may also be beneficial. UVB phototherapy is often used for patients who either do not respond to topical antipsoriatic drugs or for those who find topical treatment inconvenient.

Ultraviolet light is widely used in combination with anthralin and/or coal tar in the treatment of psoriasis. The *Goeckerman regimen* generally consists of multiple daily applications of crude coal tar ointment, a daily tar bath, and subsequent exposure to UV-B and UV-A light. Whether or not the photosensitizing action of coal tar is of therapeutic relevance in psoriasis is unclear.

Although the action spectrum of coal tar phototoxicity is in the UV-A range and the inhibitory action of coal tar on DNA synthesis is enhanced by UV-A light, coal tar is not photosensitive at commonly used doses of UV-A. Moreover, UV-B light–coal tar appears to be more effective than UV-A light–coal tar in psoriasis. Other studies suggest that the contribution of coal tar to the efficacy of Goeckerman therapy is minimal [50]. The *Ingram regimen* is a variation of the Goeckerman regimen in which an anthralin – salicylic acid paste is applied after tar baths and UV radiation.

In addition to tar and anthralin, phototherapy (UVB) effectiveness can be enhanced when combined with other topical and systemic drug treatments. These include corticosteroids, calcipotriene, methotrexate [59-05-2] and retinoids.

Systemic Therapy. Use of systemic drugs is reserved for cases of severe or widespread psoriasis. These fall into three main categories: cytotoxic agents, retinoids, and immunomodulators.

Cytotoxic compounds exert a suppressive effect on rapidly proliferating tissue by blocking nucleic acid synthesis. They are used primarily in cancer chemotherapy, but may be used to treat severe, recalcitrant psoriasis. Immunosuppression by these agents also contributes to their efficacy in psoriasis. Methotrexate, a folic acid antagonist, is considered the most effective chemotherapeutic agent for psoriasis [51]. Hepatotoxicity and immunosuppression are major drawbacks to long-term therapy with methotrexate.

Hydroxyurea [127-07-1], 6-thioguanine [154-42-7], azathioprine [446-86-6], and mycophenolic acid [24280-93-1] are other antimitotics that have been used to treat psoriasis. Like methotrexate, their use is limited by severe toxicity.

Oral retinoids, including the aromatic ester derivative etretinate [54350-48-0] and its corresponding carboxylic acid form, acitretin [55079-83-9], are highly effective in selected forms of psoriasis. The pustular and exfoliative subtypes of psoriasis are more responsive to these retinoids than plaque psoriasis. Acitretin has a more favorable elimination profile from the body than etretinate, with a half-life of about 50 hours for the acid versus 120 days for the ester. Etretinate may persist for several years because of its tendency to be stored in body fat. Systemic retinoids are sometimes combined with psoralen photochemotherapy [52].

Cyclosporine (cyclosporin A) [59865-13-3] is the first drug used to treat psoriasis that acts primarily to suppress the immune system [53]. Its major pharmacological effect is to inhibit interleukin-2 production by T-lymphocytes. Cyclosporine is highly effective for all forms of psoriasis, especially those exhibiting severe inflammation. It is commonly used to bring severe psoriasis under control, followed by a switch to other systemic or topical drugs for maintenance therapy. Long-term use of cyclosporine is limited because of its side effects of renal toxicity and hypertension.

Photochemotherapy. Although the effects of systemic chemotherapy are nonselective, the combination of a photosensitive compound and UV radiation restricts the therapeutic action of a drug to areas of epidermal hyperplasia. An ideal photosensitizer is activated by relatively low doses of UV light and is otherwise inert.

Psoralens are natural furocoumarins well-known for their photosensitizing potential. Porphyrins, including hematoporphyrin derivatives (HPD), are also potent photosensitizers. Both psoralens and porphyrins have an action spectrum in the UV-A range. Porphyrins are also activated by visible light.

Photoactivated psoralens covalently bind to DNA pyrimidines, producing monoadducts and cross-links, thereby inhibiting DNA synthesis and cell proliferation. Suppressive effects of PUVA (psoralen and UV-A) on the immune system, particularly T-lymphocytes, have been demonstrated in animals and humans. The mechanism of action of PUVA therapy in psoriasis is presumably related to its antiproliferative and immunosuppressive effects [54].

In PUVA therapy, 8-methoxypsoralen [298-81-7] is orally administered 2 h before exposure to increasing doses of UV-A. An average of 20 treatments over a 3-month period has been shown to be highly effective as initial therapy for psoriasis. Psoralen can also be administered before UVA by direct topical application to individual lesions or by soaking in a bath. In addition, PUVA therapy is valuable in treating other skin disorders, including vitiligo, mycosis fungoides, and atopic dermatitis.

Long-term side effects of PUVA primarily affect skin, immunologic and ocular systems. With PUVA, premature aging of skin is common but the major concern is increased risk of cutaneous malignancies. The risk of squamous cell carcinoma is definitely increased in patients treated over many years with PUVA [55]. It is also possible that the immunosuppressive effects associated with PUVA therapy can contribute to development of skin neoplasia. Eye protection during PUVA therapy is essential to prevent cataracts.

4.4. Antifungals

Superficial fungal infections are frequently caused by dermatophytes of the genera *Microsporum, Epidermophyton,* and *Trichophyton.* The most common dermatophyte infections are tinea cruris, tinea corporis, tinea pedis, tinea unguium, and tinea capitis. Superficial mycoses are also caused by the yeasts *Malassezia ovalis,* the etiological agent in tinea versicolor, and *Candida.*

The causative organisms are predominantly localized in keratinized areas of the skin, hair, and nails.

Topical agents are generally satisfactory for the treatment of dermatophytosis, cutaneous candidiasis, and tinea versicolor, but are often ineffective for infections of the nail and scalp. Oral agents are usually required for extensive or refractory dermatophytosis or tinea versicolor.

The following drugs are commonly employed in the treatment of superficial fungal or yeast infections. Some of these agents are also used to treat systemic mycoses.

Azoles. The broad-spectrum azole antimicrobials represent an increasingly large class of compounds, including clotrimazole [23593-75-1], miconazole [22916-47-8], econazole [27220-47-9], ketoconazole [65277-42-1], sulconazole [61318-90-9], oxiconazole [64211-46-7], itraconazole [84625-61-1], tioconazole [65899-73-2], bifonazole [60628-96-8], and fluconazole [86386-73-4].

The azoles are active against dermatophytes, yeasts, dimorphic fungi, some gram-positive bacteria, and certain protozoa. They exert their antifungal effect primarily by blocking the synthesis of ergosterol, an important regulator of cell membrane permeability [56], [57].

Topical azoles are effective in treating cutaneous candidiasis and most tinea infections [58]–[61]. They are generally considered superior to such agents as tolnaftate, haloprogin, and undecylenic acid. No significant difference in the safety or efficacy of various azoles has been established.

In recent years, combinations of topical azoles and corticosteroids (e.g., miconazole plus hydrocortisone; clotrimazole plus betamethasone dipropionate [5593-20-4]) have become increasingly popular for acute inflammatory mycoses. The steroid component appears to provide rapid symptomatic relief without interfering with antifungal efficacy [62]. Also, some topical azoles have been reported to exert an anti-inflammatory effect [63].

Ketoconazole, itraconazole, and fluconazole are the only orally active azoles in current use. Ketoconazole and itraconazole are highly effective in all superficial fungal infections. Oral ketoconazole is generally used to treat deep mycoses (e.g., paracoccidioidomycosis and histoplasmosis) or resistant superficial mycoses (e.g., chronic mucocutaneous candidiasis) and is the drug of choice in tinea versicolor (two tablets, single intake). Itraconazole and fluconazole are effective in the treatment of onychomycosis, a common fungal infection of the fingernail and toenail as well as systemic infections [64], [65]. Fluconazole primarily focuses on yeasts (*Candida albicans*).

Allylamines. Naftifine hydrochloride [65473-14-5] is a synthetic allylamine derivative effective against dermatophytes, moderately active against molds and less active against yeasts. Topical naftifine is indicated for common tinea infections. Terbinafine hydrochloride [78628-80-5] is a second generation allylamine structurally related to naftifine. Terbinafine is 10 to 100 times more potent than naftifine in vitro. Both agents inhibit squalene oxidase and decrease ergosterol, an essential component of cell membranes in fungi. Topical terbinafine is indicated for tinea infections while oral terbinafine is indicated for the treatment of onychomycosis of the toenail or fingernail due to dermaytophytes.

Butenafine hydrochloride [101827-46-7] is a benzylamine, structurally related to the allylamines, and is indicated for tinea pedis. It is active against *E. floccosum, T. mentagrophytes* and *T. rubrum*.

Ciclopirox. Ciclopirox olamine [41621-49-2] (→ Antimycotics), a pyridone–ethanolamine salt, is structurally dissimilar to other antifungal agents. Its primary mode of action appears to be inhibition of transmembrane transport of substrates and ions essential for cell metabolism and energy production. Ciclopirox is similar to the imidazoles in antimicrobial spectrum, therapeutic uses, and efficacy. Limited studies suggest superior penetration of topical ciclopirox into horny material and nails.

Acidic Compounds. Acidic compounds reduce the local pH to levels unfavorable for dermatophyte growth. Undecylenic acid [112-38-9], an 11-carbon unsaturated fatty acid, is the most widely used compound in this class; however, it is effective in only 50% of the patients. Zinc or calcium salts of undecylenic acid are also used.

Similarly, the fungistatic activity of triacetin [102-76-1] (glyceryl triacetate) results from its hydrolysis by fungal esterases to acetic acid.

Tolnaftate. Tolnaftate [2398-96-1] is a thiocarbamate derivative similar in spectrum of action and efficacy to undecylenic acid. It is most often used in dermatophyte and tinea versicolor infections.

Haloprogin. An iodinated trichlorophenol, haloprogin [777-11-7] has a broad spectrum of activity against dermatophytes, *Candida* and *Pityrosporum*. It is similar to tolnaftate in efficacy for tinea cruris, tinea corporis, and tinea pedis.

Morpholine Derivatives. Amorolfine [67467-83-8] is a morpholine derivative which possesses both fungistatic and fungicidal activity in vitro [66]. Amorolfine is an effective antifungal agent for patients with onychomycosis and dermatomycosis caused by dermatophytes, yeasts, and sensitive molds.

Griseofulvin. Griseofulvin [126-07-8] is a safe, effective, orally administered fungistatic antibiotic that inhibits cell mitosis by binding to microtubules during metaphase. Griseofulvin is active only against dermatophytes and concentrates in the keratinized outer layer of the stratum corneum. Absorption of griseofulvin is increased by fat intake, reduction in particle size, and multiple-dose regimens.

Griseofulvin is effective for superficial dermatophyte infections of the skin, hair, and nails and is generally reserved for infections that have failed to respond to topical therapy. It is considered the drug of choice for tinea capitis. Griseofulvin has no action against *C. albicans* or tinea versicolor.

Polyenes. Nystatin [1400-61-9] is a polyene macrolide antibiotic that binds to sterol constituents in the yeast cell membrane, thereby altering its permeability. Nystatin is active against various yeast and fungi, but is most often used topically in treating cutaneous and mucocutaneous candidiasis. Nystatin is ineffective in dermatophyte infections.

Like nystatin, amphotericin B [1397-89-3] exerts its antifungal activity by an effect on cell membrane permeability. It may also increase T-cell and B-cell responsiveness. Topical amphotericin is used primarily for cutaneous and mucocutaneous candidiasis and has been shown to be as effective as nystatin in intertriginous candidiasis.

Amphotericin is commonly administered intravenously for the management of various deep mycoses. It has been associated with significant toxicity, including impaired renal function and death.

Miscellaneous Topicals. Diiodohydroxyquinoline [83-73-8] (iodoquinol) and iodochlorhydroxyquinoline [130-26-7] (clioquinol) are antimicrobial agents used alone or in combination with cortisol (hydrocortisone) for treating *Candida* and dermatophyte infections, although their efficacy has not been definitely established.

Gentian violet [548-62-9] is a mixture of methylpararosanilines used for cutaneous or mucocutaneous candidiasis. Castellani's paint [8052-17-3] contains basic fuchsin, boric acid, phenol, and resorcinol in a water–ethanol–acetone solution and is of benefit in subacute and chronic dermatophytosis. Whitfield's ointment, a 2:1 combination of the fungistatic agent benzoic acid [65-85-0] and the keratolytic salicylic acid, is helpful in the treatment of chronic dermatophytosis of the palms and soles and in tinea versicolor. Sodium thiosulfate [10102-17-7], selenium sulfide [7488-56-4], and acrisorcin [7527-91-5] are primarily effective in tinea versicolor. Zinc pyrithione [13463-41-7] is effective as an anti-*Pityrosporum* agent.

4.5. Antiaging

Skin undergoes aging by two separate and distinct processes: intrinsic aging and photoaging. Chronic sun exposure of the skin leads to photoaging, defined as photodamage superimposed on the natural (intrinsic) aging process. In contrast to intrinsically aged skin, photoaging is manifested by coarse wrinkles, reduced elasticity, darkened and uneven pigmentation and roughness. The main histologic features of intrinsic aging are epidermal and dermal atrophy while the hallmark of photoaging is dermal elastosis. Photodamage is likely mediated by destruction of extracellular matrix (collagen and elastin) of the dermis by matrix degrading metalloproteinases.

Topical tretinoin is the first pharmacologic agent shown to have proven efficacy in treating photoaged skin [67], [68]. Tretinoin can enhance the appearance of photodamaged skin by effacing fine wrinkles, lightening hyperpigmented age spots, reducing roughness and improving the overall texture of the skin surface.

5. References

[1] R. J. Scheuplein, *Curr. Probl. Dermatol.* **7** (1978) 172–186.
[2] B. Idson, *J. Pharm. Sci.* **64** (1975) no. 6, 901–924.
[3] J. Ostrenga, C. Steimetz, B. J. Poulsen, *J. Pharm. Sci.* **60** (1971) no. 8, 1175–1183.
[4] T. Higuchi, *J. Soc. Cosmet. Chem.* **11** (1960) 85–97.
[5] M. F. Coldman, B. J. Poulsen, T. Higuchi, *J. Pharm. Sci.* **58** (1969) 1098–1102.
[6] J. Haleblian, B. J. Poulsen, K. H. Burdick, *Curr. Ther. Res.* **22** (1977) no. 5, 713–721.
[7] M. F. Coldman, B. J. Poulsen, T. Higuchi, *J. Pharm. Sci.* **58** (1969) no. 9, 1099–1102.
[8] A. M. Kligman, *JAMA* **193** (1965) no. 10, 140–148.
[9] W. B. Campbell, P. V. Halushka: "Lipid-Derived Autacoids: Eicosanoids and Platelet-Activating Factor," in J. G. Hardman, A. G. Gilman, L. E. Limbird (eds.): *Goodman and Gilman's The Pharmacological Basis of Therapeutics,* 9th ed., McGraw-Hill, New York 1996, pp. 601–616.
[10] J. M. Hanifin, S. C. Chan, *J. Invest. Dermatol.* **105** (1995) suppl. 1, 84S–88S.
[11] M. J. Fischer, J. J. Paulussen, R. Roozendaal et al., *Inflamm. Res.* **45** (1996) 564–573.
[12] M. M. Hamawy, S. E. Mergenhagen, R. P. Siraganian, *Cell. Signal.* **7** (1995) 535–544.
[13] K. S. Babe, W. E. Serafin: "Histamine, Bradykinin and Their Antagonists," in J. G. Hardman, A. G. Gilman, L. E. Limbird (eds.): *Goodman and Gilman's The Pharmacological Basis of Therapeutics,* 9th ed., McGraw-Hill, New York 1996, pp. 581–600.
[14] R. W. Brimblecombe et al., *Gastroenterology* **74** (1978) 339–347.
[15] O. B. Christensen, H. I. Maibach, *Semin. Dermatol.* **2** (1983) no. 4, 270–280.
[16] J. E. Bernstein, D. H. Whitney, K. Soltani, *J. Am. Acad. Dermatol.* **5** (1981) 582–585.
[17] L. A. Drake, J. D. Fallon, A. Sober, *J. Am. Acad. Dermatol.* **31** (1994) 613–616.
[18] P. A. Insel: "Analgesic-Antipyretic and Antiinflammatory Agents and Drugs Employed in the Treatment of Gout," in J. G. Hardman, A. G. Gilman, L. E. Limbird (eds.): *Goodman and Gilman's The Pharmacological Basis of Therapeutics,* 9th ed., McGraw-Hill, New York 1996, pp. 617–657.
[19] M. E. Goldyne, *J. Invest. Dermatol.* **64** (1975) 377–385.
[20] D. S. Snyder, W. H. Eaglstein, *Br. J. Dermatol.* **90** (1974) 91–93.
[21] W. H. Eaglstein, L. D. Ginsberg, P. M. Mertz, *Arch. Dermatol.* **115** (1979) 1421–1423.
[22] J. Elks, *Br. J. Dermatol.* **94** (1976) suppl. 12, 3–13.
[23] R. G. Dluhy, D. P. Lauler, G. W. Thorn, *Med. Clin. North Am.* **57** (1973) 1155–1167.
[24] S. Yawalkar et al., *J. Am. Acad. Dermatol.* **25** (1991) 1137–1144.
[25] J. Thompson, R. Van Furth: "The Effect of Glucocorticosteroids on the Kinetics of Monocytes and Peritoneal Macrophages," in Van Furth (ed.): *Mononuclear Phagocytes,* Blackwell Scientific Publ., Oxford 1970, pp. 255–281.

[26] T. F. Dougherty: "Effects of Corticosteroid Hormones on Lymphatic Tissues," in J. W. Rebuck (ed.): *The Lymphocytes and Lymphocytic Tissue*, Hoeber, New York 1960, pp. 112–124.
[27] R. C. Cornell, H. I. Maibach, "Clinical Indications: Real and Assumed," in H. I. Maibach, C. Surber (eds.): *Topical Corticosteroids*, Karger, Basel 1992, pp. 154–162.
[28] R. B. Stoughton, *Arch. Dermatol.* **106** (1972) 825–827.
[29] R. M. Fine, *Semin. Dermatol.* **2** (1983) no. 4, 250–256.
[30] J. J. Leyden, A. M. Kligman: "The Role of Bacteria in Acne Vulgaris," in R. Fleischmajer (ed.): *Progress in Diseases of the Skin*, vol. **2**, Grune & Stratton, Orlando 1984, pp. 21–29.
[31] J. Knop, T. Bossecker, P. M. Kovary, *Arch. Dermatol. Res.* **274** (1982) 267–275.
[32] G. L. Peck: "Retinoids in Clinical Dermatology," in R. Fleischmajer (ed.): *Progress in Diseases of the Skin*, vol. **1**, Grune & Stratton, Orlando 1981, pp. 227–269.
[33] J. A. Mezick, M. C. Bhatia, L. M. Shea, E. G. Thorne et al.: "Anti-Acne Activity of Retinoids in the Rhino Mouse," in H. I. Maibach, N. J. Lowe (eds.): *Models in Dermatology*, vol. **2**, Karger, Basel, Switzerland 1985, pp. 59–63.
[34] K. Nordin, T. Fredricksson, C. Rylander, *Dermatologica* **162** (1981) 104–111.
[35] J. A. Mezick, M. C. Bhatia, R. J. Capetola, *J. Invest. Dermatol.* **83** (1984) 110–113.
[36] G. J. Fisher, J. J. Voorhees, *FASEB J.* **10** (1996) 1002–1013.
[37] B. Shroot, S. Michel, *J. Am. Acad. Dermatol.* **36** (1997) part 2, S96–S103.
[38] R. A. S. Chandraratna, *J. Am. Acad. Dermatol.* **37** (1997) part 3, S12–S17.
[39] E. A. Eady, K. T. Holland, W. J. Cunliffe, *J. Am. Acad. Dermatol.* **5** (1981) 455–489.
[40] R. B. Stoughton, *Arch. Dermatol.* **115** (1979) 486–489.
[41] S. J. Adams, E. M. Cooke, W. J. Cunliffe, *J. Antimicrob. Chemother.* **7** (1981) suppl. A, 75–80.
[42] N. B. Esterly, N. L. Furey, L. E. Flanagan, *J. Invest. Dermatol.* **70** (1978) 51–55.
[43] G. L. Peck, T. G. Olsen, D. Butkus, *J. Am. Acad. Dermatol.* **6** (1982) 735–745.
[44] J. S. Strauss, A. M. Stranieri, L. N. Farrell et al., *J. Invest. Dermatol.* **74** (1980) 66–67.
[45] G. Plewig, J. Nikolowski, H. H. Wolff, *J. Am. Acad. Dermatol.* **6** (1985) 766–785.
[46] P. M. Elias, P. Fritsch, M. A. Lampe et al., *Clin. Res.* **28** (1980) 248.
[47] A. W. Lucky et al., *J. Am. Acad. Dermatol.* **37** (1997) 746–754.
[48] E. Christophers, *Int. Arch. Allergy Immunol.* **110** (1996) 199–206.
[49] K. Kragballe, *J. Am. Acad. Dermatol.* **37** (1997) part 2, S72–S76.
[50] M. J. LeVine, H. A. D. White, J. A. Parrish, *J. Invest. Dermatol.* **73** (1979) 170–173.
[51] G. D. Weinstein, *Ann. Intern. Med.* **86** (1977) 199–204.
[52] C. W. Ehmann, J. J. Voorhees, *J. Am. Acad. Dermatol.* **6** (1982) 692–696.
[53] M. A. de Rie, J. D. Bos, *Clin. Dermatol.* **15** (1997) 811–821.
[54] J. Lauharanta, *Clin. Dermatol.* **15** (1997) 769–780.
[55] R. S. Stern et al., *N. Engl. J. Med.* **310** (1984) 1156–1161.
[56] M. S. Marriott, *J. Gen. Microbiol.* **117** (1980) 253–255.
[57] H. Van Den Bossche, G. Willemsens, W. Cools, W. F. J. Lauwers et al., *Chem.-Biol. Interact.* **21** (1978) 59–78.
[58] J. E. Fulton, *Arch. Dermatol.* **111** (1975) 596–598.
[59] P. H. Spiekermann, M. D. Young, *Arch. Dermatol.* **112** (1976) 350–352.
[60] O. East, J. T. Henderson, *Dermatologica* **166** (1983) suppl. 1, 1–14.
[61] R. C. Heel, R. N. Brogden, T. M. Speight et al., *Drugs* **16** (1978) 177–201.
[62] C. Samson, E. Peets, R. Winter-Sperry, H. Wolkoff: "Clotrimazole/Betamethasone Dipropionate–Lotrisone–An Antifungal/Steroid Combination for Treatment of Cutaneous Dermatophyte Infections," in H. I. Maibach, C. Surber (eds.): *Topical Corticosteroids*, Karger, Basel 1992, pp. 318–334.

[63] J. V. Cutsem et al., *J. Am. Acad. Dermatol.* **25** (1991) 257–261.
[64] M. Brautigam, *J. Am. Acad. Dermatol.* **38** (1998) S53–S56.
[65] L. Drake et al., *J. Am. Acad. Dermatol.* **38** (1998) S87–S94.
[66] A. Polak, P. G. Hartman: "Preclinical Data of Amorolfine," in J. W. Rippon, R. A. Fromtling (eds.): *Cutaneous Antifungal Agents: Selected Compounds in Clinical Practice and Development,* Marcel Dekker, New York 1993, pp. 13–26.
[67] A. M. Kligman et al., *J. Am. Acad. Dermatol.* **15** (1986) 836–859.
[68] E. A. Olsen et al., *J. Am. Acad. Dermatol.* **26** (1992) 215–224.

Gout Remedies

LARRY J. SILVERMAN, University of Michigan, Ann Arbor, Michigan 48109, United States
THOMAS D. PALELLA, University of Michigan, Ann Arbor, Michigan 48109, United States

1.	Introduction	1737
2.	Gout and Hyperuricemia	1739
2.1.	Pathophysiology	1739
2.2.	Clinical Manifestations	1740
3.	Treatment	1741
3.1.	Drugs Used to Treat Acute Gout	1742
3.1.1.	Colchicine	1742
3.1.2.	Nonsteroidal Anti-inflammatory Drugs	1742
3.1.3.	Corticosteroids	1745
3.2.	Antihyperuricemic Drugs	1746
3.2.1.	Uricosuric Drugs	1746
3.2.2.	Xanthine Oxidase Inhibitors	1748
3.3.	Dietary Management	1749
4.	References	1749

1. Introduction

Gout is an exclusively human disease. Its cardinal manifestations are high serum urate levels (hyperuricemia) and recurrent episodes of acute arthritis. In its fully developed form, gout also results in deposition of uric acid in the kidneys (nephrolithiasis), interstitial renal disease, and deposition of monosodium urate monohydrate aggregates (tophi) in and around joints (chronic tophaceous gout).

History. The term gout was first introduced in the 13th century, although its clinical manifestations had been described by HIPPOCRATES in the 5th century B.C. [1]. In the 17th century, SYDENHAM distinguished gout from other arthritidies [2]. The importance of uric acid in the pathogenesis of this disease was elucidated in the 19th century [3].

An understanding of the pathogenesis of gout has lead to the development of specific therapies. Although colchicine has been used for centuries to treat acute gouty arthritis, newer anti-inflammatory agents are now used to manage acute attacks [4]. In addition, effective antihyperuricemic therapy has been developed. These agents decrease serum urate, reduce the total body pool of uric acid, prevent urate deposition, and promote dissolution of preformed tophi and stones.

Biochemistry. Uric acid is the ultimate degradative product of purine metabolism in humans. It is derived primarily from the breakdown of nucleic acids, nucleotides, and ingested, preformed purines. The proximate steps in uric acid formation are dephosphorylation of ribonucleotides and deoxyribonucleotides to their nucleoside forms, adenosine and guanosine. Adenosine and guanosine are both converted to inosine. Purine nucleoside phosphorylase converts inosine to hypoxanthine in a freely reversible reaction. Hypoxanthine, thus generated, is a substrate for the iron-containing flavoenzyme, xanthine oxidase. Xanthine oxidase converts hypoxanthine to xanthine, and then xanthine to uric acid.

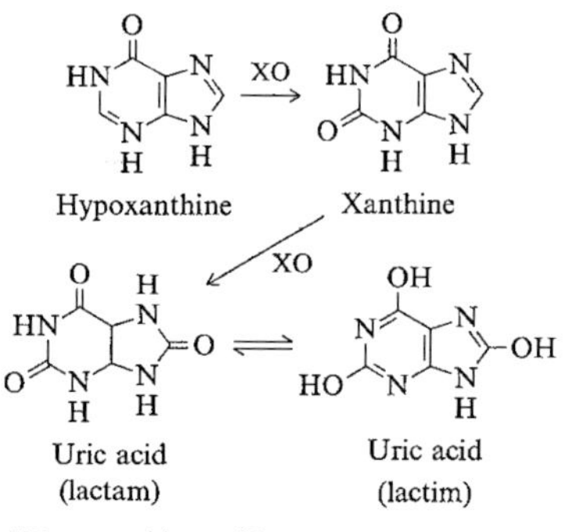

Hypoxanthine → Xanthine → Uric acid (lactam) ⇌ Uric acid (lactim)

XO = xanthine oxidase

Thus, the production of uric acid is dependent upon the rate of purine nucleotide degradation and the subsequent oxidation of hypoxanthine. The serum concentration of uric acid reflects the difference between this production rate and the net excretion rate of uric acid. Under normal conditions, two-thirds of the daily production of uric acid is excreted in the urine. The remaining fraction is excreted through the intestine where uric acid is cleaved by the action of bacterial uricases.

Epidemiology. Hyperuricemia denotes an elevation of serum urate concentration and may be defined on either a physicochemical or an epidemiologic basis. Physicochemically, serum is saturated with urate at concentrations exceeding 7 mg/dL as measured by the uricase method at 37 °C [4]. In sex- and age-matched populations, relative hyperuricemia is present when the serum urate concentration exceeds two standard deviations above the mean. In postpubertal men, this value is 7 mg/dL; in postpubertal females, 6 mg/dL [4]. Epidemiologically, an increased risk for development of gout and kidney stones exists when the serum urate concentration exceeds 7 mg/dL [4]. In the majority of subjects, however, hyperuricemia is asymptomatic. Only a small fraction of hyperuricemic individuals develop gout [5].

The serum urate concentration varies with age and sex [6]–[8]. Values range from 3 to 4 mg/dL in children. With the onset of puberty, serum urate concentration normally rises in males to 4–6 mg/dL and remains at that level throughout life. In females, the serum urate concentration typically remains unchanged until the menopause at which time it approaches values comparable to those in adult males. Factors such as blood pressure, serum creatinine, height, weight, age, and alcohol intake correlate with serum urate concentration [9]–[12].

Serum urate levels above 7 mg/dL are found in 6.5–9.3% of adult males [4]. Among women, 6.3% have serum urate values >6 mg/dL [4]. The development of gout depends on the degree and duration of serum urate elevation. Therefore, the prevalence of gout increases with age and with the degree of hyperuricemia [11], [12]. The annual incidence of gout is 0.1% for urate levels less than 7 mg/dL and 4.9% for levels greater than 9 mg/dL [13].

Symptoms and Treatment. Gout is manifested clinically by recurrent episodes of acute arthritis. Intraleukocytic crystals of monosodium urate monohydrate are found in synovial fluid from inflamed joints [14].

An acute attack of gout is generally self-limited and confined to a single joint (monoarticular). Recurrent episodes follow, which tend to become polyarticular. The frequency of attacks increases with time and leads to disabling chronic arthritis. Prompt diagnosis and implementation of therapy at the onset of an acute attack shorten the episode and relieve the disabling pain. Prophylactic therapy aimed at preventing recurrent arthritis significantly reduces the risk of chronic arthritis and joint deformity [15]. Antihyperuricemic agents lower the serum urate concentration and thus promote solubilization of tophi. Reduction of serum urate levels also decreases the risks of recurrent arthritis, uric acid nephrolithiasis, and urate nephropathy, as well as the development of chronic gouty arthritis.

2. Gout and Hyperuricemia

2.1. Pathophysiology

Hyperuricemia may result from insufficient excretion of uric acid by the kidneys, endogenous overproduction of uric acid, or a combination of both. *Primary hyperuricemia* develops independently of other diseases and is unrelated to drug use. *Secondary hyperuricemia* is either a manifestation of an underlying disease or the result of an ingested drug or toxin.

Impaired Renal Uric Acid Excretion. The most common cause of hyperuricemia is diminished uric acid excretion by the kidneys. Although this accounts for approximately 90% of cases of primary hyperuricemia [16], the reason has not been established; impaired urate clearance has been documented in these patients [17]–[21].

Underlying renal disease, drugs, or toxins may also impair the ability of the kidney to excrete uric acid and result in secondary hyperuricemia. Intravascular volume contraction resulting from diuretic therapy, adrenal insufficiency, or liver disease leads to secondary hyperuricemia due to diminished glomerular filtration and increased tubular reabsorption of uric acid [22], [23]. Drugs such as ethanol, ethambutol, and nicotinic acid may diminish renal clearance of uric acid [24]. Organic acids inhibit uric acid secretion by the kidneys. Therefore, diseases resulting in metabolic acidosis may also lead to hyperuricemia [24].

Overproduction of Uric Acid. Endogenous overproduction of uric acid accounts for approximately 10% of cases of primary hyperuricemia [16]. Overproduction is characterized by a high level of uric acid in the urine (hyperuricaciduria), which is the result of accelerated purine synthesis. The etiology of urate overproduction in primary hyperuricemia is unknown in the majority of cases, but it may be due to specific inborn errors of metabolism. Patients with 5-phosphoribosyl-1-pyrophosphate synthetase overactivity or partial hypoxanthine phosphoribosyltransferase (HPRT) deficiency demonstrate primary hyperuricemia [4].

Endogenous urate overproduction caused by the following factors may also lead to secondary hyperuricemia:

1) Inborn enzyme defects resulting in overproduction of uric acid, e.g., the Lesch–Nyhan syndrome, glucose 6-phosphatase deficiency, and the glycogen storage diseases types I, III, V, and VII [24]
2) Myeloproliferative disorders leading to accelerated nucleic acid turnover
3) Increased adenosine triphosphate (ATP) degradation caused by alcohol consumption or tissue hypoxia [24]

Urate Precipitation. The acute attack of gout is triggered by precipitation of monosodium urate monohydrate crystals in or near a joint [6], [25], [26]. The cause of this precipitation is unclear, and hyperuricemia is not essential. The inflammatory response to intra-articular urate crystals includes activation of the kallikrein–kinin and complement systems, formation of leukotrienes, platelet activation, prostaglandin synthesis [16], and leukocyte migration. The leukocytes phagocytose the urate crystals and release chemotactic factors. Intracellular urate crystals cause lysosomal disruption and cell injury. Release of lysosomal enzymes into the joint space leads to further inflammatory changes including increased capillary permeability and tissue damage. For a detailed description of inflammation, see → Anti-inflammatory – Antirheumatic Drugs.

2.2. Clinical Manifestations

Articular Disease. The clinical spectrum of gout encompasses four stages:

1) asymptomatic hyperuricemia,
2) acute gouty arthritis,
3) intercritical gout, and
4) chronic tophaceous gout [16].

Asymptomatic hyperuricemia is characterized solely by elevation of the serum urate concentration. *Acute gouty arthritis* is usually the first clinical manifestation of gout and often occurs after years of asymptomatic hyperuricemia. The first attack is usually monoarticular and develops rapidly. The most commonly affected joint is the metatarsal–phalangeal joint of the big toe [27], but other joints may be affected [16]. Pain, swelling, and redness develop and intensify within a few hours. Untreated acute gout is generally self-limited; symptoms persist from several hours to weeks in severe attacks.

In *intercritical gout* the first attack of gouty arthritis is followed by a variable, asymptomatic period. The majority of patients experience the second attack within six months to two years, but the asymptomatic (intercritical) period may be ten years or longer [16]. During the phase of intercritical gout, the asymptomatic periods between acute attacks shorten. The arthritis becomes polyarticular, and if untreated, chronic persistent arthritis may develop [16], [24].

Chronic tophaceous gout is characterized by chronic polyarticular arthritis and development of tophi in and around joints and in soft tissues [16]. Joint erosion and destruction occur, which may be accompanied by marked deformity and crippling of hands and feet.

Renal Disease. The clinical manifestations of hyperuricemia are not limited to articular disease; three types of renal disease are also found: (1) uric acid nephropathy, (2) urate nephropathy, and (3) nephrolithiases.

Uric acid nephropathy results from precipitation of uric acid within the collecting ducts of the kidneys [16]. It is typically associated with marked hyperuricemia and hyperuricaciduria. Uric acid nephropathy is seen most frequently in lymphoproliferative disorders following chemotherapy [28]–[30], when the large nucleic acid pool liberated by tumor cell lysis leads to marked hyperuricemia. Acute renal failure may occur.

Urate nephropathy is caused by deposition of monosodium urate monohydrate in the renal interstitium [16]. Unlike uric acid nephropathy, it does not cause acute renal failure but may contribute to the chronic renal insufficiency of gouty patients.

Nephrolithiasis. Gouty patients have an increased prevalence of kidney stones [16], [31], [32]. Renal calculi occur in 10–25% of patients with primary gout and in as many as 42% of patients with secondary gout. The incidence of stone formation increases with increasing serum urate concentration and urinary uric acid excretion [31], [32].

Other Disorders. Several other disorders are associated with hyperuricemia and gout; these include hypertension, obesity, hyperlipidemia, and diabetes mellitus, as well as cerebrovascular and cardiovascular disease [16]. No direct cause and effect relationship has been established between these disorders and hyperuricemia.

3. Treatment

Medical management of gout is undertaken to:

1) promote rapid resolution of acute arthritis;
2) reduce the likelihood of recurrent gouty attacks;
3) lower the serum urate concentration, thereby inhibiting urate deposition and promoting dissolution of tophi; and
4) prevent uric acid nephrolithiasis.

3.1. Drugs Used to Treat Acute Gout

3.1.1. Colchicine

Colchicine [64-86-8], (S)-N-(5,6,7,9-tetrahydro-1,2-3,10-tetramethoxy-9-oxobenzo[α] heptalen-7-yl)acetamide, is an alkaloid of the plant *Colchicum autumnale*. The plant has been used in the treatment of articular disorders since the sixth century A.D. [33] and colchicine has been used to treat gout for over two centuries. This anti-inflammatory drug is uniquely effective against gouty arthritis (see also → Anti-inflammatory – Antirheumatic Drugs).

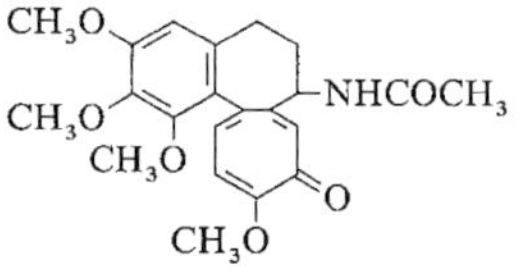

Colchicine

Colchicine impairs leukocyte migration, thereby inhibiting the inflammation responsible for acute gouty attacks [34]. Colchicine has also been shown to inhibit lysosomal degranulation, phagocytosis, and neutrophil adherence [34]–[39]. The drug binds to microtubular structures and arrests cell division [33].

Colchicine is administered orally or intravenously [33]. Peak plasma levels occur 0.5 to 2 h after an oral dose. The drug is eliminated principally through feces; ca. 10 – 20 % is eliminated in the urine.

Oral colchicine is used to treat acute gouty arthritis [16]. The drug is most effective when administered within the first few hours of an attack at a dose of 0.5 or 0.6 mg/h until symptoms subside or gastrointestinal side effects develop (maximum total dosage 6.0 mg). Symptoms usually improve within 12 h and often completely resolve within three days. As many as 50 – 80 % of patients develop gastrointestinal side effects before relief from arthritis occurs. Daily oral colchicine (0.5 – 1.0 mg) is useful in preventing recurrent episodes of acute gout [16].

Intravenous colchicine effectively terminates acute attacks of gout without causing gastrointestinal side effects [16]. It is administered intravenously as a single 2-mg dose in 20 mL of saline, followed by 1-mg dosages every 6 h up to a maximum of 4 – 5 mg. Complications include local extravasation with inflammation and skin necrosis, as well as bone marrow suppression.

3.1.2. Nonsteroidal Anti-inflammatory Drugs

Several nonsteroidal anti-inflammatory drugs have been employed in the treatment of acute gout. These drugs are thought to exert their anti-inflammatory effects by inhibiting prostaglandin synthesis, inhibiting leukotriene synthesis, and preventing free-radical formation [40].

Prostaglandins are potent mediators of inflammation. The nonsteroidal anti-inflammatory agents reversibly inhibit cyclooxygenase, an enzyme that converts arachidonic acid to prostaglandin precursors [40]. Prostaglandin synthesis is thereby suppressed.

Leukotrienes, also derived from arachidonic acid, have been implicated in leukocyte aggregation, degranulation, and free-radical formation [40]. Some nonsteroidal anti-inflammatory agents inhibit leukotriene synthesis and presumably modify the inflammatory response by interfering with these mechanisms.

Oxygen free radicals are formed during the synthesis of prostaglandins and can cause tissue injury. Inhibition of prostaglandin synthesis by nonsteroidal agents suppresses free-radical formation [40].

(For physical properties, syntheses, and trade names of the drugs treated in this section see → Anti-inflammatory – Antirheumatic Drugs).

Indomethacin [53-86-1], 1-(4-chlorobenzoyl)-5-methoxy-2-methyl-1*H*-indole-3-acetic acid (Indocin) is rapidly absorbed from the gastrointestinal tract. Peak plasma levels are obtained within 3 h of ingestion [33], and the drug is highly bound to plasma proteins. It is eliminated in both urine and feces, 10 – 20 % being excreted unchanged in the urine. Approximately 50 % of a given dose undergoes O-demethylation, and 10 % is conjugated with glucuronic acid.

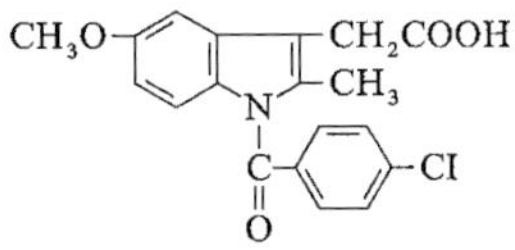

Indomethacin

For treatment of acute gout, indomethacin is administered at a 50 – 75-mg dosage, followed by 50 mg every 6 h for a maximum dose of 200 mg in the first 24 h [16]. The dosage is subsequently reduced [16].

Indomethacin may also be used for gout prophylaxis [16]. A daily dose of 25 – 50 mg is usually sufficient to prevent recurrence of acute attacks.

Phenylbutazone [50-33-9], 4-butyl-1,2-diphenyl-3,5-pyrazolidinedione (Butazolidin) has been used for over 30 years to treat acute gout [41], [42]. After oral administration, peak plasma concentration is reached within 2 h [33]. The drug is completely absorbed from the gastrointestinal tract and 98 % is bound to plasma proteins. Plasma half-life ranges from 50 to 100 h. The drug undergoes glucuronidation and hydroxylation, and is excreted primarily in the urine. Oxyphenylbutazone, an active metabolite of phenylbutazone, has a long plasma half-life and contributes to the side effects of phenylbutazone. Oxyphenylbutazone is also excreted in the urine after glucuronidation [33].

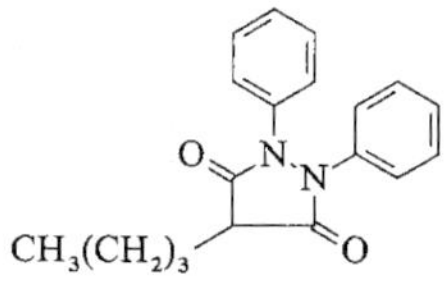

Phenylbutazone

For treatment of acute gout, phenylbutazone is administered at an initial dosage of 200 mg – 600 mg [16], followed by 100 – 200 mg every 6 – 8 h for 24 h (maximum dose = 600 mg/d). The dosage is then reduced and discontinued after two days.

Side effects of phenylbutazone include hypersensitivity reactions, gastric discomfort, peptic ulcer disease, and fluid retention [33]. Deaths from aplastic anemia have been reported [43].

Naproxen [*22204-53-1*], (S)-6-methoxy-α-methyl-2-naphthaleneacetic acid (Naprosyn) was effective in acute gout management in several clinical trials [44]. It is fully absorbed after oral administration [33]. Plasma half-life is 12 – 15 h. The drug is extensively bound to plasma protein and eliminated in the urine. Approximately 30 % undergoes 6-demethylation prior to elimination. To treat acute gout, naproxen is administered at an initial dose of 500 – 750 mg, followed by 250 mg every 8 h until the acute attack subsides [16].

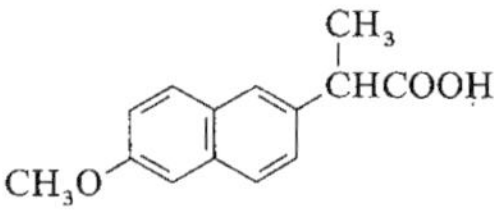

Naproxen

Ibuprofen [*15687-27-1*], α-methyl-4-(2-methylpropyl)benzeneacetic acid (Motrin) is absorbed efficiently from the gastrointestinal tract [33]. Maximum plasma levels are achieved within 1 – 2 h, and its half-life is approximately 2 h. Like naproxen, ibuprofen is almost completely bound to plasma proteins and is eliminated principally via the urine. It is metabolized prior to elimination in the urine. Ibuprofen has demonstrated efficacy in the treatment of acute gout [45]. A dosage of 800 mg every 8 h is given until symptoms are resolved [16]. The dose may then be reduced.

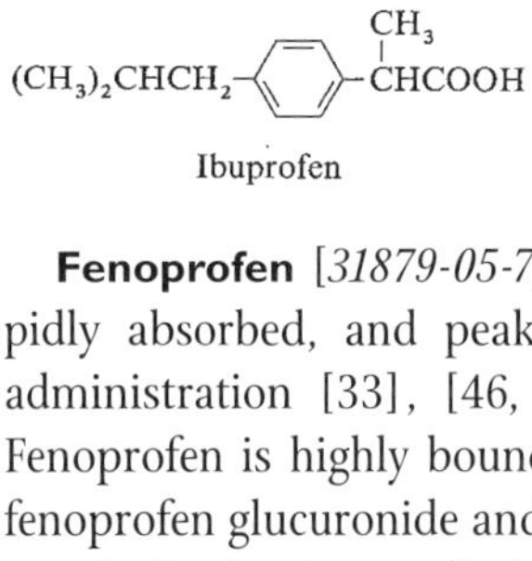

Ibuprofen

Fenoprofen [*31879-05-7*], (±)-α-methyl-3-phenoxybenzeneacetic acid (Nalfon) is rapidly absorbed, and peak plasma concentrations are achieved within 2 h of oral administration [33], [46, pp. 890 – 891]. The plasma half-life is approximately 3 h. Fenoprofen is highly bound to plasma protein, and its major urinary metabolites are fenoprofen glucuronide and 4′-hydroxyfenoprofen glucuronide. Over 90 % of the drug is metabolized prior to elimination in the urine. Successful termination of an acute gout attack may be achieved with a dosage of 800 mg every 6 h [16]. The dose is then tapered when symptomatic improvement occurs.

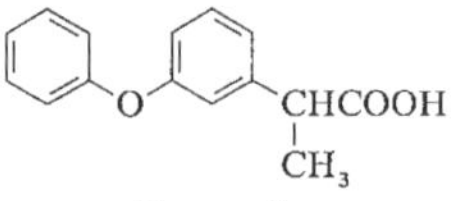

Fenoprofen

Sulindac [*38194-50-2*], (*Z*)-5-fluoro-2-methyl-1-{[4-(methylsulfinyl)phenyl]methylene}-1*H*-indene-3-acetic acid, (Clinoril) is approximately 90% absorbed after oral administration [33]. The active sulfide metabolite is formed by oxidation to the sulfone and subsequent reduction. Sulindac has a plasma half-life of 7 h; the sulfide metabolite has a half-life of 18 h. Sulindac and its metabolites are extensively protein bound. Excretion of the drug occurs in the urine and feces. Approximately 50% of the administered dose is eliminated in the urine, the major product being the conjugated sulfone. Sulindac administered at a dosage of 200 mg twice a day is effective in terminating an acute attack of gout [47].

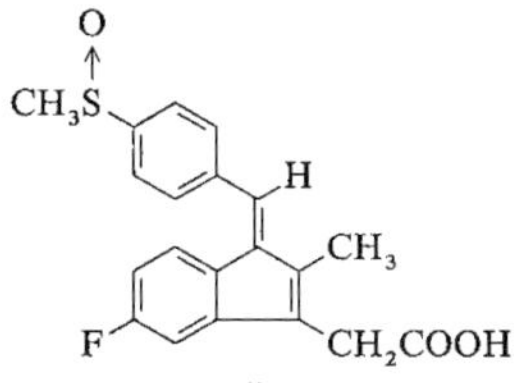

Sulindac

Piroxicam [*36322-90-4*], 4-hydroxy-2-methyl-*N*-2-pyridyl-2*H*-1,2-benzothiazine-3-carboxamide-1,1-dioxide (Feldene) is well absorbed after oral administration, and peak plasma concentrations are achieved within 3–5 h [46, pp. 1558–1559]. Mean plasma half-life is 50 h. The drug is excreted in both urine and feces, the majority being excreted as metabolites formed by hydroxylation and conjugation, cyclodehydration, and hydrolysis of the amide linkage, decarboxylation, ring contraction, or *N*-demethylation. Piroxicam administered at a daily dose of 20 mg, effectively suppresses or completely relieves symptoms of acute gouty arthritis within five days [48]–[50].

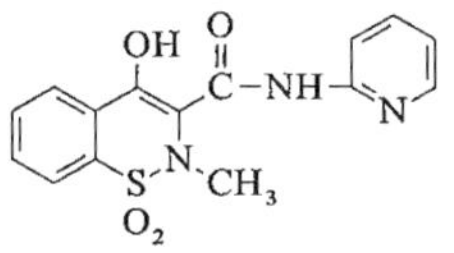

Piroxicam

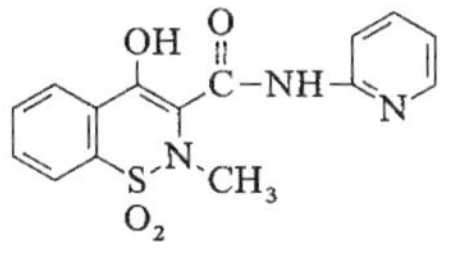

3.1.3. Corticosteroids

Corticosteroids have both anti-inflammatory and immunomodulatory properties [16]. They inhibit arachidonic acid release from phospholipids, thereby inhibiting prostaglandin synthesis. They are also purported to stabilize lysosomal membranes, inhibit cellular metabolism, decrease serum immunoglobulins, and diminish synovial permeability. Intra-articular corticosteroids are particularly effective in suppressing acute gouty arthritis [51]. Commonly used intra-articular steroids include methylprednosolone acetate (Depo-medrol), triamcinolone acetonide (Kenalog), and triamcinolone hexacetonide (Aristopan). Relief of symptoms usually occurs within 12 h. This form of treatment may be employed when one or two joints are involved and the patient is unable to tolerate other forms of therapy. Many different corticosteroid preparations are available. For structure, properties, and preparation, see → Hormones.

3.2. Antihyperuricemic Drugs

The antihyperuricemic drugs consist of (1) uriscosuric drugs which increase urinary uric acid excretion and (2) xanthine oxidase inhibitors which inhibit uric acid synthesis. The aim of antihyperuricemic therapy is to lower the concentration of serum urate below 6.4 mg/dL [16].

The selection of an antihyperuricemic agent should be based on the cause of the hyperuricemia and the condition of the patient. Patients without kidney problems who excrete less than 700 mg of uric acid per day may be treated with either uricosuric agents or xanthine oxidase inhibitors [24]. Patients who excrete more than 700 mg of uric acid a day should be treated with xanthine oxidase inhibitors. Antihyperuricemic agents should not be given until an acute gouty attack has subsided because they may exacerbate such attacks [16]. In addition, they may also precipitate the onset of acute gouty arthritis. Patients should, therefore, be treated prophylactically with colchicine or indomethacin prior to the introduction of antihyperuricemic medication.

3.2.1. Uricosuric Drugs

Uricosuric drugs increase the rate of excretion of uric acid by interfering with its reabsorption by the renal tubules [52]. They reduce serum urate levels in the majority of patients but are generally ineffective in patients with marked renal insufficiency [16]. Concomitant administration of salicylate blocks their uricosuric effect [52].

The use of uricosuric drugs should be accompanied by liberal intake of fluids to prevent crystallization of uric acid within the urinary tract. Alkalinization of urine by the administration of sodium hydrogen carbonate or potassium citrate may prevent formation of kidney stones [24].

Probenecid [57-66-9], 4-[(dipropylamino)sulfonyl]benzoic acid (Benemid) inhibits epithelial transport of organic acids [52]. It is completely absorbed after oral administration, with peak plasma levels achieved in 2–4 h. It is extensively bound by plasma proteins and has a plasma half-life of approximately 9 h. Probenecid is eliminated in the urine. Metabolism may occur by hydroxylation or conjugation to glucuronide [52].

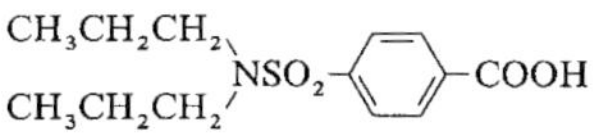

CH₃CH₂CH₂\
       NSO₂—⟨ ⟩—COOH
CH₃CH₂CH₂/
Probenecid

Probenecid is administered at a dose of 250 mg twice a day for one week; the dosage is then increased up to a maximum of 2 g per day [46, pp. 1246–1247].

Adverse side effects to probenecid include gastrointestinal irritation, hypersensitivity reactions, and central nervous system abnormalities [52].

Sulfinpyrazone [*57-96-5*], 1,2-diphenyl-4-[2-(phenylsulfinyl)ethyl]-3,5-pyrazolidine-dione (Anturane) is a strong organic acid [52]. Although sulfinpyrazone inhibits reabsorption of uric acid, like probenecid it interferes with tubular secretion of other organic acids. It is absorbed efficiently from the gastrointestinal tract and is extensively bound to plasma proteins. This drug is excreted in the urine, with 10% being the N1-*p*-hydroxyphenyl metabolite.

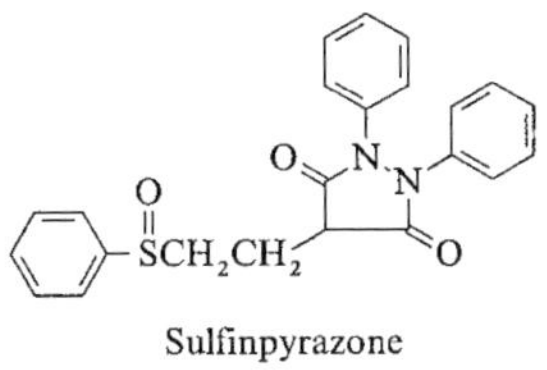

Sulfinpyrazone

Sulfinpyrazone is administered initially at a dose of 50–100 mg twice a day, which can be increased gradually to a daily dosage of up to 800 mg [46, pp. 833–834].

Side effects include gastrointestinal irritation, hypersensitivity reactions, and blood dyscrasias [52]. In addition, this drug may potentiate the action of certain sulfonamides such as sulfadiazine and sulfisoxazole [46, pp. 833–834].

Benzbromarone [*3562-84-3*] (3,5-dibromo-4-hydroxyphenyl)(2-ethyl-3-benzofuranyl)methanone, is a potent uricosuric agent with demonstrated efficacy in reducing serum urate levels [53]–[55]. Onset of uricosuric activity is rapid, with peak uric acid excretion occurring at approximately 5 h [53]. Benzbromarone does not affect tubular secretion of other organic acids. The usual maintenance dose is between 50 and 100 mg a day [53]–[56]. Side effects include diarrhea, allergy, and urolithiasis [53], [54], [56].

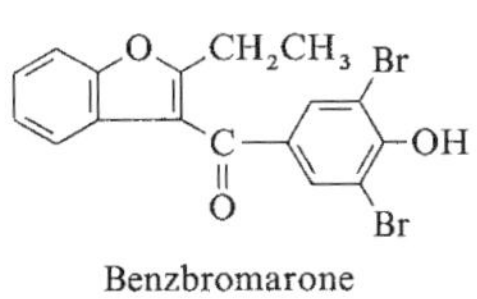

Benzbromarone

Benziodarone [*68-90-6*] (2-ethyl-3-benzofuranyl)(4-hydroxy-3,5-diiodophenyl)methanone, and *zoxazolamine* [*61-80-3*], 5-chloro-benzoxazolamine, have uricosuric activity [57], [58], but neither agent is used clinically in the United States to manage hyperuricemia.

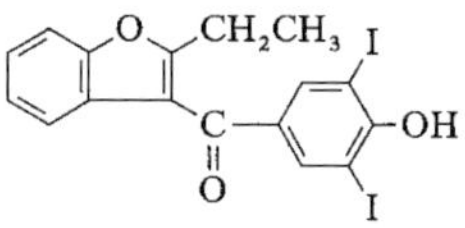

Benziodarone

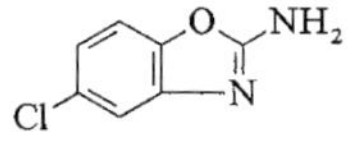

Zoxazolamine

3.2.2. Xanthine Oxidase Inhibitors

Xanthine oxidase catalyzes the oxidation of hypoxanthine to xanthine and of xanthine to uric acid [24]. Allopurinol and its oxidation product oxipurinol are structural analogues of hypoxanthine and xanthine, respectively. Both inhibit xanthine oxidase, thereby diminishing uric acid synthesis and lowering serum urate concentration.

Allopurinol [*315-30-0*], 1,5-dihydro-*H*-pyrazolo[3,4-*d*]pyrimidin-4-one (Zyloprim) is absorbed rapidly from the gastrointestinal tract [33]. Peak plasma levels occur within 30–60 min of an oral dose, and plasma half-life is 2–3 h. Allopurinol is metabolized to oxipurinol [*2465-59-0*]. Oxipurinol, 1*H*-pyrazolo[3,4-*d*]-pyrimidine-4,6(5*H*, 7*H*)-dione is cleared slowly from plasma and has a half-life of 15–20 h [46, pp. 818–820]. Both compounds are biologically active and neither is protein-bound [33]. Only 10% of a single dose of allopurinol is excreted unchanged in the urine.

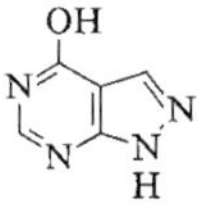

Allopurinol

The usual dosage of allopurinol is 300 mg administered once a day. The serum urate concentration falls within the first two days of therapy, and the minimum serum urate construction is achieved within two weeks [24]. The dose of allopurinol may gradually be increased to a maximum of 800 mg daily [46, pp. 818–820]. Dosages above 300 mg should be given twice daily. In patients with renal insufficiency, the half-life of oxipurinol is increased and the dose of allopurinol should be reduced.

Side effects of allopurinol therapy include acute gouty arthritis, hematologic abnormalities, gastrointestinal intolerance [33], and life-threatening hypersensitivity reactions.

Allopurinol may interfere with the hepatic metabolism of oral anticoagulant drugs and must be used with caution in this context [33]. Azathioprine and 6-mercaptopurine are potentiated when administered concomitantly with allopurinol. The combined administration of allopurinol and uricosuric drugs may improve serum uric acid control [33].

3.3. Dietary Management

The aims of dietary management in gout are to reduce body weight, alcohol intake, and ingestion of foods known to trigger acute attacks [16]. Patients with gout are often overweight [16]; weight reduction may lower the serum urate concentration [59].

Alcohol intake may lead to hyperuricemia and gout [60], therefore patients with gout should reduce or discontinue alcohol consumption.

Ingestion of certain, often purine-rich, foods may be associated with the development of acute gouty arthritis in some patients [16]. Specific foods known to precipitate acute attacks of gout should be avoided. Examples of purine-rich foods are herrings, sardines, liver, sweetbreads, kidneys, and brewer's yeast. Severe dietary restriction of purine is usually unnecessary, but patients should avoid excessive purine intake. High-protein diets should also be avoided because endogenous purine production is increased by excessive protein consumption [16].

Obesity, hyperlipidemia, and hypertension are associated with gout and hyperuricemia [16]. Management of these conditions with diet, exercise and appropriate medications is important in prolonging life and preventing the complications of these associated conditions.

This work is supported by National Institutes of Health Grant Number DK-19045. Dr. Palella is the recipient of an Arthritis Investigator Award from the Arthritis Foundation.

4. References

[1] Hippocrates: *The Genuine Works of Hippocrates*, vols. **1** and **2**, translated from the Greek with a preliminary discourse and annotations by Francis Adams, W. Wood & Co, New York 1886.
[2] T. Sydenham: *Tractatus de Podagra et Hydrope*, G. Kettilby, London 1683.
[3] A. B. Garrod, *Trans. M-Chir. Soc. Edinburgh* **31** (1848) 83–97.
[4] J. B. Wyngaarden, W. N. Kelley: *Gout and Hyperuricemia*, Grune & Stratton, New York 1976.
[5] H. E. Paulus, A. Coutts, J. J. Calabro, J. R. Klinenberg, *J. Am. Med. Assoc.* **211** (1970) 277–281.
[6] W. M. Mikkelsen, H. J. Dodge, H. Valkenburg, *Am. J. Med.* **39** (1965) 242–251.
[7] S. Akizuki, *Ann. Rheum. Dis.* **41** (1982) 272–274.
[8] W. R. Harlan, J. Cornoni-Huntley, P. E. Leaverton, *Pediatrics* **63** (1979) 569–575.
[9] W. R. Harlan, A. L. Hull, R. P. Schmouder, F. E. Thompson et al.: "Dietary intake and cardiovascular Risk Factors," part 2, *Serum Urate, Serum Cholesterol and Correlates, in Vital and Health Statistics, United States, 1971–1975*, Series 11, no. 227, Dept. of Health & Human Services (DHHS) Pub. no. (PHS) 83–1677, Public Health Service, Washington, D.C., Mar. 1983.

[10] D. Kuntz, J. M. Chretien, A. Ryckewart, *Sem. Hop. Paris* **55** (1979) 241–248.
[11] J. Zalokar, J. Lellouch, J. R. Claude, *Sem. Hop. (Paris)* **57** (1981) 664–670.
[12] K. Nishioka, K. Mikanagi, *Adv. Exp. Med. Biol.* **122 A** (1980) 155.
[13] E. W. Campion, R. J. Glynn, L. O. DeLabry, *Am. J. Med.* **82** (1987) 421–426.
[14] D. J. McCarty, J. L. Hollander, *Ann. Intern. Med.* **54** (1961) 452–460.
[15] T-F. Yu, A. B. Gutman, *Ann. Intern. Med.* **55** (1961) 179–192.
[16] W. N. Kelly, I. H. Fox in W. N. Kelley, E. D. Harris, S. Ruddy, C. B. Sledge (eds.): *Textbook of Rheumatology*, 2nd ed., W. B. Saunders Co., Philadelphia 1985, pp. 1359–1398.
[17] C. A. Nugent, W. D. MacDiarmid, F. H. Tyler, *Arch. Intern. Med.* **113** (1964) 115–121.
[18] J. T. Scott, M. L. Snaith, *Ann. Rheum. Dis.* **30** (1971) 285–289.
[19] J. B. Haupt, M. A. Ogryzla, *Arthritis Rheum.* **7** (1964) 316.
[20] J. E. Seegmiller, A. I. Grayzel, R. R. Howell, C. Plato, *J. Clin. Invest.* **41** (1962) 1094–1098.
[21] W. Latham, G. P. Rodman, *J. Clin. Invest.* **41** (1962) 1955–1963.
[22] I. H. Fox in J. A. Spittell (ed.): *Clinical Medicine*, Harper & Row, Philadelphia 1982.
[23] E. I. Feinstein, V. H. Quion, E. M. Kaptein, S. G. Massry, *Am. J. Nephrol.* **4** (1984) 77.
[24] T. D. Palella, I. H.Fox in C. R. Scriver, A. L. Beaudet, W. S. Sly, D. Valle (eds.): *Methabolic Basis of Inherited Disease*, 6th ed., McGraw-Hill, New York 1988.
[25] N. J. Zvaifler, T. J. Pekin, *Arch. Intern. Med.* **111** (1963) 99–102.
[26] J. E. Seegmiller, L. Laster, R. R. Howell, *N. Engl. J. Med.* **268** (1963) 712–716.
[27] C. Scudamore: *A Treatise on the Nature and Cure of Gout and Rheumatism, including General Consideration on Morbid States of the Digestive Organs: Some Remarks on Regimen and Practical Observations on Gravel*, 2nd ed., Longman, London 1817.
[28] E. Frei, C. J. Bentzel, R. R. Rieselbach, J. B. Block, *J. Chronic Dis.* **16** (1963) 757–776.
[29] R. E. Rieselbach et al., *Am. J. Med.* **37** (1964) 872–884.
[30] C. M. Kjellstrand, D. C. Campbell, B. von Hartitzsch, T. J. Buselmeier, *Arch. Intern. Med.* **133** (1974) 349–359.
[31] A. B. Gutman, T.-F. Yu, *Am. J. Med.* **45** (1968) 757–779.
[32] T-F. Yu, A. B. Gutman, *Ann. Intern. Med.* **67** (1967) 1133–1148.
[33] R. J. Flower, S. Moncada, J. R. Vane in A. G. Gilman L. S. Goodman, A. Gilman (eds.): *The Pharmacological Basis of Therapeutics*, 6th ed., Macmillan Publ. Co., New York 1980, pp. 682–728.
[34] S. E. Malawista, *Arthritis Rheum.* **18** (1975) 835–846.
[35] R. B. Zurier, S. Holfstein, G. Heissman, *J. Cell Biol.* **58** (1973) 27–41.
[36] Y. H. Cheng, *Arthritis Rheum.* **11** (1968) 473.
[37] R. I. Lehrer, *J. Infect. Dis.* **127** (1973) 40–48.
[38] R. Penney, D. A. G. Galton, J. T. Scott, V. Eifen, *Br. J. Haematol.* **12** (1966) 623–632.
[39] E. Dellaverde. P. T. Fan, Y. H. Chang, *J. Pharmacol. Exp. Ther.* **223** (1982) 197–202.
[40] H. E. Paulus, D. E. Furst in D. J. McCarty (ed.): *Arthritis and Allied Conditions*, 10th ed., Lea & Febiger, Philadelphia 1985, pp. 453–486.
[41] R. H. Freyberg, *Arthritis Rheum.* **5** (1962) 624–632.
[42] A. B. Gutman, *Arthritis Rheum.* **8** (1965) 911–920.
[43] M. F. Cuthbert, *Curr. Med. Res. Opin.* **2** (1974) 600–610.
[44] R. F. Willkins, J. B. Case, F. J. Huix, *J. Clin. Pharmacol.* **15** (1975) 363–366. R. A. Sturge et al., *Adv. Exp. Med. Biol.* **76 B** (1977) 290–296.
[45] M. C. Schweitz, D. J. Nashel, F. P. Alepa, *J. Am. Med. Assoc.* **239** (1978) 34–35.
[46] E. R. Barnhart (ed.): *Physicians Desk Reference*, 41st ed., Medical Economics Co., Oradell, N. J., 1987, pp. 890–891.

[47] G. N. Karachalios, G. Donas, *Int. J. Tissue React.* **4** (1982) 297–299.
[48] E. Murpy, *J. Int. Med. Res.* **7** (1979) 507–510.
[49] G. N. Karachalios, *Clin. Exp. Rheum.* **3** (1985) 63–65.
[50] R. H. Bluestone, *Am. J. Med.* **72** (1982) 66–70.
[51] J. L. Hollander, J. Bone, *Joint Surg.* **35 A** (1953) 983–990.
[52] G. H. Mudge in A. G. Gilman, L. S. Goodman, A. Gilman (eds.): *The Pharmacological Basis of Therapeutics*, 6th ed., Macmillan Publ. Co., New York 1980, pp. 929–934.
[53] A. Masbernard, C. P. Giudicelli, *S. Afr. Med. J.* **59** (1981) 701–706.
[54] W. A. M. Scott, *J. Int. Med. Res.* **9** (1981) 82–87.
[55] G. W. Schepers, *J. Int. Med. Res.* **9** (1981) 511–515.
[56] R. Bluestone, J. Klinenberg, I. K. Lee, *Adv. Exp. Med. Biol.* **122 A** (1980) 283–286.
[57] G. Lemieux, P. Vinay, A. Gougoux, G. Michaud, *Am. J. Phys.* **224** (1973) 1440–1449.
[58] J. J. Burns, T.-F. Yu, L. Berger, A. B. Gutman, *Am. J. Med.* **25** (1958) 401–408.
[59] B. T. Emmerson, *Aust. N. Z. J. Med.* **3** (1973) 410.
[60] J. Faller, I. H. Fox, *N. Engl. J. Med.* **307** (1982) 1598–602.

Immunotherapy and Vaccines

STANLEY J. CRYZ, JR., Swiss Serum and Vaccine Institute Berne, Switzerland (Chaps. 1 and 2)

MARTA GRANSTROM, Departments of Clinical Microbiology and of Vaccine Production, National Bacteriological Laboratory, Karolinska Hospital, Stockholm, Sweden (Chap. 3)

BRUNO GOTTSTEIN, Institut für Parasitologie, Universität Zürich, Zürich, Switzerland (Section 4.1)

LUC PERRIN, Division d'Hematologie, Hôpital Cantonal Universitaire, Genève, Switzerland (Section 4.2)

ALAN CROSS, Department of Bacterial Diseases, Walter Reed Army Institute of Research, Washington D.C. 20307-5100, United States (Chap. 5)

JAMES LARRICK, Genelabs Incorporated, Redwood City, California, United States (Chap. 6)

1.	Introduction	1755
1.1.	Historical Aspects	1755
1.2.	Principles and Definitions	1757
1.2.1.	Antigens	1757
1.2.2.	Antibodies	1758
1.2.3.	Immune Response	1760
1.2.4.	Active Immunization	1762
1.2.5.	Passive Immunization	1763
1.2.6.	Genetic Engineering	1764
2.	Bacterial Vaccines	1764
2.1.	Diphtheria Vaccine	1764
2.2.	Tetanus Vaccine	1766
2.3.	Pertussis Vaccine	1768
2.4.	Typhoid Fever Vaccine	1770
2.5.	*Streptococcus pneumoniae* Vaccine	1772
2.6.	Shigella Vaccines	1774
2.7.	Cholera Vaccine	1775
2.8.	Vaccines Against Nosocomial Pathogens	1776
2.9.	Meningococcal Meningitis Vaccine	1779
2.10.	Tuberculosis Vaccine	1780
2.11.	*Escherichia coli* Vaccines	1782
2.12.	*Neisseria gonorrhoeae* Vaccine	1784
2.13.	*Hemophilus influenzae* Type b Vaccines	1785
3.	Viral Vaccines	1787
3.1.	Measles Vaccine	1787
3.2.	Mumps Vaccine	1788
3.3.	Rubella Vaccine	1790
3.4.	Combined Measles – Mumps – Rubella Vaccine	1791
3.5.	Polio Vaccine	1792
3.6.	Hepatitis B Vaccine	1794
3.7.	Rabies Vaccine	1796
3.8.	Influenza Vaccine	1798
3.9.	Varicella Vaccine	1799
3.10.	Yellow Fever Vaccine	1801
3.11.	Tick-Borne Encephalitis Vaccine	1802
3.12.	Japanese Encephalitis Vaccine	1803
3.13.	Smallpox Vaccine	1804
3.14.	Rift Valley Fever Vaccine	1805
4.	Vaccines against Parasites	1805
4.1.	Vaccines against Helminths	1805
4.1.1.	Vaccines against Schistosoma	1806
4.1.2.	Vaccines against Nematodes	1808
4.1.2.1.	Gastrointestinal Nematodes	1808
4.1.2.2.	Tissue-Invading Nematodes (Filariidae)	1810

4.1.3.	Vaccines against Cestodes	1811	5.3.	**Prophylaxis with Hyperimmune Globulins** ... 1828
4.2.	**Malaria Vaccine**	1812	5.4.	**Therapy with Immune Serum Globulin** ... 1829
4.2.1.	Strategy for Malaria Vaccine Development	1813	5.5.	**Prophylaxis and Therapy with Intravenous Immunoglobulin (IVIG)** ... 1830
4.2.2.	Sporozoite Vaccines	1815		
4.2.3.	Asexual Blood Stage Vaccine	1816		
4.2.3.1.	Merozoite Surface Antigens	1817	5.5.1.	Viral Infection ... 1830
4.2.3.2.	Rhoptry Antigens	1818	5.5.2.	Bacterial Infection ... 1831
4.2.3.3.	Antigens Associated with the Membrane of Infected Erythrocytes	1819	5.5.3.	Noninfectious Diseases ... 1833
			5.5.3.1.	Therapeutic Effect of IVIG ... 1833
			5.5.3.2.	Mechanism of Action ... 1834
4.2.3.4.	Other Proteins and Synthetic Peptides	1819	5.6.	**Prophylaxis and Therapy with Plasma and Other Blood Products** ... 1835
4.2.4.	Sexual Stages—Transmission Blocking Immunity	1820		
5.	**Immunotherapy**	1821	5.7.	**Adverse Effects of Gamma Globulin Preparations** ... 1836
5.1.	**Gamma Globulin Preparations**	1822	5.8.	**Future Prospects** ... 1838
5.1.1.	Standard Immune Serum Globulin	1822	**6.**	**Immunotherapeutic Uses of Monoclonal Antibodies** ... 1839
5.1.2.	Immunoglobulin for Intravenous Use	1824	6.1.	Introduction ... 1840
5.1.3.	Hyperimmune Globulins and Antitoxins	1825	6.2.	**Bacterial Targets** ... 1841
5.1.4.	Production Requirements	1825	6.3.	**Viral and Chlamydial Targets** ... 1842
5.2.	**Prophylaxis with Immune Serum Globulin**	1826	6.4.	**Parasite Targets** ... 1843
			7.	**References** ... 1844

Abbreviations used in this article:

AIDS	acquired immune deficiency syndrome
BCG	acillus Calmette-Guerin
CMV	Cytomegalovirus
CPS	capsular polysaccharide
CS	circumsporozoite
Da	dalton
DNA	deoxyribonucleic acid
DPT	diphtheria–pertussis–tetanus
DPT-Pol.	diphtheria–pertussis–tetanus–polio
DT	diphtheria–tetanus
FHA	filamentous hemagglutinin
HBIG	human anti-HBV immune globulin
HBsAg	hepatitis B surface antigen
HBV	hepatitis B virus
HIV	human immunodeficiency virus

HRIG	human rabies immune globulin
humab	human monoclonal antibodies
Ig	immunoglobulin
IPV	inactivated polio vaccine
ISG	immune serum globulin
ITP	idiopathic thrombocytopenic purpura
IU	international units
IVIG	intravenous immunoglobulin
Lf	limit of flocculation
LPS	lipopolysaccharide
MMR	measles–mumps–rubella
NANP	asparagine–alanine–asparagine–proline
NVDP	asparagine–valine–aspartic acid–proline
OPV	oral polio vaccine
PFU	plaque forming units
PT	pertussis toxin
RESA	ring-infected erythrocyte surface antigen
RNA	ribonucleic acid
TCID	tissue culture infectious dose

1. Introduction [1]–[9]

1.1. Historical Aspects

Immunization is the most efficient, cost-effective means of preventing infectious diseases. The concept of preventing disease by vaccination is old: in China and India, the practice of "variolation," whereby small quantities of material from disease pustules were used to immunize people against smallpox, was practiced before 1000 B.C. The first "rational" approach to vaccination was taken by JENNER in 1798, who used naturally attenuated cowpox to immunize against smallpox. About 100 years later, PASTEUR introduced vaccines against anthrax and rabies based upon attenuated virulent organisms. The discovery by VON BEHRING in 1890, that serum antibodies could neutralize diphtheria toxin opened the door for a new avenue of vaccine development and passive therapy, whereby preformed antibodies were transferred to at-risk patients. By the beginning of the 20th century, serum obtained from immunized animals was used to treat a variety of diseases including diphtheria and tetanus. The use of human serum followed shortly in 1907.

Since the turn of the century, a wide range of vaccines and antisera have been introduced to manage infectious and noninfectious diseases (Table 1). The development of such agents has required expertise from a variety of disciplines including microbiology, biochemistry, immunology, and molecular biology.

Table 1. Currently available vaccines and immunoglobulins

Type of vaccine or immunoglobulin	Disease
Bacterial vaccines	cholera
	diphtheria
	Hemophilus influenzae b
	meningococcal meningitis
	pertussis
	Streptococcus pneumoniae
	tetanus
	tuberculosis
	typhoid fever
Viral vaccines	hepatitis B
	influenza
	Japanese encephalitis
	measles
	mumps
	polio
	rabies
	Rift Valley fever
	smallpox
	tick-borne encephalitis
	varicella
	yellow fever
Immunoglobulins against bacterial diseases	diphtheria
	Hemophilus influenzae, meningococcal meningitis, *Streptococcus pneumoniae* (polyvalent preparation)
	pertussis
	tetanus
Immunoglobulins against viral diseases	cytomegalovirus
	hepatitis A
	hepatitis B
	human immunodeficiency virus (HIV) (normal intravenous immunoglobulin preparation administered to HIV-positive infants)
	measles
	mumps
	rabies
	rubella
	vaccinia
	varicella
Immunoglobulins against noninfectious diseases	hypogammaglobulinemia
	rhesus factor
	idiotypic thrombocytopenia purpura

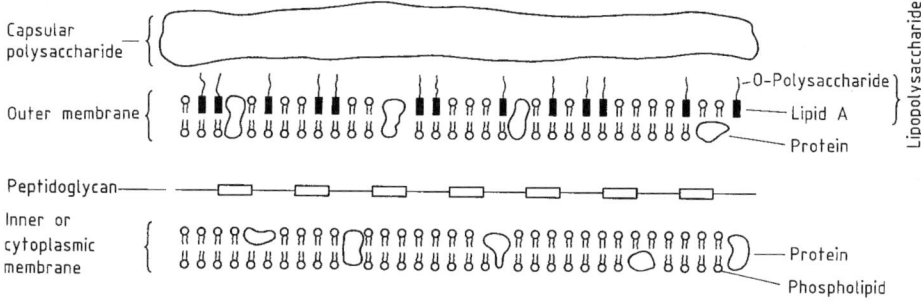

Figure 1. Schematic representation of the gram-negative bacterial cell envelope

1.2. Principles and Definitions

1.2.1. Antigens

An antigen is a molecule that can elicit an immune response (either humoral, i.e., antibody-mediated, or cellular, i.e., cell-mediated) or an immune reaction, such as an allergic reaction. An antigen that evokes an immune response is commonly referred to as an *immunogen*. Only foreign or "non-self" molecules are immunogenic. Usually the larger and more complex a molecule is, the more immunogenic it will be. For example, the gram-negative bacterial cell envelope shown in Figure 1 contains many different somatic (cell-associated) antigens, such as lipopolysaccharide (LPS), outer-membrane proteins, and phospholipids. Numerous factors determine the immunogenicity of a purified molecule. Size is of critical importance: proteins with a molecular mass of $\leq 20\,000$ are poorly immunogenic. Similarly, simple polysaccharides composed of a limited number of repeating monosaccharides are not immunogenic unless their molecular mass exceeds 500 000. Small molecules can be rendered immunogenic by covalently coupling them to larger molecules, forming conjugates.

A single antigen may contain many *epitopes*, which are specific areas of the molecule with a three-dimensional configuration that induces an immune response. Complex molecules, such as large proteins composed of many different amino acids, contain more epitopes than a comparatively simple polysaccharide composed of two or three monosaccharide repeats. The immune response to a given antigen can vary greatly among species and individuals within a species due to immune regulation genes.

Table 2. Immunoglobulin G subclasses *

Property	IgG 1	IgG 2	IgG 3	IgG 4
Percentage of total IgG (adult)	60–70	20–30	5–10	<5
Molecular mass (daltons)	146 000	146 000	170 000	146 000
No. of interchain disulfide bonds	2	4	11	2
No. of amino acids in hinge region	15	12	62	12
Half-life (days)	21–23	20–23	7–8	21–23
Fc receptor binding				
mononuclear cells **	++	+	++	±
neutrophils	++	±	++	+

* Adapted from [239].
** ++ strong binding; + detectable binding; ± binding detected in some, but not all studies.

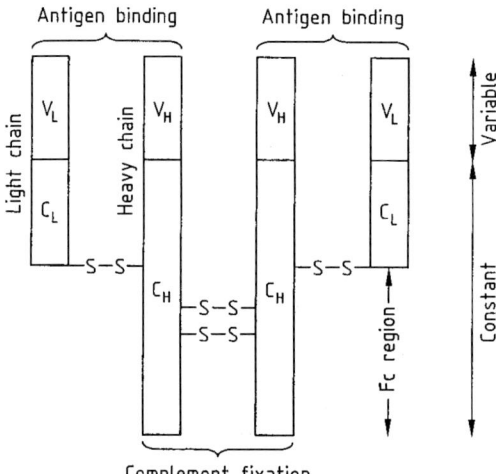

Figure 2. Schematic representation of an immunoglobulin G (IgG) molecule

1.2.2. Antibodies

Antibodies are proteins found primarily in the serum which are produced by B cells in response to contact with a foreign antigen. There are several different classes of antibodies with characteristic functions (see Chap. 5, Table 2). A schematic of an *immunoglobulin G* (IgG) molecule is shown in Figure 2. Immunoglobulins are composed of light and heavy chains held together by disulfide bonds. Each chain has a variable and a constant region. The tertiary structure of the variable region accounts for the specificity of antibody binding. An antibody produced from a given clone of B cells (Section 1.2.3) recognizes and binds to a given epitope or closely related epitopes. Antibodies which recognize more than one epitope are termed *cross-reactive*. The strength with which an antibody binds to an antigen is termed *affinity* and is determined by the "fit" between the immunoglobulin binding site and the epitope.

Antibodies produced by a single clone of B cells are termed *monoclonal antibodies* and recognize only a single epitope (see Chap. 6). *Polyclonal antibodies* are produced by several B cell clones which recognize the same antigen but bind to different epitopes.

Immunoglobulins are divided into five classes termed IgG, IgM, IgA, IgD, and IgE based upon physical and structural differences; they are all glycoproteins. The vast majority of immunoglobulins circulate in the plasma fraction, but certain cells of the immune system can express immunoglobulin on their surface.

Immunoglobulin G (IgG) has a molecular mass of 150 000 and has two antigen binding sites; it represents approximately 80% of all immunoglobulins in normal serum (8–16 mg/mL). Four IgG subclasses (IgG 1–IgG 4) are found which account for ca. 70, 19, 8, and 3% of total IgG, respectively; they differ with respect to their antigenic properties in the constant region of the heavy chains. The functional attributes of these subclasses are detailed elsewhere (see Section 5.1). Immunoglobulin G readily crosses the placenta and provides protection against a variety of infectious diseases in the neonate. For example, immunization of the mother against tetanus shortly before delivery ensures a protective level of antibody for the infant. Immunoglobulin G also diffuses into the extravascular tissue more readily than other immunoglobulin classes. The IgG antibody is thought to be responsible for neutralizing the majority of bacterial toxins (e.g., tetanus and diphtheria toxin) formed during an infection. Upon binding with invading bacteria, IgG activates the complement system (a group of interacting serum proteins) which attracts phagocytic cells. The binding of complement components to the Fc region of IgG permits the uptake and killing of the bacteria by phagocytes.

Immunoglobulin M (IgM) is a 900 000 dalton pentamer whose five IgG-like units are held together by intramolecular disulfide bonds and stabilized by a polypeptide "anchor" termed the "J-chain." IgM comprises 5–10% of normal circulating immunoglobulins (1–2 mg/mL). Due to its large size, IgM is confined to the intravascular space. It binds complement but, unlike IgG, does not bind directly to phagocytic cells. Immunoglobulin M is usually the first antibody class formed in response to foreign antigen exposure. Due to its multivalency, IgM is extremely efficient at agglutinating bacteria. This phenomenon is important in the control of bacteremia and the neutralization of LPS released by gram-negative bacteria.

Immunoglobulin A (IgA) occurs as 160 000 dalton monomers or 320 000 dalton dimers. Monomers are found primarily in the intravascular space and comprise approximately 10–15% (1.4–4 mg/mL) of total serum immunoglobulins. There are two subclasses, IgA 1 and IgA 2. Dimeric IgA, termed *secretory IgA*, is formed by noncovalent interaction with the "secretory piece," a glycoprotein of about 60 000 daltons. Secretory IgA is the predominant immunoglobulin found in mucous secretions, such as tears, colostrum, pulmonary, intestinal, and urinogenital fluids; IgA can bind complement and react with phagocytic cells.

Although parenteral immunization stimulates a vigorous serum IgA response, secretory IgA is considered to be of greater importance due to its "first line" defensive role in body secretions. Secretory IgA plays a critical role in providing protection against respiratory, intestinal, and urinogenital tract bacterial pathogens.

Immunoglobulin D (IgD) is a monomer of ca. 180 000 daltons; concentrations vary widely in humans ranging from 0 to ca. 0.4 mg/mL. Compared to other immunoglobulin classes, IgD is very susceptible to proteolysis and possesses a short half-life (about 3 days). Since IgD does not fix complement or bind to phagocytic cells, it is not considered to be a "protective" immunoglobulin as are IgA, IgG, and IgM. However, IgD is abundant on the surface of B lymphocytes and may play a central role in the activation process which leads to the development of antibody-secreting plasma cells.

Immunoglobulin E (IgE) is a monomer of 200 000 daltons; its average serum concentration (ca. 250 ng/mL) is the lowest of any immunoglobulin. Although IgE does not fix complement or react directly with phagocytic cells, it has a very high affinity for most cells and basophils. Immunoglobulin E appears to mediate both a protective and a detrimental immune response. At the mucoid surface, pathogens binding to IgE stimulate an acute inflammatory response by triggering the release of potent mediators from mast cells and basophils. A protective role for IgE is indicated in several chronic parasitic infections, most notably schistosomiasis (see Chap. 4) where high levels (> 100 µg/mL) of serum IgE have been noted. Immunoglobulin E induces an inflammatory response and "recruits" effector cells to the area. In contrast, IgE-mediated degranulation of effector cells leads to the release of vasoreactive molecules. Individuals suffering from certain allergies can also have elevated serum IgE levels.

1.2.3. Immune Response

The chain of events leading to an immune response is extremely complicated and not yet completely understood. A simple model is shown in Figure 3. The human is capable of forming an immune response to thousands of foreign antigens. The humoral (antibody-mediated) response depends on the *B cells* which are lymphocytes derived from bone marrow stem cells. Upon maturation, B cells form antibody-secreting *plasma cells*.

Each clone of B cells has an immunoglobulin molecule with a recognition site for a specific antigen on its surface. Binding of the appropriate antigen to the B cell causes proliferation of the clone whereby the progeny cells secrete antibody whose specificity is identical to that of the cell-surface immunoglobulin. A foreign antigen can be taken up by macrophages, processed, and presented upon the cell surface, or may remain in a soluble state. All antigens can be termed T-dependent or T-independent depending on whether they require or do not require the interaction of T cells for antibody synthesis (T cells are lymphocytes derived from the thymus). In the case of a *T-independent antigen* (usually polymers, such as bacterial capsular polysaccharides), the antigen can cross-link the B cell surface antibody molecules of the B cell; this initiates proliferation and antibody synthesis. A *T-dependent antigen* requires, in addition to binding to B-cell surface immunoglobulin, the release of T cell factors which act upon the B cell to initiate proliferation.

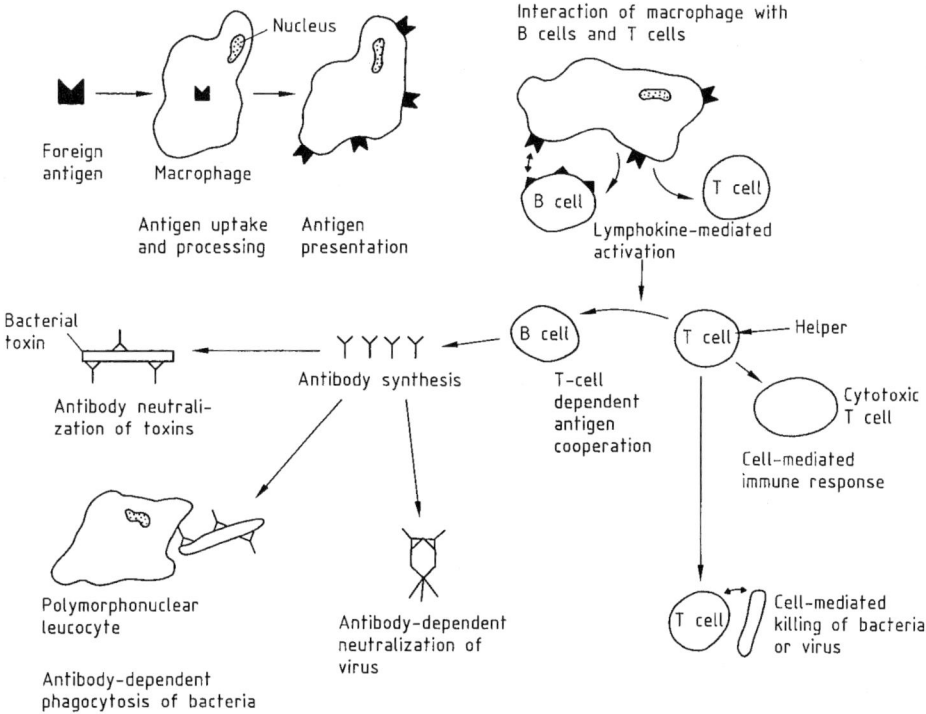

Figure 3. Immunological response cascade to a foreign antigen or vaccine, leading to antibody response or cell-mediated response

A human exposed to an antigen in this way is considered *primed*. This is a critical factor in immunization. After initial exposure to a T-dependent antigen by immunization, the induced serum antibody concentration is low and returns to near-basal levels comparatively quickly (the average half-life of IgG is ca. 22 days, whereas IgM is ca. 5 days). Upon revaccination (boosting) or natural exposure to the pathogen or toxin, a vigorous antibody response occurs in which high antibody levels are synthesized over a longer period of time. This is termed an *anamnestic response*. Therefore, high levels of serum antibodies do not have to be present for the individual to be protected by prior vaccination. A T-independent antigen does not evoke an anamnestic response.

The induction of a *cell-mediated immune (CMI) response* is rather more complicated and less well defined. The CMI response plays a critical role in immunity to pathogens able to live and proliferate within host cells. The first step is "activation" of T cells. Like B cells, T cells also recognize specific antigens. Recognition of an antigen on the surface of a macrophage by T cells results in the release of *interleukin 1* (IL-1), a lymphokine which activates them. One subpopulation of T cells synthesizes interleukin 2 (IL-2), also known as T-cell growth factor, which causes a second subpopulation of activated T cells expressing the IL-2 receptor to proliferate and become cytotoxic T cells. Cytotoxic

T cells recognize the specific antigen they are activated against when expressed on the surface of infected cells; they thus kill these infected cells by lysis. A third subpopulation of T cells evolve to be "primed" memory cells which undergo rapid proliferation upon reexposure to the same antigen.

Immunization with a given vaccine may induce a CMI, a humoral antibody response, or both depending on the infecting pathogen. Both types of immunity are usually desirable and may act synergistically. For example, vaccine-induced immunity to many viral diseases, such as rabies, measles, mumps, and rubella, is confirmed by measuring serum antibody levels. However, infected cells can only be destroyed by cytotoxic T cells. Circulating antibody probably prevents the spread of the virus, while cytotoxic T cells eliminate already infected cells. Tuberculosis vaccines stimulate a good antibody response but are less proctective due to the suppression of the critical CMI response.

1.2.4. Active Immunization

Active immunization entails the administration of one or more antigens to a host in the form of a vaccine in an attempt to elicit a protective immune response. In this way an individual can be rendered immune to a variety of diseases. Most vaccines are administered to infants or young children. The trend is to combine several *monovalent vaccines* (vaccines containing a single antigen or vaccine strain) to form *multivalent vaccines* capable of simultaneously inducing immunity to several diseases. Infants are routinely immunized against diphtheria, tetanus, pertussis, and, in some countries, polio by administering the vaccines in a single injection. Similarly, young children are immunized simultaneously with a multivalent measles, mumps, and rubella vaccine. Administration of multivalent vaccines reduces the number of visits to healthcare centers, a critical point in developing countries.

Historically, *parenteral vaccines* have usually been used (i.e., injected with a syringe and needle). This approach is extremely successful for systemic diseases such as diphtheria or measles. However, in localized diseases (e.g., intestinal infections such as cholera), parenteral vaccines are of limited use since the immune system needs to be stimulated at the site of infection (e.g., local immune system). Therefore, effort is now being directed at controlling intestinal infectious diseases with orally administered vaccines. *Oral vaccines* can consist of

1) killed intact bacteria,
2) toxoids (i.e., toxins that are still immunogenic but are rendered biologically inactive by treatment with a chemical, heat, or mutation)
3) subunit vaccines, in which only the nontoxic portion of a molecule is used, or
4) live-attenuated vaccines, in which a viral or bacterial strain is rendered nonpathogenic (e.g., by passaging the virus in cell culture or deletion of bacterial genes), but is still able to multiply to a limited degree, thereby eliciting a protective immune response in the absence of disease symptoms.

One major problem associated with highly purified antigens or subunit vaccines is reduced immunogenicity. Unfortunately, as protective antigens are purified from either bacteria, viruses, or parasites in order to free them from nontoxic substances such as LPS, they lose their ability to evoke an immune response. Their immunogenicity can, however, be improved by using an *adjuvant*. Adjuvants function either to present the antigen to the immune system for a prolonged period of time, or to nonspecifically stimulate the immune system by releasing immune modulators such as lymphokines. To date, only aluminum gels have been licensed as adjuvants for human use. Tetanus or diphtheria toxoid is absorbed onto the gel which allows the toxoid to persist longer at the injection site. Biologically active peptides that can nonspecifically stimulate the immune system via lymphokine release or downregulate the suppression of the immune system, are being evaluated.

1.2.5. Passive Immunization

Passive immunization entails the transfer of preformed immunoglobulins to a host. Passive immunization is usually employed after known or suspected exposure to a given pathogen (for example, after being bitten by a rabid animal). It is used in cases were disease progress is rapid if the pathogen or toxin is not neutralized. The time needed for the host to mount a protective immune response following active immunization (7 – 14 days) would be too long. In some instances, globulin may be passively administered as a prophylactic measure, as in the case of travellers entering an area where hepatitis A is endemic.

Immunoglobulin preparations are assayed to confirm that they contain a high titer of neutralizing antibodies against the disease in question. Such *hyperimmune globulins* must contain at least five-fold higher levels than normal globulin and are obtained by screening plasma units (human) for a given antibody. Alternatively, volunteers may be vaccinated to enrich their plasma for a given antibody.

Immunoglobulin is administered either intramuscularly or intravenously. Only a limited volume of globulin can be given intramuscularly. Intravenous administration allows for comparatively large quantities (400 mg of immunoglobulin per kilogram body weight) of immunoglobulin to be administered quickly.

Several problems are associated with the preparation of immunoglobulin for passive therapy. Firstly, the identification of plasma donors who are "hyperimmune" to a given antigen requires expensive and tedious screening programs. Secondly, the administration of immunoglobulins entails the possibility of transmitting viral diseases, such as hepatitis B and non-A, non-B hepatitis. Additional screening procedures are therefore required.

An alternative to antibody obtained from donors is the use of human monoclonal antibodies (humabs, see Chap. 6). Humabs can be synthesized by hybridoma cell lines that are produced by fusing a human B cell secreting a desired antibody to a non-secreting cell line. Hybridomas can be grown in fermentors containing up to 1000 L of serum-free medium. Antibody yields can reach 100 mg per liter of medium. The use of humabs can circumvent the need to obtain plasma from many donors and the risk of viral disease transmission, all at a lower cost. Humabs have been produced against a variety of infectious agents (e.g., cytomegalovirus, diphtheria and tetanus toxins) and will be tested in the near future.

1.2.6. Genetic Engineering

The application of recombinant DNA technology to the field of vaccine development has led to remarkable progress since the late 1970s. Critical protective antigenic determinants can be defined, cloned, produced on a large scale by using an appropriate expression system (vector and host), and purified. The first vaccine for human use to be produced by genetic engineering is that against hepatitis B; the surface antigen was cloned, then expressed in yeast, and purified to homogeneity. These vaccines have virtually replaced the first-generation, plasma-derived antigen vaccines due to increased safety and lower costs.

Numerous licensed, live-attenuated vaccines (measles, polio, mumps, rubella, varicella, and typhoid fever) have been obtained by rather empirical techniques, such as passage on tissue culture or chemical mutagenesis. Surprisingly, the precise genetic alterations responsible for avirulence are unknown. Several live-attenuated bacterial vaccine strains have been developed by using genetic engineering. The most advanced, in regards to clinical testing, are for cholera and typhoid fever. These strains have been attenuated by inactivating a gene or genes essential for virulence. A related approach has been to clone a given protective antigen and express it in a suitable, nonvirulent "carrier" strain. For example, surface glycoprotein from HIV-1 and rabies virus have been introduced into vaccinia virus (smallpox vaccine). In addition, antigens from several enteric pathogens such as *Shigella sonnei, S. dysenteriae*, and *Vibrio cholerae* have been introduced into the licensed live-oral typhoid vaccine strain, Ty21a. Such recombinant strains may prove useful as "bivalent" vaccines, conferring protection against two enteric pathogens.

2. Bacterial Vaccines

The chemotherapy of bacterial disease is treated elsewhere (→ Antibiotics; → Synthetic Chemotherapeutic Agents).

2.1. Diphtheria Vaccine

Etiological Agent and Pathogenesis. The causative agent of diphtheria is *Corynebacterium diphtheriae*, first isolated by LOEFFLER in 1884. The disease is spread by inhalation of infected droplets and is characterized by the formation of a pseudomembrane in the nasopharynx caused by localized bacterial replication. Disease symptoms are caused exclusively by the production of a lethal toxin which spreads from the site of infection via the bloodstream. Death is due to the inhibition of protein synthesis by the toxin in vital organs [10].

History of Immunization. Studies by Roux and Yersin in 1888 demonstrated that diphtheria is a "toxicosis." Two years later, Behring and Kitasato established that the disease could be prevented in animals by immunization with a crude diphtheria toxoid. Attempts to extend these findings to humans used a "toxoid" prepared by combining diphtheria toxin and antitoxin. The first large-scale immunization program with such a product was performed on New York school children in 1922. Soon after, crude toxoids prepared by formaldehyde treatment of *C. diphtheriae* filtrates replaced toxin-antitoxin toxoids. Routine mass vaccination against diphtheria was initiated in Western Europe and North America shortly after World War II.

Production and Properties of Diphtheria Vaccine. Diphtheria toxoid is prepared by detoxification of diphtheria toxin with formaldehyde. Derivatives of the hypertoxinogenic Park Williams 8 strain are used by most manufacturers. Large quantities of toxin are produced in fermentors (ca. 200 – 5000 L) and yields approach 500 mg/L. Although manufacturing processes vary, most employ the following steps. Formaldehyde [*50-00-0*] (0.4 – 0.6%) is added to cell-free culture supernatants which are then stored at 35 – 37 °C for 3 – 5 weeks. Detoxification is accomplished first by reaction of formaldehyde with the ε-amino groups of lysine and then by formation of irreversible methylene bridges between aromatic amino acids. Upon confirmation of detoxification by animal testing, the toxoid is purified by diafiltration, ammonium sulfate or ethanol fractionation, or anion-exchange chromatography.

Immunization Recommendations. Diphtheria toxoid is rarely used as a monovalent vaccine, but is usually combined with tetanus toxoid (DT, see Section 2.2), pertussis vaccine (DPT, see Section 2.3), or inactivated polio vaccine (DPT-Pol, see Section 3.5) absorbed onto an aluminum salt adjuvant. Toxoid content is usually expressed in limits of flocculation, Lf (1 Lf $\approx$ 2 µg of toxoid). For primary immunization of infants, ca. 25 Lf of toxoid is administered intramuscularly starting at 6 – 12 weeks of age. Three doses are usually administered at 4 – 8 week intervals with a fourth dose 4 – 12 months later. To maintain life-long immunity, booster doses, together with tetanus toxoid, are recommended every 7 – 10 years.

Adverse Reactions. The absolute reactogenicity of diphtheria toxoid is difficult to determine since it is usually administered with tetanus toxoid or pertussis vaccines. Overall, the vaccine is considered to be safe; it primarily evokes mild, transient, local reactions.

Vaccine Efficacy. No large-scale field trials have been performed to determine vaccine efficacy. This is because it would be unethical to withhold the vaccine from control subjects in view of the dramatic decrease in disease following immunization in the 1920s. Furthermore, overwhelming data indicate that proper vaccination can provide absolute protection. For example, mass vaccination with diphtheria toxoid in Romania in 1958 showed that within seven years morbidity and mortality due to diphtheria declined by more then 99%. In Western Europe and North America, where vaccination is universal, diphtheria has been virtually eradicated [11]. Rare cases occur almost exclusively in individuals lacking a proper history of vaccination.

Future Prospects. Given the combination of vaccine safety, efficacy, and cost, there has been little impetus to develop and test a "second generation" diphtheria toxoid. Any such vaccine would have to completely eliminate the possibility of toxic reversion and be more economical to produce. In a recent approach, synthetic peptides that express key epitopes of diphtheria toxin are conjugated to carrier proteins; these conjugates can elicit neutralizing antibodies [12]. The drawback with this approach is the cost of purifying the carrier protein.

The most promising approach to toxoid development is the use of nontoxic mutant proteins [13], obtained by recombinant DNA technology. Such proteins can be constructed by deletion of specific regions responsible for toxicity [14]. In addition, these "toxoids" can be synthesized in yields equal to those obtained by the current production strains of *C. diphtheriae*. The efficacy of such proteins is being evaluated.

2.2. Tetanus Vaccine

Etiological Agent and Pathogenesis. The causative agent of tetanus is the spore-forming bacillus, *Clostridium tetani*. The disease, first described by HIPPOCRATES, is spread via contamination of wounds or abrasions with soil containing *C. tetani* spores. Bacteria multiplying at the wound site release a potent neurotoxin which enters the bloodstream and acts upon the central nervous system evoking the "spastic paralysis" characteristic of tetanus [15]. The majority of tetanus cases occur in neonates due to the nonsterile severing of the umbilical cord.

History of Immunization. The identification of tetanus as a toxicosis by FABER in 1890 led to the development of crude tetanus toxoids as early as 1893. Initial attempts to vaccinate against tetanus were carried out during World War I. By 1926, parenteral immunization of humans with a safe, effective formalin toxoid was achieved. Until the mid-1960s, only such crude toxoids were available.

Production and Properties of Tetanus Vaccine. The presently used tetanus toxoid is a partially purified preparation. The hypertoxinogenic Harvard strain of *C. tetani* is most widely used for production of tetanus toxin. The bacteria are grown in fermentors until the cells autolyse, thereby releasing the toxin into the medium. Intact cells are removed by filtration. Formaldehyde is added to the filtrate to a final concentration of 0.4–0.6%, the pH adjusted to 7.4–7.6, and the mixture held at 35–37 °C for about four weeks. After confirmation of detoxification by animal testing, tetanus toxoid is purified by diafiltration and ammonium sulfate fractionation.

Immunization Recommendations. Tetanus toxoid is most frequently adsorbed to an aluminum salt adjuvant and administered together with diphtheria toxoid (DT, Section 2.1), pertussis vaccine (DPT, Section 2.3), or inactivated polio vaccine (DPT-Pol, Section 3.5). For primary immunization of infants, each dose of vaccine contains 20–30 Lf's of tetanus toxoid as DPT or DPT-Pol. Immunization starts at 6–12 weeks of age and consists of three doses given at 4–8 week intervals with a fourth dose 4–12 months later. To maintain life-long immunity (≥ 0.01 IU/mL of serum), booster doses, often combined with diphtheria toxoid, are recommended every 7–10 years.

Tetanus vaccine is often administered as a routine prophylactic measure following puncture wound trauma. This practice can lead to hyperimmunization and attendant reactions. Alternatively, an individual may be "tolerized" by repeated vaccination and, therefore, unable to mount a protective immune response. Currently, vaccination of pregnant women is recommended with the final dose given about three weeks before expected delivery. The immune response of humans to tetanus vaccine varies widely [16]. The immunization schedule used appears to influence the magnitude of the immune response [17].

Adverse Reactions. Serious reactions following vaccination with tetanus toxoid are infrequent: swelling, pain, and/or redness at the injection site are the most common. However, immediate and delayed hypersensitivity, an Arthuslike reaction, and in rare instances (< 1 case in 2×10^6), neurological sequelae have been noted [15]. The majority of severe reactions are associated with high levels of pre-existing antitetanus antibody. Therefore, attention should be paid to the interval between booster doses to avoid these occurrences.

Vaccine Efficacy. Data supporting the efficacy of tetanus toxoid vaccine comes primarily from retrospective analysis of attack rates in immune versus nonimmune populations. The first such study was the evaluation of soldiers during World War II, where tetanus occurred primarily in nonimmune individuals. Immunization of pregnant women completely prevented neonatal tetanus [18]. Recently, cases of tetanus observed in the United States occurred exclusively in nonimmune individuals. Numerous studies have also shown that abbreviated one- or two-dose immunization regimens are able to induce long-lasting immunity [15]. This is of particular importance in developing countries where a multidose immunization regimen may not be feasible due to an inadequate health care system or economic constraints.

Future Prospects. As with diphtheria toxoid, the current tetanus vaccine is an extremely economic, effective, safe vaccine. Efforts directed to a "new generation" of tetanus vaccine are aimed at providing a toxoid more amenable to mass vaccination. The main thrust has been to produce a safer, more immunogenic toxoid capable of inducing long-lasting immunity after one or two administrations without reactogenic adjuvants. The substitution of glutaraldehyde for formaldehyde yields a safe, immunogenic vaccine not requiring adjuvants [19]. Production of toxoids by recombinant DNA techniques similar to those described for diphtheria toxin is also feasible [14].

2.3. Pertussis Vaccine

Etiological Agent and Pathogenesis. *Bordetella pertussis*, the cause of pertussis or "whooping cough," is a gram-negative rod that was first isolated by BORDET and GENGOU in 1906. Humans are the only known host for *B. pertussis*. The disease is spread by infectious droplets with the vast majority of cases occurring in young children. *B. pertussis* shows a marked trophism for the ciliated epithelial cells of the upper respiratory tract. Initial symptoms are similar to those of the common cold, but worsen within 10–20 days. The paroxysmal stage, lasting an average of 15–20 days, is characterized by severe bouts of coughing. The convalescent stage can last for many months when secondary infections can present a major problem. *B. pertussis* remains localized within the initial disease stages which implies that the severe symptoms are toxin-induced [20]. This is supported by the fact that *B. pertussis* can synthesize a large number of toxic extracellular factors including pertussis toxin (also referred to as lymphocytosis promoting factor), adenylate cyclase, dermonecrotic toxin, and tracheal cytotoxin [21].

History of Immunization. Attempts at immunization against pertussis were initiated in the 1920s using killed whole-cell vaccines [22]. These studies demonstrated that vaccination could not only prevent a substantial proportion of disease, but that symptoms in vaccinated individuals who became infected were milder than in unvaccinated controls. Routine large-scale immunization against pertussis began shortly after World War II using standardized whole-cell vaccines of known potency. Even though the safety of these vaccines has been under attack, they are still used in most areas of the world except Japan, where acellular vaccines have been used since 1980 (see below).

Production and Properties of Pertussis Vaccine. At present, there is no standardized method, culture medium, or bacterial strain for producing pertussis vaccine. A given manufacturer's production procedure is designed to yield a vaccine which will meet the minimal requirements of the appropriate regulatory agency. *B. pertussis* is usually grown in fermentors on a synthetic or semisynthetic medium. The cells are harvested by centrifugation and resuspended to a given opacity using a reference standard. Inactivation is accomplished by heating and/or adding formaldehyde or thimerosal. In view of the diversity of production techniques employed, toxic components which escape inactivation can vary considerably from manufacturer to manufacturer.

Immunization Recommendations. Pertussis vaccine is used almost exclusively in combination with diphtheria and tetanus vaccines (see Sections 2.1 and 2.2). The vaccine is adsorbed onto an aluminum salt adjuvant. Immunization usually commences at 6–12 weeks of age and consists of three doses of vaccine given intramuscularly at 4–8 week intervals with a fourth dose given 4–12 months later. Vaccination with the acellular vaccine used in Japan consists of 2–3 doses of vaccine given at 4–8 week intervals starting at two years of age.

Adverse Reactions. The majority of children receiving pertussis vaccine experience an adverse reaction. Approximately 40–50% of children have a local reaction and 30–40% a systemic reaction (anorexia, vomiting, fretfulness, fever, or persistent crying) [23]. Of greater concern is the infrequent occurrence of convulsions and hypotonia. As a result of concern regarding vaccine safety, a large-scale study was initiated in England to determine the incidence of serious reactions in children aged 2–36 months [24]. The risk of vaccine-induced neurological illness was estimated to be 1/110 000 vaccinations and that for encephalopathy, 1/310 000 vaccinations. It is extremely difficult to establish a causal relationship in individual cases in such a study; incidence rates were evaluated on a temporal basis. Furthermore, neurological symptoms believed to be vaccine-related constitute only a small proportion of similar cases seen in this age group. Review of these data by health authorities has led to the conclusion that the benefit of vaccination outweighs the attendant risk in view of the fact that clinical pertussis can often result in neurological sequela.

Vaccine Efficacy. The first indication of vaccine efficacy came from a trial conducted in 1929 during an epidemic on the Faroe Islands; vaccination afforded 73% protection against clinical disease [22]. Extensive information has been obtained in Japan, England, and Sweden where disease incidence correlates inversely with the overall immune status of the general population.

Routine immunization against pertussis in Japan was initiated in 1947–1949 with an acceptance rate of roughly 90%. The number of pertussis cases declined from 152 072 in 1947 to less than 400 by 1971. In 1975, controversy concerning vaccine safety briefly halted vaccine usage. When it was resumed, the acceptance rate declined to ca. 25–30%. Pertussis returned to epidemic proportions by 1979 with more than 13 000 cases reported nationally. At the urging of federal health authorities, vaccine acceptance rates increased to 65–70% by 1982 with a concomitant decline in cases.

Similarly, concern about vaccine safety in England resulted in a dramatic decline in vaccine acceptance from ca. 80% in 1973 to ca. 30% in 1978. As in Japan, the disease became epidemic with more than 100 000 cases reported between 1977 and 1980. In Sweden, routine vaccination against pertussis was initiated in the early 1950s. One decade later, >90% of children were considered to be immune. Due to changes in the manufacturing technique, the pertussis vaccine used in the mid-1970s was not potent. Shortly after, a marked increase of pertussis cases was seen. When an effective vaccine was reintroduced, disease incidence was dramatically reduced.

Case-contact studies have shown the whole-cell vaccine to be 63–95% effective at preventing overt disease. Efficacy of acellular vaccines is discussed below.

Future Prospects. Recent efforts to develop a "second generation" pertussis vaccine have centered on using purified detoxified antigen preparations containing a minimum amount of lipopolysaccharide. Several such acellular vaccines have been developed and clinically evaluated. They are composed of detoxified pertussis toxin (PT) alone or in combination with filamentous hemagglutinin (FHA), two key protective antigens [25]. The first generation acellular vaccines were produced in Japan using supernatant from static cultures as a source of antigen. Cell-free supernatants were subjected to ammonium sulfate precipitation followed by sucrose density gradient ultracentrifugation to simultaneously enrich PT and FHA and eliminate lipopolysaccharide; PT was inactivated by formaldehyde treatment. Preliminary testing in Japan showed the vaccine to be far better tolerated than whole-cell vaccine; case-contact studies revealed that it was ca. 90% effective.

Second generation acellular vaccines of a greater purity have been produced on a large scale using fermentor-grown cultures. A monovalent formalin PT toxoid and a bivalent PT toxoid – FHA vaccine have been evaluated in a placebo-controlled trial in 5 – 11 month old children in Sweden [26]. Children received 2 doses of vaccine 8 – 12 weeks apart. The small number of children vaccinated (ca. 3000) does not allow evaluation of vaccine safety regarding the rare neurological reactions. However, the vaccine evoked far fewer local reactions in comparison to the whole-cell vaccine. After 15 months of observation, the PT toxoid vaccine was 54% effective whereas the PT toxoid-FHA vaccine was 69% effective against all forms of clinical pertussis. Both vaccines were equally effective (ca. 80%) against severe disease.

2.4. Typhoid Fever Vaccine

Etiological Agent and Pathogenesis. *Salmonella typhi*, the causative agent of typhoid fever, is a gram-negative bacillus first isolated by GAFFKY in 1884. Infection is due to ingestion of the organisms in contaminated food or water. The bacteria penetrate the epithelium of the small intestine and are ingested by reticuloendothelial cells. Unlike most bacteria, the typhoid bacillus can survive and multiply within phagocytic cells and then spreads to the spleen, liver, lymph nodes, and gallbladder. Symptoms appear 1 – 2 weeks after exposure when the bacteria enter the bloodstream. Despite appropriate treatment, 1 – 2% of those infected will become chronic asymptomatic carriers serving as an infectious reservoir. Typhoid primarily occurs in developing countries or in areas of poor sanitation. In endemic areas, the disease is primarily contracted by school-aged children, but travellers of all ages entering an endemic area are at risk.

History of Immunization. Attempts at immunizing against typhoid fever began in 1896. The first mass vaccination was conducted by the British Army during World War I using a killed whole-cell vaccine. The attack rate among vaccinated soldiers was far less than for nonimmunized soldiers. Consequently, the use of whole-cell vaccines became common practice in the armed forces.

In the 1960s and 1970s, attempts were made to use killed oral immunization. Killed bacteria were administered to volunteers at 1×10^{11} per dose; twelve doses resulted in a modest, but significant protection rate of 30% [27]. Field trials in India showed that such killed preparations were ineffective at preventing disease.

An alternative approach to killed oral vaccines against typhoid was to use attenuated live-oral vaccine strains. The first candidate was a streptomycin-dependent mutant developed in 1967 [28]. When orally administered to volunteers in doses of up to 10^{11} bacteria, this strain did not evoke adverse reactions. Good levels of protection (66 – 78%) were achieved with freshly harvested cultures. However, lyophilized preparations afforded little or no protection [28], which diminished interest in this strain.

The second live-oral vaccine candidate was a gal ε mutant termed Ty21a which lacked glucose 1,4-epimerase activity and therefore could not synthesize complete lipopolysaccharide [29]. Freshly harvested cultures of this strain were safe and effective in volunteer challenge studies [30]. A lyophilized Ty21a preparation afforded > 90 % protection for three years in a field trial in Egypt [31].

Production and Properties of Typhoid Vaccines. Killed whole-cell vaccines are usually produced from the Ty2 strain of *S. typhi*. Fermentor-grown cultures are inactivated either by heating in combination with phenol or formaldehyde or by acetone drying. Although the latter method yields a more efficacious vaccine (see below), only the vaccine produced by the former method is readily available. Each 0.5 mL dose of vaccine contains $(1-3) \times 10^9$ killed organisms. Vaccine potency is estimated by a mouse-protection test.

The attenuated live-oral Ty21a vaccine strain is produced from fermentor-grown cultures. The harvested cells are suspended in a cryoprotective medium consisting of sugar and amino acids and lyophilized. The lyophilizate is placed in gelatin capsules which are then coated with an acid-resistant layer. Each capsule contains $(1-5) \times 10^9$ viable bacteria. Vaccine potency is based upon the number of viable organisms per capsule [32].

Immunization Recommendations. People living in or travelling to an endemic area are vaccinated with two doses of whole-cell vaccine administered subcutaneously or intramuscularly 14 days apart. A single booster dose is advised for individuals travelling to an endemic area if 2–3 years have elapsed since primary immunization.

Three doses of attenuated live-oral vaccine are ingested, each dose on an alternate day. An additional complete immunization course is advised when entering an endemic area if more than 1–2 years have elapsed since primary immunization. The vaccine is for use in children six years of age or older and in all ages of non-immunocompromised adults.

Adverse Reactions. Reactions following immunization with the parenteral whole-cell vaccine are frequent and consist of local pain, swelling, and redness, often accompanied by fever, chills, and malaise. Transient debilitating reactions occur in ca. 10 % of vaccine recipients. Due to the high rate of adverse reactions, parenteral typhoid vaccine is not widely used.

Reactions following ingestion of the live-oral Ty21a vaccine are rare (< 1 per 100 000 doses). Reactions are usually mild, transient, and resolve of their own accord. Gastrointestinal disturbances and rash are most frequently reported.

Vaccine Efficacy. To accurately assess the efficacy of parenteral typhoid vaccine, several field trials were conducted in the 1960s [33]. In most instances, two doses of either heat–phenol or acetone-dried vaccine were administered. Protection rates ranged from 51–77 % and 79–93 %, respectively, for up to seven years. A further trial was conducted in Tonga in 1966–1973 where school-aged children received one or two doses of acetone-dried vaccine. Surprisingly, one dose gave no protection whereas two doses afforded only 40% protection.

Efficacy of the live oral Ty21a vaccine administered in enteric-coated capsules has been evaluated in field trials in school-aged children in Chile [34]. Three doses of vaccine provided about 70% protection over a three-year period.

Future Prospects. Two additional typhoid vaccines are undergoing clinical evaluation. The first is a double auxotrophic mutant of *S. typhi* requiring 4-aminobenzoate and adenine [35]. Although ingestion of up to 10^{10} organisms only evoked a poor humoral immune response to *S. typhi* cellular antigens, most vaccines manifested a specific cell-mediated immune response to *S. typhi* lipopolysaccharide.

The second candidate vaccine is a capsular polysaccharide termed V_i. Purified V_i antigen is safe upon parenteral immunization. A single dose of vaccine provided ca. 70% protection for ca. one year in Nepal [36].

2.5. *Streptococcus pneumoniae* Vaccine

Etiological Agent and Pathogenesis. *Streptococcus pneumoniae*, often referred to as "pneumococcusP," is a gram-positive diplococcus first isolated by STERNBERG and PASTEUR in 1881. The pneumococcus usually causes disease subsequent to a viral infection or in a compromised individual, i.e., is an opportunistic pathogen. Pneumonia is the most common disease syndrome caused by *S. pneumoniae*; bacteremia and meningitis are often sequelae to pneumococcal pneumonia. *Streptococcus pneumoniae* is also a leading cause of otitis media in young children. The pneumococcus often colonizes the nasopharynxes of healthy individuals which then serve as infectious foci. The organisms can then spread to another person by infected droplets or translocate to the inner ear or lungs following a viral infection.

History of Immunization. Early attempts at immunization against pneumococcal pneumonia employed heat-killed, whole-cell vaccines (reviewed in [37]). These trials and related studies led to the conclusion that immunity is mediated by serospecific antibody produced against the polysaccharide capsule surrounding the pneumococcus. A large-scale trial performed in the mid-1930s with a bivalent capsular polysaccharide (CPS) vaccine showed a modest decrease in disease. Conclusive proof of vaccine efficacy came from trials conducted with polyvalent vaccines; protection was serospecific. Further studies showed that vaccines comprising up to six distinct capsular antigens could be effectively used [38]. The first pneumococcal polysaccharide vaccines were hexavalent and were licensed in the early 1950s. However, owing to the belief that antibiotics could effectively control pneumococcal disease, vaccine acceptance was low, leading to a halt in production.

Production and Properties of *Streptococcus pneumoniae* Vaccine. The vaccine licensed for current use is composed of 23 types of purified CPS that have been selected on the basis of seroepidemiological surveillance studies of bacteremic isolates [39].

Pneumococcal CPSs are purified by a variety of techniques, depending upon serotype, which include precipitation with organic solvents, treatment with detergent, digestion with DNAse, RNAse, and protease, and ultracentrifugation. The vaccine contains only traces of protein and nucleic acids (≤ 2 wt%) and is analyzed for its CPS constituents, serological purity, and pyrogenicity. Vaccine potency is based upon the determination of the molecular mass of the CPS, because a minimum antigen size is required for immunogenicity in humans. One human dose consists of 25 µg of each antigen in 0.5 mL administered intramuscularly or subcutaneously.

Immunization Recommendations. Pneumococcal vaccine is currently recommended for immunocompetent individuals with underlying clinical conditions that increase their risk of acquiring pneumococcal bacteremia or pneumonia. These conditions include alcoholism, asplenia, sickle cell anemia, reduced pulmonary function, congestive heart failure, diabetes, reduced renal function, and cirrhosis [40]. Routine vaccination of the elderly is also warranted due to the increased incidence of disease in this group. Patients with underlying conditions (leukemia, Hodgkin's disease, myeloma, and treatment with immunosuppressive drugs) that render them susceptible to pneumococcal disease may be considered candidates for vaccination even though a proportion may not respond to immunization [40]. Revaccination is not advised due to severe reactions.

Adverse Reactions. Pneumococcal vaccine is very safe [40]. Mild local reactions occur in ca. 30–40% of vaccine recipients, fever and severe local reactions in ca. 1%. Anaphylactic reactions are extremely rare with a frequency of $5/1 \times 10^6$ doses administered. Severe local and systemic reactions are more likely to occur upon revaccination.

Vaccine Efficacy. Considerable controversy exists concerning the degree of protection provided by pneumococcal vaccine. The vaccine was highly effective in reducing the incidence of pneumococcal pneumonia among South African gold miners [41]. However, this population differs substantially in age and in general health from those patients for whom the vaccine is now recommended. Several studies have attempted to determine vaccine efficacy among debilitated and/or elderly patients. The distribution of *S. pneumoniae* serotypes among vaccinated and nonvaccinated individuals with invasive pneumococcal disease has been compared [42]; efficacy was estimated to be 61–65%. In contrast, a vaccine trial on high-risk patients, aged >55, did not demonstrate any significant benefit [43], probably due to the inability of these patients to mount or maintain a protective antibody response. Similar findings were reported in a study on patients suffering from chronic pulmonary disease [44]. In spite of these divergent findings, health authorities still strongly recommend immunization based upon risk–benefit analysis.

Future Prospects. The current pneumococcal vaccine is poorly immunogenic in children under two years of age and in certain debilitated patient populations [40]. Attempts have been made to improve immunogenicity by covalently coupling pneumococcal CPS to carrier proteins [45]. Such conjugates evoked a more vigorous response in mice and rhesus monkeys when compared to native CPS [46]. However, a type 6A-tetanus toxoid conjugate administered to young adult volunteers was only slightly more immunogenic than native 6A-CPS, and a booster dose did not significantly increase anti CPS antibody levels [47]. However, these volunteers had been previously immunized with tetanus toxoid and possessed low, but detectable, levels of antibody to 6A-CPS.

2.6. Shigella Vaccines

Etiological Agent and Pathogenesis. Shigellosis or bacillary dysentery is caused by *Shigellae*, most notably *S. sonnei*, *S. flexneri* 2 a and 3, and *S. dysenteriae* 1. Disease results from ingestion of as few as 10 bacteria, making it the most infective bacterial enteric pathogen. In developing countries, where the attack rate can exceed $100\,000/1 \times 10^6$ population, shigellosis is primarily a disease of young children aged six months to six years.

Once ingested, the organisms penetrate the epithelium of the colon and rapidly multiply. This leads to localized inflammation and ulceration, possibly due to synthesis of a potent cytotoxin [48]. Symptoms include fever, bloody diarrhea, cramps, tenesmus, and shock. The mortality rate for untreated disease can exceed 10%.

History of Immunization. Immunization of humans with killed whole-cell vaccines administered parenterally does not provide significant protection against shigellosis [49]. Subsequent attempts at immunization have therefore centered on use of attenuated live-oral vaccine strains. Vaccines based on streptomycin-dependent derivatives of *S. flexneri* and *S. sonnei* are well tolerated and provide significant immunity for 6–12 months [50], [51]. However, these vaccines were not pursued due to the necessity of administering four doses for primary immunization, yearly boosting to maintain efficacy, and genetic instability.

Attempts have been made to produce live-oral shigella vaccines by conjugal transfer of genes coding for *S. flexneri* surface antigens into *E. coli*. Multiple doses evoked no significant adverse reactions when fed to volunteers, but failed to protect against subsequent challenge [52].

The identification of critical *Shigellae*-protective antigens has allowed their expression in the attenuated live-oral typhoid vaccine strain, *S. typhi* Ty21a [53]. One strain, *S. typhi* Ty21a-5076-1C, carries a plasmid coding for the form I (O-polysaccharide) antigen of *S. sonnei*. This strain is safe and has afforded significant protection in volunteer studies [54].

Future Prospects. At present, no vaccines are available for use against shigellosis. As noted above, *S. typhi* 5076-1C that expresses *S. sonnei* form I antigen has shown promise in volunteer studies. However, efficacy varies from lot to lot and probably depends on the degree of flagellation of the bacteria. The construction of a *S. typhi* Ty21a strain expressing *S. flexneri* 2a type and group antigens has been described [55], but this strain has not yet been clinically evaluated.

Genes coding for the production of the O-antigen of *S. dysenteriae* 1 have been cloned and inserted in a plasmid [56]. The *S. dysenteriae* O-antigen can be expressed in *E. coli* K12 and *S. typhi* Ty21a by introduction of the recombinant plasmid. Volunteer studies to determine their safety are scheduled.

2.7. Cholera Vaccine

Etiological Agent and Pathogenesis. The causative agent of cholera is *Vibrio cholerae*, a gram-negative motile bacillus which is transmitted via the fecal–oral route. Upon passage through the stomach acid barrier, *V. cholerae* colonizes the ileum where it rapidly multiplies. The release of a potent enterotoxin (cholera toxin) causes massive watery diarrhea [57]. In most cases, oral or intravenous fluid replacement is sufficient treatment.

History of Immunization. Parenteral whole-cell cholera vaccines were employed by FENAN as early as 1885. Starting in the 1960s, several field trials were performed using inactivated whole-cell vaccines or cholera antigens (see "Vaccine Efficacy").

Numerous attempts have been made to orally vaccinate against cholera. Multiple oral doses of heat-killed *V. cholerae* evoked both a local and humoral antibody response [58]. Ten daily oral doses of 1.6×10^{10} killed vibrios provided protection against a homologous challenge [59]. However, parenteral vaccine gave slightly better protection. Although promising, the massive quantities of organisms used and the need for multiple doses made the oral approach expensive and cumbersome.

Several environmental isolates of *V. cholerae* with reduced virulence have been evaluated as live-oral vaccines. These strains failed to provide significant protection due to their inability to colonize the small intestine and evoke a protective immune response [60]. A hypotoxinogenic mutant of *V. cholerae* isolated by chemical mutagenesis afforded significant protection against diarrhea. However, due to its propensity to revert, it was not studied further [61]. Chemical mutagenesis has provided a strain of *V. cholerae* 3083 (Texas Star-SR), that produces only the B subunit of cholera toxin responsible for binding to cells and lacks the enzymatically active A subunit. Although vaccination with this strain afforded good levels of protection against cholera, about 25% of vaccinees experienced mild to moderate diarrhea [62].

Production and Properties of Cholera Vaccine. The current licensed cholera vaccine is composed of inactivated whole cells of *V. cholerae* and is administered parenterally. Fermentor-grown cultures are inactivated by a combination of heat and phenol or formaldehyde. Usually, two cultures of Ogawa-classical and Inaba-El Tor serotypes-biotypes are produced and then mixed. Cell concentration is adjusted to $(5-8) \times 10^9$ cells/dose (0.5 mL). Vaccine potency is estimated by an intraperitoneal mouse challenge test. Two doses of vaccine given at two-week intervals are recommended. A booster dose is recommended whenever entering an endemic area.

Adverse Reactions. Approximately 20–30% of vaccinees have a mild to moderate local reaction consisting of pain, redness, and/or swelling. Systemic reactions such as fever, headache, or malaise occur in ca. 5% of vaccinees.

Vaccine Efficacy. Various parenteral vaccines, including inactivated whole-cell, purified lipopolysaccharide, cell-free Inaba antigen, and a whole-cell vaccine in adjuvant have been evaluated in field trials [58]. The results can be summarized as follows:

1) two doses of vaccine are superior to one;
2) protection afforded by vaccination is greater in adults than children;
3) vaccination is only moderately effective in nonendemic areas; and
4) protection lasts for only 3–6 months.

Therefore, the currently available cholera vaccine is not an effective tool to control endemic cholera, it is best employed by travellers entering an endemic area for a short time.

Future Prospects. Due to advances in understanding the pathogenesis and molecular genetics of *V. cholerae*, one or more safe and effective cholera vaccines will probably be introduced in the near future.

Several attenuated live-oral vaccine strains have been developed by deleting the cholera toxin gene or its enzymatically active A subunit [63], [64]. A derivative of *V. cholerae* 16961, in which the cholera toxin gene was deleted afforded 89% protection against clinical disease although 50% of the volunteers had mild to moderate diarrhea. A second vaccine strain was produced by deletion of the A subunit of cholera toxin from strain 395. Again, good protection was afforded by a single dose, but mild diarrhea was observed in 60% of the subjects. The discovery, that strains of *V. cholerae* can produce a Shiga-like cytotoxin suggested that at least part of the diarrhea could be due to this toxin [65]. Therefore, KAPER and coworkers (personal communication) deleted the A subunit of cholera toxin from strain 569B which is naturally cytotoxin negative. This strain, called CVD-103, is far less reactinogenic than previously evaluated strains, eliciting mild diarrhea in ~10% of subjects. Vaccination with CVD-103 afforded excellent protection (87%) against a homologous challenge. Furthermore, significant protection (67–78%) was seen even when the challenge was of a different serotype or biotype.

Orally administered inactivated vaccines consisting of 1×10^{11} killed vibrios and 1 mg of the B subunit of cholera toxin or 1×10^{11} killed vibrios alone per dose have been evaluated [66]. After six months of surveillance, three doses of the combined vaccine afforded 85% protection, whereas the cells alone were 58% effective. After one year, protection had declined to 62% for the combined vaccine and 53% for the cells alone.

2.8. Vaccines Against Nosocomial Pathogenes

The routine use of broad-spectrum antibiotics, more frequent invasive surgery, increasing use of immunosuppressive agents in the management of cancer, and the ability to prolong the lives of critically ill patients have led to a dramatic increase of hospital-acquired (nosocomial) bacterial infections [67]. In the United States alone, nosocomial infections are a contributing factor in ca. 70 000 deaths per year; their treatment costs $> 1 \times 10^9$ \$ [68], [69]. The mortality rate for nosocomial bacteremia and pneumonia has remained unacceptably high ($\geq 25\%$) despite the introduction of numerous antibacterial agents [68]. Therefore, much effort has been devoted to developing immunological agents for controlling these infections [70].

Etiological agents and pathogenesis. *Pseudomonas aeruginosa, Escherichia coli, Klebsiella* spp., and *Staphylococcus aureus* account for the majority of serious nosocomial infections, such as bacteremia and pneumonia [71]. Infection with these agents usually arises due to the translocation of bacteria from the skin, intestine, or nasopharyngeal cavity following disruption of the host's defense systems by trauma, invasive surgical procedures, treatment with antibiotics, or immunosuppressive agents. Foci of infection are usually formed from which the bacteria enter the bloodstream. Death is most likely caused by shock.

Future Prospects. No vaccines are licensed against the four pathogens listed above. Attempts to control bacterial nosocomial infections by immunological means are complicated because of the diverse patient populations who are at high risk and the substantial number of bacterial pathogens of varying serotypes involved. Furthermore, vaccination of at-risk patients may not be feasible because of the time needed to mount a protective immune response and because a substantial portion of these patients are unable to mount a significant antibody response. A more effective approach is to passively immunize patients by using a hyperimmune globulin for intravenous use (see Chap. 5): vaccines developed against nosocomial pathogens could then be used to vaccinate healthy donors whose plasma would be processed into such a globulin [70].

Two different approaches are being taken to develop appropriate hyperimmune products. The first is based upon developing vaccines against the relevant serospecific bacterial antigens. The second is to evoke an antibody response to a common epitope expressed by the lipopolysaccharide of gram-negative bacteria.

Serospecific Vaccine against *Pseudomonas aeruginosa*. Human immunity to *P. aeruginosa* depends on the presence of humoral antibody directed against serospecific lipopolysaccharide (LPS) determinants and toxin A [72]. Several lipopolysaccharide-containing vaccines have been developed and clinically evaluated [73]. These vaccines have not gained wide acceptance due to their toxicity, poor immunogenicity, and poorly characterized antigenic content. Two serotypes of *P. aeruginosa* high molecular mass polysaccharides have been purified and found to be safe and immunogenic in humans [74]. In an attempt to produce both an antitoxin A and an antiLPS antibody response, toxin A – O-polysaccharide conjugate vaccines were synthesized. Such conjugates are nontoxic, nonpyrogenic, safe when administered to humans, and evoke an immune response to both vaccine moieties [75]. A polyvalent vaccine based upon O-polysaccharide serotypes is undergoing clinical evaluation; this vaccine would "cover" > 90 % of *P. aeruginosa* bacteremic isolates.

Serospecific Vaccine against *Klebsiella spp*. Antibody to *Klebsiella* capsular polysaccharide (CPS) is highly protective against experimental infections [76]. Several experimental vaccines composed of purified *Klebsiella* CPS were found to be safe and immunogenic in human volunteers [70]. Based upon the seroepidemiology of *Klebsiella* blood isolates, a 24-valent CPS vaccine has been developed which is safe and immunogenic in humans. It also elicits production of antibody to 11 "cross-reactive" CPS serotypes not included in the vaccine and therefore, offers potential "coverage" against ca. 75% of all *Klebsiella* bacteremic isolates.

Serospecific Vaccine against *Escherichia coli*. Antibodies to both O (LPS) and K (capsular) antigens can provide protection against experimental *E. coli* infections [77], [78]. However, ca. 40% of *E. coli* bacteremic isolates do not have a polysaccharide capsule [79]. In addition, the two most frequently encountered K serotypes among blood isolates, K1 and K5, are nonimmunogenic in humans. Approximately 90% of *E. coli* bacteremic isolates can be typed according to their O-antigen and most of them can be grouped within 11 serotypes, making a polyvalent formulation feasible [79]. The toxicity of native LPS precludes its use as a vaccine. However, nontoxic, serologically reactive O-polysaccharide can be isolated from *E. coli* LPS and O-polysaccharide–protein conjugates are being constructed in a manner similar to that described for *P. aeruginosa* (see above).

Serospecific *Staphylococcus aureus* Vaccine. *Staphylococcus aureus* produces numerous somatic antigens and extracellular toxins which have been implicated as virulence factors. These include teichoic acid, lipoteichoic acid, exopolysaccharide, capsular polysaccharide, α-toxin, and β-toxin. Patients with deep-seated *S. aureus* disease usually mount an antibody response to teichoic acid, α-toxin, and β-toxin [80]. However, the relative protective capacities of antibodies against these antigens are unknown.

A serotyping scheme has been described for *S. aureus* based upon cell-surface capsular polysaccharides [81]. Approximately 80–90% of *S. aureus* clinical isolates belong to one of these serotypes [82]. Serospecific CPS is produced during experimental *S. aureus* infections and can be isolated from blood [83]. The protective capacity of antibody to CPS is being evaluated.

Crossreactive Anticore Glycolipid Vaccine. Antibody produced against the "core" glycolipid of *E. coli* strain J5 crossreacts with the LPS of virtually all gram-negative bacteria [84]. Polyclonal or monoclonal antiJ5 antibody (see also Section 6.1) can afford significant protection against several gram-negative pathogens [85]. Recent findings indicate that protective antibody is directed against the lipid A moiety of LPS and neutralizes the toxic activities of LPS. When used prophylactically or therapeutically, human plasma enriched with antiJ5 antibody provides significant protection against death due to "gramnegative shock" [86], [87].

2.9. Meningococcal Meningitis Vaccine

Etiological Agent and Pathogenesis. The causative agent of meningococcal disease is *Neisseria meningitidis*, a gram-negative diplococcus. Humans are the only known reservoir for *N. meningitidis*. The meningococcus colonizes the nasopharynx; infrequently, it enters the bloodstream and infects the meninges. Meningococcal disease is most often seen in children less than 18 months of age and in settings such as military recruit camps where individuals from different geographical areas come into constant close contact [88]. The disease is endemic worldwide with an annual incidence of about 3/100 000 population. Numerous epidemics occurred in the 1970s and 1980s with attack rates exceeding 500/100 000. Eleven meningococcal capsular serotypes are known, with the majority of disease (ca. 95%) caused by groups A, B, C, W 135, and Y [88], [89].

History of Immunization. The first attempts at immunizing against meningococcal disease used vaccines made of killed bacteria and were unsuccessful [90]. Attention then centered on using purified capsular antigens. In 1945 KABAT demonstrated that up to 1 mg of purified antigen could be administered to humans with no untoward reactions. However, these studies were not continued because the prophylactic treatment of meningococcal meningitis with antibiotics appeared to be highly effective. However, the appearance of resistant strains in the early 1960s revived interest in vaccine development. In the late 1960s, highly purified group A and group C capsular polysaccharides with a high molecular mass were shown to be safe and immunogenic in humans [91]. Human immune sera contained elevated levels of antibody capable of lysing meningococci in the presence of complement.

Production and Properties of Meningococcal Vaccines. Available vaccines contain serogroup A, C, W 135, and Y capsular antigens. The most widely used formulations are the tetravalent A, C, W 135, and Y, and the bivalent A and C vaccines. Capsular antigens are purified from fermentor-grown cultures by coprecipitation with detergents followed by ethanol fractionation, extraction with organic solvents, and ultracentrifugation. Preservation of the high molecular mass of the capsular antigens is of critical importance since this governs the immune response. Each vaccine dose contains 50 μg of capsular antigen per serotype in lyophilized form. Potency of the vaccine is based upon various physicochemical properties (size and purity). No animal test is required.

Immunization Recommendations. Meningococcal vaccine can be administered to any immunocompetent individual over 18 months of age. A single dose is given either subcutaneously or intramuscularly. The group A antigen is immunogenic in children ≥ 6 months of age, but two doses must be administered within a 2–3 month interval. Due to the low incidence of the disease in the general population, vaccine acceptance has been poor, it is usually used to immunize military recruits or travellers.

Adverse Reactions. The meningococcal vaccines are very safe—about 250×10^6 doses of vaccine have been administered worldwide with no vaccine-associated fatalities reported. Mild, transient local reactions occur in about 25% of vaccinees.

Vaccine Efficacy. Conclusive proof that group C meningococcal vaccine prevents disease was presented in 1970 [92]. Subsequently, the group A vaccine was found to be effective in controlling endemic and epidemic disease in adults and children [93], [94]. Group C vaccine elicited a protective antibody response only in children ≥ 18 months of age, whereas the two doses of group A vaccine were immunogenic in children ≥ 6 months of age [95]. Efficacy for the W 135 and Y capsular antigens has not been clinically demonstrated, but is assumed because they engender appropriate levels of relevant functional antibodies in humans.

Future Prospects. Current meningococcal vaccines have two major limitations. First, they do not provide protection against group B organisms which may account for up to 50% of endemic cases. Secondly, the group C vaccine is nonimmunogenic in children ≤ 18 months of age. The major difficulty in producing a group B meningococcal vaccine is that the group B capsular antigen, a homopolymer of $\alpha (2\rightarrow 8)$-linked scialic acid, is poorly immunogenic in humans, probably because similar structures are found in human gangliosides and fetal proteins. Even when coupled to carrier proteins [96] or mixed with group B outer membrane proteins [97], the group B antigen is a poor immunogen. An alternative approach is to use serotype-specific outer membrane proteins, 18 such serotypes have been identified. Outer membrane protein vaccines containing group B or C capsular antigens were found to be safe and immunogenic in humans [98], [99]. A large-scale field trial to determine the efficacy of these vaccines is being conducted.

In an attempt to construct vaccines capable of engendering a protective immune response in young children, capsular antigens have been coupled to carrier proteins to form conjugate vaccines [96]. Group A and C antigens coupled to tetanus toxoid are more immunogenic in animals than native capsular antigens. Studies are now in progress to determine the safety and immunogenicity of group A and C conjugates in humans.

2.10. Tuberculosis Vaccine

Etiological Agents and Pathogenesis. Human pulmonary tuberculosis is caused by the bacilli, *Mycobacterium tuberculosis* and *M. bovis*, the latter causing disease primarily in children. *M. tuberculosis* is spread by infected droplets, whereas *M. bovis* is disseminated by ingestion of contaminated milk. The vast majority of clinical disease occurs in middle-aged individuals in developed countries and in children under 10 years of age in underdeveloped countries. The incidence of tuberculosis ranges from ca. 10/100 000 population in Europe and North America to 500/100 000 in parts of Africa. Although 95 – 98% of people exposed to *M. tuberculosis* become infected, i.e., become tuberculin-positive, they do not manifest signs of clinical disease and are immune [100]. Only a small percentage of infections progress to a disease state. Once the inhaled bacilli reach the lower respiratory tract, they are ingested by alveolar macrophages. Rapid intracellular growth with inflammatory cell recruitment leads to granuloma formation. In the majority of cases, this is as far as the disease progresses. Active disease presents clinical symptoms ranging from self-limiting pulmonary involvement to acute disseminated disease which rapidly leads to death.

History of Immunization. Efforts to protect humans against tuberculosis have almost exclusively employed *Bacillus Calmette–Guérin* (BCG) vaccine (see below). Oral BCG vaccine was first evaluated in the 1920s in newborns. Subsequent studies showed BCG vaccine to be safe when given orally or parenterally. In the early 1930s, large-scale trials with BCG vaccine were started. Subsequent trials involving American Indians and Eskimos showed that intradermally administered BCG conferred protection against disease in infants. Additional proof of vaccine efficacy was obtained in a small-scale trial with Canadian Indians [101]. Even in the face of a high disease transmittance rate and a mortality rate of 800/100 000, the vaccine was more than 80% effective over a 9-year observation period.

Production and Properties of Tuberculosis Vaccine. The original BCG vaccine strain was developed at the Pasteur Institute over 50 years ago by in vitro cultivation of *M. bovis* for 10 years. Numerous substrains are now used to produce commercial products. The bacteria are grown on Sauton's liquid media either in bottles with a high surface area or in submerged culture. They are then collected, suspended in a stabilizer solution, homogenized, and lyophilized. The bulk material is standardized by opacity, dry mass, and number of viable bacteria per unit mass. The vaccine is usually filled into glass ampules which are sealed under vacuum.

Controls of BCG vaccine are stringent. First, the culture is examined for purity, viability, and identity. Innocuity is tested by injection of vaccine into guinea pigs. The number of live organisms are determined by viability counting. General safety tests to confirm absence of unexpected toxicity are also performed. Resistance of the bacteria to elevated temperature is also documented. Vaccine potency is based upon the total viable bacteria per dose. However, this is not totally satisfactory because (1) different BCG strains may vary in their ability to evoke a protective cell-mediated immune response, and (2) the ratio of viable to nonviable bacteria can influence the magnitude of both the humoral and cell-mediated immune response to vaccination [102].

Immunization Recommendations. A single intradermal dose containing $(4-8) \times 10^5$ viable organisms in 0.1 mL is normally administered. The preferred use of BCG vaccine depends on the disease incidence in the area in question. In developed areas where disease is infrequent, vaccination should be limited to individuals at high risk to exposure, i.e., family members of tuberculosis patients, health care workers expected to come into contact with tuberculosis patients or infected clinical specimens, or travellers entering an area of high endemicity for a prolonged period of time. In developing areas with a high rate of disease incidence, routine immunization of infants is a cost-effective method of controlling tuberculosis. In addition, all tuberculin-negative children upon entry into school and tuberculin-negative adults should be vaccinated.

Adverse Reactions. Correctly administered BCG vaccine is considered to be safe; one vaccination in 100 000 will result in a noticeable reaction [103]. A significant proportion of these reactions are ulcers caused by inadvertent subcutaneous administration of the vaccine. Of greater concern are the reports of mycobacterial disease following immunization [104]. Great care must be taken to insure that a potential vaccinee is not immunocompromised as the vaccine organisms can then rapidly multiply and lead to death [105].

Vaccine Efficacy. In numerous field trials the efficacy of BCG vaccine ranged from 0–80% [100]. Although the three most recent trials failed to document significant vaccine-induced protection, considerable support still exists for large-scale vaccination in developing countries [106].

Future Prospects. At present, no "second-generation" tuberculosis vaccines are being clinically evaluated. Although the tuberculin skin test offers a practical method to measure "immunity" to disease, the molecular mechanism responsible for inducing a tuberculin-positive state is not known.

Recently, several *M. tuberculosis* cell surface antigens have been identified which can induce a cell-mediated response in experimental animals [107]. The cloning of such antigens and their high-level expression in bacteria or yeast would permit economical large-scale purification. Alternatively, such cloned genes could be inserted into a suitable vector such as vaccinia virus [108]. Since most tuberculosis vaccine are used in underdeveloped nations, any new vaccine must not only be safe and effective, but also economical to produce and control.

2.11. *Escherichia coli* Vaccines

Etiological Agent and Pathogenesis. *Escherichia coli*, a gram-negative motile rod, is a major cause of diarrheal disease, urinary tract infections, neonatal meningitis, and nosocomial infections (see Section 2.8). *Escherichia coli* produces a variety of virulence factors including a polysaccharide capsule antigen (K-antigen), lipopolysaccharide antigen (O-antigen), fimbriae, enterotoxins, and cytotoxins. The majority of *E. coli* strains causing a given disease fall within a limited number of K or O serotypes.

Escherichia coli is the major cause of diarrhea worldwide. Three "categories" are recognized, i.e., enterotoxigenic, enteroinvasive, and enteropathogenic *E. coli* strains.

1) *Enterotoxigenic E. coli* invariably causes watery diarrhea which may be accompanied by cramps, fever, or vomiting. The disease is acquired by ingesting contaminated food and affects infants, young children, and travellers to underdeveloped countries. The strains produce either a heat-stable enterotoxin and/or a heat-labile enterotoxin; they also possess fimbriae which mediate attachment to the ileal epithelium. Three distinct fimbriae (CFA/I, CFA/II, and E 8775) have been identified on enterotoxigenic *E. coli* strains isolated from human disease.
2) *Enteroinvasive E. coli* causes a dysentery-like disease characterized by fever, diarrhea, cramps, and stools containing blood and mucus. The disease is acquired by ingestion of contaminated food. The bacteria readily penetrate the ileal and colonic epithelia where they rapidly multiply resulting in tissue destruction. They can then infect adjacent cells or penetrate the underlying lamina propria.

3) *Enteropathogenic E. coli* affects primarily infants and young children after ingestion of contaminated food causing watery diarrhea often accompanied by fever and/or vomiting. These *E. coli* strains rarely produce a heat-labile or heat-stable enterotoxin and are unable to invade eukaryotic cells. They colonize the ileal epithelium with destruction of the surrounding brush border in the absence of invasion. Enteropathogenic *E. coli* strains can synthesize a cytotoxin similar to that produced by *Shigella dysenteriae* 1. This so-called Shiga-like toxin is thought to elicit the clinical symptoms.

Approximately 50% of neonatal meningitis is due to *E. coli* transmitted from mother to infant. The bacteria apparently penetrate the intestinal epithelium of the neonate, enter the bloodstream, and are then translocated to the meninges. The majority of strains (>80%) causing meningitis express the K1 capsular antigen. However, the O-antigen also appears to play a role in virulence. *Escherichia coli* meningitis has a high attendant fatality rate.

Escherichia coli is the causative agent of ca. 90% of nonobstructive urinary tract infections which are much more common among women than men. The clinical syndromes can include bacteriuria, cystitis, and pyelonephritis. Renal scarring associated with pyelonephritis occurs most often in children ≤ 5 years of age. The infecting organisms are thought to be acquired from the flora colonizing the vagina and periurethral areas. P and type I fimbriae are believed to play an important role by mediating tissue colonization [109].

History of Immunization. Initial attempts to vaccinate against *E. coli* diarrhea used crude cellular extracts derived from O types 111, 55, and 86 [110]. In one study, hospitalized infants up to one year of age received multiple doses of oral vaccine. Overall efficacy was 41%. Current efforts to develop vaccines against *E. coli* diarrhea center around the use of purified antigens or live-oral attenuated strains to elicit an antitoxic or antiadhesion response. Calves or neonatal piglets reared by mothers immunized with purified K88, K99, or 987P fimbriae, procholeragnoid (heat-aggregated cholera toxin with reduced toxicity) or heat-labile enterotoxin are protected against *E. coli* bacillosis [111]–[113].

Graded doses (45–1800 µg) of purified type I fimbriae have been parenterally administered to human volunteers [114]. The vaccine was well tolerated and elicited both a humoral IgG and an intestinal secretory IgA antibody response. A moderate level of serospecific protein was obtained in volunteer challenge studies.

Oral administration of 1 mg of purified CFA/I evoked a significant secretory IgA response in about 50% of vaccinated humans [115]. Responders were protected against challenge with an St^+, LT^+, CFA/I^+ *E. coli* strain.

Future Prospects. No vaccines are available to prevent diarrhea urinary tract infections, or meningitis caused by *E. coli*. At present, both attenuated live-oral and inactivated vaccines are being evaluated as to their ability to prevent *E. coli* diarrhea. The production of a killed, inactivated, oral, whole-cell vaccine from a CFA/I^+ strain has been described [115]. The majority of vaccinees responded with a good secretory IgA antibody response to CFA/I. The diarrhea attack rate for volunteers challenged with a toxinogenic CFA/I^+ *E. coli* strain was 89% for those in the placebo group and 20% for those in the vaccine group.

A spontaneous mutant of *E. coli* has been isolated which has lost the genes for production of the heat-labile and heat-stable enterotoxins but still expresses CFA/II. This strain has been proposed as a likely carrier for cloned genes expressing relevant protective antigens such as colonization factors or "enterotoxoids" [116]. Several purified antigens show promise as oral vaccine candidates including conjugates of the two enterotoxins [117], the B subunit of the heat-labile toxin [118], and toxoid termed "procoligenoid" produced by heating the heat-labile enterotoxin [119].

Since the majority of *E. coli* strains associated with neonatal meningitis possess the K 1 capsule antigen, this would appear to be an excellent vaccine candidate. However, the K 1 antigen (which is identical to the group B meningococcal capsule) is not immunogenic in humans (see Section 2.9).

The most promising vaccines to prevent serious *E. coli* urinary tract infections (pyelonephritis) contain type I fimbrial antigens. Two fimbrial amino acid sequences have been shown to prevent colonization by a homologous *E. coli* strain in a murine pyelonephritis model [120]. Antibody to one sequence also afforded protection against an *E. coli* strain expressing a distinct fimbrial serotype.

2.12. *Neisseria gonorrhoeae* Vaccine

Etiological Agent and Pathogenesis. *Neisseria gonorrhoeae* was identified as the causative agent of gonorrhea by BUMM in 1885. It is a gram-negative diplococcus with humans serving as its only known natural host. Gonorrhea is a sexually transmitted disease of epidemic proportions worldwide. More than 500 000 cases are reported annually in the United States. The infection is usually restricted to the entry site, i.e., the genitalia, pharynx, and/or rectum. More than 90% of infected males display symptoms characterized by pain upon urination and an urethral exudate [121]. However, up to 50% of infected women remain asymptomatic and can serve as carriers. In approximately 10% of women with genital gonorrhea, the bacteria invade the bloodstream and result in pelvic inflammatory disease. Disseminated gonococcal infections are far less frequent in male patients.

When cultured on solid medium, four colonial variants (T 1 – T 4) can be discerned [122]; however, only the piliated T 1 and T 2 variants can cause disease [123]. The first step in the infectious process is pili-mediated attachment of the gonococcus to mucus-secreting epithelial cells. Shortly thereafter, the bacteria are internalized and enter the submucosa [121]. Treatment of gonorrhea has recently been complicated by the occurrence of multiple antibiotic-resistant clinical isolates.

History of Immunization. The first attempt to prevent gonorrhea by immunization used a killed whole-cell vaccine administered parenterally [124], but did not provide protection in Eskimos. Subsequent human vaccine trials have used purified intact pili: antipili antibody blocks attachment of the gonococcus to epithelial cells, and, an antipili secretory IgA antibody response is mounted during natural infection [125]. Vaccination of human volunteers with purified *N. gonorrhoeae* pili elicited a humoral and a local vaginal antibody response. These antibodies blocked the attachment of *N. gonorrhoeae* to epithelial cells in vitro [125]. Two 2 mg doses of a monovalent pili vaccine resulted in significant protection against challenge with the homologous strain [126]. However, no protection was observed when the challenge strain possessed a pilus serotype modestly cross-reactive with the vaccine [121]. A subsequent field trial with this monovalent vaccine failed to show any protection [125].

Future Prospects. No vaccine has been licensed for the prevention of gonorrhea. The above human volunteer studies showed that protection against gonorrhea mediated by antipili antibody is serospecific [125]. Two different approaches are being taken to circumvent the problem of antigenic variation among pili:

1) Use of a polyvalent vaccine containing those types of pili which predominate in a given area [126].
2) Use of conserved antigenic domains from the antigenic pili molecule. A cyanogen bromide cleavage fragment can induce production of an antibody that can recognize various pili types [127]. Peptides corresponding to this conserved region block attachment of the gonococcus to epithelial cells in vitro [128]. If such peptides can induce a broadly protective immune response in humans, synthetic peptides containing the proper sequence could be coupled with carrier proteins to produce a conjugate vaccine.

A possible alternative to pili as vaccine antigens are the outer membrane proteins of *N. gonorrhoeae*, the most promising being the PI protein. The PI protein has a sufficiently restricted antigenic variation to make construction of a polyvalent vaccine feasible. Furthermore, antiPI antibody is bactericidal and may block entry of the gonococcus into epithelial cells [129]. A parenterally administered, outer membrane preparation enriched with PI protein elicits an antibody response in humans [130].

2.13. *Hemophilus influenzae* Type b Vaccines

Etiological Agent and Pathogenesis. *Hemophilus influenzae* is a gram-negative bacillus first isolated by PFEIFFER in 1892. The vast majority (> 90%) of human disease is caused by type b capsular serotype organisms in children between the ages of three months to two years [131]. *Hemophilus influenzae* type b can cause several serious disease syndromes predominated by meningitis which is endemic worldwide and is one of the three leading causes of bacterial meningitis in developed countries. Infection is by inhalation of contaminated aerosols. After localized multiplication, the organisms enter the bloodstream and translocate to the meninges. Even though effective antibiotic treatment against *H. influenzae* type b has been available for many years, meningitis due to this organism still has a high attendant morbidity and mortality rate.

History of Immunization. Intensive efforts to understand human immunity to *H. influenzae* type b disease have largely replaced clinical trials with crude, poorly characterized vaccines. The capsular polysaccharide is a critical virulence factor because it confers resistance to lysis by serum [132]. Antibody-dependent, complement-mediated bacteriolysis is the basis for immunity against *H. influenzae* type b [132]. Antibodies directed against the type b capsular polysaccharide are bactericidal in the presence of complement and are believed to be the predominant protective antibody [131].

Production and Properties of *H. influenzae* Type b Vaccines. Two vaccines are licensed for use against *H. influenzae* type b: a purified capsular polysaccharide and a conjugate vaccine composed of capsular polysaccharide covalently coupled to diphtheria toxoid. The capsular antigen is purified by coprecipitation of the capsule with detergent, digestion with nucleases and protease, extraction in phenol, and ultracentrifugation. In the conjugate vaccine, the capsule is covalently linked to diphtheria toxoid by a bifunctional spacer molecule. The potency of *H. influenzae* type b vaccines is not determined in animals but by analysis of physicochemical characteristics which confirm that the capsular antigen is in an immunogenic form.

Immunization Recommendations. The capsular polysaccharide vaccine was first licensed for use in the United States in 1985. A single 25 µg dose, administerd parenterally is recommended for children 2–6 years of age; no booster dose is needed.

For the conjugate vaccine licensed in 1987, a single dose is recommended for children 18 months–6 years of age; no booster dose is needed.

Vaccine Efficacy. The purified capsular polysaccharide vaccine was evaluated in a trial on Finnish children [133]. Vaccine efficacy was approximately 90% for children who were vaccinated at 2–5 years of age or older. Vaccination provided no protection in children 3–18 months of age due to poor immunogenicity [132]. However, retrospective studies performed in the United States have found vaccine efficacy to range from 55–88% [133]. In contrast, a capsular polysaccharide–diphtheria toxoid conjugate vaccine evoked protective levels of antibody in a majority of 18-month old children [135].

Future Prospects. A *H. influenzae* type b vaccine is clearly needed which is immunogenic in children <18 months of age, where approximately 75% of serious disease occurs. Several candidate vaccines including the diphtheria toxoid conjugate mentioned above and a *Neisseria meningitides* outer membrane protein–capsular polysaccharide conjugate are being evaluated in clinical trials [135], [136]. A large-scale Finnish trial is being performed to evaluate the diphtheria toxoid conjugate vaccine. Vaccine efficacy was 83% in children vaccinated at 3, 4, 6, and 14 months of age [137]. In contrast, this vaccine afforded insignificant protection when used to immunize Alaskan Eskimo children at 2, 3, and 4 months of age.

3. Viral Vaccines

3.1. Measles Vaccine

Etiological Agent and Pathogenesis. Measles (rubeola) is caused by a paramyxovirus first isolated by ENDERS and PEEBLES in 1954. This highly contagious disease is transmitted in an aerosol of virus-infected droplets from the respiratory tract. During the prevaccination era, children in industrialized countries generally had a self-limiting disease characterized by conjunctivitis, bronchitis, fever, and a rash; complications included pneumonia (mostly bacterial) and encephalitis. A late complication of measles in early childhood is the development of a progressive, fatal neurological disease, termed subacute sclerosing panencephalitis. Death due to complications following measles infection remains one of the leading causes of child mortality in developing countries.

History of Immunization. Two approaches to vaccine development were adopted, one was based on killed (inactivated) strains and the other on live, attenuated strains [138]. The formaldehyde-inactivated, alum-precipitated vaccine was licensed in the United States in 1963 and withdrawn in 1967 because its protective effect was insufficient. In addition, the vaccinees exposed to the wild virus developed atypical measles with fever and rash but also pneumonitis.

The attenuated live vaccine with the Edmonston B strain was also licensed in 1963 in the United States. A large-scale field trial in the United Kingdom showed high seroconversion rates, a long antibody response, and a 84–94% decrease in attack rate among the 36 000 immunized children. Further attenuation of the vaccine strain in chick embryo cells yielded the Schwartz and the Moraten strains. Both strains are included in current measles (and combined measles–mumps–rubella) vaccines. Attenuation of the original Edmonston strain in human diploid cells yielded the Edmonston-Zagreb strain, also in current use [139]. With the introduction of general immunization against measles, the disease has almost disappeared in many countries (see below).

Production and Properties of Measles Vaccines. The attenuated measles vaccine strain is usually grown in primary culture using chick embryo cells or human diploid cells. The WHO test requirements for the seed virus and cell substrates are described in [140, pp. 52–73], [141, p. 179]. Tests for monitoring measles vaccine grown in chick embryo cells include tests for nonadsorbing viruses and avian leukosis viruses. For vaccines grown in human diploid cells, chromosomal monitoring requirements have been formulated. Potency is checked by titration of the virus in tissue culture, (minimum requirement, 1000 $TCID_{50}$ per human dose). To test for stability, a sample of the final freeze-dried vaccine is incubated at 37 °C for seven days: the sample must then contain at least 1000 $TCID_{50}$ in each human dose. The vaccine is lyophilized; reconstituted vaccine should be used immediately or stored at 0–10 °C for not more than 8 h.

Immunization Recommendations. The vaccine is given at the age of 15–24 months [142] or even younger in developing countries [139]. In the United States and many European countries, a combined measles–mumps–rubella (MMR) is usually given (see Section 3.4). As for all live vaccines, immunization of immuncompromised individuals is not generally recommended. Immunization of pregnant women should be avoided although no increased risks have been documented for measles vaccine.

Adverse Reactions. Current attenuated measles strains give a few mild reactions, mostly consisting of fever and rash 7–10 days after immunization. The neurologic reaction of the Guillian-Barré syndrome has been reported but is extremely rare.

Vaccine Efficacy. The rapid decrease of measles morbidity in the United States after introduction of general immunization is a good example of vaccine efficacy [142], [143]. Over a period of 20 years the rate of notified cases fall by more than 99%. Subacute sclerosing panencephalitis also decreased. The overall protective efficacy of the current strains is 90% but is lower in children of 12 months of age due to the presence of maternal antibody. Life-long immunity after one dose of vaccine has not been proven. Increased disease incidence in the United States has raised the question of the possible need for a two-dose schedule [142].

Future Prospects. Given the efficacy and acceptability of the currently used attenuated measles strains, there has been little incentive for vaccine development. However, current vaccines are not very stable at high temperature, which is a concern in many developing countries. Furthermore, a large proportion of morbidity and mortality in children in developing countries occurs before the recommended age for immunization. Research is centered on solving these two problems. More immunogenic strains and/or higher doses of attenuated vaccines are being investigated [139]. Another approach would be to renew work on the development of a killed (inactivated) vaccine. Inactivated vaccines have so far failed, this is probably the result of a lack of neutralizing antibodies to the viral fusion surface protein that is destroyed in the inactivation process [144]. Subunit vaccines based on the hemagglutinin and the fusion proteins of measles virus have successfully been tested in animals and may provide a vaccine especially suited for developing countries [145].

3.2. Mumps Vaccine

Etiological Agent and Pathogenesis. Mumps is caused by a member of the paramyxovirus group and is also known as (epidemic) parotitis due to its predominant symptom—infection of the salivary (parotid) gland. Transmission occurs via infected aerosol droplets. The disease is generally characterized by moderate fever and swelling of the salivary glands. Subclinical infection occurs in ca. one third of infected individuals. Most cases occur in children 5–10 years of age. The most frequent complication is meningoencephalitis, giving clinical symptoms in >10% of patients. Orchitis, a less common but feared complication in postpubertal males, results in impairment of fertility in 10–20% of cases but absolute sterility is rare.

History of Immunization. Isolation of the causative agent for mumps by HABEL in 1945 from chick embryo was followed by attenuation studies for vaccine development. Development of an inactivated (killed) vaccine, was also initiated; a killed vaccine was licensed in the US in 1950–1978 but had low long-term protective efficacy [146]. An inactivated mumps vaccine was also produced in Finland for immunizing military recruits; it decreased disease incidence by 94% [147]. Immunization with an attenuated mumps vaccine (Jeryl Lynn strain cultured in chick embryonic cells), licensed in 1967 in the United States resulted in a 97% decline of disease incidence in the general population by 1981 [146], [148], [149]. Chick embryonic cells are used to propagate another highly attenuated strain, Urabe Am 9 developed in Japan [150]. The Rubini strain, developed in Switzerland, is used in a vaccine in which the virus is grown in human diploid cells [151]. Vaccines based on the Leningrad-3 strain have been used in the Soviet Union and other countries since 1974.

Production and Characterization of Mumps Vaccines. The WHO requirements for the above-mentioned live mumps vaccine have recently been formulated [152, pp. 139–164]. Controls, thermostability tests, and requirements are as for measles vaccines, i.e., the minimum potency should be retained after inoculation for one week at 37 °C (see Section 3.1). No minimum requirement of potency has been established by the WHO but a commonly used minimum dosage is 5000 TCID$_{50}$ per human dose. The vaccine is lyophilized, reconstituted vaccine should be used without delay.

Immunization Recommendations. Mumps vaccine is recommended to be given to all susceptible individuals over the age of 12 months [153]. It is usually given in combination with measles and rubella vaccines in industrialized countries (see Section 3.4).

Vaccine Efficacy. Clinical efficacy is 75–90% for the live vaccine containing the Jeryl Lynn strain [149]. Similar results have been shown for the other strains or inferred from serologic comparisons. The long-term protective efficacy of the inactivated mumps vaccine used in Finland has not been established but antibody responses indicate that it may give a shorter-term immunity than a live vaccine due to a lack of antibody response to the fusion protein [154] as in the case of the inactivated measles vaccine (see Section 3.1).

Adverse Reactions. Mumps vaccine is one of the least reactogenic attenuated vaccines. Side effects are rare and mild. No cases of atypical mumps were found after immunization of Finnish recruits with killed mumps vaccine [154].

Future Prospects. The present vaccine is highly satisfactory and little effort has been devoted to further development. Subunit vaccines based on the hemagglutinin neuraminidase and the fusion surface proteins of mumps virus are protective in animal models [155].

3.3. Rubella Vaccine

Etiological Agent and Pathogenesis. Rubella, also known as German measles, is caused by a member of the togavirus group. The virus was isolated in 1962 by two independent American groups. In 1941 GREGG reported that this mild disease could cause severe congenital cataracts in children born to mothers who were infected during the first trimester of pregnancy. Transmission of the virus occurs by the respiratory route. The clinical picture is a discrete rash and low-grade fever, 25–50% of cases remain subclinical. Arthritis is a common complication; encephalitis is less common than in measles or varicellae (1/6000 cases). The congenital rubella syndrome is usually characterized by hearing impairment, ocular lesions, cardiac malformation, microcephaly, and mental retardation. The severity of defects is related to fetal age at the time of maternal infection, with the most severe damage seen after infection during the first month of pregnancy.

History of Immunization. Live attenuated rubella vaccines were licensed in several countries in 1969–1970. One of the first vaccines contained the Cendehill strain, grown in rabbit kidney cells [156]. Other vaccines were based on the HPV-77 strain, grown in either duck embryo cultures or in dog kidney cultures. The latter vaccine produced arthritis symptoms; 50–60% of adult female vaccinees as compared to 10–20% for the other two vaccines. A vaccine based on the RA 27/3 strain, isolated and propagated in human diploid cells, was licensed in the mid-1970s in Europe and in 1979 in the USA [157]. This vaccine gave a better immune response and had less side effects than the other vaccines; it has replaced other rubella strains in the current vaccines.

Production and Properties of Rubella Vaccines. The attenuated RA 27/3 strain is grown in human diploid cells. The WHO requirements for rubella vaccine include control of normal karology of the human diploid cells [158, pp. 54–84], [159, pp. 313–316]. The minimal potency requirement, determined by titration in tissue culture, is 1000 $TCID_{50}$ per human dose. The vaccine is lyophilized; reconstituted vaccine should be used immediately or stored at 2–8 °C for not more than 8 h.

Immunization Recommendations. Vaccination strategies range from protection of the individual as in the U.K. to indirect protection as in the US [160], [161]. Protection is achieved by general immunization of adolescent girls, usually combined with post partum vaccination of seronegative women. Indirect protection of adult women by immunization of young children is often combined with vaccination of seronegative women.

A combination of these strategies is used in some European countries, e.g., Sweden [162]. This program includes routine screening of all pregnant women, post partum vaccination of seronegatives, and two-dose immunization of children with combined measles–mumps–rubella vaccine. Although the risks for the fetus seem small, the vaccine should not be given to pregnant women and contraceptive measures are recommended for three months after immunization.

Adverse Reactions. The currently used RA 27/3 strain has few and mild side effects. The joint manifestations, mainly arthralgia, are more common in adults. Intrauterine infections during pregnancy have been documented for all three strains (Cendehill, HPV-77, and RA 27/3) [163]. No abnormalities related to congenital rubella infections have been documented for any of the strains.

Vaccine Efficacy. The efficacy of the current strain is ca. 90%. The incidence of congenital rubella virus infection decreased in countries with a routine immunization program against the disease. The duration of immunity has not yet been determined. The risks to the fetus upon a maternal reinfection are also unknown and may be very small.

Future Prospects. The current vaccine is considered safe and efficient and no efforts are devoted to further development.

3.4. Combined Measles–Mumps–Rubella Vaccine

History of Immunization. The first combined measles–mumps–rubella (MMR) vaccine was licensed in the US in 1971. It contains the Moraten measles strain (Section 3.1), the Jeryl Lynn mumps strain (Section 3.2), both grown in chick embryo cultures, and the RA 27/3 rubella strain (Section 3.3), grown in human diploid cells. Three other combined vaccines were licensed later and are also in current use. The triple vaccine has replaced the monovalent vaccines used in general immunization programs for children in the United States and in many European countries.

Production and Properties of Combined Vaccines. The minimum potencies for the attenuated strains of measles, mumps, and rubella virus, are the same as in the monovalent vaccines, i.e., 1000 $TCID_{50}$ for measles, > 5000 $TCID_{50}$ for mumps, and > 1000 $TCID_{50}$ for rubella. The vaccine is lyophilized and should be stored at 2–8 °C.

Immunization Recommendations. In many countries, one dose of vaccine is administered to children at the age of 15–24 months. Two doses are given in some European countries [162]. In Sweden, a first dose is administered at 18 months and a second at 12 years of age, replacing the monovalent programs for measles and rubella immunization respectively. In the United States, the combination of MMR with oral polio vaccine and diphteria–tetanus–pertussis (DTP) at 15 months has been recommended [164], [165].

Adverse Reactions. The side effects are the same as for the monovalent vaccines, the measles component being the most reactogenic. The rate of adverse reactions is < 0.5–4% [166].

Vaccine Efficacy and Future Prospects. The efficacy of the combined preparation is the same as for the monovalent vaccines with an overall efficacy of 90 – 95 % [143]. The life-long protective efficacy of a single injection has not been proven. Further combinations with an attenuated varicella component are currently under investigation [167].

3.5. Polio Vaccine

Etiological Agent and Pathogenesis. Polio (infantile paralysis) is caused by a picornavirus of the genus enteroviruses. A poliovirus strain was first isolated in cell culture by ENDERS, WELLER, and ROBBINS in 1949. In 1951, polio virus isolates were officially grouped into three serotypes, type 1 (Brunhilde), type 2 (Lansing), and type 3 (Leon). Transmission is mainly via the oral – fecal route and the virus reaches the central nervous system by way of the blood stream. The vast majority of infections are subclinical. The nonparalytic disease has a mild or minor form with fever and general malaise and a more severe or major form with additional symptoms of meningitis/ meningoencephalitis. Paralytic polio, with its most severe bulbar form, is estimated to represent 5 – 10 % of clinical cases. The mortality rate is 5 – 10 % of clinical cases, i.e., 1 – 2 % of all infections. Severe paralysis is seen in 10 – 20 % of clinical cases, and mild or moderate paralysis in 30 %.

With improved sanitation and standards of living, a shift towards infection at a higher age was observed in industrialized countries in the prevaccination era. During the first half of this century, peak incidence was noted in the age group 5 – 14 years. A large proportion of cases occurred in young adults. In the developing countries, polio has maintained its character of infantile paralysis with the majority of children being infected during the first few years of life. Cases of paralytic polio occur only in the youngest age groups.

History of Immunization. The first large-scale field trial of an inactivated (killed) polio vaccine was launched in the US in the early 1950s, only a few years after successful propagation of the virus in tissue culture. Inactivated polio vaccine (IPV) was licensed for general use in the United States in 1955 [168], [169]. In 1955, cases of atypical paralytic polio were reported, most of them were associated with two lots of vaccine from CUTTER [170]. Although live poliovirus was recovered from vaccine supplied by other manufacturers, this failure of inactivation is known as the Cutter incident. The total toll was 269 cases, of which 192 were paralytic (with ten deaths). Clinical trials with a live attenuated oral polio vaccine (OPV) were started in 1958 and the vaccine was licensed in the United States in 1962 [171], [172]. Immunization with OPV alone has been used from the late 1950s – early 1960s in Europe. One of the largest immunization campaigns was launched in the Soviet Union in 1960 when OPV was given to ca. 77×10^6 people. A few European countries have used only IPV [173] – [175]. In Sweden, clinical trials with IPV were started in 1955 and general immunization was introduced in 1957 [173]. All three strategies were effective in reducing the incidence of paralytic polio. With IPV alone, an 80 % reduction rate was achieved in the US with less than 50 % of the population immunized [168], [169], [171]. The decrease of polio continued at the same rate after introduction of OPV. Introduction of the massive IPV campaign in Sweden resulted in a 93% decrease of paralytic disease after six years of immunization and elimination of the disease by 1964 [173].

Production and Properties of Polio Vaccines. Oral polio vaccines are manufactured with the attenuated Sabin strains. Inactivated polio vaccines are mostly produced with the type 1 Mahoney strain, type 2 MEF-1 strain, and type 3 Saukett strain. Primary and secondary monkey (Cynomolgus) kidney cultures or human diploid cells are most commonly used for culture. The WHO requirements for both OPV and IPV produced in primary monkey kidney cells include tests

1) for the absence of cytopathogenic viruses with the exception of some foamy viruses [141, pp. 40–84], [159, pp. 157–173], [176], [182, pp. 108–110]
2) the absence of simian B virus in rabbits, and
3) the absence of SV 40 in sensitive cells.

(usually primary green monkey *Cercopitecus* kidney cultures). Minimal potency requirements for OPV were formulated by the WHO in 1987[152pp. 165–166]. A single human dose of trivalent oral vaccine should contain approximately $10^{5.5} - 10^{6.5}$ infectious units of type 1, $10^{4.5} - 10^{5.5}$ of type 2, and $10^{5.0} - 10^{6.0}$ of type 3. For IPV, recommendations of 40:8:32 D-antigen units per human dose for types 1, 2, and 3 respectively were issued in 1981 [141]. The potency tests for OPV (and IPV prior to inactivation) are performed by titration in tissue culture. In vivo potency tests are also required for IPV but neither the animal species nor the number of injections is specified. Inactivation of polio vaccine by formaldehyde is controlled by titration of polio-sensitive cells (usually *Cynomolgus* or *Cercopitecus*) in tissue culture. Oral polio vaccines are supplemented with a stabilizer, usually magnesium chloride or sorbitol. Both OPV and IPV contain antibiotics, usually neomycin; OPV is best stored at −20 °C but can be stored at 2–4 °C for a variable length of time; IPV is stored at 2–8 °C.

Immunization Recommendations. Recommendations vary but at least three doses of OPV are given. In the US, five doses are recommended, in Sweden four, and in Finland six. The WHO recommends four doses for infants in developing countries.

Adverse Reactions. Serious adverse reactions of paralytic polio have only been reported for OPV with a predominance for type 3 [169], [177] (incidence = $1/1 \times 10^6$ vaccine recipients, including both nonimmune and immune individuals). No cases of IPV-induced paralytic polio have been reported after 30 years of use (with the exception of the Cutter incident). IPV is considered to be the least reactogenic of all the vaccines used for childhood immunization.

Vaccine Efficacy. Both OPV and IPV have proven highly efficient in eliminating polio from industrialized countries with general immunization programs. The protective efficacy is close to 100% after at least three doses of vaccine with adequate immunogenicity. Serologic data reported for OPV from developing countries have been less than encouraging. Both vaccines have eliminated the circulation of wild type strains in industrialized countries with high immunization rates. Outbreaks among groups refusing immunization have occurred [168]. In 1984, nine cases of paralytic polio caused by type 3 occurred in Finland [177]. The low immunogenicity of the IPV used in Finland has been known since the late 1960s [178]. The new IPV developed in the Netherlands with its high antigen content is given in European countries in at least three doses [179]. Two doses or even one dose of the new IPV preparations have been claimed to be highly protective. A two dose-schedule has been used in Africa and protective efficiency was ca. 89% [180]. Inactivated polio vaccine is more heat-stable and can be combined with diphtheria–tetanus–pertussis vaccines.

Future Prospects. Further development of OPV is aimed at reducing production costs by use of microcarrier cultures (cell cultures on beads etc.) in fermentors [179]. Use of continuous cell lines is being investigated. An IPV produced in a continuous cell line is currently licensed in France [181]. The WHO requiremet [182 pp. 93–107].

3.6. Hepatitis B Vaccine

Etiological Agent and Pathogenesis. Hepatitis B is caused by the hepatitis B virus (HBV), a member of the hepadna virus group. The presence of a new antigen called the Australia (Au) antigen or hepatitis B surface antigen (HBsAg) in the blood of patients with leukemia, Down's syndrome, and hepatitis was shown by BLUMBERG in 1964–67. Other antigens (HBeAg and HBcAg) were described later and correlated with infectivity; they are found in the 42-nm Dane particle, i.e., the infectious virion.

The HBV is a pathogen only for humans, but certain monkeys, in particular chimpanzees, are also susceptible to infection. Transmission in humans occurs mainly by inoculation with infected blood. The virus is excreted in body fluids, causing infection through saliva and sexual contact. A chronic carrier state is established in 5–10% of adult cases; chronic active hepatitis develop in 25–30% of the carriers, often leading to cirrhosis of the liver: 90% of infants infected at birth become carriers. Hepatitis B occurs throughout the world; the incidence varies from <0.5% in Western Europe and the US to 5–15% in Southeast Asia and Southern Africa [183], [184].

Hepatitis B virus plays an important role in hepatocellular carcinoma although the exact mechanism has not yet been determined [183].

History of Immunization. Heat-inactivated HBsAg-positive serum was used by KRUGMAN in 1971 to immunize children. A 70% protection rate was found upon challenge with active virus. The high protective efficacy of hepatitis B vaccine derived from purified, inactivated HBsAg from positive plasma was demonstrated in 1980 [185]. The vaccine, developed in the US, was licensed for general use in 1981. Studies in 1980–1983 also showed that the HBV vaccine in combination with human antiHBV immune globulin (HBIG) prevented development of the carrier state in infants born to carrier mothers. Hepatitis B vaccine was the first vaccine to be routinely produced by DNA recombinant technology in yeast. Serologic studies indicate that the protective efficacy of the recombinant vaccines is the same as that of the plasma-derived vaccine. Genetically engineered vaccines were licensed in the US in 1986 and subsequently in several European countries.

Production and Properties of Hepatitis B Vaccines. The WHO requirements for plasma-derived HBV specify guidelines for selection of donors of HBsAg-positive plasma [186, pp. 70–101]. The plasma pool must be subjected to extensive tests in animals, fertile eggs and cell cultures for extraneous viruses and *Mycobacterium tuberculosis*. The purified HBsAg (>95% pure) is inactivated by treatment with pepsin followed by urea (8 mol/L), and formaldehyde or by heat with or without formaldehyde. After controls for purity and antigen content, an alum adjuvant is added; the vaccine must be stored at $5 \pm 3\,°C$. No minimal requirement of antigen content has been formulated but a commonly used vaccine contains 20 µg HBsAg per human dose.

The WHO requirements for HBV made by recombinant DNA technology state that such vaccines may contain the S gene products or the S/pre-S combination [152, pp. 106–138]; licensed recombinant vaccines contain only the S gene product. A full description of the host cell (currently *Saccharomyces cerevisiae*) and the expression vector is required.

The HBsAg is commonly purified by precipitation, ultrafiltration, and chromatography. Tests for HBsAg and residual cell or plasmid DNA are required before addition of adjuvant. No minimal antigen requirements have been formulated but a potency assay in mice has been outlined. Commonly used recombinant yeast vaccines contain 10 or 20 µg HBsAg per human dose.

Immunization Recommendations. Most Western European countries and the US have issued recommendations for vaccination of risk groups [187]–[189]. In general, pre-exposure prophylaxis is recommended for medical and other staff in frequent contact with high-risk groups or with blood from such groups. Immunization of patient groups frequently receiving blood or blood products, as well as patients in certain institutions is also recommended.

For pre-exposure prophylaxis, three 20 µg doses of plasma-derived vaccine is given at day 0, 30 and 6 months intramuscularly in the deltoid region. Children <11 years of age should receive 5 µg at each injection. Postexposure prophylaxis for infants to HBsAg, HBeAg-positive mothers is 0.5 mL intramuscular injection of human antiHBV immune globulin (HBIG, see also Section 5.3). Vaccine (10 µg) should be given at birth and then at one and six months of age. For postexposure prophylaxis of adults, 1 mL HBIG should be given immediately together with three doses of vaccine administrated as for preexposure prophylaxis.

Adverse Reactions. Side effects are mild and mainly local at the site of injection.

Vaccine Efficacy. The overall vaccine efficacy is 90% with no response upon immunization in 5–10% of the vaccines. In individuals with seroconversion to HBsAg, protective efficacy is almost 100%. Studies on the duration of protection beyond 5 – 10 years are not yet available. The protective efficacy of HBIG and vaccine administered to newborns born to carrier mothers has been estimated to be more than 90% [183].

Future Prospects. Current HBV vaccines are safe and have a high efficacy but also high production costs. Research is aimed at the development of cheaper, more immunogenic vaccines. The inclusion of the pre-S (P 31) region of the viral genome in addition to the present S (P 25) region is being investigated in yeast-derived recombinant vaccines [190], [191]. Other antigens such as HBcAg are also being studied. Polypeptide vaccines, based on the major determinants of HBsAg, are undergoing clinical trials. Synthetic peptides may provide the ultimate solution to the problem of immunogenic, readily available HBV vaccines.

3.7. Rabies Vaccine

Etiological Agent and Pathogenesis. Rabies is a lethal disease caused by a neutropic rhabdovirus affecting humans and warm-blooded animals. The most common route of transmission to humans is by infected saliva through the bite from a rabid animal. The virus ascends along peripheral nerves to the central nervous system. The incubation period varies from ten days to several months. Symptoms include a nonspecific prodromal stage followed by an acute neurological phase ending in coma and death. The disease is endemic in animals in most parts of the world with some exceptions such as the United Kingdom and parts of Scandinavia [192]. In the urban form, stray dogs and cats act as vectors, whereas the sylvatic form involves wild animals. Urban rabies is well-controlled in most countries, but progressive spread of the sylvatic form by foxes is a major problem in Europe. In the US, other animal species are involved including bats. In Europe and Africa, bats are the vector for an antigenic variant of the virus [193].

History of Immunization. In 1885 PASTEUR administered the first ever vaccination to a young boy who had been bitten by a rabid dog [194]. Serial injections containing a virus strain that had been propagated and attenuated in the spinal cord of rabbits ("fixed" virus) were given in an increasingly virulent form. The treatment was rapidly adopted as standard postexposure prophylaxis. The attenuated vaccine was later replaced by an inactivated (killed) vaccine also produced in neural tissue [194]. However, in some patients the myelin content of the vaccine caused sensitization resulting in neurological disease. An inactivated vaccine produced in duck embryos decreased the risk of neural tissue sensitization. The most widely used rabies vaccine in Europe and in the United States was developed in the 1960s and is produced in human diploid cells. A highly purified, concentrated duck embryo vaccine is also available. In Europe, other vaccines are produced in primary animal cells and continuous cell lines; non human primate diploid cells are used in the US.

Production and Properties of Rabies Vaccines. Attenuated strains of rabies virus are grown in tissue culture or embryonated duck eggs (see above). The virus suspension is usually concentrated and in some cases purified. Only inactivated vaccines are allowed for human use. Inactivation is mostly achieved by treatment with β-propiolactone, but phenol, formaldehyde, or ultraviolet irradiation are also used. The WHO has established requirements for controls of the cells used for virus propagation [152, pp. 167–194], [159, pp. 54–95]. The potency of the vaccine is determined in a mouse challenge assay (minimal requirement 2.5 IU per human dose). The vaccine is then usually lyophilized although adjuvanted preparations are also in use.

Immunization Recommendations. The recommendations vary by vaccine and country. As pre-exposure prophylaxis, three 1 mL doses are usually recommended on days 0, 7, and 28 or on days 0, 28, and 90 (or 365) as intramuscular injections in the deltoid or suprascapular region [192], [195]. As postexposure prophylaxis, recommended measures include local wound care, administration of vaccine, and administration of human rabies immune globulin (HRIG, 20 IU per kilogram of body weight; see also Section 5.3). Purified equine rabies immune globulin (40 IU/kg) is used in many developing countries with few side effects [196]. Postexposure vaccine prophylaxis is given in five 1 mL doses on days 0, 3, 7, 14, and 28 by intramuscular injection in previously unimmunized individuals. In individuals who have received pre-exposure prophylaxis, two doses are given on days 0 and 3. Children under four years of age receive 0.5 mL injections.

Vaccine Efficacy. The protective efficacy of the human diploid cell vaccine given (with HRIG) as postexposure prophylaxis was first documented in field trials conducted in Iran in 1974–75; the survival rate was 100%. Other studies have confirmed the protective effect of pre- and postexposure immunization with this vaccine. The general opinion that postexposure vaccine prophylaxis in combination with immune globulin conveys 100% protection has been challenged by two cases of vaccine failure [197]. In other vaccines, a high protective efficacy has been proven in field trials or inferred from comparative serologic studies.

Adverse Reactions. The most common side effects are local redness and induration at the injection site and fever. Desquiting episodes of a serum sickness-like reaction upon repeated immunization have been reported [198] and are possibly due to a sensitizing complex formation between β-propiolactone and human serum albumin [199] or to the high content of bovine serum residues in the vaccine [200].

Future Prospects. Work is centered on the development of large-scale vaccine production techniques to decrease the high costs and thereby increase availability in poorer countries. One such approach is the use of microcarrier culture systems in fermentors [179]. A vaccine produced in this manner in a continuous cell line is currently being evaluated. A subunit vaccine based on the surface glycoprotein of the virus is a further possibility. The peptide segment of the glycoprotein that induces the production of neutralizing antibodies has been identified and synthesized. A synthetic peptide vaccine may therefore be the most attractive future alternative [201].

3.8. Influenza Vaccine

Etiological Agent and Pathogenesis. Influenza is caused by two antigenically distinct members of the orthomyxovirus group—influenza viruses A and B. In 1934 ANDREWES managed to transfer the influenza A virus from human material to ferrets and later to mice. In 1940, FRANCIS and MAGILL independently isolated influenza B by transmission to ferrets. The virus was subsequently propagated in embryonated hen's eggs and tissue culture. The disease is transmitted by droplet infection from the respiratory tract; it is characterized by high fever, muscle pain, and a dry cough. Pneumonia is the main cause of mortality in elderly people and persons with chronic underlying diseases. Influenza occurs worldwide with regular epidemics during the winter months. The recurrent epidemics are caused by small changes (antigenic drift) in the main pathogenic determinants of the virus, i.e., hemagglutinin and neuraminidase. Large pandemics occur at 10–20 year intervals and are due to substantial changes (antigenic shift) in the pathogenic determinants. Antigenic drift is seen in both influenza A and B whereas antigenic shift has only been noted in influenza A [202].

History of Immunization. The first influenza vaccines produced in hens' eggs were tested in humans in the early 1940s. These inactivated *whole virus vaccines* had a low content of viral antigens and a high content of contaminating egg protein. Consequently, the protective effect was low and the rate of adverse reactions high. Whole virus vaccines purified by ultracentrifugation showed clearly improved immunogenicity and a decreased rate of side reactions. Zonal centrifugation further improved the vaccine for adults but still caused adverse reactions in children. Disruption of influenza vaccine with detergent was used to develop *split vaccines* in the mid 1960s. *Subunit vaccine* introduced in the mid 1970s contain concentrated, purified hemagglutinin and neuraminidase [202].

The development of live, attenuated virus strains was pursued in the Soviet Union by serial passages in hens' eggs [202]. However, doubt was cast on their stability. More stable attenuated vaccines were obtained by isolation of temperature-sensitive and cold-adopted mutants [203]. These strains were successfully tested in humans in the late 1970s but are not in general use. Attenuated vaccines, based on avian–human recombinant strains, are being evaluated in clinical trials.

Production and Properties of Influenza Vaccines. The most commonly used influenza vaccines are of the whole or split virus type. The strains used are specified in annual WHO recommendations. The virus is grown in embryonated hens' eggs (usually pathogen-free) and the allantoic fluid is harvested. According to the WHO requirements, the inactivation method used (usually addition of formaldehyde or β-propiolactone) should inactivate avian leukosis viruses and mycoplasma [204, pp. 148–170]. The virus is concentrated and purified by high-speed centrifugation either before or after inactivation. Effective inactivation is controlled by inoculation of embryonated hens' eggs. Hemagglutinin content is usually checked by single radial immunodiffusion against a WHO standard. Whole virus and split virus vaccines usually contain 10–15 µg hemagglutinin per human dose. The inactivated vaccines are usually supplemented with a preservative and stored at 2–8 °C.

The WHO requirements for attenuated (live) vaccines stipulate that pathogen-free eggs must be used; the absence of other pathogens must be controlled in tissue cultures and animals [204, pp. 171–194]. Vaccine potency is determined by titration in embryonated eggs. No minimal requirements of infective dose have been formulated.

Immunization Recommendations. Most countries have recommendations for yearly immunization of high-risk groups. Some countries and the WHO recommend vaccination of all individuals over 65 years. In the United States, immunization of children with chronic pulmonary or cardiac disorders or with other chronic diseases residing in institutional care is recommended; 0.25 and 0.5 mL of split virus vaccine are given in the age groups 6–35 months and 3–12 years, respectively [205]. For individuals older than 12 years, 0.5 mL of vaccine is recommended.

Adverse Reactions. The incidence of systemic (febrile) and local reactions is <10% and <20% for the whole virus vaccines and split vaccines, respectively. An increased rate of the neurological reaction of Guillain–Barre's syndrome was reported in the United States after large-scale immunization against swine influenza in 1976.

Protective Efficacy. The protective efficacy of influenza vaccines is controversial. Discrepancies are probably due to different vaccines, number of injections given, the age groups under study, and occurrence of antigenic drift during the study period. The newer purified whole virus and split virus vaccines are considered to be 70–90% effective in healthy adults. The duration of protection is largely dependent on the degree of antigenic drift. In the case of antigenic shift, little or no protection can be expected. The immune response and protective efficacy in children receiving chemotherapy and in debilitated elderly persons are lower [206]. The protective efficacy in preventing death has been estimated at 74% [207].

Future Prospects. Research on influenza vaccines is directed toward the synthesis of the hemagglutinin and the neuraminidase antigens and to improvement of their immunogenicity [201]. Another line of development is the attenuated live vaccine approach [203].

3.9. Varicella Vaccine

Etiological Agent and Pathogenesis. The varicella-zoster virus, a member of the herpes virus group, causes two distinct clinical manifestations [208]—varicella and herpes zoster. *Varicella*, also known as chickenpox, is the primary infection. *Herpes zoster* is caused by the reactivation of the latent varicella virus in ganglion tissue. Varicella is usually a mild infection characterized by a vesicular rash. Transmission occurs by droplet infection from the respiratory tract and by direct contact with vesicle fluid. Complications are encephalitis (1/1000–1/3000 cases) and pneumonia. The disease occurs worldwide, with 90–95% of cases occurring before the age of 15 years.

Suspected cases of congenital varicella have been described mainly after varicella during the first trimester of pregnancy. Neonatal varicella with a 30% mortality rate occurs with maternal varicella within one week prior to term without prophylaxis. Varicella in the compromised host is a severe disease with a 10–20% mortality rate. Herpes zoster, a localized, often painful vesicular rash is most common in the elderly; it occurs in 10% of the population.

History of Immunization. The varicella-zoster virus was first isolated and propagated in tissue culture by WELLER and STODDARD in 1952. Attenuated, live varicella vaccine was developed in Japan in 1970 using the Oka strain [209]. After serial passages, the attenuated strain was adapted to human diploid cell cultures. The immunogenicity and protective efficacy of the vaccine in healthy children has been demonstrated [209], [210]. Efficacy of the vaccine in children with malignant disease, in particular leukemia, has also been shown [211]. The Oka strain is currently used for vaccine production in Japan, Europe, and the United States.

Production and Properties of Varicella Vaccines. The attenuated, live Oka strain is propagated in human diploid cell cultures. The WHO requirements include control for the absence of adventitious agents and the usual conditions for culture in human diploid cells [186, pp. 102–133]. No minimal potency requirements have yet been formulated. Varying doses have been used in clinical trials. The most common formulation is ≥ 2000 plaque-forming units per human dose. The vaccine is lyophilized; reconstituted vaccine should be used without delay.

Immunization Recommendations. Varicella vaccine is not yet recommended for general immunization in Europe or the United States. It is given to children with leukemia during remission or when chemotherapy is withheld for one week prior to and after vaccination [211]. A second dose is often given to children who remain seronegative after the first injection.

Adverse Reactions. In healthy children and adults, local swelling and pain occurred in 1–5% of healthy children and 20% of adults. Systemic reactions with fever and a rash are reported in 5–10% of both healthy children and adults. Fever and rash occurred in 40% of vaccine recipients with chemotherapy suspended for two weeks.

Vaccine Efficacy. Protective efficacy in healthy children is 95–100% with persistence of immunity over a 5–10 year period. In adults, protective efficacy is 60–80% and possibly of shorter duration. In children with malignancies, protective efficacy is 60–92%. Spread from vaccine-induced vesicular rash has been documented in household contacts.

Future Prospects. The currently investigated, attenuated Oka strain vaccine is effective but long-term immunity and zoster incidence remain to be established. The problems of latency and of vaccine-induced rash could be overcome by the development of a subunit vaccine produced either by a recombinant DNA technique or by using synthetic peptides.

3.10. Yellow Fever Vaccine

Etiological Agent and Pathogenesis. Yellow fever is a hemorrhagic fever caused by a member of the toga virus group [212], [213]. The virus was transmitted to rhesus monkeys by MATHIS and coworkers in 1927. The virus strain was then propagated by serial passages of intracerebral inoculations in white mice. Yellow fever is transmitted by mosquitoes of the genus *Haemagogus* in South America and of the genus *Aedes* in Africa. The animal reservoir is mainly monkeys. Subclinical infections are common. The incubation period in clinical cases is 3–6 days; symptoms range from transient fever and headache to high fever with meningoencephalitis, followed by jaundice and hemorrhagic manifestations. In the malignant form all these symptoms are present and death occurs within one week. Mortality rates are 40–50% in the severe forms of the disease.

History of Immunization. Two types of attenuated vaccine were developed in the early 1930s cultured in neural tissue (Dakar vaccine) and in chick embryo [214]. Initially, both types were given together with human immune serum. In the late 1930s, the vaccine cultured in neutral tissue was used for mass immunization in Senegal and the chick embryo vaccine was tested in Brazil. The vaccines were administered by scarification using normal human serum as stabilizer. The chick embryo cultured vaccine (first the 17 E vaccine and later the 17 D vaccine) has been most widely used. Numerous problems and accidents were associated with both vaccines. The human serum used as stabilizer caused hepatitis affecting many vaccinees in the armed forces during World War II. Systemic reactions were common. Severe postvaccination encephalitis with a high mortality rate occurred mainly with the neural tissue cultured vaccine. This reaction (mostly in children < one year of age) was also reported with some lots of the 17 D vaccine with increased neurotropism; furthermore, loss of protective efficacy was noted in tropical climates due to low thermostability. The 17 D vaccine in current use is the result of developments aimed at careful definition of the properties of the seed virus and at improved thermostability [215].

Production and Properties of Yellow Fever Vaccines. Only certain institutes are approved by the WHO for production of yellow fever vaccine [159, pp. 34–53]. The seed lot virus, usually a substrain of 17 D-204, has to be shown to be free from neurotropism by testing in monkeys [159, pp. 34–53], [182, pp. 113–141], [216]. Most producers use seed lots that are free of leukosis virus for production but their use is not mandatory. Virus-infected embryos are harvested, homogenized, and the supernatant is used as vaccine. Several tests for adventitious agents are performed. Virus titrations are performed in a mouse assay with a minimal potency requirement of 1000 LD_{50} per human dose. The vaccines are lyophilized in the presence of stabilizer. Most current vaccines retain the minimal requirement for two weeks at 22 °C. A vaccine stable for two weeks at 37 °C is requested by WHO for use in tropical areas.

Immunization Recommendations. Vaccination of visitors to endemic areas in equatorial Africa and northern parts of South America is recommended. Immunization is mandatory in several countries for visitors from endemic areas. One 0.5 mL injection is given to both adults and children for both primary and booster immunization. Booster injections are recommended every ten years. The vaccine is not recommended for children under one year of age or for pregnant women.

Adverse Reactions. Yellow fever vaccines are safe and induce only minor local reactions. Transient headache can occur.

Protective Efficacy. Current vaccines are estimated to be 90–95% protective. Immunization has decreased or eliminated the disease in many endemic areas.

3.11. Tick-Borne Encephalitis Vaccine

Etiological Agent and Pathogenesis. The tick-borne encephalitis virus, a member of the toga virus group, was first isolated in the Soviet Union in 1937 [212]. Two antigenically distinct forms of the virus cause the disease in Europe and in the eastern Soviet Union. The main vector for the European form is *Ixodes ricinus* and for the Eastern form *I. perulcatus*. Many wild and domestic animals can be infected; the main animal reservoirs are small mammals such as field mice. Transmission is usually by tick bite but infection can occur by drinking untreated cow milk. The incubation period is 7–14 days before onset of fever and malaise. After a 1–2 week recovery period, a second stage with fever and neurological symptoms can follow: meningitis (40% of cases), meningoencephalitis (40%), and severe meningoencephalomyelitis (20%). Mortality rates are 1–2% in the European form and 20–25% in the Eastern form; neurological sequelae are seen in 15–40% of cases. The disease is subclinical or abortive with only the first stage in 75% of infections. Tick-borne encephalitis is endemic in Central Europe, in the Balkan countries, and in Finland and Sweden.

History of Immunization. The first inactivated (killed) vaccine was developed and used in humans in the Soviet Union in 1939 followed by an attenuated, live vaccine in the 1960s [217]. An inactivated vaccine, developed in Europe with a virus strain isolated from a tick in Austria, was subjected to clinical trials in 1973 [218]. The vaccine currently used in Western Europe is a purified, concentrated version of this.

Production and Properties of Tick-Borne Encephalitis Vaccine. The inactivated vaccine is produced by propagation of the virus in hens' eggs, followed by purification by continuous flow zonal ultracentrifugation, and inactivation with formaldehyde. No WHO requirements have been formulated. The vaccine contains not less than 25 protective doses per human dose assayed in a mouse protection test. Human albumin is used as stabilizer and aluminum hydroxide as adjuvant.

Immunization Recommendations. In Austria and Bavaria, immunization of children older than one year is recommended. Other endemic countries recommend immunization of forest workers and other high-risk groups. Primary immunization consists of two 0.5 mL intramuscular injections at 1–3 month interval, followed by a third 0.5 mL dose 9–12 months after the second. Booster injections are recommended every three years.

Adverse Reactions. Local and systemic reactions are rare and mild. Low-grade fever is occasionally seen, mainly in children after the first injection.

Protective Efficacy. The vaccine is at least 95% protective against all European virus strains in children and young adults. Seroconversion rates of about 90% are reported for persons over 65 years of age.

3.12. Japanese Encephalitis Vaccine

Etiological Agent and Pathogenesis. The Japanese encephalitis virus, belonging to the toga virus group, was first isolated in 1935 in Japan [212]. The disease is transmitted by the mosquito *Culex tritaeniorhynchus*; the main animal reservoirs are pigs and birds. The incubation period is 5–15 days. The disease is subclinical in at least 95% of infections. In clinical cases symptoms vary from mild febrile disease with headache to severe encephalitis. Paralytic forms more commonly affect the upper extremities. In endemic areas the disease affects mainly younger children but also elderly people with mortality rates of 50%. Neurological sequelae have been reported in 30–40% of survivors of the severe clinical forms.

History of Immunization. A formaldehyde-killed vaccine, consisting of a 5% suspension of infected mouse brain tissue, was introduced for human use in Japan in 1954 [219]. The vaccine (Nakayama strain) has been purified by protamine sulfate precipitation since the late 1950s and by absorption with charcoal or kaolin since the early 1960s. A protective efficacy of 80–90% for the vaccine was shown. The highly purified, mouse brain cultured vaccine produced in Japan is also used for immunizing travellers to endemic areas. In China, a vaccine produced in primary hamster kidney cells has been extensively used since the late 1950s.

Production and Properties of Japanese Encephalitis Vaccines. Several Japanese manufacturers produce the purified vaccine from culture of mouse neural tissue by similar methods. No WHO requirements have been formulated. A vaccine used for immunizing travellers to endemic areas is prepared by infecting mouse brain with the Nakayama strain. The brain homogenate is purified by protamine sulfate treatment and then inactivated with formaldehyde. Further purification involves ultracentrifugation on a sucrose density gradient. The vaccine is lyophilized; the reconstituted vaccine must be used immediately.

Immunization Recommendations. Primary immunization consists of subcutaneous injection of two 1 mL doses at a 1–2 week interval. A third 1 mL dose is recommended one month later as is a regular booster injection every 1–3 years. Extensive immunization in endemic areas has been considered by the WHO. Children less than three years of age should receive 0.5 mL doses. Immunization is generally recommended for health care workers and other people with extended stay in endemic areas.

Adverse Reactions. Only a few mild, local and systemic reactions have been reported.

Protective Efficacy. Immunization has drastically reduced disease incidence in Japan. The protective efficacy of the current vaccines is estimated to be 90–95% from serologic studies.

3.13. Smallpox Vaccine

Smallpox (variola) was caused by a member of the pox virus group, which also includes the vaccinia virus used for immunization. The disease was one of the most devastating infections in human history. Eradication of smallpox is the success story of immunization. In 1967 the WHO launched a massive eradication program—smallpox was still reported from 42 countries. In May 1980, WHO officially declared the world free from smallpox. No proven cases of variola have occurred in the past decade.

Two types of vaccine were manufactured, calf lymph vaccine and egg vaccine [220]. Both liquid and lyophilized forms were used. Requirements for manufacturing, control, and potency were last formulated by the WHO in 1966. Immunization by multiple puncture with a bifurcated needle was most commonly used. Severe adverse reactions included postvaccination encephalitis and disseminated vaccinia.

General immunization against smallpox was withdrawn in most European countries in the mid 1970s. Requirement for smallpox vaccination was abandoned for international travel in 1982.

Recommendations for civilian immunization in the United States include only laboratory workers handling variola virus or other closely related orthopox viruses. Military personnel in the United States and the Soviet Union are routinely vaccinated against smallpox [221].

3.14. Rift Valley Fever Vaccine

Rift Valley fever is an arthropod-borne disease known only in Africa [212], [213]. It is caused by a member of the Bunya virus group. The major vectors are mosquitoes (*Culex theileri* and *Aedes cabbalus*). The main natural hosts are sheep, cattle, and goats. The incubation period is 2–6 days. Rift Valley fever is a febrile disease with headache and abdominal pain lasting less than one week. Hemorrhagic fever with liver necrosis and encephalitis are the severe manifestations causing mortality and sequelae. Outbreaks occurred in Africa during the 1970s, with the largest outbreak in Egypt in 1977–1978.

An inactivated (killed) vaccine (NDBR-103) was produced by the US army in 1967 [222]. The Entebbe strain of the virus was grown in primary monkey kidney cells, inactivated with formaldehyde, and lyophilized. The adverse reactions are few and mild. One case of Guillain–Barré occurred in a Swedish military vaccinee [223]. Seroconversion rates after subcutaneous injection of three 1 mL doses given at 1–2 week intervals were over 95%. A new vaccine (GSD-200) based on a cloned version of the original seed virus (Entebbe strain) and grown in diploid rhesus monkey cells is currently undergoing clinical studies. The WHO has formulated requirements for inactivated Rift Valley fever vaccines produced in primary monkey kidney cells and in human or non-human primate diploid cells [141, pp. 104–143]. No minimal potency requirements have been formulated.

4. Vaccines against Parasites

4.1. Vaccines against Helminths

Helminths represent one of the major causes of infectious diseases affecting humans and domestic animals. This results not only in a deleterious effect to the health of the hosts, but also in great economic losses. Although major advances have been made in the chemotherapy and epidemiology of diseases caused by helminths (→ Anthelmintics) immunotherapy has produced only minor breakthroughs in the field of veterinary parasitology.

As a result of a long evolutionary development and a close parasite–host relationship, helminths have evolved strategies for circumventing complete elimination by the host immune response. Nevertheless, the immune response usually exerts deleterious effect upon their growth and proliferation. Thus, the use of a vaccine resulting in partial or complete protection might be one of the most cost-effective means of controlling helminth diseases.

Only a few reliable vaccines are available on a commercial scale for immunoprophylaxis of helminthoses in livestock. A vaccine against lungworms in cattle is a commercial success. Another vaccine against hookworms in dogs, although immunologically efficient, has failed commercially.

Important in the development of vaccines is the identification, isolation, and testing of putative protective antigens. Successful vaccination against helminth infections requires the priming of those responses which may subsequently be triggered during natural infection. Since antigen presentation plays a central role in the acquired immune response, the development of accessory cells and the activation of T and B lymphocytes has to be taken into account [224]. Recent developments have focussed on vaccines produced by recombinant DNA techniques. In vitro cultivation methods have been used for producing helminth antigens for immunoprophylaxis and combined with recombinant DNA technology [225].

The helminthoses described in this chapter were chosen on the basis of the parasites' importance to human health or because of their interesting biology.

4.1.1. Vaccines against Schistosoma

Etiologic Agents, Pathogenesis, and Epidemiology. Human schistosomiasis occurs primarily in tropical countries where it is one of the most threatening diseases. The infection affects about $(200-300) \times 10^6$ people throughout the world; more than 600×10^6 people live in *Schistosoma*-endemic areas. The importance of schistosomiasis has also increased in industrialized countries with intensive tourism and influx of high numbers of refugees from endemic areas [226].

The main causative agents of schistosomiasis are helminths of the genus *Schistosoma* which use aquatic or amphibious snails as intermediate hosts:

1) *Schistosoma haematobium* uses aquatic snails of the genus *Bulinus* as intermediate hosts and occurs mainly in Africa and some middle-eastern countries.
2) *S. mansoni* uses aquatic snails of the genus *Biomphalaria* as intermediate hosts; it occurs in Africa, parts of Arabia, northern and eastern parts of South America, and some Carribean islands.
3) *S. japonicum* uses amphibious snails of the genus *Oncomelania* as intermediate hosts; it occurs in Japan, the Philippines, and parts of China, Thailand, and Indonesia.

The life cycle of all three *Schistosoma* species is similar. The fully developed *miracidium* hatches from the egg in water and infects the intermediate host (i.e., the snail) where it multiplies asexually to produce numerous *cercariae*. The cercariae are released into the water and infect humans by penetration through the skin. They develop into immature worms (schistosomula) which migrate to the lungs and liver. The adult, sexually mature worms mate and migrate to their final destination which varies according to the species: *S. haematobium* migrates to the veins of the vesical plexus, *S. mansoni* and *S. japonicum* to the mesenteric veins. The adult paired worms produce eggs (300–3000 per pair per day) which pass through the vessels and tissues into the lumen of the gut and bladder. The eggs escape from the host in the feces and urine and the cycle is repeated.

Eggs from *S. haematobium* are mainly found in the bladder and urogenital tract causing hematuria and fibrosis of the bladder. In severe cases malignancies may develop. Eggs from *S. mansoni* and *S. japonicum* are trapped in the liver and bowels causing hepatomegaly. Subacute disease is probably due to the passage of worms through lungs leading to cough, pulmonary infiltrates, and fever [227]. Mainly in *S. japonicum*, acute schistosomiasis occurs 5–7 weeks after heavy primary infection. The illness if often associated with diarrhea, fever, hepatosplenomegaly, resulting in liver fibrosis and ascites.

Immunity and Vaccine Design. An age-dependent resistance to reinfection after chemotherapy was demonstrated with *S. haematobium* and *S. mansoni* [228], [229]. This is associated with an increased lymphocyte proliferation in response to egg, cercarial, and adult worm antigen [230], [231]. Lack of reinfection is also related to the eosinophil count [228], [232]; human eosinophils mediate antibody-dependent damage to the schistosomula of *S. mansoni* [233]. The main immune mechanisms involved in schistosomiasis are summarized in Table 3 and reviewed in [234].

The immune response may result in resistance to reinfection but not to simultaneous expulsion of an established parasite population from primary exposure [235]. In this way, the parasites evade the immune response. An alternative escape mechanism is that the worms may be covered by bound host IgG [236]. The immunoglobulin is partially cleaved by a parasite protease to produce peptides which may inhibit macrophage activation and thus depress macrophage-mediated, IgE-dependent destruction (cytotoxicity) of the schistosomula.

In the development of schistosomiasis vaccines, the findings regarding age-dependent host resistance will require application and efficacy at a very early age, before the child is exposed to natural infections [237]. The search for antigens that mediate protective immunity has concentrated on the exposed outer surface of the young schistosomula [238]. About 90% of exposed epitopes consist of carbohydrates that crossreact with *Schistosoma* egg antigen. Antibodies to surface polypeptide antigen do not generally crossreact with egg antigen but are present on the surface membrane of adult worms. Monoclonal antibodies to some of these molecules seem to confer partial resistance when passively administered to animals and could thus be potentially protective antigens [239]–[241]. A range of candidate vaccine antigens of *S. mansoni* have been identified, several have been cloned and expressed in *Escherichia coli*, yeast, or vaccinia virus [238]. This important achievement will facilitate the production of large amounts of polypeptides for vaccination trials.

One of the most promising candidates is a schistosomula surface polypeptide with a molecular mass of 28 000. The gene coding for this polypeptide has been cloned [242] and expressed in *E. coli*. Immunization with this recombinant antigen induced a high level of serum cytotoxicity towards schistosomulas in the rat, hamster, and monkey. Significant protection against a natural challenge infection with live cercariae was obtained in rats and hamsters [242]. These recent results can be viewed with optimism for the development of a schistosomiasis vaccine.

Table 3. Main cells and antibodies active against *Schistosoma* in experimental animal models or in vitro

Developmental stage of parasite	Cells and antibodies involved	Mechanisms
Egg	effector T lymphocytes	initiation of granuloma formation (lymphokines)
Schistosomula	neutrophils, IgG	tegumental damage and killing of young larvae under certain conditions
Schistosomula	eosinophils, IgG	killing of young larvae in the presence of complement
Schistosomula	eosinophils, IgE	killing of young larvae
Schistosomula	macrophages, IgE	IgE-dependent cytotoxicity

4.1.2. Vaccines against Nematodes

4.1.2.1. Gastrointestinal Nematodes

The most prevalent and pathogenic gastrointestinal nematode parasites of humans belong to the genera *Ascaris, Strongyloides, Trichinella, Trichuris, Ancylostoma,* and *Necator*. Although many species parasitize deeper tissues of the body, the majority are intestinal. The intestine has been maintained as a site for adult stage development, whereas larval stages often invade other host tissues. The intestine consists of a series of distinct parasite habitats (gut sections, lumen, mucosa, etc.) each having its own characteristics. Large worms such as *Ascaris* must live within the lumen, smaller species such as hookworms are associated with the mucosa. For a long time research on immunity against intestinal worms was given low priority due to the lack of knowledge about local immune responses in the gut. Recent studies have shown that intestinal worms are indeed subject to protective immune responses, although these responses differ somewhat from classical immunity in the body because the worms live in the gut lumen. Special features are (1) macromolecular antigen uptake across the intact mucosal epithelium or by specialized epithelial cells overlying Peyer's patches and (2) complexation of the antigen by dimeric IgA secreted from the mucosa or intestinal IgG Fab fragments. Cells from the underlying lamina propria participate in cytotoxicity and hypersensitivity reactions and thus affect mucosal structure and function. Furthermore, a large variety of nonlymphoid effector cells occur within the intestine, including natural killer cells, macrophages, neutrophils, eosinophils and basophils; their numbers increase during parasite infections.

No vaccine against gastrointestinal nematodes is presently available for human application. However, promising, successful trials in veterinary parasitology have initiated interesting work in human gastrointestinal nematode infections and will probably result in vaccine supply in the near future.

Hookworm Disease. Various species of the family Ancylostomatidae are responsible for hookworm disease in humans, the most important being *Ancylostoma duodenale* and *Necator americanus*. Over a fifth of the world's population is afflicted by this disease, mainly in tropical and subtropical regions. Adult hookworms have a length of about 0.7 – 1.8 cm and a hooklike anterior end with a distinct mouth area. Females release eggs in the small intestine which are excreted in the feces of the host. In humid surroundings first-stage larvae hatch from the eggs and develop into infective larvae, which penetrate through the skin into new hosts. After migration through lymph or blood vessels, the larvae finally develop into mature adult worms in the small intestine, completing the parasite's life cycle. Hookworms suck blood from microlesions in the host's small intestine, thereby causing chronic gastrointestinal blood loss, anemia, and hypoalbuminemia. With a large worm burden death may occur. Patients usually suffer from extreme weakness, pallor, secondary respiratory tract infections, skin irritations, heart palpitations, and gastrointestinal distress.

Immunity and Vaccine Design. Nematode parasites present special problems for the host's protective immune response, because they possess a tough, protective, external cuticle. The cuticle is both antigenic and immunogenic, but it is doubtful whether responses directed against its surface play a major role in immunity against intestinal species. Protective responses are more likely initiated by antigens released through the orifices of the parasite [243]. One such antigen with a potential vaccine function is a secreted proteolytic enzyme [244]. Hookworms attached to the mucosa secrete the enzyme from glands in their mouths. The enzyme degrades the host proteins and inhibits blood coagulation which permits the hookworms to feed for an indefinite period of time. Immunizing the host against this enzyme would result in inhibition of any enzyme secreted by worms after a subsequent challenge infection. The worms would then be unable to feed. The gene coding for the enzyme in question has been cloned. Future investigations will have to demonstrate the applicability of this kind of vaccine.

A vaccine has been developed for the control of hookworm infections in dogs [245]. It was based upon irradiation-attenuated infective larvae and was administered parenterally. Although the vaccine was highly effective in preventing hookworm disease, it was withdrawn because it did not completely prevent infection, and effective anthelminthic chemotherapy was available. This example suggests that protective immunity may also occur in human hookworm infections.

4.1.2.2. Tissue-Invading Nematodes (Filariidae)

Many nematode species which live as adults in the intestine, (e.g., *Ascaris*, hookworms, and *Trichinella*) undergo development in parenteral tissues. Other species are wholly confined to these tissues and have no contact with the intestine (tissue-invading nematodes). This closed habitation site requires special conditions in order to obtain biological contact with the outside world, especially for reproduction. In the major group of tissue-invading nematodes, the Filariidae, the worms overcome this problem by using bloodfeeding arthropods as intermediate hosts. The female worms release microfilariae larvae which circulate in the blood or accumulate in the skin of the host. The arthropod feeds on the blood and takes up the microfilariae which develop into infective larvae. At a following blood meal, infective larvae are reinoculated into new human hosts.

Filariasis. The human disease filariasis comprises an extremely heterogeneous group of diseases. The main filarial parasite species in humans are *Wuchereria bancrofti* and *Brugia malayi*, which are both transmitted by mosquitoes. The disease is widespread in tropical and subtropical regions affecting ca. 90×10^6 persons. Its early symptoms are fever, lymphangitis, and lymphadenitis. A following chronic stage is frequently characterized by more serious clinical manifestation including elephantiasis, hydrocele, and pulmonary eosinophilia. Adult worms are found in the lymphatic system, microfilariae may be found in blood.

Filarial infections caused by *Onchocerca volvulus* occur in Africa and South and Central America, currently affecting some 20×10^6 patients, 250 000 of them being already blind. Adult worms are generally located under the skin, forming typical nodules; less often they penetrate deeply into the tissues. Pathology in onchocerciasis is due entirely to the microfilarial stage. Microfilariae are present in the skin and may penetrate into the eye, thus leading to severe eye lesions and blindness.

Loa loa is prevalent in the forest areas of West Africa. Adult worms penetrate into the tissues, provoking transient edema. Severe clinical symptoms are rare, loiasis being generally regarded as a benign infection [246]. *Dipetalonema perstans*, *D. streptocerca*, and *Mansonella ozzardi* infect humans (Africa, South America) but usually asymptomatically and rarely cause significant diseases [247].

Many of the changes associated with filarial infection are immunopathological in origin and hypersensitivity reactions are important in their development.

Immunity and Vaccine Design. Although filarial parasites provoke a strong immune response in the human host, the chronicity of these infections implies the absence of a protective response or the evasion of such responses by the worms. There is also no direct evidence that filarial infections in nature confer resistance to reinfection with the same parasite species.

No vaccines against filarial parasites are available. The development of new strategies for immunological control depends on a thorough understanding of immunological host–parasite relationship. Many studies on protective immunity in animals have concentrated upon responses directed against larval stages. A vaccine against microfilariae would inhibit transmission of the disease, a vaccine against infective larvae would provide protective immunity against primary infection of hosts. Inoculation with infective larvae attenuated by irradiation [248], [249], only confers partial protection to a subsequent challenge infection. Filarial vaccines based on irradiated larvae cannot be used in humans without first determining whether these attenuated larvae induce pathological changes [250].

The target antigens of antimicrofilarial immune responses are probably located on the surface of the microfilariae; they are currently being characterized [251]. Antibodies to surface antigens mediate adherence of host cells to the microfilariae; this can result in worm killing [243]. Using *Dipetalonema vitae* as a model, IgE was found to be the primary immunoglobulin involved in cell adherence to the worm cuticle. The first cell type to adhere is the eosinophil. Subsequent, adherence of macrophages is followed by release of lysosomal enzymes which degrade the cuticle. In humans, a comparatively long time is needed to establish such immunity. This may be due to pronounced immunosuppression or because the microfilariae cover their surface with host components and thus render their antigenic surface epitopes less accessible to the host's immune system. The search for filarial antigens that can safely and successfully be used in humans is still continuing; it is still not known how restricted series of antigens can be used to protect natural hosts against first or persistent infections with a complex, adaptable, genetically diverse parasite. Recombinant DNA technology and hybridoma technology may provide potential candidates for successful filarial vaccines.

4.1.3. Vaccines against Cestodes

Etiologic Agents, Pathogenesis, and Epidemiology. This section deals with the hydatidosis and cysticercosis disease complexes, which are caused by the larval stages of tapeworms belonging to the family Taeniidae. The most important causative agents of *hydatidosis* (echinococcosis) are *Echinococcus granulosus* and *E. multilocularis,* whose life cycles involve a definitive and an intermediate mammalian host. The definitive hosts are carnivores (mainly dogs for *E. granulosus* and foxes for *E. multilocularis*) in whose intestines the adult stage worms occur. Intermediate hosts are herbivorous and omnivorous species in which the larvae (metacestodes) develop. Humans and other intermediate hosts become infected by ingesting eggs passed in the feces of definitive hosts. The diseases caused by the metacestodes are referred to as cystic echinococcosis for *E. granulosus* and alveolar echinococcosis for *E. multilocularis. Echinococcus granulosus* is prevalent throughout the world and is a public health and economic problem in many areas. *Echinococcus multilocularis* only exists in the northern hemisphere and is relatively frequently seen in the Soviet Union (Siberia), central Europe, northern China, Japan, and Alaska. The fully developed metacestode of *E. granulosus* is a typically unilocular, fluid-filled cyst, which is located in the liver, lungs, and other organs. *Echinococcus multilocularis* metacestode conforms a vesiculated parasitic mass in the liver of the host, it proliferates by continuous exogenous budding with possible metastasis formation in other organs.

The causative agent of *cysticercosis* is the tapeworm *Taenia solium*. Humans are the obligatory definitive hosts, pigs act as intermediate hosts. Metacestode stage infection can also occur in humans and may result in infection of the central nervous system by parasite larvae (neurocysticercosis). Morbidity includes intracranial hypertension, basal arachnoiditis, focal neurological deficits, and dementia. Hyperendemic areas are found in Latin America, Africa, and Asia; areas of lower endemicity occur in southern and eastern Europe.

Immunity and Vaccine Design. Host protective immunity is a striking feature of repeated infection with cestodes in mammalian intermediate hosts [225]. It plays a major role in regulating natural transmission of these parasites, and substantial research efforts have been undertaken towards development of vaccines against cestodes of veterinary importance [252]. For human cestode infections, it is debatable whether there is sufficient importance to warrant the research required to develop a vaccine. In areas of high egg contamination (e.g., the Turkana district (Kenya) for *E. granulosus*, St Lawrence Island for *E. multilocularis*, or Mexico for *T. solium*), vaccination should be given to very young persons, as patients usually become infected at very young age. An ideal human vaccine requires complete, long-lasting protection. Experiments in veterinary parasitology, however, demonstrated exactly opposite results. Vaccination of animals against *Taenia* species resulted only in marked reduction of cyst numbers which persisted only for a maximum of one year. In addition, strong adjuvants had to be employed which are not tolerated by humans. Other criteria may be important; a review of protective immune mechanisms is given in [253].

Little attention has been paid to the vaccination of definitive hosts (dogs and foxes for *Echinococcus*, and humans for *T. solium*). Experiments [254], [255] demonstrated an immune response after infection as well as highly significant suppression of egg production by *E. granulosus* after immunization of dogs with secretory antigens derived from adult tapeworms. This approach seems to be the most likely control measure for reducing infection risk in humans.

4.2. Malaria Vaccine

Malaria remains a major health problem in many tropical and subtropical countries and affects hundreds of million of people each year. A major effort was made to control malaria from 1950–1970 with insecticides and antimalarial drugs (→ Synthetic Chemotherapeutic Agents). The initial remarkable results have been difficult to maintain. The advent of drug-resistant parasite strains and of insecticide-resistant mosquito vectors are major obstacles in the effort to control malaria. Since the early 1970s new approaches have been explored such as vector control through biological agents and control of malaria infection through vaccines.

The development of malaria vaccine has received considerable impetus: first, because immunization of animals with whole parasites can induce a degree of protection equal or superior to that induced following natural infection [256]–[258]; second, because in vitro culture systems have been developed for the blood and hepatic stages of the malarial parasite *Plasmodium falciparum* [259], [260]; and third, because monoclonal antibody and recombinant DNA techniques have been used to identify and produce the parasite polypeptides possibly involved in the development of protective immunity.

4.2.1. Strategy for Malaria Vaccine Development

More than 100 species of malarial plasmodia are known, but only four infect humans: *Plasmodium falciparum*, which is responsible for the majority of human deaths, *P. vivax*, *P. malariae*, and *P. ovale*.

The life cycle of the plasmodia is complex (Fig. 4). The female anopheline mosquitoes inoculate *sporozoites* into the blood of the vertebrate host. Within minutes the sporozoites invade the liver parenchymal cells (hepatocytes) where they divide asexually and develop into *merozoites* which rupture the hepatocytes and reenter the blood. In the subsequent *erythrocytic cycle*, the merozoites invade the red blood cells and mature into *schizonts* within 48–72 h depending on the species. The mature schizonts release merozoites which invade new erythrocytes. The erythrocytic cycle is responsible for the clinical manifestations of malaria. Some merozoites differentiate into sexual stages called *gametocytes* which are ingested by the mosquito. Fertilization of the gametes occurs solely in the midgut of mosquito. The resulting zygotes develop into *ookinetes* and then into *oocysts*. Sporozoites are released from mature oocysts and migrate to the mosquito salivary glands. The cycle is then repeated.

This complex life cycle involves continuous morphologic, enzymatic, and antigenic changes which are linked to the parasite's environmental adaptation to the host. The invasive stages of the parasite (sporozoites, merozoites, gametes) have unique, stage-specific surface determinants. Furthermore, immunologic crossreactivity exists between plasmodia species and between the developmental stages of a given species. However, immunization experiments in animal models have demonstrated that the antigenic determinants involved in protective responses are species- and stage-specific. Three types of malaria vaccine can therefore be devised, based on:

1) sporozoites,
2) asexual stages (merozoites, schizonts), and
3) sexual stages (gametes).

In addition, recent data indicate that stagespecific parasitic antigens are expressed on liver cells containing merozoites [261] and may also be candidates for vaccine development. A favored approach is the development of a multivalent vaccine containing components of several malaria stages. The selection of defined parasite components

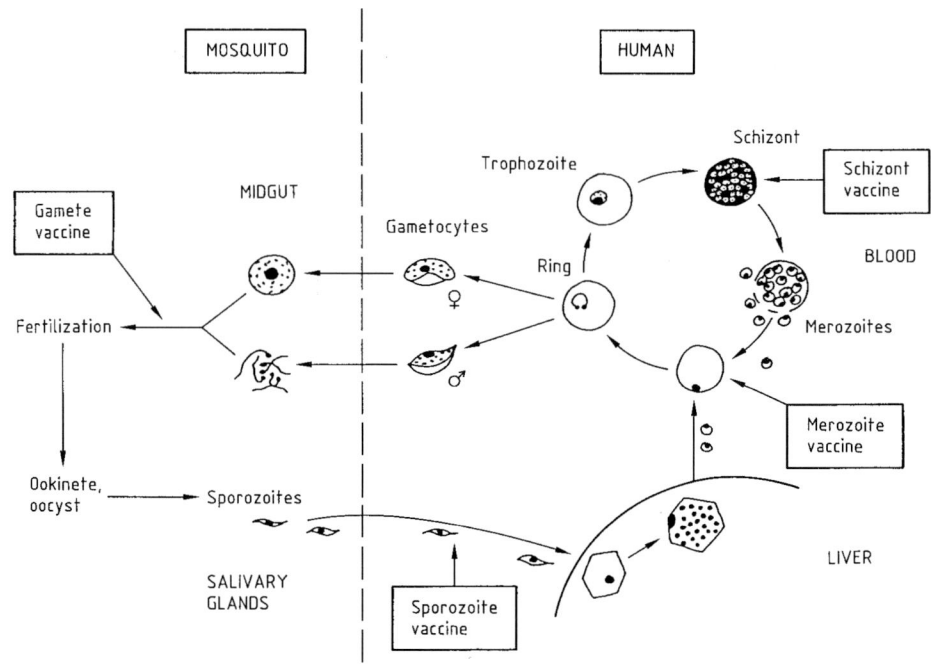

Figure 4. *Plasmodium falciparum* life cycle showing targets for vaccine development

versus whole parasites (sporozoites, merozoites, gametes) for vaccine development is indicated by the following reasons:

1) Large-scale production and purification of whole parasites is not feasible.
2) Parasites cannot be obtained free of host components (e.g., mosquito salivary glands, erythrocyte membranes) which may induce adverse autoimmune reactions.
3) Most of the parasite components are irrelevant to the induction of protective responses and their inclusion may impair truly protective responses or induce immunopathologic lesions in the host.

The current strategy for the development of malaria vaccine is as follows:

1) Identification and selection of malaria antigens that are the target of protective immune responses.
2) Functional, biochemical, and immunological characterization of these antigens, including identification of B and T cell epitopes.
3) Cloning of the genes coding for protective antigens, determination of DNA and amino acid sequences—what is the level of antigenic diversity?
4) Production of candidate protective antigens or epitopes by genetic engineering or chemical synthesis.
5) Evaluation of candidate protective antigens in terms of production of antigens, safety tests, adjuvants, carriers, etc.
6) Immunization trials in monkeys and human volunteers.

4.2.2. Sporozoite Vaccines

Following invasion of the hepatocytes, the sporozoites develop into thousands of merozoites, each of which may invade an erythrocyte. Obviously a vaccine which neutralizes sporozoites before their entry into liver cells or within liver cells would optimally prevent malaria infection. Vaccination with attenuated, irradiated sporozoites in rodents, monkeys, and humans induced complete protection against malaria [262], [263]. The induced immunity is species- and stage-specific but not strain-specific. It is at least partially mediated by antibodies as is shown by the protection against sporozoite challenge afforded by the passive transfer of antisporozoite monoclonal antibody [268] and by neutralization of sporozoite infectivity following incubation with the serum of protected animals. Deposition of specific antibodies on the sporozoite surface results in the formation of a tail-like precipitate; this is called the circumsporozoite (CS) reaction. Indirect evidence suggests that antisporozoite antibodies may play a role in human malaria [264], [265].

The control of sporozoite-induced infection also involves T cell-dependent mechanisms. In animal models, effector T cells can mediate antisporozoite immunity via antibody-independent mechanisms; for example, immunization with irradiated sporozoites can protect B cell-deficient mice against subsequent challenge infection with viable sporozoites [266]. Activation of malaria-specific T cells by malaria antigens induces the secretion of lymphokines which may act directly on the malaria parasite or indirectly by activating host effector systems [267].

The CS protein has been identified in several plasmodia species. Passive transfer of monoclonal antibodies directed against the CS protein protects mice from sporozoite challenge infection [268]. Indirect evidence suggests that CS protein is involved in the binding and penetration of sporozoites into liver cells. Similarly, fragments of monoclonal antibodies against CS prevent the attachment of sporozoites to hepatocytes in vitro [269].

The gene coding for the CS protein has been cloned in several plasmodial species [270], [271]. The protein contains a central block of tandemly repeated amino acids which vary in number and sequence among the different malaria species [272], [273]. For example, in *P. falciparum* the central area consists of 37 copies of the tetrapeptide asparagine – alanine – asparagine – proline (NANP) and four copies of the tetrapeptide asparagine – valine – aspartic acid – proline (NVDP). The regions flanking the repeats are more highly conserved between species than the repeats. Within a species, limited variations also occur outside the repeats [274].

The repeats cover the surface of mature sporozoites; for *P. falciparum* ca. 10^8 molecules of NANP are expressed on the membrane of mature sporozoites. The NANP repeats are the target of protective monoclonal antibodies passively transferred in vivo. The antibody response against sporozoites in humans is also mainly directed against the NANP repeats [275], [276].

In view of these findings, and because the NANP repeats are present on all the isolates of *P. falciparum* tested, two malaria vaccines based on NANP repeats have been prepared and tested on human volunteers. The first, produced by DNA recombinant technology, consisted of 32 repeats of NANP and NVDP fused to a 32 amino acid tail. The second was composed of three NANP repeats (NANP 3) conjugated to tetanus toxoid. Both formulations use aluminum hydroxide as adjuvant [277], [278]. The vaccines were safe and did not induce adverse reactions. Volunteers with high antisporozoite antibody titers were challenged with sporozoites and some were protected or presented a delay in appearance of parasitemia. Protection was shown in individuals with the highest antibody titers.

An antisporozoite vaccine must induce high antibody titers for a prolonged period and ideally a boosting effect should occur following exposure to sporozoites. Higher antibody responses can be obtained by changing the formulation and concentration of the immunogens and by using other adjuvants or live-attenuated vectors (vaccinia virus, salmonella) that carry and express the gene coding for the CS protein. Optimal antibody formation is dependent on the collaboration of T helper cells and B cells. In the two sporozoite vaccines tested, T cell help was provided by foreign protein (tetanus toxoid) but was unable to boost the antiNANP antibody response following exposure to sporozoites. More efficient sporozoite vaccines should contain T cell epitopes derived from sporozoites and ideally from the CS protein; in this context, proper help is provided by sporozoite-specific, primed T cells and can lead to optimal antiNANP antibody production by B cells. In a mouse model the response to some of the T cell epitopes of the CS protein is restricted by antigens of the major histocompatibility complex (MHC class II) [279]–[282]. In human populations, only three immunodominant epitopes are located in polymorphic regions of the CS protein outside the repetitive area. Since T cells have exquisitely specific reactivity, it follows that the polymorphism of T cell determinants may be responsible for a lack of proper help following exposure to sporozoites with T cell areas on CS that are different from those present on sporozoites responsible for previous infections in the same individual.

Therefore, it seems that more efficient vaccines should be based on either native malaria polypeptide(s) or cocktails of synthetic polypeptides containing multiple B and T cell epitopes. These epitopes should be selected in relation to constant and variant parasite components and in relation to epitope binding and recognition by components of the major histocompatibility complex of the human host.

4.2.3. Asexual Blood Stage Vaccine

The multiplication of asexual blood stages (merozoites and schizonts) is responsible for the morbidity and mortality associated with malaria. The level of parasitemia usually correlates with the severity of malaria infection. Immunity to malaria is mostly acquired but natural immunity also plays a role. Several single-gene disorders affecting erythrocytes, (e.g., sickle cell anemia, the thalassemias, and glucose phosphate deficiency) reduce the severity of malaria infection. Another genetic characteristic, the lack of the Duffy blood group antigens, is associated with complete resistance to *P. vivax* infection [283]; the Duffy blood group antigen or a closely associated antigen may be the receptor for *P. vivax* merozoites at the surface of erythrocytes.

The development of acquired resistance to malaria depends on the frequency and duration of the exposure to the parasite [284]. In endemic areas, babies born to immune mothers are resistant to malaria during the first three months of life due to the presence of maternal antibodies transferred during gestation. Later they suffer from severe, recurrent attacks; most deaths due to malaria occur in young children. From adolescence to adulthood there is a decrease in the severity and frequency of malaria attacks but sterile immunity is probably never achieved. In this context two types of vaccine based on asexual blood stages can be envisaged; (1) a vaccine which is more efficient than nature and leads to sterile immunity (i.e., infection no longer detectable) or (2) a vaccine capable of attenuating the parasite load by transforming the immune system of a non immune individual into that of an adult living in an endemic area.

The immune response to blood stages is complex and is directed against several antigens. Both antibody-mediated responses and cell-mediated, antibody-independent responses control asexual blood-stage infection. In humans, passive transfer of immunoglobulins purified from the sera of immune adults abort malaria infection in nonimmune infected children [285]. The antibodies may react with the surface of the merozoites and provoke their lysis upon addition of complement, or enhance their phagocytosis by mononuclear cells, or simply inhibit the binding of merozoites to erythrocytes. Other targets for antibodies are antigens on the surface of erythrocytes containing schizonts; binding of antibodies to schizonts may also lead to their destruction by phagocytosis [286] or induce the endothelial release of schizonts which may be later destroyed in the spleen [287]. Immunity to asexual blood stages also operates through a variety of antibody-independent mechanisms: T cell-dependent release of lymphokines, induction of oxidizing radicals leading to intracellular death of the malaria parasites, and activation of mononuclear cells in the spleen.

Immunization with merozoites and/or schizonts in a variety of plasmodia–host systems resulted in partial to almost complete protection [288]. Subsequent investigations were aimed at the characterization of components capable of inducing immunity. Several hundreds of asexual blood stage components can raise an immune response but only very few of the evoked responses are helpful to the host. Characteristics of some of the candidate antigens for asexual bloodstage vaccines are discussed in Sections 4.2.3.1–4.2.3.4.

4.2.3.1. Merozoite Surface Antigens

A protein with a molecular mass of 190–200 kDa has been identified at the surface of *P. falciparum* schizonts and merozoites [289], [290]. During maturation of schizonts this polypeptide is processed into several components, one of them (M_r 83 000) being the main surface component of the merozoites [290]. An important feature of the 190–200 kDa protein is its genetic polymorphism [291]–[294]. The gene coding for the protein can be divided into blocks ranging in homology among different *P. falci-*

parum isolates from 10–87% at the amino acid level. A relatively short region of variable tripeptide repeats is found close to the N terminus. The blocks encoding for the N and C terminal sequences are highly conserved.

Immunization with the 190–200 kDa protein derived from *P. falciparum* in monkeys [295]–[297] and with an analogous protein from *P. yoelii* [298] can induce at least partial protection. Immunization with synthetic polypeptides corresponding to defined parts of the molecule (for example, the N terminus and amino acids 277–287) also confer partial protection in monkeys [299], [300].

An antigen with a molecular mass of 51 kDa is also expressed at the surface of *P. falciparum* merozoites and is the target of inhibitory monoclonal antibodies. It contains variant and constant epitopes for various *P. falciparum* isolates.

There is considerable antigenic diversity among *P. vivax* isolates as regards the components exposed at the surface of merozoites.

4.2.3.2. Rhoptry Antigens

Rhoptries are apical organelles of the merozoites which release their contents onto the erythrocyte membrane during invasion. A monoclonal antibody directed against a rhoptry protein of a rodent malaria, *P. yoelii*, reduced the virulence of the infection and a monoclonal antibody directed against 82 and 41 kDa components of *P. falciparum* inhibited the growth of *P. falciparum* in vitro [301]–[303]. The 82 kDa component is processed into 82 and 65 kDa components [304].

The 82 kDa polypeptide is membrane-bound through a glycosyl–phosphatide–inositol anchor; hydrolysis of its anchor activates the proteolytic activity of the 76 kDa polypeptide and may play a role in the invasion of erythrocytes by merozoites [305]. The 41 kDa polypeptide displays aldolase activity [306]. Interestingly, both the 76 kDa and the 41 kDa components can induce at least partial protection against *P. falciparum* infection in monkeys [307], [308]. The gene coding for the 41 kDa protein has been cloned and presents two interesting characteristics in terms of vaccine development; absence of variable amino acid repeats and almost complete conservation of the amino acid sequences among isolates from *P. falciparum* [306].

Another rhoptry antigen of *P. falciparum* with a molecular mass of 225 kDa has been identified in the peduncle of the rhoptries. It is synthesized as a 240 kDa polypeptide which is processed into a 225 kDa protein during schizogony and is quantitatively recovered in the culture supernatant following merozoite invasion [309].

A third set of rhoptry-associated proteins is the 105–130–140 kDa complex composed of three coprecipitating but unrelated proteins [310]. The 225 kDa proteins and the 105–130–140 kDa complex have not been evaluated in immunization trials.

4.2.3.3. Antigens Associated with the Membrane of Infected Erythrocytes

The ring-infected erythrocyte surface antigen (RESA) is a *P. falciparum* antigen with a molecular mass of 155 kDa. It is synthesized in trophozoites, accumulates in the merozoite, and following invasion becomes associated with the membrane of erythrocytes containing ring forms of the parasite [311], [312] but is not directly accessible on the external erythrocyte surface. AntiRESA antibodies inhibit the multiplication of asexual blood stages in vitro and may interfere with the invasion process [313]. The gene coding for RESA has been cloned and sequenced [314]. It contains two blocks of repetitive amino acid sequences which are the immunodominant regions of the molecules in terms of antibody response. Antibodies directed against the RESA repeats crossreact with at least six other asexual blood stage components. Aotus monkeys have been immunized with fusion proteins corresponding to various areas of the RESA and with synthetic polypeptides corresponding to the repetitive sequences [315]. Partial protection was observed in some groups of animals; work is in progress to optimize the efficacy of immunization based on RESA-derived molecules.

Erythrocytes containing mature asexual blood stages of *P. falciparum* attach to endothelial cells lining the venules of deep tissues. This cytoadherence of mature parasites prevents their passage through the spleen and thus their exposure to localized destructive mechanisms. Electron-dense protuberances (knobs) on the plasma membranes of infected erythrocytes are implicated in cytoadherence [316]. The genes coding for two knob components (knob-associated histidine-rich protein M_r 85 – 105 kDa) and mature parasite-infected erythrocyte surface antigen (M_r 240 – 300 kDa) have been cloned [317], [318]. The two proteins differ antigenically among isolates and contain repeated amino acid sequences. Cytoadherence can be inhibited by antisera in a strain-specific manner [319]. However, an antigenically invariant epitope has also been identified on the surface of infected erythrocyte isolates and may be an important antigen for vaccine development [320].

4.2.3.4. Other Proteins and Synthetic Peptides

A number of other antigens are also candidates for vaccine development. *P. falciparum* requires exogenous iron in the form of ferrotransferrin. A malaria transferrin receptor at the surface of infected erythrocytes transports bound ferrotransferrin to the parasite and may be used as a target for the vaccine [321]. Glycophorins exposed at the erythrocyte surface may act as ligands for *P. falciparum* merozoites and *P. falciparum* proteins have been identified which either bind to glycophorins or to human erythrocytes [322], [323]. A prominent antigen of *P. falciparum* with an apparent molecular mass of 126 – 140 kDa is associated with merozoite release. The gene coding for this protein has been cloned and contains at least two stretches of amino acid repeats, one being composed of polyserine repeats [324]. Monkeys immunized with this protein are protected from a lethal challenge infection [295].

Several immunization trials have been conducted in monkeys using synthetic peptides derived from asexual blood stages of *P. falciparum* coupled to carrier proteins [298], [299], [315]. A partial protective response was observed with peptides corresponding to various areas of the 190–200 kDa protein, to RESA, and to fragments of parasite components identified by their molecular mass of 55 and 35 kDa [299]. Recently synthetic hybrid polymer–proteins containing several peptides corresponding to epitopes of 195–200 kDa, RESA, 55 kDa, 35 kDa, and CS protein have been used for immunization of human volunteers [324]. The vaccine was well tolerated and no adverse effects were observed. All the immunized and control volunteers had patent parasitemia but the majority of the immunized volunteers were able to control their parasitemia in the absence of drug therapy. The immune response was low in terms of specific antibody production, and cell mediated responses as measured by proliferation assays was undetectable.

4.2.4. Sexual Stages—Transmission Blocking Immunity

The transmission of malaria from the vertebrate host to the mosquito vector is effected by sexual parasite stages—the gametocytes—which develop from merozoites. Within the vertebrate host the gametocytes are surrounded by the erythrocyte membrane; following ingestion by the mosquito vector, the gametes become extracellular. The female gametes are fertilized by the male gametes in the midgut of the mosquito to produce zygotes which develop into ookinetes. The ookinetes penetrate the midgut wall where they remain to form oocysts in which the sporozoites develop.

In the vertebrate host, the sexual stages do not produce illness, and their intracellular localization prevents direct attack by host effector mechanisms. Within the midgut of the mosquito the extracellular gametes are exposed to antigamete and/or antizygote antibodies from the vertebrate host that are ingested with the mosquito's blood meal. The antibodies partially or completely prevent the development of sexual stages and subsequent production of sporozoites. Transmission of the parasite is therefore blocked. This phenomenon is termed *transmission blocking immunity*.

The development of vaccines based on sexual blood stages could have an important impact on the epidemiology of malaria in endemic areas by reducing the level of malaria transmission. Ideally, transmission blocking vaccines have to be used in combination with vaccines based on sporozoite and/or asexual blood stages (see Sections 4.2.2 and 4.2.3).

Transmission blocking immunity has been induced by immunization with extracellular gametes in several animals [325]–[328]. The induced antigamete response is long lasting and in some cases is boosted by malaria infection, probably due to the gametocyte antigens in the circulation of the vertebrate host [328], [329]. There is also evidence that in *P. vivax* malaria in humans the antigamete response is boosted during natural infection [330]. Addition of sera of previously infected individuals to gametes can prevent fertilization and development of oocysts in mosquitos.

Specific targets for antigamete immunity have been identified using monoclonal antibodies in species including the human parasites *P. falciparum* and *P. vivax* [331], [332]. The antibodies act against the gametes by preventing fertilization and against the zygotes and ookinetes by preventing further development. In *P. falciparum* the target antigens for inhibition of fertilization are polypeptides with a molecular mass of 230 kDa and 45–48 kDa [331], [332]. Some of the epitopes on the 45–48 kDa antigen have been defined and are the targets of inhibitory monoclonal antibodies [333]. New antigens are expressed at the surface of the zygote and ookinete and one of them (M_r = 25 kDa) is a probable target of inhibitory monoclonal antibodies [331].

5. Immunotherapy

Since the late nineteenth century considerable progress has been made in our concepts of passive immunotherapy and in the development of preparations safe for human use; however, the proper role for such therapy in clinical medicine still needs to be defined.

By 1900 immune serum from various animal species had been used to treat pneumonia, tetanus, diphtheria, and rabies. Human serum was first used in 1907 by CENCI for the modification of measles and later for mumps and pertussis [334]. Placental extracts prepared by ammonium sulfate precipitation [335], [336] were also employed and may be considered as the first immunoglobulins prepared for human therapy [336]. Placental material is still used as a source of immunoglobulin.

The serious hypersensitivity reactions associated with animal serum proteins and the risk of viral hepatitis with convalescent human serum limited the use of serum therapy to life-threatening infections. In the 1920s attempts were made to separate the immune substances from animal serum by alcohol or acetone treatment. One such preparation, Huntoon's antibody solution, was administered intravenously to over 400 patients without the occurrence of anaphylaxis or serum sickness; however, pyrexia, cyanosis, and dyspnea did occur and were implicated in the deaths of three patients [337]. The introduction of antibiotics in the 1930s decreased the demand for serum therapy [338]. During this period, however, the experimental basis for combination therapy with antimicrobials and hyperimmune animal serum was established [339], [340].

A wide variety of biological products is now available for immunotherapy, the most important being purified gamma globulin for intramuscular injection (standard immune serum globulin, ISG) and globulin for intravenous injection (standard intravenous immunoglobulin, IVIG). Hyperimmune globulins with a high antibody titer against specific pathogens are also used. Additional preparations include antitoxins, plasma, and other blood products.

5.1. Gamma Globulin Preparations

5.1.1. Standard Immune Serum Globulin

Historical Aspects. In 1936 ARNE TISELIUS separated serum proteins into four major fractions by electrophoresis; subsequently he and KABAT found that immunoglobulin occurred predominantly in the gamma electrophoretic fraction [341]. COHN and colleagues devised a procedure for recovering immunoglobulins (gamma globulins) from serum on a large scale. The serum proteins were fractionated by precipitation with ethanol under carefully controlled conditions of pH, temperature, protein concentration, and ionic strength [336], [342]. This process enriched and stabilized the gamma globulin from plasma at a relatively uniform antibody content while also denaturing most viruses [334]. Such gamma globulin prepared from large pools (>500 donors) of donor plasma were used in the treatment of infectious diseases during World War II [343].

Shortly after World War II gamma globulin was shown to contain antibody titers adequate for the prevention or attenuation of measles [344], infectious hepatitis [345], and polio [346]. Since the report of hypogammaglobulinemia in 1952, and the demonstration that gamma globulin administered on a monthly basis decreased the incidence of infection [347], antibody replacement of this deficiency has been routine.

Properties of Gamma Globulins. Gamma globulins occur at a serum concentration of 600–1200 mg/100 mL in adults, they represent approximately 11–14% of total serum proteins [341] and 80% of serum antibody [348]. Immunoglobulin G has a half-life in the normal circulation of approximately 25 days (35–40 days in patients with agammaglobulinemia) and is synthesized by adults at a daily rate of 35 mg/kg body weight [348]. Its rate of synthesis is regulated by serum IgG levels. Immunoglobulin M and IgA have lower serum concentrations and shorter half-lives than IgG. The molecular mass of IgG has been estimated to be 145 000 [341] with 2.5 wt% being carbohydrate that is associated with the heavy chain (see also Fig. 5).

The IgG isotype can be divided into four subclasses (Table 2). The most abundant is IgG 1 (60–70% of total IgG) which binds to the Fc receptors of neutrophils and mononuclear cells and to the first component of complement [349]. The IgG 2 subclass (20–30% of total IgG) activates the classical complement pathway poorly but can activate complement by the alternate pathway. It is more resistant to proteolysis than the other subtypes [348]. The antibody response to pneumococcal and hemophilus polysaccharides may be related to the preimmune levels of this subclass [350]. The subclasses IgG 3 and IgG 4 bind to the Fc receptors of phagocytes (mononuclear cells and neutrophils) and basophils, respectively [349]. The IgG 3 has a short half-life (nine days); IgG 4 is unable to bind complement.

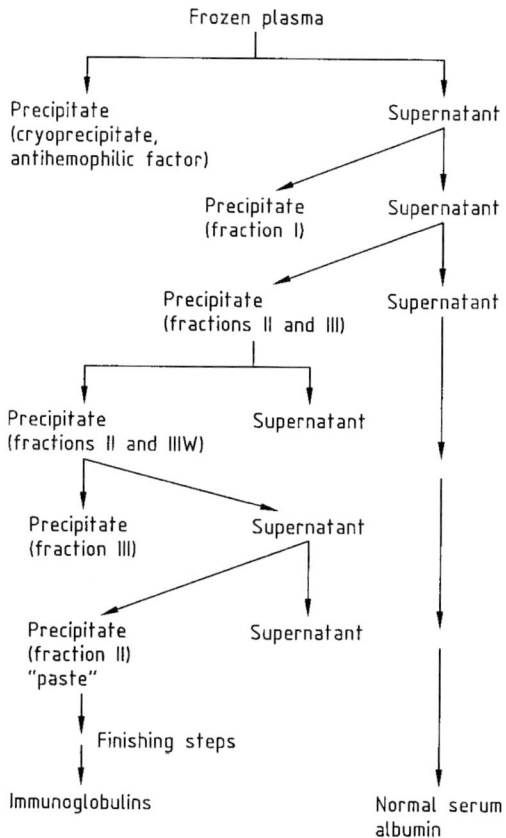

Figure 5. Preparation of immunoglobulin by Cohn–Oncley fractionation ("6/9 method") Cohn fractionation of plasma by cold–alcohol procedures yields a fraction II precipitate or "paste". This is followed by applying the Oncley procedure to Cohn fraction II material to give immunoglobulin. The procedure is based on solubility differences that depend on ethanol concentration, ionic strength, pH, temperature, and protein concentration. Precipitates are usually collected by centrifugation.

Preparation of Immune Serum Globulin. Standard immune serum globulin (ISG, gamma globulin) is prepared from the plasma of pools of >1000 donors (see Fig. 5); → Blood). This minimizes differences between individual antibody levels to specific antigens and ensures a broad range of antibody specificities.

Standard immune serum globulin is usually prepared as a 16.5% injectable solution (165 mg per mL) and represents a ca. 20–25-fold concentration of plasma IgG [336], [343]. This provides an effective dose of antibody in a relatively small volume [343]. Because of the viscosity of the preparation, it can only be given intramuscularly or subcutaneously [343], usually through a large gauge needle (16–18 ga). It has been estimated that 1 g of ISG contains 4×10^{18} antibody molecules with $> 10^7$ specificities [334]. The product, although highly stable at 4 °C, can still undergo proteolysis, presumably due to plasmin contamination [336]. The ISG may also contain blood group substances, IgA, and IgG dimers. The presence of IgA can result in anaphylactic reactions in patients who lack IgA [336].

5.1.2. Immunoglobulin for Intravenous Use

Intravenously administered gamma globulin is needed to allow greater patient comfort, increase acceptability, and provide larger quantities of antibody. This is particularly important for patients who have a small muscle mass (children) or insufficient skin surface (burns), who are at risk from uncontrollable bleeds (Wiskott-Aldrich or other bleeding diatheses), or who need either large repeated doses (immunodeficient) or rapid onset of peak levels (intoxicated patients). Furthermore, intravenous administration avoids the local degradation of ISG antibody that may occur at the injection site [351].

In 1948 COHN administered 25–50 mL of ISG intravenously and saw no adverse effects [336], [352]. JANEWAY and COHN repeated the experiment with a new preparation and induced severe reactions after only 2 mL. This reaction was attributed to contamination of the preparation by staphylococcal enterotoxin [352]. The intravenous administration of ISG was repeated by JANEWAY in 1970 [336], and induced severe cardiovascular (tachycardia, arrhythmias, hypotension, severe chest pain), tachypneic, and pyrogenic (fever, chills, malaise) reactions [343]. These adverse effects were attributed to the presence of immunoglobulin aggregates that activated complement [353], [354].

To avoid the above-mentioned severe reactions, methods were introduced to rid the preparations of the high molecular mass immunoglobulin aggregates. Although they could be removed by centrifugation, reaggregation usually occurred and this method was not practical on a large scale [355]. Consequently, attention was turned to chemical modification.

Various methods of chemical modification have been tried. Degradation with the enzymes pepsin and plasmin resulted in the formation of antibody fragments with immunologic activity [336], [356]. In addition, since the proteolytic enzyme was not removed, it may have continued to be active in the absence of lyophilization; porcine pepsin also induced antibody formation [354]. Reaggregation was also prevented by acidification with hydrochloric acid [357] and reduction–oxidation with dithiothreitol, iodoacetamide, sulfite, or tetrathionite. Alkylation and acylation of immunoglobulin has also been accomplished with β-propiolactone.

The IVIG preparations that have been reduced and alkylated or otherwise modified may have impaired complement binding and altered subclass distribution [358], impaired opsonic activity in vitro [359], shortened serum half-life, and decreased protective efficacy [360]. Consequently, a new generation of native or intact IVIG preparations has been produced without modification by methods which include adjustment to pH 4, use of poly(ethylene glycol), ethanol precipitation, ultra- or diafiltration, and ion-exchange chromatography [353]. Currently, four unmodified products are licensed in the United States: one prepared by adjustment to pH 4 in the presence of traces of pepsin, one prepared by ion-exchange chromatography and ultrafiltration, and a third by diafiltration, ultrafiltration, and adjustment of pH to 4–4.5. A fourth product (Venoglobulin-1 from Alpha Therapeutic Corporation) licensed in the United States but produced in Japan is prepared by poly(ethylene glycol) fractionation and ion-exchange adsorption.

Commercially prepared IVIG is usually stabilized with a mono- or disaccharide, such as 10% maltose [338] or glucose. This considerably decreases the incidence of side effects that accompany the infusion of IVIG [336], [357], [361], perhaps by minimizing precipitation or aggregation. Maltose can cause a mild diuretic effect [362].

5.1.3. Hyperimmune Globulins and Antitoxins

Hyperimmune globulins are high-titered preparations of ISG against a specific antibody. They are prepared from the plasma of patients who have recently recovered from the disease or who have a high titer as a result of a previous vaccination or natural infection. Antitoxins for use in intoxication with botulinus or diphtheria toxin are prepared from snake venom or in horses [363].

5.1.4. Production Requirements

The WHO requirements stipulate that current immunoglobulin products

1) are prepared from pools of >1000 donors;
2) are free of kinins, plasmin, and prekallikrein activity;
3) have a low IgA content;
4) are as free as possible from immunoglobulin aggregates;
5) contain at least 90% intact IgG without fragments;
6) are as unmodified as possible so that opsonic, complement, and other biologic activities are maintained;
7) contain all IgG subclasses;
8) have high levels of antibody to at least two bacterial species or toxins and two viruses (to be ascertained by neutralization tests), and
9) contain at least 0.1 International Units (IU) of antibody to hepatitis B and have a 1:1000 titer to hepatitis A [364].

In addition, manufacturers should specify the diluent and any chemical modifications used. Finally a product can only be classified as a hyperimmune immunoglobulin if the antibody level is five times that of standard ISG preparations.

There is no consensus on which laboratory tests should be used to predict the safety of IgG (contact activation, anticomplementary activity) [336]. Safety requirements should, however, consider the IgG half-life and virus transmission (particularly of human immunodeficiency virus HIV, but also of hepatitis B and non-A, non-B hepatitis viruses) [353].

5.2. Prophylaxis with Immune Serum Globulin

Passive immunization with ISG has been recommended for the short-term prevention of disease when vaccines for active immunization are unavailable or when active immunization was not given before disease exposure; it should be given before expected contact or early in the disease incubation. In these situations active immunization is always preferable because the antibodies subsequently formed by the vaccine provide longterm protection. Immune serum globulin is also used for antibody replacement in patients who lack adequate serum levels of immunoglobulin (i.e., hypogammaglobulinemia) [335].

The efficacy of ISG for the short-term prophylaxis of specific infections was established shortly after its widespread availability in the 1940s. These include:

Hepatitis. STOKES and NEEFE showed that ISG could prevent or modify the course of hepatitis A [335], [336], [345], but KRUGMAN found that ISG modified, but did not prevent this disease [365]. The ISG dose for treating hepatitis A is 0.02 mL/kg body weight [336]. A hyperimmune antibody preparation is also available commercially. In the case of hepatitis B, ISG is not recommended. Instead, hyperimmune globulin is given (see Section 5.3). Although ISG can decrease the incidence of post-transfusion hepatitis [366], its use for non-A, non-B hepatitis is considered optional [364].

Measles. The ability of convalescent serum to modify the course of measles was demonstrated in 1907 [335]. The ISG has also been shown to prevent or attenuate the disease [343]. The introduction of active immunization with a measles vaccine in 1963 (see Section 3.1) significantly decreased the incidence of this disease in the United States. Standard ISG is now recommended for infants under one year of age or for immunodeficient patients within six days of acute exposure to a case of measles [364].

Polio. If given early, ISG can modify the paralytic complications of polio [335], [346]. An unexposed individual should receive 0.15 mL/kg body weight.

Prevention of Infection in Patients with Hypogammaglobulinemia. An intramuscular ISG dose of 100 mg/kg given every 3–4 weeks is currently recommended for patients with hypogammaglobulinemia. This maintains a serum level of circulating IgG above 200 mg/mL [336] and confers protection against a wide variety of infections [342]. This minimum recommended dose is based on a study by the British Medical Research Council conducted between 1956–66 on 176 patients [367]. Although serum IgG levels at this dose rarely rise to the normal range [368], total replacement of IgG does not appear necessary for preventing infection [369]. Although a monthly dose of 200 mg/kg was found superior to 100 mg/kg, most patients did not tolerate more than the 100 mg/kg [361]. Two other studies also showed that higher doses of ISG decreased the incidence of acute infections [334]. Individualization of doses and their frequency has also been suggested [368].

Rubella. Standard ISG is unreliable in the modification of rubella [335], [337], [365]. High titers of antibody are needed [365]. Nevertheless, the use of ISG is recommended for women exposed to the disease during early pregnancy [364].

Clinical Studies. On the basis of data available in 1980, a committee of the WHO observed that it was "inappropriate" to use standard immunoglobulin for the prevention of infection in premature infants, during the physiologic hypogammaglobulinemia of infancy, or for malnutrition. Its use is contraindicated in patients with selective IgA deficiency [364]. Previous attempts in the 1960s showed that ISG did not prevent infection in a variety of clinical situations [370], [371] including multiple myeloma [372]. Although ISG was unable to prevent upper or lower respiratory tract infections in children or institutionalized, elderly adults, a significant decrease was seen in the incidence of mumps in children and fevers of unknown origin in adults [373]. Its use in the prevention of infection in burned patients has yielded conflicting data [335]. In a study on Peruvian children with burns over at least 10% of their body surface area, there was a 41% incidence of septicemia and 15% mortality in control patients compared to a 21% incidence of septicemia and 6% mortality among those who received either plasma or ISG [374]. However, in another study in the United States administration of ISG did not beneficially affect either the rate of septicemia or mortality in burned patients [375]. More recently, patients admitted to a burn unit in India were randomized into groups that received active immunization against *Pseudomonas aeruginosa*, passive immunization with a cold ethanol precipitate of plasma from normal immunized volunteers, both immunologic treatments, or neither. In children, but not adults, there was a significant reduction in mortality following a daily dose of just over 20 mg protein for three days. There was a decrease in the incidence of not only *P. aeruginosa*, but also other gram-negative bacilli among those treated immunologically [376].

The ability of ISG to prevent infections has been most extensively studied in high-risk, premature infants. This population has a hypogammaglobulinemia that exposes them to a high incidence of bacterial infection [377]. The more premature the infant the lower the IgG, since most of the transplacental transfer of IgG occurs during the last six weeks of gestation [370]. Children do not attain adult levels of immunoglobulin until two years of age. In two studies using 0.5 – 3 mL/kg, no prophylactic effect was observed [371], [378], although there was a suggestion that the high dose regimen may have had a beneficial effect.

5.3. Prophylaxis with Hyperimmune Globulins

Several hyperimmune globulins are commercially available or undergoing development. Hyperimmune IVIG preparations are discussed in Section 5.5.

Diphtheria [342]. Diphtheria antitoxin is obtained from the blood of horses immunized against diphtheria toxin. Individuals should be tested for sensitivity to horse serum before being given the product. For prophylaxis, 1000–5000 IU of antitoxin is administered to Schick-positive individuals exposed to diphtheria. Higher doses (20 000–80 000 IU) are given for treatment.

Hepatitis. A hyperimmune globulin is available for hepatitis A. In the case of hepatitis B, a hyperimmune globulin is administered after mucosal or percutaneous exposure (including sexual contact) to an antigen-positive individual [336]. This is also recommended for newborns of antigen-positive mothers [365].

Mumps. Immune serum globulin has no efficacy in the modification of this disease, but hyperimmune globulin does [343].

Pertussis. Hyperimmune globulin modifies the disease. For example, 2.5 mL of hyperimmune antipertussis gamma globulin with a followup dose at 5–7 days led to a 75 % reduction in disease among exposed, nonimmune individuals [343]. However, use of hyperimmune globulin has been superseded by antibiotics [379].

Rabies. Hyperimmune globulin is recommended in addition to active immunization following exposure to a possible or proven case of rabies [364] (see Section 3.7). Combined active and passive immunization against rabies have been shown to be superior to active immunization alone [336].

Rho(D) immune globulin. This hyperimmune globulin is recommended for rhesus-negative mothers who deliver rhesus-positive infants [364].

Tetanus. The efficacy of tetanus immune globulin in either the prophylaxis or treatment of tetanus has not been clearly shown in clinical study. Current recommendations are for 250 IU injected intramuscularly for individuals whose tetanus immunization is not known and whose wound is of a sufficiently serious nature. For treatment of clinical tetanus, doses of 3000–6000 units are recommended.

Vaccinia. Hyperimmune globulin is used for prophylaxis against smallpox and for treatment of the dermal complications of vaccination [343].

Varicella. The efficacy of standard ISG in the prophylaxis of this disease is not well-established [335], [365]. Use of the hyperimmune product, however, is indicated in individuals who have never had chicken pox, who are exposed to acute cases and belong to a high risk group (newborns or immunocompromized patients), and women who are pregnant [364].

Hyperimmune Globulins under Development. Globulins with high titers to cytomegalovirus, and *Pseudomonas aeruginosa* are currently being tested. A product (bacterial polysaccharide immune globulin) has been tested that is prepared from the sera of donors immunized with licensed vaccines against *Hemophilus influenzae* (type b), pneumococci, and meningococci.

Certain subpopulations of children, such as native American Indians and Eskimos, are at particularly high risk of acquiring serious infection with encapsulated bacteria. Standard ISG preparations have failed to show a beneficial prophylactic effect for many types of infection. Immune serum globulin made hyperimmune to polysaccharide antigens by immunizing volunteers with licensed vaccines against pneumococci, *H. influenzae*, and meningococci has been used to immunize Apache infants. There was a significant reduction in the incidence of systemic disease caused by *H. influenzae* and pneumococci during the first six followup months as well as a significant decrease in the incidence of bacteremia [380].

5.4. Therapy with Immune Serum Globulin

Data suggest ISG is active in bacterial infections in animal models [379], [381] and may be synergistic with antibiotics [339], [340], [382]. However in 1968 SCHLESS and HARRELL [379] and others [336] observed that there was little evidence to support its therapeutic efficacy in established infection in humans; they suggested the need for a controlled clinical trial of ISG in the treatment of systemic infection in patients that were not deficient in antibody [379]. Indeed, since the most functionally-active antibody to gram-negative bacteria is IgM and not the IgG found in ISG, little benefit was to be expected [342]. The ISG is active against a wide variety of human pathogens in animal models of infection [379]. When used with antimicrobial agents, ISG has possible benefits in experimental infection with a wide variety of organisms [382], [383] and in clinical infections in humans [339], [340], [384], [385]. Large daily doses of ISG administered intravenously for 10 days to patients with leukemia and fever was well tolerated [386]. However, no benefit was found for patients who received the ISG in addition to antibiotics (compared to antibiotics alone). The administration of ISG to shorten the course of infection in children under two years of age was ineffective [387].

5.5. Prophylaxis and Therapy with Intravenous Immunoglobulin (IVIG)

In studies on the prophylactic or therapeutic efficacy of ISG, the possibility that higher doses of standard immunoglobulin might improve efficacy was a recurring theme. With the availability of IVIG, larger volumes of immunoglobulin could be administered directly into the bloodstream (see Section 5.1.2). In addition, with the ability to screen large numbers of samples for antibody levels or to immunize volunteers with an increasing number of vaccines, an increasing number of publications have examined the efficacy of passive immunotherapy, particularly with hyperimmune IVIG preparations, in the treatment and prophylaxis of infectious diseases.

Intravenous immunoglobulin G is effective in the prevention of infection in patients with hypogammaglobulinemia [358]. It has also been shown to be effective for treating chronic infection in such patients who developed sinopulmonary infection despite ISG maintenance therapy [351].

5.5.1. Viral Infection

Since cytomegalovirus (CMV) is a frequent cause of infection in patients undergoing organ transplantation and effective antiviral therapy is lacking, there has been considerable interest in the use of IVIG that is hyperimmune in CMV antibody for both the prophylaxis and treatment of CMV infections.

Prophylaxis. Hyperimmune CMV-IVIG (total dose 550 mg/kg) was shown to decrease the attack rate of symptomatic CMV infection among CMV-negative patients who received kidneys from CMV antibody-positive donors from 60–21% [388]. This treatment also decreased the incidence of fungal and parasitic infection. Hyperimmune CMV-IVIG (3×200 mg/kg doses) also prevented mortality and interstitial pneumonia from CMV for 120 days in leukemic patients who underwent bone marrow transplantation [389]; however in another study it did not prevent acquisition of infection [390]. In another trial, nonimmune IVIG with a high antiCMV titer did not decrease the incidence of CMV seroconversion in bone marrow transplantation patients [391]; however, the incidence of symptoms and interstitial pneumonia decreased. In contrast, CMV-IVIG given to CMV-negative patients undergoing bone marrow transplantation had no effect on either the prevention of new disease or the amelioration of established disease [392]. Therapy of bacterial infections following infection with human immunodeficiency virus are discussed in Section 5.5.2.

Therapy. Immunoglobulin has been used sporadically in the treatment of viral disease. High-titered CMV-IVIG from screened donors did not show efficacy in the treatment of bone marrow transplant patients with documented CMV infection [393].

IVIG has been shown to alter the course of echovirus encephalitis infection in three patients with hypogammaglobulinemia. Although no benefit occurred from giving the IVIG intravenously in two of these patients, intraventricular administration resulted in clinical cures [394], [395]. The use of IVIG was unable to alter the lethal course of polymyositis secondary to echovirus in another patient with hypogammaglobulinemia [396]. IVIG has also been used experimentally in the successful treatment of herpes infection in mice [397].

5.5.2. Bacterial Infection

Numerous studies have demonstrated the efficacy of hyperimmune IVIG in the prophylaxis and, if used early after infection, the treatment of bacterial infection. In 1943 ALEXANDER showed that the combination of a sulfa drug and animal hyperimmune sera was more effective in reducing mortality than either agent alone [339], [340]. Recently a number of studies have demonstrated the efficacy of IVIG in both the prevention and treatment of experimental infection with *H. influenzae* in neonatal rat models [398]; with *E. coli* [399] and *Klebsiella* [400] in mouse models; and with *P. aeruginosa* in neutropenic and burned rodent models [401], [402]. Intravenous immunoglobulin hyperimmune to group B streptococcal surface antigens can prevent and treat serious bacteremia in experimental infection in monkeys [403].

Prophylaxis. Earlier data with ISG in both experimental bacterial infection in animals and in clinical infection in humans indicated that in some situations (e.g., patients with hypogammaglobulinemia), exogenous standard gamma globulin could prevent the acquisition of serious bacterial infection. Similar studies with IVIG have established the efficacy of these preparations [358].

Since earlier investigators believed that larger doses of ISG might be effective in the prevention of bacterial infection in high-risk neonates, it is not surprising that similar studies have now been reported with IVIG. In one nursery with a high rate of infection, a single dose of IVIG (120 mg/kg within 2 h of birth) decreased acquisition of infection and mortality in preterm, lowbirthweight neonates [404]. A second dose at eight days, conferred no further advantage. Prophylaxis with IVIG (0.5 mg/kg/week for four weeks) was significantly better in preventing infection and death in neonates, but only in the subpopulation that weighed less than 1500 g and had a gestational age of less than 34 weeks [370]. In another study, antibiotics were given either alone or with IVIG to women 27–36 weeks pregnant who had chorioamnionitis. Only high doses (24 g/d for five days) of IVIG given after the 32nd week of pregnancy prevented infection in the delivered babies [405]. The investigators concluded that little transplacental transfer of IgG occurred before the 32nd week of gestation.

Data on the use of standard IVIG for the prevention of bacterial infections in patients without hypogammaglobulinemia is limited. High levels of specific antibody are needed; these may not be found in standard, nonimmune IVIG, despite the possibility of delivering larger amounts of IVIG than was the case with ISG. Hyperimmune products have been shown to prevent specific infections in experimental models (see Sections 5.4, 5.5.2, and 5.6). The administration of nonimmune IVIG (1000 mg/kg) before bone marrow transplantation and weekly for 17 weeks thereafter had no effect on the acquisition of either bacterial or fungal infection [406]. Clinical trials are being performed to test the efficacy of IVIG in the prophylaxis of bacterial infection in adults with chronic lymphocytic leukemia, a condition which may be complicated by hypogammaglobulinemia [407].

Therapy. Patients with human immunodeficiency virus (HIV) infection (i.e., acquired immune deficiency syndrome, AIDS) have a dysfunction in their humoral immune system which includes both a decreased antibody response to bacterial antigens and an altered distribution of IgG subclasses [408]. Unlike adults, who tend to acquire opportunistic infections (infections in which immunodeficient individuals are infected by organisms that are normally withstood by immunocompetent individuals), young children and particularly infants infected with HIV often resemble patients with primary humoral immunodeficiency and tend to suffer from bacterial infections. Consequently, prophylaxis with monthly doses of IVIG has been used in the management of these patients. In one pilot study, a decreased incidence in episodes of fever and bacteremia was noted. This was accompanied by clinical improvement, prolongation of life, and improvements in other immunologic parameters [408]. In a 37 month old child with AIDS, monthly doses of IVIG produced increases in IgG 2 antibody and antibody to 12 pneumococcal serotypes, and prevented subsequent episodes of bacteremia [409]. On the basis of these preliminary data, IVIG prophylaxis has been advocated in the treatment of childhood AIDS [408], although a well-controlled, prospective study has yet to be done. Periodic administration of IVIG decreased lactic dehydrogenase activity (a proposed indicator of pulmonary interstitial inflammation) in adults and children with HIV infection [410].

Attempts to demonstrate a significant effect in the therapy of bacterial infections with IVIG have been much less successful than with its use in prophylaxis. This may be attributable to the shorter half-life of IVIG in the blood during infection [368], [381]. The addition of IVIG to standard regimens of antibiotics decreased the mortality from documented bacteremia, particularly among preterm infants. The number of subjects studied was too small for statistical analysis, however [411].

In adults, little data is available to support the use of IVIG in the treatment of bacterial infections. This may be due to the need for high levels of antibody specific for the invading organism (rather than simply high levels of non-specific antibody) as well as the need for prompt initiation of therapy. In most experimental studies with hyperimmune IVIG, little benefit can be shown if the exogenous immunoglobulin is given more than 8 h after infection (see [381], [383]). In clinical medicine, however, identification of the time of onset of infection is often difficult.

5.5.3. Noninfectious Diseases

5.5.3.1. Therapeutic Effect of IVIG

In 1981 it was reported that a patient who received IVIG for hypogammaglobulinemia also recovered from bleeding secondary to a coincidental platelet deficiency (*thrombocytopenia*) following the infusion [412]. Following this fortuitous observation, experiments showed that a high dose of IVIG could indeed reverse thrombocytopenia in the absence of hypogammaglobulinemia; furthermore, a regimen of 400 mg/kg/d for five days was effective treatment for idiopathic thrombocytopenic purpura (ITP) in children [413]. The use of IVIG and oral steroids was also compared. Among those who responded rapidly to treatment (62%), IVIG was as efficacious as steroids; however the slower responders responded better to the IVIG treatment. Administration of IVIG was accompanied by a doubling of serum IgG; IgM levels rose under both treatment regimens. The IVIG may possibly reduce clearance of antibody-coated platelets via an Fc-mediated mechanism (see Section 5.5.3.2) [414]. The dosage regimen used in the above study (400 mg/kg/d for five days) has been used in many other studies on the noninfectious uses of IVIG. However, in another study 800–1000 mg/kg was given as a single infusion to children with ITP with similar results and no reported untoward effects [415]. Experience with IVIG for ITP has not been uniformly successful however [416].

IVIG has been used in the treatment of *immunologically-mediated blood disorders* in both pediatric and adult populations. These include autoimmune hemolytic anemia [417], [418], autoimmune neutropenia [419], posttransfusion purpura [420], ITP in adults [421], [422], chronic ITP in adults and children [423], [424], thrombocytopenia secondary to alloimmunization in leukemic patients receiving platelet transfusions who became refractory to subsequent platelet transfusions [425], thrombocytopenia secondary to transplacental passage of antiplatelet antibodies [426], ITP of pregnancy [427], antibody-mediated red cell aplasia [428], during pregnancy in women with severe rhesus immunization [429], and in conjunction with cyclophosphamide used to treat antibody to factor VIII in hemophilia [430].

In addition to these hematologic disorders, two large trials conducted in patients with *Kawasaki's disease* have compared the use of aspirin alone to that of aspirin and IVIG (400 mg/kg/d for 4 and 3 days) [431], [432]. The IVIG reduced the fever and the incidence of coronary artery disease [431], [432]. It suppressed the T and B cell activation characteristic of patients with this disease and decreased the levels of spontaneous immunoglobulin synthesis in vitro [433].

Placentally derived immunoglobulin has been administered to patients with severe *rheumatoid arthritis* because the symptoms of some patients improved during pregnancy [434]. None of five patients given IVIG derived from control plasma improved whereas 3/5 improved under the placentally derived globulin. An antihuman leukocyte antigen–DR surface antigen (anti HLA-DR) antibody in the placental preparation was presumed to be a possible mechanism for this improvement.

High dose immunoglobulin treatments have also been used in patients with *Felty's syndrome* [435], *myasthenia gravis* [436], *epilepsy* [437], and *multiple sclerosis* [438]. In the latter study one-third of patients worsened with IVIG therapy.

5.5.3.2. Mechanism of Action

Many immunologic mechanisms have been invoked to explain the beneficial effects of IVIG in noninfectious illnesses. IVIG is a potent immune modulator; the role of the Fc portion of the immunoglobulin molecule in this respect is still a subject of active investigation.

Antibody-coated material is removed by phagocytes in a process known as *Fc-receptor-mediated clearance*. The Fc portion of the antibody coating the bacterium cell, protein etc. binds to the Fc receptor of the phagocyte. The material is then taken up by the phagocyte and digested. Initially, the Fc receptors in the reticuloendothelial system were thought to become saturated by the high dose of immunoglobulins in IVIG resulting in a decreased clearance of IgG-coated particles due to competitive inhibition [417], [439]. This is consistent with studies in which radiolabelled, autologous erythrocytes were cleared more slowly after IVIG infusion than before [414], [423]; this treatment altered Fc receptor affinity, not receptor number [440]. Other suggested mechanisms include decreased synthesis of autoantibodies [441], the clearance of persistent, occult viral infections [396], [441], the protection of platelets or megakaryocytes against antiplatelet antibodies [441], antiidiotypic suppression of antibody synthesis [436], [442], blocking of the Fc receptor by antibodies, and the production of antilymphocyte antibody [443]. However, patients with ITP have responded to IVIG without demonstrable alteration in Fc-receptor mediated clearance [441]. In addition, preincubation of erythrocytes in IVIG failed to inhibit the phagocytosis of sensitized erythrocytes by cultured macrophages [444]. Thus, IVIG may lead to improvement in such patients by multiple mechanisms.

1IVIG is a potent modulator of antibody production both in vitro and in vivo. It inhibits lectin-driven B cell differentiation in vitro [445], [446] and immunoglobulin production by peripheral blood mononuclear cells stimulated with pokeweed mitogen [446]. For these effects the Fc portion of immunoglobulin must be present: the Fc portion alone was 100 fold more effective than the intact IgG. The observation that IgM antibody rises after IVIG infusion has raised the possibility that it may stimulate some immunoglobulin-producing cell populations [423]. Interestingly, the monthly administration of ISG to premature infants during the first year of life resulted in a significantly lower level of immunoglobulin compared to the control group [371], [447].

In adults with ITP, the infusion of IVIG (400 mg/kg) led to a decrease in T 4 (helper) lymphocytes and elevation in T 8 (suppressor) lymphocytes (and decrease in T 4/T 8 ratio) [421]. Similarly, the administration of IVIG to patients with hypogammaglobulinemia led to increased suppressor T cell activity, decreased total T cells, a significant decrease in the T 4/T 8 ratio, and a decrease in lectin-induced immunoglobulin syntheses in vitro [448].

The ability of IVIG to decrease antibody production may be desirable in patients producing autoantibody, but dangerous in those with infection [445]. In patients with acute otitis, repeated monthly [379] infusions of IVIG resulted in higher IgG levels if the initial antipneumococcal antibodies were low. However, in the presence of high initial levels of antibody, specific antipneumococcal antibody levels decreased [449]. The suggestion has also been made that high levels of nonspecific antibody could lead to a reduction in the survival of specific antibody [450], [451]. Evidence suggests that IVIG may even exacerbate infections. A patient with neutropenia died due to acceleration of the yeast infection shortly after infusion of IVIG; the IVIG may have blocked the normal Fc-mediated clearance mechanism [452].

Immunoglobulin G can bind both native C 3 and C 3 b during complement activation by soluble immune complexes and by bacteria [453], [454]. Finally since immunoglobulin has unique antigenic determinants, antibodies to the immunoglobulin may develop [455], [456]. These antibodies may play a role in the anti-idiotype network.

In summary, exogenous immunoglobulin has a diverse, potent effect on a wide variety of immune regulatory mechanisms. These interactions may often work to the patient's benefit, but may also result in previously unsuspected adverse effects.

5.6. Prophylaxis and Therapy with Plasma and Other Blood Products

Although the regular, large-scale, clinical use of plasma and other blood products is considered impractical (both for production and safety reasons), passive immunotherapy with these agents may provide some efficacy in the prophylaxis and treatment of infections in humans.

Plasma has been used to treat infections, particularly those caused by *Pseudomonas* [374], [457]. Plasma was superior to ISG in the prevention of infection among patients who suffered > 30 % burns [457]. Similar observations were made during experimental infection [458]. One possible explanation for the higher efficacy of plasma is that it contains IgG, IgA, and IgM (ISG contains only IgG). This may improve the distribution of antibody at different anatomic sites. Since the isotypes have functional differences, the antibacterial activity is also increased (natural antibody against gram-negative bacilli is thought to be predominantly of the IgM isotype). However, plasma infusions carry the risk of transmitting hepatitis (there is no manufacturing process for plasma that inactivates the virus as is the case with ISG).

Further efforts in the use of passive immunotherapy have also focused on septic shock. The rapid infusion of large volumes of lyophilized human plasma from blood with > 40 µg/mL of antibody to a mixture of lipopolysaccharide serotypes resulted in nearly 7-fold decrease in mortality in South African women treated in an obstetrical–gynecological ward for septic shock [459]. Equine plasma similarly screened for antilipopolysaccaride antibodies has also been used routinely in veterinary practice in South Africa [460].

Opsonins. The defective neonatal antibody (opsonic) response that mediates the uptake and killing of bacteria by phagocytes can be partially corrected with gamma globulin [461]. The ability passively administered opsonins to prevent infection was first suggested in one study in which fresh whole blood was administered to infants to prevent group B streptococcal sepsis. All nine infants transfused with blood having antibody to group B streptococci lived versus 3/6 transfused with blood having undetectable antibody titers. Protection was correlated with opsonic antibody titers in the infants' blood and with having >40% of their blood volume replaced [462].

Postimmune Serum. In a recent study, donors were immunized with a J5 mutant of *E. coli* to elicit antibodies to widely shared core epitopes in the lipopolysaccharides of gram-negative bacteria. The passive administration of postimmune serum from these individuals to patients in bacterial endotoxic shock decreased the incidence of death as well as the need for pressor therapy in patients with profound shock. However, since this protection did not correlate with the presence of antibody to the core epitopes, it is not clear whether the protection was conferred by the antibody [463]. In a followup study, the prophylactic administration of postimmune plasma appeared to significantly prevent shock and death from gram-negative infections in patients admitted to a surgical intensive care unit. Such treatment had no effect on infection rate, however [464]. In both studies the protective moiety in the plasma and serum was presumed to be antibody to the endotoxin. In any event, for this to be a practical therapy, the protective portion of the postimmune product must be mass produced in the form of a safe, standardized preparation (e.g., made into an IVIG from a large pool of donors immunized with the J5 vaccine or into a monoclonal antibody preparation). Attempts to show a protective effect from an IVIG prepared from immunized donors have been unsuccessful to date [465]. Nevertheless, monoclonal antibody preparations directed toward a J5 epitope are currently being evaluated in human volunteers (see also Section 6.1).

5.7. Adverse Effects of Gamma Globulin Preparations

Immune serum globulin G is one of the safest biological products available with an overall incidence of reactions of 3–12% [334]. Rare systemic (anaphylactic) reactions may occur during or within minutes of administration of ISG (ca. 1/500–1/1000 injections). Late-occurring systemic reactions (within hours or days) include arthralgias, pyrexia, and diarrhea and are not uncommon in immunodeficient patients [364]. Local reactions occur with intramuscular administration of ISG and are related primarily to the large volume administered. Local reactions to ISG can be prevented by using small volumes or by pretreatment with analgesics. Systemic reactions may be modified by the use of aspirin, hydrocortisone, or antihistamines.

Systemic reactions with IVIG are more common than with ISG and often depend on the rate of infusion and the type of preparation. Patients receiving IVIG for hypogammaglobulinemia have a reported incidence of 2.5% reactions which are usually mild [358]; the incidence of these reactions decreases after the first two months of therapy. The cause of reactions to IVIG are not clear but may depend on aggregate formation, IgA contamination, and activated enzymes that initiate the release of inflammatory mediators [364].

The experimental work discussed above and clinical experience give reason for caution in the use of immunoglobulin preparations. Clinical experience primarily involves concern about viral transmission, anaphylactic reactions among patients with IgA deficiency, and isolated case reports of unusual occurrences. The indications for the use of gamma globulin should be well founded [363].

Viral Transmission. Immune serum globulin G has had an outstanding safety record with regard to lack of transmission of virus to recipients [466]. For example, ISG is free of serum hepatitis, even if the virus occurs in the original plasma. Ten children who received plasma for prophylaxis against measles in England became jaundiced and three died. In contrast, when ISG prepared from the same plasma was given to 56 children, only one child became jaundiced [343]. In the United States, 4 out of 15 volunteers who were inoculated subcutaneously with hepatitis-positive plasma developed hepatitis. Five further volunteers received large doses of ISG and none developed hepatitis [343]. The ability of the Cohn–Oncley fractionation (see Fig. 5) to inactivate virus may in part be due to the high concentration of antivirus antibody [353]. Non-A, non-B hepatitis has been reported in pastes produced in this procedure: in one, the lyophilization step was omitted, and in two others an ion-exchange step was added to lower IgA but it lowered IgG 4 levels as well [353].

In contrast, non-A, non-B (NANB) hepatitis has been reported following IVIG infusion [467]–[469]. In one such episode 16 out of 77 patients who received IVIG for immunodeficiency acquired hepatitis with death occurring in 5 cases. Interestingly, no hepatitis was observed following administration of ISG prepared from the same serum pool [467]. Since neither of the IVIG products was subjected to a recognized virucidal finishing step, such as treatment with acid, β-propiolactone, or ultraviolet radiation, the virus may still have been viable. Thus ethanol fractionation by itself might not fully inactivate putative virus [470], though it is virucidal for enveloped viruses [471]. Supplementary virucidal procedure could make IVIG safe from NANB hepatitis transmission [471]. Unlike plasma, IVIG has not been known to transmit hepatitis B [472].

There have been no confirmed cases of transmission of HIV virus in IVIG but antibody to HIV has been isolated from two patients with primary hypogammaglobulinemia [471]. When HIV was added to plasma and then processed into IVIG, $> 10^4$ PFU/mL (PFU denotes plaque-forming units) of HIV were inactivated during alcohol fractionation and poly-(ethylene glycol) fractionation to IVIG [473]. Antibody to HIV was detected by an enzyme-linked immunoabsorbent assay (ELISA) in all 10 lots of a reduced, alkylated IVIG and in 4/8 lots of a pH 4/pepsin IVIG; 8/10 and 3/8 lots were also positive by Western blot analysis. The HIV antibody-positive lots, which were negative for HIV by culture and reverse transcriptase activity, were infused into patients. Comparison of pre- and post-infusion sera demonstrated that antibody to HIV was detectable for up to one month before converting to a negative antibody status [472].

Recently one woman who had received hyperimmune Rho (D) immunoglobulin subsequently was found to be infected with HIV but belonged to a group with an increased risk of acquiring infection with HIV [474].

Anaphylaxis. Anaphylaxis has been documented following the administration of IVIG to some patients with hypogammaglobulinemia. Indeed, patients with antibody deficiency have an increased incidence of adverse reactions to IVIG [342]. Low levels of IgA are associated with an increased risk of such anaphylactic reactions [475]; currently available IVIG products differ markedly in their IgA content [476]. In two patients with common variable immunodeficiency, IgE antibody to IgA in the infused IVIG was felt to account for these reactions [475]. In addition, IgE has been detected in commercial preparations of IVIG [477].

Other clinical problems have been described which involve the use of IVIG. IVIG prophylaxis precipitated cryoglobulinemic nephropathy in a patient with hypogammaglobulinemia secondary to B cell neoplasm [478].

The passive transfer of antibodies against specific blood types in IVIG may also be associated with hemolytic anemia or cause problems in the crossmatching of blood prior to surgery [479]. Immunologically normal patients may receive antibody to immunoglobulin that is of a different genotype.

5.8. Future Prospects

Large amounts of immunoglobulin can now be safely delivered directly into the bloodstream by using IVIG. This need was recognized by COHN over 40 years ago. Since immunoglobulin is now known to be a potent modulator of the immune system, IVIG has been used as a form of therapy for many noninfectious diseases. However, the impact of such treatment on the ability to respond to infectious diseases often complicates these conditions and still has to be assessed. Studies to date do not clearly show that the indications for the use of IVIG in infectious diseases should be expanded beyond those already shown for ISG. In some conditions involving functional hypogammaglobulinemia (e.g., HIV infection in young children, chronic lymphocytic leukemia), prophylaxis with standard IVIG may ultimately demonstrate efficacy. In certain subpopulations that are at risk of infection with specific organisms over a finite period of time, immunoglobulin preparations enriched in antibody to those organisms (i.e., hyperimmune preparations) may prove useful. The most promising use appears to be in the prophylaxis of specific infections; this usually involves hyperimmune serum and, as shown by studies in neonates, the substantial replacement of blood volume.

The use of IVIG to treat established infections with pharmacological (not replacement) doses of immunoglobulin, (with the possible exception of antitoxin treatment of septic shock), still has no proven role. This may be related to the need for prompt intervention and a large amount of a specific antibody, requiring rapid identification of the microorganisms involved. Preparations that are known to be immunotherapeutically effective are less efficacious when given 8 h or more after the onset of infection. Substantial evidence suggests that passive immunotherapy may provide an enhanced therapeutic benefit; however, this still remains to be proved in a prospective, controlled therapeutic trial.

Finally, the most recent studies that show some benefit from passive immunotherapy have not been performed with IVIG preparations, e.g., [376], [463], [464], [459]. Thus despite the substantial progress made in the development of IVIG preparations for human use, their firm recommendation for widespread use in infectious disease requires further well-designed studies.

6. Immunotherapeutic Uses of Monoclonal Antibodies

When a vertebrate is vaccinated with a foreign antigen, it produces a mixture of antibodies that can bind to the antigen. Each antibody binds to a different epitope on the immunogen and is made by an individual clone of B lymphocytes. Industrial quantities of a single (i.e., monoclonal) antibody recognizing a specific epitope can be made by fusing a B lymphocyte producing the antibody of interest with an immortal tumor cell line to give hybridomas.

Infectious diseases continue to cause much morbidity and mortality despite the discovery of a wide variety of novel antimicrobial agents and improved methods for immunizing at-risk populations. Therapy with human or animal serum is an effective adjunct to conventional drug therapy and, prior to the development of antimicrobials, was the only effective therapy for many infectious diseases (see Chap. 5). The efficacy of serum treatment for various infectious diseases suggests that this approach could be a useful adjunct to drug therapy if safety, efficacy, and production problems could be overcome. Human monoclonal antibodies (humabs) may solve many of the problems associated with serum therapy. Therapeutic humabs have been developed against many target organisms and are being tested in clinical trials.

6.1. Introduction

Despite the fact that the first reports describing the generation of antigen-specific murine monoclonal antibodies were published in the mid 1970s, routine production of humabs has been more difficult. Methods to generate and produce these molecules have improved and serious efforts are underway to develop them as therapeutic products for which passively administered human antisera against specific target antigens are currently in use in a number of clinical settings:

1) Red cell antigens: Rh (hemolytic disease of the newborn)
2) White cell antigens: antilymphocyte/thymocyte globulin
3) Viral antigens: hepatitis A and B, rabies, CMV, HSV, and varicella zoster
4) Bacterial antigens: tetanus, endotoxins, and pneumococcus
5) Elimination of circulating drugs (overdoses)
6) Antisnake venom
7) Fertility control (e.g., anti-β-human chorionic gonadotropin)

Humabs that recognize many of these antigens have been produced, and will probably augment or replace pooled antisera in the near future. Concerns, which include the possible contamination of pooled globulins by infectious agents (e.g., human immunodeficiency virus HIV, various hepatitis viruses, and cytomegalovirus) and the diminished availability of serum donors, as well as the relative ease of reproducible humab manufacture, will accelerate this trend.

Humabs will be preferred for the diagnosis and treatment of a variety of diseases because they minimize the problems encountered when a foreign animal monoclonal antibody is administered (e.g., anaphylaxis, clinical manifestations of immune complex formation, and reduction of efficacy by anti-antibodies). In well over half of the patients treated to date with murine monoclonal antibodies, the human antimouse immunoglobulin response has limited their usefulness [480]. Only a fraction of the antimouse immune response is directed to the variable region (idiotype) of the rodent immunoglobulins. This suggests that humabs will be more effective than murine monoclonal antibodies. Furthermore, humabs are more likely to have species-specific carbohydrates which may be important in a number of effector functions, such as Fc receptor-mediated, antibody-dependent, cellular cytotoxicity, complement activation, and phagocytosis [481]. Serum half-life and effector functions of immunoglobulin subclasses are very important for designing the optimal anti-infectious agent therapeutic. There are a number of reasons why passively administered humabs will be preferred for therapeutic use. A good summary and technical background are to be found in [482], [483].

6.2. Bacterial Targets

Various bacterial targets are recognized by humabs:

Tetanus toxoid [484]–[488]
Diphtheria toxoid [489]
Gram-negative endotoxins [491]–[494]
Pseudomonas aeruginosa lipopolysaccharide [496] and exotoxin A [497]
Hemophilus influenzae [498], [499]
Mycobacterium leprae [500]
Neisseria meningitides [501]
Pneumococcus [502][503]

Tetanus-neutralizing humabs have been frequently generated because of the ease of obtaining immune B cells from vaccinated persons [484]–[488].

Because immunization against tetanus in the United States is universal and most individuals have relatively high titers, it is unlikely that any of these monoclonals will be scaled up and used in this country. The same is true of antidiphtheria monoclonal antibodies [489]. In other countries, vaccination against tetanus and diphtheria is not widespread and administration of immune human or animal sera is still practiced. Recent concern about HIV-contaminated serum has stimulated efforts to generate humabs against tetanus for use in India, South America, and southern Europe.

Gram-negative bacterial infections account for 1–2% of hospital admissions and up to 100 000 deaths each year in the USA. Despite currently available therapies including antibiotics and various support measures, mortality rates remain as high as 50–70%. Much interest has been generated by the idea that lipopolysaccharide, the lethal component of gram-negative bacteria, might have antigenic determinants that are shared by many bacterial species. The outer carbohydrate domain of lipopolysaccharide varies highly from one species to the next whereas the inner carbohydrate core and lipid A regions appear to be much more conserved. In a clinical trial antisera were obtained from firemen vaccinated with a rough mutant *E. coli* strain called J 5 that lacks the outer carbohydrate. This antiserum gave significant protection to patients with gram-negative sepsis [490]. Several laboratories have reported the generation of humabs that are highly cross-reactive with bacterial lipopolysaccharide [491]–[494]. They are all of the IgM class and protect mice against infection when given at high doses. Two of these and a murine monoclonal of the same specificity are in clinical trials for the treatment of gram-negative sepsis. At present it is too early to determine whether they are successful, however several points should be made regarding this approach:

Firstly, it is unclear whether antibody was the active factor in the antisera in the original study [490]. Secondly, no documentation on the binding of the antisera to the bacteria causing a specific case of sepsis was made, i.e., no correlation could be made between response and nonresponse. Thirdly, immune serum is known to contain many other factors that might have benefited the treated patients. Fourthly, although cross-reactive antisera and monoclonal antibodies appear to bind to a wide number of bacteria and blotted lipopolysaccharide, questions still exist regarding the accessibility of the core and lipid A region for antibody binding. Fifthly, several trials with slightly different design have given inconsistent results [495]. Thus for a number of reasons the so-called core concept is questionable.

Type-specific horse antibodies recognizing typhoid and pneumococcus immunotypes were used from the turn of the century up to the beginning of the antibiotic era in the 1940s. The generation of humabs recognizing type-specific determinants on gram-negative bacteria is a reasonable alternative to the above-mentioned highly cross-reactive monoclonals. Although the diversity of possible immunotypes of gram-negative bacteria is large, a particular subset is probably responsible for most of the invasive bacteremias. Several groups are trying to generate a number of type-specific humabs and administer a cocktail of antibodies. This approach has already been tried for *Pseudomonas aeruginosa* [496]. In model systems, murine monoclonal antibodies recognizing a single immunotype were much more effective than those recognizing a core determinant that was shared between Pseudomonas species. If the cocktail approach is to prove effective, it may be necessary to add humabs that neutralize various virulence factors. In the case of *Pseudomonas* infections, humabs have also been generated to exotoxin A [497].

Humabs have been reported that recognize *Hemophilus influenza* type B capsular polysaccharide [498]. The successful introduction of an *H. influenza* type B capsular polysaccharide vaccine [499] (see Section 2.13) has made the clinical use of these humabs less attractive than originally anticipated. However, a safe humab for this life-threatening disease may still be developed for clinical use in conjunction with the vaccines.

Anti-mycobacterium leprae humabs have been reported [500]. In view of the central role of cell-mediated immunity in resistance and recovery from this disease, these reagents are unlikely to find clinical application. Nevertheless, these antibodies and those made by patients suffering from other infectious diseases can be used to probe the human humoral immune response, to clone antigens recognized by the humabs, and to investigate antibody-mediated autoimmunity initiated by microorganisms.

6.3. Viral and Chlamydial Targets

Antiviral humabs are especially attractive therapeutic targets. Immunoprophylaxis of cytomegalovirus (CMV) infections in immunosuppressed patients (eg., transplantation) is of major clinical interest. Several groups have generated CMV-neutralizing humabs [504], [505] and clinical trials are scheduled to begin in 1989.

Humabs that neutralize hepatitis B virus [506]–[508] will replace the antisera currently used after acute exposure to this agent. These humabs may also be useful for "active–passive" immunization of at-risk infants. A very large population of infants, particularly in less developed countries, become chronically infected in the perinatal period. Evidence suggests [509] that administration of immune sera will prevent lifelong infection and could have an impact on the prevalence of the most common human cancer in the world, hepatitis-B-positive hepatoma [509]. The widespread use of anti-hepatitis A antisera suggests that the humabs generated to this target may also find clinical application.

A minor yet important target for humabs is the treatment of varicella zoster infections [510]. These humabs may have to be developed as an orphan drug given the small numbers of patients. (The development of drugs for rare diseases is supported by the US government; these pharmaceuticals are called orphan drugs.)

The acquired immune deficiency syndrome (AIDS) is caused by the human immunodeficiency viruses (HIV) and is currently the major infectious viral disease problem. Many groups are generating humabs that recognize one or more HIV strains. Many of these humabs recognize gene product gp 160 that is found in the viral envelope [511], [512]. At present, there is no evidence that a humoral immune response can prevent or change the pattern of HIV infection in people or animals. The viruses undergo rapid mutation (five times the rate of influenza). Furthermore, they can spread by cell to cell contact and reside permanently out of reach of humoral immunity inside the mononuclear phagocytes scattered throughout the body. Trials are underway to test the efficacy of a high-titer anti-HIV antiserum in the prevention of neonatal AIDS. The dismal failure of all attempted vaccines and passive serum to delay the spread of the virus in subhuman primate models is very discouraging. Antiviral humabs attached to toxins or engineered to bring cytotoxic cells into contact with infected cells might improve the potency of the antibody approach.

Several humabs have been generated against chlamydia [513] and other viruses:

Epstein Barr Virus [514]
Herpes simplex [515]–[517]
Human T cell leukemia virus I [518]
Measles (SSPE) virus [519], [520]
Rabies [521]
Rubella [522]
X 31 influenza virus [523]

However, clinical interest in the therapeutic purposes and use of these antibodies is limited.

6.4. Parasite Targets

The major therapeutic target of interest among parasitic diseases is malaria (see Section 4.2). Much less is known about the humoral immune response of humans suffering from other parasitic diseases (Section 4.1). Passively administered humabs may have therapeutic potential in acute *Falciparum malaria*. Immunoglobulin M and IgG humabs have been generated [524], [525]. Monoclonal antibodies are also useful research tools in the development of malaria vaccines (Chap. 4).

7. References

[1] I. Roitt: *Essential Immunology*, 5th ed., Blackwell Scientific Publications, Oxford, United Kingdom 1984.

[2] B. D. Davis, R. Dubecco, H. N. Eisen, H. S. Ginsberg: *Microbiology: Including Immunology and Molecular Genetics*, 3rd ed., Harper and Row, Hagerstown, MD, 1980.

[3] R. H. Schwartz, A. Yano, W. E. Paul: "Interaction between antigen-presenting cells and primed T-lymphocytes," *Immunol. Rev.* **40** (1978) 153.

[4] W. Stroler, L. A. Hanson, K. W. Sell (eds.): *Recent Advances in Mucosal Immunity*, Raven Press, New York 1982.

[5] T. Yamamura, T. Tada: *Progress in Immunology V*, Academic Press, Tokyo 1984.

[6] J. B. Robbins, J. C. Hill, J. C. Sadoff (eds.): *Bacterial Vaccines, Seminars in Infectious Disease*, vol. IV, Thieme-Stratton, New York 1982.

[7] U. E. Nydegger (ed.): *Immunotherapy, A Guide to Immunoglobulin Prophylaxis and Therapy*, Academic Press, London 1981.

[8] R. E. Black, M. M. Levine, M. L. Clements, G. Losonsky et al.: "Prevention of shigellosis by a Salmonella typhi–Shigella sonnei bivalent vaccine," *J. Infect. Dis.* **155** (1987) 1260–1265.

[9] M. M. Levine, J. B. Kaper, D. Herrington, J. Ketley et al.: "Safety, immunogenicity, and efficacy of recombinant live oral cholera vaccines, CVD 103 and CVD 103-HgR," *Lancet* **2** (1988) 467–470.

[10] R. J. Collier: "Diphtheria toxin: mode of action and structure," *Bacteriol. Rev.* **39** (1975) 54–85.

[11] A. M. Pappenheimer, Jr. in R. Germanier (ed.): "Diphtheria," *Bacterial Vaccines*, Academic Press, Orlando, FL 1984, pp. 1–36.

[12] F. Audibert et al.: "Active antitoxic immunization by a diphtheria toxin synthetic oligopeptide," *Nature (London)* **289** (1981) 593–594.

[13] T. Uchida, D. M. Gill, A. M. Pappenheimer, Jr.: "Mutation in the structural gene for diphtheria toxin carried by temperate phage β," *Nature (London)* **283** (1971) 8–11.

[14] L. Greenfield, H. L. Dovey, F. C. Lawyer, D. H. Gelfand: "High level expression of diphtheria toxin peptides in Escherichia coli," *Bio/Technology* **4** (1986) 1006–1011.

[15] B. Bizzinin in R. Germanier (ed.): "Tetanus," *Bacterial Vaccines*, Academic Press, Orlando, FL 1984, pp. 37–68.

[16] M. C. Hardegree et al.: "Titration of tetanus toxoids in international units; relationship to antitoxin responses of rhesus monkeys," *6th Proc. Int. Conf. Tetanus 1981*, 1982, pp. 409–423.

[17] M. Huet: "La standardisation des vaccins tetaniques," *6th Proc. Int. Conf. Tetanus 1981*, 1982 pp. 425–433.

[18] K. W. Newell, A. Dueñas-Lehmann, D. R. Leblanc, N. Garces-Osorio: "The use of toxoid for the prevention of tetanus neonatorum; final report of a double-blind controlled field trial," *Bull. W.H.O.* **35** (1966) 863–871.

[19] N. Guerin, C. Fillastre: "Vaccin D.T.C. Polan," *6th Proc. Int. Conf. Tetanus 1981*, 1982, pp. 477–479.

[20] M. Pittman: "Pertussis toxin; the cause of the harmful effects and prolonged immunity of whooping cough," *Rev. Infect. Dis.* **1** (1979) 401–412.

[21] C. R. Manclark, J. L. Cowell in R. Germanier (ed.): "Pertussis," *Bacterial Vaccines*, Academic Press, Orlando, FL 1984, pp. 69–106.

[22]   T. Madsen: "Vaccination against whooping cough," *JAMA J. Am. Med. Assoc.* **101** (1933) 187–188.

[23]   C. L. Cody et al.: "Nature and rates of adverse reactions associated with DTP and DT immunizations in infants and children," *Pediatrics* **68** (1981) 650–660.

[24]   R. Aldersdale et al.: *Whooping Cough. Reports from the Committee on Safety of Medicines and the Joint Committee on Vaccination and Immunization*, vol. **4**, H. M. Stationary Office, London, England 1981, pp. 79–169.

[25]   M. Oda, J. L. Cowell, D. G. Burstyn, R. Manclark: "Protective activities of the filamentous hemagglutinin and the lymphocytosis-promoting factor of Bordetella pertussis in mice," *J. Infect. Dis.* **150** (1984) 823–833.

[26]   Ad Hoc Group for the study of Pertussis Vaccines: "Placebo controlled trial of two acellular pertussis vaccines in Sweden—protective efficacy and adresse events." *Lancet* **1** (1988) 955–960.

[27]   C. S. Chuttani et al.: "Controlled field trial of a high-dose oral killed typhoid vaccine in India," *Bull. W.H.O.* **55** (1977) 643–644.

[28]   M. Reitman: "Infectivity and antigenicity of streptomycin-dependent Salmonella typhosa," *J. Infect. Dis.* **117** (1967) 101–107.

[29]   R. Germanier, E. Fürer: "Isolation and characterization of gal ε mutant Ty 21 a of Salmonella typhi; a candidate strain for a live, oral typhoid vaccine," *J. Infect. Dis.* **131** (1975) 553–558.

[30]   R. H. Gilman, R. B. Hornick, W. E. Woodward, H. L. Dupont et al.: "Evaluation of a UDP-glucose-4-epimeraseless mutant of Salmonella typhi as a live oral vaccine," *J. Infect. Dis.* **136** (1977) 717–723.

[31]   M. H. Wahdan, C. Série, Y. Cerisier, S. Sallam et al.: "A controlled field trial of live Salmonella typhi Ty 21 a oral vaccine against typhoid; three year results," *J. Infect. Dis.* **145** (1982) 292–295.

[32]   World Health Organization Committee on Biological Standardization: Requirement for typhoid vaccine (live-attenuated, Ty21 a, oral). World Health Organization Technical Report, series 700; Geneva, Switzerland: 34th report, 1984, pp. 48–68.

[33]   R. Germanier in R. Germanier (ed.): "Typhoid fever," *Bacterial Vaccines*, Academic Press, Orlando, FL 1984, pp. 137–165.

[34]   M. M. Levine, C. Ferreccio, R. E. Black, R. Germanier, Chilean Typhoid Committee: "Large-scale field trial of Ty21 a live oral typhoid vaccine in enteric-coated capsule formulation," *Lancet* **ii** (1987) 1049–1052.

[35]   M. M. Levine, D. Herrington, J. R. Murphy, J. G. Morris et al.: Safety, infectivity, immunogenicity, and in vivo stability of two auxotrophic mutant strains of *Samonella typhi*, 541 Ty and 543 Ty, as live oral vaccine in humans, *J. Clin. Invest.* **79** (1987) 888–902.

[36]   I. L. Acharya, C. V. Lowe, R. Thapa, V. L. Gurubacharya et al.: "Prevention of typhoid fever in Nepal with the Vi capsular polysaccharide of *Salmonella typhi*," *N. Engl. J. Med.* **317** (1987) 1101–1104.

[37]   R. Austrian: "Of gold and pneumococci; a history of pneumococcal vaccines in South Africa," *Trans. Am. Clin. Climatol. Assoc.* **89** (1977) 141–161.

[38]   C. M. MacLeod, M. R. Krauss: "Stepwise intratype transformation of pneumococcus from R to S by way of a variant intermediate in capsular polysaccharide production," *J. Exp. Med.* **86** (1947) 439–453.

[39]   J. B. Robbins, R. Austrian, C.-J. Lee, S. C. Rastogi et al.: "Considerations for formulating the second-generation pneumococcal capsular polysaccharide vaccine with emphasis on the cross-reactive types," *J. Infect. Dis.* **148** (1983) 1136–1159.

[40] Health and Public Policy Committee, American College of Physicians: "Pneumococcal vaccine," *Ann. Intern. Med.* **104** (1986) 118–120.

[41] P. Smit, D. Oberholzer, S. Hayden-Smith, H. J. Koornhof et al.: "Protective efficacy of pneumonococcal polysaccharide vaccines," *JAMA J. Am. Med. Assoc.* **238** (1977) 2613–2616.

[42] G. Bolan, C. V. Broome, R. R. Facklam, B. D. Plikaytis et al.: "Pneumococcal vaccine efficacy in selected populations in the United States," *Ann. Intern. Med.* **104** (1986) 1–6.

[43] M. S. Simberkoff, A. P. Cross, M. Al-Ibrahim, A. L. Baltch et al.: "Efficacy of pneumococcal vaccine in high-risk patients," *N. Engl. J. Med.* **315** (1986) 1318–1327.

[44] J. A. Leech, A. Gervais, F. L. Ruben: "Efficacy of pneumococcal vaccine in severe chronic obstructive pulmonary disease," *Can. Med. Assoc. J.* **136** (1987) 361–365.

[45] C. Chu, R. Schneerson, J. B. Robbins, S. C. Rastogi: "Further studies on the immunogenicity of Haemophilus influenzae type b and pneumococcal type 6A polysaccharide-protein conjugates," *Infect. Immun.* **40** (1983) 245–256.

[46] R. Schneerson, J. R. Robbins, C. Chu, A. Sutton et al.: "Serum antibody response of juvenile and infant rhesus monkeys injected with Haemophilus influenzae type b and pneumococcus type 6A capsular polysaccharide-protein conjugates," *Infect. Immun.* **45** (1984) 582–591.

[47] R. Schneerson, J. B. Robbins, J. C. Parke, Jr., C. Bell et al.: "Quantitative and qualitative analyses of serum antibodies elicited in adults by Haemophilus influenzae type b and pneumococcus type 6A capsular polysaccharide-tetanus toxoid conjugates," *Infect. Immun.* **52** (1986) 519–528.

[48] A. D. O'Brien, M. R. Thompson, P. Gemski, B. P. Doctor et al.: "Biological properties of Shigella flexneri 2A toxin and its serological relationship to Shigella dysenteriae 1 toxin," *Infect. Immun.* **15** (1977) 796–798.

[49] A. R. Higgins, T. M. Floyd, M. A. Kader: "Studies in shigellosis. III. A controlled evaluation of a monovalent Shigella vaccine in a highly endemic environment," *Am. J. Trop. Med. Hyg.* **4** (1955) 281–288.

[50] D. M. Mel, E. J. Gangarosa, M. L. Radovanović, B. L. Arsić et al.: "Studies on vaccination against bacillary dysentery. 6. Protection of children by oral immunization with streptomycin-dependent Shigella strains," *Bull. W.H.O.* **45** (1971) 457–464.

[51] M. M. Levine, P. A. Rice, E. J. Gangarosa, G. K. Morris et al.: "An outbreak of Sonnei shigellosis in a population receiving oral attenuated shigella vaccines," *Am. J. Epidemiol.* **99** (1973) 30–36.

[52] S. B. Formal, M. M. Levine in R. Germanier (ed.): "Shigellosis," *Bacterial Vaccines*, Academic Press, Orlando, FL 1984, pp. 167–186.

[53] S. B. Formal, L. S. Baron, D. J. Kopecko, O. Washington et al.: "Construction of a potential bivalent vaccine strain: introduction of Shigella sonnei Form I antigen genes into the gal ε Salmonella typhi Ty 21a typhoid vaccine strain," *Infect. Immun.* **34** (1981) 746–750.

[54] R. E. Black, M. M. Levine, M. L. Clements, G. Losonsky et al.: "Prevention of shigellosis by a Salmonella typhi–Shigella sonnei bivalent vaccine," *J. Infect. Dis.* **155** (1987) 1260–1265.

[55] L. S. Baron, D. J. Kopecko, S. B. Formal, R. Seid et al.: "Introduction of Shigella flexneri 2a type and group antigen genes into oral typhoid vaccine strain Salmonella typhi Ty 21a," *Infect. Immun.* **55** (1987) 2797–2801.

[56] S. Sturm, K. Timmis: "Cloning of the rfb gene region of Shigella dysenteriae 1 and construction of an rfb-rfp gene cassette for the development of lipopolysaccharide-based live anti-dysentery vaccines," *Microb. Pathog.* **1** (1986) 289–297.

[57] R. A. Finkelstein: "Immunology of cholera," *Immunol.* **69** (1975) 137–195.

[58] R. A. Finkelstein: "Cholera," *CRC Crit. Rev. Microbiol.* **2** (1973) 553–623.

[59] R. A. Cash, S. I. Music, J. P. Libonati, J. P. Craig et al.: "Response of man to infection with Vibrio cholerae. II. Protection from illness afforded by previous disease and vaccine," *J. Infect. Dis.* **130** (1974) 325–333.

[60] R. A. Cash, S. I. Music, J. P. Libonati, A. R. Schwartz et al.: "Live oral cholera vaccine; evaluation of the clinical effectiveness of two strains in humans," *Infect. Immun.* **10** (1974) 762–764.

[61] W. E. Woodward, R. H. Gilman, R. B. Hornick, J. P. Libonati et al.: "Efficacy of a live oral cholera vaccine in human volunteers," *Dev. Biol. Stand.* **33** (1976) 108–112.

[62] M. M. Levine, R. E. Black, M. L. Clements, C. Lanata et al.: "Evaluation in humans of attenuated *Vibrio cholerae* El Tor Ogawa strain Texas Star-SR as a live oral vaccine, *Infect. Immun.* **43** (1984) 515–522.

[63] J. B. Kaper, M. M. Levine: "Cloned cholera enterotoxin genes in study and prevention of cholera," *Lancet* **ii** (1981) 1162–1163.

[64] J. J. Mekalanos, D. J. Suartz, G. D. N. Pearson, N. Harford et al.: "Cholera toxin genes: nucleotide sequence, deletion analysis, and vaccine development," *Nature* (London) 306 (1983) 551–557.

[65] A. D. O'Brien, M. E. Chen, R. K. Holmes et al.: Environmental and human isolates of *Vibrio cholerae* and *Vibrio parahaemolyticus* produce a *Shigella dysenteriae* 1 (Shiga)-like cytotoxin," *Lancet* **i** (1984) 77–78.

[66] J. D. Clements, J. R. Harris, M. R. Khan, B. A. Kay et al.: "Field trial of oral cholera vaccine in Bangladesh," *Lancet* **ii** (1986) 124–127.

[67] J. E. McGowan, M. W. Barnes, M. Finland: "Bacteremia at Boston City Hospital; occurrence and mortality during 12 selected years (1935–1972), with special reference to hospital-acquired cases," *J. Infect. Dis.* **132** (1975) 316–335.

[68] C. S. Bryan, K. L. Reynolds, E. R. Brenner: "Analysis of 1186 episodes of gram-negative bacteremia in non-university hospitals; the effects of antimicrobial therapy," *Rev. Infect. Dis.* **5** (1983) 629–638.

[69] R. W. Haley: *Preliminary cost-benefit analysis of hospital infection control programs (The SENIC Project)*, Proc. Internat. Workshop, Baiersbromm, Germany (1977) pp. 93–95.

[70] S. J. Cryz, Jr.: "Prospects for the prevention and control of gram-negative nosocomial infections," *Vaccine* **5** (1987) 261–265.

[71] W. R. Jarvis, J. W. White, V. P. Munn, J. L. Mosser et al.: "Nosocomial infection surveillance," 1983. Centers for Disease Control Surveillance Summaries **33**, no. 299, pp. 955–2155 (1985).

[72] M. S. Pollack, L. S. Young: "Protective activity of antibodies to exotoxin A and lipopolysaccharide at the onset of Pseudomonas aeruginosa septicemia in man," *J. Clin. Invest.* **63** (1979) 276–286.

[73] S. J. Cryz, Jr. in R. Germanier (ed.): "Pseudomonas aeruginosa infections," *Bacterial Vaccines*, Academic Press, Orlando, FL 1984, pp. 317–351.

[74] G. B. Pier, S. E. Bennett: "Structural analysis and immunogenicity of Pseudomonas aeruginosa immunotype 2 high molecular weight polysaccharide," *J. Clin. Invest.* **77** (1986) 491–495.

[75] S. J. Cryz, Jr., E. Fürer, A. S. Cross, A. Wegmann et al.: "Safety and immunogenicity of a Pseudomonas aeruginosa O-polysaccharide-toxin A conjugate vaccine in humans," *J. Clin. Invest.* **80** (1987) 51–56.

[76] S. J. Cryz, Jr., A. S. Cross, E. Fürer, N. Chariatte, et al.: "Activity of intravenous immune globulin against *Klebsiella*," *J. Lab. Clin. Med.* **108** (1986) 182–189.

[77] W. D. Welch, W. J. Martin, P. Stevens, L. S. Young: "Relative opsonic and protective activities of antibodies against Kl, O and lipid A antigens of *Escherichia coli*," *Scand. J. Infect. Dis.* **11** (1979) 291–298.

[78] B. Kaijser, S. Ahlstedt: "Protective capacity of antibodies to Escherichia coli O and K antigens," *Infect. Immun.* **17** (1977) 286–289.

[79] A. S. Cross, P. Gemski, J. C. Sadoff, F. Ørskov et al.: "The importance of the Kl capsule in invasive infections caused by *Escherichia coli*," *J. Infect. Dis.* **149** (1984) 184–193.

[80] M. Granström, I. Julander, R. Möllby: "Serological diagnosis of deep Staphylococcus aureus infections by enzyme-linked immunosorbent assay (ELISA) for staphylococcal hemolysin and teichoic acid," *Scand. J. Infect. Dis. Suppl.* **41** (1983) 132–139.

[81] W. W. Karakawa, J. M. Fournier, W. F. Vann, R. Arbeit et al.: "Method for the serological typing of the capsular polysaccharides of *Staphylococcus aureus*," *J. Clin. Microbiol.* **22** (1985) 445–447.

[82] H. K. Hochkeppel, D. G. Braun, W. Visher, A. Imm et al.: "Serotyping and electron microscopy studies of Staphylococcus aureus clinical isolates with monoclonal antibodies to capsular polysaccharide types 5 and 8," *J. Clin. Microbiol.* **25** (1987) 526–530.

[83] R. D. Arbeit, J. M. Nelles: "Capsular polysaccharide antigenemia in rats with experimental endocarditis due to *Staphylococcus aureus*," *J. Infect. Dis.* **155** (1987) 242–246.

[84] L. M. Mutharia, G. Crockford, W. C. Bogard, Jr., R. E. W. Hancock: "Monoclonal antibodies specific for Escherichia coli J5 lipopolysaccharide: cross-reaction with other gram-negative bacterial species," *Infect. Immun.* **45** (1984) 631–636.

[85] E. J. Ziegler, J. A. McCutchan, H. Douglas, A. I. Braude: "Treatment of E. coli and Klebsiella bacteremia in agranulocytopenic animals with antiserum to a UDP-GAL epimerase-deficient mutant," *J. Immunol.* **111** (1973) 433–438.

[86] E. J. Ziegler, J. A. McCutchan, J. Fierer, M. P. Glauser et al.: "Treatment of gramnegative bacteremia and shock with human antiserum to a mutant of *Escherichia coli*," *N. Engl. J. Med.* **307** (1982) 1225–1230.

[87] J. D. Baumgartner, M. P. Glauser, J. A. McCutchan, E. J. Ziegler et al.: "Prevention of gramnegative shock and death in surgical patients by antibody to endotoxin core glycolipid," *Lancet* **ii** (1985) 59–63.

[88] J. D. Band, M. E. Chamberland, T. Platt, R. E. Weaver et al.: "Trends in meningococcal disease in the United States, 1975–1980," *J. Infect. Dis.* **148** (1983) 754–758.

[89] E. C. Gotschlich in R. Germanier (ed.): "Meningococcal meningitis," *Bacterial Vaccines*, Academic Press, Orlando, FL 1984, pp. 237–255.

[90] L. Lapeyssonnie: "La meningite cerebro-spinale en Afrique," *Bull. W.H.O.* **28** (1963) 1–114.

[91] E. C. Gotschlich, I. Goldschneider, M. S. Artenstein: "Human immunity to the meningococcus. IV. Immunogenicity of group A and group C meningococcal polysaccharides in human volunteers; *J. Exp. Med.* **129** (1969) 1367–1384.

[92] M. S. Artenstein, R. Gold, J. G. Zimmerly, F. A. Wyle et al.: "Prevention of meningococcal disease by group C polysaccharide vaccine," *N. Engl. J. Med.* **282** (1970) 417–420.

[93] P. H. Mäkelä et al.: "Effect of group-A meningococcal vaccine in army recruits in Finland," Lancet **ii** (1975) 883–886.

[94] M. H. Wahdan, S. A. Sallam, M. N. Hassan, A. A. Gawad et al.: "A second controlled field trial of a serogroup A meningococcal vaccine in Alexandria," *Bull. W.H.O.* **55** (1977) 645–651.

[95] H. Peltola, P. H. Mäkelä, H. M. Käyhty, H. Jousimies et al.: "Clinical efficacy of meningococcus group A capsular polysaccharide vaccine in children three months to five years of age," *N. Engl. J. Med.* **297** (1977) 686–691.

[96] H. J. Jennings, C. Lugowski: "Immunochemistry of groups A, B and C meningococcal polysaccharide-tetanus toxoid conjugates," *J. Immunol.* **127** (1981) 1011–1018.

[97] C. E. Frasch: "Prospects for the prevention of meningococcal disease: special reference to group B," *Vaccine* **5** (1987) 3–4.

[98] W. D. Zollinger, R. E. Mandrell, J. M. Griffis in J. B. Robbins, J. C. Hill, J. C. Sadoff (eds.): "Enhancement of immunological activity by monovalent complexing of meningococcal group B polysaccharide and outer membrane protein," *Seminars in Infectious Disease*, vol. **4**, Thieme-Stratton, New York 1982, pp. 254–262.

[99] C. E. Frasch, G. Coetzee, D. M. Zahradrik, L. Y. Wang in G. K. Schoolnik (ed.): "New developments in meningococcal vaccines," *Pathogenic Neisseria*, American Society for Microbiology, Washington, D.C. 1985, pp. 633–640.

[100] F. M. Collins in R. Germanier (ed.): "Tuberculosis," *Bacterial Vaccines*, Academic Press, Orlando, FL 1984, pp. 373–418.

[101] R. G. Ferguson, A. B. Simes: "BCG vaccination of Indian infants in Saskatchewan," *Tubercle* **30** (1949) 5–11.

[102] S. V. Boyden: "The effect of previous injections of tuberculoprotein on the development of tuberculin sensitivity following B.C.G. vaccination in guinea pigs," *Br. J. Exp. Pathol.* **38** (1957) 611–617.

[103] A. Lotte, O. Wasz-Höckert, F. Lert, N. Dumitrescu et al.: "Balance of risks and balance of cost regarding tuberculosis and antituberculosis vaccination," *Dev. Biol. Stand.* **43** (1979) 111–119.

[104] R. Torriani, A. Zimmermann, A. Morell: "Die BCG-Sepsis als letale Komplikation der BCG-Impfung," *Schweiz. Med. Wochenschr.* **109** (1979) 708–713.

[105] W. Hennessen, H. Freudenstein, H. Engelhardt: "Observations on a BCG vaccine causing adverse reactions in newborns," *J. Biol. Stand.* **5** (1977) 139–146.

[106] D. B. Travers: "BCG vaccination," *Lancet* **i** (1981) 1001–1002.

[107] M. C. Lu, M. H. Lien, R. E. Becker, H. C. Heine et al.: "Genes for immunodominant protein antigens are highly homologous in Mycobacterium tuberculosis, Mycobacterium africanum, and the vaccine strain Mycobacterium bovis BCG," *Infect. Immun.* **55** (1987) 2378–2382.

[108] B. Moss, C. Flexner: "Vaccinia virus expression vectors," *Ann. Rev. Immunol.* **5** (1987) 305–324.

[109] H. Leffler, C. Svanborg-Eden: "Glycolipid receptors for uropathogenic E. coli attaching to human urinary tract epithelial cells and agglutinating human erythrocytes," *FEMS Microbiol. Lett.* **8** (1980) 127–134.

[110] K. Rauss, I. Kétyi, E. Matusovits, L. Szendrei et al.: "Specific oral prevention of infantile enteritis. III. Experiments with corpuscular vaccine," *Acta. Microbiol. Acad. Sci. Hung.* **19** (1972) 19–28.

[111] E. Fürer et al.: "Protection against colibacillosis in neonatal piglets by immunization of dams with procholeragenoid," *Infect. Immun.* **35** (1982) 887–894.

[112] L. Doberescu, C. Huygelen: "Protection of piglets against neonatal *E. coli* enteritis by immunization of the sow with a vaccine containing heat-labile enterotoxin (LT), *Zentralbl. Veterinärmed. Reihe A* **23** (1976) 79–88.

[113] R. L. Morgan, R. E. Isaacson, H. W. Micon, C. C. Brinton et al.: "Immunization of suckling pigs against enterotoxigenic Escherichia coli-induced diarrheal disease by vaccinating dams with purified 987 or K 99 pili; protection correlates with pilus homology of vaccine and challenge," *Infect. Immun.* **22** (1978) 771–777.

[114] M. M. Levine, R. E. Black, C. C. Brinton, Jr., M. L. Clements et al.: "Reactogenicity, immunogenicity and efficacy studies of *Escherichia coli* type 1 somatic pili parenteral vaccine in man, *Scand. J. Infect. Dis. Suppl.* **33** (1982) 83–95.

[115] D. G. Evans, D. J. Evans, Jr., A. R. Opekun, D. Y. Graham in A. Tagliabue, R. Rappuoli, S. E. Piazzi (eds.): "Oral whole cell vaccine protective against enterotoxigenic Escherichia coli diarrhea," *Bacterial Vaccines and Local Immunity*, Edita Da Sclavo Sp.A., Siena, Italy 1986, pp. 155–156.

[116] M. M. Levine in R. Germanier (ed.): *Escherichia coli* infections," Bacterial Vaccines, Academic Press, Orlando, FL 1984 pp. 187–235.

[117] F. A. Klipstein, R. F. Engert, R. A. Houghton: "Protection in rabbits immunized with a vaccine of Escherichia coli heat-stable toxin cross-linked to the heat-labile toxin B subunit," *Infect. Immun.* **40** (1983) 888–893.

[118] J. Holmgren: "Actions of cholera toxin and the prevention and treatment of cholera," *Nature (London)* **292** (1981) 413–417.

[119] R. A. Finkelstein, C. V. Sciortino, L. C. Rieke, M. F. Burks et al.: "Preparation of procholigenoids from Escherichia coli heat-labile enterotoxins," *Infect. Immun.* **45** (1984) 518–521.

[120] M. A. Schmidt, P. O'Hanley, G. K. Schoolinik in A. Tagliabue, R. Rappuoli, S. E. Piazzi (eds.): Synthetic peptides protect against infection in a model of experimental pyelonephritis in mice," *Bacterial Vaccines and Local Immunity*, Edita Da Sclavo Sp.A., Siena, Italy 1986, pp. 389–398.

[121] E. C. Gotschlich in R. Germanier (ed.): "Gonorrhea," *Bacterial Vaccines*, Academic Press, Orlando, FL 1984, pp. 353–371.

[122] D. S. Kellogg, Jr., I. R. Cohen, L. C. Norins, A. L. Schroeter et al.: "Neisseria gonorrhoeae. II. Colonial variation and pathogenicity during 35 months in vitro," *J. Bacteriol.* **96** (1968) 596–605.

[123] J. Swanson, S. J. Kraus, E. C. Gotschlich: "Studies on gonococcus infection. I. Pili and zones of adhesion; their relation to gonococcal growth patterns," *J. Exp. Med.* **134** (1971) 886–906.

[124] L. Greenberg, B. B. Diena, F. A. Ashton, R. Wallace et al.: "Gonococcal vaccine studies in Inuvik," *Can. J. Public Health* **65** (1974) 29–33.

[125] E. C. Tramont, J. W. Boslego: "Pilus vaccines," *Vaccine* **3** (1985) 3–10.

[126] C. C. Brinton, S. W. Wood, A. Brown, A. M. Labik et al. in: J. B. Robbins, J. Hill, J. C. Sadoff (eds.): "The development of a Neisseria pilus vaccine for gonorrhea and meningococcal meningitis," *Seminars in Infectious Disease*, vol. **4**, Bacterial Vaccines, Thieme-Stratton, Inc., New York 1982, pp. 140–159.

[127] G. K. Schoolinik, J. Y. Tai, E. C. Gotschlich: "A pilus peptide vaccine for prevention of gonorrhea," *Prog. Allergy* **33** (1983) 314–331.

[128] J. B. Rothbard, R. Fernandez, L. Wang, N. N. H. Teng et al.: "Antibodies to peptides corresponding to conserved sequence of gonococcal pilins block bacterial adhesion," *Proc. Nat. Acad. Sci. U.S.A* (2) (1985) 915–919.

[129] M. S. Blake in G. G. Jackson, H. Thomas (ed.): "Functions of the outer membrane proteins of *Neisseria gonorrhoeae*," *The Pathogenesis of Bacterial Infections*, Springer-Verlag, Berlin 1985, pp. 51–66.

[130] T. M. Buchanan, M. S. Siegel, K. C. S. Chen, W. A. Pearce in J. B. Robbins, J. Hill, J. C. Sadoff (ed.): "Development of a vaccine to prevent gonorrhea," *Seminars in Infectious Disease*, vol. **4**, Bacterial Vaccines, Thieme-Stratton, Inc., New York 1982, pp. 160–164.

[131] J. B. Robbins, R. Schneerson, M. Pittman in R. Germanier (ed.): *Haemophilus influenzae type b infections*," *Bacterial Vaccines*, Academic Press, Orlando, FL 1984, pp. 289–316.

[132] A. Sutton, R. Schneerson, S. Kendall-Morris, J. B. Robbins: "Differential complement resistance mediates virulence of Haemophilus influenzae type b," *Infect. Immun.* **35** (1982) 95–104.

[133] H. Peltola, H. Käyhty, A. Sivonen, P. H. Mäkelä: "Haemophilus influenzae type b capsular polysaccharide vaccine in children; a double-blind field study of 100 000 vaccines 3 months to 5 years of age in Finland," *Pediatrics* **60** (1977) 730–737.

[134] E. A. Mortimer, "Efficacy of Haemophilus b Polysaccharide Vaccine: an Enigma," *J. Am. Med. Assoc.* **260** (1988) 1454–1455.

[135] H. Käyhty, J. Eskola, H. Peltola, M. G. Stout et al.: "Immunogenicity in infants of a vaccine composed of Haemophilus influenzae type b capsular polysaccharide mixed with DPT or conjugated to diphtheria toxoid," *J. Infect. Dis.* **155** (1987) 100–106.

[136] A. A. Lenoir, P. D. Granoff, D. M. Granoff: "Immunogenicity of Haemophilus influenzae type b polysaccharide-Neisseria meningitidis outer membrane protein conjugate vaccine in 2- to 6-month-old infants," *Pediatrics* **80** (1987) 283–287.

[137] J. Eskola, H. Peltola, A. K. Takala, H. Käyhty et al.: "Efficacy of Haemophilus influenzae type b polysaccharide-diphtheria toxoid conjugate vaccine in infancy," *N. Engl. J. Med.* **317** (1987) 717–722.

[138] C. D. Mitchell, H. H. Balfour, Jr.: "Measles control: So near and yet so far," *Prog. Med. Virol.* **31** (1985) 1–42.

[139] L. E. Markowitz, R. H. Bernier: "Immunization of young infants with Edmonston-Zagreb measles vaccine," *Pediatr. Infect. Dis. J.* **6** (1987) 809–812.

[140] WHO Expert Committee on Biological Standardization: *WHO Tech. Rep. Ser.* **329** (1966) 52–73.

[141] WHO Expert Committee on Biological Standardization: *WHO Tech. Rep. Ser.* **673** (1982).

[142] Centers for Disease Control: "Measles – United States, 1986," *Morb. Mort. Wkly. Rep.* **36** (1987) 301–305.

[143] K. J. Bart, W. A. Orenstein, A. R. Hinman: "The virtual elimination of rubella and mumps from the United States and the use of combined measles, mumps and rubella vaccines (MMR) to eliminate measles," *Dev. Biol. Stand.* **65** (1986) 45–52.

[144] E. Norrby, G. Enders-Ruckle, V. ter Meulen: "Differences in the appearance of antibodies to structural components of measles virus after immunization with inactivated and live virus," *J. Infect. Dis.* **132** (1975) 262–269.

[145] T. M. Varsanyi, B. Morein, A. Löve, E. Norrby: "Protection against lethal measles virus infection in mice by immune-stimulating complexes containing the hemagglutinin or fusion protein," *J. Virol.* **61** (1987) 3896–3901.

[146] M. R. Hilleman: "Mumps vaccination," in: *Modern Trends in Medical Virology*, vol. **2**, Butterworths, London (1970) pp. 241–261.

[147] K. Penttinen, K. Cantell, P. Somer, A. Poikolainen: "Mumps vaccination in the Finnish defense forces," *Am. J. Epidemiol.* **88** (1968) 234–244.

[148] M. R. Hilleman, R. E. Weibel, E. B. Buynak, J. Stokes, Jr. et al.: "Live, attenuated mumps-virus vaccine; 4. Protective efficacy as measured in a field evaluation," *N. Engl. J. Med.* **276** (1967) 252–258.

[149] Centers for Disease Control: "Efficacy of mumps vaccine – Ohio," *Morb. Mort. Wkly. Rep.* **32** (1983) 391–398.

[150] F. E. Andre, J. Peetermans: "Effect of simultaneous administration of live measles vaccine on the "take rate" of live mumps vaccine," *Dev. Biol. Stand.* **65** (1986) 101–107.

[151] R. Glück, J. M. Hoskins, A. Wegmann, M. Just et al.: "Rubini, a new live attenuated mumps vaccine virus strain for human diploid cells," *Dev. Biol. Stand.* **65** (1986) 29–35.

[152] WHO Expert Committee on biological standardization: *WHO Tech. Rep. Ser.* **760** (1987).

[153] Recommendation of the Immunization Practices Advisory Committee (ACIP): "Mumps vaccine," *Morb. Mort. Wkly. Rep.* **31** (1982) 617–625.

[154] K. Penttinen, E.-P. Helle, E. Norrby: "Differences in antibody response induced by formaldehyde inactivated and live mumps vaccines," *Dev. Biol. Stand.* **43** (1979) 265–268.

[155] A. Löve, R. Rydbeck, G. Utter, C. örvell et al.: "'Monoclonal antibodies against fusion protein are protective in necrotizing mumps meningoencephalitis," *J. Virol.* **58** (1986) 220–222.

[156] F. T. Perkins: "Licensed vaccines," *Rev. Infect. Dis.* **7** (1985) Suppl. 1, 73–76.

[157] S. A. Plotkin, F. Buser: "History of RA 27/3 rubella vaccine," *Rev. Infect. Dis.* **7** (1985) Suppl. 1, 77–78.

[158] WHO Expert Committee on Biological Standardization: *WHO Tech. Rep. Ser.* **610** (1977) 54–84.

[159] WHO Expert Committee on Biological Standardization: *WHO Tech. Rep. Ser.* **658** (1981).

[160] J. A. Dudgeon: "Selective immunization; Protection of the individual," *Rev. Infect. Dis.* **7** (1985) Suppl. 1, 185–190.

[161] Recommendation of the Immunization Practices Advisory Committee (ACIP): "Rubella prevention," *Morb. Mort. Wkly. Rep.* **33** (1984) 301–318.

[162] B. Christenson, M. Böttiger, L. Heller: "Mass vaccination program aimed at eradicating measles, mumps, and rubella in Sweden; first experience," *Br. Med. J.* **287** (1983) 389–391.

[163] Centers for Disease Control: "Rubella vaccination during pregnancy – United States, 1971–1986," *Morb. Mort. Wkly. Rep.* **36** (1987) 457–461.

[164] Recommendation of the Immunization Practices Advisory Committee (ACIP): "New recommended schedule for active immunization of normal infants and children," *Morb. Mort. Wkly. Rep* **35** (1986) 577–579.

[165] A. Deforest, S. S. Long, H. W. Lischner, J. A. C. Girone et al.: "Simultaneous administration of measles-mumps-rubella vaccine with booster doses of diphtheria-tetanus-pertussis and poliovirus vaccines," *Pediatrics* **81** (1988) 237–246.

[166] H. Peltola, O. P. Heinonen: "Frequency of true adverse reactions to measles-mumps-rubella vaccine," *Lancet* **i** (1986) 939–942.

[167] A. M. Arbeter, L. Baker, S. E. Starr, S. A. Plotkin: "The combination measles, mumps, rubella and varicella vaccine in healthy children," *Dev. Biol. Stand.* **65** (1986) 89–93.

[168] L. B. Schonberger, J. Kaplan, R. Kim-Farley, M. Moore et al.: "Control of paralytic poliomyelitis in the United States," *Rev. Infect. Dis.* **6** (1984) Suppl. 2, 424–426.

[169] J. Salk: "Commentary: Poliomyelitis vaccination – choosing a wise policy," *Pediatr. Infect. Dis. J.* **6** (1987) 889–893.

[170] G. S. Wilson: *The hazards of immunization*, University of London, The Athlone Press 1967.

[171] A. B. Sabin: "Oral poliovirus vaccine: History of its development and use and current challenge to eliminate poliomyelitis from the world," *J. Infect. Dis.* **151** (1985) 420–436.

[172] A. B. Sabin: "Commentary: Is there a need for a change in poliomyelitis immunization policy?" *Pediatr. Infect. Dis. J.* **6** (1987) 887–889.

[173] M. Böttiger: "Experiences of vaccination with inactivated poliovirus vaccine in Sweden," *Dev. Biol. Stand.* **47** (1981) 227–232.

[174] K. Lapinleimu, M. Stenvik: "Experiences with polio vaccination and herd immunity in Finland," *Dev. Biol. Stand.* **47** (1981) 241–246.

[175] H. Bijkerk: "Poliomyelitis in the Netherlands," *Dev. Biol. Stand.* **47** (1981) 233–240.

[176] WHO Expert Committee on Biological Standardization: *WHO Tech. Rep. Ser.* **687** (1983) 107–174.

[177] A. M. Mcbean, J. F. Modlin: "Rationale for the sequential use of inactivated poliovirus vaccine and live attenuated poliovirus vaccine for routine poliomyelitis immunization in the United States," *Pediatr. Infect. Dis. J.* **6** (1987) 881–887.

[178] K. Lapinleimu: "Sero-immune pattern of poliomyelitis type I and III in a population vaccinated with inactivated polio vaccine," in: *Proceedings of the 11th Symposium of the European Association Against Poliomyelitis and Allied Diseases,* Rome 1966, Brussels; European Association Against Poliomyelitis, 1967, pp. 119–25.

[179] A. L. van Wezel, G. van Steenis, Ch. A. Hannik, H. Cohen: "New approach to the production of concentrated and purified inactivated polio and rabies tissue culture vaccines," *Dev. Biol. Stand.* **41** (1978) 159–168.

[180] S. E. Robertson, H. P. Traverso, J. A. Drucker, E. Z. Rovira et al.: "Clinical efficacy of a new, enhanced-potency, inactivated poliovirus vaccine," *Lancet* **i** (1988) 897–899.

[181] B. J. Montagnon, B. Fanget, A. J. Nicolas: "The large-scale cultivation of vero cells in microcarrier culture for virus vaccine production. Preliminary results for killed poliovirus vaccine," *Dev. Biol. Stand.* **47** (1981) 55–64.

[182] WHO Expert Committee on Biological Standardization: *WHO Tech. Rep. Ser.* **745** (1987).

[183] Prevention of liver cancer: *WHO Tech. Rep. Ser.* **691** (1983).

[184] A. A. McLean: "Development of vaccines against Hepatitis A and Hepatitis B," *Rev. Infect. Dis.* **8** (1986) 591–598.

[185] W. Szmuness, C. E. Stevens, E. J. Harley, E. A. Zang et al.: "Hepatitis B vaccine; Demonstration of efficacy in a controlled clinical trial in a high-risk population in the United States," *N. Engl. J. Med.* **303** (1980) 833–841.

[186] WHO Expert Committee on Biological Standardization: *WHO Tech. Rep. Ser.* **725** (1985) 102–133.

[187] Recommendation of the Immunization Practices Advisory Committee (ACIP): "Recommendations for protection against viral hepatitis," *Morb. Mort. Wkly. Rep.* **34** (1985) 313–335.

[188] Recommendation of the Immunization Practices Advisory Committee (ACIP): "Update on Hepatitis B prevention," *Morb. Mort. Wkly. Rep.* **36** (1987) 353–366.

[189] G. Eder, J. L. McDonel, F. Dorner: "Hepatitis B vaccine," in: *Progress in Liver Diseases,* vol. **8,** Grune and Stratton, New York 1986, pp. 367–394.

[190] A. Zuckerman: "Tomorrow's hepatitis B vaccines," *Vaccine* **5** (1987) 165–167.

[191] D. N. Standring, W. J. Rutter: "The molecular analysis of hepatitis B virus," in: *Progress in Liver Diseases,* vol. **8,** Grune and Stratton, New York 1986, pp. 311–333.

[192] WHO Expert Committee on Rabies: *WHO Tech. Rep. Ser.* **709** (1984).

[193] Centers for Disease Control: "Bat-rabies – Europe," *Morb. Mort. Wkly. Rep.* **35** (1986) 430–432.

[194] P. Sureau: "Rabies vaccine production in animal cell cultures," *Adv. Biochem. Eng. Biotechnol.* **34** (1987) 111–128.

[195] B. D. Perry: "Rabies," *Vet. Clin. North. Am. Small Anim. Pract.* **17** (1987) 73–89.

[196] H. Wilde, P. Chomchey, S. Prakongsri, P. Punyaratabandhu: "Safety of equine rabies immune globulin," *Lancet* **ii** (1987) 1275.

[197] Centers for Disease Control: "Human rabies despite treatment with rabies immune globulin and human diploid cell rabies vaccine – Thailand," *Morb. Mort. Wkly. Rep.* **36** (1987) 759–765.

[198] Centers for Disease Control: "Systemic allergic reactions following immunization with human diploid cell rabies vaccine," *Morb. Mort. Wkly. Rep.* **33** (1984) 185–187.

[199] M. C. Swanson, E. Rosanoff, M. Gurwith, M. Deitch et al.: "IgE and IgG antibodies to β-propiolactone and human serum albumin associated with urticarial reactions to rabies vaccine," *J. Infect. Dis.* **155** (1987) 909–913.

[200] M. Granström, M. Eriksson, G. Edevåg: "A sandwich ELISA for bovine serum in viral vaccines," *J. Biol. Stand.* **15** (1987) 193–197.

[201] B. Morein, K. Simons: "Subunit vaccines against enveloped viruses; virosomes, micelles and other protein complexes," *Vaccine* **3** (1985) 83–93.

[202] C. H. Stuart-Harris, G. C. Schild: "Influenza," *The Viruses and Disease*, E. Arnoldt Ltd, London 1976.

[203] P. F. Wright, D. T. Karzon: "Live Attenuated Influenza Vaccines," *Prog. Med. Virol.* **34** (1987) 70–88.

[204] WHO Expert Committee on Biological Standardization: *WHO Techn. Rep. Ser.* **638** (1979).

[205] Recommendations of the Immunization Practices Advisory Committee (ACIP): "Prevention and Control of Influenza," *Morb. Mort. Wkly. Rep.* **36** (1987) 373–387.

[206] P. A. Gross, A. L. Gould, A. E. Brown: "Effect of cancer chemotherapy on the immune response to influenza virus vaccine: Review of published studies," *Rev. Infect. Dis.* **7** (1985) 613–618.

[207] M. A. Strassburg, S. Greenland, F. J. Sorvillo, L. E. Lieb et al.: "Influenza in the elderly: report of an outbreak and a review of vaccine effectiveness reports," *Vaccine* **4** (1986) 38–44.

[208] J. M. Ostrove, G. Inchauspe: "The biology of varicella-zoster virus" in: S. E. Straus "Varicella-zoster virus infections: biology, natural history, treatment, and prevention," *Ann. Intern. Med.* **108** (1988) 221–237.

[209] M. Takahashi: "Clinical overview of varicella vaccine; Development and early studies," *Pediatrics* **78** (1986) Suppl., 736–741.

[210] R. E. Weibel, B. J. Neff, B. J. Kuter, H. A. Guess et al.: "Live attenuated varicella virus vaccine. Efficacy trial in healthy children," *N. Engl. J. Med.* **310** (1984) 1409–1415.

[211] A. A. Gershon, S. P. Steinberg, L. Gelb, and The National Institute of Allergy and Infectious Diseases Varicella Vaccine Collaborative Study Group: "Live attenuated varicella vaccine use in immunocompromised children and adults," *Pediatrics* **78** (1986) Suppl., 757–762.

[212] Arthropod-borne and rodent-borne viral diseases: *WHO Tech. Rep. Ser.* **719** (1985).

[213] Viral haemorrhagic fevers: *WHO Tech. Rep. Ser.* **721** (1985).

[214] K. C. Smithburn, C. Durieux, R. Koerber, H. A. Penna et al.: *Yellow Fever Vaccination*, World Health Organization, Geneva 1956.

[215] J. C. Roche, A. Jouan, B. Brisou, R. Rodhain et al.: "Comparative clinical study of a new 17 D thermostable yellow fever vaccine," *Vaccine* **4** (1986) Suppl., 163–165.

[216] WHO Expert Committee on Biological Standardization: *WHO Tech. Rep. Ser.* **594** (1976) 23–49.

[217] A. A. Smorodintsev, A. V. Dubov, V. I. Ilyenko, V. G. Platonov: "A new approach to development of live vaccine against tick-borne encephalitis," *J. Hyg.* **67** (1969) 13–20.

[218] C. Kuntz, F. X. Heinz, H. Hofmann: "Immunogenicity and reactogenicity of a highly purified vaccine against tick-borne encephalitis," *J. Med. Virol.* **6** (1980) 103–109.

[219] A. Oya in H. Fukumi (ed.): "Japanese encephalitis vaccine," *The vaccination. Theory and practice*, International Medical Foundation of Japan (1975) pp. 69–82.

[220] WHO Expert Committee on Biological Standardization: *WHO Tech. Rep. Ser.* **323** (1966) 56–71.

[221] Recommendation of the Immunization Practices Advisory Committee (ACIP): *Smallpox Vaccine Morb. Mort. Wkly. Rep.* **34** (1985) 341–342.
[222] R. Randall, L. N. Binn, V. R. Harrison: "Immunization against Rift Valley fever virus," *J. Immunol.* **93** (1964) 293–299.
[223] B. Niklasson: "Rift Valley fever virus vaccine trial; Study of side effects in humans," *Scand. J. Infect. Dis.* **14** (1982) 105–109.
[224] P. M. Kaye, *Parasitol. Today* **3** (1987) 293–299.
[225] M. D. Rickard, M. J. Howell in E. E. R. Taylor, J. R. Baker (ed.): *In Vitro Methods for Parasite Cultivation*, Academic Press, London 1987, pp. 407–451.
[226] G. Piekarski: *Medizinische Parasitologie*, Springer Verlag, Berlin 1987.
[227] V. Houba in V. Houba (ed.): *Immunological Investigation of Tropical Parasitic Diseases*, Churchill Livingstone, Edinburgh 1980, pp. 130–147.
[228] P. Hagan, P. J. Moore, A. B. Adjukiewicz, B. M. Greenwood et al.: *Parasite Immunol.* **7** (1985) 617–624.
[229] A. E. Butterworth, M. Capron, J. S. Cordingley, P. R. Dalton, *Trans. R. Soc. Trop. Med. Hyg.* **79** (1985) 393–408.
[230] G. Gazzinelli, J. R. Lambertucci, N. Katz, R. S. Rocha, *J. Immunol.* **135** (1985) 2121–2127.
[231] C. W. Todd, R. W. Goodgame, D. Colley, *J. Immunol.* **122** (1979) 1440–1446.
[232] R. F. Sturrock, R. Kimani, B. I. Cottrell, A. E. Butterworth, *Trans. R. Soc. Trop. Med. Hyg.* **77** (1983) 363–371.
[233] A. E. Butterworth, R. F. Sturrock, V. Houba, A. A. F. Mahmoud, Nature (*London*) **256** (1975) 727–729.
[234] A. E. Butterworth, *Adv. Parasitol.* **23** (1984) 143–235.
[235] S. R. Smithers, R. J. Terry, *Adv. Parasitol.* **14** (1976) 399–422.
[236] A. Capron, *Fortschr. Zool.* **27** (1982) 259–264.
[237] A. E. Butterworth, P. Hagan, *Parasit. Today* **3** (1987) 11–16.
[238] A. J. G. Simson, D. Cioli, *Parasit. Today* **3** (1987) 26–28.
[239] C. Dissous, I. M. Crzych, A. Capron, *J. Immunol.* **129** (1982) 2232–2234.
[240] Q. D. Bickle, B. I. Andrews, M. G. Taylor, *Parasit Immunol.* **8** (1986) 95–107.
[241] C. Kelly, A. I. G. Simpson, E. Fox, S. M. Phillips et al.: *Parasit Immunol.* **8** (1986) 193–198.
[242] J. M. Balloul, P. Sondermeyer, D. Dreyer, M. Capron, *Nature (London)* **326** (1987) 149–153.
[243] D. Wakelin: *Immunity to Parasites*, Edward Arnold Ltd., London 1984.
[244] S. Friedmann, *Gen Eng. News* (1985) 24–25.
[245] V. K. Vinayak, N. K. Gupta, A. K. Chopra, G. L. Sharma et al., *Parasitology* **82** (1981) 375–382.
[246] P. Ambroise-Thomas in V. Houba (ed.): *Immunological Investigation of Tropical Parasitic Diseases*, Churchill Livingstone, Edinburgh 1980, pp. 84–103.
[247] B. M. Ogilvie, M. J. Worms in S. Cohen, E. H. Sadun (eds.): *Immunology of Parasitic Infections*, Blackwell Scientific Publ., Oxford 1976, 397–402.
[248] M. M. Wong, H. I. Fredericks, C. P. Ramachandran, *Bull. W.H.O.* **40** (1969) 493–501.
[249] C. P. Ramachandran, *Southeast Asian J. Trop. Med. Public Health* **1** (1970) 78–92.
[250] A. Haque, A. Capron in T. W. Pearson (ed.): *Parasite Antigens*, Marcel Dekker Inc., New York 1986, pp. 317–402.
[251] R. M. Maizels, F. Partono, Sri Oemijati, D. A. Denham et al., *Parasitology* **87** (1983) 249–263.
[252] M. D. Rickard in C. Arme, P. W. Pappas (eds.): *Biology of the Eucestoda*, Academic Press, London 1983, pp. 539–579.
[253] M. D. Rickard, F. F. Williams, *Adv. Parasitol.* **21** (1982) 229–296.

[254] R. P. Herd, R. I. Chappel, D. Biddell, *Int. J. Parasitol.* **5** (1975) 395–399.
[255] D. J. Jenkins, M. D. Rickard, *Aust. Vet. J.* **63** (1986) 40–42.
[256] R. S. Nussenzweig, J. Vanderberg, H. Most, C. Orton, *Nature (London)* **216** (1967) 160–162.
[257] W. A. Siddiqui, *Science* **197** (1977) 388–389.
[258] G. H. Mitchell, W. H. G. Richards, G. A. Butcher, S. Cohen, *Lancet (i)* (1977) 1335–1358.
[259] W. Trager, J. B. Jensen, *Science* **193** (1976) 673–675.
[260] D. Mazier, R. L. Beaudoin, S. Mellouk, P. Druilhe et al., *Science* **227** (1985) 440–442.
[261] C. Guerin-Marchand, P. Druilhe, B. Galey, A. Londono et al., *Nature (London)* **329** (1987) 164–167.
[262] A. H. Cochrane in J. P. Kreier (ed.): *Malaria, Immunology and Immunization* vol. **3,** Chap. 4, Academic-Press, New-York, 1980.
[263] D. F. Clyde, *Am. J. Trop. Med. Hyg.* **24** (1975) 397–401.
[264] E. H Nardin, R. S. Nussenzweig, I. A. McGregor, J. H. Bryan, *Science* **206** (1979) 597–599.
[265] G. Del Giudice, H. D. Engers, C. Tougne, S. S. Biro et al., *Am. J. Trop. Med. Hyg.* **36** (1987) 203–212.
[266] D. H. Chen, R. E. Tigelaar, F. I. Weinbaum, *J. Immunol.* **118** (1977) 1322–1327.
[267] A. Ferreira, L. Schofield, V. Enea, H. Schellekins et al., *Science* **232** (1986) 881–884.
[268] P. Potocnjak, N. Yoshida, R. S. Nussenzweig, V. Nussenzweig, *J. Exp. Med.* **151** (1980) 1504–1513.
[269] D. Mazier, S. Mellouk, R. L. Beaudoin, B. Texier et al., *Science* **231** (1986) 156–159.
[270] J. B. Dame, J. L. Williams, T. F. McCutchan, J. L. Weber et al., *Science* **225** (1984) 593–599.
[271] V. Enea, J. Ellis, F. Zavala, D. E. Arnot et al., *Science* **225** (1984) 628–630.
[272] D. J. Kemp, R. L. Coppel, R. F. Anders, *Ann. Rev. Microbiol.* **41** (1987) 181–208.
[273] V. F. De la Cruz, A. A. Lal, T. F. McCutchan, *J. Biol. Chem.* **262** (1987) 11935–11941.
[274] M. J. Lockyer, R. T. Schwarz, *Mol. Biochem. Parasitol.* **22** (1987) 101–107.
[275] G. Del Giudice, A. S. Verdini, M. Pinori, J. P. Verhave et al., *J. Clin. Microbiol.* **25** (1987) 91–96.
[276] S. L. Hoffman, R. Wistar, Jr., W. R. Ballou, M. R. Hollingdale et al., *New-Engl. J. Med.* **315** (1986) 601–603.
[277] D. A. Herrington, D. F. Clyde, G. Losonsky, M. Cortesia et al., *Nature (London)* **328** (1987) 257–259.
[278] W. R. Ballou, S. L. Hoffman, J. A. Sherwood, M. R. Hollingdale et al., *Lancet* **I** (1987) 1277–1279.
[279] G. Del Giudice, J. A. Cooper, J. Merino, A. S. Verdini et al., *J. Immunol.* **137** (1986) 2952–2955.
[280] M. F. Good, J. A. Berzofsky, W. L. Maloy, Y. Hayashi et al., *J. Exp. Med.* **164** (1986) 655–660.
[281] A. R. Togna, G. Del Giudice, A. S. Verdini, F. Bonetti et al., *J. Immunol.* **137** (1986) 2956–2960.
[282] M. F. Good, D. Pombo, I. A. Quakyi, E. M. Riley et al., *Proc. Natl. Acad. Sci. USA* **85** (1988) 1199–1203.
[283] L. H. Miller, S. J. Mason, D. F. Clyde, M. H. McGinis, *New-Eng. J. Med.* **295** (1976) 302–304.
[284] M. J. Miller, *Trans. Roy. Soc. Trop. Med. Hyg.* **52** (1958) 152–158.
[285] S. Cohen, I. A. McGregor, S. C. Carrington, *Nature (London)* **192** (1961) 733–735.
[286] A. Celada, A. Cruchau, L. H. Perrin, *Clin. Exp. Immunol.* **47** (1982) 635–641.
[287] P. H. David, *Proc. Natl. Acad. Sci. USA* **80** (1983) 5075–5081.
[288] L. H. Miller, R. J. Howard, R. Carter, M. F. Good et al., *Science* **234** (1986) 1349–1356.
[289] L. H. Perrin, E. Ramirez, L. Er-Hsiang, P.-H. Lambert; *Clin. Exp. Immunol.* **41** (1980) 91–106.

[290] A. A. Holder, R. R. Freeman, *J. Exp. Med.* **156** (1982) 1528–1538.
[291] J. S. McBride, D. Walliker, G. Morgan, *Science* **217** (1982) 254–256.
[292] A. A. Holder, M. J. Lockyer, K. G. Odink, J. S. Sandhu et al., *Nature (London)* **317** (1985) 270–273.
[293] K. G. Oding, M. J. Lockyer, S. C. Nicholls, Y. Hillman et al., *FEBS Lett.* **173** (1984) 108–112.
[294] M. Mackay, M. Goman, N. Bone, J. E. Hyde et al., *EMBO J.* **4** (1985) 3823–3829.
[295] L. H. Perrin, B. Merkli, M. Loche, C. Chizzolini et al., *J. Exp. Med.* **160** (1984) 441–447.
[296] R. Hall, J. E. Hyde, M. Goman, D. L. Simmons et al., *Nature (London)* **311** (1984) 379–382.
[297] W. A. Siddiqui, L. Q. Tam, K. J. Kramer, G. S. N Hui et al., *Proc. Natl. Acad. Sci. USA* **84** (1987) 3014–3018.
[298] A. A. Holder, R. R. Freeman, *Nature (London)* **294** (1981) 361–363.
[299] M. E. Patarroyo, P. Romero, M. L. Torres, P. Clavijo et al., *Nature (London)* **328** (1985) 629–631.
[300] A. Cheung, J. Leban, A. R. Shaw, B. Merkli et al., *Proc. Natl. Acad. Sci. USA* **83** (1986) 8323–8332.
[301] L. H. Perrin, R. Dayal, *Immunol. Rev.* **61** (1982) 245–268.
[302] R. R. Freeman, A. J. Trejosiewicz, G. A. Cross, *Nature (London)* **284** (1980) 366–369.
[303] L. H. Perrin, E. Ramirez, P.-H. Lambert, P. A. Miescher, *Nature (London)* **289** (1981) 301–303.
[304] C. Braun-Breton, M. Jendoubi, E. Brunet, L. Perrin et al., *Mol. Bioch. Parasitol.* **20** (1986) 33–38.
[305] C. Braun-Breton, T. L. Rosenberg, L. Pereira Da Silva, *Nature (London)* **332** (1988) 457–459.
[306] U. Certa, P. Ghersa, H. Dobeli, H. Matile et al., *Science* **240** (1988) 1036–1038.
[307] P. Dubois, J. P. Dedet, T. Fandeur, C. Roussilhon et al., *Proc. Natl. Acad. Sci. USA* **81** (1984) 229–335.
[308] L. H. Perrin, B. Merkli, M. S. Gabra, J. Stocker et al., *J. Clin. Invest.* **75** (1985) 1718–1725.
[309] J. F. Dubremetz, P. Delplace, B. Fortier, G. Tronchin et al., *Mol. Biochem. Parasitol.* **27** (1988) 135–142.
[310] R. L. Coppel, J. G. Culvenor, A. E. Bianco, P. E. Crewther et al., *Mol. Biochem. Parasitol.* **25** (1987) 73–81.
[311] H. Perlamnn, K. Berzins, M. Wahlgren, J. Carlsson et al., *J. Exp. Med.* **159** (1984) 1686–1693.
[312] G. V. Brown, I. G. Culvenor, P. E. Crewther, A. E. Bianco et al., *J. Exp. Med.* **162** (1985) 774–779.
[313] B. Wahlin, M. Wahlgren, H. Perlamnn, H. Berzins et al., *Proc. Natl. Acad. Sci. USA* **81** (1984) 7912–7916.
[314] J. M. Favarolo, R. L. Coppel, L. M. Corcoran, S. J. Foote et al., *Nucl. Acid. Res.* **14** (1986) 8265–8277.
[315] W. E. Collins, R. F. Anders, M. Pappaioanou, G. H. Campbell et al., *Nature (London)* **323** (1986) 259–262.
[316] S. A. Luse, L. H. Miller, *Am. J. Trop. Med. Hyg.* **20** (1971) 655–670.
[317] T. Triglia, H. D. Stahl, P. E. Crewther, D. Scanlon et al., *EMBO J.* **6** (1987) 1413–1419.
[318] M. Koenen, A. Scherf, O. Mercereau, G. Langsley et al., *Nature (London)* **311** (1984) 382–385.
[319] I. J. Udeinya, L. H. Miller, I. A. McGregor, J. B. Jensen, *Nature (London)* **231** (1983) 429–431.
[320] K. Marsh, R. J. Howard, *Science* **231** (1986) 150–153.
[321] S. Pollack, J. Fleming, *Brit. J. Haematol.* **58** (1984) 289–294.
[322] M. E. Perkins, *J. Exp. Med.* **160** (1984) 788–797.
[323] D. Camus, T. J. Hadley, *Science* **230** (1985) 553–557.

[324] P. Delplace, B. Fortier, G. Tronchin, J. F. Dubremetz et al., *Mol. Biochem. Parasitol.* **23** (1987) 193–202.
[325] M. E. Patarroyo, R. Amador, P. Lavijo, A. Moreno et al., *Nature (London)* **332** (1988) 158–161.
[326] R. W. Gwadz, *Science* **193** (1976) 1150–1153.
[327] K. N. Mendis, G. A. T. Targett, *Nature (London)* **277** (1979) 289–391.
[328] R. W. Gwadz, L. C. Koontz, *Infect. Immun.* **44** (1984) 137–145
[329] N. Kumar, R. Carter, *Mol. Biochem. Parasitol.* **13** (1984) 333–341.
[330] K. Mendis, P. Udayama, R. Carter, P. H. David, *J. Cell Biochem. Suppl.* **10 A** (1986) 149–156.
[331] A. N. Vermuelen, T. Ponnudurai, P. J. A. Beckers, J. P. Verhave et al., *J. Exp. Med.* **162** (1985) 1460–1476.
[332] J. Rener, P. M. Graves, R. Carter, J. L. Williams et al., *J. Exp. Med.* **158** (1983) 976–980.
[333] P. M. Graves, R. Carter, T. R. Burkot, J. Rener et al., *Infect. Immun.* **48** (1985) 611–619.
[334] E. R. Stiehm, E. Ashida, K. S. Kim, D. J. Winston et al., *Ann. Intern. Med.* **107** (1987) 367–82.
[335] E. R. Stiehm, *Pediatrics* **63** (1979) 301–19.
[336] J. A. Finlayson, in C. S. F. Easmon, J. Jeljaszewics (eds.): *Medical Microbiology*, vol. **1**, Academic Press, London 1982, pp. 129–82.
[337] R. Heffron: *Pneumonia, With Special Reference to Pneumococcus Lobar Pneumonia*, The Commonwealth Fund, New York 1939.
[338] R. H. Rousell, M. S. Collins, M. B. Dobkin, R. E. Louie et al., *Am. J. Med. Suppl.* **76** (1984) 40–5.
[339] H. E. Alexander, *Am. J. Dis. Child.* **66** (1943) 172–87.
[340] H. E. Alexander, *Am. J. Dis. Child.* **66** (1943) 160–71.
[341] E. Merler, F. S. Rosen, *N. Engl. J. Med.* **275** (1966) 480–86;536–542.
[342] C. A. Janeway, F. S. Rosen, *N. Engl. J. med.* **275** (1966) 826–31.
[343] P. A. M. Gross, D. Gitlin, C. A. Janeway, *N. Engl. J. Med.* **260** (1959) 170–78.
[344] J. Stokes, E. P. Maris, S. S. Gellis, *J. Clin. Invest.* **23** (1944) 531–40.
[345] J. Stokes, J. R. Neefe, *JAMA J. Am. Med. Assoc.* **127** (1945) 531–40.
[346] W. M. Hammon, L. L. Coriell, J. Stokes et al., *JAMA J. Am. Med. Assoc.* **150** (1950) 739–49;750–56.
[347] C. A. Janeway, L. Apt, D. Gitlin, *Trans. Assoc. Am. Physician* **66** (1953) 200–202.
[348] P. A. M. Gross, D. Gitlin, C. A. Janeway, *N. Engl. J. Med.* **260** (1959) 121–25.
[349] R. G. Hamilton, *Clin. Chem. (Winston Salem N.C.)* **33** (1987) 1707–25.
[350] G. R. Siber, P. H. Schur, A. C. Weitzman, G. Schiffman, *N. Engl. J. Med.* **303** (1980) 178–82.
[351] C. M. Roifman, H. M. Lederman, S. Lavi, L. D. Stein et al., *Am. J. Med.* **79** (1985) 171–4.
[352] S. Barandun, H. Isliker, *Vox Sang.* **51** (1986) 157–60.
[353] A. Hassig, *Vox Sang.* **51** (1986) 10–17.
[354] D. D. Schroeder, M. L. Dumas, *Am. J. Med. Suppl.* **76** (1984) 33–9.
[355] C. A. Janeway, E. Merler, F. S. Rosen, S. Salmon et al., *N. Engl. J. Med.* **278** (1968) 919–23.
[356] J. Passwell, F. S. Rosen, E. Merler in B. Alving, J. Finlayson (eds.): *Immunoglobulins; characteristics and uses of intravenous preparations*, U.S.Dept. H.H.S., Washington 1979, pp. 139–42.
[357] H. D. Ochs, S. H. Fischer, R. J. Wedgwood, *J. Clin. Immunol Suppl.* **2** (1982) 22 S–29 S.
[358] C. Cunningham-Rundles, F. P. Siegal, E. M. Smithwick, A. Lion-Boule et al., *Ann. Intern. Med.* **101** (1984) 435–9.
[359] M. S. Collins, J. H. Dorsey, *J. Infect. Dis.* **151** (1985) 1171–3.
[360] J. R. Schreiber, V. A. Barrus, G. R. Siber, *Infect. Immun.* **47** (1985) 142–48.

[361] H. D. Ochs et al., *Lancet* **2** (1980) 1158–9.
[362] *Medical Letter* **24** (1982) 81–82.
[363] R. H. Johnson, R. J. Ellis, *Ann. Intern. Med.* **81** (1974) 61–67.
[364] C. Cunningham-Rundles, L. A. Hanson, W. H. Hitzig, W. Knapp et al., *Bull. W.H.O.* **60** (1982) 43–47.
[365] S. Krugman, *N. Engl. J. Med.* **269** (1963) 195–201.
[366] E. B. Grosman, S. G. Stewart, J. Stokes, *JAMA J. Am. Med. Assoc.* **129** (1945) 991–94.
[367] Medical Research Council Workingparty, *Lancet* **1** (1969) 163–8.
[368] C. L. S. Leen, P L. Yap, D. B. L. McClelland, *Vox Sang.* **51** (1986) 278–86.
[369] *Lancet* **1** (1983) 105–6.
[370] G. Chirico, G. Rondini, A. Plebani, A. Chiara et al., *J. Pediatr.* **110** (1987) 437–42.
[371] J. Amer, E. Ott, F. A. Ibbott, D. O'Brien et al., *Pediatrics* **32** (1963) 4–9.
[372] A. A. Hertler, S. C. Ross, *J. Clin. Lab. Immunol.* **21** (1986) 177–81.
[373] S. Baron, E. V. Barnet, R. S. Goldsmith, S. Silbergeld et al., *Am. J. Hyg.* **79** (1964) 186–95.
[374] N. A. Kefalides, J. A. Arana, A. Bazan, M. Bocanegra et al., *N. Engl. J. Med.* **267** (1962) 317–24.
[375] H. H. Stone, C. D. Graber, J. D. Martin, L. Kolb, *Surgery St. Louis* **58** (1965) 810–14.
[376] R. J. Jones, E. A. Roe, J. L. Gupta, *Lancet* **2** (1980) 1263–5.
[377] R. L. Wasserman, *Pediatric Infectious Diseases* **5** (1986) 620–21.
[378] J. A. Steen, *Arch. Pediatr.* **77** (1960) 291–4.
[379] A. P. Schless, G. S. Harell, *Amer. J. Med.* **44** (1968) 325–329.
[380] M. Santosham, R. Reid, D. M. Ambrusino, M. C. Wolff et al., *Engl. J. Med.* **317** (1987) 923–9.
[381] S. M. Rosenthal, R. C. Millican, J. Rust, *Proc. Soc. Exp. Biol. Med.* **94** (1957) 214–217.
[382] M. W. Fisher, *Antibiot. Chemother. (Washington D.C.)* **7** (1957) 315–21.
[383] M. W. Fisher, M. C. Manning, *Antibiot. Chemother. (Washington D.C.)* **8** (1958) 29–31.
[384] B. A. Waisbren, *Antibiot. Chemother. (Washington D.C.)* **7** (1957) 322–33.
[385] B. A. Waisbren, D. Lepley, *Arch. Intern. Med.* **109** (1962) 712–6.
[386] G. P. Bodey, B. A. Nies, N. R. Mohberg, E. J. Freireich, *JAMA J. Am. Med. Assoc.* **190** (1964) 1099–1102.
[387] K. C. Finkel, J. C. Haworth, *Pediatrics* **25** (1960) 798–806.
[388] D. R. Snydman, B. G. Werner, B. Heinzer-Lacey, V. P. Berardi et al., *N. Engl. J. Med.* **317** (1987) 1049–54.
[389] R. M. Condie, R. J. O'Reilly, *Am. J. Med.* **76** (Suppl. 30 March 1984) 134–41.
[390] A. Hagenbeek, H. G. J. Brummelhuis, A. Donkers, A. M. Dumas et al., *J. Infect. Dis.* **155** (1987) 897–902.
[391] D. J. Winston, W. G. Ho, C.-H. Lin, M. D. Budinger et al., *Am. J. Med.* **76** (Suppl. 30 March 1984) 128–33.
[392] R. A. Bowden, M. Sayers, N. Flournoy, B. Newton et al., *N. Engl. J. Med.* **314** (1986) 1006–1010.
[393] E. C. Reed et al., *J. Infect. Dis.* **156** (1987) 641–44.
[394] K. Erlendsson, T. Swartz, J. M. Dwyer, *N. Engl. J. Med.* **312** (1985) 351–3.
[395] P. J. Mease, H. D. Ochs, R. J. Wedgood, *N. Engl. J. Med.* **304** (1981) 1278–81.
[396] J. M. Crennan, R. E. Van Scoy, C. H. McKenna, T. F. Smith, *Am. J. Med.* **81** (1986) 35–42.
[397] K. S. Erlich, J. Mills, *Rev. Infect. Dis.* **8** (1986) S 439–45.
[398] D. Ambrosino, J. R. Schreiber, R. S. Daum, G. R. Siber, *Infect. Immun.* **39** (1983) 709–14.
[399] T. E. Harper, R. D. Christensen, G. Rothstein, *Pediatr. Res.* **22** (1987) 455–60.
[400] S. J. Cryz, A. S. Cross, E. Furer, N. Chariatte et al., *J. Lab. Clin. Med.* **108** (1986) 182–89.

[401] I. A. Holder, J. G. Naglich, *Am. J. Med.* **76** (1984) 161–67.
[402] J. E. Pennington, G. J. Small, *J. Infect. Dis.* **155** (1987) 973–78.
[403] V. G. Hemming, W. T. London, G. W. Fischer, B. L. Curfman et al., *J. Infect. Dis.* **156** (1987) 655–658.
[404] K. N. Haque, M. H. Zaidi, S. K. Haque, H. Bahakim et al., *Pediatric Infectious Diseases* **5** (1986) 622–25.
[405] D. Sidiropoulos, U. Herrmann, A. Morell, G. von Muralt et al., *J. Pediatr. St. Louis* **109** (1986) 505–508.
[406] W. G. Ho, D. J. Winston, K. Bartoni, R. E. Champlin et al., Abstr. 456, 1983 Interscience Confer. Infect. Dis. Chemother., Las Vegas, Nevada.
[407] C. Bunch, H. Chapel, K. R. Rai, R. P. Gale, *Blood* **70** (1987) 224 a (Abstr); *N. Engl. J. Med.* **319** (1988) 902–907.
[408] H. D. Ochs, *Pediatr. Infect. Dis.* **6** (1987) 509–11.
[409] C. C. Wood, J. G. McNamara, D. F. Schwarz, W. W. Merrill et al., *Pediatr. Infect. Dis.* **6** (1987) 564–66.
[410] B. A. Silverman, A. Rubinstein, *Am. J. Med.* **78** (1985) 728–36.
[411] D. Sidiropoulos, U. Boehme, G. von Muralt, A. Morell et al., *Pediatr. Infect. Dis.* **5** (1986) 193–94.
[412] S. Barandun, P. Imbach, A. Morrell, H. P. Wagner in U. E. Nydegger (ed.): *Immunohemotherapy. A guide to immunoglobulin prophylaxis and therapy*, Academic Press, New York 1981, pp. 275–82.
[413] P. Imbach, H. P. Wagner, W. Berchtold, G. Gaedicke et al., *Lancet* **2** (1985) 464–68.
[414] J. Fehr, V. Hoffmann, U. Kappeler, *N. Engl. J. Med.* **306** (1982) 1254–58.
[415] S. Rosthoj, G. K. Steffensen, T. K. Guld, *Acta Paediatr. Scand.* **76** (1987) 631–5.
[416] R. Bohm, C. Hofstaetter, R. C. Briel, *Blut* **48** (1984) 469–70.
[417] F. E. Leickly, R. H. Buckley, *Am. J. Med.* **82** (1987) 159–62.
[418] G. W. Richmond, I. Ray, A. Korenblitt, *J. Pediatr. (St. Louis)* **110** (1987) 917–9.
[419] M. W. Hilgartner, J. Bussel, *Am. J. Med.* **83** (1987) 35–9.
[420] T. Becker, S. Panzer, D. Maas, V. Kiefel et al., *Br. J. Haematol.* **61** (1985) 149–55.
[421] F. Dammacco, G. Iodice, N. Campobasso, *Br. J. Haematol.* **62** (1986) 125–35.
[422] A. C. Newland, J. G. Treleaven, R. M. Minchinton, A. H. Waters, *Lancet* **1** (1983) 84–7.
[423] J. B. Bussel, R. P. Kimberly, R. D. Inman, I. Schulmann et al., *Blood* **62** (1983) 480–6.
[424] J. R. Duran-Suarez, A. Martin, M. C. Botella, S. de la Torre et al., *Haematologica* **68** (1983) 564–6.
[425] C. A. Schiffer et al., *Blood* **64** (1984) 937–40.
[426] G. Chirico, M. Duse, A. G. Ugazio, G. Rondini, *J. Pediatr. (St. Louis)* **103** (1983) 654–5.
[427] E. C. Besa, M. W. McNab, A. J. Solan, M. J. Lapes et al., *Am. J. Hematol.* **18** (1985) 373–9.
[428] W. A. McGuire, H. H. Yang, E. Bruno, J. Brandt et al., *N. Engl. J. Med.* **317** (1987) 1004–8.
[429] C. De la Camara, R. Arrieta, A. Gonzalez, E. Iglesias et al., *N. Engl. J. Med.* **318** (1988) 519–20.
[430] I. M. Nilsson, E. Berntorp, O. Zettervall, *N. Engl. J. Med.* **318** (1988) 947–50.
[431] J. W. Newburger, M. Takahashe, J. C. Burns, A. S. Beiser et al., *N. Engl. J. Med.* **315** (1986) 341–7.
[432] M. Nagashima, M. Matsushima, H. Matsuoka, A. Ogawa et al., *J. Pediatr. (St. Louis)* **110** (1987) 710–12.
[433] D. Y. M. Leung, J. C. Burns, J. W. Newburger, R. S. Geha, *J. Clin. Invest.* **79** (1987) 468–72.
[434] B. Combe, B. Cosso, J. Clot, M. Bonneau et al., *Am. J. Med.* **78** (1985) 920–8.

[435] F. C. Breedveld, A. Brand, W. G. van Aken, *J. Rheumatol.* **12** (1985) 700–2.
[436] E. L. Arsura, A. S. Bick, N. G. Brunner, T. Namba et al., *Arch. Intern. Med.* **146** (1986) 1365–8.
[437] K. Kawada, P. I. Terasaki, *Exp. Hematol (N.Y.)* **15** (1987) 133–6.
[438] E. Schuller, A. Govaerts, *Eur. Neurol.* **22** (1983) 205–12.
[439] A. Salama, C. Mueller-Eckhardt, V. Kiefel, *Lancet* **2** (1983) 193–5.
[440] R. P. Kimberly, J. E. Salmon, J. B. Bussel, M. K. Crow et al., *J. Immunol.* **132** (1984) 745–50.
[441] U. Budde et al., *Scand. J. Haematol.* **37** (1986) 125–129.
[442] A. Etzioni, S. Pollack, A. Benderly, *N. Engl. J. Med.* **318** (1988) 994.
[443] G. P. Sandilands, H. I. Atrah, G. Templeton, J. E. Cocker et al., *J. Clin. Lab. Immunol.* **23** (1987) 109–15.
[444] T. W. Jungi, S. Barandun, *Vox Sang.* **49** (1985) 9–19.
[445] W. Stohl, *Clin. Exp. Immunol.* **62** (1985) 200–7.
[446] F. Hashimoto, Y. Sakiyama, S. Matsumoto, *Clin. Exp. Immunol.* **65** (1986) 409–15.
[447] H. H. Hodes, *Pediatrics.* **32** (1963) 1–3.
[448] W. B. White, C. R. Desbonnet, M. Ballow, *Am. J. Med.* **83** (1987) 431–4.
[449] K. Prellner, P. Christensen, O. Kalm, K. Offenbartl, *Acta Pathol. Microbiol. Scand. Sect.* **94 C** (1986) 207–11.
[450] K. K. Christensen, P. Christensen, *Pediatric Infectious Diseases* **5** (1986) S 189–92.
[451] J. G. Kelton, C. J. Carter, C. Rodger, G. Bebenek et al., *Blood* **63** (1984) 1434–8.
[452] A. S. Cross, B. M. Alving, J. C. Sadoff, P. Baldwin et al., *Lancet* **1** (1984) 912.
[453] M. Berger, P. Rosencranz, C. Y. Brown, *Clin. Immunol. Immunopathol.* **34** (1985) 227–36.
[454] J. Kulics, E. Rajnavolgyi, G. Fust, J. Gergely, *Mol. Immunol.* **20** (1983) 805–10.
[455] E. F. Ellis, C. S. Henney, *J. Allergy* **43** (1969) 45–54.
[456] R. C. Williams, O. J. Mellbye, G. Kronvall, *Infect. Immun.* **6** (1972) 316–23.
[457] D. S. Feingold, F. Oski, *Arch. Intern. Med.* **116** (1965) 226–8.
[458] R. C. Millican, J. D. Rust, *J. Infect. Dis.* **107** (1960) 389–394.
[459] E. Lachman, S. B. Pitsoe, S. L. Gaffin, *Lancet* **1** (1984) 981–3.
[460] M. A. Thomson, *J. S.Afr. Vet. Assoc.* **54** (1983) 279–81.
[461] M. L. Forman, R. Stiehm, *N. Engl. J. Med.* **281** (1969) 926–31.
[462] A. S. Shigeoka, R. T. Hall, H. R. Hill, *Lancet* **1** (1978) 636–8.
[463] E. J. Ziegler, J. A. McCutchan, J. Fierer, M. P. Glauser et al., *N. Engl. J. Med.* **307** (1982) 1225–30.
[464] J.-D. Baumgartner, M. P. Glauser, J. A. McCutchan, E. J. Ziegler et al., *Lancet* **2** (1985) 59–63.
[465] T. Calandra, J. Schellekens, J. Verhoef, M. P. Glauser, *Program and Abstracts for the 4th International Symposium on Infections in the Immunocompromised Host*, Ronneby Brunn, Sweden, 1986. Abstract No. 128.
[466] A. H. Levy, *J. Chronic Dis.* **15** (1962) 589–98.
[467] J. Bjorkander, C. Cunningham-Rundles, P. Lundin, R. Olsson et al., *Am. J. Med.* **84** (1988) 107–111.
[468] R. S. Lane, *Lancet* **2** (1983) 974–5.
[469] H. D. Ochs, S. H. Fischer, F. S. Virant, M. L. Lee et al., *Lancet* **1** (1985) 404–5.
[470] C. L. S. Leen, P. L. Yap, G. Neill, D. B. L. McClelland et al., *Vox Sang.* **50** (1986) 26–32.
[471] B. Cuthbertson, R. J. Perry, P. R. Foster, K. G. Reid, *J. Infect.* **15** (1987) 125–33.
[472] C. C. Wood, A. E. Williams, J. G. McNamara, J. A. Annunziata et al., *Ann. Intern. Med.* **105** (1986) 536–8.
[473] Y. Hamamoto, S. Harada, N. Yamamoto, Y. Uemura et al., *Vox Sang.* **53** (1987) 65–9.

[474] Morb. Mort. Wkly. Rep. **36** (1987) 728–9.
[475] A. W. Burks, H. A. Sampson, R. H. Buckley, *N. Engl. J. Med.* **314** (1986) 560–4.
[476] R. Apfelzweig, D. Piskiewicz, J. A. Hooper, *J. Clin. Immunol.* **7** (1987) 46–50.
[477] R. Paganelli, I. Quinti, G. P. D'Offizi, C. Papetti et al., *Vox Sang.* **51** (1986) 87–91.
[478] J. C. Barton, G. A. Herrera, J. H. Galla, L. F. Bertoli et al., *Am. J. Med.* **82** (1987) 624–9.
[479] A. G. Brox, D. Cournoyer, M. Sternbach, G. Spurll, *Am. J. Med.* **82** (1987) 633–5.
[480] R. Schroff, K. Foon, S. Beatty, R. Oldham et al.: "Human anti-murine immunoglobulin responses in patients receiving monoclonal antibody therapy," *Can. Res.* **45** (1985) 879–885.
[481] M. Nose, H. Wigzell: "Biological significance of carbohydrate chains on monoclonal antibodies," *Proc. Natl. Acad. Sci. U.S.A.* **80** (1983) 6632–6636.
[482] E. Englemann, S. Foung, J. Larrick, A Raubitschek (eds.): *Human Hybridomas and Monoclonal Antibodies*, Plenum Press, New York 1985.
[483] A. Strelkelkaus (ed.): *Human Hybridomas; Diagnostic and Therapeutic Applications*, Marcel Dekker, New York 1986.
[484] D. Kozbor, J. Roder, T. Chang, Z. Stplewski et al.: "Human antitetanus toxoid monoclonal antibody secreted by EBV-transformed human B cells fused with murine myeloma," *Hybridoma* **1** (1982) 323–8.
[485] J. Larrick, K. Truitt, A. Raubitschek, G. Senyk et al., "Characterization of human hybridomas secreting antibody in tetanus toxoid," *Proc. Natl. Acad. Sci. U.S.A.* **80** (1983) 6376–6380.
[486] N. Chiorazzi, R. Wasserman, H. Kunkel: "Use of Epstein-Barr virus-transformed B cell lines for the generation of immunoglobulin-producing human B cell hybridomas," *J. Exp. Med.* **156** (1982) 930–935.
[487] F. Gigliotti, R. Insel: "Protective human hybridoma antibody to tetanus toxin," *J. Clin. Invest.* **70** (1982) 1306–1309.
[488] L. Olsson, T. Mazauric, J. Vincent-Falquet, J. Armand: "A human monoclonal antibody specific for tetanus toxoid," *Dev. Biol. Stand.* **57** (1984) 87–91.
[489] S. Tsuchiya, S. Yokoyama, O. Yoshie, Y. Ono: "Production fo diphtheria antitoxin antibody in Epstein-Barr virus induced lymphoblastoid cell lines." *J. Immunol.* **124** (1980) 1970–1976.
[490] E. Ziegler, J. McCutchan, J. Fierer et al.: "Treatment of Gram-negative bacteremia and shock with human anti-serum to a mutant Escherichia coli," *N. Engl. J. Med.* **307** (1982) 1225–1230.
[491] W. Bogard, E. Hornberger, P. Kung in E. Englemann, S. Foung, J. Larrick, A. Raubitscheck (eds.): "Production and characterization of human monoclonal antibodies against Gram-negative bacteria," *Human Hybridomas and Monoclonal Antibodies*, Plenum Press, New York 1985, pp. 95–112.
[492] N. Teng, H. Kaplan, J. Hebert et al.: "Protection against Gram-negative bacteremia and endotoxemia with human monoclonal IgM antibodies," *Proc. Natl. Acad. Sci. U.S.A.* **82** (1985) 179–184.
[493] J. Larrick, M. Jahnsen, G. Senyk, S. Weiss et al. in H. Friedman (ed.): Production of human monoclonal antibodies recognizing cross-reactive determinants on lipopolysaccharides," *The Immunology and Immunopharmacology of Endotoxins*, 1986, pp. 75–81.
[494] M. Pollack, A. Raubitschek, J. Larrick: "Cross-reactive human monoclonal antibodies that recognize conserved epitopes in the core-lipid A region of endotoxin," *J. Clin. Invest.* **79** (1987) 1421–1430.
[495] J. Baumgartner, M. Glauser, J. McCutchan, E. Zeigler et al.: "Prevention of gramnegative shock and death in surgical patients by antibody to endotoxin core glycolipid." *Lancet* **2** (1985) 59–63.

[496] J. Larrick, S. Hart, D. Lippman, M. Glembourtt et al. in A. Strelkelkaus (ed.): "Generation and characterization of human monoclonal anti-Pseudomonas aeruginosa antibodies," *Human Hybridomas; Diagnostic and Therapeutic Applications*, Marcel Dekker, New York 1986, pp. 65–80.

[497] J. Larrick, B. Dyer, G. Senyk et al. in E. Engleman, S. Foung, J. Larrick, A. Raubitschek (eds.): "In vitro expression of human B cells for the production of human monoclonal antibodies," *Human Hybridomas and Monoclonal Antibodies*, Plenum Press, New York 1985, pp. 149–165.

[498] K. Hunter, Jr., G. Fischer, V. Hemming, S. Wilson et al.: "Antibacterial activity of a human monoclonal antibody to Haemophilus influenzae type B capsular polysaccharide," *Lancet* **2** (1982) 789–799.

[499] H. Peltola, H. Kayhty, M. Virtanen, P. Makela: "Prevention of Hemophilus influenza type b bacteremic infections with the capsular polysaccharide vaccine," *N. Engl. J. Med.* **310** (1984) 1561–1565.

[500] T. Atlaw, D. Kozbor, J. Roder: "Human monoclonal antibodies against Mycobacterium Leprae," *Infect. Immun.* **49** (1985) 104–110.

[501] B. Brodeur, L. Lagace, Y. Larose, M. Martin et al. in B. Schook (ed.): "Mouse-human myeloma partner for the production of heterohybridomas," *Monoclonal Antibodies*, Marcel Dekker, 1986, p. 51.

[502] M. Steinitz, S. Tamir, A. Goldfarb: "Human anti-pneumococci antibody produced by an Epstein-Barr virus (EBV)-immortalized cell line," *J. Immunol.* **132** (1984) 877–882.

[503] J. Schwaber, M. Posner, S. Schlossman, H. Lazarus: "Human-human hybrids secreting pneumococcal antibodies," *Hum. Immunol.* **9** (1984) 137–142.

[504] D. Emanuel, J. Gold, J. Colacino, C. Lopez et al.: "A human monoclonal antibody to cytomegalovirus (CMV)," *J. Immunol.* **133** (1984) 2202–2205.

[505] C. Amadei, S. Michelson, J. Frot, M. Fruchart et al.: "Human anticytomegalovirus (CMV) immunoglobulins secreted by EBV-transformed B-lymphocytes cell lines," *Dev. Biol. Stand.* **57** (1984) 283–286.

[506] K. Burnett, J. Leung, J. Marinis in E. Engleman, S. Foung, J. Larrick, A. Raubitschek (eds.): "Human monoclonal antibodies to defined antigens," *Human Hybridomas and Monoclonal Antibodies*, Plenum Press, New York 1985, pp. 113–133.

[507] Y. Ichimori, K. Sasano, H. Itoh, S. Hitosumachi et al.: "Establishment of hybridomas secreting human monoclonal antibodies against tetanus toxin and hepatitis B virus surface antigen," *Biochem. Biophys. Res. Commun.* **129** (1985) 26–38.

[508] E. Stricker, R. Tiebout, P. Lelie, W. Zeijlemaker: "A human monoclonal IgG antihepatitis B surface antibody; production, properties and applications," *Scand. J. Immunol.* **22** (1985) 337–345.

[509] R. Beasley et al.: "Hepatitis B immune globulin (HBIG) efficacy in the interruption of perinatal transmission of hepatitis B virus carrier state," *Lancet* **2** (1981) 388–393.

[510] S. Foung et al.: "Human monoclonal antibodies neutralizing Varicella Zoster virus," *J. Infect. Dis.* **52** (1985) 280–285.

[511] L. Evans, J. Homsy, W. Morrow, I. Gaston et al.: "Human monoclonal antibody directed against gag gene products of the human immunodeficiency virus," *J. Immunol.* **140** (1988) 941–943.

[512] B. Banapour et al.: "Characterization and epitope mapping of a human monoclonal antibody reactive with the envelope glycoprotein of human immunodeficiency virus," *J. Immunol.* **138** (1987) 4027–33.

[513] A. Rosen, K. Persson, G. Klein: "Human monoclonal antibodies to a genus-specific chlamydial antigen, produced by EBV-transformed B cells," *J. Immunol.* **130** (1983) 2899–2902.

[514] S. Koizumi, S. Fujiwara, H. Kikuta et al.: "Production of human monoclonal antibodies against Epstein-Barr virus specific antigens by the virus immortalized lymphoblastoid cell lines," *Virology* **150** (1986) 161–170.

[515] J. Seigneurin et al.: "Herpes simplex virus glycoprotein D; human monoclonal antibody produced by bone marrow cell line," *Science* **221** (1983) 173–175.

[516] Y. Masuho et al.: "Generation of hybridomas producing human monoclonal antibodies against herpes simplex virus after in vitro stimulation," *Biochem. Biophys. Res. Commun.* **135** (1986) 495–505.

[517] L. Evans, C. Maragos, J. May: "Human lymphoblastoid cell lines established from peripheral blood lymphocytes secreting immunoglobulins directed against herpes simplex virus," *Immunol. Lett.* **8** (1984) 39–50.

[518] S. Matsushita, M. Robert-Gurhoff, J. Trepel, J. Cossman et al.: "Human monoclonal antibodies directed against an envelope glycoprotein of human T-cell leukemia virus type 1," *Proc. Natl. Acad. Sci. U.S.A.* **83** (1986) 2672–2677.

[519] C. Croce, A. Linnenbach, W. Hall, Z. Steplewski et al.: "Production of human hybridomas secreting antibodies to measles virus," *Nature (London)* **288** (1980) 488–489.

[520] R. Ritts, Jr., A. Ruiz-Arguelles, K. Weyl et al.: "Establishment and characterization of a human non-secreting plasmacytoid cell line and its hybridization with human B cells," *Int. J. Cancer* **31** (1983) 133–151.

[521] J. Hilfenhaus, E. Kanzy, R. Kohler, W. Willems: "Generation of human anti-rubella monoclonal antibodies from human hybridomas constructed with antigen-specific Epstein-Barr virus transformed cell lines," *Behring Inst. Mitt.* **80** (1986) 31–40.

[522] F. van Meel, P. Steenbakkers, J. Oomen: "Human and chimpanzee monoclonal antibodies," *J. Immunol. Methods* **80** (1985) 267–280.

[523] D. Crawford et al.: "Production of human monoclonal antibody to X 31 influenze virus nucleoprotein," *J. Gen. Virol.* **64** (1983) 697–700.

[524] R. Schmidt-Ullrich, J. Brown, R. Whittle, P. Lin: "Human-human hybridomas secreting monoclonal antibodies to the M.W. 195 000 Plasmodium falciparum blood stage antigen," *J. Exp. Med.* **163** (1986) 179–189.

[525] R. Udomsangpetch et al.: "Human monoclonal antibodies to Pf-155, a major antigen of malaria parasite Plasmodium falciparum," *Science* **231** (1986) 55–59.

Cancer Chemotherapy

BERNHARD KUTSCHER, ASTA Medica AG, Frankfurt am Main, Federal Republic of Germany
GREGORY A. CURT, National Cancer Institute, Bethesda, Maryland 20205, United States
CARMEN J. ALLEGRA, National Cancer Institute, Bethesda, Maryland 20205, United States
ROBERT L. FINE, National Cancer Institute, Bethesda, Maryland 20205, United States
HAMZA MUJAGIC, National Cancer Institute, Bethesda, Maryland 20205, United States
GRACE CHAO YEH, National Cancer Institute, Bethesda, Maryland 20205, United States
BRUCE A. CHABNER, National Cancer Institute, Bethesda, Maryland 20205, United States

1.	Introduction	1866
2.	Antimetabolites	1868
2.1.	Methotrexate	1868
2.1.1.	Mechanism of Action and Mechanisms of Resistance	1869
2.1.2.	Analogs	1871
2.2.	Fluoropyrimidines	1871
2.2.1.	Mechanism of Action	1873
2.2.2.	Mechanisms of Resistance	1874
2.2.3.	Other Fluoropyrimidines	1875
2.3.	5-Azacytidine	1875
2.3.1.	Mechanism of Action	1876
2.3.2.	Mechanism of Resistance	1876
2.3.3.	New Analogs	1877
2.4.	Cytosine Arabinoside (Ara-C)	1877
2.4.1.	Mechanisms of Resistance	1878
2.4.2.	New Analogs	1879
2.5.	Deoxycytidine and Analogs	1879
2.6.	2-Halopurines and Analogs	1880
2.7.	6-Mercaptopurine and 6-Thioguanine	1882
2.7.1.	Mechanism of Action	1883
2.7.2.	Mechanism of Resistance	1883
2.7.3.	New Analogs	1885
3.	Alkylating Agents	1885
3.1.	Nitrogen Mustard	1886
3.1.1.	Mechanism of Action	1886
3.1.2.	Mechanisms of Drug Resistance	1887
3.2.	Melphalan	1887
3.2.1.	Mechanism of Action	1887
3.2.2.	Mechanism of Resistance	1888
3.3.	Cyclophosphamide	1888
3.3.1.	Mechanism of Action	1889
3.3.2.	Mechanism of Resistance	1889
3.4.	Chlorambucil	1889
3.5.	Thio-TEPA	1889
3.6.	Ifosfamide	1890
3.7.	Estramustine	1890
3.8.	Nitrosoureas	1891
3.8.1.	Mechanism of Action	1892
3.8.2.	Mechanisms of Resistance	1893
3.8.3.	Analogs	1893
3.9.	Procarbazine	1894
3.9.1.	Mechanism of Action	1895
3.9.2.	Mechanisms of Resistance	1895
3.10.	Dacarbazine	1895
3.11.	Hexamethylmelamine	1896
3.12.	Mitomycin-C	1896
4.	Anthracyclines	1897
4.1.	Mechanism of Action	1899
4.2.	Mechanism of Resistance	1901
4.3.	Analogs	1901
5.	Intercalating Anthracenes and Analogs	1902
5.1.	Mitoxantrone	1902
5.2.	Analogs	1903

6.	Antitumor Antibiotics Other than Anthracyclines	1904	8.3.	Analogs	1921
6.1.	Actinomycin D	1904	9.	Hormonally Active Anticancer Drugs/Antihormones	1922
6.2.	Bleomycin	1905	9.1.	Antiestrogens	1922
6.2.1.	Analogs	1906	9.1.1.	Antagonists	1922
6.2.2.	Mechanism of Action	1906	9.1.2.	Tamoxifen, Toremifene	1922
6.3.	DNA Interactive Natural Products	1907	9.1.3.	Analogs	1924
			9.2.	Aromatase Inhibitors	1924
7.	Antitubulin Agents	1909	9.3.	Antiandrogens	1928
7.1.	Vinca Alkaloids	1910	9.3.1.	Flutamide	1928
7.1.1.	Vincristine and Vinblastine	1910	9.3.2.	Nilutamide	1929
7.1.2.	Vindesine	1911	9.3.3.	Bicalutamide	1930
7.1.3.	Vinorelbine	1912	9.4.	LHRH Analogs	1931
7.2.	Podophyllotoxin and Its Derivatives	1912	9.4.1.	LHRH Agonists	1932
			9.4.1.1.	Leuprorelin Acetate	1933
7.3.	Camptothecin and Analogs	1913	9.4.1.2.	Goserelin	1933
7.4.	Taxoids	1915	9.4.2.	LHRH Antagonists	1933
			9.4.2.1.	Receptor Assays	1935
7.5.	Epothilone A and B	1918	9.4.2.2.	Peptidomimetics	1935
8.	Heavy-Metal Complexes	1918	10.	Signal Transduction Inhibitors	1937
8.1.	cis-Platinum	1918	10.1.	Enzyme Inhibitors	1937
8.1.1.	Mechanism of Action	1919	10.2.	Phospholipid – Based Antineoplastics	1938
8.1.2.	Mechanisms of Resistance	1920			
8.2.	Carboplatin	1920	11.	Economic Aspects	1939
			12.	References	1940

1. Introduction

Malignant tumors represent one of the most common human diseases worldwide. Based on an estimation made in the United States, cancer will become the leading cause of death in the year 2000 [1].

Unfortunately, the subset of human cancer types that are amenable to curative treatment today still is rather small. Although there is a tremendous progress in understanding the molecular events that lead to malignancy and many agents are known that effectively kill cancer cells, progress in development of clinically innovative drugs that can cure humans is slow [2], [3].

The heterogeneity of malignant tumors with respect to their genetics, biology, and biochemistry as well as primary or treatment-induced resistance to therapy hamper curative treatment [4], [5].

Searching for antineoplastic agents with improved selectivity to malignant cells remains the central task for drug discovery and development [5], [6].

According to a survey published in 1997 more then 315 drugs are under development in the United States for the treatment of cancer. This figure includes 42 drugs for treatment of lung cancer, 58 for breast cancer [8], [9], 60 for treatment of skin tumors (60), 36 for prostate cancer, and 35 for colon cancer.

In 1997 more than 1500 Americans are expected to die of cancer each day and more than a million new cases will be diagnosed with overall medical costs of $ 35×10^9. The total disease costs are estimated to sum up to more than $ 100×10^9 per year in treatment expenses and lost wages.

Increase in the incidence of cancer, mostly as a result of an aging population, is the driving growth in the marketplace. However, new medicines like hormone-analogs have lead to decrease in severity of the side effects of cancer therapy and have spurred wider use. According to a most recent review [11] of ca. 90 approved anticancer drugs, more than 60% are of natural origin or modeled on natural products parents. Cancer chemotherapeutics can be grouped according to their pharmacological and mechanistic profiles into

Antimetabolites
Alkylating agents
DNA-intercalating (agents) antibiotics
Mitose inhibitors
Signaltransduction-inhibitors

Two primary events in cell proliferation are DNA replication and cell division [12], [13]. The cell cycle has been divided into four sequential phases (G_0, G_1, S, M) and the cytostatics vary in the way they interfere with the cell cycle. Phasespecific agents/drugs, e.g., those interacting in G_1, S- or M-phase, are the mitose-inhibitors vincristine or vinblastine or the antimetabolites methotrexate or cytarabine, attacking in the S-phase. Alkylating drugs such as cyclophosphamide, cisplatin or carboplatin inhibit and damage the cell in all phases and are thus phase unspecific. In general cells in the resting phase (G_0) are insensitive.

Approximately 70% of patients diagnosed as having cancer have metastatic disease, i.e., disease that has spread beyond the primary site at the time of diagnosis [15]. However, the steady progress made in the treatment of cancer with drugs has contributed to the curing of an increasing proportion of patients with metastatic disease. The greatest change and improvements in cancer treatment have occurred because of the discovery and clinical development of drugs and the demonstration that metastatic cancer can be cured by these agents. Drugs, such as cytoxan, adriamycin, vincristine, *cis*-platinum, and bleomycin, all developed since 1960, are now regularly used by physicians to treat patients who would have been considered incurable a short time ago.

This article details the pharmacology and clinical use of the major classes of anticancer agents, including (1) antimetabolites, (2) alkylating agents, (3) anthracyclines and analogs, (4) other antitumor antibiotics, (5) antitubulin agents, (6) platinum complexes, (7) antihormones, and (8) signaltransduction-inhibitors. In each case, special emphasis is given to the progress made in developing clinically useful drugs and analogs that retain antitumor activity while decreasing host toxicity.

2. Antimetabolites

2.1. Methotrexate

Folic acid analogs comprise a class of antineoplastic agents of which methotrexate has gained the most widespread clinical use. These agents were the first to produce impressive remissions in acute leukemia [16] and cures in choriocarcinoma in women [17].

Reduced folates (tetrahydrofolates) are the biologically active form of folates required as cosubstrates in one-carbon transfer reactions. Included in these reactions are several important enzymatic steps in the de novo synthesis of purines and pyrimidines.

Methotrexate (NSC-740) [59-05-2], N-(4-{(2,4-diamino-6-pteridinyl)methyl]methylamino}benzoyl)-L-glutamic acid, $C_{20}H_{22}N_8O_5$, M_r 454.46, is a 2,4-diamino,N^{10}-methyl analog of folic acid that is capable of inhibiting certain folate-requiring reactions.

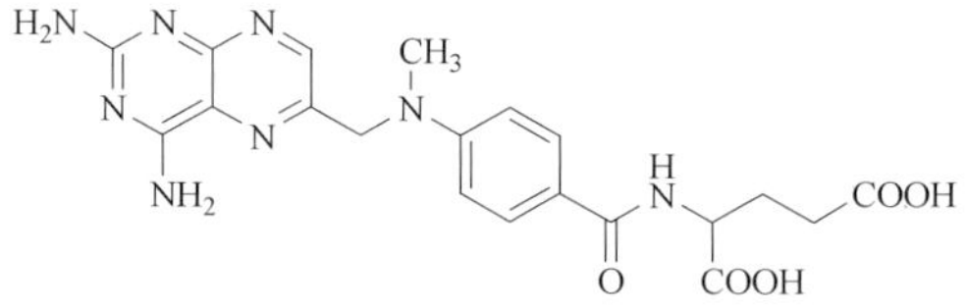

Methotrexate

Most importantly, methotrexate can inhibit dihydrofolate reductase (DHFR, $K_i = 10^{-11}$ M), a key enzyme for the maintenance of biologically active intracellular reduced folate pools. The folate-requiring reactions utilize reduced folates, and all reactions except that catalyzed by thymidylate synthase maintain folates in a reduced state during carbon transfer. Thymidylate synthase, which catalyzes the methylation of deoxyuridylate to thymidylate (required for DNA synthesis), requires the transfer of a carbon group from the folate cofactor N^{5-10}-methylene tetrahydrofolate with resultant oxidation of the folate to dihydrofolic acid. Oxidized folates must be reduced to the tetrahydro form by DHFR to be useful for intracellular metabolism. Inhibition of DHFR following methotrexate exposure ultimately leads to depletion of intracellular reduced folates. Cessation of first thymidylate and then purine nucleotide synthesis occurs as an indirect effect of methotrexate on reduced-folate levels. Methotrexate metabolites (polyglutamates) may also have direct inhibitory effects on folate-requiring enzymes, e.g., thymidylate synthase [18], and these effects may be important in inducing cytotoxicity.

2.1.1. Mechanism of Action and Mechanisms of Resistance

Transport. At concentrations less than 10 µM, methotrexate and reduced folates enter cells via an energy-dependent, temperature-sensitive carrier mechanism [19]. The affinity for this carrier has been variously reported to fall between 1 and 10 µM for tumor cell lines [19]–[21] and to be 87 µM for normal intestinal epithelial cells [22]. These differences in efficiency of transport may account for some of the selectivity of methotrexate for neoplastic cells. In addition to this carrier-mediated transport system, there exists a second, low-affinity transport mechanism that is poorly understood but appears to play a role in the transport of drug when concentrations exceed 20 µM [21], [23]. Methotrexate and reduced folates do not compete for uptake by this process, and it may represent a means for drug entry in cells resistant to low doses of drug by virtue of defective transport by the high-affinity mechanism. In some models, the sensitivity of a cell to methotrexate can be directly correlated with efficiency of drug transport, i.e., sensitive cells have a greater capacity for drug transport and longer intracellular retention of drug when compared to methotrexate-resistant cells [24]. Because decreased membrane transport may play a role in clinical drug resistance, a number of analogs with high lipid solubility have been developed that can circumvent a transport deficit. Methotrexate esters, diaminopyrimidines, and triazenates have been synthesized and used with success against experimental cell lines with defective methotrexate transport.

Intracellular Metabolism. Once inside the cell, naturally occurring folates may be metabolized to *polyglutamates;* that is, additional glutamyl moieties are added to the terminal glutamate present on the parent compound. This process allows for selective intracellular retention of the polyglutamated forms and an increased affinity for certain folate-requiring enzymes, such as thymidylate synthase, and for enzymes required for the de novo production of purine nucleotides [25]. Like the naturally occurring folates, intracellular methotrexate is also polyglutamated with the addition of from one to four additional glutamyl residues [26]. This process has been demonstrated in a variety of tissues, including human breast cancer cell lines [26], normal human liver [27], and murine leukemia cells [28]. The polyglutamates of methotrexate are selectively retained by the cells and appear to have an enhanced inhibitory potential for certain enzymes [18]. The inhibitory capacity of methotrexate polyglutamates for dihydrofolate reductase appears to be somewhat greater than that of the parent drug. The selective retention of methotrexate polyglutamates may be critical for the delayed cytotoxicity exhibited by cells capable of polyglutamate synthesis. In vitro experiments using MCF-7 human breast cancer cells have demonstrated that the intracellular retention and duration of binding to dihydrofolate reductase are directly related to the length of the polyglutamate tail [29].

Interaction with Dihydrofolate Reductase. The binding of methotrexate to dihydrofolate reductase has been extensively investigated by X-ray crystallographic and amino acid sequencing studies [30]–[34]. Methotrexate binds in a stoichiometric fashion to a hydrophobic pocket in the target enzyme DHFR [35]. The binding affinity depends on multiple factors, including pH, salt concentration, and NADPH concentration, and has been reported to be ca. 10 pM [36]. In the cell, methotrexate is a reversible inhibitor capable of being displaced by high concentrations of substrate. Thus, free intracellular drug in excess of the cellular dihydrofolate reductase binding capacity is required to maintain complete inhibition of the enzyme and thereby produce and maintain a state of reduced-folate depletion. If an excess of intracellular drug is not maintained, the intracellular reduced-folate pool recovers through enzymatic reduction of oxidized folates and cellular metabolism resumes.

Cellular resistance to methotrexate has been most commonly associated with an increase in dihydrofolate reductase activity. In general, the amplified enzyme is identical to the native protein in its affinity for methotrexate; however, altered methotrexate affinity has been reported to correlate with sensitivity to methotrexate in a series of murine leukemias [36]. Increased reductase activity and resistance to methotrexate has also been demonstrated in a number of cell lines made resistant in vitro by stepwise increases in drug concentration [37], [38], and in human tumor samples from clinically resistant tumors [39]. The increased enzyme levels can be correlated to gene amplification that may take the form of small new pieces of chromosomal material, called double minutes, or of large chromosomes, referred to as homogeneously staining regions (HSRs). The former variety of amplification imparts relatively unstable resistance, which requires the ongoing selective pressure of drug presence to be maintained [40], whereas the HSRs represent a more durable form of amplification and thus resistance. Several investigators [41]–[43] have successfully transvected amplified reductase genes into normal hematopoietic cells, allowing greater marrow resistance to methotrexate, an important dose-limiting toxicity of the drug.

Determinants of Cytotoxicity. Methotrexate is an S-phase-specific agent whose cytotoxic effects are determined by drug concentration and duration of cell exposure. These effects may be altered by the cellular milieu. The toxic effects of methotrexate can be completely reversed by exogenous administration of the end products (purines and thymidine) whose de novo synthesis is inhibited by methotrexate treatment. Also, the synthesis of these products may resume if an exogenous source of reduced folates is provided. These data provide the rationale for the treatment of patients with high-dose methotrexate and subsequent administration of a reduced folate in the form of leucovorin calcium (N^5-formyl tetrahydrofolic acid) as "rescue." The reversal of methotrexate cytotoxicity by reduced folates is a competitive process. The reasons for the competitive nature of this relationship are unclear but may be the result of a shared membrane transport system. In addition, methotrexate may have direct inhibitory effects on enzymes other than dihydrofolate reductase that require competitive levels of the folate cosubstrate to overcome the inhibition.

2.1.2. Analogs

Many new analogs to methotrexate have been developed in an effort to circumvent the cellular resistance that occurs with prolonged methotrexate exposure. As mentioned, drugs with increased lipid solubility have been successful in treating transport-resistant cells in vitro [44]. The lipophilic derivative metoprine and variations of the 10-deazaaminopterin series are the most interesting recent additions. Of the 10-deaza series, an ethyl sub-stitution at the 10-position imparts a marked increase in cytotoxicity when compared to methotrexate [45]. The analog possesses an improved membrane transport ability while retaining a high affinity for dihydrofolate reductase. A new antineoplastic agent is piritrexim isothionate [*79483-69-5*], 6-[(2,5-dimethoxyphenyl)methyl]-5-methylpyrido[2,3-d]pyrimidine-2,4-diamine mono-2-hydroxyethanesulfonate, $C_{17}H_{19}N_5O_2$ [46], that inhibits dihydrofolate reductase (DHFR). Further lipophilic DHFR-inhibitors are trimetrexate [*82952-645*] [47], and edatrexate [48] which are clinically studied and have shown activity, e.g., in non-small cell lung cancer. Thymidylate synthase (TS) is the rate-limiting enzyme in the anabolism of thymidine resulting in the incorporation into DNA. Raltitrexed (company codes: ZN-1694, D-1694, ICI-D1694) [*112887-68-0*] [49], is currently under clinical investigation with response rates in colon and breast cancers of up to 30% [50], [51]. Myelosuppression seems to be the predominant dose limiting toxicity.

Crystallographic data and computer-assisted drug design led to the development of thymidylate synthase (TS) inhibitors of the type of AG 331 [52].

Finally, CB 3717, a potent inhibitor of thymidylate synthase, is toxic for cell lines with altered dihydrofolate reductase; and homofolate, a de novo purine inhibitor requiring dihydrofolate reductase for activation, is effective in reductase-amplified lines [44].

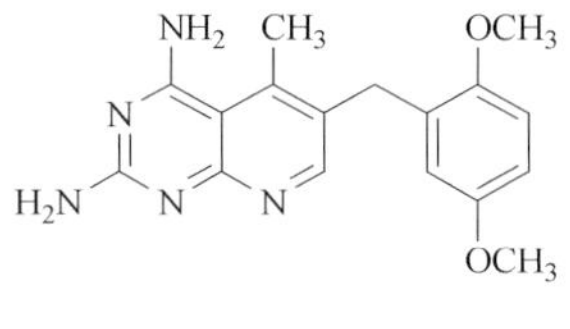

Piritrexim

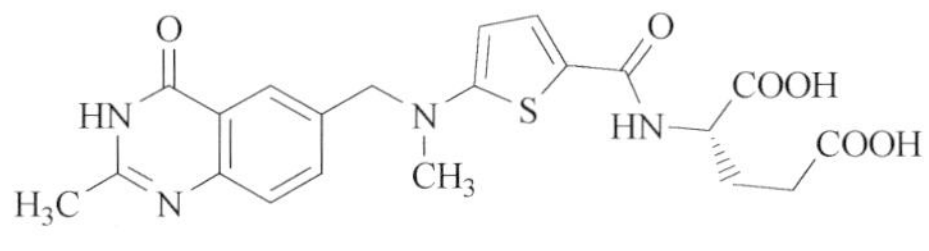

Raltitrexed

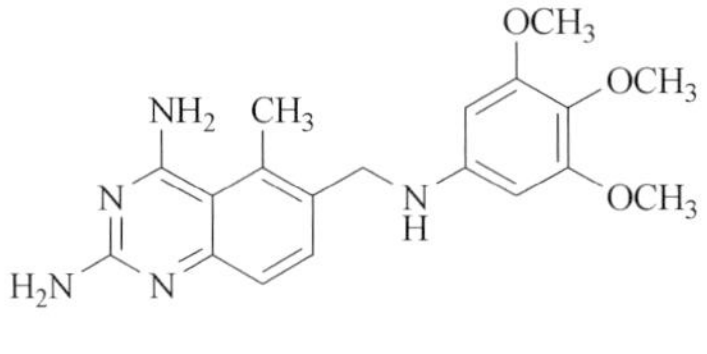

Trimetrexate

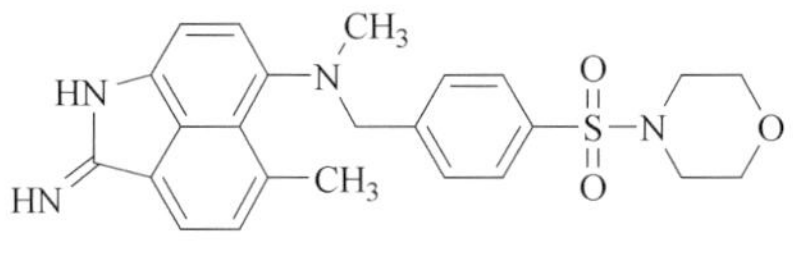

AG 331

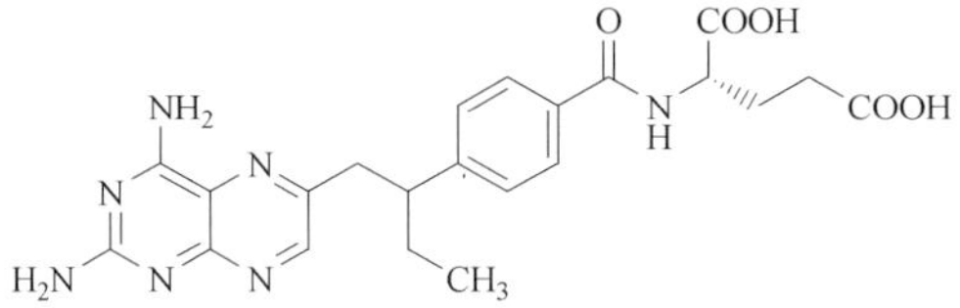

Edatrexate

2.2. Fluoropyrimidines

5-Fluorouracil (NSC-19893) [51-21-8], 5-fluoro-2,4-(1H, 3H)-pyrimidinedione, 5-FU, $C_4H_3FN_2O_2$, M_r 130.08, is a fluorinated pyrimidine whose structural formula resembles thymine; the hydrogen in the 5-position of the naturally occurring pyrimidine being replaced by fluorine. The *synthesis* of 5-FU in 1957 [15] represents the first successful effort in the rational design of anticancer drugs [53], and was predicated on the earlier observation that malignant cells selectively utilized uracil (and possibly toxic uracil analogs) in vivo [16], [54].

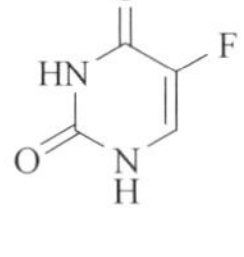

5-Fluorouracil

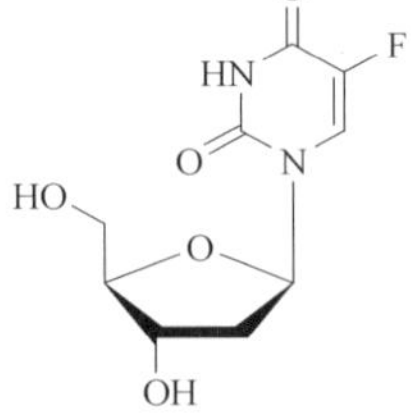

5-FUDR

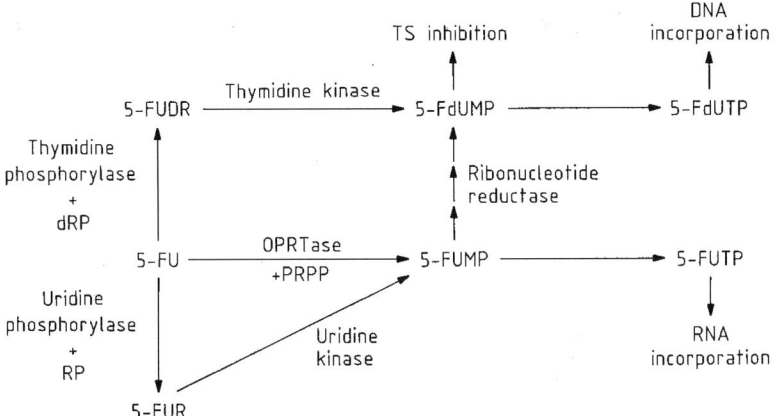

Figure 1. Pathways for fluoropyrimidine activation

Since the original synthesis of 5-FU and its nucleoside 5-FUDR (NSC-27640) [50-91-9], 2′-deoxy-5-fluorouridine, floxuridine, $C_9H_{11}FN_2O_5$, M_r 246.21, much has been learned about the mechanism of action of the fluoropyrimidines. These drugs are useful in the treatment of a wide range of human malignancies.

2.2.1. Mechanism of Action

Both 5-FU and 5-FUDR are prodrugs that require intracellular metabolism to their respective nucleotides for cytotoxicity. The pathways for fluoropyrimidine activation are shown in Figure 1. Each drug is enzymatically activated by different routes to FdUMP, FUMP, or FUTP, and each of these fluorinated nucleosides has different mechanisms of cytotoxicity.

Thymidine phosphorylase converts 5-FU to the deoxyribonucleotide 5-FUDR, which is then phosphorylated by thymidine kinase to yield 5-FdUMP. In the presence of methylene tetrahydrofolate, 5-FdUMP forms a stable ternary complex with thymidylate synthetase (TS), inhibiting this critical enzyme to cause "thymineless death." As expected, cytotoxicity is prevented in the presence of exogenous thymidine in these cells with intact salvage pathways. Inhibition of TS has long been considered the principal mechanism of 5-FU cytotoxicity, but it has also been demonstrated that 5-FU can be converted to 5-FUMP, either by orotic acid phosphoribosyl-transferase (OPRTase) in the presence of phosphoribosyl pyrophosphate (PRPP), or by stepwise conversion to the ribonucleotide 5-FUR (by uridine phosphorylase) followed by formation of 5-FUMP by uridine kinase. This intermediate can be converted to 5-FdUMP by ribonucleotide reductase to inhibit TS. Alternatively, 5-FUMP can be phosphorylated to 5-FUTP, which may be fraudulently incorporated into RNA to induce cytotoxicity. In a number of tumor models, loss of clonogenic capacity is directly correlated with the extent of incorporation of 5-FUTP into RNA [55], [56]. This RNA-specific toxicity is not reversed by thymidine. However, the precise mechanism of RNA-induced cell kill is speculative. The most consistent structural effect of 5-FU exposure is impaired processing of ribosomal RNA [57]. This concept is not supported by current evidence, however, as neither the synthesis nor the translation of messenger RNA (mRNA) appears affected by documented 5-FUTP incorporation into mRNA. Using human colon carcinoma cells propagated in vitro, neither quantitative nor qualitative differences in the translational products (polypeptides) of 5-FUTP-containing mRNA could be demonstrated [58]. However, small nuclear RNA species responsible for exon recognition during RNA splicing do contain significant quantities of uridylic acid [59]. Specific substitution of 5-FUTP into this RNA fraction may be critical for 5-FU toxicity.

As shown in Figure 1, 5-FdUMP can be further phosphorylated to 5-FdUTP, which can be incorporated into tumor cell DNA [60], [61]. This mechanism of drug action has been particularly difficult to appreciate, since the fraudulent base is quickly excised from DNA by the enzymes uracil-DNA glycosylase and dUTP nucleotidohydrolase. Thus, when tumor cells are incubated at low (0.1 µM) concentrations of 5-FUDR, single-strand DNA shifts to lower molecular mass species, suggesting excision of the fluorinated base [62]. However, the actual presence of 5-FdUTP in cellular DNA cannot be detected until tumor cells are exposed to higher drug concentrations. Apparently, the importance of 5-FU incorporation into DNA varies from tumor to tumor, with human promyelocytic leukemia cells incorporating nearly 100-fold more 5-FdUTP into DNA than mouse leukemia cells [63]. The importance of this pathway both to tumor cell cytotoxicity and therapeutic index remains to be elucidated.

2.2.2. Mechanisms of Resistance

Tumor cells selected for in vitro resistance may demonstrate a deletion of critical drug-activating enzymes, including uridine kinase [64], orotic acid phosphoribosyltransferase [65], [66], and uridine phosphorylase [67]. Methylene tetrahydrofolate is required for 5-FdUMP inhibition of TS, and decreased availability of intracellular folates has been involved as a mechanism of 5-FU resistance [68].

In addition, alterations in the target enzyme TS can result in drug resistance. Resistant cells have been described with altered thymidylate synthetase having decreased affinity for FdUMP [69]–[71]. In addition, increased specific activity of TS has been reported in drug-resistant fibroblasts [72] and tumor cells [73], [74]. Whether elevated levels of target protein represent the end result of specific gene amplification (as has been documented for methotrexate resistance) remains to be determined.

2.2.3. Other Fluoropyrimidines

Attempts to develop fluoropyrimidines with improved therapeutic indexes have resulted in the synthesis and clinical trial of so-called *masked fluoropyrimidines*. The masked fluoropyrimidine 5′-deoxy-5-fluorouridine is a nontoxic prodrug that is converted to 5-FU by pyrimidine nucleoside phosphorylase. Because this enzyme may be present to a greater degree in some tumor cells than in normal human bone marrow, an improved therapeutic index can be demonstrated in vitro using breast, sarcoma, leukemia, and colon carcinoma cells [75]. Since 5′-deoxy-5-fluorouridine requires conversion to 5-FU and is in itself nontoxic, tumor cells that lack phosphorylase activity are resistant to the masked compound, while remaining cross-sensitive to 5-FU [76].

Ftorafur is a second masked fluoropyrimidine that is less myelosuppressive than 5-FU. However, the drug has a higher incidence of gastrointestinal and nervous system toxicity, which is probably due to organ-specific localization of activating enzymes [77]. In early clinical studies, ftorafur has shown antitumor activity in patients with 5-FU refractory colorectal and breast cancer [78].

2.3. 5-Azacytidine

5-Azacytidine (NSC-102816) [*320-67-2*], 4-amino-1-β-D-ribofuranosyl-1,3,5-triazine-2(1*H*)-one, 5-azacitidine, $C_8H_{12}N_4O_5$, M_r 244.21, is a pyrimidine analog that was first isolated as a fermentation product from *Streptoverticillium* cultures [79] and was chemically synthesized in Czechoslovakia in 1964 [80]. Structurally, 5-azacytidine differs from cytidine by the substitution of nitrogen in the 5 position of the pyrimidine nucleus.

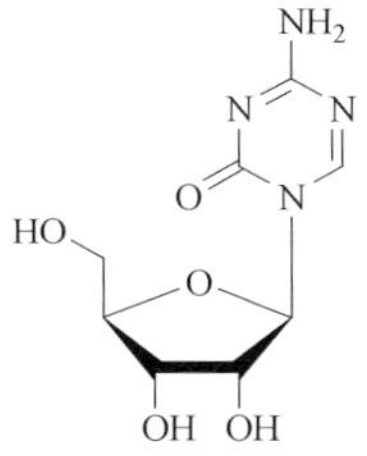

5-Azacytidine

2.3.1. Mechanism of Action

5-Azacytidine shares the facilitated transport system for cytidine for entry into cells [81] and must be phosphorylated to exert cytotoxic effects. Conversion to the monophosphate is catalyzed by the enzyme uridine–cytidine kinase [82], and this is likely the rate-limiting step for drug activation [83]. 5-Azacytidine monophosphate inhibits the enzyme orotidylate decarboxylase and interferes with de novo pyrimidine biosynthesis [84]. Subsequent metabolism of the monophosphate to azacytidine di- and triphosphate, catalyzed by cytidine monophosphate kinase and nucleoside diphosphate kinase, occurs rapidly and does not appear to be a rate-limiting step in drug activation [81].

The diphosphate of azacytidine is a substrate for ribonucleotide reductase and dAzaCTP (deoxyribonucleotide triphosphate of azacytidine) for DNA polymerase, allowing direct incorporation of drug into DNA [85]. This pathway may be critical for cytotoxicity because active DNA synthesis correlates with drug sensitivity in vitro [86]. In addition, incorporation of 5-azacytidine into DNA may affect gene expression. Mammalian DNA contains ca. 5% of incorporated cytosine methylated in the 5 position [87]; methylation appears to inhibit gene expression. Thus the globin gene is hypomethylated in bone marrow as compared to other tissues [88]–[91]. The DNA containing even low levels of 5-azacytidine is a potent inhibitor of the enzyme responsible for cytosine methylation, i.e., DNA cytosine methyltransferase [92], [93]. The enzyme inhibition is disproportionately great compared with the small amount of incorporated fraudulent base, and appears to result from formation of a stable complex between 5-azacytidine residues and the methyltransferase, similar to the complex formed with thymidylate synthase and FdUMP [94]. Thus, in vitro treatment with 5-azacytidine can induce DNA hypomethylation and differentiation of murine cells [95], [96]. More recently these observations have been extended to clinical medicine. Azacytidine treatment in a patient with severe β-thalassemia could stimulate gamma-globin synthesis by inducing hypomethylation and expression of the gamma-globin gene [97].

After conversion to a triphosphate, azacytidine also competes with CTP for incorporation into RNA [98] and inhibits maturation of ribosomal and transfer RNA [99]. This causes disassembly of polyribosomes [100] and interferes with protein synthesis [101].

2.3.2. Mechanism of Resistance

Cellular resistance to 5-azacytidine may be the result of either decreased drug activation or possibly increased drug degradation. Drug metabolism by cytidine kinase appears to be rate-limiting in drug activation; deletion of this enzyme has been reported in 5-azacytidineresistant cells in vitro [81], [102]. Cytidine deaminase degrades 5-azacytidine to 5-azauridine; however, the role of this enzyme in drug resistance remains uncertain. For example, drug toxicity may be dependent on deamination; bacteria incapable of forming 5-azauridine from 5-azacytidine are resistant to the drug [103].

2.3.3. New Analogs

In clinical trials 5-azacytidine has demonstrated consistent antileukemic activity, inducing complete remissions in a significant number of heavily pretreated patients with acute myelogenous leukemia [104]. Acute dose-limiting toxicities associated with bolus drug administration (severe gastrointestinal symptoms, fever, life-threatening hypotension) may be ameliorated by administering the drug via constant intravenous infusion [105]. However, 5-azacytidine is chemically unstable, undergoing ring opening at the 5,6-imino double bond to form *N*-formylamidinoribofuranosylguanylurea, which further decomposes to ribofuranosylurea [106]. The halflife of this decomposition is 4 h, complicating the task of strict dosage control of prolonged infusions.

To circumvent the problem of aqueous instability, the hydrolytically susceptible 5,6-imino bond of 5-azacytidine was reduced to produce *dihydro-5-azacytidine* (NSC-264880) [*62402-31-7*], 4-amino-5,6-dihydro-1-β-D-ribofuranosyl-1,3,5-triazine-2(1*H*)one · monohydrochloride, $C_8H_{14}N_4O_5$ · HCl, M_r 282.7 [107]. This compound has recently completed phase I trials with preliminary evidence of activity in lymphoma; in addition it exhibits the unusual dose-limiting toxicity of chest pain at the maximally tolerated dose [108].

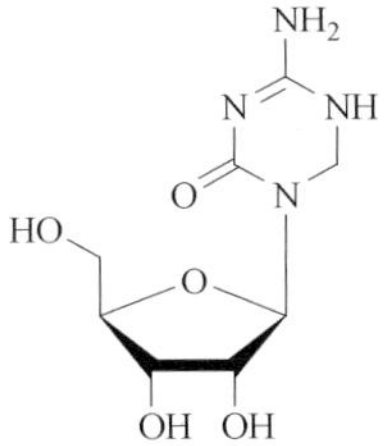

Dihydro-5-azacytidine

2.4. Cytosine Arabinoside (Ara-C)

Ara-C (NSC-63878) [*147-94-4*], 4-amino-1-β-D-arabinofuranosyl-2(1*H*)-pyrimidone, cytarabine, cytosine arabinoside, $C_9H_{13}N_3O_5$, M_r 243.22, an antimetabolite that is a structural analog of cytidine, differs from the physiologic nucleoside by the epimeric configuration of the β-*trans*-hydroxyl group at the 2′ position of the sugar. The drug is transported into cells by a carrier-mediated process with shared affinity for deoxycytidine [109]. Once ara-C has entered the cell, cytotoxicity is dependent on formation of the triphosphate ara-CTP, which is responsible for inhibition of DNA synthesis.

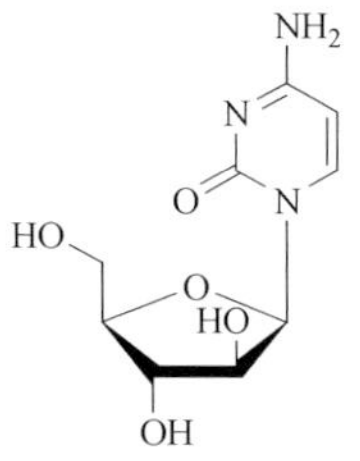

Cytosine arabinoside

The precise *mechanism* by which ara-CTP inhibits DNA synthesis remains uncertain. Ara-CTP does inhibit both DNA polymerase α [110] and β [111]. The former is essential for DNA synthesis, the latter for DNA repair. Recently, studies have shown that ara-CTP can be directly incorporated into DNA as well, and this pathway correlates strongly with cytotoxicity. The extent of drug incorporation into DNA is proportional to cell kill in both acute myelocytic and promyelocytic leukemia cells [112]; incorporation and cytotoxicity can be modulated by thymidine [113]. Also ara-C-substituted DNA is unstable under conditions of alkaline elution, suggesting drug-induced strand fragility and breakage [114]. Ara-CTP incorporation also directly blocks strand elongation [115] and causes premature strand termination [116]. This results in accumulation of DNA peaks of small sizes, suggesting that preexisting DNA may be nicked following exposure to ara-C [117]. A further effect of ara-C exposure is inhibition of DNA repair, as determined by alkaline elution studies in L 1210 cells [118]. As might be expected, cells pretreated with ara-C are more sensitive to ionizing radiation, suggesting potentiation of radiation damage by inhibition of DNA repair [119], [120]–[123].

2.4.1. Mechanisms of Resistance

A number of mechanisms of resistance to ara-C have been described. Ara-C itself is a prodrug that is metabolized first to ara-CMP by deoxycytidine kinase, then to ara-CDP by pyrimidine nucleoside monophosphate kinase, and finally to ara-CTP by nucleoside diphosphate kinase. Resistant cell lines lack the initial, rate-limiting activating enzyme [115], [124].

Since the 1980s, HPLC has been used to separate and semipurify deoxycytidine kinase, pyrimidine nucleoside monophosphate kinase, and pyrimidine nucleoside diphosphate kinase from cell extracts [125]. In murine leukemia cells selected for ara-C resistance, deoxycytidine kinase was found in lower specific activity, and similar results have been reported for other systems [126].

In addition, expansion of the intracellular pool of the physiologic substrate deoxycytosine triphosphate can inhibit ara-C activation by feedback inhibition of the initial activating enzyme [127]. Moreover, increased dCTP pools may also compete directly with ara-CTP for DNA polymerase. Experiments have shown a relationship between duration of leukemia remission and the ability of tumor cells obtained from patients to form and retain ara-CTP in vitro [128], but the mechanisms underlying this relationship remain to be established.

Although some workers have reported that increased drug catabolism by cytidine deaminase may underlie both de novo and acquired drug resistance in patients with leukemia [129], a definite clinical correlation between response and levels of this enzyme has not been established in several studies [130]–[132].

2.4.2. New Analogs

Several ara-C analogs have been rationally designed with the goal of overcoming specific mechanisms of resistance. Compounds with lipophilic side chains are relatively resistant to inactivation by cytidine deaminase. Another analog with lipophilic modifications in the side chain is enocitabine (NSC-239336, BHAC) [*055726-47-1*] [133], [134]. N^4-Behenoyl-ara-C (BHAC) has undergone clinical trial in patients with acute leukemia [135]. Despite its lipophilicity, the drug does not enter the cerebrospinal fluid and concentrates in bone marrow and red blood cells. N^4-Palmitoyl-ara-C can be administered orally and appears to be more active than the parent compound in preclinical models [136]. Although developed as an antiviral agent, 2'-fluoro-5-iodo-1-β-D-arabinofuranosylcytosine (FIAC) has significant antitumor activity [137]. Interestingly, this compound may be relatively cytotoxic for cells with high levels of cytidine deaminase because the catabolism product, FIA-uracil, is more toxic than the parent compound [138].

2.5. Deoxycytidine and Analogs

The pyrimidine antimetabolite gemcitabine (LY-188011, dFdC) [*095058-81-4*], 2'-deoxy-2',2'-difluorocytidine, $C_9H_{11}F_2N_3O_4$ is an analog of deoxycytidine and a result of a program initiated at Lilly Research to synthesize fluorinated D-ribose and fluorinated nucleosides [139]–[142]. The difluorinated analog of cytarabine, gemcitabine was identified as novel antimetabolite with a broad spectrum of antitumor activity.
Trade Names: Gemicitabine monohydrochloride [*122111-03-9*] is on the market as Gemzar in the indication of palliative treatment of locally advanced or metastatic non-small cell lung cancer.
Mechanism of Action. Gemcitabine shows good activity against human leukemic cell lines, a number of murine solid tumors, and human tumor xenografts [143]–[146]. Gemcitabine was significantly more cytotoxic than cytarabine in Chinese hamster ovary cells. The major cellular metabolite is the 5'-triphosphate of gemcitabine. The cytotoxicity was competitively reversed by deoxycytidine, suggesting that the biological activity required phosphorylation by deoxycytidine kinase [145].

Tumor-bearing mice were treated with either gemcitabine or cytarabine (20 mg/kg). DNA synthesis reached 1 % of control levels upon administration of gemcitabine. The greater accumulation of gemcitabine-5'-triphosphate compared with cytarabine-5'-triphosphate may cause greater cytotoxicity and therapeutic activity [146].

Further gemcitabine may enhance its own cytotoxic effects by self-potentiation mechanisms that act on, e. g., deoxycytidine monophosphate deaminase, deoxycytidine kinase or on DNA-synthesis [147], [148].

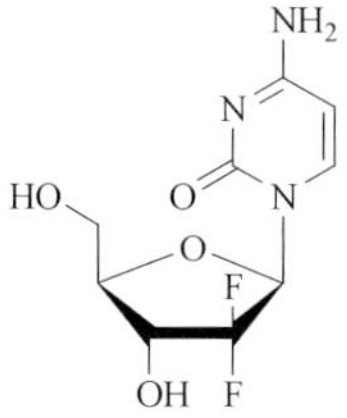

Gemcitabine

2.6. 2-Halopurines and Analogs

Cladribine (2-CdA, RWJ-26251) [*004291-63-8*], is a purine deoxyribonucleoside with remarkable antileukemic activity. It represents a significant advance over existing therapy because it is given as a single 7-day continuous treatment, thus minimizing the side effects observed with multiple treatments. Remission rates of up to 89 % lasting for up to 25 months were observed in clinical trials in patients with hairy cell leukemia [149]–[152]. The antimetabolite used for first line treatment of hairy cell leukemia [153].
Trade Name: Leustatin.

Fludarabine phosphate (NSC-312887, 2-F-ara-AMP) [*075607-67-9*] [154], 2-fluoro-9-(5-*O*-phosphono-*β*-D-arabino-furanosyl)-9H-purin-6-amine, $C_{10}H_{13}FN_5O_7P$, M_r 365.21, another cytotoxic purine antimetabolite, acts via inhibition of DNA synthesis. The product is used for treatment of patients with chronic lymphocytic leukemia.
Trade Names: Fludara (Berlex), Benefluor (Schering AG).

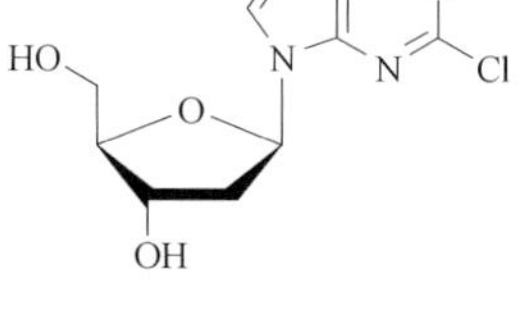

Cladribine

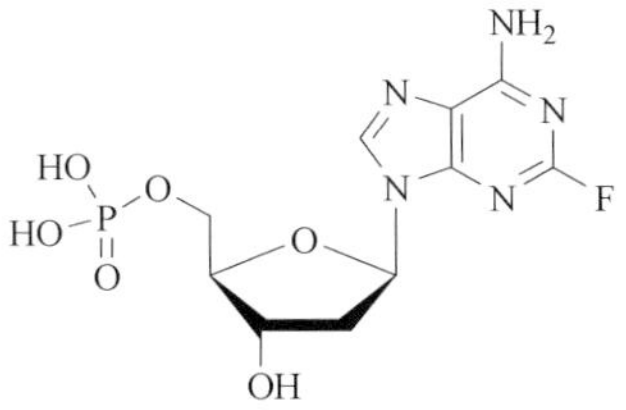

Fludarabine

Mechanism of Action. Fludarabine and its soluble derivatives interfere with phosphorylation, e.g., in L 1210 cells. Fludarabine behaves more like an analog of deoxycytidine than adenine or deoxyadenine as indicated by reports [155] demonstrating that the presence of fluorine in the 2-position of the adenine ring alters its function as a substrate for deaminase and nucleoside kinases. This results in differences in biological activity and metabolism. Halogenation does not simply block deamination, but also influences the enzyme that carries out the phosphorylation, as a result cytotoxicity is increased [156]. Fludarabine phosphate may selectively inhibit the incorporation of thymidine and uridine into the DNA molecule by inhibiting both ribonucleotide reductase [157] and DNA polymerase [158]. The maximum tolerated dose (MTD) in heavily pretreated patients with advanced malignancy/solid tumors on the daily regimen was about 15 mg/m^2. Granulocytopenia and thrombocytopenia were dose-limiting [159]–[161].

Pentostatin. The cytotoxic and immunosuppressant pentostatin (NSC-218321, CI-825, PD-81565, YK-176, 2-deoxycoformycin, 2'-dCF) [*063677-95-2*], (*R*)-3-(2-deoxy-*β*-D-erythro-pentofuranosyl)-3,6,7,8-tetrahydroimidazo[4,5-d][1,3]diazepin-8-ol, $C_{11}H_{16}H_4O_4$, M_r 268.13, *mp* 220–225 °C, can be isolated from the fermentation broth of *Streptomyces antibioticus* NRRL 3238 [162], [163] or *Aspergillus nidulanus* Y 176-2 or *Emericella* [164]. The adenosine nucleoside analog pentostatin, is the most potent inhibitor of adenosine deaminase, which is an important and ubiquitous cellular enzyme. The inhibition of this enzyme leads to accumulation of dATP which inhibits ribonucleotide reductase and thus DNA synthesis. Pentostatin was launched for treatment of hairy-cell leukemia refractory to α-interferon.
Trade Names: Nipent (Parke-Davis, Lederle), Coforin (Katetsuken).

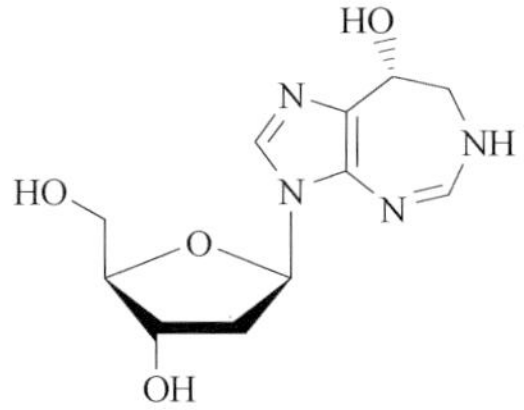

Pentostatin

Mechanism of Action. The highest activity of adenosine deaminase is found in lymphoid tissue as well as in the malignant cells of acute lymphoblastic leukemia [165]. Since pentostatin is the most potent inhibitor of adenosine deaminase it was expected to possess antitumor properties against certain malignancies, especially acute leukemias, chronic myelogenous leukemia, and lymphomas. Surprisingly, when tested against murine tumor models and also against various tumor cell cultures no antitumor effect was found [166]. However, in a phase I clinical trial the compound was found to produce a drop in lymphoblast count and showed antitumor activity in acute leukemia and lymphoma. The toxicity observed consisted mostly of CNS side effects including nausea, hepatic and renal dysfunction. In combination with vidarabin (Ara-A) (→ Synthetic Chemotherapeutic Agents) superiority to monotherapy was demonstrated in various studies [167], [168]. Durable complete remission was observed after pentostatin treatment in patients with hairycell leukemia resistant to α-interferon [169]–[174].

2.7. 6-Mercaptopurine and 6-Thioguanine

The purine antimetabolites, 6-mercaptopurine (NSC-755) [*50-44-2*], 1,7-dihydro-6*H*-purine-6-thione, 6-MP, $C_5H_4N_4S$, M_r 152.19; and 6-thioguanine (NSC-63878) [*154-42-7*], 2- amino-1,7-dihydro-6*H*-purine-6-thione, 6-TG, $C_5H_5N_5S$, M_r 167.21, have been used for several decades in the treatment of leukemia and certain other neoplastic diseases. The 6-thiopurine analog, 6-MP, was first synthesized in 1952 [175]; subsequently the same workers synthesized 6-TG in 1955 [176]. The clinical evaluation of 6-MP in the treatment of acute leukemia and chronic myelocytic leukemia was carefully studied in 1953 [177]. Shortly thereafter, the 2-amino analog of 6-MP, 6-TG, was submitted for clinical evaluation as an antitumor agent [178].

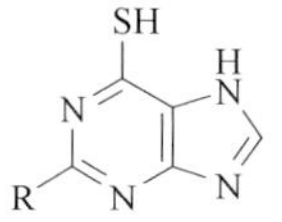

R = H, 6-Mercaptopurine
R = NH$_2$, Thioguanine

6-Mercaptopurine is used for maintenance therapy of acute lymphocytic and acute myelogenous leukemia, and 6-TG is used primarily for remission induction in acute myelogenous leukemia. Although 6-MP and 6-TG have an important role in chemotherapy of leukemia patients, the purine analogs produce low-response rates in patients with solid tumors, lymphomas, and chronic lymphocytic leukemia.

2.7.1. Mechanism of Action

6-Mercaptopurine has been used as an antineoplastic and immunosuppressive agent for decades, but the precise mechanism by which it exerts its cytotoxic effects has not yet been established. Both 6-MP and 6-TG must be converted to their thiol nucleotide form, which is the active cytotoxic moiety. The conversion is catalyzed by hypoxanthine guanine phosphoribosyltransferase (HGPRT) [179]–[181] and the reaction is dependent on the phosphoribosylpyrophosphate (PRPP) level in the cells [182]–[186]. 6-Thiouric acid (6-TU) is the major catabolic product of 6-MP. The rapid conversion of 6-MP to 6-TU by xanthine oxidase [187], [188] in leukemic cells may be a possible mechanism of 6-MP resistance [182], [189], [190]. The antineoplastic effect of 6-TG is similar to that of 6-MP. In its nucleotide form, 6-TG inhibits de novo purine biosynthesis and purine interconversions [191]–[195].

The cytotoxicity of 6-MP and 6-TG has been linked to (1) the interference with de novo purine biosynthesis and purine interconversions, (2) the inhibition of in vitro RNA synthesis, and (3) the incorporation into DNA during S phase, resulting in a deformation of the DNA.

The interference with de novo purine biosynthesis by 6-MP is regulated by 6-MP nucleotides. These nucleotides inhibit the enzyme 5-phosphoribosylpyrophosphate amidotransferase that catalyzes the initial reaction in the purine biosynthetic pathway [196]. The 6-MP nucleotides also inhibit conversion of inosine monophosphate (IMP) to adenine monophosphate (AMP) and to xanthine monophosphate (XMP), and limit the availability of XMP to form guanine monophosphate (GMP), thereby interfering with the supply of purine precursors for nucleic acid synthesis [181]. In the 1980s two new findings pertaining to the cytotoxicity mechanism in tumor cells have been reported. Studies of human lymphoma revealed that 6-MP was a potent inhibitor of cellular RNA synthesis and that 6-thioITP inhibited both the RNA polymerase I and RNA polymerase II activities of these cells [197]. These data suggested that direct inhibition of the enzymes mediating transcription by 6-thio-IMP may be one of the mechanisms for the cytotoxic action of 6-MP in human tumor cells. Using 6-TG as a cytotoxic agent resulted in severe chromosome damage in wild-type CHO cells [198]. Gross unilateral chromatid damage resulted, and the unilateral nature of this damage was probably due to malfunction of 6-TG-containing DNA as a replication template.

2.7.2. Mechanism of Resistance

Several mechanisms of resistance to these agents have been described in experimental tumors and relate to the pathways of antimetabolite activation and degradation. A decreased HGPRT activity in tumor cells diminishes antimetabolite activation, and this resistance pathway has been reported by several workers [199]–[202]. However, the HGPRT-regulated mechanism as a basis for drug resistance in human leukemic cells is relatively uncommon [203]. The resistance of 6-MP is related to an increase in alkaline phosphatase in sarcoma cells [204]. Increased alkaline phosphatase, a membrane-bound enzyme that converts the active mononucleotide to 6-thioinosine and inorganic phosphate, has also been reported in human leukemia patients resistant to drug treatment [205], [206]. Another enzyme in the degradation pathway of 6-MP that needs to be considered for drug efficacy and bioavailability is xanthine oxidase, which is responsible for converting 6-MP to 8-OH-6-MP and subsequently to thiouric acid, which is excreted through the urine.

To understand further the bioavailability and pharmacokinetics of thiopurines, the effect of allopurinol on 6-MP catabolism was studied. Allopurinol, an analog of hypoxanthine, enhances the therapeutic efficacy of 6-MP by inhibiting xanthine oxidase. The urinary excretion of 6-MP metabolites is markedly reduced in patients treated with allopurinol. Furthermore, allopurinol increased the plasma level of 6-MP in rabbits [207]. The data suggest that inhibition of 6-MP catabolism by allopurinol may contribute to a greater availability of 6-MP to tissues. Studies of the effect of allopurinol on the kinetics of oral and intravenous 6-MP in Rhesus monkeys and in humans demonstrated that allopurinol pretreatment resulted in a nearly 400 % increase in peak plasma concentration of 6-MP in monkeys and a 500 % increase in humans, but only when 6-MP was administered orally [208]. Allopurinol pretreatment had no effect on the kinetics of intravenously administered 6-MP. This difference is due to the action of allopurinol on liver or intestinal xanthine oxidase and inhibition of first-pass metabolism of oral 6-MP. This finding may explain the low and variable plasma levels of mercaptopurine in patients with acute lymphoblastic leukemia treated with oral 6-MP [209]. Although these studies emphasize the catabolic pathway for purines, all the purine and pyrimidine metabolic enzymes may be important to the bioavailability and activation of antimetabolites.

To clarify the mechanisms of resistance to thiopurines and potential drug interactions in tumor cells, current studies have focused on the thiopurine-resistant cell lines deficient in HGPRT (L1210) and the regulation of PRPP formation in thiopurine-resistant cell lines.

The thiopurines are inactive in the base form and must be converted to their respective nucleotides. This activation step requires PRPP as the cofactor and HGPRT as the enzyme to catalyze the conversion to nucleotide. A major biochemical effect of methotrexate is the suppression of purine biosynthesis and expansion of the PRPP pool. Studies of the cytotoxic and biochemical interaction of methotrexate and 6-TG in L1210 mouse leukemia cells demonstrated that methotrexate can markedly enhance 6-TG activity [210]. Preexposure of cells to methotrexate resulted in a large increase in cytotoxic potency of 6-TG, whereas simultaneous exposure caused an antagonism of 6-TG cytotoxic activity. Although PRPP pools were not measured quantitatively, the effect of methotrexate preexposure was to increase PRPP pools, enhance the activation of 6-TG to 6-TG monophosphate, and thereby increase its incorporation into RNA.

2.7.3. New Analogs

Alkyl disulfide derivatives have been used as masked compounds of 6-mercaptopurines. The antitumor effect of these derivatives has been measured by their ability to decrease the degradation of 6-MP monophosphate. Of the seven 6-alkyl disulfide derivatives of 6-MP and 6-TG in L 1210 leukemia cells tested, decyl derivatives of both 6-MP and 6-TG were the most effective with therapeutic ratios as high as 50 and 48, while those of parent compounds were 6.2 and 5.0, respectively [211]. Partial circumvention of thiopurine resistance may have resulted from cellular uptake of intact acylated bis(6-MP-9-β-D-ribofuranoside-5′)pyrophosphate derivatives in 6-MP-resistant human cell lines deficient in HGPRT (L 1210) [212].

The 6-MP-resistant sublines of P 388 and L 1210 leukemia are also sensitive to two new *purine antagonists:* 5-carbamoyl-1*H*-imidazol-4-yl piperonylate and 4-carbamoylimidazolium-5-olate [213]. These two new purine analogs kill 6-MP-resistant cells by suppressive de novo purine synthesis. The activation of those new purine analogs is mediated by adenine phospho-ribosyltransferase.

3. Alkylating Agents

Historically, alkylating agents were important in the early development of cancer chemotherapies. Victims of sulfur mustard gas exposure in World War I were found to have severe lymphoid aplasia as well as pulmonary irritation [214]. This led to clinical trials of the related, but less toxic, nitrogen mustard derivative, which produced tumor regressions in lymphoma patients [215].

Clinical use of nitrogen mustard today is mostly limited to the treatment of lymphomas, especially Hodgkin's disease, where it is used in a multidrug regimen called MOPP (nitrogen mustard, vincristine, procarbazine, and prednisone). Three widely used derivatives of nitrogen mustard used in patients with malignancies today include melphalan, cyclophosphamide, and chlorambucil (Sections 3.2, 3.3, 3.4).

The mechanism by which alkylating agents act can be classified as either S_N1 or S_N2. In the S_N1 *reaction,* a highly reactive intermediate forms initially and quickly reacts with a nucleophile to produce an alkylated product. This reaction follows first-order kinetics because the rate-limiting step is formation of the intermediate. The S_N2 *reaction* is a second-order reaction and thus is dependent on the concentration of both the alkylating agent and its target nucleophile [216]. In general, alkylating agents that react by an S_N1 mechanism, such as nitrogen mustard, are less selective in their reactions than S_N2 agents, but this rule is not always true. Selectivity also depends upon membrane permeability, charge, and reactivity of the drug.

3.1. Nitrogen Mustard

3.1.1. Mechanism of Action

Nitrogen mustard (NSC-762) [*51-75-2*], mechloroethamine, HN2, $CH_3N(CH_2CH_2Cl)_2$, $C_5H_{11}Cl_2N$, M_r 156.07, is activated through loss of one of its chlorines. The α carbon then reacts with the nucleophilic nitrogen to form the positively charged, highly reactive, cyclic aziridinium compound, which is attacked by nucleophiles to give the initial alkylated product. The second chlorine can also leave, initiating a second alkylation, which produces a cross-linked alkylation between two nucleophiles.

Because HN2 bonds covalently to many biologic molecules, such as DNA, RNA, and proteins, the alkylated sites responsible for its cytotoxicity are difficult to determine. However, studies have shown that cytotoxicity is likely to result from inhibition of DNA synthesis by damaging the DNA template [217]–[219]. The DNA molecule is rich in potential sites for alkylation, including the phosphate groups in the sugar phosphate backbone structure, and the oxygen and nitrogen sites in the purine and pyrimidine bases. However, the tendency for alkylation to occur in the N-7 position of guanine is enhanced. This may be mediated by the increased nucleophilic characteristics of the N-7 deoxyguanosine due to base stacking and charge transfer [220]. Other preferred sites of alkylation in decreasing order are the N-1 of adenosine, the N-3 of cytidine, and the N-3 of thymidine [221].

Bifunctional alkylating agents, such as HN2, produce intra- and interstrand cross-linking between DNA in the double-helix structure preferentially at the N-7 guanosine site. Thus, these bifunctional compounds are more effective antitumor agents than their monofunctional analogs; however, increasing the number of alkylating sites on the agent beyond two does not appear to increase antitumor activity [222]. This evidence suggests that DNA cross-linking is critical for alkylator activity. Further evidence for the importance of cross–linking to cytotoxicity comes from alkaline elution studies that can detect low levels of cross-linking in cells exposed to minimal doses of bifunctional alkylating agents [223].

In contrast, monofunctional agents, such as procarbazine and dacarbazine, do not produce DNA interstrand cross-links and appear to exert toxicity by producing single-strand DNA breaks. The increased carcinogenicity seen with some monofunctional alkylating agents may be due to incorrect base pair substitution by DNA repair enzymes that could result in malignant transformation.

3.1.2. Mechanisms of Drug Resistance

Several mechanisms have been elucidated for nitrogen mustard and other bifunctional alkylating agents. First, resistant cells with defective drug transport have been described. Nitrogen mustard enters cells by an active transport system that is physiologically utilized for choline transport. Lymphoma cells resistant to nitrogen mustard demonstrate decreased drug uptake by this specific active transport site, which also decreases its uptake of choline [224]. Second, cytosolic increase in nonprotein sulfhydryl levels [225] and higher nonprotein-bound thiol compounds that could inactivate the drugs before they reach the nucleus have been found in nitrogen mustard-resistant cells [226]. Third, enhanced repair of DNA cross-linking by repair enzymes has been demonstrated in vitro [227] and in vivo [228].

3.2. Melphalan

Melphalan (NSC-8806) [148-82-3], 4-[bis(2-chloroethyl)amino]-L-phenylalanine, L-phenylalanine mustard, $C_{13}H_{18}Cl_2N_2O_2$, M_r 305.20, was rationally designed and synthesized as a phenylalanine derivative of nitrogen mustard with the aim of obtaining increased specificity against melanoma tumor cells that utilize phenylalanine or tyrosine to produce melanin.

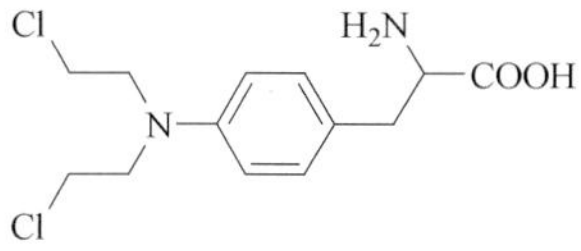

Melphalan

Although this compound does not exhibit specific antimelanoma activity, it has a broad-spectrum cytotoxicity in multiple myeloma, breast cancer, and lymphomas.

3.2.1. Mechanism of Action

Melphalan is a bifunctional alkylating agent but differs from HN2 mainly by the presence of an aromatic ring. This ring reduces the nucleophilicity of the nitrogen atom by withdrawing electrons, making the drug less reactive. Thus, it can be taken orally. It retains its alkylating activity but is more selective than nitrogen mustard because it is less likely to form the unstable and highly reactive aziridine intermediate indiscriminately. Like nitrogen mustard, it forms DNA cross-links that are critical for cytotoxic effect. Its cellular uptake is also mediated by an active, energy-dependent transport mechanism shared with leucine and glutamine uptake [218]. Thus, high concentrations of leucine or glutamine can reduce the cytotoxicity of melphalan in a marrow colony-forming unit assay [229]. Another transport mechanism, although less active, can be used by melphalan, and this mechanism is shared by the neutral amino acids alanine, cysteine, and serine [218].

3.2.2. Mechanism of Resistance

Decreased transport of melphalan into drug-resistant leukemia cells has been demonstrated and correlated to the melphalan-resistant phenotype. Specifically, a mutation in the higher velocity transport system has been suggested that results in a decreased affinity of the carrier protein for melphalan and leucine [230]. Evidence for the other two mechanisms of resistance seen with nitrogen mustard, i.e., increased intracellular thiol compounds and increased DNA crosslinking repair also, exists.

3.3. Cyclophosphamide

Cyclophosphamide (NSC-26271) [*50-18-0*], *N,N*-bis(2-chloroethyl)tetrahydro-2*H*-1,3,2-oxazaphosphorin-2-amine 2-oxide, $C_7H_{15}Cl_2N_2O_2P$, M_r 261.10.

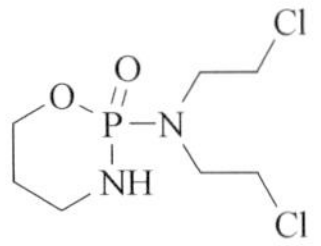

Cyclophosphamide

This widely used bifunctional, cyclic alkylating agent has important clinical use in lymphomas, leukemias, sarcomas, carcinomas of breast and ovary, as well as childhood malignancies. The compound was rationally designed based on data that tumor cells possess high concentrations of enzymes capable of cleaving the P–N bond. This reaction would activate drug by release of the potent antitumor agent phosphoramide mustard. Cyclophosphamide requires hepatic activation by oxidase enzymes. First, it is metabolized by liver microsomes to hydroxycyclophosphamide, which is spontaneously tautomerized to aldophosphamide. Aldophosphamide reaches peripheral tissues and tumors where it is hydrolyzed to yield the active antitumor agent phosphoramide mustard and acrolein. Acrolein has very weak antitumor activity and, when concentrated within the bladder by excretion, can cause hemorrhagic cystitis [216].

3.3.1. Mechanism of Action

Phosphoramide mustard can undergo similar bifunctional alkylation as nitrogen mustard. Also, because of its need for metabolic conversion for biologic activity, cyclophosphamide can be given orally or intravenously.

3.3.2. Mechanism of Resistance

Similar mechanisms of resistance occur for cyclophosphamide as for nitrogen mustard. Also, defective metabolic conversion by the hepatic microsomal system could serve to decrease the bioavailability of aldophosphamide to tumor tissue, but it is not known whether this is significant.

3.4. Chlorambucil

Chlorambucil (NSC-3088) [305-03-3], 4-[bis(2-chloroethyl)amino]benzenebutanoic acid, $C_{14}H_{19}Cl_2NO_2$, M_r 304.23. This drug is a close congener of melphalan and exhibits similar stability because of the electron-withdrawing properties of the aromatic ring. It is given orally and has proven efficacy in treating chronic lymphocytic leukemias, multiple myeloma, and lymphomas.

Chlorambucil is thought to have activation properties and mechanisms of resistance similar to melphalan.

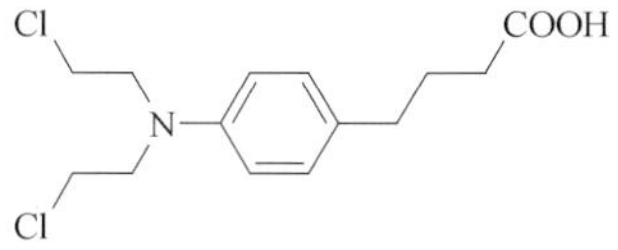

Chlorambucil

3.5. Thio-TEPA

Triethylenethiophosphoramide (NSC-6396) [52-24-4], 1,1′,1″-phosphinothioylidyne-trisaziridine, thio-TEPA, $C_6H_{12}N_3PS$, M_r 189.23. This agent is representative of alkylating agents that have two or more aziridine rings. It has clinical activity against the same tumors as nitrogen mustard and has been used clinically in carcinomas of the breast and ovary. Thio-TEPA is also indicated for intrathecal therapy of meningeal carcinomas. The reactivity of the aziridine groups is increased by protonation; thus thio-TEPA is most active at low pH and has been used to cause sclerosis of malignant pleural effusions that often have an acidic pH.

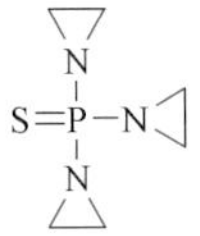

Thio-TEPA

The mechanisms of action and resistance are similar to those of nitrogen mustard.

3.6. Ifosfamide

Ifosfamide, or isofosphamide (NSC-109724) [*3778-73-2*], *N*,3-bis(2-chloroethyl)-tetrahydro-2*H*-1,3,2-oxazaphosphorin-2-amine 2-oxide, $C_7H_{15}Cl_2N_2O_2P$, M_r 261.07, is an analog of cyclophosphamide [231]. It has approximately one-third the alkylating activity of cyclophosphamide and requires hepatic microsomal conversion to its active form. Also, the rates of conversion by metabolism are similar in both drugs, but less biologically active alkylating moieties are formed in ifosfamide [232]. It has shown promising results in refractory pediatric bone and soft tissue sarcomas, refractory testicular tumors, and Wilms' tumor in children [233], [235].
Trade Name: Ifex, Holoxan (ASTA Medica)

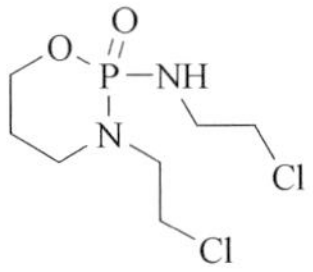

Ifosfamide

3.7. Estramustine

From the numerous analogs few found a way into the clinic or to the market. However, estramustine [*489-15-0*] was successfully launched for treatment of prostate cancer [234].

Trade Name: Estracyt (Pharmacia/Upjohn)

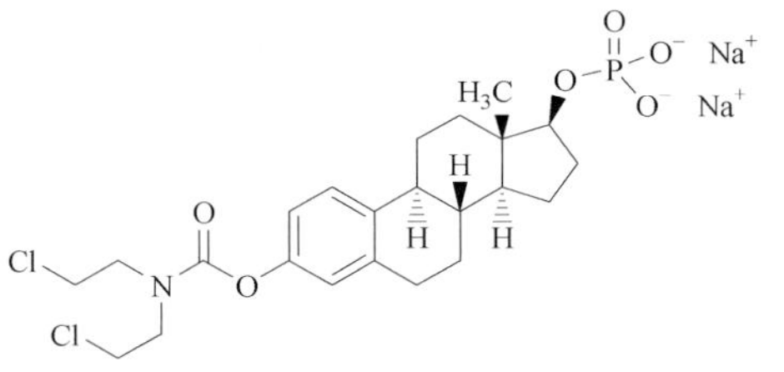

Estramustine

3.8. Nitrosoureas

The clinically useful nitrosoureas include carmustine (NSC-409962) [*154-93-8*], N,N'-bis(2-chloroethyl)-N-nitrosourea, BCNU, $C_5H_9Cl_2N_3O_2$, M_r 214.04; lomustine (NSC-79037) [*13010-47-4*], N-(2-chloroethyl)-N'-cyclohexyl-N-nitrosourea, CCNU, $C_9H_{16}ClN_3O_2$, M_r 233.69; methylcyclohexylchloroethylnitrosourea (NSC-94941) [*52662-76-7*], 1-(2-chloroethyl)-3-(4-methylcyclohexyl)-1-nitrosourea, methyl-CCNU, $C_{10}H_{18}ClN_3O_2$, M_r 248; streptozotocin (NSC-85998) [*18883-66-4*], 2-deoxy-2{[(methylnitrosoamino)carbonyl]amino}-D-glucopyranose, $C_8H_{15}N_3O_7$, M_r 265.22; and chlorozotocin (NSC-178248) [*54749-90-5*], 2-({[(2-chloroethyl)nitrosoamino]carbonyl}amino)-2-deoxy-D-glucose, DCNU, $C_9H_{16}ClN_3O_7$, M_r 313.69. Streptozotocin is a naturally occurring nitrosourea derived from *Streptomyces acromogenes*.

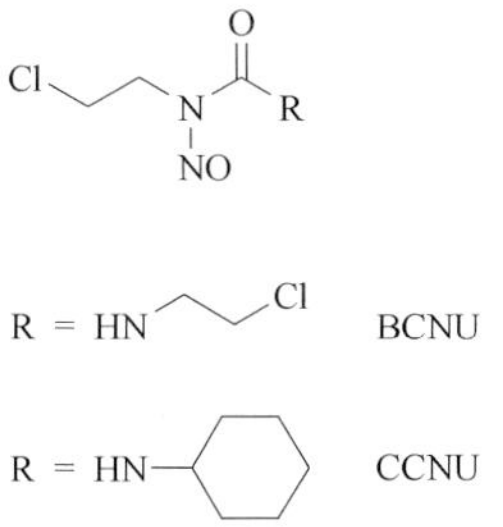

R = HN–CH₂CH₂–Cl   BCNU

R = HN–cyclohexyl   CCNU

This group of agents was developed by careful structure–function studies based on the antitumor activity of methyl-CCNU [235]–[237]. The chloroethyl derivatives were found to possess increased activity and a capacity to cross the blood–brain barrier because of their lipophilic nature. Each of these agents is capable of undergoing alkylating reactions with biologic molecules in a manner similar to the classic mustards through the formation of highly reactive chloroethyl carbocations. With BCNU, each molecule of drug may undergo two such reactions to produce nucleic acid strand breaks, and DNA–DNA and DNA–protein cross-links [238]. Monofunctional CCNU and methyl-CCNU cross-link by initial carbocation formation and alkylation followed by loss of the chloride substituent and formation of a second reactive carbocation. These alkylation reactions appear to be the major mode of cytotoxicity for these agents [239], [240]. With the exception of chlorozotocin, the nitrosoureas may also form an isocyanate compound that may play a role in the toxic side effects of these agents [241], [242], but has little importance in the antitumor effect. In support of this view is the fact that chlorozotocin retains its potent cytotoxic capacity while producing little or no isocyanate compound and reportedly has less marrow toxicity. Recent experimental evidence raises additional questions concerning the role of isocyanates, which may enhance the antitumor activity of these compounds. Methylnitrosourea, which cannot alkylate DNA to produce cross-linking, produces alterations of nuclear protein in a manner similar to BCNU, and isocyanates are considered a possible explanation for these effects from both agents [243]. Also, sensitivity of a Walker tumor line made resistant to bifunctional alkylating agents can be restored by simultaneous treatment with an isocyanate-producing agent [244].

3.8.1. Mechanism of Action

Because of their lipophilicity, nitrosoureas enter cells by passive diffusion as opposed to an active transport mechanism common to the classic alkylating agents [245]. Once inside the cell, alkylating agent exposure results in pancellular covalent binding of drug to proteins, nucleic acids, and to a variety of smaller intracellular molecules. Which of these reactions is critical for cytotoxicity remains uncertain, but the majority of evidence points to interaction directly with DNA as the focal point of cytotoxicity [246]–[248]. The 7 position of guanine is particularly susceptible to alkylation, and accounts for the majority of the total alkylation of DNA [249], [250]. Since the chloroethyl nitrosoureas are each capable of two independent alkylations [238], DNA can be cross-linked by either interstrand or intrastrand processes [251]. Multiple studies using a variety of alkylating agents have confirmed that DNA cross-linking, which leads to inactivation of the DNA template, may well be the key mechanism of cytotoxicity [252]–[254]. Monofunctional DNA alkylations also occur following nitrosourea exposure and these must also be considered cytotoxic as monofunctional alkylating agents, incapable of cross-linking, retain cytotoxic activity. Monofunctional alkylations may produce single-strand DNA breaks by endonuclease cleavage at apurinic sites produced by the alkylation and repair process [255].

3.8.2. Mechanisms of Resistance

The mechanisms of cellular resistance to nitrosoureas remain unclear, although defective drug transport, as has been demonstrated for the classic alkylating agents, would be unlikely given the lack of need for an active membrane transport system. In human glioblastoma cells, increased activity of a specific excision enzyme, guanine-O^6-alkyltransferase, is correlated with in vitro resistance to nitrosoureas [256], and repair mechanisms would seem a likely mechanism since mammalian cells are capable generally of such repair [257].

3.8.3. Analogs

New agents have been synthesized that contain a sugar moiety similar to streptozotocin. Analogs that contain mannose, glucose, ribose, maltose, and galactose have all been produced, with the maltosyl derivative being exceptionally active in a variety of tumors tested [258], [259]. Sugar alcohols, such as mannitol, linked to nitrosoureas retain their cytotoxicity but appear to protect against the marrow toxicity induced by the parent nitrosourea compounds. A number of derivatives with di- and tripeptides containing alkyl nitrosoureas have also been introduced but have met with only moderate preclinical success [260].

There is continuous effort to obtain new nitrosoureas with higher efficiency and/or lower toxicity and several new compounds are undergoing clinical trials. In 1987 ranimustine (NSC-270561, MCNU) [*058994-96-0*], was marketed as Cymerin (Tokyo Tanabe) [261].

The phosphonoalanine derivative fotemustine (S-10036) [*092118-27-9*], (±)-diethyl[1-[3-(2-chloroethyl)-3-nitrosoureido]ethyl]phosphonate, $C_9H_{19}ClN_3O_5P$, M_r 315.69, *mp* 85 °C was approved as Muphoran (Servier) for treatment of disseminated malignant melanoma [262], [263]. Side effects of the related compound BCNU (carmustine) have been linked to the inhibitory effect on a major enzyme of the glutathione pathway, the cytosolic glutathione reductase. Fotemustine has no inhibitory effect on cytosolic glutathione reductase, indicating that fotemustine has a lower toxicity than carmustine [264]–[266]. In addition it has been shown to have lower mutagenicity in the Ames and micronucleus tests compared to BCNU [267]. In the clinical studies the major toxic effects of fotemustine were thrombocytopenia and leukopenia, which were delayed and reversible, nausea and vomiting being mild [268].

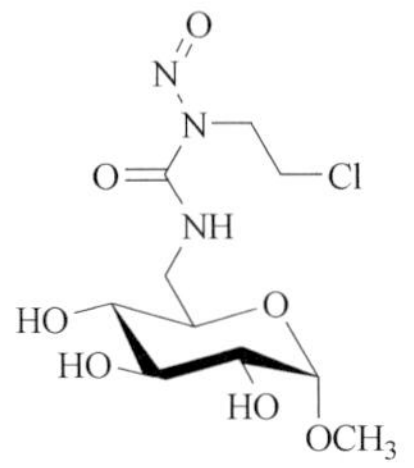

Ranimustine

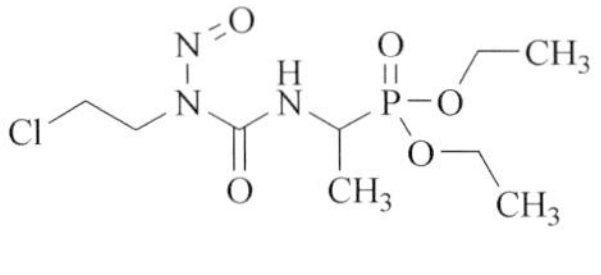

Fotemustine

3.9. Procarbazine

Procarbazine (NSC-77213) [671-16-9], N-(1-methylethyl)-4-[(2-methylhydrazino)-methyl]-benzamide, $C_{12}H_{19}N_3O$, M_r 221.30, is one of a number of substituted hydrazine compounds originally synthesized as monoamine oxidase inhibitors in the early 1960s [269], and found to possess antineoplastic activity, particularly in the treatment of Hodgkin's disease [270]. Procarbazine is nontoxic as the parent compound but undergoes rapid chemical and metabolic degradation to intermediate products that, through a variety of mechanisms, are capable of cytotoxicity [271].

A crucial step in the activation of procarbazine appears to be the production of the azo analog N-isopropyl-α-(2-methyldiazeno)-p-toluamide, which is catalyzed by hepatic microsomal cytochrome P-450 [272], [273]. Further metabolism by hepatic microsomes leads to the production of methyl- and benzyl-azoxy metabolites [272], [273]. Cytotoxic alkylating compounds may be formed from these metabolites via hydroxylation reactions, and these may play the major role in cytotoxicity. This sequence of metabolism and formation of the alkylating intermediates is consistent with the time course of appearance of the active species isolated in the serum and then excreted [274], [275]. In addition to the formation of these alkylating intermediates, methyl- and benzyl-azoxy metabolites may result in free radicals through the formation of diazenes [276], which in the presence of oxygen form free radicals and nitrogen. However, free-radical formation is not likely a major cause of cytotoxicity, because drug activity is preserved when cells are exposed to procarbazine under conditions that do not support the formation of free radicals [277]. The degree of toxicity induced by alkylation and by free-radical formation is unknown.

3.9.1. Mechanism of Action

The exact mechanism by which procarbazine produces cytotoxicity is unknown; however, the effects of its action have been well studied at the cellular level. Chromosome breaks and translocations have been demonstrated in vivo in Ehrlich ascites and L 1210 leukemia cells [278]. Inhibition of nucleic acid and protein synthesis and of a variety of enzymes has also been documented following procarbazine treatment. Inhibition of transfer and nuclear RNA synthesis occurs 2 h after procarbazine exposure and lasts for up to 24 h [279]. Thymidine incorporation into DNA is inhibited concomitantly with protein synthesis, reaching maximum inhibition in 12–16 h [271], [280]. Although many potentially cytotoxic events have been associated with procarbazine administration, it is not clear which of these effects causes cell death.

3.9.2. Mechanisms of Resistance

Resistance mechanisms for procarbazine are poorly understood, and no detailed studies have been reported that illustrate the typical cellular resistance encountered with alkylating agents. Since the drug enters cells by simple diffusion, resistance is unlikely to involve altered transport mechanisms [281]. New pieces of chromosomal material were found in Ehrlich ascites cells made resistant to procarbazine [271], and these may represent gene amplification, perhaps encoding for a target enzyme or detoxifying enzyme.

3.10. Dacarbazine

Dacarbazine (NSC-45388) [4342-03-4], 5-(3,3-dimethyl-1-triazenyl)-1H-imidazole-4-carboxamide, DTIC, $C_6H_{10}N_6O$, M_r 182.18, was synthesized in the late 1950s as an analog of 5-aminoimidazole-4-carboxamide, an intermediate in de novo purine synthesis. The drug is a product of rational synthesis, designed as a false intermediate capable of inhibiting de novo purine synthesis. Despite this theoretical basis for antitumor action, DTIC does not function as a purine analog; instead, it is extensively metabolized to a methylating agent [282]. Similar to procarbazine, DTIC must undergo activation by a microsomal oxidase to form a compound that can spontaneously produce a methyl diazonium ion intermediate that is probably the active metabolite. However, recent evidence suggests a further metabolism to N-hydroxymethyl diazonium ion may be responsible for the selective antitumor effect of the drug [283]. Evidence for methylation of nucleic acids has been demonstrated in tissue culture [284]. An additional metabolic pathway involves the light sensitivity of the drug. Exposure to ultraviolet energy converts the parent compound to metabolites with moderate cytotoxicity in vitro [285], but this probably does not represent an important pathway for cytotoxicity in vivo.

Mechanism of Action. The mechanism of action of DTIC has not been systematically investigated, but it appears that the drug may act during any phase of the cell cycle [286] and can produce inhibition of RNA, DNA, and protein synthesis.

3.11. Hexamethylmelamine

Hexamethylmelamine (NSC-13875)[*645-05-6*], N,N,N',N',N'',N''-hexamethyl-1,3,5-triazine-2,4,6-triamine, altretamine, $C_9H_{18}N_6$, M_r 210.27, was synthesized in 1951 [287]. It has consistent antitumor activity in a variety of solid tumors, including ovarian, lung, and breast cancer.

The compound is almost insoluble in water, and thus, must be administered orally. Neither the mechanism of action nor the products of metabolic breakdown have been firmly established. Following administration of hexamethylmelamine, a spectrum of N-demethylation species has been isolated in the urine. The triazine ring appears unaffected by metabolism, as is evident by almost complete recovery of the intact ring in the urine using ring-labeled compound [288]. Existing evidence suggests possible formation of an alkylating species through N-demethylation [289] or the formation of N-methylol intermediates by hydroxylation of the parent compound [290] to account for the cytotoxic effects of the drug. N-Methylol derivatives have cytotoxic effects in vitro, but whether or not these derivatives are formed in vivo is unclear.

Analogs. Pentamethylmelamine (NSC-118742) [*35832-09-8*], N,N,N',N',N''-pentamethyl-1,3,5-triazine-2,4,6-triamine, $C_8H_{17}N_6$, M_r 196, is the most commonly used analog of hexamethylmelamine and differs by the absence of a single methyl group. Its major advantage is aqueous solubility, allowing an intravenous formulation. Its metabolism, toxicity, and antitumor activity parallel those of hexamethylmelamine [291], [292]. Other metabolites of hexamethylmelamine with varying numbers of methyl groups also possess antitumor activity that, in general, is directly proportional to the number of methyl groups on the triazene ring [292], [293].

3.12. Mitomycin-C

The mitomycins are a family of antibiotics isolated from *Streptomyces caespitosus*.
Mitomycin-C (NSC-26980, MIT-C) [*50-07-7*] [1aR-(1aα,8β,8aα,8bα)]-6-amino-8-[(aminocarbonyl)oxymethyl]-1,1a,2,8,8a,8b-hexahydro-8a-methoxy-5-methylazirino[2',3':3,4]pyrrolo[1,2-a]indole-4,7-dione, $C_{15}H_{18}N_4O_5$, has DNA-alkylating properties

[294]. Mitomycin is applied for treatment of stomach, breast, and gynecological cancers [295], [296]. Dose-limiting are leucopenia and thrombopenia.

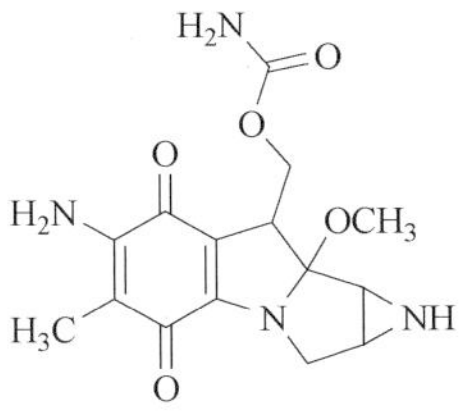

Mitomycin

4. Anthracyclines (→ Antibiotics)

Doxorubicin (NSC-123127) [*23214-92-8*], 10-[(3-amino-2,3,6-trideoxy-α-L-lyxo-hexopyranosyl)oxy]-7,8,9,10-tetrahydro-6,8,11-trihydroxy-8-(hydroxyacetyl)-1-methoxy-5,12-naphthacenedione, adriamycin, $C_{27}H_{29}NO_{11}$, M_r 543.54; and daunorubicin (NSC-82151 for HCl salt) [*20830-81-3*], 8-acetyl-10-[(3-amino-2,3,6-trideoxy-α-L-lyxo-hexopyranosyl)oxy]-7,8,9,10-tetrahydro-6,8,11-trihydroxy-1-methoxy-5,12-naphthacenedione, daunomycin, $C_{27}H_{29}NO_{10}$, M_r 527.51. Doxorubicin is used in the treatment of breast cancer, sarcoma, lymphoma, and small-cell lung cancer. Daunorubicin is used more commonly in acute myelocytic and lymphocytic leukemias.

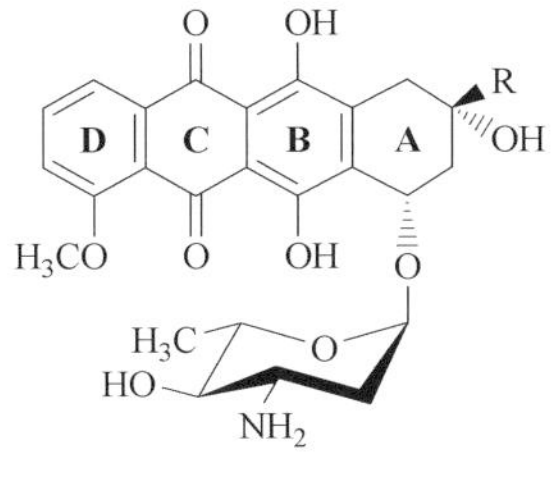

R = COCH$_2$OH, Doxorubicin
R = COCH$_3$, Daunorubicin

Meanwhile epirubicin (4'-epiadriamycin, pidorubicin, IMI-28) [*56420-45-2*], (8S-*cis*)-10-[(3-amino-2,3,6-trideoxy-β-L-arabino-hexopyranosyl)oxy]-7,8,9,10-tetrahydro-6,8,11-trihydroxy-8-(hydroxyacetyl)-1-methoxy-5,12-naphthacenedione, is the most commonly used antineoplastic antibiotic for breast cancer treatment. The total turnover exceeded $ 200 × 10^6 in 1996. The compound can be synthesized by classical chemical synthesis [297] – [299].

Trade Names: Farmorubicin and Pharmorubicin (Farmitalia).

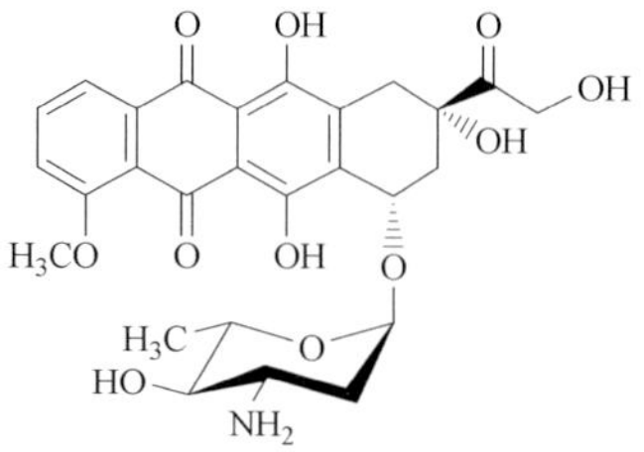

Epirubicin

In addition idarubicin (NSC-256439, IMI-30, DMDR) [*058957-92-99*] and pirarubicin (THP-ADM) [*072496-41-4*], (8S-*cis*)-10-[[3-amino-2,3,6-trideoxy-4-*O*-(tetrahydro-2*H*-pyran-2-yl]-α-L-lyxo-hexopyranosyl]oxy]-7,8,9,10-tetrahydro-6,8,11-trihydroxy-8-(hydroxyacetyl)-1-methoxy-5,12-naphthacenedione, $C_{32}H_{37}NO_{12}$, M_r 627,64 are available as idamycin (Adria) and pinorubicin (Nippon Kayaku), respectively.

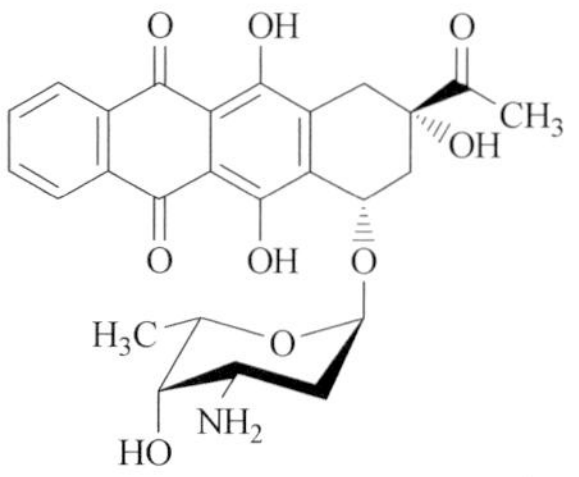

Idarubicin

Pirarubicin is the 4'-*O*-tetrahydropyranyl analog of adriamycin and can be synthesized from adriamycin [300], [301]. The acute cardiac toxicity was significantly less than that of adriamycin and general toxicity lower than that of other analogs [302]–[303], [303]. The main indications for pirarubicin are cancer of the bladder, head and neck, and cervix.

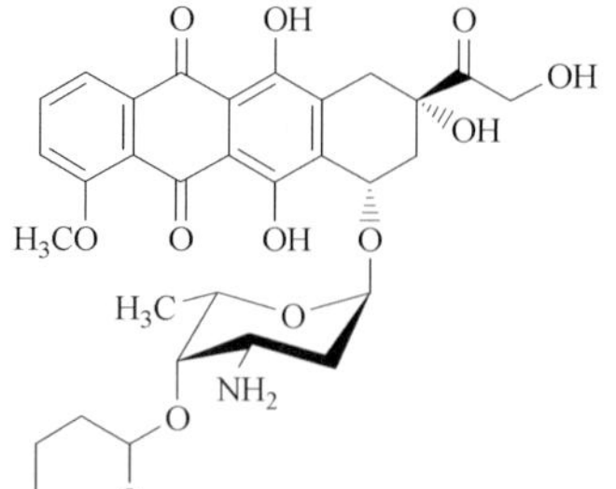

Pirarubicin

The anthracyclines are derived from *Streptomyces* species and are structurally tetracyclic chromophore antibiotics [305]. They are classified by their chromophore, otherwise known as aglycone, structure. The sugar most commonly attached to the aglycones is daunosamine, but other sugars may be involved, and these are mentioned in Section 4.3. The basic tetracyclic aglycone structure of the anthracyclines shares many characteristics with the hydroxyanthraquinones, which are ubiquitous in nature.

4.1. Mechanism of Action

DNA–RNA Binding. The exact mechanism of cytotoxicity by anthracyclines is unknown, but they do have multiple and distinct toxic effects that may kill a tumor cell in one or more ways [306].

Initially these drugs were found to bind DNA by intercalation between base pairs perpendicular to the long axis of the double helix, with the major binding occurring between the B and C rings of the drugs with the bases above and below [307]. The daunosamine sugar is thought to bind ionically with the sugar-phosphate backbone of DNA. The binding association constant is between 10^5 and $10^6 \, M^{-1}$. Intercalation of the DNA causes a partial unwinding of the helix and thus disrupts DNA polymerases and transcription. However, these experiments were done with DNA in vitro; DNA in vivo is organized into chromatin, which is DNA wrapped around a series of histone core particles. Also, it has been established that the anthracyclines can affect every DNA function, including initiation, chain elongation, DNA synthesis, DNA repair, and RNA synthesis [308], [309]. Experimentally these compounds can cause sister chromatid exchanges, single- and double-strand breaks, and alkylation of DNA [310], [311]. Interestingly, some results strongly suggest that inhibition of DNA synthesis is not essential for cell kill. New anthracycline analogs, such as aclacinomycin A, which selectively inhibits preribosomal RNA synthesis and not DNA synthesis, retain cytotoxicity. This suggests other non-DNA-mediated mechanisms of cytotoxicity, such as preribosomal RNA synthesis [309].

Free-Radical Generation. Free-radical formation (highly reactive compounds with an unpaired electron) occurs during the metabolism of anthracyclines. When the microsomal enzyme P 450 reductase or xanthine oxidase interacts with and reduces the ketone oxygen in ring B to O^-, a semiquinone radical intermediate is formed. This interacts with oxygen to produce the superoxide radical with regeneration of the original anthracycline structure [312]. The superoxide radical can serve as a substrate for superoxide dismutase to form hydrogen peroxide, which can interact with the superoxide molecules to form hydroxyl radicals [313]. The superoxide and hydroxyl radicals can interact with and damage cells, especially the hydroxyl radical, which is one of the most reactive substances known. Hydroxyl radicals can also react with purine or pyrimidine bases, amines, and thiols. This free-radical formation is responsible for cardiac damage seen with chronic use of the anthracyclines in the treatment of human malignancies. Evidence exists in many animal models, as well as in humans, that superoxide and hydroxyl radical formation occurs in cardiac tissue, leading to lipid peroxidation of mitochondria and sarcosomes [314], [315]. Since mitochondria account for more than 40 % of cardiac muscle mass, as well as being the major source for ATP needed for contraction, and are coupled to calcium release during the action potential, one can easily visualize how these agents mediate cardiac toxicity. However, it now seems apparent that cardiac tissue lacks catalase and that doxorubicin destroys glutathione peroxidase [315], [316]. Agents that can scavenge free-radicals are under active investigation. Interestingly, there is no evidence to date to link anthracycline free-radical formation to its antitumor activity so that analogs incapable of free-radical formation may demonstrate improved therapeutic index.

Membrane Interactions. Anthracycline binding to cell membranes appears to be an important mechanism for cytotoxicity. Changes in membrane glycoproteins, transmembrane flux of ions, and membrane morphology have been demonstrated in a variety of cells. Doxorubicin binds most tightly to cardiolipin, a phospholipid found in high concentration in mitochondrial and tumor cell membranes but in low concentration in normal cell membranes [317]. Also, membrane redox potential changes occur with drug binding, and may promote free-radical generation. Perhaps the most interesting and important finding to date on this subject is that doxorubicin, covalently linked to beads to prevent cell entry of the drug, retains cytotoxic effects [318]. This suggests that doxorubicin does not need to enter cells or interact with DNA to mediate cell kill.

Another potential mechanism of action may be the dissociation of the cytochrome oxidase electron transport chain for ATP generation [319]. Cytochrome c oxidase requires cardiolipin for activity, and doxorubicin can remove cardiolipin from the enzyme complex, thus inactivating the enzyme.

Metal Ion Chelation. The anthracyclines are capable of chelating ions, including copper, calcium, magnesium, zinc, and iron. Of the metal ions, the tightest complex seems to be with iron(III). Doxorubicin–iron chelates can act as a redox catalyst for electron transfer from glutathione to oxygen, which leads to formation of cytotoxic oxygen radicals. This reaction can also utilize hydrogen peroxide and superoxide, leading to hydroxyl radical formation. Thus, evidence exists that the anthracycline–iron complex can mediate free-radical formation capable of lipid peroxidation and cell damage [320]. Whether this is a mechanism of tumor cell kill is unknown, but the phenomenon may be important to cardiac toxicity.

4.2. Mechanism of Resistance

Probably the most common and important mechanism by which tumor cells become resistant to the anthracyclines is decreased net intracellular accumulation. Many studies have shown that doxorubicin-resistant tumor cells are capable of effluxing the drug more efficiently than their parent sensitive cells [355]. In fact, this mechanism of drug resistance may be responsible for resistance to a variety of structurally unrelated compounds with different modes of antitumor action, such as the vinca alkaloids and actinomycin D. This has been termed pleiotropic drug resistance, and recently several laboratories have shown the reversal of this resistant phenotype by co-incubation with various calcium channel blockers with doxorubicin [356]. These compounds increase the net intracellular drug accumulation in resistant cells, but the precise mechanism of this action is uncertain.

4.3. Analogs

As mentioned in Section 4.1, anthracycline tumor toxicity is probably related to intercalation with DNA, preribosomal RNA inhibition, and/or membrane binding effects, while cardiac toxicity may be due to free-radical formation by drug and drug–metal complex. Thus, analogs have been developed with less potential for free-radical formation. One such drug in early clinical development is aclacinomycin A (NSC-208734), which has an aklavinone aglycone structure and is derived as a fermentation product of *Streptomyces* [321]. This drug lacks the 11-hydroxyl, and therefore has about 10% the potential to generate free-radicals in the P 450 reductase system. Although it does not bind DNA as well as doxorubicin, it retains significant antitumor activity.

The National Cancer Institute (USA) has screened hundreds of anthracycline analogs for antitumor activity. From these studies it can be concluded that (1) the amino sugar is not required for activity, (2) disaccharide analogs are generally more active than the parent saccharide in inhibiting RNA synthesis, and (3) in the aglycone, substituents in the 7 and 9 positions are important for activity. Studies have also shown that if the 4'-hydroxyl group is removed, cardiac toxicity is lessened significantly [322]–[324]. Thus, it seems that the amino sugar and/or the 4'-hydroxyl group are major determinants of cardiac toxicity in the anthracyclines.

5. Intercalating Anthracenes and Analogs

5.1. Mitoxantrone

Mitoxantrone hydrochloride (NSC-301739, DAD, CL-232315) [*070476-82-3*], 1,4-dihydroxy-5,8-bis[(2-(2-hydroxyethyl)amino)ethyl]amino-9,10-anthracenedione, is a new type of antineoplastic agent. Mitoxantrone is active against breast cancer, acute leukemia, lymphoma, cervix carcinoma, and liver cell cancer. Unlike the anthracyclines that have a red color, the anthracenediones are deep blue. Mitoxantrone is structurally similar to adriamycin but without the aminosugar at C9 and can be synthesized starting from 1,8-dihydroxy-anthrachinone [325] – [329].

Trade Names: Novanthrone (Lederle / American Cyanamide); Onkotrone (ASTA Medica AWD).

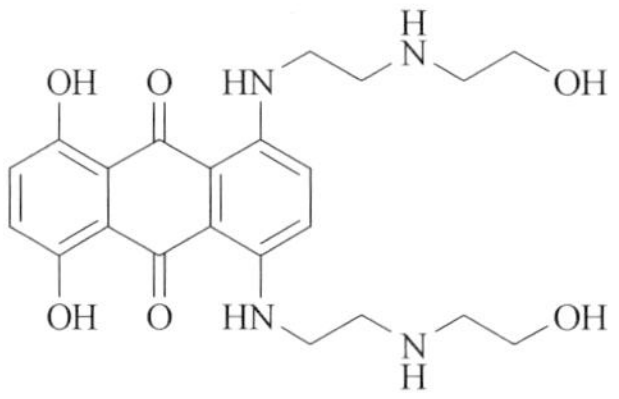

Mitoxantrone

Mechanism of Action. The quinone structure of mitoxantrone was recognized as being similar to that of adriamycin, having a lower cardiotoxic potential. However, its discovery was both a result of serendipity and of rational drug development [325]. The exact mechanism of action by which mitoxantrone exerts its cytotoxic effects has not been fully defined. The cytotoxicity is most likely associated with the action on chromosomal elements, resulting in DNA damage and leading to inhibition of nucleic acid synthesis and the eventual death of the cell. When compared to doxorubicin on an equimolar basis mitoxantrone proved to be six to seven times more potent in inhibiting the incorporation of uridine and thymidine into DNA [330]. Mitoxantrone is a cell phase nonspecific agent and has wide spectrum of activity against several experimental animal tumors. Cross-resistance to adriamycin was not always seen [331] – [333]. Adriamycin-like cardiac toxicity was not found in comparative studies using rats, dogs, rabbits, and monkeys maybe partially due to inhibition of free radical formation and due to the lack of the amino sugar moiety [334], [335]. Although mitoxantrone is not entirely free of cardiac toxicity in humans, minimal nausea and vomiting was observed [336] – [341].

5.2. Analogs

Bisantrene hydrochloride (NSC-337766, ADD, CL-216942) [*071439-68-4*], 9,10-anthracenedicarboxaldehyde bis[(4,5-dihydro-1H-imidazol-2-yl]] dihydrochloride, $C_{22}H_{22}N_8 \cdot 2\,HCl$, M_r 471.39, is an intercalating anthracenebishydrazine cytostatic [342]–[344]. The product was launched for treatment of acute non-lymphocytic leukemia [345].
Trade Names: Zantrene and Cyabin (Lederle).

As solubility is a problem, bisantrene prodrugs with enhanced water solubility (e.g., 199344, see below) were developed [346]–[348].

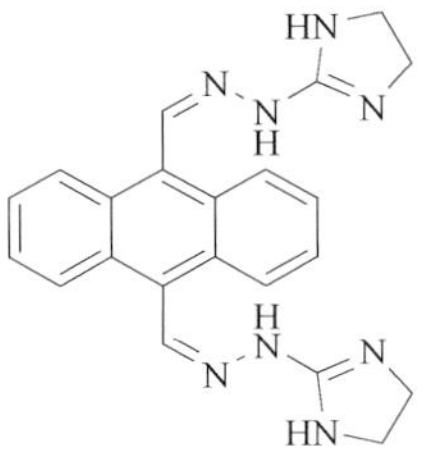

Bisantrene

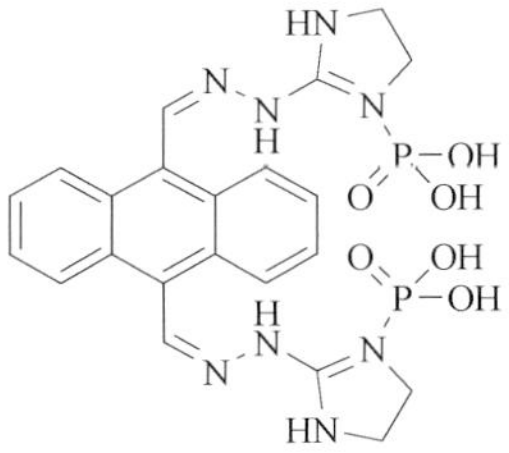

199344

Further intercalating agents such as amsacrine (NSC-156303/NSC-249992, m-AMSA, SN-11841) [*051264-14-3*] [349], [350], or nitracin [*4533-39-5*] [351] have the acridine structure in common.

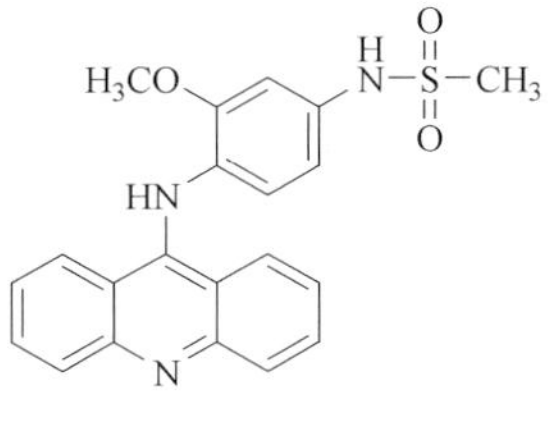

Amsacrine

Trade Names: Amsacrine is marketed as Amsidine, Amecrin, and Ansidyl (Parke-Davis) [352]–[354].

6. Antitumor Antibiotics Other than Anthracyclines

The antitumor antibiotics are a group of microbial products capable of inhibiting tumor growth. This class of antitumor agents has been extensively studied as to mechanism of action and has a rather broad spectrum of activity. In contrast to antibacterial antibiotics, the therapeutic index of these drugs tends to be narrow and toxicity to normal host tissues is considerable.

6.1. Actinomycin D (→ Antibiotics)

The actinomycins are a family of antibiotics derived from *Actinomyces* broths during the 1940s [357]. Structurally, compounds in this class share a common phenoxone ring and two cyclic pentapeptides. The natural products differ among themselves in the amino acid composition of the peptide chains, but only actinomycin D is used in the clinical treatment of cancer, where it demonstrates reproducible activity against gestational choriocarcinoma and Wilms' tumor.

Actinomycin D (NSC-3053) [*50-76-0*], dactinomycin, $C_{62}H_{86}N_{12}O_{16}$, M_r 1255.47.

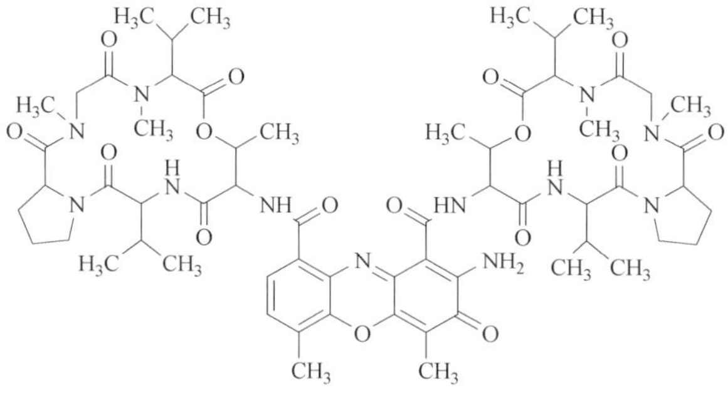

Actinomycin D

Mechanism of Action. Actinomycin interacts with DNA through "pseudo-intercalation" at deoxyguanylyl-3′,5′-deoxycytidine sequences [358], [359], resulting in inhibition of DNA-directed RNA synthesis and inhibition of protein synthesis [360]. Resistance to actinomycin D is linked to cross-resistance against other structurally unrelated amphiphilic drugs with dissimilar mechanisms of action as part of the phenomenon of pleiotropic drug resistance (see Section 4.2). Certainly it had been previously demonstrated that the accumulation of drug is greater in sensitive cells [361], [362], suggesting that alterations in transport at the membrane level might be important to resistance.

6.2. Bleomycin (→ Antibiotics)

Bleomycins (NSC-125066) are a group of antitumor antibiotics initially isolated from broths of *Streptomyces verticillus* [363].

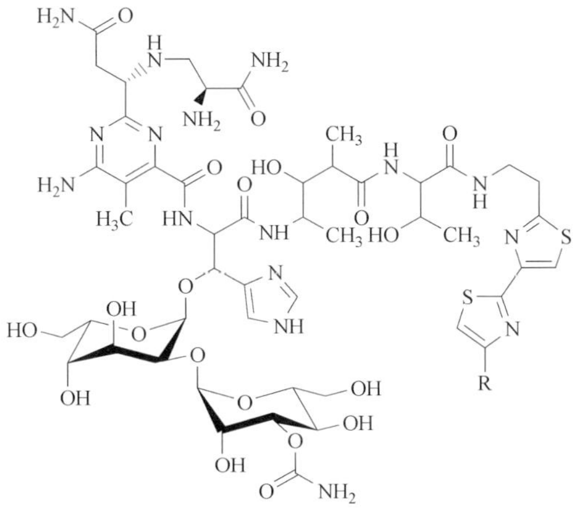

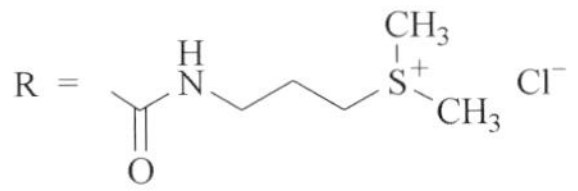

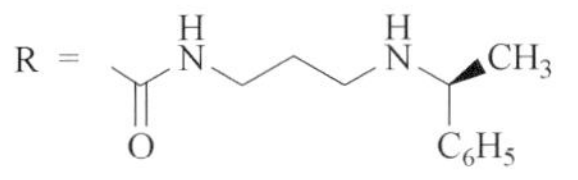

The fermentation product consists of approximately 12 different components clinically marketed as bleomycin. Each is a peptide with a low relative molecular mass (ca. 1500), all containing bleomycinic acid but differing in terminal alkylamine groups. The clinical product is approximately 60–70 % bleomycin A_2 (NSC-146842), N^1-[3-(dimethylsulfonio)propyl]bleomycinamide, $C_{55}H_{84}N_{17}O_{21}S_3$, M_r 1416; and 20–30 % bleomycin B_2. Other analogs comprise about 5 % of the total. The drug is highly active against germ cell neoplasm of the testis.

6.2.1. Analogs

Peplomycin [68247-85-8], N^1-{3-[(1-phenylethyl)amino]propyl}bleomycinamide, $C_{61}H_{88}N_{18}O_{21}S_2$, M_r 1473.62, a biosynthetic bleomycin analog, demonstrates significantly reduced pulmonary toxicity in rodents [364]. The toxicology studies are supported by clinical trials in Japan (initiated in 1981) and France (1983), and suggest that peplomycin has a greater therapeutic index than the parent compound and may replace bleomycin in the clinic [365]. In addition, total synthesis of bleomycin described in the 1980s will likely lead to other analogs and a further understanding of the drug's mechanism of action [367]. A new drug delivery system which comprises peplomycin absorbed on to small activated carbon particles was introduced as slow release formulation suggesting to decrease the systemic toxicity [366].

6.2.2. Mechanism of Action

Bleomycin enters cells slowly. Labeling studies demonstrate that the drug is first detected on the cell membrane and reaches the nucleus only after several hours [368]. Bleomycin kills cells by producing single- and double-strand DNA breaks. Bleomycin first binds guanine bases in DNA through the amino terminal peptide of the drug [369]. Free-radical formation occurs through oxidation of a bleomycin – Fe(II) complex to Fe(III), which catalyzes the reduction of molecular oxygen to superoxides and free hydroxyl radicals [370]. In vitro resistance to the drug appears to be mediated either by defective accumulation [371] or by increased intracellular drug degradation by a specific bleomycin hydrolase [372] – [374].

The clinical toxicity of bleomycin is unusual. The drug has little hematopoietic toxicity; its major dose-limiting toxicities are to the lungs and skin. Acute and chronic pneumonitis followed by progressive pulmonary fibrosis appears to be dose dependent, with risk increasing significantly in patients receiving a cumulative dose of more than 450 mg. This toxic endpoint may be due to the terminal amines of the parent compound. This hypothesis is supported by preclinical studies [375] as well as by early clinical studies of peplomycin. The maximum-tolerated cumulative dose of this terminal amino acid-substituted bleomycin analog has yet to be determined, although during phase I analysis, pulmonary toxicity was not observed until weekly doses exceeded 10 mg/m^2 [376].

6.3. DNA Interactive Natural Products

Analogs related to the natural product CC-1065 like adozelesin, bizelesin, and carzelesin bind in the minor grove of the DNA and form a covalent adduct with adenine [377]. Adozelesin [*110314-48-2*] [378] is the clinically farthest advanced agent, while the prodrug carzelesin [*119813-10-4*] [379] and the dimer bizelesin [*129655-21-6*] [380] demonstrated impressive preclinical activity.

Further modified cyclopropylpyrroloindoles (CPI) forming covalent adducts with DNA are duocarmycin A [*118292-35-6*] [381], [382], pyrindamycin [*118292-36-7*] and FCE 24517 [383], [384].

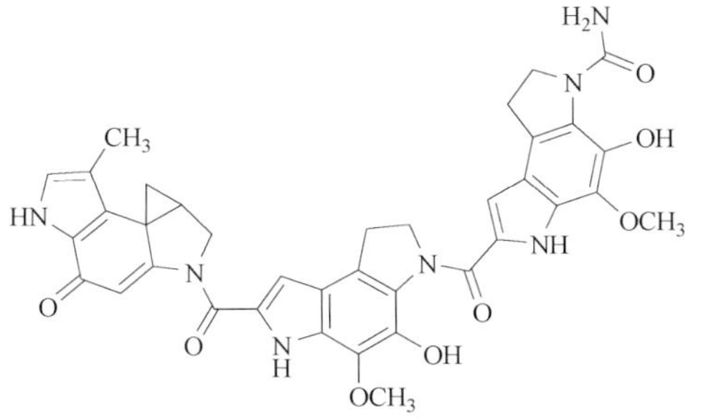

CC-1065

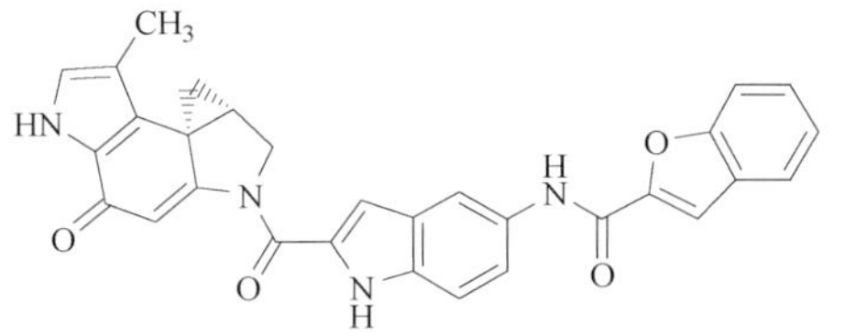

Adozelesin

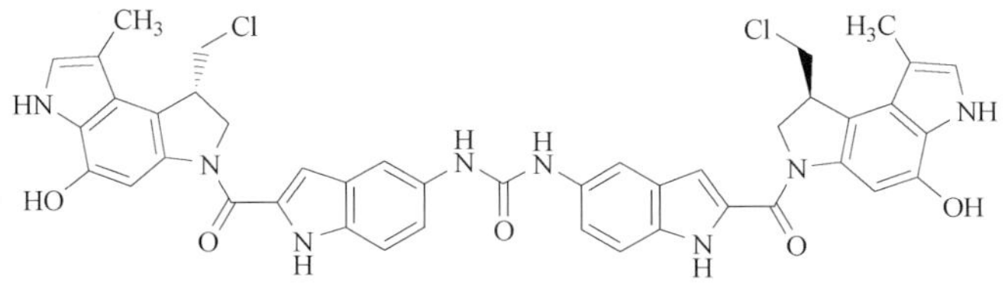

Bizelesin

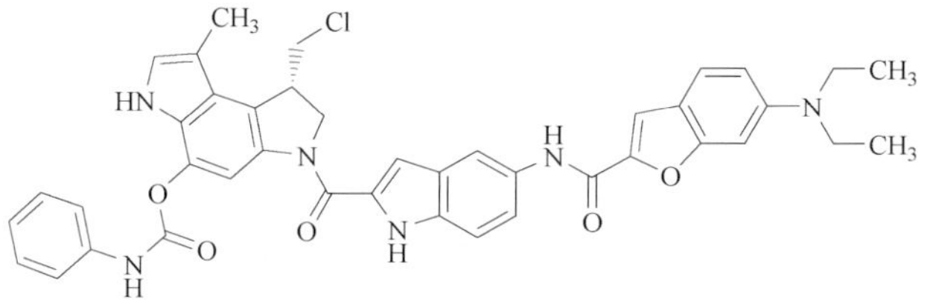

Carzelesin

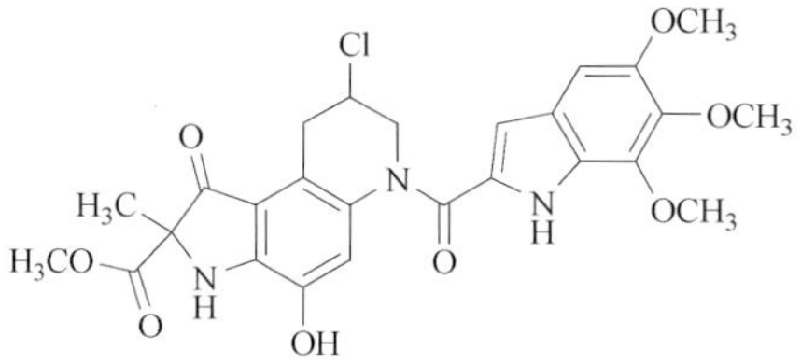

Duocarmycin C1

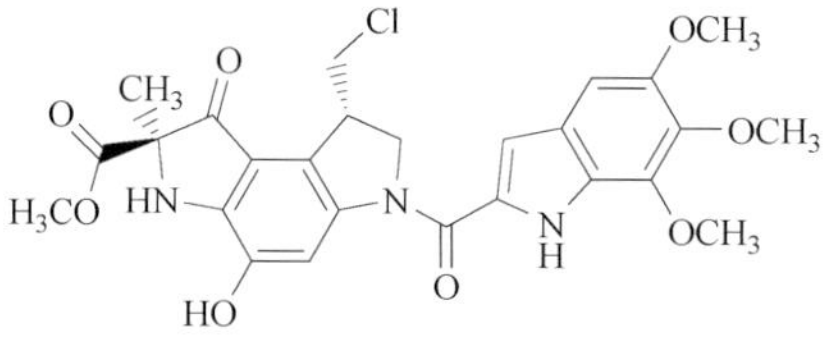

Pyrindamycin A

A series of new DNA cleaving molecules based on the reactive enediyne moiety are the anticancer antibiotics calicheamicin [*113440-58-7*] [385], esperamicin [*114797-28-3*] [386], and dynemicin.

Simplified enediyne-type compounds damage DNA upon activation by chemical or biological means and are extremely potent cytotoxic agents in vitro [387], [388].

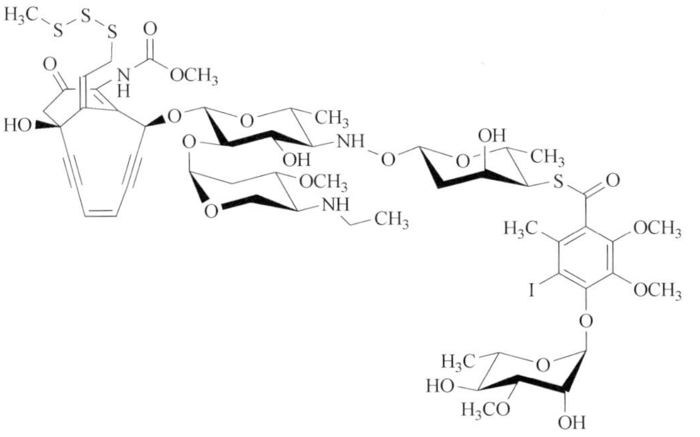

Esparamicin A1

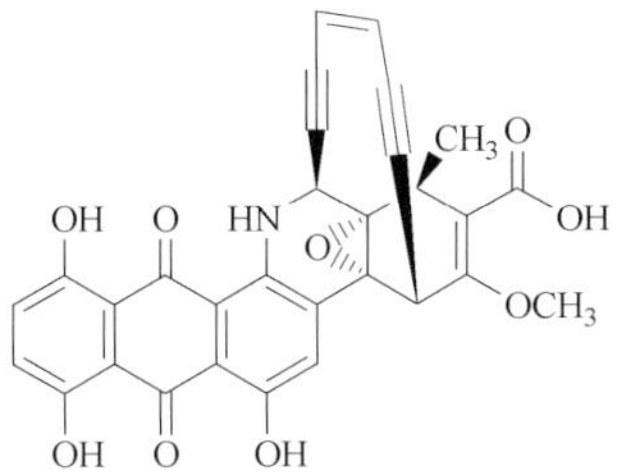

Dynemicin A

7. Antitubulin Agents

Tubulin-containing structures are important for diverse cellular functions, including chromosome segregation during cell division, intracellular transport, development and maintenance of cell shape, cell motility, and possibly distribution of molecules on cell membranes. The drugs that interact with tubulin are heterogeneous in structure. A common characteristic of these agents is binding to tubulin, causing its precipitation and sequestration to interrupt many important biologic functions that depend on the microtubular class of subcellular organelles. The tubulin–drug aggregates can be visualized by the indirect immunofluorescence technique as brightly stained cytoplasmic paracrystals. Of the tubulin binders, those that are important in cancer medicine include vinca alkaloids, podophyllotoxins, and taxoids.

7.1. Vinca Alkaloids

Vinca alkaloids are dimeric indole derivatives isolated from the periwinkle plant, *Catharantus roseus*. Of the whole family of more than 70 naturally occurring alkaloids, only a few have cytotoxic activity.

7.1.1. Vincristine and Vinblastine

The molecular structures of the dimeric *Catharantus roseus* alkaloids vincristine (NSC-67574) [*57-22-7*], 22-oxovincaleukoblastine, $C_{46}H_{56}N_4O_{10}$, M_r 824.94, Oncovin (sulfate); and vinblastine (NSC-49842) [*865-21-4*], vincaleukoblastine, $C_{46}H_{58}N_4O_9$, M_r 811.00, Velban (sulfate hydrate), are very similar. Both are formed of multiringed units, vindoline and catharantine, linked by a carbon–carbon bridge. They differ only in the nature of the substituent on the vindoline nitrogen atom.

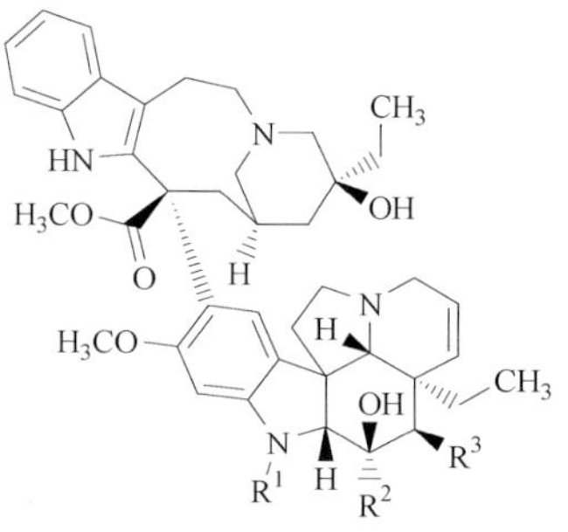

R^1 = CHO, R^2 = CO(OCH$_3$), R^3 = CO(CH$_3$), Vinristine
R^1 = CH$_3$, R^2 = CO(OCH$_3$), R^3 = CO(CH$_3$), Vinblastine
R^1 = CH$_3$, R^2 = CO(NH$_2$), R^3 = OH, Vindesine

Cellular Pharmacology. As yet it is not clear how vinca alkaloids cross cell membranes. Some data suggest an energy-dependent transport system [389], while other data suggest simple diffusion across membranes [390]. However, passive diffusion is important only at drug concentrations exceeding 100 µM. Transport of vincristine is completely inhibited by vinblastine, suggesting a common carrier.

Vincristine and vinblastine exert their biologic effect through binding to tubulin. This occurs in interphase (late S and G_2), producing a transient G_2 block [391]. They do not affect DNA synthesis directly. The metabolic consequences of tubulin binding include polyploidy, nuclear fragmentation, and inhibition of cytokinesis, which occurs after prolonged drug exposure. Present studies suggest that sensitivity to these drugs increases progressively as cells approach mitosis and that cells at the end of the cycle at the time of exposure are likely to exhibit the greatest degree of mitotic disorganization [391].

Vincristine and vinblastine share a common binding site on each tubulin monomer, with the binding affinity (K_d) of about 1.6×10^{-6} M. The drug concentrations necessary to produce 50% cell kill in vitro are ca. 4×10^{-8} M [392]. Malignant lymphocytes appear more susceptible than normal lymphocytes, presumably because of high tubulin content expressed on the surfaces of leukemic cells [393]. Other metabolic effects include inhibition of DNA and RNA and protein synthesis. However, these effects are exerted only at very high drug concentrations (1000 times greater than those achieved in vivo) and probably are secondary phenomena.

Drug Interactions. Both vincristine and vinblastine potentiate the effect of methotrexate through their blockade of methotrexate exit from cells [394], [395]. Some amino acids (glutamine, aspartic acid, ornithine, citrulline, and arginine) completely reverse the cytotoxic effect of vinblastine in tissue culture [396].

Mechanisms of Resistance. Vinca resistance may arise through mutations in tubulin, leading to decreased drug binding. Vinca-resistant cells may also share cross-resistance with antitumor antibiotics through a separate mechanism of resistance. The precise mechanism of this pleiotropic drug resistance remains to be clarified. However, tubulin is not responsible for that phenomenon because there is very little difference in affinity binding of colchicine to tubulin isolated from drug-sensitive and drug-resistant cells [397]. The appearance of a novel glycoprotein on the membrane of resistant cells and the accelerated drug efflux — leading to impaired drug accumulation — seem, at present, to be the possible mechanisms responsible for resistance [398].

7.1.2. Vindesine

Vindesine (NSC-245467) [*53643-48-4*], 3-(aminocarbonyl)-O^4-deacetyl-3-de(methoxycarbonyl)vincaleukoblastine, deacetylvinblastineamide, $C_{43}H_{55}N_5O_7$, M_r 753.95 (structure, Section 7.1.1), is a vinblastine metabolite. It possesses antitumor activity that is similar to vincristine's rather than that of its parent compound, vinblastine. Vindesine demonstrates better activity than vinblastine in some tumor models (Gardner lymphosarcoma, Ridgway osteogenic sarcoma, and mammary carcinoma). In the past several years vindesine has attracted considerable attention among clinical investigators and has been intensively investigated. Response rates vary, with the highest responses observed in the highest lymphatic malignancies [399]. The drug is less neurotoxic than vincristine and appears to be active in vincristine-resistant tumors [400], [401].

7.1.3. Vinorelbine

Vinorelbine (KW-2307, NVB) [071486-22-1] 3',4'-didehydro-4'-deoxy-C'-norvincaleukoblastine, $C_{45}H_{54}N_4O_8$, M_r 778.94, is a semisynthetic vinca alkaloid. Starting from anhydrovinblastine the compound is obtained as 5'-noranhydrovinblastine in three steps [402], [403]. The product was launched for the treatment of non-small cell lung cancer.

Trade Names: Navelbine (Pierre-Fabre/Glaxo), Ennades (Farmitalia)

Mechanism of Action. Vinorelbine was selected for drug development due to its high affinity for tubulin and its ability to prevent tubulin polymerization [403]–[405]. The compound induced total depolymerization of microtubules in P 388 murine leukemia cells, possibly via stimulation of microtubular protein synthesis [403]. Vinorelbine and vincristine were equally active against L 1210 leukemia in mice, while vinblastine had no significant effect. Vinorelbine exerted significant antitumor activity in the vincristine-resistant cell line P 377/VCR and low cross-resistance to other vinca alkaloids was observed [406], [407].

In clinical studies leukopenia was dose-limiting, no thrombocytopenia was observed [408]. Efficacy in non-small cell lung cancer was demonstrated [409]–[411].

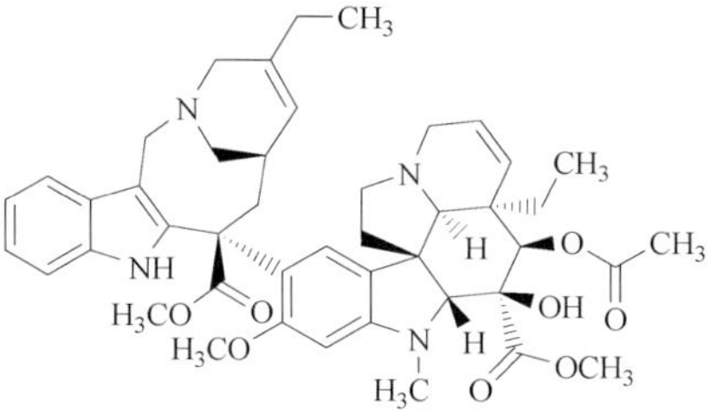

Vinorelbine

7.2. Podophyllotoxin and Its Derivatives

Podophyllin is a complex mixture of crystalline compounds derived from the mayapple plant. The active agent derived from this plant product, podophyllotoxin [518-28-5], shares a common binding to tubulin with colchicine, and the morphological effects of podophyllotoxin and colchicine exposure are indistinguishable [412]. The semisynthetic glycoside derivatives of podophyllotoxin, VP-16-213 (NSC-141540) [33419-42-0], etoposide, and VM-26 (NSC-122819) [29767-20-2], teniposide, have reproducible clinical activity against testicular cancer, small-cell lung cancer, and lymphomas [413].

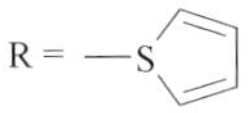

R = —S⟨thiophene⟩

R = CH₃

H₃CO   OH

OCH₃

Etoposide (VB-16-213)
Teniposide (VM-26)

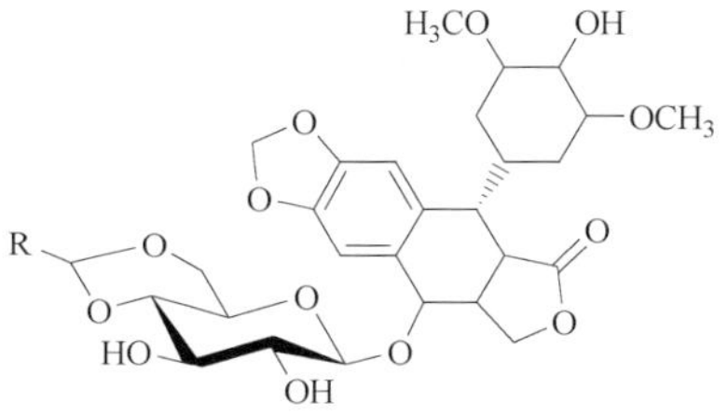

Despite structural similarities between the semisynthetic derivatives and parent natural product, neither VP-16 nor VM-26 binds tubulin. Recent evidence from a number of laboratories suggests that VP-16 exerts cytotoxic effects by causing DNA breakage [414], [415]. Selective double-strand DNA breaks, which are particularly lethal, appear to be caused by these agents [416].

The precise mechanism for VP-16 and VM-26 DNA damage is unknown, but the fact that the presence of a 4′-hydroxyl group is critical for cytotoxicity suggests formation of free-radical intermediates. Indeed, free-radical scavengers appear to be able to protect from podophyllin derivative cytotoxic effects in vitro [417]. Formation of DNA breaks may also be due to specific inhibition of the DNA repair enzyme topoisomerase II by both VP-16 and VM-26 [418]. It is likely that further characterization of topoisomerase II and the effects of its inhibition will result in the development of new leads in antitumor drug development.

7.3. Camptothecin and Analogs

Camptothecin (CPT) [7689-03-4], an alkaloid isolated from the Chinese plant *Camptotheca accuminata* in 1966 [419] was proved to have antineoplastic activity in various tumor systems [420]. The clinical use of CPT was, however, limited because of its high toxicity and low solubility in water. New derivatives have been synthesized carrying substitutents in 7,9,10- or 11-position of the ring A.

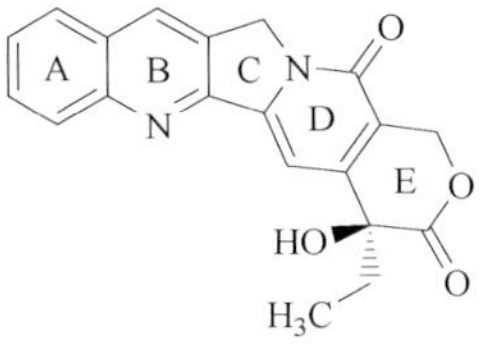

Campothecin

Irinotecan (NSC-616348, CPT-11, DQ-2805) [*097682-44-5*] [421], (+)-7-ethyl-10-[4-(1-piperidyl)-1-piperidyl]carbonyloxycamptothecin, $C_{33}H_{38}N_4O_6$, M_r 586.69 and topotecan (NSC-609699, SKF-S-104864-A, E-89/001) [*119413-54-4*] [422], (S)-10-dimethylaminomethyl-9-hydroxycamptothecin, $C_{23}H_{23}N_3O_5$, M_r 457.91 are camptothecin derivatives with topoisomerase I-inhibitory activity. As antineoplastic alkaloids, they can be synthesized semisynthetically starting from CPT [423]. In addition both agents have intercalating properties and are water soluble. Irinotecan hydrochloride was launched for the treatment of small cell and non-small cell lung cancer and cancers of the uterine, cervix, and ovaries.

Trade Names: Campto (Rhône-Poulenc Rorer/Yakult Honsha) Topotecin (Daiichi Seiyaku).

Unlike irinotecan, topotecan is not a prodrug and does not require metabolic activation.

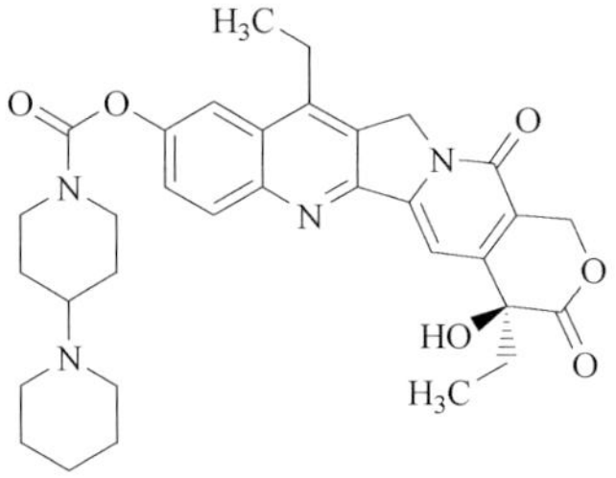

Irinotecan

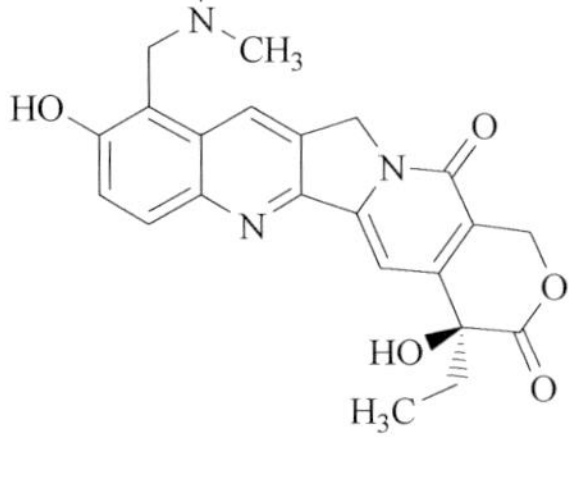

Topotecan

Mechanism of Action. Topoisomerase (topo I) is an ubiquitous nuclear enzyme involved in the regulation of essential cellular functions by relieving torsional DNA strain. Relaxation of supercoiled DNA is achieved through a series of topo I-mediated reactions. The main physiological role of topo-I is in sensing and subsequently releasing the positive supercoiling generated ahead of the moving transcription apparatus. A number of malignancies including acute leukemia, blasts, colon, esophageal, and ovarian cancers contain increased topo-I levels as compared to normal tissues [424], [425]. Topotecan binding to topo-I-DNA adducts results in markedly decreased rates of nick resealing and in delayed enzyme release, leading to increased numbers of strand breaks [426].

In animal tumor models including xenografts of human cancer lines topotecan and irinotecan exhibit a wide spectrum of antineoplastic activity [427]–[429]. Cell lines overexpressing P-glycoprotein display low levels of resistance to topotecan through decreased drug retention, more pronounced resistance is found in cell lines containing low levels of topo I [430], [431].

Dose-limiting toxicity in patients were stomatitis and esophagitis as well as neutropenia [432].

7.4. Taxoids

Paclitaxel (NSC-125973, BMS-181339) [*33069-62-4*], $C_{47}H_{51}NO_{14}$, M_r 853.9, is a natural antineoplastic taxane derivative originally isolated from the plant *Taxus brevifolia* [433] and subsequently synthesized due to limited supply from the bark of endangered yew trees. Starting from 10-deacetyl baccatin III (isolated from the renewable twigs and needles of *Taxus baccata*) the compound can be prepared in three steps utilizing protected *N*-benzoyl-(2R,3S)-3-phenylisoserine as key building block [434]–[436]. Various synthetic approaches are described and patented, including three total synthesis of paclitaxel [437]–[440].

Paclitaxel has a novel mechanism of action. Unlike vincristine or vinblastine, which bind tubulin to inhibit tubulin polymerization, paclitaxel stabilizes microtubular structures [441]. Indicated for the treatment of primary ovarian cancer in combination with cisplatin and for metastatic ovarian cancer where standard therapy has failed, paclitaxel has been marketed since 1993 as Taxol (Bristol-Myers Squibb) and was the best-selling anticancer agent in 1996. Taxol does not share a common binding site with other antitubulin agents and may instead bind to and stabilize polymerized tubulin [442].

A semisynthetic analog of paclitaxel, docetaxel (NSC-628503, RP-56976) [*114977-28-5*], (2R,3S)-*N*-carboxy-3-phenylisoserine, *N*-*tert*-butyl ester, 13-ester with 5β,20-epoxy-1,2α-4,7β,10β,13α-hexahydroxytax-11-en-9-one 4-acetate 2-benzoate, $C_{43}H_{53}NO_{14}$, M_r 807,9, is a derivative with an *N*-*tert*-butyloxycarbonyl-(2R,3S)phenylisoserine C-13 side chain. Docetaxel is more potent than paclitaxel in vitro [443], [444]. Impressive clinical results have been reported for the treatment of ovarian, breast, and bronchial cancers

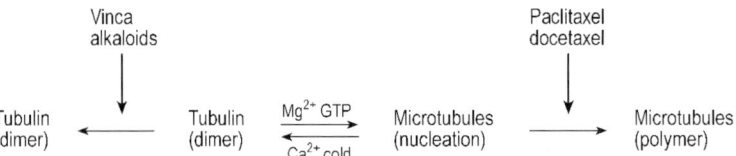

Figure 2. Mechanism of action of paclitaxel and vinca alkaloids

with docetaxel [445]. The compound was launched in 1996 for the treatment of locally advanced breast cancer or relapse during anthracycline therapy of NSCLC and breast cancer.

Trade Name: Taxotere (Rhône-Poulenc Rorer).

Paclitaxel

Docetaxel

Mechanism of Action. More than two decades after its isolation and the elucidation of its complex structure and cytotoxic activity [433] interest in paclitaxel raised again when S. HORWITZ et al. [446] reported on the novel mechanism of action. As spindle poison paclitaxel promotes the polymerization of tubulin to microtubules and stabilizes them against depolymerization, whereas vinca alkaloids induce microtubule disassembly (see Fig. 2). Thus, with paclitaxel the dynamic equilibrium of assembly and disassembly of microtubules is shifted in favor of the polymer, preventing cell division [447]. Paclitaxel binds preferentially to the β-tubulin subunit. This binding is reversible and the site is different from the binding sites of vinca alkaloids, colchicine and podophyllotoxin [448]–[450]. If paclitaxel is present, tubulin polymerizes without exogenous GTP and these stabilized, rigid microtubules cannot be disassembled. As a result, the dynamic organization of the cell is interrupted which leads to irreversible damage in rapidly dividing cells. Further evidence has been reported that the antiproliferative activity of paclitaxel is caused by additional effects [451]–[453].

A major impediment in the development of taxol as a drug was its poor water solubility [454]. With the solubility enhancer Cremophor EL hypersensitivity reactions, including hypotension, urticaria, and dyspnea, occurred in patients during rapid infusion. To cope these allergic reactions, 24-hour infusions and pretreatment with dexamethasone, diphenhydramine, or cimetidine was recommended [455].

The dose-limiting *toxicity* of paclitaxel is neutropenia. Several other toxic effects such as diarrhea, nausea, and emesis are less common. Docetaxel shares many toxic effects with paclitaxel such as dose-limiting neutropenia, alopenia, myalgias, and mucositis. In addition fluid retention and cutaneous toxicities are observed [445], [460], [461]. Because of the low water solubility of paclitaxel and docetaxel, the synthesis of more soluble taxoid prodrugs and smaller analogs has become an interesting area of research (see Section 7.5).

Structure–Activity Relationship. A large number of taxoid analogs has been synthesized with emphasis to enhance biological activity and to improve the water solubility [436]. C-13 side chain depleted analogs such as baccatin III and its derivatives [456] as well as N-benzoyl-(2R,3S)-phenylisoserine are inactive. Simplified side chains at C-13 (like acetic, crotonic, or phenylacetic acid) possess reduced activity.

3'-Cyclohexyl-3'-dephenylpaclitaxel has a similar cytotoxicity to paclitaxel. Further modifications in the aromatic 3'-phenyl group gave compounds that were equipotent with paclitaxel. The compound with a 3'-(p-methoxyphenyl) has a slightly increased activity. Compounds with different substituents in the N-benzoyl part were similar to paclitaxel if these substituents were aromatic, aliphatic substituents reduced cytotoxicity.

As the 2'-hydroxyl group is essential for maximal biological activity [457] esterification leads to a total loss of activity in the microtubule assay, whereas cytotoxicity remains unchanged. Thus various amino acid esters were produced as prodrugs.

Notable loss of cytotoxicity was observed with A-ring modified analogs and oxidation at C-10 or C-7. 7-Acetylpaclitaxel and the C-7 epimer were similar in their ability to inhibit cell proliferation. All derivatives without the intact oxetane moiety are inactive [458]. Further, all C-4 modified analogs were devoid of activity underscoring the vital importance of these functional groups [459].

7.5. Epothilone A and B

The high cytotoxicity and good stabilization of microtubule raised interest in the natural products epothilone A [*152044-53-6*] and B [*152044-54-7*] originally isolated from myxobacteria *Sorangium cellulosum* [462]. Since BOLLAG et al. [463] reported on the mechanism of action, which resembles that of paclitaxel numerous reports were published on total synthesis [464]–[468] and biology of epothilones. Their unique capability to inhibit taxol resistant tumor cell lines [470] and their good solubility in water are the biggest advantages as compared to paclitaxel. Their in vivo activity is similar to that of paclitaxel [469].

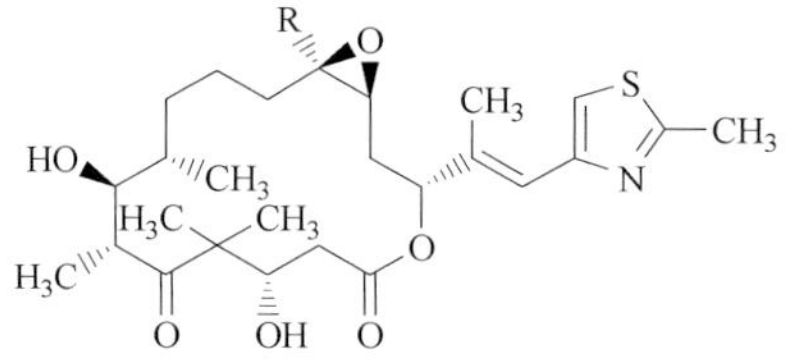

R = H, Epothilone A
R = CH$_3$, Epothilone B

8. Heavy-Metal Complexes

8.1. *cis*-Platinum

cis-Platinum (NSC-119875) [*15663-27-1*], *cis*-diamminedichloroplatinum, cisplatin, PtCl$_2$(NH$_3$)$_2$, M_r 300.05, was the first heavy-metal compound to be introduced into clinical cancer chemotherapy. The discovery of this unique agent was predicated on the serendipitous observation that bacterial growth was inhibited when culture medium was subjected to an alternating current using platinum electrodes [471]. Moreover, the spent medium itself developed bacteriocidal characteristics, even in the absence of electrical current. Detailed analysis confirmed that, of the several platinum species produced by electrolysis, it was the cis isomer of PtCl$_2$(NH$_3$)$_2$ that had antibacterial activity. In 1969, ROSENBERG and co-workers reported that *cis*-platinum also had potent antitumor activity in murine tumors [472], and phase I clinical studies with the drug were initiated two years later [473], [474]. Today *cis*-platinum, in combination with other agents, has led to highly active and often curative regimens in patients with testicular, ovarian, and head and neck cancer [475], [476].

8.1.1. Mechanism of Action

As with the alkylating agents, one of the major factors mediating the cytotoxicity of *cis*-platinum is probably the formation of cross-links between opposing strands of DNA (interstrand cross-links), linkage within a single strand of DNA (intrastrand cross-links), or the formation of linkages dependent on the hydrolysis of *cis*-platinum in solution. While the covalent stability of Pt–NH$_3$ bonds is high, both chlorides are good leaving groups and can be displaced by water or hydroxyl ions to form positively charged, aquated platinum species that avidly react with nucleophilic sites on macromolecules, especially the N^7 position of guanine and the N^3 position of cytosine [477]. Formation of the active intermediate is inhibited in the presence of Cl$^-$; in plasma the Cl$^-$ concentration is sufficient to inhibit aquation of the drug, which has an in vitro plasma half-life of several hours [478]. However, this process occurs rapidly in the intracellular milieu, where Cl$^-$ concentrations are low.

The cis configuration is central to cytotoxicity; the trans isomer is devoid of antitumor effects. This fact suggests that of the reactions caused by the aquated platinum species, formation of intrastrand cross-links is most important (intrastrand cross-links cannot be formed by the inactive trans compound).

As a result of nucleophilic attack on macromolecules, *cis*-platinum causes changes in the structural conformation of DNA [479] as well as intra- and interstrand cross-links. These changes inhibit RNA transcription from the DNA template [480] and, probably more important, directly inhibit DNA synthesis itself [481]. In addition, *cis*-platinum can react with tumor cell membrane to cause presentation of new antigenic determinants [482]. Whether this mechanism is related to tumor cell recognition and immune response remains to be determined.

As a heavy-metal-based compound, *cis*-platinum has unique clinical toxicities, including renal tubular damage, severe nausea and vomiting, high-tone hearing loss, and peripheral neuropathy. Less common are myelosuppression, hemolytic anemia, hypomagnesemic tetany, allergic reactions, and hepatotoxicity. Nephrotoxicity was dose limiting in early clinical trials and appeared to be due to tubular reabsorption of active platinum species causing proximal and distal tubule necrosis [483]–[485]. Tubular damage has been reported to cause defective reabsorption of magnesium, resulting in hypomagnesemia (which may result in tetany) [486], [487], as well as inappropriate renal loss of calcium, potassium, and phosphorus [488]. The acute renal toxicities of *cis*-platinum appear to be secondary to activation of the renin–angiotensin system, resulting in reduced renal blood flow and glomerular filtration [484], [489]. Thus, this toxicity can be mitigated by high-volume diuresis [490]. In 1984 it has been demonstrated that even high doses of *cis*-platinum are well tolerated when administered with high-volume chloresis, which not only dilutes urinary platinum levels but also prevents leaving of the chloride groups to form the toxic aquated molecule [491]. When administered by this regimen, the limiting side effects of *cis*-platinum become neurotoxicity and myelosuppression, while renal function is remarkably spared.

8.1.2. Mechanisms of Resistance

Although bacterial resistance to *cis*-platinum appears to be due to increased efficiency of DNA repair [492], the mechanisms of tumor cell resistance are less clear. Alkaline elution studies have demonstrated a direct relation between DNA cross-linking and tumor cell resistance [493]–[495], and a *cis*-platinum-resistant murine leukemia line has been described in which cross-links are formed at a reduced rate [496]. However, it remains uncertain whether this resistance is secondary to accumulation, impaired activation, or altered DNA repair processes. As with the alkylating agents in general, high levels of metallothionein have been reported in *cis*-platinum-resistant cells [497]. This sulfhydryl-rich protein is known to protect from metal toxicity by specifically binding to platinum, cadmium, and other heavy metals [498].

Because the cellular mechanisms of resistance to *cis*-platinum are so poorly characterized, new platinum analogs in clinical trial have been selected in an attempt to reduce host toxicity with the aim of improving therapeutic index [499].

8.2. Carboplatin

Carboplatin (NSC-241240, CBDCA) [*839805-03-3*], *cis*-diamine[1,1-cyclobutanedicarboxylato-(2)-*O*,*O'*]platinum(II), $C_6H_{12}N_2O_4Pt$, M_r 371.3, is a second generation cisplatin analog without significant nephrotoxicity or neurotoxicity and with less emetic potential than the parent compound [500]–[503]. Clinical trials show activity against several tumor types. Carboplatin is especially effective in treatment of ovarian and small-cell lung cancer [504], [505]. The minimal emetogenic dose of *cis*-platinum in dogs is 9 mg/m^2, whereas for carboplatin the dose is 624 mg/m^2. In addition, this compound retains antitumor activity in *cis*-platinum-resistant murine leukemia [506], demonstrating consistent antitumor activity in patients with ovarian cancer in the absence of either ototoxicity, nephrotoxicity, or neuropathy [507], [508]. Carboplatin entered the market in 1989 and is the leading platinum complex cancer drug with sales of \$ 373 × 10^6 worldwide in 1996. Carboplatin is indicated for the treatment of ovarian cancer and sales have benefited from the drug's use in combination with Taxol [509]–[512].

Trade Names: Paraplatin (Bristol-Myers Squibb), Carboplat (BMS).

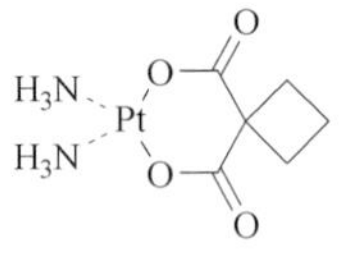

Carboplatin

8.3. Analogs

The synthesis and development of "second-generation" platinum compounds such as carboplatin has modified the problem of nephrotoxicity. Iproplatin (NSC-256927, CHIP) [34348-60-2], dichlorodihydroxy-bis-(2-propanamine)platinum $C_6H_{20}Cl_2N_2Pt$, M_r 418 [514], nedaplatin (NSC-375101D, 254-S), [095734-82-0], cis-diamine(glycolato-O^1,O^2)platinum, $C_2H_8N_2O_3Pt$, M_r 303.19 [515], and oxaliplatin (OHP) [61825-94-3], [Sp-4-2-(1R-trans)](1,2-cyclohexanediamine-N,N')[ethanedioxato(2⁻)-O,O']-platinum, $C_8H_{14}N_2O_4Pt$ [516] were clinically investigated in depth. Nedaplatin entered the market in 1997 for the treatment of head and neck, small-cell lung, non-small cell lung, esophageal, bladder, testicular, ovarian, and uterine cervical cancers [517]–[520]. *Trade Name:* Agupla (Shionogi).

Further studies of additional "third-generation" compounds are centering on the elimination of toxicity, enhanced therapeutic activity, non-cross-resistance and selective drug delivery [521]. Ormaplatin (NSC-363812, U-77233) [62816-98-2] [±(trans)]-tetrachloro(1,2-cyclohexanediamine-N,N']platinum, $C_6H_{14}Cl_4N_2Pt$ and Lobaplatin (D-19466) [135558-11-1] are representatives of a series of platinum complexes in clinical development [522]–[524].

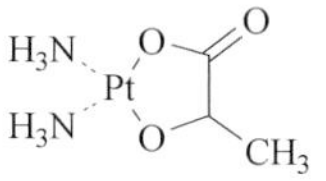

Nedaplatin

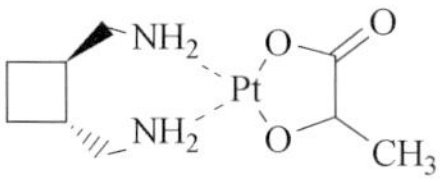

Lobaplatin

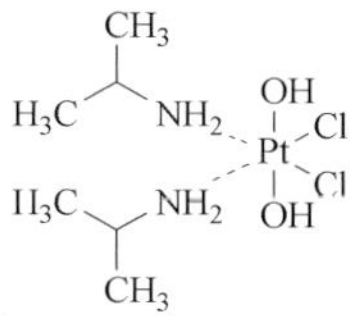

Iproplatin

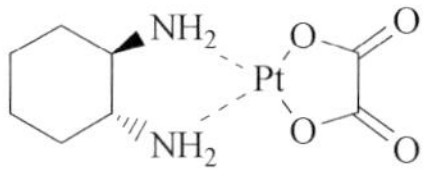

Oxaliplatin

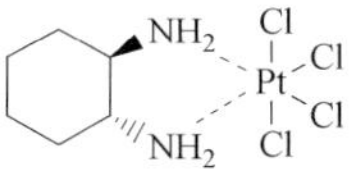

Ormaplatin

Orally active cisplatin-analogs with higher therapeutic ratio and specifically high anticancer activity are the challenge in the search for new "fourth-generation" derivatives. Platinum(IV)-complexes like JM 216, drug targeting approaches, e.g., the use of ligands with hormone receptor-binding affinity [525] or intercalating structures [526], and special formulation techniques are the major focus.

9. Hormonally Active Anticancer Drugs/Antihormones

Hormones and in particular, the sex hormones were the first growth factors discovered to be involuntary helpers of cancer. Female breast cancer and male prostate cancer are the best known examples of tumors acknowledged to be hormone-dependent. Shutting down the main production site of the sex hormones estrogen and testosterone either by removing the ovaries or by castration is a well-known and often effective therapy; however, these procedures can be problematic due to the concomitant psychological stress. Modern hormone therapy for advanced breast cancer and prostate cancer attempts to spare the patient such irreversible operative procedures for as long as possible by using hormone antagonists. Examples are the antiestrogens or LHRH antagonists, which hinder deployment of the hormone itself and thus its growth-promoting activity.

9.1. Antiestrogens (→ Hormones)

9.1.1. Antagonists

Estrogens can induce hormone-dependent human breast carcinoma and stimulate tumor growth. Reduced estrogen production is correlated with a lower risk of breast cancer and in particular with tumor regression. Hormonally active drugs are considered to be the treatment of first choice for advanced breast cancer, unless metastatic complications require immediate aggressive chemotherapy.

9.1.2. Tamoxifen, Toremifene

The antiestrogen tamoxifen (ICI 46474) [10540-29-1], (Z)-2-[p-(1,2-diphenyl-1-butenyl)phenoxy]-N,N-dimethylethylamine, $C_{26}H_{29}NO$, has become the standard first-line agent in postmenopausal patients [527]–[529]. The product is indicated for the treatment of breast cancer with worldwide sales of $ 561×10^6 in 1996.
Trade Name: Nolvadex (Zeneca).

However, primary and secondary resistance to tamoxifen treatment requires the introduction of new analogs with improved therapeutic activity for hormon-dependent neoplasia. Toremifene (NK-622, Fc-1157a) [089778-26-7] [530], [531], 2-[p-[(Z)-4-chloro-1,2-diphenyl-1-butenyl]phenoxy]-N,N-dimethylethylamine, $C_{26}H_{28}ClNO$, M_r 405.97, raloxifene [532], and droloxifene [533] were investigated clinically.

Toremifene is a novel triphenylethylene derivative structurally related to tamoxifen [527]–[529]. Almost all compounds related to tamoxifen contain an alkylaminoethoxy side chain, which seems to be essential for their binding to the estrogen-receptor (ER) and antiestrogenic activity [536].

Trade Names: Toremifene has been launched as Fareston (Farmos, ASTA Medica) and Estrimex (Adria) for treatment of postmenopausal breast cancer.

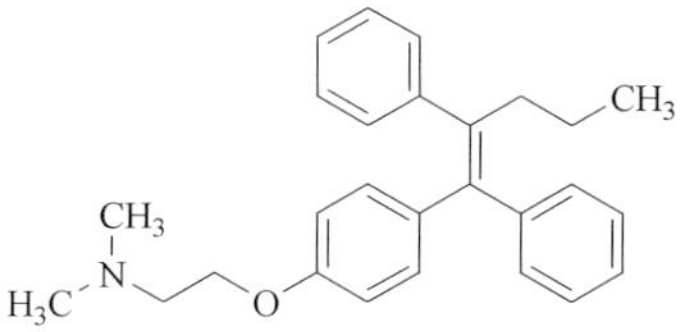

Tamoxifen

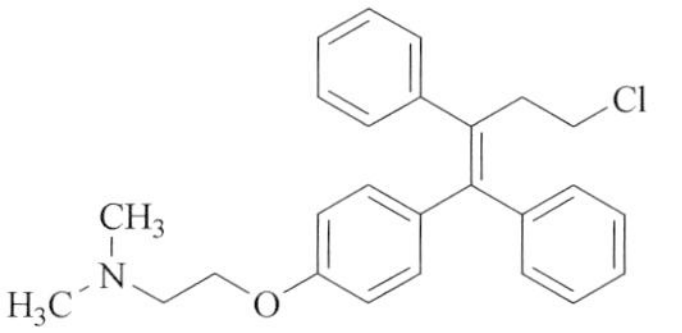

Toremifene

Mechanism of Action. Tamoxifen and toremifene bind to estrogen receptors in the cytosol, are translocated to the nucleus, and block estrogen-induced cell proliferation. However, specific differences in the drug profiles exist, indicating improved therapeutic properties of toremifene compared with tamoxifen. These differences include lower intrinsic estrogenic activity, longer nuclear retention, no retinal damage or neoplastic liver changes. Toremifene is more active against dimethylbenz[a]anthracene (DMBA)-induced rat mammary cancer and unlike tamoxifen, it inhibited the growth of an ER-negative transplantable mouse uterus sarcoma, although the antitumor effect of this compound was preferentially directed against estrogen-dependent tumors of the mammary gland and endometrium [534], [535]. In clinical trials no concrete side effects or pathological clinical chemistry values were observed in most patients [537], [538]. Some patients complained of light hot waves, sweating, nausea, and transient vertigo.

Cytotoxicity in vitro, dose related activity in ER-positive (i.e., cancer cells containing estrogen receptors) and ER-negative tumor models suggest that additional mechanism like growth-factor production may be triggered by toremifene [539].

9.1.3. Analogs

Raloxifene (LY-139481) [*82640-04-8*] is a benzo(b)thien-3yl-antiestrogen formerly under development for treatment of breast cancer. It mimics the effects of estrogen on the skeleton and is therefore effective in the prevention of postmenopausal osteoporosis [540]. Faslodex (ICI-182780, ZD-182780) [*129453-61-8*] is a pure, steroidal estrogen antagonist with oral anticancer activity [541], [542]. The in vivo antitumor activity of faslodex in xenografts of MCF-7 and Br 10 human breast cancers in mice was equivalent to that of tamoxifen. Miproxifene phosphate (Tat-59) [*115767-74-3*] [543], panomifene (Gyki-13504) [*77599-17-8*], idoxifene [544] (SB-223030, CB-7432) [*116057-75-1*] and droloxifene [*83647-29-4*] are more potent than tamoxifen against estrogen-dependent tumors in mice.

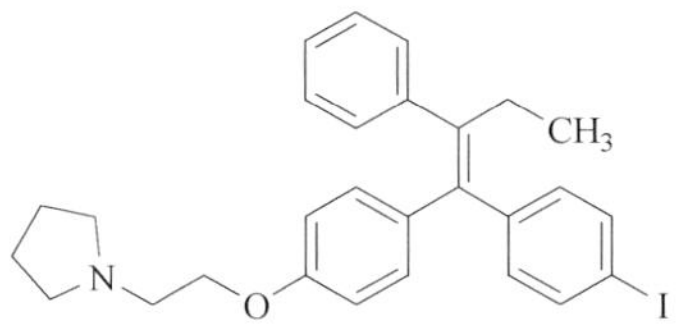

Idoxifene

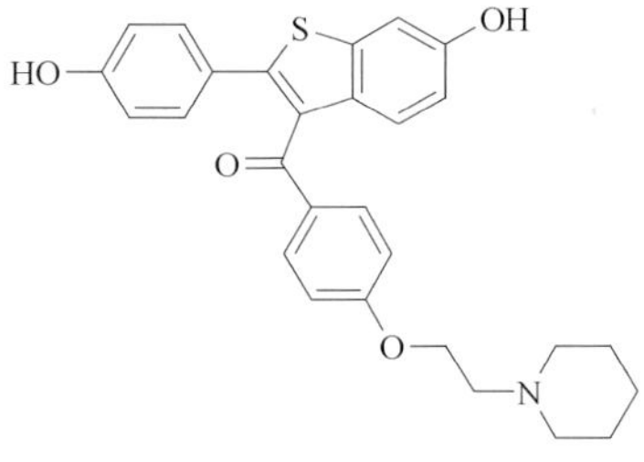

Raloxifene

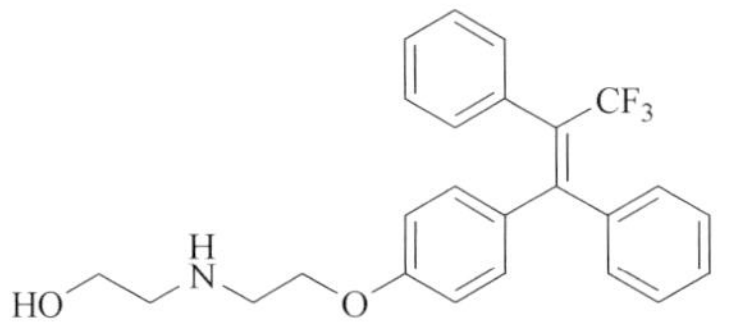

Panomifene

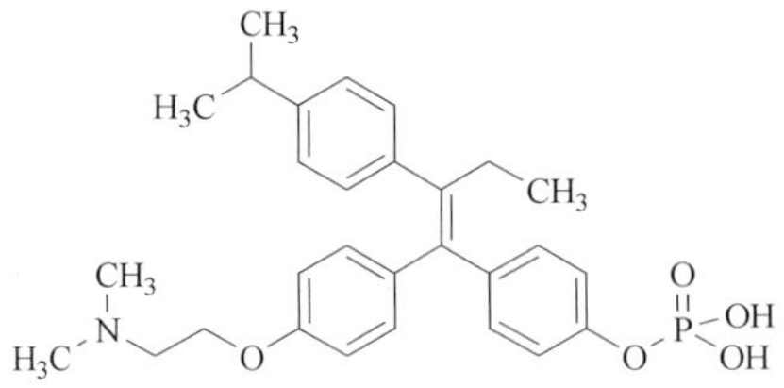

Miproxifene phosphate

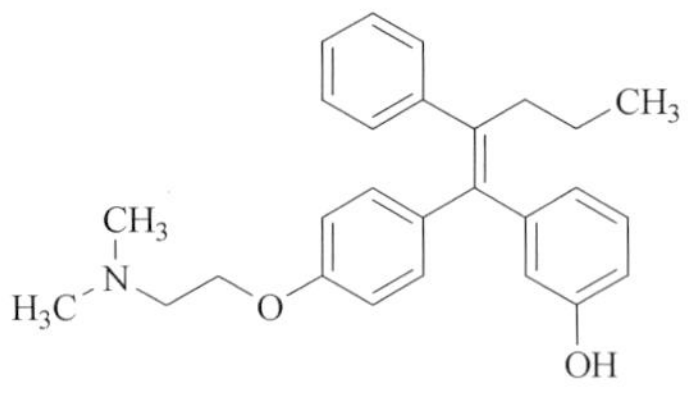

Droloxifene

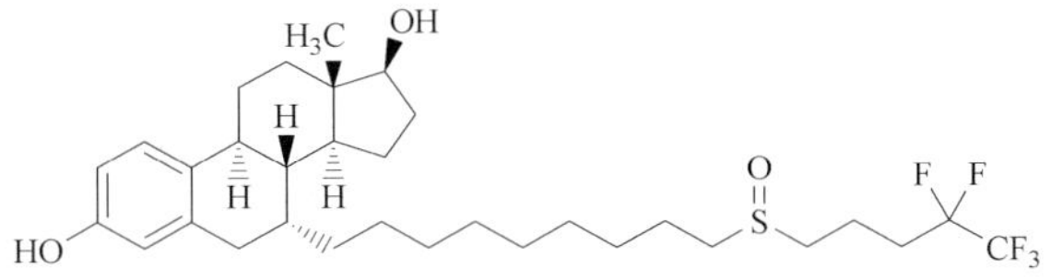

Faslodex

9.2. Aromatase Inhibitors (→ Hormones)

All endocrine therapies inhibit endogenous estrogen production or the interaction between estrogens and cellular estrogen receptors. Both ablative and additive hormone therapies are equal in efficiency.

Aromatase inhibitors block cellular estrogen synthesis and thus induce, particularly during postmenopause, a marked decrease in estrogen production. The inhibition of estrogen synthesis is caused by a suppression of the enzyme aromatase which converts androstenedione to estrone [545].

Aminoglutethimide [125-84-8], (±)-2-(4-aminophenyl)-2-ethylglutarimide, an unspecific aromatase inhibitor which also suppresses adrenal desmolase and 11-β-hydrolase, was the first aromatase inhibitor in the clinic and on the market.

Trade Name: Orimeten (Ciba-Geigy).

Aminoglutethimide induces a decrease in cortisol, followed by an increase in ACTH [546]. Therefore new powerful specific aromatase inhibitors without influence on adrenal steroid synthesis were developed.

Formestane (4-OHA; CGP-32349) [000566-48-3], 4-hydroxyandrost-4-ene-3,17-dione, $C_{19}H_{26}O_3$, M_r 302.41, is an androstane derivative with highly specific aromatase inhibition [547]. The substance is used for the treatment of advanced breast cancer in postmenopausal women [548]–[553].

Trade Name: Lentaron (Ciba-Geigy).

Atamestane, exemestane, and NKS-01 are orally active steroids.

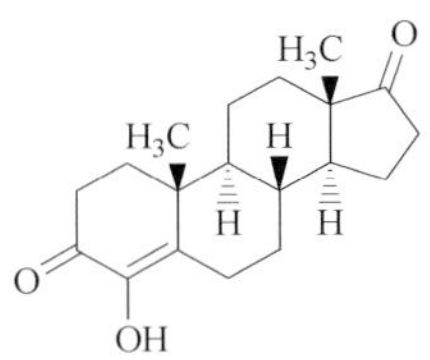

Formestane

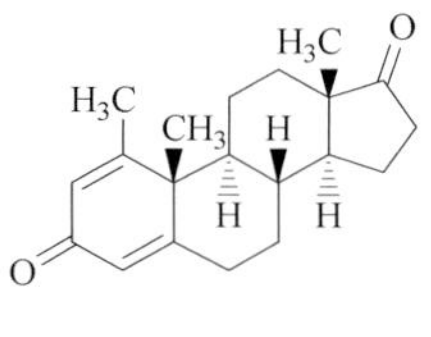

Atamestane

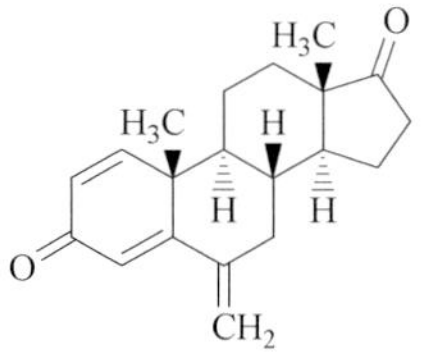

Exemestane

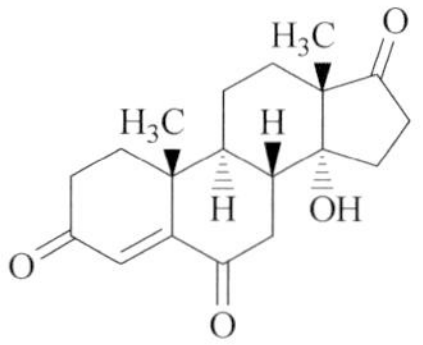

NSK-O1

In addition to steroidal also nonsteroidal, in particular, imidazole/triazole derivatives were investigated as aromatase inhibitors. Anastrazole (ICI-D1033, ZD-1033) [120511-73-1] [554], 2,2'-[5-(1H-1,2,4-triazol-1-ylmethyl)-1,3-phenylene]-bis(2-methylpropionitrile), $C_{17}H_{19}N_5$, M_r 293.37, fadrozole (CGS-16949A) [102676-96-0] [555], (±)-4-(5,6,7,8-tetrahydroimidazo[1,5-a]pyridin-5-yl)benzonitrile, $C_{14}H_{13}N_3$, M_r 259.74, letrozole (CGS-20267) [112809-51-5] [556], 4,4'-(1H-1,2,4-triazol-1-ylmethylene)bis[benzonitrile], $C_{17}H_{11}N_5$, and vorozole (R83842), [129731-10-8] [557], (+)-6-[(4-chlorophenyl)-1H-1,2,4-triazol-1-ylmethyl]-1-methyl-1H-benzotriazole, $C_{16}H_{13}ClN_6$, are highly selective nonsteroidal aromatase inhibitors without intrinsic androgenic or estrogenic properties.

Trade Names: Anastrazole and Fadrozole are marketed as Arimedex (Zeneca) and Arensin (Ciba-Geigy), respectively, for treatment of postmenopausal breast cancer.

Another aminoglutethimide analog in development is rogletimide [558] (pyridoglutethimide) [121840-95-7], (±)-3-ethyl-3(4-pyridinyl)-2,6-piperidinedione, $C_{12}H_{14}N_2O_2$.

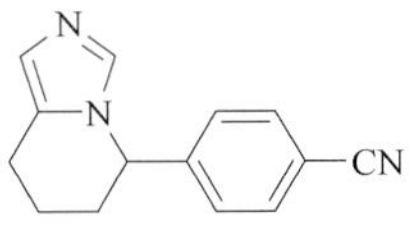

Fadrazole

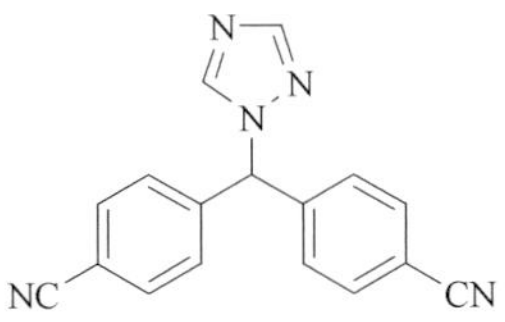

Letrozole

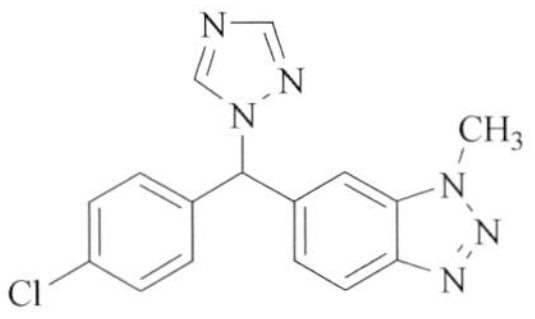

Vorozole

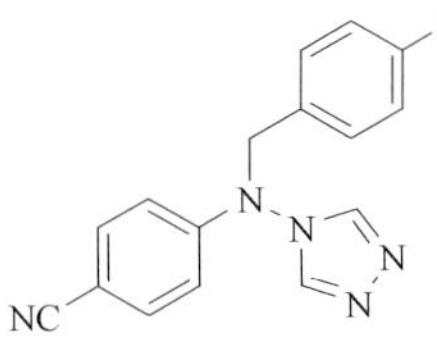

YM-511

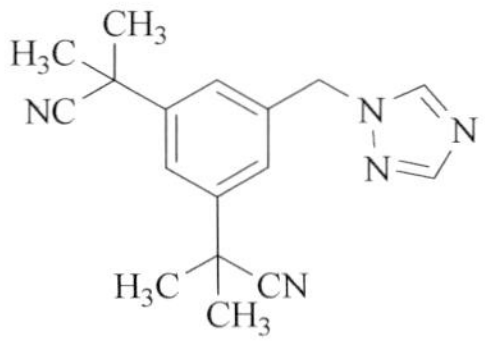

Anastrazole

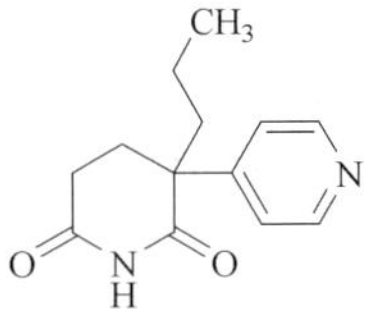

Rogletimide

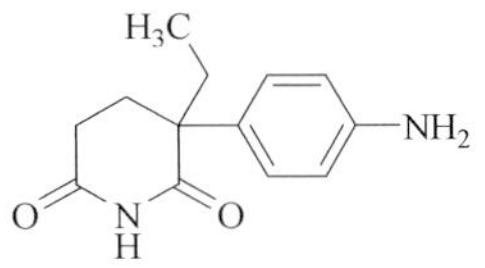

Aminoglutethimide

Mechanism of Action. Fadrazole [559] and anastrazole [560]–[567] for example, were found to be potent and specific aromatase inhibitors with neither androgenic nor estrogenic activity. Fadrazole was 180 times more potent than aminoglutethimide as an aromatase inhibitor. Anastrazole inhibited human placental aromatase with an $IC_{50} = 15$ nM (IC_{50} = inhibitory concentration). In animal studies fadrazole was found to lower serum estrogen, raise luteinizing hormone (LH) level, and reduce uterine weight as a result of aromatase inhibition. In addition, administration of 2 mg per kilogram body weight caused regression of DMBA-induced mammary tumors in female rats. In humans half-life of, e.g., anastrazole was more than 30 h. No serious side effects were reported and no significant effects on cortisol or aldosterone secretion was observed.

9.3. Antiandrogens

Prostate cancer is primarily a disease of the elderly and the second most common cancer in men in the United States. Ever since HUGGINS and HODGES demonstrated the partial androgen dependence of most prostatic tumors more than 50 years ago, androgen deprivation has become the commonly used initial treatment for prostate cancer [568], [569].

The major circulating androgen in man is testosterone, 90 % of which is produced in the testis. In addition, a small amount of androgen is produced by the adrenal gland under the control of ACTH. (see Fig. 3).

Four ways for androgen deprivation exist:

1) Removal of organs by surgery
2) Interference with control mechanism
3) Inhibition of biosynthesis
4) Competitive inhibition of androgens at the receptor site

The group 2 approach with, e.g., LHRH agonists and antagonists will be discussed in Section 9.4. Examples of group 3 include compounds (e.g., ketoconazole), which inhibit the synthesis of adrenal androgens [570] as well as inhibitors of the enzymes 5α-reductase [571] and aromatase [572]. "True" antihormones are compounds of group 4, based on the definition that an antiandrogen is a substance which binds to the target tissue androgen receptor and prevents the stimulatory effects of androgens. Although the exact mechanism of action of antiandrogens is not totally understood, an important feature is the competitive inhibition of the binding to the cytosol receptor. The first antiandrogen to be used clinically was the synthetic steroidal antiandrogen cyproterone acetate [*427-51-0*] [573]. However, cyproterone acetate also exhibits other steroidal activities, it is, e.g., a potent progestin, exhibits weak antigonadotrophic activity and has glucocorticoid like properties. Beyond the success in treating prostate cancer the steroidal properties are largely responsible for the fluid retention and thrombosis seen in patients [574]. Therefore, nonsteroidal antiandrogens are expected to have advantages by avoiding steroid-related side effects [575].

9.3.1. Flutamide

Flutamide (Sch-13521, NK-601, FTA) [*013311-84-7*], 2-methyl-*N*-[4-nitro-3(trifluormethyl)phenyl]-propanamide, $C_{11}H_{11}F_3N_2O_3$ was discovered in the early 1970s. Unlike the steroids it can easily be synthesized [576], [577] and is devoid hormonal activities. Total sales of the product were $ 271 × 10^6$ in 1996.
Trade Names: Eulexin and Drogenil (Schering-Plough).

Mechanisms of Action. In a comparative study in castrated rats flutamide was shown to be equipotent to cyproterone acetate as an antiandrogen [578]. Several groups had observed that flutamide was a more potent antiandrogen in vivo than in vitro and suggested the involvement of an active metabolite [579], [580]. The major metabolite was identified as 2-hydroxyflutamide analog (Sch-16423) [52806-53-8] and high levels of 2-hydroxyflutamide in the plasma led to the conclusion that this was the active form of flutamide [581]. Although the plasma levels of testosterone increased on flutamide treatment, the levels of testosterone and dihydrotestosterone in androgen target tissue were reduced [582]. Flutamide exerts its antiandrogenic action by blocking the binding of androgens to the cytosolic androgen receptor and/or inhibiting the nuclear binding of androgens in the target tissue [581].

In patients improvements were seen in pain relief, prostatic enlargement and induration, reduction of metastases, and increase in body weight and phosphatase. Most common side effects of flutamide therapy are gynecomastia and breast tenderness [584]. More recent trials use a combination of flutamide with LHRH agonists (see Section 9.4). This is the concept of maximal androgen withdrawal in which the LHRH agonist wipes out androgens of testicular origin and the antiandrogen blocks the action of androgens of adrenal origin at the androgen receptor [585]–[587]. The therapeutic benefits seem to be greatest in patients with minimal disease at the start of treatment.

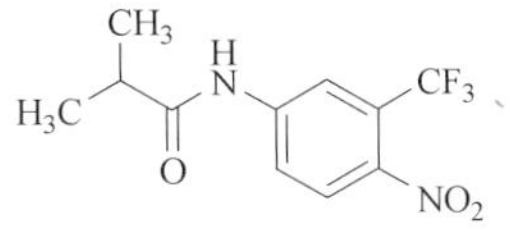

Flutamide

9.3.2. Nilutamide

Nilutamide has been discovered bei Roussel-Uclaf.
Trade Name: Anandron (Roussel-Uclaf).

Mechanism of Action. Nilutamide competitively inhibits binding of androgens to the cytosolic androgen receptor. Administration over 7 days to immature, castrated male rats, nilutamide inhibited the increase of prostate weights induced by testosterone in a dose-dependent manner [588]. In rat pituitary cells nilutamide reverses the inhibition of LHRH-induced LH release elicited by dihydrotestosterone. It is probably because of these effects that nilutamide is recommended for use only in surgically or medically castrated males [589]. Nilutamide has demonstrated an antiandrogenic action in several animal tumor models [590].

Single dose kinetics of nilutamide in volunteers (100 mg) indicate a half-life of 43 ± 3 h, compared to 5.2 h for flutamide (200 mg) [591].

Clinical trials with nilutamide have concentrated on combination therapy with surgical or medical castration. The side effects observed in patients include hot flushes, nausea, vomiting, and visual problems [592].

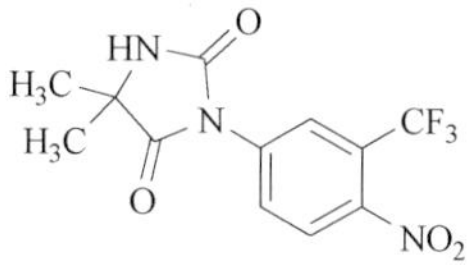

Nilutamide

9.3.3. Bicalutamide

Bicalutamide (ICI-176334) [*090357-06-5*], (±)-4-[3-(4-fluorophenylsulfonyl)-2-hydroxy-2-methylpropionamido]-2-(trifluoromethyl)benzonitrile, $C_{18}H_{14}F_4N_2O_4S$, M_r 430.37 was discovered at ICI/Zeneca and selected from more than 1000 compounds as having the desired properties of a pure, nonsteroidal, peripherally selective antiandrogen [593], [594]. The product was launched for the treatment of advanced prostate cancer in combination with LHRH analogs or surgical castration.
Trade Name: Casodex (Zeneca).

Mechanism of Action. Bicalutamide inhibits the binding of the synthetic androgen [^{3}H]-R-1881 to both rat prostate and pituitary cytosol androgen receptors. The substance binds some 50 times less effectively than dihydrotestosterone and about 100 times less effectively than R-1881 to the prostate androgen receptor, Its affinity for the prostate receptor is about four-fold higher, that for the pituitary receptor ten times higher than that of hydroxyflutamide [595]. In vivo studies revealed that bicalutamide is about five times as potent as flutamide after oral application. Bicalutamide did not cause a significant elevation in LH or testosterone at any of the doses tested, whereas flutamide elicited increases. Half-life of bicalutamide in prostate cancer patients who received 10, 30 or 50 mg/d bicalutamide was around 6 d. The compound was well tolerated in all doses [596]–[598].

Nilutamide and bicalutamide offer advantages over flutamide because of their long half-lives and sustained serum levels on once-daily dosing. The latter is essential to prevent androgen stimulation.

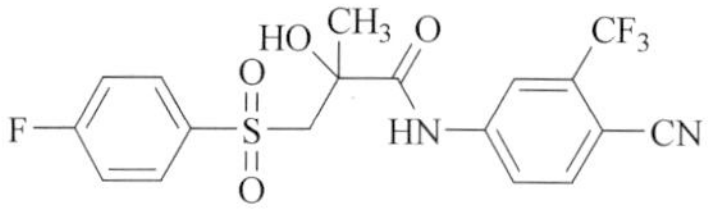

Bicalutamide

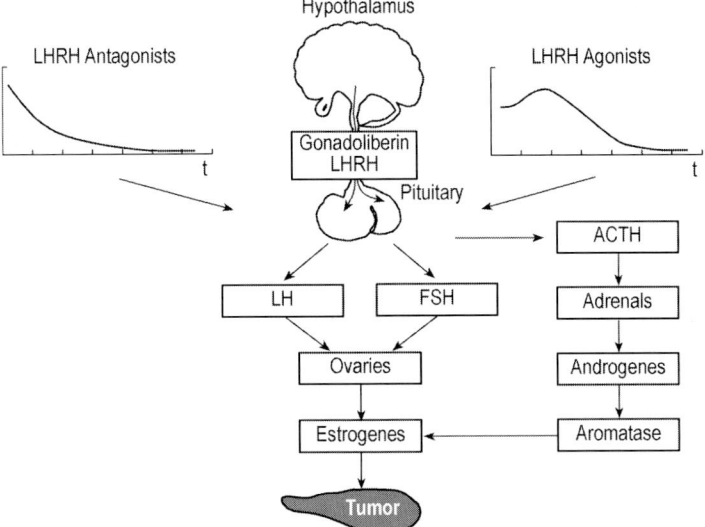

Figure 3. Antitumor activity of LHRH agonists and antagonists in female FSH = follicle stimulating hormone; ACTH = adrenocorticotropic hormone

9.4. LHRH Analogs (→ Peptide and Protein Hormones)

The releasing hormone gonadorelin (GnRH, synonymous with LHRH, luteinizing hormone-releasing hormone or gonadoliberin), together with its specific receptor, plays a central role in neuroendocrinology [599]. The decapeptide LHRH is formed in the cell bodies of hypothalamic neurons and is secreted in pulses into the blood stream [600] – [603]. Ultimately it stimulates secretion of the sex-specific hormones in the testes and ovaries. Specific receptors for LHRH and synthetic analogs are also present in the pituitary gland and other tissues (for example, tumor cells) [604] and organs.

Three concepts for therapeutic application have emerged. The first is the restoration of normal physiology by administration of LHRH by infusion pump to promote fertility in men and women who are infertile due to defective endogenous LHRH secretion. Second, long-lasting LHRH agonists (so-called superagonists) are used in a depot form, which bring about desensitization of the pituitary receptors and thus interrupt the signal cascade. This results in a biochemical "castration", which opens up new therapeutic possibilities for hormone-dependent diseases such as prostate cancer, breast cancer, and endometriosis. Although superagonists are generally well tolerated, they have the disadvantage that hormone secretion (estrogen, testosterone) is initially stimulated before the depletion of receptors or "down" regulation can take place, and thus the illness temporarily worsens [605]. This has led to the development of the third concept: use of LHRH antagonists [606] (see Fig. 3). In the late 1980s about 5000 LHRH analogs had been synthesized worldwide and tested in vitro or in vivo

Table 1. Structure of LHRH agonists on the market (given are only those amino acid residues that are different in LHRH).

Name (Co.)	1	2	3	4	5	6	7	8	9	10
LHRH	Glp[a]	His	Trp	Ser	Tyr	Gly	Leu	Arg	Pro	Gly-NH$_2$
Buserelin (Hoechst)						D-Ser(tBu)				Gly-NHEt
Nafarelin						D-(2)Nal				
Leuprorelin (Abbott, Takeda)						D-Leu				Gly-NHEt
Goserelin (Zeneca)						D-Ser(tBu)				Azagly-NH$_2$
Histrelin (Ortho)						D-His(Bzl)				Gly-NHEt
Triptorelin (Ferring)						D-Trp				

[a] Glp = pyroglutamic acid.

[607]–[609]. Whilst LHRH agonists have been on the market for about ten years, the LHRH antagonists that have been developed farthest are still in clinical testing [610]–[614].

9.4.1. LHRH Agonists

The first years after the discovery of the gonadorelins were marked by the search for more active agonists, since the therapeutic potential of gonadorelins as, for example, antitumor agents or in gynecology, was apparent [606], [615]. Such superagonists bring about a very effective reversible inhibition of the release of steroidal sex hormones. The exchange of glycine at postion 6 (glycine$_6$) of the native LHRH for other, always D-configured, amino acids, is common to all modern superagonists; some have a C-terminal ethylamide (buserelin, leuprorelin) or azaglycinamide (goserelin) residue instead of glycinamide. Eight to ten amino acids of the LHRH sequence are thus conserved in all clinically relevant superagonists; by exchange at a maximum of two positions, the biological activity or hormone suppression in tumor patients, can be increased by a factor of up to 100 on subcutaneous application [606], [615]. Table 1 summarizes the most important derivatives currently on the market.

Buserelin (Profact, Suprecur) [57982-77-1], leuprorelin (Lupron, Carcinil, Enatone) [53714-56-0], triptorelin (Decapeptyl) [57773-63-4], and goserelin (Zoladex) [65807-02-5] (trade names in Germany in parantheses) are the products on the market with the highest turnover. The Lupron line of products is the leading anticancer hormone drug with worldwide sales in 1996 of $ 810 \times 10^6$ followed by Zoladex with sales of $ 563 \times 10^6$.

Annual production is less than 100 kg for buserelin and significantly over 100 kg for the market leader, leuprorelin. At this order of magnitude, substances are only produced by classical organic preparative synthesis (fragment condensation in solution). Hoechst, for example, synthesizes the nonapeptide buserelin from the units pyroglutamic acid-histidine (Glp-His), tryptophan-serine-tyrosine (Trp-Ser-Tyr), and D-Serine-tert-butyl ether-leucine-arginine-proline-NHEt (D-Ser(tBu)-Leu-Arg-Pro-NHEt); the tri-

and tetrapeptide units are coupled to form the corresponding C-terminal heptapeptide, and then the N-terminal dipeptide is condensed with this to form the complete sequence. Control of the physicochemical process parameters, such as concentrations, precipitation, separations, reaction temperature profiles, and purification techniques, is important for successful scale-up of peptide synthesis to a technical scale [617].

9.4.1.1. Leuprorelin Acetate

Because leuprorelin inhibits the synthesis of androgen and estrogen, the drug blocks the growth of hormone-dependent tumors by shutting down testosterone production. Leuprorelin is indicated for the treatment of advanced prostate cancer, as an alternative to castration for the treatment of endometriosis, and for the presurgical management of patients with anemia caused by benign fibroid tumors.
Trade Name: Lupron (Abbott, Takeda).

9.4.1.2. Goserelin

As certain prostate tumors grow in response to testosterone, goserelin prevents the production of testosterone in testes and is therefore indicated for the treatment of prostate cancer, advanced breast cancer, and endometriosis.
Trade Name: Zoladex (Zeneca).

9.4.2. LHRH Antagonists

Common to the intrinsic activity of all superagonists is the initial temporary stimulation of gonadotropin release. Soon after the use of highly active agonists became an established therapy, a search began for corresponding antagonists, which do not bring about an initial hormone release, to avoid this therapeutically counterproductive effect. A final big hurdle for clinical use of highly active antagonists was the inherent anaphylactic potential of these peptides. Starting from the sequence of native LHRH, the individual positions of the peptide chain were examined in rapid succession for their contribution to biological activity. Particular attention was paid to side effects. The most effective early improvements in antagonistic activity were achieved by using D-phenylalanine (D-Phe) instead of histidine at position 2, by D-amino acids at position 6 instead of Gly, and the exchange of C-terminal glycine for D-alanine (D-Ala^{10}). Further stepwise optimization led to the sequence scheme now usual for all modern antagonists of D-Nal^1-D-Cpa^2-D-Pal^3 (Nal = 2-naphthylalanine, Cpa = Phe(4-Cl), Pal = 3-pyridylalanine) as hydrophobic cluster, a D-configured aromatic or aliphatic, yet hydrophilic, aminocarboxylic acid at position 6, and the C-terminal hydrophilic sequence Xxx^8-Pro^9-D-Ala^{10} where Xxx is either arginine or isopropyllysine. Excellent documentation of the stepwise optimization can be found in [618], [619].

Antagonists of the second generation caused formation of temporary edemas of the face and extremities in animal experiments, due in part to massive histamine release by mast cell degranulation. Cyanosis and respiratory impairment were also observed [620]. The cause of these intolerable side effects is thought to be the combination of D-arginine at position 6 with the three aromatic amino acids at the N-terminus of the sequence [621]. For the desired biological antagonist potency, a D-configured basic amino acid is necessary at position 6. A. V. SCHALLY et al. achieved the breakthrough to highly active antagonists free of side effects with the derivatives SB-75 (cetrorelix) [120287-85-6] and SB-88 [120287-93-6], by incorporating hydrophilic, nonbasic amino acids with side-chain carbamoyl functions at position 6 (D-Citrulline^6, D-homocitrueline) [621], [622].

Today, e.g., cetrorelix is *manufactured* exclusively by classical fragment condensation on a kilogram scale. Two synthesis strategies proceed via either the N-terminal tripeptide (D-Nal^1-D-Cpa^2-D-Pal^3) or the C-terminal tripeptide (Arg^8-Pro^9-D-Ala^{10}) and the complementary heptapeptide (Ser^4-D-Ala^{10}) or (Nal^1-Leu^7) to the protected decapeptide with *tert*-butyl side chain protection. The deprotection with hydrochloric acid is followed by final purification by preparative HPLC. The C- and N-terminal functionalization of the acetylated decapeptide amide, necessary for biological activity and to avoid rapid enzymatic degradation, is introduced at the level of the terminal tripeptide. Functionalization is achieved by acetylation of the free α-amino group of naphthylalanine^1 with acetylhydroxysuccinimide, or amminolysis of the resulting alanine^{10} methyl ester in alcoholic ammonia [623].

FOLKERS et al. were able to improve active antagonists successively by consistent and systematic modifications both at the relevant sequence positions and the side chain substituents; the most successful were complex substitutions at the positions 5, 6 and, in part, 8 [624]–[628]. An important contribution was made by RIVIER et al., who synthesized the decapeptide "azaline" with novel modifications at positions 5 and 6, where aminotriazole-substituted *p*-aminophenylalanine or lysine are positioned [629]–[631]. Azaline B [134457-28-6] is probably one of the most active antagonists presently available worldwide.

DEGHENGHI published a highly active decapeptide sequence with minimal histamine release and good water solubility. This structure, known as antarelix [151277-78-5], differs from SB-75 (cetrorelix) in that it has homocitrulline^6 instead of citrulline^6 and isopropylysine^8 instead of arginine^8 [633]. By using Lys(*i*Pr)^8, residual potential for histamine release can be further reduced. Organon is developing the antagonist ganirelix (RS26306) [124904-93-4] under license from Syntex; this is a decapeptide with novel alkyl-modified D- and L-homoarginine units at positions 6 and 8 [634]. With ramorelix (HOE 013) [127932-90-5], Hoechst (HMR) has a peptide antagonist with a sugar–amino acid unit (*O*-α-L-rhamnosyl-D-serine^6), that has improved water solubility [632]. Schering has also synthesized peptide antagonists, using nonproteinogenic amino acids, such as ε-dialkylated lysine or benzodiazepine aminocarboxylic acids [635], [636]. Abbott's A-76154 [136989-30-5] is an octapeptide antagonist with LHRH receptor affinity of a similar order of magnitude to that of active decapeptides such as A-75998 [*135215-95-1*] or "Nal-Glu" [*103733-02-4*] [644].

9.4.2.1. Receptor Assays

At present, selected peptides are being tested in vivo by measuring the testosterone concentration in male rat serum. Here, a single subcutaneous application of a potent LHRH antagonist leads to persistent testosterone suppression. According to current knowledge, the activity in animals generally correlates well with the binding affinity determined in vitro on human receptors. In cases where no in vivo activity was observed despite high binding affinity, the peptide probably had pharmacokinetic characteristics that played a decisive role. In the future, transgenic animals will also be very important for in vivo testing [645] – [647].

9.4.2.2. Peptidomimetics

For some years, increased efforts have also been made to find substances with affinity to the LHRH receptor that do not have the characteristic substance-specific properties and also disadvantages of peptides (short half-life, lack of bioavailability), yet have a high binding affinity. Ideally, such substances should be able to be administered orally, be sufficiently stable in the organism, and possess favorable pharmacological parameters comparable to peptide antagonists. Clinical investigations of the influence of the antimycotic ketoconazole on prostate cancer and testosterone suppression indicate that it may have a LHRH-mimetic effect [648], [649]. Further investigations by Abbott showed weak antagonistic activity both in vitro and in vivo for ketoconazole and its modified analogs; however, the biological activity of such derivatives is most probably not mediated mainly by LHRH antagonism [650], [651]. Complex, highly substituted nitrogen heterocycles seem to have LHRH antagonistic potential [652]. In patent literature benzodiazepines, benzodiazepinones, heterocyclic benzo-substituted alkylamines, and recently, thienopyridine carboxylic acid derivatives are described as LHRH receptor antagonists with receptor-binding inhibition at submicromolar concentrations, suitable amongst other uses as antitumor agents for hormone-dependent tumors [647], [653]. A lead structure from the latter substance class is currently in extensive pharmacological trials [654] – [656].

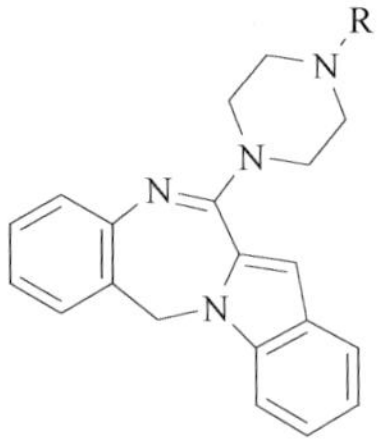

McNeil, US 4 678 784

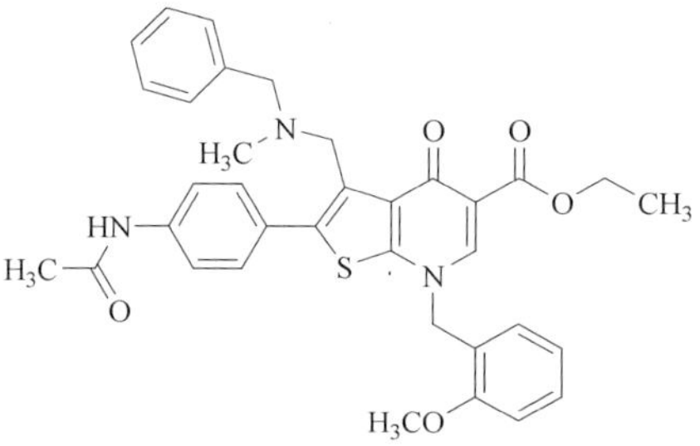

Takeda, WO 95/28405

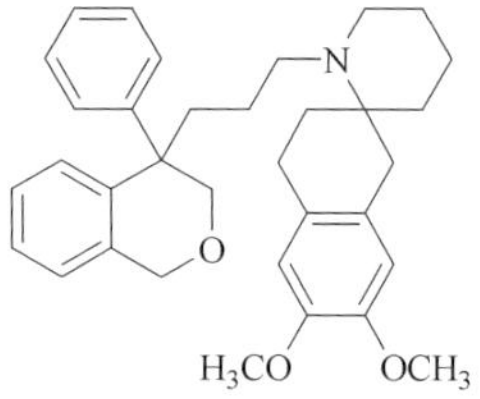

Takeda, EP 679 642 A1

10. Signal Transduction Inhibitors

Recently anticancer drug discovery has been directed away from agents that affect cells by producing DNA damage towards modulators of signal transduction pathways that have become unregulated or aberrant in malignant transformation [657].

10.1. Enzyme Inhibitors

Binding of growth factors [658], such as epidermal growth factor (EGF) [659] to a membrane bound receptor tyrosine kinase results in dimerization and autophosphorylation of tyrosine residues on the protein surface. As result the GTP- bound form of *ras* [660] undergoes a conformational change on its surface, enabling it to bind to several effector molecules leading to the activation of transcription factors involved in DNA synthesis [661]. An overexpression of the EGF-receptor is observed in various types of human cancers.

Mechanism-based screens have identified the natural products bryostatin [662] [*83314-01-6*] as partial agonist of protein kinase C and fumagillin [663] [*23110-15-8*] as interfering with tumor-induced neovascularization thus inhibiting angiogenesis. FR-111142 as analog of fumagillin is reported to be less toxic and more active than the parent compound [664]. A potent inhibitor of *ras* protein farnesylation [665] is pepticinnamin E [666], and peptides related to the CAAX-tetrapeptide (where C is cysteine, A is valine, isoleucine, or leucine, and X is methionine or serine). In order to prepare completely nonpeptide and potentially more stable inhibitors a series of potent peptidomimetic inhibitors were designed [667].

Specific natural and synthetic kinase inhibitors are in clinical development as anticancer agents [668]–[670].

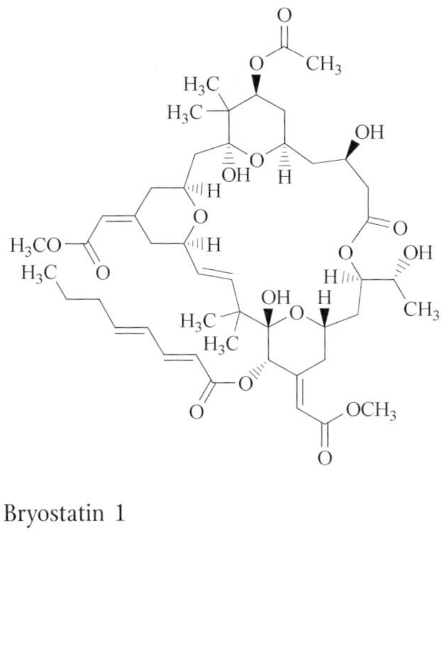

Bryostatin 1

10.2. Phospholipid – Based Antineoplastics

Ether phospholipids and lysophospholipids are naturally occurring derivatives of phospholipids of the cell membrane with many interesting properties. For example, the ether phospholipid 1-O-alkyl-2-acetyl-sn-glycero-3-phosphocholine (platelet activating factor, PAF) [671] causes platelet aggregation and dilation of blood vessels and lysophosphatidic acid (LPA), the simplest natural phospholipid, is a potent mitogen. The observation that alkyllysophospholipids exerted experimental antitumor activity in vitro and in vivo [672], [673] led to systematic modification of the structures and subsequent clinical investigation of, e.g., ilmofosine [83519-04-4] [675]. In the course of modification it was found that an analog without glycerol backbone was equally effective.

Miltefosine (D-18506, hexadecylphosphocholine), [058066-85-6] [676], 2-[[(hexadecyloxy)-hydroxyphosphinyl]oxy]-N,N,N-trimethylethanaminium hydroxide (inner salt); $C_{21}H_{46}NO_4P$, M_r 407.57 is a new phospholipid-based antineoplastic agent which exerts substantial antitumor activity in appropriate models. Because of its special pharmacology — high activity against mammary carcinoma and low toxicity to normal tissue — it is an ideal candidate for topical treatment of cutanous breast cancer metastases. The substance was launched in 1993 as a topical formulation for palliative treatment of refractory skin metastases of breast cancer [676].
Trade Name: Miltex (ASTA Medica).

A variety of derivatives were synthesized and characterized, some of which appeared to be considerably less toxic to the gastrointestinal tract [677]. One of them, perifosine (D-21266) [157716-52-4] is currently in clinical trials.

Mechanism of Action. Miltefosine shows highly selective antineoplastic activity in DMBA- and N-methyl-N-nitrosourea (MNU)- mammary carcinoma of the rat. The activity is not due to an antiestrogenic effect in these estrogen-dependent tumor models [678]. The treatment was tolerated and no overt toxic symptoms were observed. The mechanism of action of miltefosine is most probably different from that of other chemotherapeutic agents [679]. Incorporation of miltefosine and metabolites into biological membranes was demonstrated, affecting the interactions of receptor proteins or membrane associated enzymes involved in growth control and cellular signalling [680]. The reduction in tumor mass was accompanied by morphological changes compatible with the induction of differentiation [675], [681]. Furthermore, the decrease of tumors was mainly due to cell loss by apoptosis.

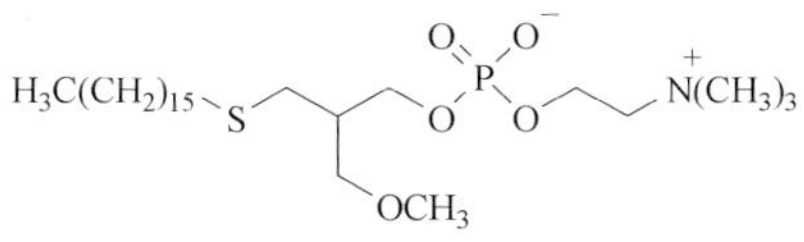

H₃C(CH₂)₁₅−S−CH(CH₂OCH₃)−CH₂−O−P(=O)(O⁻)−O−CH₂CH₂−N⁺(CH₃)₃

Ilmofosine

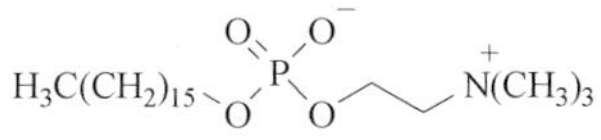

H₃C(CH₂)₁₅−O−P(=O)(O⁻)−O−CH₂CH₂−N⁺(CH₃)₃

Miltefosine

In clinical trials mild dryness and flaking of the skin were noted, but no systemic toxicity was reported after topical administration. Several phase I and II studies in patients that were treated with oral capsules were terminated early due to gastrointestinal intolerance [682].

11. Economic Aspects

The world cancer drug market is expected to reach $ 14×10^9 by 2000 [10], [14]. Total sales of the top 100 anticancer drugs generated a turnover of $ 4.66×10^9 in 1996. Of the worlds eight top-selling anticancer drugs, four — the prostate cancer-drugs flutamide (Eulexin), bicalutamide (Casodex), leuprorelin (Lupron) and goserelin (Zoladex) — are merely palliative, yet have combined annual sales of $ 1.7×10^9, whereas sales of the breast cancer drugs tamoxifen (Nolvadex) and paclitaxel (Taxol) are approaching $ 500×10^6 and $ 850×10^6, respectively.

Taxol, BMS's drug for treating ovarian and breast cancers, was the biggest selling anticancer agent in 1996, with sales of $ 813×10^6.

12. References

[1] I. B. Weinstein, *Cancer Res. (Suppl.)* **51** (1991) 5080.
[2] R. Doll, *Eur. J. Cancer* **26** (1990) 5007.
[3] C. LaVecchina, F. Levi, F. Lucchini, S. Garattini, *Anti-Cancer-Drugs* **2** (1991) 215.
[4] D. Kessel, *In Vivo* **8** (1994) 829.
[5] T. Tsurno, A. Toninda, *Anti-Cancer-Drugs* **6** (1995) 213.
[6] G. Eisenbrand, S. Lauch-Birkel, W. C. Tang, *Synthesis* (1996) 1246.
[7] G. B. Elion, *Cancer Res.* **45** (1985) 2943.
[8] *Scrip* 1997, July 15, p. 20.
[9] *Chem. Marketing Report*, July 14, 1997, p. 19.
[10] *Med. Ad-News*, May 1996, p. 56.
[11] G. M. Cragg, D. J. Newman, K. M. Snader, *J. Nat. Prod.* **60** (1997) 52–60.
[12] K. Nasmyth, *Science* **274** (1996) 1643–1645.
[13] G. Dratta, H. Pagano, *Annu. Rep. Med. Chem.* **31** (1996) 241–248.
[14] *Pharma Business*, July/August 1997, p. 62.
[15] S. A. Rosenberg: "Principles of Surgical Oncology," in V. T. DeVita, S. Hellman, S. A. Rosenberg (eds.): *Cancer: Principles and Practice of Oncology*, J. B. Lippincott Co., Philadelphia 1982, pp. 93–102.
[16] S. Farber, L. K. Diamond, R. D. Mercer, R. F. Sylvester, V. A. Wolff, *N. Engl. J. Med.* **238** (1948) 787–793.
[17] R. Hertz, G. T. Ross, M. B. Lipsett, *Am. J. Obstet. Gynecol.* **86** (1963) 808–814.
[18] R. L. Kisliuk, Y. Gaumont, C. M. Baugh, J. M. Galivan, G. F. Maley, F. Maley: "Inhibition of Thymidylate Synthetase by Poly-gamma-glutamyl Derivatives of Folate and Methotrexate," in R. L. Kisliuk, G. M. Brown (eds.): *Chemistry and Biology of the Pteridines*, Elsevier/North Holland, New York 1979, pp. 431–435.
[19] I. D. Goldman, N. S. Lichtenstein, V. T. Oliverio, *J. Biol. Chem.* **243** (1968) 5007–5017.
[20] F. M. Sirotnak, R. C. Donsbach, *Cancer Res.* **36** (1976) 1151–1158.
[21] R. D. Warren, A. P. Nichols, R. A. Bender, *Cancer Res.* **38** (1978) 668–671.
[22] P. L. Chello, F. M. Sirotnak, D. M. Dorck et al., *Cancer Res.* **37** (1977) 4297–4303.
[23] B. T. Hill, B. D. Bailey, J. C. White et al., *Cancer Res.* **39** (1979) 2440–2446.
[24] F. M. Sirotnak, R. C. Donsbach, *Cancer Res.* **35** (1975) 1737–1744.
[25] J. J. McGuire, J. R. Bertino, *Mol. Cell Biochem.* **38** (1981) 19–48.
[26] R. L. Schilsky, B. D. Bailey, B. A. Chabner, *Proc. Natl. Acad. Sci. USA* **77** (1980) 2919.
[27] S. A. Jacobs, C. J. Derr, D. G. Johns, *Biochem. Pharmacol.* **26** (1977) 2310–2313.
[28] V. M. Whitehead, *Cancer Res.* **37** (1977) 408–412.
[29] J. Jolivet, B. A. Chabner, *J. Clin. Invest.* **72** (1983) 773–778.
[30] S. V. Gupta, N. J. Greenfield, M. Poe et al., *Biochemistry* **16** (1977) 3073–3079.
[31] H. Nakamura, J. Littlefield, *J. Biol. Chem.* **247** (1972) 179–187.
[32] D. A. Matthews, R. A. Alden, J. T. Bolin et al.: "X-ray Structural Studies of Dihydrofolate Reductase," in R. L. Kisliuk, G. M. Brown (eds.): *Chemistry and Biology of the Pteridines*, Elsevier/North Holland, New York 1979, pp. 465–470.
[33] P. A. Charlton, D. W. Young, B. Birdsall et al., *Chem. Commun.* **20** (1979) 922–924.
[34] D. A. Matthews, R. A. Alden, J. T. Bolin et al., *Science* **197** (1977) 452–455.
[35] W. L. Werkheiser, *Cancer Res.* **23** (1963) 1277–1285.
[36] B. A. Kamen, W. Whyte-Bauer, J. R. Bertino, *Biochem. Pharmacol.* **32** (1983) 1837–1841.

[37] R. J. Kaufman, J. R. Bertino, R. T. Schimke, *J. Biol. Chem.* **253** (1978) 5852–5860.
[38] E. W. Alt, R. E. Kellems, J. R. Bertino et al., *J. Biol. Chem.* **253** (1978) 1357.
[39] G. A. Curt, J. Jolivet, B. D. Bailey, D. N. Carney, B. A. Chabner, *Biochem. Pharmacol.* **33** (1984) 1682–1685.
[40] G. A. Curt, D. N. Carney, K. H. Cowan, J. Jolivet et al., *N. Engl. J. Med.* **308** (1982) 199.
[41] M. J. Cline, H. Stang, K. Mercola et al., *Nature (London)* **284** (1980) 922–925.
[42] M. Bar-Eli, H. D. Stang, K. E. Mercola, M. J. Cline, *Somatic Cell Genet.* **9** (1983) 55–67.
[43] F. Carr, W. D. Medina, S. Dube, J. R. Bertino, *Blood* **62** (1983) 180–185.
[44] H. Diddens, D. Niethammer, R. C. Jackson, *Cancer Res.* **43** (1983) 5286–5292.
[45] F. M. Sirotnak, J. I. Degraw, F. A. Schmid, L. J. Goutas et al., *Cancer Chemother. Pharmacol.* **12** (1984) 26–30.
[46] R. P. Hertzberg, R. K. Johnson, *Annu. Rep. Med. Chem.* **28** (1993) 167–176.
[47] G. F. Fleming, R. L. Schilsky, *Semin. Oncol.* **19** (1992) 707.
[48] K. Y. Shum et al., *J. Clin. Oncol.* **6** (1988) 446.
[49] A. L. Lackman et al., *Cancer Res.* **51** (1991) 5579.
[50] D. Cunningham, *Cancer Res.* **53** (1991) 810.
[51] A. L. Jackman, D. C. Farrugia, W. Gibson, *Eur. J. Cancer* **13A** (1995) 1277–1282.
[52] M. D. Varney et al., *J. Med. Chem.* **35** (1992) 663.
[53] C. Heidelberger, N. K. Chandhavi, P. Dannenberg et al., *Nature (London)* **179** (1957) 663–666.
[54] R. J. Rutman, A. Cantarow, K. E. Paschkis, *Cancer Res.* **14** (1954) 119–126.
[55] R. L. Glazer, L. S. Lloyd, *Molec. Pharmacol.* **21** (1982) 468–473.
[56] D. W. Kufe, P. P. Major, *J. Biol. Chem.* **256** (1981) 9802–9806.
[57] K. V. Hadjiolova, Z. G. Naydenova, A. A. Hadjiolova, *Biochem. Pharmacol.* **30** (1981) 1861–1866.
[58] R. I. Glazer, K. D. Hartman, *Molec. Pharmacol.* **23** (1983) 540–546.
[59] Y. Oshima, M. Itoh, N. Okada, T. Miyata, *Proc. Natl. Acad. Sci. USA* **78** (1981) 4471–4474.
[60] D. W. Kufe, P. P. Major, E. M. Egan, E. Leh, *J. Biol. Chem.* **256** (1981) 8885–8889.
[61] P. V. Danenberg, C. Heidelberger, M. A. Mulkins, A. R. Peterson, *Biochem. Biophys. Res. Commun.* **102** (1981) 654–659.
[62] Y. C. Cheng, K. Nakayama, *Molec. Pharmacol.* **23** (1983) 171–174.
[63] M. Tanaka, K. Kimura, S. Yoshida, *Cancer Res.* **43** (1983) 5145–5150.
[64] P. Reichard, O. Skold, G. Klein et al., *Cancer Res.* **22** (1962) 235–243.
[65] P. Reyes, T. C. Hall, *Biochem. Pharmacol.* **18** (1969) 1587–1590.
[66] D. K. Kasbeker, D. M. Greenberg, *Cancer Res.* **23** (1963) 818–825.
[67] P. Reichard, O. Skold, G. Klein, *Nature (London)* **183** (1959) 939–941.
[68] J. A. Houghton, S. J. Masada, J. O. Phillips et al., *Cancer Res.* **42** (1982) 144–149.
[69] C. Heidelberger, G. Kaldos, K. L. Mukherjee et al., *Cancer Res.* **20** (1960) 903–909.
[70] M. M. Lastieboff, B. Kedzioska, W. Rode, *Biochem. Pharmacol.* **32** (1983) 2259–2267.
[71] A. R. Bapat, C. Zarow, P. V. Dannenberg, *J. Biol. Chem.* **258** (1983) 4130–4136.
[72] C. C. Rossana, L. G. Rao, L. F. Johnson, *Mol. Cell. Biol.* **2** (1982) 1118–1125.
[73] F. Baskin, S. C. Carlin, P. Kraus et al., *Molec. Pharmacol.* **11** (1975) 105–117.
[74] D. G. Priest, B. E. Ledford, M. T. Doig, *Biochem. Pharmacol.* **29** (1980) 1549–1553.
[75] R. D. Armstrong, E. Cadman, *Cancer Res.* **43** (1983) 2525–2528.
[76] J. L. Au, Y. M. Rustum, J. Minonda, B. I. Sriva-stava, *Biochem. Pharmacol.* **32** (1983) 541–545.
[77] Y. M. El Sayed, W. Sadee, *Cancer Res.* **43** (1983) 4039–4044.
[78] E. J. Ansfield, G. J. Kallas, J. P. Simpson, *J. Clin. Oncol.* **1** (1983) 107–110.
[79] F. Sorm, A. Piskala, A. Cihak et al., *Experientia* **20** (1964) 202–203.

[80]   A. Piskala, F. Sorm, *Collect. Czech. Chem. Commun.* **29** (1964) 2060–2075.
[81]   P. G. W. Plagemann, M. Behrens, D. Abraham, *Cancer Res.* **38** (1978) 2458–2466.
[82]   J. C. Drake, R. G. Stoller, B. A. Chabner, *Biochem. Pharmacol.* **26** (1977) 64–66.
[83]   T. Lee, M. Karon, R. L. Momparler, *Cancer Res.* **34** (1974) 2481–2488.
[84]   A. Cihak, *Collect. Czech. Chem. Commun.* **39** (1974) 3782–3792.
[85]   J. Doskel, V. Paces, F. Sorm, *Biochim. Biophys. Acta* **145** (1967) 771–779.
[86]   L. H. Li, E. J. Olin, T. J. Fraser et al., *Cancer Res.* **30** (1970) 2770–2775.
[87]   B. F. Vanyosbin, S. G. Tkacheug, A. W. Belozinsky, *Nature (London)* **225** (1976) 948–949.
[88]   J. L. Mandell, P. Chambon, *Nucleic Acids Res.* **7** (1979) 2081–2103.
[89]   J. D. McGhee, G. D. Gineler, *Nature (London)* **280** (1979) 419–420.
[90]   C. Scheir, T. Maniatis, *Proc. Natl. Acad. Sci. USA* **77** (1980) 6634–6638.
[91]   L. H. Van der Ploeg, R. A. Flavill, *Cell* **19** (1980) 947–958.
[92]   O. Niwa, T. Sugarhar, *Proc. Natl. Acad. Sci. USA* **78** (1981) 6290–6294.
[93]   S. Friedman, *Molec. Pharmacol.* **19** (1980) 314–320.
[94]   D. V. Santi, C. E. Garrett, P. J. Barr, *Cell* **25** (1983) 9–10.
[95]   P. G. Constantinides, P. A. Jones, W. Geness, *Nature (London)* **267** (1977) 364–366.
[96]   P. A. Jones, S. M. Taylor, *Cell* **20** (1980) 85–93.
[97]   T. J. Ley, J. DeSimone, N. P. Anagnov et al., *N. Engl. J. Med.* **307** (1982) 1469–1475.
[98]   J. Vesely, A. Chiak, *Pharmacol. Therap.* **2** (1978) 813–840.
[99]   T. Lee, M. R. Kovan, *Biochem. Pharmacol.* **25** (1976) 1737–1742.
[100] A. Cihak, H. Vessela, F. Sorm, *Biochim. Biophys. Acta* **166** (1968) 277–279.
[101] A. Cihak, J. Vesely, *Biochem. Pharmacol.* **21** (1972) 3257–3265.
[102] J. Vesely, A. Cihak, F. Sorm, *Int. J. Cancer* **2** (1967) 639–646.
[103] J. Doskacil, F. Sorm, *FEBS Lett.* **2** (1974) 30–32.
[104] D. D. Von Hoff, M. Slavik, F. M. Muggia, *Ann.Intern. Med.* **85** (1976) 237–245.
[105] Z. H. Israili, W. R. Vogler, E. S. Mingidi et al., *Cancer Res.* **36** (1971) 1453–2461.
[106] J. A. Beisler, *J. Mol. Chem.* **21** (1978) 204–208.
[107] J. A. Beisler, M. M. Abbasi, J. S. Driscoll, *Cancer Treat. Rep.* **60** (1976) 1671–1674.
[108] G. A. Curt, J. A. Kelley, R. L. Fine et al., *Proc. Am. Assoc. Clin. Oncol.* **3** (1984) 37.
[109] P. G. W. Plagemann, R. Marz, R. M. Wolhvete, *Cancer Res.* **38** (1978) 978–989.
[110] A. Iwagaki, T. Nakamura, G. Wakisaka, *Cancer Res.* **29** (1969) 2169–2176.
[111] M. Y. Chu, G. A. Fisher, *Biochem. Pharmacol.* **11** (1962) 423–430.
[112] P. P. Major, E. M. Egan, G. P. Beardsley, M. D. Minden et al., *Proc. Natl. Acad. Sci. USA* **78** (1983) 3235–3239.
[113] P. P. Major, L. Sargent, E. M. Egan, D. W. Kufe, *Biochem. Pharmacol.* **30** (1981) 2221–2224.
[114] P. P. Major, E. M. Egan, D. Herrick, D. W. Kufe, *Biochem. Pharmacol.* **31** (1982) 861–866.
[115] M. Y. Chu, G. A. Fisher, *Biochem. Pharmacol.* **14** (1965) 333–341.
[116] M. R. Atkinson, M. Deutscher, A. Kornberg, A. Russel et al., *Biochemistry* **8** (1969) 4897–4904.
[117] P. A. Diskwel, F. Wanka, *Biochim. Biophys. Acta* **520** (1978) 461–471.
[118] R. J. Frann, C. M. Egan, D. W. Kufe, *Leuk. Res.* **7** (1983) 243–249.
[119] J. F. Ward, E. I. Jones, W. F. Blakely, *Cancer Res.* **44** (1984) 59–63.
[120] S. R. Bähring-Kuhlmey, *Drugs of Today* **13** (1977) 475.
[121] A. A. Claesen et al., *Tetrahedron Lett.* (1966) 3499.
[122] P. Major et al., *Proc. Natl. Acad. Sci. USA* **78** (1981) 3235.
[123] R. Preston, *Treat. Carcinogen Mutagen* **1** (1980) 147.
[124] P. Drahovsky, W. Kreis, *Biochem. Pharmacol.* **19** (1970) 940–944.

[125] W. Kreis, J. Graham, L. A. Damin, *Biochem. Pharmacol.* **31** (1982) 3831–3837.
[126] J. Balzarini, E. De Cleireg, *Mol. Pharmacol.* **23** (1983) 175–181.
[127] A. W. Harris, E. C. Reynolds, L. R. Finch, *Cancer Res.* **39** (1979) 538–541.
[128] Y. Rustum, H. Priesler, *Cancer Res.* **39** (1979) 42–49.
[129] C. D. Stuart, P. J. Burke, *Nat. New Biol.* **233** (1971) 109–110.
[130] P. Chang, P. H. Wiernik, S. D. Reich et al.: "Prediction of Response to Cytosine Arabinoside and Daunorubicin in Acute Nonlymphocytic Leukemia," in F. Mandelli (ed.): *Therapy of Acute Leukemias,* Lombardo-Editore, Rome 1977, pp. 148–159.
[131] J. F. Smyth, A. B. Robins, C. L. Leese, *Eur. J. Cancer* **12** (1976) 567–573.
[132] M. N. H. Tattersall, U. K. Ganeshaguru, A. V. Hoffbrand, *Br. J. Haematol.* **27** (1974) 39–46.
[133] M. Aoshima et al., *Cancer Res.* **37** (1977) 1481.
[134] F. Cabaninas, *Drugs of the Future* **5** (1980) 12, 603.
[135] T. Ueda, T. Nakamura, S. Ando et al., *Cancer Res.* **43** (1983) 3412–3416.
[136] K. Hoie, T. Tsuro, K. Naganuma, S. Tsukagoshi et al., *Cancer Res.* **44** (1984) 172–177.
[137] C. W. Young, R. Schneider, B. Leyland-Jones et al., *Cancer Res.* **43** (1983) 5006–5009.
[138] Y. C. Cheng, R. S. Tan, J. L. Ruth, G. Dutschman, *Biochem. Pharmacol.* **32** (1983) 726–729.
[139] L. W. Hertel, J. S. Kroin, J. W. Misner, J. M. Tustin, *J. Org. Chem.* **53** (1988) 11, 2406–2409.
[140] Eli Lilly, EP 122707, 1984 (L. W. Hertel).
[141] Eli Lilly, EP 184365, 1986 (G. B. Grindey, L. W. Hertel).
[142] Eli Lilly, EP 306190, 1989T. S. Chou, P. Heath).L. E. Patterson).
[143] S. Chubb et al., *Proc. Am. Assoc. Cancer Res.* **28** (1987) Abst. 1282.
[144] D. Y. Bouffard, L. F. Momparler, R. L. Momparler, *Eur. J. Pharmacol.* **183** (1990) 2, Abst. 032.
[145] V. Heinemann, L. W. Hertel, G. B. Grindley, W. Phukett, *Cancer Res.* **48** (1988) 14, 4024–4031.
[146] M. N. Serradel, J. Castañer, *Drugs of the Future* **15** (1990) 8, 794–797.
[147] V. Heinemann et al., *Proc. Am. Assoc. Cancer Res.* **30** (1989) Abst 2204.
[148] H. H. Hansen, *Ann. Oncol.* **120** (1994) 519–528.
[149] E. Beutler, *Lancet* **340** (1992) 8825, 952.
[150] H. M. Bryson, E. M. Sorkin, *Drugs* **46** (1993) 5, 872.
[151] E. H. Estey et al., *Blood* **79** (1992) 4, 882.
[152] S. P. Smith, *Hosp. Formul.* **28** (1993) 7, 621.
[153] C. P. Robinson, *Drugs Today* **29** (1993) 6, 379.
[154] US Dept. Of Health US 4357324, 1982 (J. A. Montgomery, A. T. Shortnacy).
[155] R. W. Brockman, Y. C. Cheng, F. M. Schabel, J. M. Montgomery, *Cancer Res.* **40** (1980) 3610–3615.
[156] J. A. Montgomery, *Cancer Res.* **42** (1982) 3911–3917.
[157] W. C. Tseng et al., *Mol. Pharmacol.* **21** (1982) 474–477.
[158] S. J. Hopkins, *Drugs of the Future* **10** (1985) 1, 20.
[159] *Annu. Drug Data Rep.* (1992) 633–634.
[160] J. S. Whelan et al., *Brit. J. Cancer* **64** (1991) 1, 120.
[161] H. S. Hochster et al., *J. Clin. Oncol.* **10** (1992) 1, 28.
[162] H. W. Dion et al., *Ann. NY Acad. Sci.* **284** (1977) 21–29.
[163] Parke Davis, US 3923785, 1975 (A. Ryder et al.).
[164] P. W. Woo et al., *J. Heterocycl. Chem.* **11** (1974) 4, 641–643.
[165] J. F. Smith et al., *Cancer Chemother. Pharmacol.* **1** (1978) 49–51.
[166] J. F. Smith et al., *Proc. Am. Assoc. Cancer Res.* **20** (1979).
[167] G. A. Le Page et al., *Cancer Res.* **36** (1976) 1481–1485.

[168] M. N. Seradel, J. Castañer, *Drugs of the Future* **6** (1981) 419–420.
[169] M. Blick et al., *Am. J. Hematol.* **33** (1990) 3, 205–209.
[170] S. Bernhard et al., *Med. Pediatr. Oncol.* **19** (1991) 4, 276.
[171] A. D. Ho et al., *J. Natl. Canver Inst.* **82** (1990) 17, 1416–1420.
[172] E. H. Kraut et al., *J. Clin. Oncol.* **7** (1989) 2, 168–172.
[173] R. S. Witte et al., *Invest. New Drug* **10** (1992) 1,49.
[174] B. J. Kane, J. G. Kuhn, M. K. Roush, *Ann. Pharmacother.* **26** (1992) 939–947.
[175] G. B. Elion, E. Burgi, G. H. Hitchings, *J. Am. Chem. Soc.* **74** (1952) 411.
[176] G. B. Elion, G. H. Hitchings, *J. Am. Chem. Soc.* **77** (1955) 1676.
[177] J. H. Burchenal, M. L. Murphy, R. R. Ellison et al., *Blood* **8** (1953) 965.
[178] G. H. Hitchings, G. B. Elion, *Ann. N.Y. Acad. Sci.* **60** (1954) 195.
[179] R. W. Brockman, C. S. Debavadi, P. Stutts, D. J. Hutchison, *J. Biol. Chem.* **236** (1961) 1471–1479.
[180] J. D. Davidson, *Cancer Res.* **20** (1960) 225–232.
[181] G. B. Elion, *Fed. Proc.* **26** (1967) 898–904.
[182] J. F. Henderson, M. K. Y. Khoo, *J. Biol. Chem.* **240** (1965) 2349–2357.
[183] F. F. Snyder, J. F. Henderson, *Can. J. Biochem.* **51** (1973) 943–948.
[184] P. C. L. Wong, J. F. Henderson, *Biochem. J.* **129** (1972) 1085–1094.
[185] T. Fields, L. Brox, *Can. J. Biochem.* **52** (1974) 441–446.
[186] C. D. Green, D. W. Martin Jr., *Proc. Natl. Acad. Sci. USA* **70** (1973) 3698–3702.
[187] G. B. Elion, S. Callahan, W. Rundles, G. H. Hitchings, *Cancer Res.* **23** (1963) 1207–1217.
[188] G. H. Hitchings, *Cancer Res.* **23** (1963) 1218–1225.
[189] T. Nakamura, *Blood & Vessel (Jpn.)* **2** (1971) 237–248.
[190] T. Higuchi, T. Nakamura, G. Wakisaka, *Blood &Vessel (Jpn.)* **3** (1972) 313–318.
[191] P. W. Allan, L. L. Bennett Jr., *Biochem. Pharmacol.* **20** (1971) 847–852.
[192] A. Hampton, *J. Biol. Chem.* **238** (1963) 3068–3074.
[193] R. J. McCollister, W. R. Gilbert, D. M. Ashton, J. B. Wyngaarden, *J. Biol. Chem.* **239** (1964) 1560–1563.
[194] R. P. Miech, R. E. Parks Jr., J. H. Anderson Jr., A. C. Sartorelli, *Biochem. Pharmacol.* **16** (1967) 2222–2227.
[195] A. R. P. Paterson, D. M. Tidd: "6-Thiopurines," in A. C. Sartorelli, D. G. Johns (eds.): *Handbook of Experimental Pharmacology*, vol. **38**, Springer-Verlag, Berlin 1975, Part 2, pp. 384–403.
[196] D. L. Hill, L. L. Bennett Jr., *Biochemistry* **8** (1969) 122.
[197] R. T. Kavahata, L. F. Chuang, C. A. Holmberg, B. I. Osburn et al., *Cancer Res.* **43** (1983) 3655–3659.
[198] J. Maybaum, H. G. Mandel, *Cancer Res.* **43** (1983) 3852–3856.
[199] J. S. Lazo, K. M. Huang, A. C. Sartorelli, *Cancer Res.* **37** (1977) 4250.
[200] C. K. Carrico, A. C. Sartorelli, *Cancer Res.* **37** (1977) 1868.
[201] K. Kim, W. J. Blechman, V. G. H. Riddle, A. B. Pardee, *Cancer Res.* **41** (1981) 4529.
[202] J. D. Strobel-Stevens, S. M. El Dareer, M. W. Trader, D. L. Hill, *Biochem. Pharmacol.* **31** (1982) 3133.
[203] M. Rosman, H. E. Williams, *Cancer Res.* **33** (1973) 1202.
[204] M. K. Wolpert, S. P. Damle, J. E. Brown et al., *Cancer Res.* **31** (1971) 1620.
[205] M. Rosman, M. L. Lee, W. A. Creasey et al., *Cancer Res.* **34** (1974) 1952.
[206] E. M. Scholar, P. Calabresi, *Biochem. Pharmacol.* **28** (1979) 445.
[207] L. Tterlikkis, J. L. Day, D. A. Brown, E. C. Schroeder, *Cancer Res.* **43** (1983) 1675–1679.

[208] S. Zimm, J. M. Collins, D. O'Neill, B. A. Chabner et al., *Clin. Pharmacol. Therap.* **34** (1983) 810–817.
[209] S. Zimm, J. M. Collins, R. Riccardi, D. O'Neill et al., *N. Engl. J. Med.* **308** (1983) 1005–1009.
[210] R. D. Armstrong, R. Vera, P. Snyder, E. Cadman, *Biochem. Biophys. Res. Commun.* **109** (1982) 595.
[211] M. Inomata, F. Fukuoka, A. Hoshi, K. Kuretani et al., *J. Pharmacobiodyn.* **4** (1981) 928.
[212] D. M. Tidd, I. Gibson, P. D. G. Dean, *Cancer Res.* **42** (1982) 3769.
[213] M. Inaba, M. Fukui, N. Yoshida, S. Tsukagoshi et al., *Cancer Res.* **42** (1982) 1103.
[214] C. P. J. Adair, H. J. Bragg, *Ann. Surg.* **93** (1931) 190.
[215] L. P. Jacobson, C. L. Spurr, E. S. O. Barron et al., *J. Am. Med. Assoc.* **132** (1946) 126.
[216] M. Colvin: "The Alkylating Agents," in B. A. Chabner (ed.): *Pharmacologic Principles of Cancer Treatment*, W. B. Saunders, Philadelphia 1982, pp. 276–308.
[217] N. O. Goldstein, R. J. Rutman, *Cancer Res.* **24** (1964) 1363.
[218] R. W. Ruddon, J. M. Johnson, *Mol. Pharmacol.* **4** (1968) 258.
[219] G. P. Wheeler, J. A. Alexander, *Cancer Res.* **29** (1969) 98.
[220] P. Brookes, P. D. Lawley, *Biochem. J.* **80** (1961) 486.
[221] C. C. Price, G. M. Gaucher, P. Koneru et al., *Biochim. Biophys. Acta* **166** (1968) 327.
[222] A. Loveless, W. C. J. Ross, *Nature (London)* **166** (1950) 1113.
[223] K. W. Kohn, L. C. Erickson, R. A. G. Ewig et al., *Biochemistry* **15** (1976) 4629.
[224] R. J. Rutman, E. H. C. Chun, F. A. Lewis, *Biochem. Biophys. Res. Commun.* **32** (1968) 650.
[225] I. Hirono, *Nature (London)* **186** (1960) 1059.
[226] G. Calcutt, T. A. Connors, *Biochem. Pharmacol.* **12** (1963) 839.
[227] R. A. G. Ewig, K. W. Kohn, *Cancer Res.* **37** (1977) 2114.
[228] L. C. Ericson, G. Laurent, N. A. Sharkey et al., *Nature (London)* **288** (1980) 727–729.
[229] D. T. Vistica, J. N. Toal, M. Rabinowitz, *Biochem. Pharmacol.* **27** (1978) 2865.
[230] W. R. Redwood, M. Colvin, *Cancer Res.* **40** (1980) 1144.
[231] W. P. Brade, J. Engel, *Drugs Today* **20** (1984) 491–496.
[232] P. J. Creaven, L. M. Allen, D. A. Alford et al., *Clin. Pharmacol. Therap.* **16** (1974) 77.
[233] T. A. Conners: "Alkylating Agents, Nitrosourea, and Alkyltriazines," in H. M. Pinedo, B. A. Chabner (eds.): *Cancer Chemotherapy Annual* **5**, Elsevier, Amsterdam 1983, pp. 30–65.
[234] A. Mittelman et al., *J. Urol. (Baltimore)* **115** (1976) 409.
[235] W. A. Skinner, H. F. Gram, M. O. Green et al., *J. Med. Pharmaceut. Chem.* **2** (1960) 299.
[236] K. A. Hyde, E. Acton, W. A. Skinner et al., *J. Med. Pharmaceut. Chem.* **5** (1962) 1.
[237] F.M. Schabel, T. P. Johnston, G. S. McGaleb et al., *Cancer Res.* **23** (1963) 226.
[238] K. W. Kohn, *Cancer Res.* **37** (1977) 1450.
[239] R. J. Weinkam, D. F. Deen, *Cancer Res.* **42** (1982) 1008.
[240] J. Mendel, R. Thust, H. Schwartz, *Arch. Geschwulstforsch.* **52** (1982) 371.
[241] B. J. Bowdon, J. Grimsley, H. H. Lloyd, *Cancer Res.* **34** (1974) 194.
[242] L. C. Panasci, D. Green, R. Nagourney et al., *Cancer Res.* **37** (1977) 2615.
[243] J. M. Dornish, I. Smith-Kielland, *FEBS Lett.* **139** (1981) 41.
[244] K. D. Tew, A. L. Wang, *Molec. Pharmacol.* **21** (1982) 729.
[245] A. Begleiter, H. Y. P. Lam, G. J. Goldenberg, *Cancer Res.* **37** (1977) 1022.
[246] N. G. Goldstein, R. J. Rutman, *Cancer Res.* **24** (1964) 1363.
[247] R. W. Ruddon, J. M. Johnson, *Molec. Pharmacol.* **4** (1968) 258.
[248] J. J. Roberts, T. P. Brent, A. R. Crathorn, *Eur. J. Cancer* **7** (1971) 515.
[249] W. P. Tong, D. B. Ludlam, *Biochim. Biophys. Acta* **608** (1980) 174.
[250] C. C. Price, G. M. Gaucher, P. Koneru et al., *Biochim. Biophys. Acta* **166** (1968) 327.

[251] P. Brookes, P. D. Lawley, *Biochem. J.* **80** (1961) 486.
[252] C. B. Thomas, R. Osieka, K. W. Kohn, *Cancer Res.* **38** (1978) 2448.
[253] C. C. Erickson, M. O. Bradley, S. M. Ducore et al., *Proc. Natl. Acad. Sci. USA* **77** (1980) 467.
[254] R. A. G. Ewig, K. W. Kohn, *Cancer Res.* **38** (1978) 3197.
[255] W. G. Verly, Y. Paquette, *Cancer J. Biochem.* **50** (1972) 217.
[256] L. C. Ericson, G. Laurent, N. A. Sharkey et al., *Nature (London)* **288** (1980) 727–729.
[257] A. R. Crathorne, J. J. Roberts, *Nature (London)* **211** (1966) 150.
[258] K. Tsujihara, M. Ozeki, T. Morikawa, M. Kawamori et al., *J. Med. Chem.* **25** (1982) 441.
[259] Y. Akaike, Y. Arai, H. Takuchi, H. Satoh, *Gann* **73** (1982) 480.
[260] W. J. Zeller, M. Berger, G. Eisenbrand, W. Tang et al., *Arzneim.-Forsch.* **32** (1982) 484.
[261] J. R. Prous, *Drug News Perspect* **1** (1988) 1, 35.
[262] ADIR, FR 2536075, 1984 (C. Cudennec, G. Lavielle).
[263] M. N. Serradel, J. Castañer, *Drugs of the Future* **14** (1989) 11, 1042–1046.
[264] S. Filippeschi et al., *Anticancer Res.* **14** (1988) 11, 1351–1354.
[265] D. Khayat et al., *Cancer Res.* **47** (1987) 6782–6785.
[266] J. A. Boutin et al., *Eur. Cancer Oncol.* (1989).
[267] P. Deloffre, C. A. Cudennec, G. Lavielle, J. P. Bizzari, *15th Int. Cong. Chemother.*, Istanbul 1987, Abst. 755.
[268] J. Jacquillat et al., *Proc. Am. Assoc. Cancer Res.* **30** (1989) Abst. 1088.
[269] P. Zeller, H. Gutmann, B. Hegedus et al., *Experientia* **19** (1963) 129.
[270] V. T. DeVita, A. Serpick, P. Carbone, *Ann. Intern. Med.* **73** (1970) 542.
[271] J. Gutterman, A. Huang, P. Hochstein, *Proc. Soc. Exp. Biol. Med.* **130** (1979) 797.
[272] R. J. Weinkam, D. A. Shiba, *Life Sci.* **22** (1978) 937.
[273] D. L. Dunn, R. A. Lubet, R. A. Prough, *Cancer Res.* **39** (1979) 4555.
[274] D. E. Schwartz, W. Bollag, P. Obrecht, *Arzneim.-Forsch.* **17** (1967) 1389.
[275] D. Reed, F. Dost, *Proc. Am. Assoc. Cancer Res.* **7** (1965) 57.
[276] M. Baggiolini, B. Dewald, H. Aebi, *Biochem. Pharmacol.* **18** (1969) 2187.
[277] C. Pueyo, *Mutat. Res.* **67** (1979) 189.
[278] E. Therman, *Cancer Res.* **32** (1972) 1111.
[279] W. Kreis, *Proc. Am. Assoc. Cancer Res.* **7** (1966) 39.
[280] A. Sartorelli, S. Tsunamura, *Proc. Am. Assoc. Cancer Res.* **6** (1965) 55.
[281] H. Lam, A. Begleiter, W. Stein et al., *Biochem. Pharmacol.* **27** (1978) 1883.
[282] J. L. Skibba, D. D. Beal, G. Ramirez et al., *Cancer Res.* **30** (1970) 147.
[283] P. Farina, A. Gescher, J. A. Hickman, J. K. Horton et al., *Biochem. Pharmacol.* **31** (1982) 1887.
[284] T. L. Loo, G. E. Housholder, A. H. Gerulath et al., *Cancer Treat. Rep.* **60** (1976) 149.
[285] D. D. Beal, J. L. Skibba, K. K. Whitnable et al., *Cancer Res.* **36** (1976) 2827.
[286] A. H. Gerulath, T. L. Loo, *Biochem. Pharmacol.* **21** (1972) 2335.
[287] D. W. Kaiser, I. T. Thurston, J. R. Dudley et al., *J. Am. Chem. Soc.* **73** (1951) 2984.
[288] J. F. Worzalla, B. D. Kaima, B. M. Johnson et al., *Cancer Res.* **34** (1974) 2669.
[289] J. F. Worzalla, B. D. Kaima, B. M. Johnson et al., *Cancer Res.* **33** (1973) 2810.
[290] B. C. V. Mitchley, S. A. Clarke, T. A. Connors et al., *Cancer Treat. Rep.* **61** (1977) 3.
[291] M. Ames, G. Powis, J. S. Kovach et al., *Cancer Res.* **39** (1979) 5016.
[292] L. M. Lake, E. E. Grunden, B. M. Johnson, *Cancer Res.* **35** (1975) 2858.
[293] A. J. Cumber, W. C. J. Ross, *Chem. Biol. Interact.* **17** (1977) 349.
[294] V. Iver, W. Szybalski, *Science (Washington)* **145** (1964) 55–58.
[295] L. Littlefield et al., *Mut. Res.* **81** (1981) 377.
[296] J. Mac Donald et al., *Ann. Intern. Med.* **93** (1980) 533.

[297] *Drugs Today* **20** (1984) 489; **21** (1985) 420.
[298] *Drugs of the Future* **8** (1983) 402; **9** (1984) 371; **10** (1985) 420.
[299] F. Arcamone et al., *J. Med. Chem.* **18** (1975) 703.
[300] Microbiochem. Res. Found., US 4303785, (H. Naganawa, T. Takenchi, H. Umezawa).
[301] H. Umezawa et al., *J. Antibiot.* **32** (1979) 1082–1085.
[302] M. N. Serradel, J. Castañer, *Drugs of the Future* **8** (1983) 7, 610–611.
[303] T. Nishimura et al., *J. Antibiot.* **33** (1980) 737–743.
[304] D. Dantchev et al., *J. Antibiot.* **32** (1979) 1085–1086.
[305] H. Brockmann, *Fortschr. Chem. Org. Naturst.* **50** (1973) 121.
[306] C. E. Myers: "Anthracyclines," in B. A. Chabner (ed.): *Pharmacologic Principles of Cancer Treatment*, W. B. Saunders, Philadelphia 1982, pp. 416–434.
[307] R. Phillips, A. DiMarco, F. Zunino, *Eur. J. Biochem.* **85** (1978) 487.
[308] R. B. Painter, *Cancer Res.* **38** (1978) 4445.
[309] G. Daskal, C. Woodard, S. T. Crooke et al., *Cancer Res.* **38** (1978) 467.
[310] W. E. Ross, L. A. Zwelling, K. W. Kohn, *Int. J. Radiat. Oncol., Biol. Phys.* **5** (1979) 1221.
[311] B. Sinha, R. H. Sik, *Biochem. Pharmacol.* **29** (1980) 1867.
[312] K. Handa, S. Sato, *Gann* **66** (1975) 43.
[313] N. R. Bachur, S. L. Gordon, M. V. Gee, *Cancer Res.* **38** (1977) 1745.
[314] R. Ogura, H. Toyama, T. Shimada et al., *J. Appl. Biochem.* **1** (1979) 325.
[315] J. H. Doroshow, G. Y. Locker, C. E. Myers, *J. Clin. Invest.* **65** (1980) 128.
[316] N. W. Revis, N. Marusic, *J. Mol. Cell. Cardiol.* **10** (1978) 945.
[317] T. R. Tritton, S. A. Murphree, A. C. Sartorelli, *Biochem. Biophys. Res. Commun.* **84** (1978) 802.
[318] T. R. Tritton, G. Yeh, *Science* **217** (1982) 248.
[319] E. Goormaghtigh, R. Brasseur, J. M. Ruysschaert, *Biochem. Biophys. Res. Commun.* **104** (1982) 314.
[320] C. E. Myers, C. Simone, L. Gianni et al., *Proc. Am. Assoc. Cancer Res.* **22** (1981) 112.
[321] T. Oki, *Jpn. J. Antibiot.* **30** (1977) 570.
[322] R. K. Y. Zee-Cheng, C. C. Cheng, *J. Med. Chem.* **21** (1978) 291.
[323] L. H. Patterson, B. M. Gandecka, J. R. Brown, *Biochem. Biophys. Res. Commun.* **110** (1983) 399.
[324] E. D. Kharasch, R. F. Novak, *Biochem. Biophys. Res. Commun.* **108** (1982) 1346.
[325] S. S. Legha, *Drugs Today* **20** (1984) 12, 629–638.
[326] *Drugs of the Future* **5** (1980) 234; **6** (1981) 316; **7** (1982) 357; **8** (1983) 64; **9** (1984) 380; **10** (1985) 429.
[327] American Cyanamid, USP 4197249, 1980 (F. E. Durr, K. C. Murdock).
[328] R. K.-Y. Zee-Cheng, C. C. Cheng, *J. Med.-Chem.* **22** (1979) 1024.
[329] K. C. Murdock et al., *J. Med. Chem.* **22** (1979) 9, 1024–1030.
[330] F. I. Durr, R. E. Wallace, R. V. Citarella, *Cancer Treat Rev.* **10** (1983) 3–11.
[331] R. E. Wallace, K. C. Murdock, R. B. Angier, F. E. Durr, *Cancer Res.* **39** (1979) 1570–1574.
[332] R. K. Johnson et al., *Cancer Treat. Rep.* **63** (1979) 425–439.
[333] D. D. Von Hoff, C. A. Coltuan, B. Forseth, *Cancer Res.* **41** (1981) 1853–1855.
[334] B. M. Henderson et al., *Cancer Treat. Rep.* **66** (1982) 1139–1143.
[335] B. M. Sparano, G. Gordon, C. Hall, M. J. Iatropoulos, J. F. Noble, *Cancer Treat. Rep.* **66** (1982) 1145–1158.
[336] K. C. Anderson et al., *Cancer Treat. Rep.* **67** (1983) 435–438.
[337] F. C. Schell et al., *Cancer Treat. Rep.* **66** (1982) 1641–1643.
[338] C. B. Pratt et al., *Cancer Treat. Rep.* **67** (1983) 85–88.
[339] D. V. Unverferth et al., *Cancer Treat. Rep.* **67** (1983) 343–350.

[340] M. S. Aapro, D. S. Alberts, *Invest. New Drugs* **2** (1984) 329–330.
[341] S. A. Taylor, B. T. Tranum, D. D. Von Hoff, J. J. Constanzi, *Invest. New Drugs* **3** (1985) 67–69.
[342] J. A. Elliott et al., *Anti-Cancer Drug Res.* **3** (1989) 4, 271–282.
[343] R. F. Marschke et al., *Med. Pediat. Oncol.* **16** (1988) 4, 269–270.
[344] K. Hillier, *Drugs of the Future* **6** (1981) 12, 762; **7** (1982) 896.
[345] K. C. Murdock et al., *J. Med. Chem.* **25** (1982) 505.
[346] American Cyanamid, EP 338372, 1989 (V. J. Lee, K. C. Murdock)
[347] American Cyanamid, US 4900838, 1990 (K. C. Murdock)
[348] K. C. Murdock et al., *J. Med. Chem.* **36** (1993) 15, 2098.
[349] US Dept. Of Health, USP 25157, 1981 (H. Dubicki, J. L. Parsons, F. W. Starks).
[350] B. F. Cain et al., *J. Med. Chem.* **18** (1975) 1110.
[351] *Annu. Drug Data Report* (1981) 177.
[352] *Drugs of the Future* **5** (1980) 277; **8** (1983) 535.
[353] L. Steinherz et al., *Cancer Treat. Rep.* **66** (1982) 483.
[354] J. Kolwas, *Drugs Today* **15** (1979) 200.
[355] G. A. Curt, N. J. Clendeninn, B. A. Chabner, *Cancer Treat. Rep.* **68** (1984) 87–99.
[356] T. Tsuruo, H. Lida, M. Nojiri, *Cancer Res.* **43** (1983) 2905.
[357] S. Waksman, H. B. Woodruff, *Proc. Soc. Exp. Biol. Med.* **45** (1940) 609–611.
[358] H. M. Sobell, S. C. Jam, T. D. Sakore et al., *Nat. New Biol.* **231** (1971) 200.
[359] F. Takusagawa et al.: *12th Congress of the IUCR Associated Meeting on Molecular Structure and Biologic Activity*, Buffalo, New York, August 26–28, 1981, p. 55.
[360] E. Reich, R. M. Franklin, A. J. Shatkin et al., *Proc. Natl. Acad. Sci. USA* **48** (1982) 1238.
[361] H. S. Schwartz, E. Godoyien, R. Y. Ambaye, *Cancer Res.* **287** (1968) 192.
[362] M. N. Goldstein, K. Hamm, E. Amrod, *Science* **151** (1966) 1555–1556.
[363] H. Umezawa, K. Meada, T. Takeuchi et al., *J. Antibiot. Ser. A.* **19** (1966) 200–206.
[364] B. T. Sikic, Z. H. Siddik, T. E. Gram, *Cancer Treat. Rep.* **64** (1980) 659–667.
[365] T. Takita, Y. Muraoka, H. Umezawa: "Bleomycin and Peplomycin," in H. M. Pinedo, B. A. Chabner (eds.): *Cancer Chemotherapy Annual 6*, Elsevier, Amsterdam 1984, pp. 85–90.
[366] A. Hagiwara et al., *Anti-Cancer Drug Res.* **2** (1988) 319–324.
[367] S. Saito, Y. Umezawa, T. Yoshioka et al., *J. Antibiot. (Tokyo)* **36** (1983) 92–95.
[368] J. Fugimito, H. Higashi, G. Kosaki, *Cancer Res.* **36** (1976) 2248–2251.
[369] M. Chien, A. P. Grollman, S. B. Horwitz, *Biochemistry* **16** (1977) 3641–3645.
[370] W. J. Caspay, C. Niziak, D. A. Lanzo et al., *Mol. Pharmacol.* **16** (1979) 256–260.
[371] S. Biables, J. R. Warr, *Genet. Res.* **34** (1979) 269–279.
[372] M. Mayaki, T. Ono, S. Hori et al., *Cancer Res.* **35** (1975) 2015–2018.
[373] S. Akiyama, M. Kuwano, *J. Cell Physiol.* **107** (1981) 147–153.
[374] S. Akiyama, K. Ikezaki, H. Kunamochi et al., *Biochem. Biophys. Res. Commun.* **101** (1981) 55–60.
[375] W. E. G. Mueller, M. Geisort, R. K. Zahn et al., *Eur. J. Cancer Clin. Oncol.* **19** (1983) 665–669.
[376] P. G. Sorensen, M. Dorth, H. H. Hansen, *Eur. J. Cancer Clin. Oncol.* **19** (1983) 319–323.
[377] R. K. Johnson, R. P. Hertzberg, *Annu. Rep. Med. Chem.* **25** (1990) 129.
[378] P. A. Aristoff et al., *Invest. New Drugs* **7** (1989) 364.
[379] J. P. Mc Govren et al., *Invest. New Drugs* **7** (1989) 448.
[380] M. A. Mitchell, P. D. Johnson, M. G. Williams, P. A. Aristoff, *J. Am. Chem. Soc.* **11** (1989) 6428.
[381] K. Goni et al., *Jpn. J. Cancer Res.* **83** (1992) 113.
[382] D. L. Boger, M. S. S. Palanki, *J. Am. Chem. Soc.* **114** (1992) 9318.

[383]   M. Fontana et al., *Anti-Cancer Drug Res.* **7** (1992) 131.
[384]   D. Volpe et al., *Invest. New Drugs* **10** (1992) 255.
[385]   N. Zein, M. Poncin, R. Nilakantan, G. A. Ellestad, *Science* **244** (1989) 697.
[386]   B. H. Long et al., *Proc. Natl. Acad. Sci. USA* **86** (1989) 2.
[387]   K. C. Nicolaou, *Science* **256** (1992) 1172.
[388]   K. C. Nicolaou, W. M. Dai, *J. Am. Chem. Soc.* **114** (1992) 8908.
[389]   W. A. Bleyer, S. A. Frisby, V. T. Oliverio, *Biochem. Pharmacol.* **24** (1979) 633–639.
[390]   R. A. Bender, W. D. Kornreich, *Proc. Am. Assoc. Cancer Res.* **22** (1981) 227.
[391]   H. Mujagic, S. S. Chen, R. Geist et al., *Cancer Res.* **43** (1983) 3591–3597.
[392]   D. V. Jackson, R. A. Bender, *Cancer Res.* **39** (1979) 4341–4349.
[393]   R. Schrek, *Am. J. Clin. Path.* **62** (1974) 1–7.
[394]   R. A. Bender, W. A. Bleyer, S. A. Frisby et al., *Cancer Res.* **35** (1975) 1305–1308.
[395]   R. F. Zager, S. A. Frisby, V. T. Oliverio, *Cancer Res.* **33** (1973) 1670–1676.
[396]   I. S. Johnson, H. F. Wright, G. H. Svoboda et al., *Cancer Res.* **20** (1980) 1016–1022.
[397]   G. A. Curt, B. D. Bailey, H. Mujagic et al., *Proc. Am. Assoc. Cancer Res.* **25** (1984) 337.
[398]   W. T. Beck, M. C. Cirtain, J. L. Lefko, *Cancer Treat. Rep.* **67** (1983) 875–879.
[399]   R. B. Sklaroff, D. Straus, C. Young, *Cancer Treat. Rep.* **63** (1979) 793–794.
[400]   G. Mathe, J. L. Misset, F. deVassal et al., *Cancer Treat. Rep.* **62** (1978) 805.
[401]   M. Bayssas, J. Gouveia, F. deVassal et al., *Cancer Res.* **74** (1980) 91–99.
[402]   R. Z. Andriamialisoa, N. Langlois, Y. Langlois, P. Potier, *Tetrahedron* **36** (1980) 3053–3060.
[403]   M. N. Serradel, J. Castañer, *Drugs of the Future* **11** (1986) 7, 575–577.
[404]   M. R. Paintrand, I. Pignot, *J. Electron. Microsc.* **32** (1983) 2, 115–124.
[405]   P. Mangeney et al., *J. Org. Chem.* **44** (1979) 3765–3768.
[406]   P. Maral, C. Bourut, E. Chenn, G. Mathe, *Cancer Chemother. Pharmacol.* **5** (1981) 3, 197–199.
[407]   R. Maral, C. Bourut, E. Chenn, G. Mathe, *Cancer Lett.* **22** (1984) 49–54.
[408]   G. Mathe, P. Reizenstein, *Cancer Lett.* **27** (1985) 285–293.
[409]   M. Besenval et al., *Semin. Oncol.* **16** (1989) 2, 37.
[410]   A. Depierre et al., *Semin. Oncol.* **16** (1989) 2, 26.
[411]   J. B. Sorensen, *Drugs* **44** (1992) 60.
[412]   P. B. Schiff, A. S. Kende, S. B. Horwitz, *Biochem. Biophys. Res. Commun.* **85** (1978) 737–740.
[413]   R. A. Bender, B. A. Chabner: "Tubulin Binding Agents: Epipodophyllotoxin," in B. A. Chabner (ed.): *Pharmacologic Principles of Cancer Treatment*, W. B. Saunders, Philadelphia 1982, pp. 263–266.
[414]   J. D. Loike, S. B. Horwitz, *Biochemistry* **15** (1976) 5435–5438.
[415]   D. K. Kalwinsky, A. T. Look, J. Ducore, A. Fridland, *Cancer Res.* **43** (1983) 1592–1596.
[416]   A. J. Wozniak, W. E. Ross, *Cancer Res.* **43** (1983) 130–135.
[417]   A. J. Wozniak, B. S. Glisson, K. R. Hande, W. E. Ross, *Cancer Res.* **44** (1984) 626–629.
[418]   B. H. Long, A. Minocha, *Proc. Am. Assoc. Cancer Res.* **24** (1983) 321.
[419]   M. E. Wall, M. C. Wani, C. E. Cook, K. H. Palmer, *J. Am. Chem. Soc.* **88** (1966) 3888.
[420]   J.-C. Cai, C. R. Hutchinson, *Chem. Heterocyl. Compd.* **25** (1983) 753.
[421]   Yakult Honsha, EP 137145, 1985 (T. Miyasaka et al.).
[422]   Smith Kline Beecham Corp., EP 321122, (J. C. Boehm, S. M. Hecht, K. G. Holden, R. K. Johnson, W. D. Kingsbury).
[423]   W. D. Kingsbury et al., *J. Med. Chem.* **34** (1991) 98–107.
[424]   N. Osheroff, *Pharmacol. Ther.* **41** (1989) 223–41.
[425]   L. Liu, J. Wang, *Proc. Natl. Acad. Sci. USA* **84** (1987) 7024–7027.
[426]   A. J. Ryan, S. Squires, H. L. Strutt, R. T. Johnson, *Nucl. Acid Res.* **19** (1991) 12, 3295–3300.

[427] M. N. Serradel, J. Castañer, R. M. Castañer, *Drugs of the Future* **12** (1987) 3, 207.
[428] C. H. Spiridonis, *Drugs of the Future* **20** (1995) 5, 483–489.
[429] W. J. Slichenmyer, E. K. Rowinsky, R. C. Donehower, S. H. Kaufmann, *J. Natl. Cancer Inst.* **85** (1993) 4, 271–291.
[430] C. B. Hendricks et al., *Cancer Res.* **52** (1992) 8, 2268–2278.
[431] W. K. Eng et al., *Mol. Pharmacol.* **38** (1990) 4, 471–480.
[432] D. Abigerges et al., *J. Natl. Cancer Inst.* **86** (1994) 446.
[433] M. C. Wani et al., *J. Am. Chem. Soc.* **93** (1971) 2325–2327.
[434] J.-N. Denis et al., *J. Am. Chem. Soc.* **110** (1988) 5917–5919.
[435] J.-N. Denis, A. Correa, A. E. Greene, *J. Org. Chem.* **55** (1990) 1957–1959.
[436] M. Hepperle, G. I. Georg, *Drugs of the Future* **19** (1994) 573–584.
[437] K. C. Nicolaou, W. M. Dai, R. K. Guy, *Angew. Chem.* **106** (1994) 38–69.
[438] K. C. Nicolaou et al., *Nature* **367** (1994) 630–634.
[439] R. A. Holton et al., *J. Am. Chem. Soc.* **116** (1994) 1599–1600.
[440] J. J. Masters et al., *Angew. Chem.* **107** (1995) 1883.
[441] P. B. Schiff, S. B. Horwitz, *Proc. Natl. Acad. Sci. USA* **77** (1980) 1561–1564.
[442] P. B. Schiff, S. B. Horwitz, *J. Cell Biol.* **91** (1981) 479–483.
[443] M. A. Bissery, D. Guenard, F. Gueritte-Voegelein, F. Lavelle, *Cancer Res.* **51** (1991) 4845–4852.
[444] I. Ringel, S. B. Horwitz, *J. Natl. Cancer Inst.* **83** (1991) 288–291.
[445] R. Pazdur et al., *Cancer Treat. Rev.* **19** (1993) 351–386.
[446] P. B. Schiff, J. Faut, S. B. Horwitz, *Pharmacol. Ther.* **25** (1984) 83–124.
[447] S. Rao, J. J. Manfredi, S. B. Horwitz, I. Ringel, *J. Natl. Cancer Inst.* **84** (1992) 785–788.
[448] S. Rao, S. B. Horwitz, I. Ringel, *J. Natl. Cancer Inst.* **84** (1992) 785–788.
[449] J. Parness, S. B. Horwitz, *J. Cell. Biol.* **91** (1981) 479–487.
[450] P. B. Schiff, S. B. Horwitz, *Biochemistry* **20** (1981) 3247–3252.
[451] A. H. Ding, F. Porten, E. Sanchez, C. F. Nathan, *Science* **248** (1990) 370–372.
[452] C. Bogdan, A. Ding, *J. Leukocyte Biol.* **52** (1992) 119–121.
[453] M. E. Stearns, M. Wang, *Cancer Res.* **52** (1992) 3776–3781.
[454] D. M. Vyas et al., *Biorg. Med. Chem. Lett.* **3** (1993) 1357–1360.
[455] E. K. Rowinsky et al., *Semin. Oncol.* **20** (1993) 3, 1–15.
[456] H. Lataste et al., *Proc. Natl. Acad. Sci. USA* **81** (1984) 4090–4094.
[457] J. Kant et al., *Biorg. Med. Chem. Lett.* **3** (1993) 2471–2474.
[458] G. Samaranayake, N. F. Magri, C. Jitransgri, D. G. Kingston*J. Org. Chem.* **56** (1991) 5114–5119.
[459] A. Datta, L. Jayasinghe, G. I. Georg, *J. Med. Chem.* (1994) 4258–4260.
[460] J. L. Fabre, D. Lolli-Tonelli, L. H. Spiridonidis, *Drugs of the Future* **20** (1995) 5, 464.
[461] *Annual Data Report* (1995) 765.
[462] G. Höfle et al., *Angew. Chem.* **108** (1996) 1671; *Int. Ed. Engl.* **35** (1996) 1567.
[463] D. M. Bollag et al., *Cancer Res.* **55** (1995) 2325–2333.
[464] A. Bolag et al., *Angew. Chem.* **108** (1996) 2976; *Int. Ed. Engl.* **35** (1996) 2801.
[465] D. Meng et al., *J. Amer. Chem. Soc.* **119** (1997) 2733.
[466] K. C. Nicolaou et al., *Angew. Chem.* **108** (1996) 2534.
[467] K. C. Nicolaou et al., *Nature* **387** (1997) 268–272.
[468] D. Schinzer et al., *Angew. Chem.* **109** (1997) 543–544.
[469] D.-S. Su et al., *Angew. Chem.* **109** (1997) 2178.
[470] R. J. Kowalski et al., *J. Biol. Chem.* **272** (1997) 2534.

[471] B. Rosenberg, L. Van Camp, T. Krigas, *Nature (London)* **205** (1965) 698–700.
[472] B. Rosenberg, L. Van Camp, J. E. Troska et al., *Nature (London)* **222** (1969) 385–386.
[473] A. H. Rossof, R. E. Slayton, C. P. Perlia, *Cancer* **30** (1972) 1451–1455.
[474] D. J. Higby, H. J. Wallace Jr., J. F. Holland, *Cancer Chemother. Rep.* **57** (1973) 459–463.
[475] L. H. Einhorn, S. D. Williams, *N. Engl. J. Med.* **300** (1979) 289–292.
[476] A. W. Prestayko, J. C. D'Aousst, B. F. Issel et al., *Cancer Treat. Rev.* **6** (1979) 17–24.
[477] W. M. Scovell, T. O'Connor, *J. Am. Chem. Soc.* **99** (1977) 120–126.
[478] A. F. LeRoy, R. J. Lutz, R. L. Dedrick et al., *Cancer Treat. Rep.* **63** (1979) 59–64.
[479] G. L. Cohen, W. R. Bauer, J. K. Berton et al., *Science* **203** (1979) 1014–1016.
[480] R. C. Srivastava, J. Froelich, G. L. Eichhorn, *Biochimie* **60** (1979) 879–881.
[481] J. J. Roberts, A. J. Thomson, *Prog. Nucleic Acid Res. Mol. Biol.* **22** (1979) 71–133.
[482] B. Rosenberg, *Naturwissenschaften* **60** (1973) 399–408.
[483] T. F. Slater, M. Ahmed, S. A. Ibrahim, *J. Clin. Hematol. Oncol.* **7** (1977) 534–539.
[484] N. E. Madias, J. T. Harrington, *Am. J. Med.* **65** (1978) 307–311.
[485] D. C. Dobyan, J. Levi, C. Jacobs et al., *J. Pharmacol. Exp. Therap.* **213** (1980) 551–556.
[486] R. L. Schilsky, T. Anderson, *Ann. Intern. Med.* **90** (1979) 929–931.
[487] F. A. Hayes, A. A. Green, N. Jenzen et al., *Cancer Treat. Rep.* **63** (1979) 547–549.
[488] S. Davis, W. Kessler, B. M. Haddad et al., *J. Med.* **11** (1980) 133–137.
[489] M. Dentino, F. L. Luft, M. N. Yum et al., *Cancer* **41** (1978) 1274–1279.
[490] K. K. Chang, D. J. Higby, E. S. Henderson et al., *Cancer Treat. Rep.* **61** (1977) 367–371.
[491] R. F. Ozols, B. Corden, J. Jacobs et al., *Ann. Intern. Med.* **100** (1984) 19–24.
[492] J. Drobnik, M. Urbankova, A. Krekulova, *Mut. Res.* **17** (1973) 13–20.
[493] L. A. Zwelling, S. Michaels, H. Schwartz, *Cancer Res.* **41** (1981) 640–649.
[494] J. Ducore, L. Zwelling, K. Kohn, *Proc. Am. Assoc. Cancer Res.* **21** (1980) 267.
[495] L. C. Erikson, L. A. Zwelling, J. M. Ducore, *Cancer Res.* **4** (1981) 2791–2794.
[496] K. Micetich, S. Michaels, G. Jude et al., *Proc. Am. Assoc. Cancer Res.* **22** (1981) 252.
[497] A. Bakka, L. Endresen, A. B. S. Johnson et al., *Toxicol. Appl. Pharmacol.* **61** (1981) 215–226.
[498] L. R. Beach, R. D. Palmiter, *Proc. Natl. Acad. Sci. USA* **78** (1981) 2110–2114.
[499] C. R. Wilkenson, P. J. Cox, M. Jones et al., *Biochimie* **60** (1978) 851–857.
[500] *Drugs of the Future* **8** (1983) 489; **9** (1984) 463; **10** (1985) 497; **11** (1986) 499.
[501] *Drugs Today* **22** (1986) 255.
[502] Research Corp., DE 2 329 485, 1973 (M. J. Cleare, J. D. Hoeschele, B. Rosenberg).
[503] R. C. Harrison et al., *Inorg. Chim. Acta* **46** (1980) 215.
[504] A. H. Calvert et al., *Cancer Chemother. Pharmacol.* **9** (1982) 140.
[505] B. D. Evans et al., *Proc. Ass. Cancer Res.* **24** (1983) 154.
[506] CBDCA, Clinical Brochure, Investigational Drug Branch, Division of Cancer Treatment, National Cancer Institute, Bethesda, Md., 1980.
[507] B. D. Evans, K. S. Raju, A. H. Calvert, S. U. Harland et al., *Cancer Treat. Rep.* **67** (1983) 997–1001.
[508] P. J. Creaven, S. Madajewicz, L. Pendyala et al., *Cancer Treat. Rep.* **67** (1983) 795–798.
[509] C. Sternberg et al., *Cancer Treat. Rep.* **69** (1985) 1305–1307.
[510] R. Canetta et al., *Cancer Treat. Rep.* **63** (1985) 2107–2109.
[511] I. N. Olver et al., *Cancer Treat. Rep.* **70** (1986) 421–422.
[512] A. P. Kyriazis et al., *Cancer res.* **45** (1985) 2012–2015.
[513] K. R. Harrap, M. Jones, C. R. Wilkenson et al.: "Antitumor Toxic and Biochemical Properties of *cis*-Platinum and Eight Other Platinum Analogs," in A. W. Prestayko, S. T. Crooke, S. K.

[514] Carter (eds.): *cis-Platinum: Current Status and New Developments (Pap Symposium)*, Academic Press, New York 1980, pp. 193–212.
[514] P. Chang, *Drugs of the Future* **8** (1983) 364; **10** (1985) 7, 561.
[515] *Drugs of the Future* **12** (1987) 11, 1029–1031.
[516] *Annu. Drug Data Rep.* **8** (1986) 6, 590.
[517] Shionogi & Co., EP 216362 1987 (H. Kagawa, K. Shima, T. Tsukada).
[518] K. Hirabayashi, E. Okada, *Cancer* **71** (1993) 9, 2769.
[519] M. Koenuma et al., *Clin. Rep.* **29** (1995) 12, 259.
[520] N. Uchida et al., *Clin. Rep.* **29** (1995) 12, 269.
[521] I. H. Krakoff, 5th *Intl. Symp. Platinum Cancer Chemother.*, Abano Therme, 1987, Abst. L7.
[522] B. K. Bhuyan et al., *Cancer Commun.* **3** (1991) 2, 53.
[523] ASTA Medica AG, EP 324154 1989 (J. Engel et al.).
[524] R. Voegeli, E. Günther, P. Aulenbacher, J. Engel, P. Hilgard, *Drugs of the Future* **17** (1992) 883–886.
[525] A. M. Otto et al., *Pharm. Pharmacol. Lett.* **1** (1992) 103–106.
[526] W. A. Denny et al., *J. Med. Chem.* **35** (1992) 2983–2987.
[527] *Drugs Today* **10** (1974) 74.
[528] ICI, GB 1013907 1962.
[529] G. R. Bedford, D. N. Richardson, *Nature* **212** (1966) 733.
[530] M. N. Serradel, J. Castañer, *Drugs of the Future* **11** (1986) 5, 398–400.
[531] Farmos Group, EP 95875 (R. J. Tiovola et al.).
[532] J. T. Pento et al., *Drugs of the Future* **9** (1984) 7, 5.
[533] R. Löser, P.-St. Jamak, K. Seibel, *Drugs of the Future* **9** (1984) 3, 186.
[534] S. Kallio et al., *Cancer Chemother. Pharmacol.* **17** (1986) 103–108.
[535] L. Kangas et al., *Cancer Chemother. Pharmacol.* **17** (1986) 109–113.
[536] V. C. Jordan, B. Gosden, *Mol. Cell. Endocrinol.* **27** (1982) 27, 921.
[537] S. P. Robinson et al., *Eur. J. Cancer Clin. Oncol.* **24** (1988) 12, 1817.
[538] R. Valavaara et al., *Eur. J. Cancer Clin. Oncol.* **24** (1988) 4, 785.
[539] S. R. Ebbs et al., *Lancet II* (1987) 621.
[540] R. F. Kauffmann et al., *J. Pharmacol. Exp. Ther.* **280** (1997) 146–153.
[541] M. A. Ferreira, M. M. Caramona, L. M. Celeste, *Proc. Br. Pharmacol. Soc.* (1995) Abs. 267 P.
[542] M. Dukes, *Pharm. J.* **257** (1996) 176.
[543] A. Hoshi, *Drugs of the Future* **16** (1991) 3, 217.
[544] *Annu. Drug Data Report* **15** (1993) 4, 378.
[545] A. Brodie, *J. Steroid Biochem. Mol. Biol.* **40** (1991) 1–3, 255.
[546] S. A. Wells et al., *Cancer Res.* Suppl. **42** (1982) 3454.
[547] R. D. Burnett, D. N. Kirk, *J. Chem. Soc. Perkin Trans.* **I** (1973) 1830.
[548] R. C. Coombes, *Eur. J. Cancer* **28A** (1992) 12, 1941.
[549] D. Cunningham et al., *Brit. J. Cancer* **55** (1987) 331.
[550] J. H. Davis et al., *Brit. J. Cancer* **66** (1992) 1, 139.
[551] M. Dowsett et al., *Cancer Res.* **49** (1989) 1306.
[552] M. Dowsett et al., *Eur. J. Cancer* **28** (1992) 2–3, 415.
[553] R. C. Stein et al., *Cancer Chemother. Pharmacol.* **26** (1990) 75.
[554] ICI, US 4935437, 1990 (P. N. Edwards, M. S. Lange).
[555] Ciba Geigy AG, US 4617307, 1986 (L. J. Browne).
[556] J. Prous, *Drugs of the Future* **19** (1994) 4, 335–337.
[557] J. Castañer, *Annu. Drug Data Rep.* **13** (1991) 8, 716.

[558] A. B. Foster et al., *J. Med. Chem.* **28** (1985) 200.
[559] J. T. Pento, *Drugs of the Future* **14** (1989) 9, 843–845.
[560] J. Prous et al., *Drugs of the Future* **20** (1995) 1, 30–32.
[561] R. E. Steele et al., *Steroids* **50** (1987) 147.
[562] M. Dukes et al., *Proc. Am. Assoc. Cancer Res.* **33** (1992) Abst. 1677.
[563] P. V. Plourde et al., *Breast Cancer Res. Treat.* **30** (1994) 1, 103–111.
[564] K. Schieweck, A. S. Bhatnagar, A. Matter, *Cancer Res.* **48** (1988) 834.
[565] K. Schieweck et al., *Proc. Am. Assoc. Cancer Res.* **29** (1988) Abst. 968.
[566] M. Dowsett et al., *Breast Cancer Res. Treat.* **27** (1993) 1–2, Abst. 87.
[567] P. V. Plourde et al., *Proc. Am. Soc. Clin. Oncol.* **12** (1993) Abst. 165.
[568] C. Huggins, C. V. Hodges, *Cancer Res.* **1** (1941) 293.
[569] J. A. Smith, *J. Urol.* **137** (1987) 1–10.
[570] J. Trachtenberg, *J. Urol.* **132** (1984) 61.
[571] G. H. Rasmusson et al., *J. Med. Chem.* **27** (1984) 1690–1701.
[572] M. R. Robinson, B. S. Thomas, *Brit. Med. J.* **4** (1971) 391–394.
[573] F. Neumann, *J. Steroid Biochem.* **19** (1983) 391–402.
[574] B. J. Furr, *Clin. Oncol.* **2** (1988) 581–590.
[575] H. Tucker, *Drugs of the Future* **15** (1990) 3, 255–265.
[576] *Drugs Today* **20** (1984) 296.
[577] *Drugs of the Future* **1** (1976) 108; **8** (1983) 270.
[578] R. O. Neri, E. A. Peets, *J. Steroid Biochem.* **6** (1975) 815–819.
[579] W. I. P. Mainwaring, F. R. Mangan, P. A. Feherty, M. Freifeld, *Med. Cell. Endocrinol.* **1** (1974) 113–128.
[580] S. Liao, D. K. Howell, T. M. Chang, *Endocrinology* **94** (1974) 1205–1209.
[581] B. Katchen, S. Buxbaum, *J. Clin. Endocrinol. Metab.* **41** (1975) 373–379.
[582] J. Geller et al., *Prostate* **2** (1981) 309–314.
[583] E. A. Peets, M. F. Henson, R. O. Neri, *Endocrinology* **94** (1974) 532–540.
[584] R. Neri, N. Kassem, *Prog. Cancer Res. Ther.* **31** (1984) 507–518.
[585] F. Labrie et al., *Oncol.* **2** (1988) 597–619.
[586] F. Labrie, A. Dupont, M. Ciguere, *Brit. J. Urol.* **61** (1988) 341–346.
[587] A. Belanger, A. Dupont, F. Labrie, *J. Clin. Endocrinol. Metab.* **59** (1984) 422–426.
[588] J.-P. Raynaud, C. Bonne et al., *Prostate* **5** (1984) 299–311.
[589] T. Ojasoo, *Drugs of the Future* **12** (1987) 763–770.
[590] L. Proulx, F. Labrie, *Prostate* **5** (1984) Abst. 429.
[591] F. Labrie, A. Dupont, A. Belanger, *Prostate* **4** (1983) 579.
[592] C. Harnois, A. Dupont, F. Labrie, *Brit. J. Opthalmol.* **70** (1986) 471–473.
[593] H. Tucker, G. J. Chesterson, *J. Med. Chem.* **31** (1988) 885–887.
[594] H. Tucker, J. W. Crook, G. J. Chesterson, *J. Med. Chem.* **31** (1988) 954–959.
[595] B. J. A. Furr et al., *J. Endocrinol.* **113** (1987) R7–9.
[596] S. N. Freeman, W. I. P. Mainwaring, B. J. A. Furr, *J. Endocrinol.* (1986) Suppl. III, 155.
[597] C. J. Tyrell, *Prostate* (Suppl. 4) **20** (1992) 97.
[598] G. Wilding et al., *Proc. Am. Assoc. Clin. Oncol.* **10** (1991) Abst. 593.
[599] A. V. Schally, S. M. McCann, *Fertil. Steril.* **64** (1995) 452–453.
[600] A. V. Schally, A. Arimura, A. J. Kastin, *Science* **173** (1971) 1036–1038.
[601] H. Matsuo, Y. Baba, M. Nair, A. Arimura, A. V. Schally, *Biochem. Biophys. Res. Commun.* **43** (1971) 1334–1339.
[602] K. Amoss et al., *Biochem. Biophys. Res. Commun.* **44** (1971) 205–210.

[603] Y. Baba, H. Matsuo, A. V. Schally, *Biochem. Biophys. Res. Commun.* **44** (1971) 459–463.
[604] G. Emons, A. V. Schally, *Hum. Reprod.* **9** (1994) 1364–1370.
[605] P. M. Conn, W. Crowley, *New Engl. J. Med.* **324** (1991) 93–103.
[606] G. F. Weinbauer, E. Nieschlag in K. Höffgen (ed.): *Peptides in Oncology* Springer, Heidelberg, 1992, pp. 113–136.
[607] A. S. Dutta, *Drugs of the Future* **13** (1988) 43–57.
[608] M. Filicori, C. Flamingi, *Drugs* **35** (1988) 63–82.
[609] J. J. Nestor, B. H. Vickery, *Annu. Rep. Med. Chem.* **23** (1988) 211–220.
[610] M. T. Goulet, *Annu. Rep. Med. Chem.* **30** (1995) 169–178.
[611] T. Reissmann et al., *Hum. Reprod.* **20** (1995) 1974–1981.
[612] M. J. Karten in W. F. Crowley, P. M. Conn (eds.): *Modes of Action of GnRH and GnRH Analogs*, Springer, Heidelberg 1992, pp. 277–297.
[613] G. Flouret et al., *Pept. Sci.* **1** (1995) 89–105.
[614] P. M. Conn, W. F. Crowley, *Annu. Rev. Med.* **45** (1994) 391.
[615] A. V. Schally in J. F. Holand et al. (eds): *Cancer Medicine* 3rd ed., Lee & Febiger, Philadelphia, PA 1993, pp. 827–840.
[616] A. V. Schally in P. Belfort, J. Pinotti, T. K. Eskes (eds.): *Advances in Gynecology and Obstetrics* Vol. **6**, Parthenon, Cornforth 1989, pp. 3–22; b) Scrip 22, 1995, 2066.
[617] A. Friedrich, G. Jaeger, K. Radscheit, R. Uhmann, *Pept. Proc. Eur. Pept. Symp.* 22nd 1992/1993, 47–49.
[618] B. Kutscher et al., *Angew. Chem. Int. Ed. Engl.* **36** (1997) 2148–2161.
[619] R. L. Barbieri, *Trends Endocrinol. Metab.* **3** (1992) 30–34.
[620] R. Schmidt, K. Sundaram, R. B. Thau, C. W. Badrin, *Contraception* **29** (1984) 283–289.
[621] S. Bajusz et al., *Int. J. Pept. Protein Res.* **32** (1988) 425–435.
[622] J. Pinski et al., *Int. J. Pept. Protein Res.* **45** (1995) 410–417.
[623] A. Kleemann et al., *Proc. Akabori Conf. Ger. Jpn. Symp. Pept. Chem.* 4th 1991, 96–101; b) F. R. Kunz, T. Müller, K. Drauz, *Proc. Akabori. Conf. Ger. Jpn. Symp. Pept. Chem.* 5th 1994, 15–16.
[624] A. Ljugquist et al., *Proc. Natl. Acad. Sci. USA* **85** (1988) 8236–8240.
[625] J. Leal et al., *Drugs of the Future* **16** (1991) 529–537.
[626] A. Janecka, T. Janecki, C. Bowers, K. Folkers, *J. Med. Chem.* **37** (1994) 2238–2241.
[627] A. Janecka et al., *Med. Chem. Res.* **1** (1991) 306–311.
[628] A. Janecka, T. Janecki, C. Bowers, K. Folkers, *Int. J. Pept. Protein Res.* **44** (1994) 19–23.
[629] P. Theobald et al., *J. Med. Chem.* **34** (1991) 2395–2402.
[630] J. Rivier et al., *J. Med. Chem.* **35** (1992) 4270–4278.
[631] J. E. Rivier et al., *J. Med. Chem.* **38** (1995) 2649–2662.
[632] K. Stoeckemann, J. Sandow, *J. Cancer Res. Clin. Oncol.* **119** (1993) 457–462.
[633] R. Deghenghi, F. Boutignon, P. WüthrichV. Lenaerts, *Biomed. Pharmacother.* **47** (1993) 107–110.
[634] J. Nester, et al., *J. Med. Chem.* **35** (1992) 3942–3948.
[635] J. Mulzer in E. Ottow, U. Schöllkopf, B. G. Schulz (eds.): *Stereoselective Synthesis*, Springer, Heidelberg, 1994, pp. 37–61.
[636] J. Mulzer et al., *Angew. Chem.* **106** (1994) 1813–1815; *Angew. Chem. Int. Ed. Engl.* **33** (1994) 1737–1739.
[637] F. Haviv et al., *J. Med. Chem.* **32** (1989) 2340–2344.
[638] F. Haviv et al., *J. Med. Chem.* **36** (1993) 928–933.
[639] Abbott Laboratories, PCT/US 95/02410, 1995 (F. Haviv).
[640] Abbott Laboratories, WO 95/04540, 1995 (F. Haviv).

[641] Abbott Laboratories, WO 94/14841, 1994 (F. Haviv).
[642] Abbott Laboratories, WO 94/13313, 1994 (J. Greer).
[643] Tap Pharmaceuticals, US-A 5300492, 1994 (F. Haviv).
[644] F. Haviv et al., *J. Med. Chem.* **37** (1994) 701–707.
[645] T. Beckers, K. Marheineke, H. Reiländer, P. Hilgard, *Eur. J. Biochem.* **231** (1995) 535–543.
[646] R. P. Millar, C. A. Flanagan, R. C. Milton, J. A. King, *J. Biol. Chem.* **264** (1989) 21007–21013.
[647] G. A. McPherson, *J. Pharmacol. Methods* **14** (1985) 213–228.
[648] J. Trachtenberg, A. Pont, *Lancet* **2** (1984) 433–435.
[649] S. Bhasin et al., *Endocrinology* **118** (1986) 1229–1232.
[650] B. De et al., *J. Med. Chem.* **32** (1989) 2036–2038.
[651] Abbott Laboratories, US-A 4992421, 1991 (B. De).
[652] McNeillab Inc., US-A 4678784, 1987 (C. Y. Ho).
[653] Takeda Chemical Industries, WO 95/28405, 1995 (S. Furuya).
[654] Takeda Chemical Industries, WO 96/34012A1, 1996 (C. Kitada).
[655] Merck, WO 97/21435, 1997 (M. Goulet).
[656] Takeda Chemical Industries, WO 97/14697, 1997 (S. Furuya).
[657] C. Unger, *Drugs of the Future* **22** (1997) 12, 1337–1345.
[658] G. Powis, *Pharmacol. Ther.* **62** (1994) 57–95.
[659] W. J. Fantl, D. E. Johnson, L. T. Williams, *Annu. rev. Biochem.* **62** (1993) 453–481.
[660] F. McCormick, *Nature* **363** (1993) 15–16.
[661] C. A. Lang-Carter et al., *Science* **260** (1993) 315–319.
[662] D. Rea et al., *Proc. 7th NCI-EORTC Symp. On New Drugs in Cancer Ther.* Amsterdam (1992), p. 62.
[663] D. Ingber et al., *Nature* **348** (1990) 555.
[664] T. Ozsuka et al., *J. Antibiotics* **45** (1992) 348.
[665] Y. Reiss et al., *Cell* **62** (1990) 81–88.
[666] S. Omura, D. van der Pyl, *Cell* **46** (1993) 222.
[667] S. M. Sebti, A. D. Hamilton, *Drug Discovery Technol.* **3** (1998) 26–33.
[668] A. Levitski, A. Gazit, *Science* **267** (1995) 1782–1788.
[669] A. Levitski, *Eur. J. Biochem.* **226** (1994) 1–13.
[670] R. T. Abraham, M. Aquarone, A. Anderson, *Biol. Cell.* **83** (1995) 105.
[671] D. J. Hanahan, *Ann. Rev. Biochem.* **55** (1986) 483–509.
[672] W. E. Berdel, *Onkologie* **13** (1990) 245–250.
[673] W. E. Berdel et al., *Anticancer Res.* **1** (1981) 345–352.
[674] G. Rodriguez et al., *Proc. Am. Assoc. Cancer Res.* **33** (1992) 262.
[675] C. Unger, H. Eibl, *Lipids* **26** (1991) 1412.
[676] P. Hilgard, J. Engel, *Drugs Today* Suppl. B, (1994) 30.
[677] P. Hilgard et al., *Cancer Chemother. Pharmacol.* **32** (1993) 90–95.
[678] P. Hilgard, J. Stekar, C. Unger, *Proc. Annu. Meet. Am. Assoc. Cancer Res.* **31** (1990) A2457.
[679] J. Engel et al., *Drugs of the Future* **13** (1988) 10, 948–951.
[680] C. Geilen et al., *Eur. J. Cancer* **27** (1991) 12, 1650–1653.
[681] R. Hass et al., *Cancer Res.* **52** (1992) 1445–1450.
[682] R. Becher et al., *Onkologie* **16** (1993) 1, 11.

Interferons

TATTANAHALLI L. NAGABHUSHAN, Schering–Plough Research, Bloomfield, New Jersey 07003, United States

PAUL P. TROTTA, Schering–Plough Research, Bloomfield, New Jersey 07003, United States

1.	Introduction	1958
2.	Classification and Origin	1959
3.	Biological Activity	1960
3.1.	Antiviral Activity	1960
3.2.	Anticellular Activity	1961
3.3.	Immunomodulatory Activity	1962
4.	Mechanism of Action	1963
5.	Physicochemical Properties	1964
5.1.	Primary Structure	1964
5.2.	Secondary and Tertiary Structure	1966
5.3.	Solution Properties	1968
6.	Purification	1968
6.1.	Natural Leukocyte IFN	1968
6.2.	Recombinant IFNα	1969
6.3.	Removal of Pyrogens	1971
7.	Quality Specifications	1972
7.1.	Purity	1972
7.2.	Pyrogenicity Testing	1975
8.	Clinical Studies	1975
9.	References	1978

Abbreviations used in this article:
AIDS   acquired immune deficiency syndrome
cAMP   cyclic adenosine 5′-monophosphate
cGMP   cyclic guanosine 5′-monophosphate
CD   circular dichroism
CPE   cytopathic effect
DEAE   diethylaminoethyl
DNA   deoxyribonucleic acid
HPLC   high-performance liquid chromatography
IFN   interferon
IU   international antiviral units
MHC   major histocompatibility complex
RIEF   recycling isoelectric focusing
RNA   ribonucleic acid
SDS–PAGE   sodium dodecyl sulfate–polyacrylamide gel electrophoresis
Tris   tris(hydroxymethyl)aminomethane

1. Introduction

In 1957, A. Isaacs and J. Lindenmann observed that mammalian cells incubated with a heat-inactivated influenza virus produced a substance that could confer on fresh cells a resistance to infection by live virus [1]. This substance, called *interferon*, was subsequently shown to consist of a system of distinct proteins, many of which are structurally related, that are products of multigene families as well as separate genes [2]. Interferons (IFNs) are unique antiviral agents because they act directly on the target cell to confer a state of resistance to viral infectivity at one or more phases of the viral replication cycle, but have no direct toxicity toward the virus itself. The biological effects of IFN are not limited to antiviral activity, however: highly purified preparations of genetically engineered IFN have potent antiproliferative and immunomodulatory effects both in vitro and in animal models [3], [4]. Many types of animal cells can produce IFN in response to a variety of external stimuli or "inducers," including certain types of double-stranded ribonucleic acid (RNA), antigens, and mitogens [5].

Purification of IFN from natural sources for structural and biological characterization, as well as for initiation of clinical trials, proved difficult because IFNs exhibit a high specific biological activity and are normally produced by cells in low absolute quantities. Thus, only relatively impure preparations containing up to 1% IFN could be partially purified from normal human leukocytes for use in clinical trials in the mid 1970s. Not until the late 1970s could sufficient quantities of naturally occurring IFNs be purified from lymphoblastoid cells for initiation of structural studies [6], [7]. However, the achievement of Weissmann and coworkers in cloning the gene for a single subtype of human leukocyte IFN, IFNα-2, in *Escherichia coli* allowed the preparation of gram quantities through genetic engineering [8]. The availability of virtually unlimited quantities of IFN has permitted detailed biological and structural characterization, as well as extensive clinical trials worldwide [9], [10].

Genetically engineered IFNα-2 b (Intron A) has now been approved for marketing in over 30 countries, including the United States, for both cancer and viral indications such as hairy cell leukemia, acquired immune deficiency syndrome (AIDS)-related Kaposi's sarcoma, renal cell carcinoma, malignant melanoma, multiple myeloma, laryngeal papillomatosis, and condyloma acuminata (genital warts). Extensive clinical trials have also been performed on IFNγ [11]–[14] and IFNβ [15], [16]. This article reviews the properties, production, and specifications of recombinant human IFNα, for which the most data are available. However, for comparative purposes selected data are also provided for human IFNβ and IFNγ, as well as natural leukocyte IFN. For further information, see [2]–[4], [9], [17]–[44].

Table 1. Classes, sources, and inducers of IFN

Class	Cell source	Inducers
Alpha	macrophages	viruses
(leukocyte, type 1)	B cells	foreign cells and their products
	null cells *	low molecular mass inducers (e.g. tilorone)
Beta	epithelial cells	viruses
(fibroblast, type I)	fibroblasts null cells *	polynucleotides
		foreign cells and their products
	lymphoblasts	metabolic inhibitors
		B-cell mitogens
Gamma	T cells	T-cell mitogens
(immune, type II)	null cells *	foreign antigens
		metabolic inhibitors

* Null cells are non-B, non-T lymphocytes.

2. Classification and Origin

The first classification of IFN subtypes was based on the observation that exposure to pH 2 inactivated the antiviral activity of some IFN preparations. The IFN stable at pH 2 was designated class I, whereas IFN labile at pH 2 was designated class II. Thus, heterogeneity exists within the IFN system. Subsequently the type I IFNs produced by leukocytes and fibroblasts were shown to differ in their antigenicity and in selected chemical properties. As a result of this observation, type I IFNs were renamed according to the cells of origin, i.e., leukocyte and fibroblast IFN, respectively. These definitions, however, were not adequate because IFN identified as fibroblast with specific antisera was also shown to be produced by leukocytes; leukocyte IFN could likewise be produced by fibroblasts. Therefore, international agreement was reached that IFNs should be categorized as *alpha* and *beta*, depending principally on their antigenic properties. A third antigenic category, referred to as *gamma* (or *immune*) IFN, was also recognized; this IFN class is produced principally by T lymphocytes stimulated by specific antigens or mitogens and is categorized as type II (i.e., acid-labile).

In addition to viruses, a number of natural and synthetic agents are capable of inducing IFN production in a wide variety of cell types [5]. A general summary of cell types and IFN inducers for the three IFN classes is presented in Table 1. Interferon-inducing agents include bacteria, protozoa, mycoplasma, rickettsia, and bacterial products (e.g., bacterial cell wall extracts, lipopolysaccharides, and endotoxins). The effect of synthetic and natural nucleic acids and polyanions, including double-stranded RNA, synthetic polynucleotides, and synthetic polycarboxylates, has been of special interest. Low molecular mass compounds such as anthraquinones, propanediamines, acridines, pyrimidines and fluorenones can also induce IFN titers in the serum of certain animals. Specific structural requirements for these low molecular mass compounds to induce IFN have not been identified yet. Metabolic inhibitors (e.g., actinomycin D), mitogens, antigens, and tumor cells have also been identified as IFN inducers.

3. Biological Activity

As previously noted, IFN was first discovered as an antiviral agent that could protect cells from viral infectivity. However, a variety of other biological effects can be elicited by both natural and recombinant IFNs. Utilizing relatively crude preparations, PAUCKER et al. [45] described the first nonantiviral activity of IFN, i.e., the ability to inhibit cell growth. Subsequently, IFN was also shown to exert potent immunomodulatory effects as measured by its ability to enhance the activity of immune effector cells (e.g., natural killer cells) [46]. The range of biological activity demonstrated for IFN has increased considerably [47]. One of the important features of IFNs is their species specificity; all classes of human IFN are most effective on human cells and are generally much less potent on animal cells. However, because many of the early studies on nonantiviral activity were performed with impure preparations of naturally occurring IFNs, these biological effects were not known to be inherent properties of IFN. Subsequent use of highly purified recombinant IFN has demonstrated that most observed biological effects can be attributed to IFN [2], [3].

These properties of IFN provide strong support for the clinical application of this class of biological response modifiers as both antiviral and antitumor agents. Aspects of the three categories of biological activity of IFN are reviewed briefly below.

3.1. Antiviral Activity

Early studies indicated that the replication of a broad spectrum of both RNA- and DNA-containing viruses, either oncogenic or nononcogenic, could be inhibited by IFN [37]. Examples of such DNA viruses are herpes virus types 1 and 2 and cytomegalovirus. Examples of RNA viruses are rhinovirus and respiratory syncytial virus. In addition, IFNα can act synergistically with other antiviral agents [48]. However, the degree of antiviral activity depends not only on the nature of the virus, but also on the characteristics of the target cell, the type of IFN, and the ratio of infecting virus to cell number. In vivo studies in nonhuman primate models have also demonstrated the antiviral activity of IFNα and IFNβ against herpesvirus [49].

The specific antiviral activities of highly purified recombinant IFNα subtypes utilizing human foreskin fibroblast cells and encephalomyocarditis virus are summarized in Table 2 [50]. Four of these (α-1, -2, -4, and -7) are naturally occurring gene products, whereas the remaining two subtypes (δ-4 α-1 and the δ-4 α-2/α-1 hybrid) are products of recombinant DNA technology in which the first four amino acids of the amino terminus are deleted. Both of these molecules lack a cysteine residue at position 1 and, hence, are predicted to contain only one disulfide bond (see Chap. 5). The hybrid molecule consists of residues 5–62 of IFNα-2 and residues 64–166 of IFNα-1. Despite the high degree of purity of these preparations and their amino acid sequence homology

Table 2. Specific antiviral activity of IFNα subtypes *

IFN subtype	Activity, IU/mg
α-1	7.1×10^6
α-2	1.7×10^8
α-4	1.0×10^8
α-7	2.0×10^7
δ-4 α-1	2.8×10^6
δ-4 α-2/α-1	1.0×10^8

* Activity was determined by the CPE-inhibition assay employing encephalomyocarditis virus and human foreskin cells; data represent the mean of at least nine assays [50].

(at least 70 %), the specific antiviral activities span approximately two orders of magnitude. Interferon α-2 has the highest antiviral activity (1.7×10^8 IU/mg).

Interferon confers resistance to viral infectivity by affecting various stages of the viral replication cycle. It may inhibit early events (e.g., attachment, receptor-mediated endocytosis, uncoating, and transcription), the translation of messenger RNA, and virus maturation, including budding of the virion at the cell membrane. Other evidence also indicates that IFN lowers the infectivity of the progeny virions.

3.2. Anticellular Activity

Interferon can inhibit the growth of a variety of normal and malignant cells in vitro. For example, it can inhibit the growth of fresh tumor cells in the clonogenic assay [51]. This assay is an in vitro cloning technique in which cancer cells form colonies in a semisolid medium. The antigrowth activity of IFN is generally cytostatic (i.e., prevents cell division) rather than cytotoxic (i.e., kills cells directly). Sensitivity to growth inhibition by IFN is variable, even within cells of the same histologic type; this may be due to differences in IFN receptor levels or in the affinity of IFN for the receptor.

Combinations of IFN and chemotherapeutic agents may have a synergistic (enhancing) or additive interaction as antitumor agents [52], [53]. Synergy may be the result of a complex interaction of factors, including the nature of the tumor, the mechanism of action of the cytotoxic agent, and the administration schedule of IFN and the cytotoxic agent. Thus, synergistic interactions between IFN and cytotoxic drugs have been observed in the human tumor clonogenic assay utilizing recombinant IFNα and vinblastine, cisplatin, or doxorubicin [54], [55]. However, synergy is highly schedule-dependent. When cells were first treated with IFN α followed by doxorubicin or vice versa, no synergy was observed. In contrast, dramatic synergy was observed when both agents were administered simultaneously. Other drugs tested in this assay (e.g., bleomycin, methotrexate, or vinca alkaloids) were not synergistic with IFNα. The mechanism of the observed synergy as well as of drug failure remains unknown.

Another important aspect of the anticellular activity of IFNα is its ability to modulate the state of cellular differentiation. Both inhibition and acceleration of differentiation have been observed [56]. Interferon also interacts synergistically with other agents that inhibit differentiation, e.g., phorbol ester tumor promoters [57]. The induction of differentiation in tumor cells mediated by IFN may account for at least a part of its antitumor activity because the degree of cellular differentiation appears to be inversely related to the rate of cell division.

3.3. Immunomodulatory Activity

A variety of immune functions can be affected by IFN [41]. Immune response may be mediated either by cells or by antibodies. Interferon can activate or inhibit both cellular and humoral (antibody) immune function: higher concentrations generally inhibit, whereas lower concentrations activate. Natural killer cell activity as well as macrophage tumoricidal and tumoristatic activities can be significantly enhanced by both IFNα and IFNγ. Other types of immune effector cells activated by IFN include cytotoxic T lymphocytes, cells involved in antibody-dependent cytotoxicity, and mast cells. Interferons of all three types induce expression of class I antigens of the major histocompatibility complex (MHC), and IFNγ induces expression of class II MHC antigens. This property is significant because class I MHC antigens play a role in the lysis of virus-infected cells and tumor cells by cytotoxic T lymphocytes. Class II MHC antigens are essential for the ability of macrophages to function as antigen-presenting cells. However, both classes of MHC antigens are fundamentally important for achieving maximal immune responsiveness. All types of IFN appear to inhibit antibody production, although under certain conditions of dosing and time of addition, an enhancement of antibody production has been observed.

The role of immune stimulation in clinical responses to IFN remains unknown. For example, measurements of natural killer cell activity after administration of recombinant IFNα have demonstrated both enhancement and depression of activity, as well as no response [58], [59]. However, stimulation by IFN of cytotoxic effector cell function and antibody production may still represent major mechanisms of its therapeutic activity.

4. Mechanism of Action

The first step in expression of the biological activity of IFN is its binding to specific cell surface receptors [60]. The number of receptors for IFNα on human cells is reported to be ca. 650–13 000 per cell. The ligand and receptor form a high-affinity complex with a dissociation constant of 10^{-10}–10^{-11} mol/L. The IFN-receptor complex has an apparent molecular mass of 140 000–150 000 deduced from chemical cross-linking of radioactively labeled IFN to the cell surface. Thus, the apparent molecular mass of the receptor is 120 000–130 000 if one molecule of IFN is bound per receptor. The failure of certain cells to respond to IFN may be due to absence of the receptor or a low affinity of IFN for the receptor.

Ultrastructural studies indicate that bound IFNα enters cells by a clustering of the receptor–ligand complex into "coated pits." These pits, which are surrounded by the protein clathrin, function in the transport of a variety of molecules into the cell. The pits pinch off to form vesicles (so-called receptosomes), which can transport their contents to a variety of targets within the cell. The role of internalization of IFNα into the cell in the mechanism of action is yet to be elucidated. Interferon alone, the IFN–receptor complex, or IFN degradation products may trigger a biological response; alternatively, binding of IFN to the cell surface may itself be sufficient.

Binding of IFNα to its receptor induces the synthesis of about a dozen different proteins ranging from 15 000 to 120 000 in molecular mass. Many of these have not been identified, and their role in the mechanism of IFN activity is unknown. Interferon also affects the activity of specific proteins or enzymes. For example, levels of ornithine decarboxylase and S-adenosylmethionine decarboxylase, enzymes involved in polyamine biosynthesis, are reduced by IFN treatment [61]. Because these enzymes are critical for cellular proliferation, their inhibition may account for the antigrowth activity of IFN. Interferon treatment can increase cyclic guanosine 5'-monophosphate (cGMP) and cyclic adenosine 5'-monophosphate (cAMP); both of these compounds have also been implicated in growth regulation [61]. However, the exact biochemical mechanism for the antiproliferative effects of IFN remains unknown.

As noted above, all three IFN classes can induce the expression of class I MHC antigens [62], which mediate the lysis of virus-infected cells and tumor cells by cytotoxic T lymphocytes. In addition, IFN can enhance the expression of the immunoglobulin G receptor on lymphocytes and macrophages; this, in turn, may enhance antibody-dependent cytotoxicity or macrophage-mediated phagocytosis [63].

An extensively characterized enzyme that is induced by IFN treatment and appears to play a critical role in interferon's antiviral and antiproliferative activities is 2',5'-oligoadenylate synthetase (E. C. 2.7.7.-) [64]. This enzyme catalyzes the synthesis of a family of novel 2',5'-oligoadenylate nucleotides that regulate the breakdown of RNA in cells. Reported levels of induction of the synthetase vary between 10- and 10 000-fold. The enzyme requires double-stranded RNA, which may be produced by the virus as a replicative intermediate, for expression of enzyme activity. Although the exact mechanism for this activation is not known, the double-stranded RNA may bind to the enzyme and convert it to its biologically active form. The 2′,5′-oligoadenylates activate a latent endoribonuclease, known as RNase L, which catalyzes the cleavage of both messenger and ribosomal RNA. Thus, the localized activation of endonuclease by double-stranded RNA present during viral replication may account for the selective cleavage of viral versus host RNA.

Interferon induces another system that also requires double-stranded RNA for activation, namely, a protein kinase that catalyzes the phosphorylation of protein initiation factor eIF-2, as well as the ribosomal protein P1 [65]. In cell-free systems, the kinase is activated by much lower concentrations of double-stranded RNA than is 2′,5′-oligoadenylate synthetase; higher concentrations of double-stranded RNA inhibit the kinase but activate the synthetase. Thus, the level of double-stranded RNA may determine which biochemical pathway is important in the expression of antiviral activity.

5. Physicochemical Properties

Only very low (picogram) quantities of IFN can normally be produced from natural sources because the biological activity is expressed at very low protein levels. Thus, characterization of its properties was for many years slow and incomplete. Studies on IFNα prepared from a variety of natural sources, including virally induced natural leukocytes, lymphoblastoid cells, and chronic myelogenous leukemic cells, suggested that IFNα consists of a family of multiple species with apparent molecular masses determined by sodium dodecyl sulfate–polyacrylamide gel electrophoresis (SDS–PAGE), ranging from 16 000 to 27 000. Amino acid analysis of the purified proteins indicated that their compositions were similar but that IFNα consisted of a family of proteins with different amino acid sequences. However, further characterization of the physicochemical properties of IFNs was severely hindered by the fact that IFNs purified from natural sources were frequently contaminated with other proteins and IFN subtypes. In addition, the extremely low levels of substance produced resulted in large losses of protein because of adsorption and surface denaturation. These problems were solved by the large-scale production of genetically engineered IFNα-2b (Intron A) in *E. coli* [3, pp. 1–12], [10].

5.1. Primary Structure

The amino acid sequence of IFNα-2 deduced from the DNA sequence of the cloned gene is depicted in Figure 1. This sequence has been confirmed by N-terminal sequencing through automated Edman degradation of the native protein and of fragments produced by either proteolytic or cyanogen bromide cleavage. The DNA sequences of

CYS	ASP	LEU	PRO	GLN	THR	HIS	SER	LEU	GLY	SER	ARG	ARG	THR	LEU
MET	LEU	LEU	ALA	GLN	MET	ARG	ARG	ILE	SER	LEU	PHE	SER	CYS	LEU
LYS	ASP	ARG	HIS	ASP	PHE	GLY	PHE	PRO	GLN	GLU	GLU	PHE	GLY	ASN
GLN	PHE	GLN	LYS	ALA	GLU	THR	ILE	PRO	VAL	LEU	HIS	GLU	MET	ILE
GLN	GLN	ILE	PHE	ASN	LEU	PHE	SER	THR	LYS	ASP	SER	SER	ALA	ALA
TRP	ASP	GLU	THR	LEU	LEU	ASP	LYS	PHE	TYR	THR	GLU	LEU	TYR	GLN
GLN	LEU	ASN	ASP	LEU	GLU	ALA	CYS	VAL	ILE	GLN	GLY	VAL	GLY	VAL
THR	GLU	THR	PRO	LEU	MET	LYS	GLU	ASP	SER	ILE	LEU	ALA	VAL	ARG
LYS	TYR	PHE	GLN	ARG	ILE	THR	LEU	TYR	LEU	LYS	GLU	LYS	LYS	TYR
SER	PRO	CYS	ALA	TRP	GLU	VAL	VAL	ARG	ALA	GLU	ILE	MET	ARG	SER
PHE	SER	LEU	SER	THR	ASN	LEU	GLN	GLU	SER	LEU	ARG	SER	LYS	GLU

Figure 1. Amino acid sequence of human IFNα-2b (Intron A)

16 other IFN genes indicate an amino acid sequence homology of 75% or more and an identity of amino acids at approximately 60% of all positions for the majority of IFNα subtypes. However, some IFNα subtypes differ by only one amino acid and yet are produced by distinct, nonallelic genes, e.g., IFNα-A and IFNα-2. The IFNα subtypes generally consist of 165 or 166 amino acid residues preceded by a 23-amino acid signal sequence. These signal sequences are not found in the mature protein because they are removed by proteolysis during secretion of IFN by the cell.

All IFNα subtypes contain four cysteine residues. As shown in Figure 1, for IFNα-2b these residues occur at positions 1, 29, 98, and 138. Chromatographic analysis of proteolytic fragments has shown that these amino acids form two disulfide bonds between residues 1 and 98, and residues 29 and 138. Pairing of the cysteine residues is probably the same in all IFNα subtypes. Studies with reducing agents indicate that the 29–138 disulfide bond is required for the full expression of biological activity, whereas the 1–98 disulfide bond may be disrupted without a significant loss in activity.

Based on examination of the biological activity of various recombinant hybrid IFNα molecules, IFNα appears to contain at least two distinct binding sites located at the amino and the carboxyl terminals, respectively. Each binding site is thought to interact with the IFN receptor, and the nature of the biological response is determined by the quality of fit between ligand and receptor [66]. However, other studies suggest that the final 13 carboxyl-terminal amino acids of IFNα are not required for biological activity.

Two forms of IFNβ have been cloned, and the amino acid sequence predicted from the complementary DNA sequence [67], [68]. The principal form of IFNβ produced by fibroblasts, IFNβ-1, contains 166 amino acids and exhibits approximately 30% homology in amino acid sequence with IFNα-1. However, the region of highest homology between IFNβ-1 and IFNα-1 lies between residues 115 and 151, which suggests an important functional role for this region.

A single form of IFNγ has been identified [69]. The gene structure predicts that this protein contains 146 amino acids, significantly fewer than IFNα or IFNβ. Virtually no homology exists at the amino acid level with either IFNα or IFNβ. Another form of IFNγ that lacks three amino acids (cysteine–tyrosine–cysteine) at the amino terminus has also been cloned. The specific antiviral activity of this species appears to be higher than that of the 146-amino acid form.

5.2. Secondary and Tertiary Structure

The secondary structure of purified recombinant IFNα-2 b has been examined by circular dichroism (CD) in both the near and the far UV range (Fig. 2).

Intense, negative bands at 218 and 208 nm are characteristic of an alpha helix. Calculations based on the amplitude of these bands indicate an alpha helicity of ca. 50%. Although the CD spectra do not indicate the presence of beta sheets, a small percentage (e.g., 0–10%) may be present but undetectable in the spectrum. Near-ultraviolet CD spectra indicate two intense negative bands at 291 and 286 nm, which can be assigned to transitions of tryptophan. The intensity and location of these bands suggest that at least one of the two tryptophan residues is present in a hydrophobic environment. Weak, negative CD bands at 255, 262, and 268 nm can be assigned to phenylalanine. Titration of IFNα-2 b solution to pH 2 results in virtually complete loss of tryptophan CD bands as well as substantial depression in the CD bands representative of the alpha helix. However, the apparent loss in native conformation that occurs on acidification is completely reversible on retitration to pH 7.4. In parallel with these findings, the antiviral activity of IFNα-2 is also retained completely after exposure to pH 2 and neutralization.

Precise information on the tertiary structure of IFNα is not yet available. However, crystallization of recombinant IFNα-2 b (Fig. 3) has provided a basis for X-ray diffraction analysis of three-dimensional structure. Needle-shaped crystals, with dimensions of ca. 0.1×0.01 mm, are prepared by incubation of a purified IFNα-2 solution at pH 6.0, which is near the isoelectric point of the protein [10]. Crystal dimensions vary depending on the conditions of crystallization.

Crystallization of a form of human IFNγ (IFNγ-D'), in which the five amino acids at the carboxyl terminus are deleted, has also been reported [70].

Crystals suitable for high-resolution X-ray diffraction have been prepared by vapor-diffusion equilibration against a solution of 30% saturated ammonium sulfate in 50 mmol/L sodium acetate, pH 5.9 [70]. Rhombohedral crystals ($0.45 \times 0.45 \times 0.40$ mm) are formed after 5-d incubation at room temperature (Fig. 4). The X-ray precession photographs indicate that the crystals are trigonal. The crystals are moderately stable to X radiation and diffracted to a resolution of 0.33 nm on a rotating anode generator. The three dominant faces of the rhombohedral crystal correspond to the (110), (101), and (011) planes of the rhombohedral unit cell. The measured density of the crystals is 1.224 g/cm^3 in a Ficoll–45% ammonium sulfate solution. The calculated mass of the protein per unit volume is 63 200, which indicates that four molecules of IFNγ-D' are present in the asymmetric unit. Diffraction data from native crystals and from several heavy-atom derivatives have been collected by oscillation photography with synchrotron radiation [70]. The intensity data sets from native crystals (resolution to 0.285 nm) are currently being analyzed to determine the three-dimensional structure at this degree of resolution.

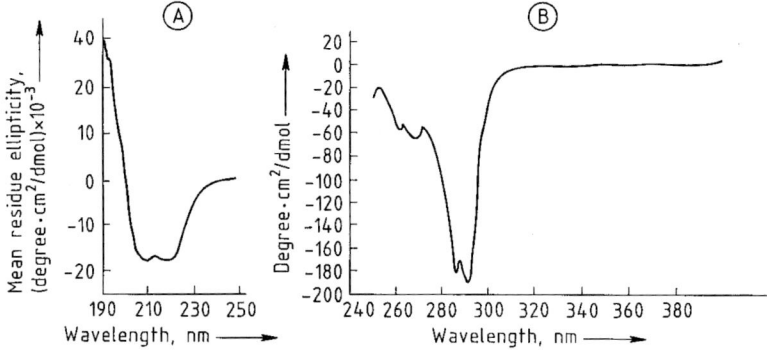

Figure 2. Circular dichroism spectra of human IFNα-2 b (Intron A)
A) Far ultraviolet; B) Near ultraviolet

Figure 3. Human IFNα-2 b (Intron A) crystals

Figure 4. Human IFNγ crystals in a suspended droplet

5.3. Solution Properties

Based on their predicted amino acid compositions, the polypeptide molecular masses for human IFNα, β, γ are 19 269, 19 990, and 17 145, respectively. However, under certain conditions the apparent molecular mass of IFNα in solution appears to be considerably higher than the polypeptide molecular mass [10]. Sedimentation velocity experiments have established that IFNα-2 undergoes self-association to yield a molecular mass at least three times that of the monomer. Lowering the pH from 7 to 2 favors monomer formation. Under the low protein concentration in a biological assay, it is anticipated that IFNα is completely monomeric; thus, the monomer is probably the active species. Even under highly concentrated conditions (e.g., 10 mg/mL or higher), IFNα-2b appears to be very soluble. At its isoelectric point (ca. pH 6.0), IFNα is relatively insoluble and can be crystallized readily. In contrast, IFNγ appears to form a stable dimer that does not dissociate on dilution and has a highly basic isoelectric point of ca. 9.5.

6. Purification

6.1. Natural Leukocyte IFN

Starting materials used for the purification of human IFN include leukocytes isolated from whole blood, neonatal fibroblasts, and lymphoblastoid and various leukemic cell lines. The principal source for purification of IFN prior to the use of genetically engineered microorganisms was human cell cultures treated with an appropriate inducer (e.g., a virus or a double-stranded polynucleotide). The principal problem with the isolation of IFN from such sources is the low level of expression of IFN due to its potent biological activity. Large starting volumes and multiple purification steps are thus required, and only small quantities of material can be purified. For example, in 1978, two laboratories in Helsinki together produced only 1.3 g of IFNα of <1% purity from 90 000 bags (45 000 L) of donated blood. Lack of precision in the antiviral assay commonly employed for monitoring the degree of purification (see Section 7.1) has also been a problem in the development of purification protocols.

Partially purified human leukocyte IFN was first developed by CANTELL and his associates in Finland and has been available in limited quantities for clinical trials since 1973 [71]. In this procedure, fresh blood from normal donors is centrifuged, and the leukocyte fraction (i.e., the buffy coat) is removed. Pooled leukocytes are treated with 0.83% ammonium chloride to lyse residual erythrocyte contaminants, suspended in a culture medium (e.g., Eagle's minimum essential medium), and then primed with high-titer crude IFN. Interferons production (primarily by monocytes) is then induced by addition of Sendai virus. After overnight incubation, the cells and debris are removed by centrifugation; the supernatant containing crude IFN is the starting material for purification.

Crude IFN is first precipitated with potassium thiocyanate at pH 3.4 and extracted with 95% ethanol. Extraction with acidified ethanol inactivates residual Sendai virus and partially purifies the IFN because not all proteins redissolve. The pH is then increased stepwise to precipitate contaminating proteins, which are removed by centrifugation. The IFN is precipitated at pH 8, concentrated, and dialyzed against phosphate-buffered saline.

The specific antiviral activity of the IFN preparation is approximately 10^6 IU/mg of protein, which represents approximately 1% purity. The net yield per unit of blood (i.e., 0.5 L) is ca. $(5-7) \times 10^6$ antiviral units. The final product contains a number of different species of IFNα, including α-1, α-2, and α-4. Other lymphokines (i.e., hormones released by leukocytes such as interleukin-1 and 2) probably also contaminate the preparation.

The partially purified leukocyte IFN preparation can be purified further to a specific activity of ca. 10^8 IU/mg of protein by chromatography on an immobilized monoclonal antibody referred to as NK-2. The IFN is eluted from the column by exposure to low pH. However, the relative composition of IFNα subtypes in this preparation differs from that in crude IFN prior to chromatography because the NK-2 antibody does not bind all IFNα species equally.

6.2. Recombinant IFNα

Purification of recombinant IFN from a genetically engineered microorganism poses unique problems distinct from those encountered in purification from natural sources [72]. The following requirements must be satisfied:

1) A procedure for efficient extraction of IFN from the cell that does not denature the protein must be developed. High concentrations of a conformation-disrupting agent such as urea or guanidine should be avoided for extraction because renaturation of the extracted protein may be difficult or impossible. This consideration is especially important for human IFNα because two disulfide bonds must be reformed in the native structure and the use of a chaotrope may lead to incorrect disulfide pairing. Furthermore, even if the final product is biologically active and appears to be properly refolded, small quantities of incorrectly folded material may be highly antigenic and remain undetected by standard analytical methods.
2) The extraction and purification procedure must yield a protein that is uncontaminated with bacterial pyrogens (gram-negative bacterial lipopolysaccharides that increase body temperature when injected parenterally). Pyrogens are not readily removed from purified IFN preparations because effective methods for their removal from solution (e.g., adsorption on activated charcoal or destruction with ionizing

radiation) may damage the biological activity or structure of the recombinant protein.
3) The purified IFN must be substantially free of nucleic acids and proteins derived from the host microorganism if it is to be employed in human clinical trials.
4) The product must be a homogeneous IFN species that is free of aggregates and proteolytic fragments. These IFN-related components are of special concern because they could render the product antigenic when injected into humans.
5) To be commercially viable, purification must be achieved with a good yield (e.g., 20%) of biological activity (e.g., antiviral units).

Purification of recombinant IFNα from *E. coli* for the IFNδ-4 α-2/α-1 hybrid that was genetically engineered to contain amino acid residues 4–62 of IFNα-2 and 54–166 of IFNα-1 is summarized in Table 3. The vector for this species was a PBR-322-based plasmid under control of the *lac* promoter [73]. The purification of IFN is monitored by following changes in specific antiviral activity (IU) expressed per milligram of protein.

The *E. coli* extract is prepared by acidification of the bacteria to pH 2, followed by adjustment to neutral pH (see Section 6.1) [74]. This procedure exploits the acid stability of type I (α and β) IFNs (see chap. 2) and avoids high concentrations of denaturants. The first purification step removes nucleic acids and contaminating *E. coli* proteins by acidification to pH 4.5 followed by centrifugation and is accompanied by a 20-fold increase in specific antiviral activity. The stability of IFNα at low pH and in organic solvents permits subsequent precipitation by addition of trichloroacetic acid and extraction from the precipitate with ethanol. This step increases the specific activity by an additional 20-fold and also concentrates the IFN solution with virtually no loss in activity. The 118% yield indicates complete recovery of antiviral activity within the experimental accuracy of the assay. The next step consists of affinity chromatography on a blue dextran affinity medium [Matrex Gel Blue A (Cibracron Blue F 3 GA) Sepharose]. This resin is used because both human and murine IFNs bind to blue dextran. Elution from this column is achieved with sodium chloride solution (2 mol/L). The final purification step can be performed in two ways: (1) repeated cycles of selective precipitation by dialyzing to low ionic strength and redissolving in 2 mol/L sodium chloride or (2) selective precipitation followed by ion-exchange chromatography on diethylaminoethyl (DEAE) Sepharose. The specific activity of the final product prepared by either procedure is 7×10^7 IU/mg, which represents a 140-fold purification.

An alternative approach for purifying recombinant IFNα employs immunosorbent chromatography with an immobilized monoclonal antibody [72]. Purification of IFNα-5 by this technique is summarized in Table 4.

The IFNα-5 is extracted by acidification of *E. coli* followed by neutralization (see Section 6.1). Extracted IFN is adsorbed on silica gel and eluted by lowering the pH to 2.0, which results in an almost tenfold increase in specific activity. Silica gel chromatography also removes the majority of contaminating nucleic acids in the unbound fraction. Matrex Gel Blue A chromatography results in a further 3.3-fold purification. Next, IFN is adsorbed onto an anion-exchange resin (i.e., DEAE Sepharose) and eluted with a pH gradient from 8.0 to 6.0, resulting in an additional tenfold increase in specific activity. The DEAE Sepharose eluate is purified further on an immobilized monoclonal antibody, YOK Sepharose. Elution is achieved with 0.1 mol/L sodium citrate, pH 2.0, to give a final specific activity of 1.9×10^8 IU/mg.

Table 3. Purification of recombinant human IFNδ-4 α-2/α-1

Purification step	Volume, L	Antiviral activity			Purification*, fold
		10^7 IU/mg	Total, 10^9 IU	Yield, %	
Escherichia coli extract**	765.0	0.05	629	100	(1.0)
Acidification (pH 4.5)	765.0	1.12	466	67	2.4
Trichloroacetic acid–ethanol extract	21.0	2.1	740	118	42
Matrex Gel Blue A Sepharose	15.3	7.7	177	28	154
Concentration by ultrafiltration followed by	1.0	8.2	192	31	164
Selective precipitation	0.069	7.0	30	5	140
DEAE Sepharose chromatography	0.031	7.0	18	3	140

* Based on antiviral units.
** Extract was prepared by acidification of the *E. coli*, followed by adjustment to neutral pH for release of the IFN.

Table 4. Purification of recombinant human IFNα-5

Purification step	Volume, L	Antiviral activity			Purification, fold
		10^7 IU/mg	Total, 10^9 IU	Yield, %	
Escherichia coli extract	200.0	0.019	42	(100)	(1.0)
Silica (Si 200)	1.8	0.18	32	76	9.5
Matrex Gel Blue A	0.8	0.6	25	60	31.6
DEAE Sepharose CL-6 B	0.2	5.8	10	24	305
YOK Sepharose	0.04	19.0	8	19	1000

6.3. Removal of Pyrogens

The final IFN preparation should be low in pyrogenic substances if it is to be employed in clinical trials (see Section 6.2). Although the purification protocols described in Sections 6.1 and 6.2 generally result in a product that is very low in pyrogens (i.e., 5–10 ng of endotoxin per milligram of protein, determined as described in Section 7.2), preparations with high pyrogen levels are occasionally encountered. An effective procedure for lowering the endotoxin level of these preparations is recycling isolelectric focusing (RIEF). Isoelectric focusing is based on the principle that charged substances migrate in an electric field and a stable pH gradient to a pH at which their net change is zero. In the RIEF apparatus designed by BIER et al. [75], the protein solution is continuously recycled between a multichannel heat exchange reservoir and a multichannel focusing cell. Thus, heat is dissipated in the heat exchange reservoir and not in the focusing cell. As shown in Figure 5, when RIEF is applied to a solution of IFNα-2 with a high pyrogen content, the majority of endotoxins and the IFNα-2 migrate to different pH values [76]. Thus, whereas endotoxins reach maximum levels at pH 5.07 (channel 5), the antiviral activity of IFNα-2 peaks at pH 6.4 (channel 10).

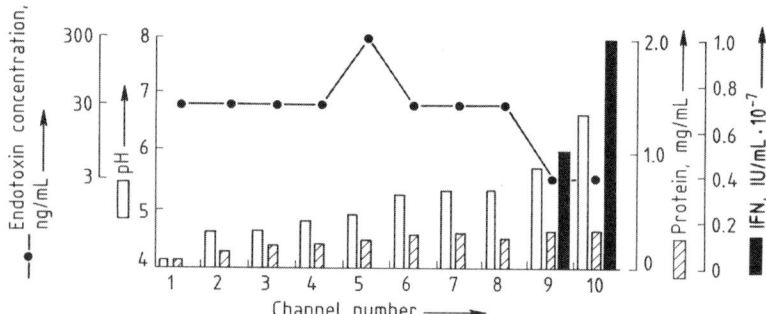

Figure 5. Recycling isoelectric focusing (RIEF) of human recombinant IFNα-2 b (Intron A) containing *Escherichia coli* pyrogens (endotoxins)

The apparatus was prefocused with 0.1% ampholines (pH range 3–10) in urea (4 mol/L) for 60 min followed by addition of the sample. Protein concentration was determined from the optical density reading at 280 nm; pyrogen levels were assayed by the limulus amebocyte lysate assay, and antiviral activity was determined by CPE-inhibition assay.

In the removal of contaminating pyrogens from IFNα-2, a 40-fold purification is obtained from a single RIEF experiment, which results in a pyrogen level of 12.5 ng/mg of protein, compared to 500 ng/mg of protein prior to RIEF. No apparent loss in specific antiviral activity is observed, and the total recovery of IFN activity is > 90%. The ampholines used to establish the pH gradient are finally removed from the preparation by crystallization of the protein, gel filtration chromatography, or ammonium sulfate precipitation.

7. Quality Specifications

7.1. Purity

Purified IFN must be shown to be biologically active, and its activity must remain constant from batch to batch. Because IFN was originally defined in terms of its antiviral activity, quantitation of antiviral activity per milligram of purified protein is routinely employed for characterization of purified product. The most commonly employed assay, which is readily adapted to screening large numbers of samples, is the *cytopathic effect- (CPE-) inhibition assay*. This assay is based on the ability of IFN to interfere with the cytopathic effects (e.g., cell lysis) caused by viruses. In a simple CPE assay, IFN is serially diluted (usually by a factor of two) in the cells of a microtiter plate, and target cells (e.g., human foreskin diploid fibroblasts) are then added to each well. Cells to which IFN has not been added are also included as controls. After an incubation period (e.g., 4 h), a challenge virus (e.g., encephalomyocarditis virus) is added, followed by a further period of incubation. In the presence of IFN, cells are protected from viral infection and can be detected by staining with a dye such as Crystal Violet.

However, in the absence of IFN, the cells lyse and are not stained by the dye because they do not adhere to the microtiter plate. The end point is arbitrarily defined (e.g., 50% protection against viral infectivity). An appropriate international reference standard provided by the World Health Organization is included in each microtiter plate and allows conversion to the international antiviral units (IU) established by international accord. Further details of the CPE-inhibition assay are described in [77].

Sodium dodecyl sulfate–polyacrylamide gel electrophoresis is a powerful technique for demonstrating the purity of the final IFN preparation [78]. The quantitation of impurities and sensitivity of the method depend on the nature of the stain employed for protein detection. Quantities of protein as low as 0.05–0.1 µg can be detected by staining with Coomassie Blue R 250, whereas silver staining may result in detection of levels of impurities up to 100-fold lower. However, silver-stained gels must be interpreted with caution because staining intensity varies widely among proteins (up to a factor of ten). A typical gel electrophoresis of purified recombinant IFNα subtypes stained with Coomassie Blue R 250 is presented in Figure 6; SDS-PAGE reveals a purity of $\geq 95\%$ for each subtype, which was confirmed by silver staining. The two components observed for the δ-4 α-2/α-1 hybrid (Fig. 6, lane e) were both demonstrated to be IFN-related. Approximate quantitation of impurities can be achieved by electrophoresis at IFN loading levels ranging from 25 to 0.05 µg. The loading level at which an impurity disappears relative to the main component can be used to calculate the approximate percentage of impurity. The SDS-PAGE technique can also be employed for estimation of apparent molecular mass by comparison to the migration of standards with known molecular mass. Thus, for example, the average apparent molecular mass of IFNα-2 was observed to be $18\,500 \pm 600$, the standard deviation representing gel-to-gel variation. The calculated molecular mass of IFNα-2 is 19 270, in good agreement with the experimental value.

Reversed-phase high-performance liquid chromatography (HPLC) is also employed to characterize the purity of the final product [80]. Reversed-phase HPLC of a clinical-grade sample of IFNα-2 b prior to formulation is presented in Figure 7. Results indicate a high degree of homogeneity, in agreement with those obtained by SDS-PAGE.

Amino acid analysis can also be used to characterize the purified preparation. Table 5 contains the average amino acid analysis of 15 batches of purified IFNα-2 b. Interferon was first hydrolyzed with 6.7 mol/L hydrochloric acid for 18 h at 160 °C, followed by chromatographic resolution of the individual amino acids on a C-18 HPLC reversed-phase column utilizing paired ion chromatography [81]. The amino acids were detected by fluorescence following postcolumn derivatization with *o*-phthalaldehyde. Excellent agreement was observed between experimental and theoretical values.

Amino acid sequencing provides a more sensitive method than amino acid analysis for detecting contaminating proteins, as well as for demonstrating that the recombinant protein has the correct amino acid sequence predicted from the nucleotide sequence of the gene. For example, N-terminal sequencing has been achieved for recombinant IFNα-2 by using automated Edman degradation in a gas-phase sequenator [82].

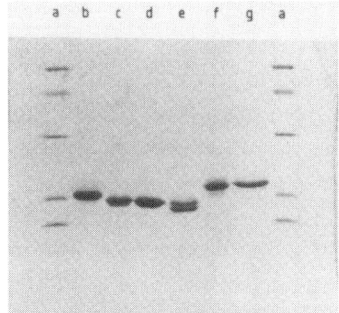

Figure 6. Polyacrylamide gel electrophoresis of purified recombinant IFNα subtypes

Before loading, 0.01 mg of each subtype was boiled for 2 min in 1% 2-mercaptoethanol, 1% sodium dodecyl sulfate, and 10% glycerol dissolved in 0.06 mol/L Tris hydrochloride, pH 6.8, containing 0.001% bromophenol blue. Discontinuous SDS–PAGE was run according to [79].

a) Standard proteins (lysozyme, β-lactoglobulin, carbonic anhydrase, ovalbumin, and bovine serum albumin); b) IFNα-1; c) IFNδ-4 α-1; d) IFNα-2; e) IFNδ-4 α-2/α-1; f) IFNα-4; g) IFNα-7

Figure 7. Reversed-phase HPLC of purified recombinant human IFNα-2 b (Intron A)

Chromatography was performed on a Water Associates μ-Bondapak C-18 column (30 cm×3.9 mm). The eluting gradient extends from 31.4 to 62.7% acetonitrile containing 0.01 mol/L trifluoroacetic acid at a flow rate of 1 mL/min for 30 min. Detection was accomplished by monitoring fluorescence after post-column derivatization with o-phthalaldehyde. The first peak is caused by the solvent.

Table 5. Amino acid analysis of recombinant human IFNα-2 b

Amino acid	Theoretical content, mol%	Measured content, mol%
Aspartic acid, asparagine	7.74	9.06 ± 0.46
Serine	9.03	10.33 ± 0.48
Glycine	3.23	3.80 ± 0.91
Glutamic acid, glutamine	16.77	19.03 ± 0.79
Threonine	6.45	5.54 ± 0.58
Alanine	5.16	5.61 ± 0.58
Valine	4.52	2.62 ± 0.23
Methionine	3.23	2.82 ± 0.59
Tyrosine	3.23	2.81 ± 0.45
Isoleucine	5.16	3.52 ± 0.39
Leucine	13.55	13.96 ± 0.46
Phenylalanine	6.45	6.51 ± 0.20
Histidine	1.94	1.95 ± 0.15
Lysine	7.10	6.53 ± 0.58
Arginine	6.45	5.99 ± 0.70

The phenylthiohydantoin–amino acids were resolved by reversed-phase HPLC on a cyanopropyl silane column, with acetonitrile–methanol (4:1) as eluent. Sequencing was performed after reduction with dithiothreitol and alkylation with iodoacetamide to permit identification of cysteine residues. The first 57 amino acids agreed completely with the amino acid sequence predicted from the complementary DNA (Fig. 1). In addition, by utilizing peptide fragments produced by cyanogen bromide treatment (which cleaves the protein at methionyl residues), 90% of the amino acid sequence was determined. Tryptic digestion of the C-terminal cyanogen bromide fragment followed by reversed-phase HPLC and amino acid analysis of the separated components was performed to identify glutamic acid as the C-terminal amino acid. The amino acid sequence was also surveyed by gas chromatography–mass spectrometry of fragments of IFNα-2 produced by subtilisin cleavage followed by trifluoroacetylation and permethylation. Selected sequences with IFNα-2 were identified and demonstrated to correspond to the gene structure.

7.2. Pyrogenicity Testing

Amebocytes circulating in the horseshoe crab (*Limulus*) hemolymph contain a coagulation system that is highly sensitive to gram-negative bacterial endotoxins [83]. The *Limulus* amebocyte lysates form a firm gel in the presence of low levels of endotoxin. The "limulus test" for detection of endotoxins, based on this gelation reaction, is extensively employed in the pharmaceutical industry. As shown in Figure 8, extremely low endotoxin levels are detectable in purified preparations of recombinant IFNα-2, in which the values generally are significantly lower than 1.0 ng/mg of protein.

An alternative, traditional method for investigating pyrogen content is to measure the *increase in body temperature* following intravenous injection into rabbits. However, as shown in Figure 8, no apparent correlation existed between temperature increase and either IFN dose or endotoxin level measured in the limulus test. The in vivo test is further complicated by the fact that IFN itself has inherent pyrogenic properties. Thus, the limulus test appears to be a convenient and easily performed in vitro test for estimation of endotoxin levels.

8. Clinical Studies

Although IFN was first discovered in 1957, not until 1981 were large enough quantities of purified recombinant IFNα available to permit the initiation of phase I/II human clinical trials. To date over 2000 patients with either advanced malignancies or viral infections have been treated with IFNα-2 b (Intron A). Despite the fact that cancer patients treated with Intron A generally displayed poor prognosis as well as resistance to standard chemotherapy, a number of patients demonstrated complete recovery of long-term duration.

Phase I trials with Intron A were generally performed on patients with advanced malignancies and on a small number of normal volunteers [9], [84]. These studies indicate that Intron A is safe in humans; its toxic effects are dose-related and completely reversible. No evidence for cumulative toxicity or drug intolerance was obtained. Three routes of administration (subcutaneous, intramuscular, and intravenous) were employed in a variety of schedules. For example, Intron A was administered in an escalating daily dose for 28 d to determine the maximum tolerable dosage. Dosages

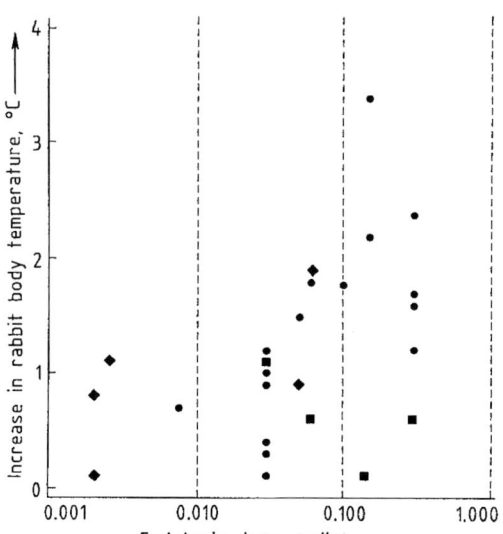

Figure 8. Pyrogenicity of recombinant human IFNα-2b (Intron A)
IFN dose, IU/kg: ■ 0.75; ● 1.5; ◆ 2.5

were also elevated on a weekly schedule. The total daily doses examined were generally in the range of $(1-100) \times 10^6$ IU. The principal toxic effect observed in 90% of all patients consisted of flu-like symptoms (e.g., fatigue, fever, headache). However, these dose-related effects reversed rapidly when the dosage was lowered or treatment discontinued. Other hematologic toxicities, including mild depression of white blood cell and platelet counts, and elevation of hepatic enzymes were also observed in some patients but were all reversible upon discontinuance of IFN administration. Single Intron A doses of up to 200×10^6 IU per square meter of body surface area could be tolerated when administered intravenously. However, for 5-d intravenous scheduling, 50×10^6 IU/m^2 was the maximum dose tolerated.

The pharmacokinetics of Intron A were also evaluated in phase I trials. Intravenous infusion for 30 min produced peak serum levels within 15–30 min. Because of a relatively short half-life, serum levels returned to base line within 4 h. However, sustained levels could be achieved by continuous intravenous infusion. Subcutaneous and intramuscular dosing demonstrated distinct pharmacokinetics with slowly rising blood levels that peaked at 4 h and maintained sustained levels for 8–12 h. Based on in vitro studies, the peak serum levels obtained by any of these three routes of administration should be sufficient to activate immune effector cells or to exert a direct antiproliferative effect on tumor cells.

Phase II clinical trials of Intron A have been conducted on a variety of malignancies. Results are as follows:

Active:	hairy cell leukemia
	non-Hodgkin's lymphoma
	mycosis fungoides
	Kaposi's sarcoma
	multiple myeloma
	malignant melanoma
	renal cell cancer
	bladder cancer (superficial tumors)
	ovarian cancer
Inactive:	breast cancer
	colon cancer
	non-small cell lung cancer
	prostate cancer
	acute myelogenous leukemia
To be determined:	chronic myelogenous leukemia
	chronic lymphocytic leukemia
	preleukemia
	Hodgkin's disease
	osteogenic sarcoma
	astrocytoma

Efficacy has been observed most often in hematological malignancies, particularly those of pre-B-cell and B-cell origin. For example, in hairy-cell leukemia, Intron A administered subcutaneously at relatively low doses for continuous periods produced hematological responses in ca. 90% of patients and objective responses in ca. 75%. Complete responses of relatively long duration have also been observed in multiple myeloma and low-grade lymphomas.

Most adult solid tumors (e.g., breast, lung, and colon) have not responded well to Intron A therapy. However, clear activity has been observed in malignant melanoma, in which the therapeutic effects were apparently schedule-dependent because continuous dosing was more efficacious than intermittent dosing. Tumor burden is probably also important in determining the response to Intron A because malignant melanoma patients with large tumor burdens appeared to have a lower incidence of response. Objective responses have also been observed in superficial bladder cancer, renal cell cancer, ovarian cancer, and AIDS-related Kaposi's sarcoma.

Intron A has also demonstrated activity in viral indications. For example, intralesional administration into genital warts (three times weekly for three weeks) can result in complete regression of the injected warts between four and twelve weeks. Therapeutic activity has also been observed in laryngeal papillomatosis, in chronic hepatitis B, and as a prophylactive treatment for the common cold.

In all these clinical trials, the incidence of human antibody formation to Intron A was very low [85]. An antibody response was elicited in <3% of more than 500 patients treated systemically and in <1% of more than 1000 patients treated intralesionally or intranasally. Significantly higher rates of neutralizing antibody production have been reported for recombinant IFNβ (ca. 65% of treated patients) and recombinant IFNα-2a (Roferon) (ca. 25–30% of treated patients).

The presence of neutralizing antibodies to Roferon has been reported in 7 out of 12 responding patients and in 9 out of 29 nonresponding patients with metastatic renal cell carcinoma [86]. A comparison of the medium duration of remission between antibody-positive and antibody-negative patients in this study suggested that antibody formation contributed to early relapse. Similarly, 17 out of 23 patients in a phase I clinical trial with a form of recombinant IFNβ containing a genetically engineered serine residue at position 17 showed evidence for antibody formation [87]. These data strongly support that Intron A is significantly less antigenic in human subjects than other recombinant IFNs.

9. References

[1] A. Isaacs, J. Lindenmann, *Proc. R. Soc. (London)* Sec. B **147** (1957) 258–267.
[2] P. P. Trotta, S. K. Narula in S. Baron, F. Dianzani, G. J. Stanton, W. R. Fleischmann, Jr. (eds.): *Interferon Genes*, Univ. of Texas Press, Austin 1987, pp. 137–147.
[3] P. P. Trotta in P. L. Kisner, J. F. Smyth (eds.): *Interferon Alpha-2: Preclinical and Clinical Evaluation*, Martinus Nijhoff Medical Division, The Hague 1985, pp. 13–27.
[4] P. P. Trotta, R. J. Spiegel in F. M. Muggia (ed.): *Concepts, Clinical Developments, and Therapeutic Advances in Cancer Chemotherapy*, Martinus Nijhoff Publishers, Boston 1987, pp. 4–159.
[5] D. A. Stringfellow (ed.): *Interferon and Interferon Inducers-Clinical Applications*, Marcel Dekker, New York 1980.
[6] K. C. Zoon, D. Miller, D. Zur Nedden, M. W. Hunkapiller, *Proc. Natl. Acad. Sci. U.S.A.* **76** (1979) 5601–5605.
[7] K. C. Zoon in E. DeMaeyer, G. Galasso, H. Schellekens (eds.): *The Biology of the Interferon System 1981*, Elsevier/North-Holland Biomedical Press, Amsterdam 1981, pp. 47–55.
[8] M. Streuli, S. Nagata, C. Weissmann, *Science* **209** (1980) 1343–1349.
[9] R. J. Spiegel in [3] pp. 43–50.
[10] T. L. Nagabhushan et al. in K. C. Zoon, P. D. Noguchi, T. Y. Liu (eds.): *Interferon: Research, Clinical Application and Regulatory Consideration*, Elsevier, New York 1984, pp. 79–88.
[11] C. L. Bennett, N. J. Vogelzang, M. J. Ratain, S. D. Reich, *Cancer Treat. Rep.* **70** (1986) 1081–1084.
[12] K. Foon et al., *Cancer Immunol. Immunother.* **20** (1985) 193–197.
[13] E. S. Kleinerman et al., *Cancer Res.* **46** (1986) 5401–5405.
[14] R. Kurzrock et al., *Cancer Res.* **45** (1985) 2866–2872.
[15] J. Rinehart, L. Malspeis, D. Young, J. Neidhart, *Cancer Res.* **46** (1986) 5364–5367.
[16] E. C. Bordon et al., *Cancer Res.* **45** (1985) 5914–5920.
[17] L. H. Kronenberg in P. N. Cheremisinoff, R. P. Quellette (eds.): *Biotechnology*, Technomic Publishing Co., Lancaster 1985, pp. 451–462.
[18] S. Pestka, J. H. Langer, K. C. Zoon, C. E. Samuel in C. C. Richardson, P. D. Boyer, I. B. David, A. Meister (eds.): *Annual Reviews of Biochemistry*, vol. **56**, Annual Reviews, Palo Alto 1987, pp. 727–777.
[19] S. Pestka, *Arch. Biochem. Biophys.* **221** (1983) 1–37.
[20] S. Baron, F. Dianzani, G. J. Stanton (eds.): *The Interferon System: A Review to 1982*, Univ. of Texas Medical Branch Publishers, Galveston 1982, Part 1. Reference 20, Part 2.
[21] S. Baron, F. Dianzani, G. J. Stanton (eds.): *The Interferon System: A Review to 1982*, Univ. of Texas Medical Branch Publishers, Galveston 1982, Part 2.

[22] S. Baron, F. Dianzani, G. J. Stanton, W. R. Fleischmann, Jr. (eds.): *The Interferon System-A Current Review to 1987*, Univ. of Texas Press, Austin 1987.
[23] P. E. Came, W. A. Carter: *Interferons and Their Applications*, Springer Verlag, New York 1984.
[24] E. DeMaeyer, H. Schellekins (eds.): *The Biology of the Interferon System 1983*, Elsevier, New York 1983.
[25] R. M. Friedman, T. M. Merigan, T. Sreevalson: *Interferons as Cell Growth Inhibitors and Antitumor Factors*, Alan R. Liss, New York 1986.
[26] I. Gresser (ed.): *Interferon 1-1979*, Academic Press, London-New York 1979.
[27] I. Gresser (ed.): *Interferon 2-1980*, Academic Press, London-New York 1980.
[28] I. Gresser (ed.): *Interferon 3-1981*, Academic Press, London-New York 1981.
[29] I. Gresser (ed.): *Interferon 4-1982*, Academic Press, London-New York 1982.
[30] I. Gresser (ed.): *Interferon 5-1983*, Academic Press, London-New York 1983.
[31] I. Gresser (ed.): *Interferon 6-1985*, Academic Press, London-New York 1985.
[32] I. Gresser (ed.): *Interferon 7-1986*, Academic Press, London-New York 1986.
[33] H. Kirchner, H. Schellekens (eds.): *The Biology of the Interferon System 1984*, Elsevier, New York 1985.
[34] T. C. Merigan, R. M. Friedman (eds.): *Interferons*, Academic Press, New York 1982.
[35] S. Pestka (ed.): *Methods in Enzymology*, vol. **78,** Academic Press, New York 1981, Part A.
[36] S. Pestka (ed.), *Methods in Enzymology*, vol. **79,** Academic Press, New York 1981, Part B.
[37] W. E. Stewart II: *The Interferon System*, 2nd ed., Springer-Verlag, New York 1981.
[38] W. E. Stewart II, H. Schellekins (eds.): *The Biology of the Interferon System 1985*, Elsevier, Amsterdam-New York 1986.
[39] K. C. Zoon, P. D. Noguchi, T.-Y. Liu (eds.): *Interferon: Research, Clinical Application and Regulatory Consideration*, Elsevier Science Publishers, New York 1984.
[40] A. Billiau (ed.): *Interferon 1. General and Applied Aspects*, Elsevier, Amsterdam 1984.
[41] J. Vilcek, E. DeMaeyer (eds.): *Interferon 2. Interferons and the Immune System*, Elsevier, Amsterdam 1984.
[42] R. M. Friedman (ed.): *Interferon 3. Mechanisms of Production and Action*, Elsevier, Amsterdam 1984.
[43] N. B. Finter, R. K. Oldham (eds.): *Interferon 4. In Vivo and Clinical Studies*, Elsevier, Amsterdam 1985.
[44] J. Taylor-Papadimitriou (ed.): *Interferons. Their Impact in Biology and Medicine*, Oxford University Press, Oxford 1985.
[45] K. Paucker, K. Cantell, W. Henle, *Virology* **17** (1962) 324–334.
[46] P. Lindahl, P. Leary, I. Gresser, *Proc. Natl. Acad. Sci. U.S.A.* **69** (1972) 721–726.
[47] J. Taylor-Papadimitriou in A. Billiau (ed.): *Interferon 1. General and Applied Aspect*, Elsevier, Amsterdam 1984, pp. 139–166.
[48] D. A. Eppstein, Y. V. Marsh, *Biochem. Biophys. Res. Commun.* **120** (1984) 66–73.
[49] E. Lvovsky et al., *J. Natl. Cancer Inst.* **66** (1981) 1013–1019.
[50] N. B. Lydon et al., *Biochemistry* **24** (1985) 4131–4141.
[51] S. E. Salmon et al., *J. Clin. Oncol.* **1** (1983) 217–225.
[52] I. Gresser, C. Maury, M. Tovey, *Eur. J. Cancer* **14** (1978) 97–99.
[53] M. Tozawa et al., *Cancer Treat. Rep.* **66** (1982) 1575–1577.
[54] M. S. Aapro, D. S. Alberts, S. E. Salmon, *Cancer Chemother. Pharmacol.* **10** (1983) 161–166.
[55] C. E. Welander, T. M. Morgan, H. D. Homesley, P. P. Trotta et al., *Int. J. Cancer* **35** (1985) 721–729.
[56] J. L. Taylor, J. L. Sabran, S. E. Grossberg, *Eur. J. Immunol.* **10** (1980) 3.

[57] P. B. Fisher, R. H. Mufson, I. B. Weinstein, *Biochem. Biophys, Res. Commun.* **100** (1981) 823–830.

[58] E. Lotzova, C. A. Saavary, J. U. Gutterman, E. M. Hersh, *Cancer Res.* **42** (1982) 2480–2488.

[59] A. E. Maluish et al., *J. Immunol.* **131** (1983) 503–507.

[60] K. C. Zoon, H. Arnheiter, *Pharmacol. Ther.* **24** (1984) 259–278.

[61] T. Sreevalson in R. M. Friedman (ed.): *Interferon-3. Mechanisms of Production and Action*, Elsevier, Amsterdam 1984, pp. 343–387.

[62] M. E. Hokland, *Acta Pathol. Microbiol. Immunol. Scand. [Suppl.]* **93** (1985) 4–35.

[63] K. Stoh, M. Inoue, S. Kataoka, K. Kumagai, *J. Immunol.* **124** (1980) 2589–2595.

[64] B. R. G. Williams, R. H. Silverman (eds.): *The 2-5 A System: Molecular and Clinical Aspects of the Interferon-Regulated Pathway*, Alan R. Liss, New York 1985.

[65] B. Lebleu, G. C. Sen, S. Shaula, B. Carrer, *Proc. Natl. Acad. Sci. U.S.A.* **73** (1976) 3107–3111.

[66] M. Streuli et al., *Proc. Natl. Acad. Sci. U.S.A.* **78** (1981) 2848–2852.

[67] T. Taniguchi, S. Ohno, Y. Fuji-Kuriyama, M. Muramatsu, *Gene* **10** (1980) 11–15.

[68] A. Zilberstein, R. Ruggieri, J. H. Horn, M. Revel, *EMBO J.* **5** (1986) 2529–2537.

[69] P. W. Gray, D. V. Goeddel, *Nature (London)* **298** (1982) 859–863.

[70] S. Vijay-Kumar et al., "Crystallization and Preliminary X-ray Investigation of a Recombinant Form of Human Gamma Interferon," *J. Biol. Chem.* **262** (1987) 4804–4805.

[71] K. E. Mogensen, K. Cantell, *Pharmacol. Ther.* **1** (1977) 369–383.

[72] P. P. Trotta, H. V. Le, B. Sharma, T. L. Nagabhushan in G. Pierce (ed.): *Developments in Industrial Microbiology*, vol. **27**, Elsevier, Amsterdam 1987, pp. 53–64.

[73] P. Leibowitz, M. Ryan, *Eur. Patent Publication No. 146* (1985) 903.

[74] Schering Corp., US 4 315 852, 1982 (P. J. Leibowitz, M. J. Weinstein).

[75] M. Bier et al.: "New Developments in Isoelectric Focusing," in E. Gross, J. Meienhofer (eds.): *Peptides: Structures and Biological Function*, Pierce Chem. Co., Rockford 1979, pp. 79–89.

[76] T. L. Nagabhushan, B. Sharma, P. P. Trotta, *Electrophoresis* **7** (1986) 552–557.

[77] S. Rubinstein, P. C. Fameletti, S. Pestka, *J. Virol.* **37** (1981) 755–758.

[78] K. Weber, M. Osborn, *J. Biol. Chem.* **244** (1969). 4406–4412

[79] U. K. Laemmli, *Nature (London)* **227** (1970) 680–685.

[80] M. Rubinstein, S. Chen-Kiang, S. Stein, S. Odenfriend, *Anal. Biochem.* **111** (1981) 184–188.

[81] M. K. Radjai, R. T. Hatch, *J. Chromatogr.* **196** (1980) 319–322.

[82] P. Edman, *Acta Chem. Scand.* **4** (1956) 761–768.P. Edman, G. Begg, *Eur. J. Biochem.* **1** (1967) 80–91.

[83] J. Levin, F. B. Bang, *Bull. Johns Hopkins Hosp.* **115** (1975) 265–274.

[84] R. J. Spiegel in K. Hellman (ed.): *Cancer Treatment Reviews*, vol. **12,** Academic Press, New York 1985, supplement B, pp. 5–16.

[85] R. J. Spiegel et al., *Am. J. Med.* **80** (1986) 223–228.

[86] R. R. Quaseda et al., *J. Clin. Oncol.* **3** (1985) 1522–1528.

[87] M. Hawkins et al., (eds.): *The Biology of the Interferon System 1984*, Elsevier, New York 1985, pp. 503–508.

Monoclonal Antibodies

GIOVANNI L. GALFRE, Celltech Ltd., Slough, United Kingdom

DAVID S. SECHER, Celltech Ltd., Slough, United Kingdom

PATRICK E. CRAWLEY, Celltech Ltd., Slough, United Kingdom (Chap. 6)

1.	Introduction..........	1981
2.	Derivation of Murine Monoclonal Antibodies by Somatic Cell Fusion......	1983
3.	Human Monoclonal Antibodies...........	1987
4.	Production...........	1989
5.	Purification..........	1991
6.	Monoclonal Antibody Patents	1993
7.	Uses................	1994
8.	References...........	1997

1. Introduction

Antibodies are proteins produced in the blood of all vertebrate species and are one of the mechanisms for defending the organism against invasion by bacteria, viruses, or other foreign material. Each animal can produce millions of different antibodies, each antibody being able to bind specifically to a particular foreign substance known as an *antigen*.

All antibodies have a similar structure based on a unit of four polypeptide chains (Fig. 1): two identical *heavy chains* (M_r ca. 50 000) and two identical *light chains* (M_r 25 000) [1]. The protein fraction containing antibodies (the immunoglobulin or Ig fraction) can be purified from blood by simple physicochemical procedures (e.g., salt precipitation, → Immunotherapy and Vaccines). This fraction contains a very large number of antibody molecules that are structurally similar but differ in their antigen-binding specificity. Further purification of a single antibody must, therefore, rely on specific antigen-binding properties. Purification methods based on this principle have been extensively used but the heterogeneity of the normal immune response to even a single, simple antigen has limited their usefulness.

In the 1960s it was shown that antibodies are made by *lymphocytes* (a type of white blood cell) and that each lymphocyte makes only a single antibody. Maintenance of clones derived from single lymphocytes in long-term culture, could therefore provide a source of single, pure antibodies. Occasionally, this happens naturally in a form of cancer (*myeloma*) that results from the uncontrolled proliferation of a clone of cells derived from a single lymphocyte. Pure antibody in large quantities can be derived from the blood of tumor-bearing animals but the rarity of the disease and the apparent randomness of the transformation event make it impossible to identify the specificity of the antibody.

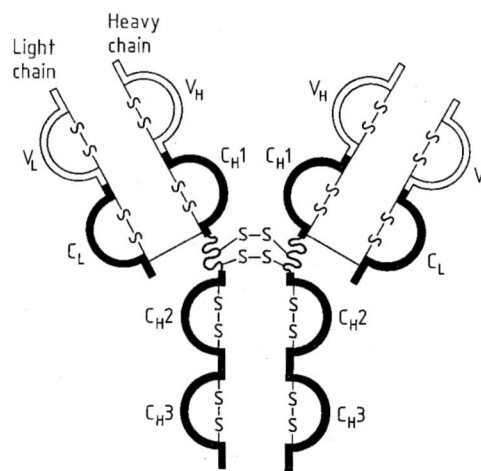

Figure 1. Schematic representation of an immunoglobulin (Ig) molecule (reproduced with permission [2])

The tertiary structure of Ig molecules is composed of interacting protein domains. The heavy chain domains (C_H2 and C_H3) interact with like domains; the light chain domains (C_L and V_L) interact with heavy chain domains (C_H1 and V_H). Inter- and intrachain disulfide bridges (–S–S–) stabilize the structure.

In 1975 GEORGES KÖHLER and CÉSAR MILSTEIN discovered a general method for the immortalization of lymphocytes making antibodies of predefined specificity [3].

They injected a mouse with a single antigen to boost the number of lymphocytes making specific antibodies. Lymphocytes removed from the mouse were then fused with cells derived from a previously isolated mouse myeloma. The resulting hybrid cell (*hybridoma*) was able to grow permanently like the parent myeloma and secreted the same specific antibody as the parent lymphocyte. Hybridomas derived from a single lymphocyte are monoclonal and the antibodies they produce are called *monoclonal antibodies*.

Since its introduction, monoclonal antibody technology has had a dramatic impact on life science research and clinical medicine. The extensive introduction of immunological methods based on monoclonal antibodies amounts to a revolution, the importance of which was recognized by the award of the Nobel Prize for medicine in 1984 to KÖHLER and MILSTEIN.

Although other methods of lymphocyte immortalization have been used to produce monoclonal antibodies, the original method of KÖH-LER and MILSTEIN (Chap. 2) remains the basis for the majority of monoclonal antibody derivation today. It has been very successfully used to produce mouse and rat hybridomas, but the derivation of human monoclonal antibodies has proven particularly difficult (Chap. 3). The derivation of hamster hybridomas is becoming a well-established method using mouse myelomas as fusion partners to derive hamster–mouse *heterohybridomas* [4]. The production of heterohybridomas of either rabbit [5] or cow origin have been reported [6] but not developed into general methods. With the rapid progress of genetic engineering techniques, direct antibody gene cloning may become the method of choice for the future derivation of monoclonal antibodies (Chap. 3).

2. Derivation of Murine Monoclonal Antibodies by Somatic Cell Fusion

Somatic cell fusion is the process by which two somatic (body) cells are fused together to give a viable hybrid cell that carries some of the genotypic and phenotypic characteristics of the two parent cells. For the derivation of hybridomas, poly(ethylene glycol) (PEG) is used to fuse a large number of spleen cells from an immunized donor with cells of a suitable myeloma line.

Spleen cells from immunized animals can produce specific antibodies, but die after a short time in culture. Myeloma cells have been adapted to grow permanently in culture, and mutants have been isolated that lack either the enzyme hypoxanthine guanine ribosyltransferase (azaguanine-resistant) or thymidine kinase (bromodeoxyuridine-resistant). Unlike normal cells, they are unable to utilize hypoxanthine or thymidine for DNA replication. Such mutants therefore cannot grow in media containing aminopterin and supplemented with hypoxanthine and thymidine (HAT-selective media); aminopterin blocks the utilization of essential amino acids for the synthesis of these purine and pyrimidine bases. Hybrids between these mutant cells and spleen cells can therefore be selected from the parents because they are the only cells that actively multiply in HAT-selective media.

From the growing hybrids, individual clones can be selected that secrete the desired antibodies. Such antibodies are thus of monoclonal origin. The selected clones, like ordinary myeloma lines, can be maintained indefinitely.

A scheme for the general procedures involved in the derivation of monoclonal antibodies is illustrated in Figure 2. Overall the process takes between three months and one year depending on the antigen and on the monoclonal antibody characteristics required. A number of well-defined steps can be identified:

1) The animal is immunized with the antigen of choice
2) The myeloma cells are prepared in sufficient quantities and proper growth phase (spinner culture)
3) Spleen cells and myeloma cells are fused and hybrids selected by the use of HAT medium
4) Antibodies of particular interest are isolated and established in culture by repeated cycles of cloning and screening
5) Frozen stocks of hybridomas are prepared throughout the procedure so that if individual cultures are intentionally or accidentally stopped it is possible to go back to earlier stocks
6) Selected clones are propagated in vitro or in vivo for large-scale production of monoclonal antibody and for the preparation of consistent banks of vials that can be kept below −80 °C for several years

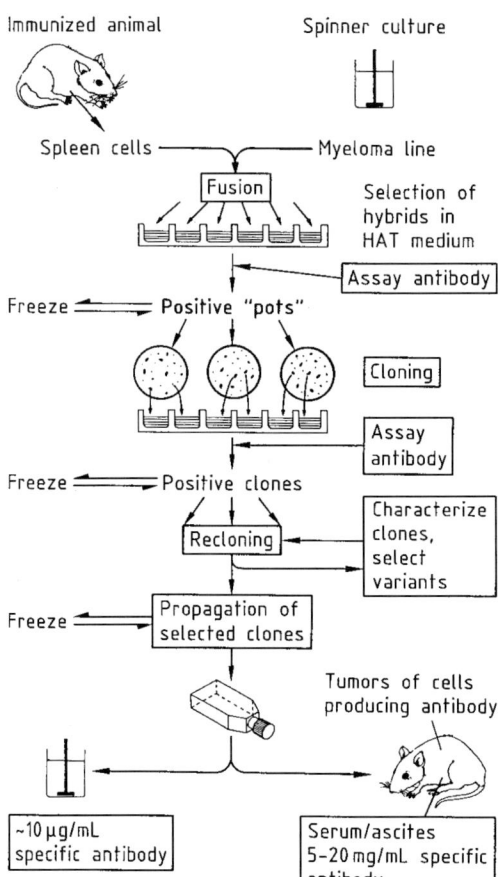

Figure 2. The process used for the derivation of monoclonal antibodies (reproduced with permission from [7])

Although this basic scheme has remained unchanged since 1976, a large number of variations have been introduced; for efficient derivation of specific monoclonal antibody the procedure must be carefully planned and prepared.

Cells and Fusion. The vast majority of monoclonal antibodies are made from immunized mice, but rats are also used. The choice of animal strains and the immunization procedure depend on the antigen to be used and on the requirements for the monoclonal antibody product.

A number of mouse and rat myeloma cell lines have been developed specifically for hybridoma derivation (Table 1). They differ in their specific drug resistance and their ability to secrete myeloma proteins. Since myelomas originate from lymphocytes, they generally secrete immunoglobulins of undefined specificity (hence the term myeloma proteins rather than antibody). Hybridomas derived from them secrete both the antibodies derived from immunized spleen cells and the myeloma protein. The two sets of immunoglobulin are difficult to separate, and it is generally a considerable

Table 1. Myeloma cell lines used for hybridoma derivation

Cell line	Strain	Immunoglobulin expression *	References
Mouse Lines			
P3-X 63 Ag.8	BALB/c	mH/mL	[3]
X 63 Ag 8.653	BALB/c	none	[8]
NSI/1 Ag 4.1	BALB/c	mL (nonsecreted)	[9]
NS 0/U	BALB/c	none	[10]
SP 2/0-AG 14	BALB/c	none	[11]
MPC 11.45.6 Tg 1.7	BALB/c	mL	[12]
MPC 11.45.6 Tg 1.7.5	BALB/c	mL	[13]
S 194/5.XX 0.BuI	BALB/c	none	[14]
Rat Lines			
IR 983 F	LOU	none	[15]
Y 3 Ag 1.2.3	LOU	rL	[16]
YB 2/0	(LOU×AO)	none	[17]

* mH, mL, and rL refer to mouse heavy, mouse light, and rat light immunoglobulin chains of myeloma origin (myeloma proteins).

advantage to use mutant myelomas that no longer make myeloma proteins but are still able to secrete antibody when fused with spleen cells. A number of protocols have been developed for cell fusion [10]. Most are based on mixing spleen cells and myeloma cells and treating the mixture for a few minutes with poly(ethylene glycol) solutions. An important alternative to the use of a chemical fusing agent is the use of a strong electric field [18].

Screening. Fusion results in a viable mixture of residual nonfused spleen cells and a few hybridomas (generally 1 for every 10 000 spleen cells). This mixture is aliquoted in a number of cultures and incubated with selective media. In 5–20 days, the only remaining cells in the cultures will be hybridomas. Intensive screening is then necessary to identify the cultures that contain a hybridoma secreting the antibody of interest. Screening consists of harvesting spent culture media and subjecting these to immunological tests to detect the presence of specific antibody.

The choice of assay used for the screening stage is extremely important. Different screening assays detect antibodies with different characteristics, e.g., different antigen specifity, cross-reactivity with molecules similar to the specific antigen, different binding affinity, and different antibody effector functions. Immunologists have developed an enormous variety of methods for detecting the presence of antibodies ranging from precipitation reactions and radioimmunoassays to assays based on the biological activity of the recognized antigen [19]. These methods can very often be adapted.

Binding assays are probably most generally used on account of their specificity, accuracy, and versatility. Insoluble antigens (either naturally insoluble or absorbed on an appropriate solid support) are allowed to react with the antibody in the culture fluid. The free antibody is washed away. The amount of monoclonal antibody bound is measured directly (direct binding assay) or by binding a second, labeled antibody capable of recognizing the first (indirect binding assay). This second antibody can be labeled in several ways: radioactively labeled, fluoresceinated, and enzyme-linked derivatives are the most common.

Although less extensively used, *assays based on the biological activity* of the antigen are of particular importance because they allow the detection of antibody to antigens that are only identified by their biological activity but cannot be chemically purified. The first example of this was the production of monoclonal antibody to human interferon [19]. A biologically active antigen preparation can be added to the culture supernatants to be screened. After incubation a decrease in biological activity is assumed to be preliminary evidence for the presence of inhibitory antibody. Alternatively, the precipitation of antibody–antigen complexes can be effected (for instance by addition of anti-mouse or anti-rat immunoglobulin) thus removing part of the biological activity from the supernatant.

Cloning. Once cultures of hybridomas secreting specific antibody are identified, they can be cloned (i.e., individual cells physically separated and propagated in individual cultures). The best way is to culture the cells embedded in a semisolid agar-containing medium [10], [20]. Clones are generally rescreened, often using a battery of screening assays to select clones meeting more stringent criteria of antibody production and specificity.

Probably the most important feature of hybridoma technology is that immortalization by cell fusion and cell cloning allow the preparation of antibody specific for an individual antigenic determinant present in a mixture, without the need for purifying the antigen.

When an animal is injected with an antigenic mixture, its usual response is the production of a highly heterogeneous population of antibody directed against the immunogen. Some of these antibodies are specific for various components of the antigenic mixture and are mixed with many others that express other properties. When monoclonal antibodies are prepared, the collection of randomly derived clones represents a cross-section of such a heterogeneous population. The strength of the response to each individual antigenic determinant is reflected in the hybridoma population producing antibody to that determinant (Fig. 3).

Thus the derivation of monoclonal antibody to antigens present in a complex mixture has become a powerful tool for the definition and identification of previously unknown components of the mixture. The most spectacular example is the derivation of a very large number of monoclonal antibodies for mammalian cell surface antigens to define or "map" the structure of the cell surface [22].

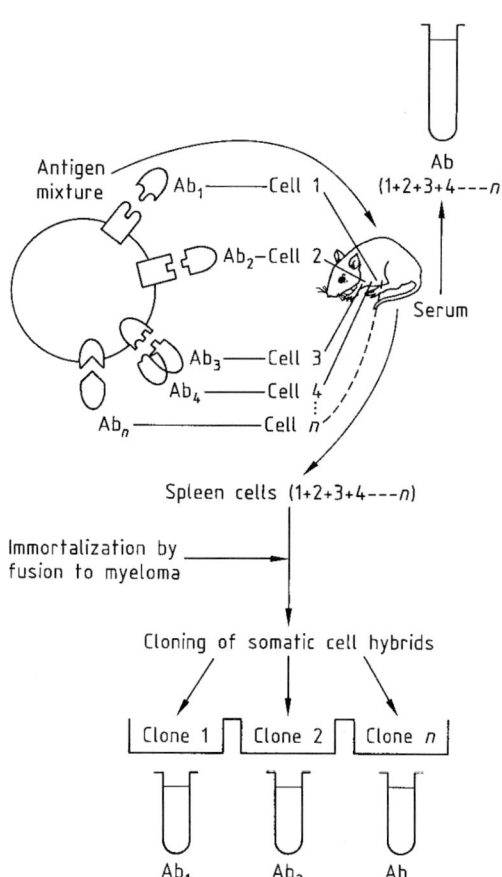

Figure 3. Antibody (Ab) production by animals and by hybridomas (reproduced with permission from [21])
Isolation of hybridoma clones allows the preparation of monospecific antibodies starting from a complex mixture of antigens.

3. Human Monoclonal Antibodies

For many potential medical applications of monoclonal antibodies the ultimate goal is a general method for the production of human monoclonal antibodies. Although several instances of human monoclonal antibody production have been reported, two main problems have been encountered in the development of a general methodology. The first is the problem of obtaining highly immunized human lymphocytes for use in the fusion. The second is the production of a monoclonal antibody in the absence of suitable myeloma partners.

A key feature of the standard technique for mouse monoclonal antibody production is the ability to repeatedly inject laboratory mice with the specific antigen. Although this is sometimes possible in humans (e.g., tetanus toxoid vaccination), it is clearly unethical to immunize humans with most antigens. The need for alternative sources of immunized human lymphocytes has led to the development of two new techniques that avoid the requirement of injection of the antigen into humans. The first consists of adding the antigen to a suspension of human blood lymphocytes in a culture medium that allows the necessary differentiation of the lymphocytes to antibody-producing cells [23]. Although not yet a routine operation, this in vitro immunization method has become more feasible with the availability of recombinant growth factors and lymphokines [23]. A more recent and promising advance is the discovery that human lymphocytes can be grown in a strain of mice suffering from Severe Combined Immune Deficiency disease (SCID mice). Human lymphocytes are used to repopulate the immune system of the mouse, and the animal responds to immunization by making human antibodies. Human cells growing in the spleen of such an animal have been fused with myeloma cells to produce a human monoclonal antibody [24].

This exciting development may open the door to the first routine method for the derivation of human monoclonal antibodies. A step beyond transfer of human lymphocytes to a mouse host would be the transfer of the genes coding for human antibodies. This may soon become reality with the development of transgenic mice [25].

However immunization is performed, a second major problem in the production of human monoclonal antibody is the absence of suitable myeloma partners. Mouse–human heterohybridomas are unstable and none of the available cell lines has become generally accepted. Despite considerable efforts, no reliable, consistent methodology based on somatic cell fusion alone has emerged. Alternative approaches to lymphocyte immortalization have therefore been explored.

Even before the advent of hybridoma technology the natural transformation of human lymphocytes by viruses had been used to derive stable cell lines. In most cases, however, these were unable to secrete useful amounts of antibody over a long period of time. The combination of viral transformation followed by somatic cell fusion has resulted in the most successful technique presently available for the derivation of human monoclonal antibodies [26].

Due to the above problems, the production of human monoclonal antibodies by hybridomas is likely to be based on genetic engineering. Antibodies of all species have relatively independent protein domains of about 110 amino acids folded in a characteristic shape (Fig. 1). Although the amino acid sequence is species-specific, the domain shape is common to all known antibodies and the domain carrying the antigen-binding specificity (V domains in Fig. 1) can be transplanted from the antibody of one species to another. Although this cannot be achieved directly on the protein, recombinant DNA technology has allowed antibody genes to be cloned and antibody gene domains to be reshuffled. The engineered genes are then expressed in a suitable mammalian host cell line that can be grown like a hybridoma. An example of a "reshuffled" antibody is the "chimeric" antibody that consists of the antigen-binding domain of the mouse monoclonal antibody B 72.3 attached to the remaining domains from a human antibody. The original (mouse) B 72.3 recognizes an antigen present on human tumor cells; the chimeric antibody has identical binding properties to the mouse antibody and is now being tested in clinical trials as a potential therapy for colon cancer [27]. Chimerization of antibodies has become a subject of considerable industrial interest.

A recent refinement of this technology has been achieved by grafting onto a human antibody molecule only those parts of the antigen-binding domain that are actually in physical contact with the antigen [28]. Whether this will prove to be a practical improvement for chimeric antibodies remains an open question.

4. Production

Monoclonal antibodies accumulate in the spent media of growing hybridomas at an average concentration of 10–100 mg/L. The productivity is a characteristic of individual clones, but is strongly affected by culture conditions. Under optimal conditions productivity may reach 500–600 mg/L.

Small volumes of supernatant can be collected either from stationary cultures ("flasks" of up to 1 L) or from slowly rotating bottles ("rollers" of up to 3 L). These conditions are generally satisfactory only for laboratory-scale production (<1 g). For larger quantities two alternatives are available: the hybridomas are either grown as tumors in animals or in an industrial-scale cell-culture plant.

Whichever culture technique is used, a number of important considerations apply:

1) The system must be effectively sterilized before introduction of cells, and maintained free of microorganism contamination for prolonged periods (months for continuous cultures). Small vessels may be sterilized by autoclaving, but large production plant must be sterilized in situ, generally by steaming at high temperature and pressure. Materials and engineering standards have to be of high quality.
2) For effective cell growth and antibody production, the system must provide adequate mixing. Heat and mass transfer characteristics must be sufficient to maintain correct temperature, pH, dissolved gas and nutrient concentrations.
3) Maximum economy of scale cannot be achieved by simply increasing the number of production units.

Production in Animals. When hybridomas are grown in animals (either as solid tumor or as ascites), the monoclonal antibody accumulates in the blood or in the peritoneal cavity at very high concentration (5–20 mg/L) but the volume that can be harvested from each animal is rather small and therefore large numbers of animals are required (around 20 000 mice to produce 1 kg of antibody).

Production in Cell Cultures. In addition to the ethical considerations, production in cell cultures has other considerable advantages over production in animals:

1) Monoclonal antibody purified from blood is always contaminated with other antibodies of the same species and class, naturally produced by the animal, and these are difficult to eliminate.
2) The risk of contamination with rodent viruses is eliminated.

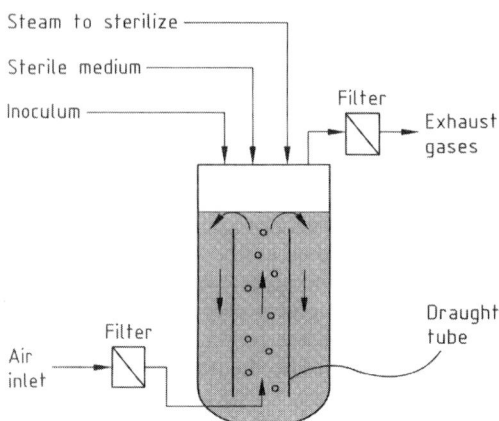

Figure 4. Schematic diagram of an airlift fermenter

3) The process can be automated and made reproducible
4) Scaling up the process by increasing the number of animals does not provide the economy of scale offered by scaling up of industrial fermentation.

With product safety legislation becoming ever more stringent it may soon even be impossible to produce antibodies for therapeutic use in animals.

A wide variety of systems have been developed for in vitro production of monoclonal antibody and are treated in depth in [29]. The most relevant methods will only be summarized here. They depend on growth in free suspension (homogeneous suspension cultures) or as immobilized cell cultures. More than 50% of current monoclonal antibody production is probably performed with homogeneous suspension cultures.

Homogeneous Suspension Culture. Two types of homogeneous suspension systems have been developed and are used extensively: continuously stirred deep-tank fermenters (originally developed for microorganisms) and airlift fermenters. Although not yet extensively used for monoclonal antibody, stirred deep-tank fermenters (up to 80 000-L capacity) have been used to grow lymphoblastoid cells for interferon production [30]. The airlift fermenter has an internal, concentric draught tube (Fig. 4). Gas mixture introduced from a sparge ring inside the base of the tube causes the culture medium to circulate. Control of gas pressure controls the circulation, and changes in gas mixture composition are used to adjust the pH and dissolved oxygen tension. Airlift fermenters of up to 2000-L capacity have been used for monoclonal antibody production [31].

Suspension cultures are usually operated in a batch mode but may also be used for continuous culture in which a feed of fresh medium is added to the culture over a prolonged period and the culture is at the same time continuously harvested at an appropriate scale to maintain a steady state [32].

Immobilized Cell Cultures. Various systems have been developed in which the cells are immobilized on a solid matrix (microcarrier), in microbeads, in hollow-fiber reactors, or on ceramic or glass matrices.

The use of *microcarriers* was first developed for bulk cultivation of anchorage-dependent cells [33] but microcarriers with surface charge properties suitable for the absorption of suspension cells are now used. Suitable microcarrier materials include cross-linked dextran, collagen, gelatin, glass, plastic, cellulose, and even droplets of a liquid that is immiscible with the culture medium (perfluorinated hydrocarbons, [34]). Cells immobilized on microcarriers can be grown in either perfusion culture systems or in fluidized-bed reactors.

In *microbead culture*, cells are encapsulated in semipermeable microbeads of agarose or alginate. The technique of alginate encapsulation, pioneered by LIM and SON [35] has been developed for industrial-scale cultivation by Damon Corporation [36]. As with microcarriers, encapsulation of cells lends itself to perfusion culture. Cells can grow to very high population densities and crude products can be retained at high concentration within the capsules.

Monoclonal antibodies have also been produced using *hollow-fiber reactors* which consist of loose bundles of semipermeable capillary fibers enclosed within a culture chamber. Cells grow within the culture chamber in the extracapillary space while the culture medium is circulated through the capillary. Gases, nutrients, and catabolites of low molecular mass diffuse between the culture chamber and the inside of the fibers along their respective concentration gradients; macromolecular products including the monoclonal antibodies are retained within the culture chamber from where they can be harvested at relatively high concentration [37].

The hollow-fiber principle is also used in the flat-bed membrane bioreactor [38] which contains three chambers: a culture chamber bounded on one side by a perfusion chamber and on the other by a harvest chamber. A semipermeable membrane separates the central culture chamber from the perfusion chamber through which a feeding medium is circulated. The other side of the culture chamber is separated from the harvest chamber by a microporous membrane, which allows the passage of macromolecules but not cells. Macromolecules, including the desired antibody product can thus be recovered from the harvest chamber.

Ceramic or glass matrices have also been developed for the immobilization of hybridomas [39]. They are used in reactors in which the culture chamber is filled with porous rigid honeycomb units or beads.

5. Purification

Antibody can be recovered from cell cultures or from ascitic fluid by a wide variety of methods depending on the purity required. For preparation of diagnostic reagents, ascitic fluids or concentrated spent culture media can often be used directly, although some simple purification is generally used. High-purity material is, however, necessary when the monoclonal antibody is to be injected into patients for diagnostic (e.g., tumor imaging) or therapeutic purposes. Levels of contaminants such as proteins, nucleic

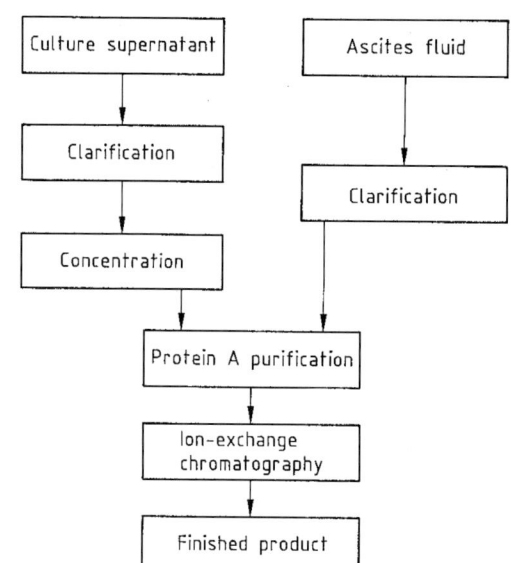

Figure 5. Flow sheet of monoclonal antibody production

acids, and pyrogens must also be reduced to acceptable levels. Furthermore the purification process must be validated to confirm that it removes contaminating substances irrespective of whether their presence can be demonstrated in the antibody source.

To achieve high purity, control of the environment in which the purification process is run is essential. Equipment used exclusively for monoclonal antibody production must be installed in a purpose-built facility.

The most important purification methods will only be summarized, more details can be found in specialized references [40]. A typical flow sheet is shown in Figure 5. In most cases the first step in antibody purification is some form of *clarification* to eliminate macroimpurities (lipids in ascites and cell debris from culture supernatants). This is generally achieved by either filtration and/or centrifugation. Culture supernatants are then usually concentrated by ultrafiltration, or (for smaller scale production) by protein precipitation.

Nearly all published antibody purification processes employ a chromatographic step. One of the most common methods is *ion-exchange chromatography* because cellulose-based ion exchangers are economical and their use in protein purification is well documented. Each monoclonal antibody has an isoelectric point in the pH range 4.7–8.3 and this dictates the operating conditions and the particular matrix for optimal purification.

Since most serum proteins are acidic, ion exchange is particularly favorable for separating the basic antibodies. Ion exchange has been used both as a single purification step to achieve relatively pure antibodies used for diagnostic or immunopurification purposes, and as a final step to clean affinity-purified antibody for therapeutic purposes.

Purification of monoclonal antibody by *affinity chromatography* using, for example, Protein A Sepharose (Pharmacia) is another very popular method. This offers the advantages of giving a high degree of purity and high yield in a single step, even from very dilute supernatants. Not all antibody subclasses and species bind equally well to protein A. However, since the interaction of antibody with protein A is primarily hydrophobic, the introduction of a buffer with a high salt concentration enhances binding considerably and effective purification processes for most monoclonal antibodies can be designed [41]. Affinity chromatography is best used as the first chromatographic purification step due to its very high purification power. Although the matrix is very expensive, the columns can be used for over 100 cycles. Large quantities of monoclonal antibody are commonly purified by using several purification cycles for each batch. Thus, with a 500-mL protein A column of 4 g capacity, up to 400 g of antibody can be purified in 100 cycles. Automation reduces costs and allows fine tuning of the purification process.

Affinity chromatography matrices are not easy to sterilize. Although Protein A Sepharose columns can be sterilized with a variety of agents (e.g., guanidine hydrochloride, hibitane and benzyl alcohol, propylene oxides) the purification process is usually designed to minimize the need for repeated sterilization. All solutions are sterilized before they come into contact with the column.

Although very high purity may be obtained in a single step by affinity chromatography, antibody for therapeutic use generally requires further purification to reduce the levels of contaminants to acceptable limits. This is often achieved by ion-exchange chromatography to bind the impurities. Because of high concentration and relatively small volumes involved, better antibody yields are obtained than if ion exchange is used at the outset.

6. Monoclonal Antibody Patents

The opportunity to obtain patent protection relating to hybridoma–monoclonal antibody technology was missed when the original publication was allowed to proceed without a patent application being filed [3]. With the benefit of hindsight, it is likely that a generic technology patent position could have been established with patent terms extending into the mid-1990s.

Subsequently, attempts were made to establish broad monoclonal antibody patents directed to classes of antigens, e.g., the Koprowski–Wistar patent applications relating to monoclonal antibodies to viral antigens and tumor-associated antigen [42]. Patents were granted on these filings in the United States, although the corresponding patents were rejected by the United Kingdom Patent Office. Broad patent applications were also filed by Hybritech directed to the use of monoclonal antibodies in immunoassays [43].

Again, patents have been granted in the United States and have now been upheld by the United States Courts in patent infringement actions. The corresponding European Hybritech patent would, however, probably be held invalid if challenged. The difference reflects the United States "First to Invent" and European "First to File" patent law.

To obtain patent protection, it is necessary to have identified a new, nonobvious invention over the "State of the Art" at the initial filing date. Furthermore once monoclonal antibody technology had been established, its application to a particular known antigen is unlikely to give rise to a patentable invention. Thus, following the early broad patents of Koprowski and Hybritech, the scope for other broad monoclonal patents was somewhat limited. Early exceptions, however, were the patents relating to T-cell markers (i.e., Ortho OKT series patents [44]) where the application of hybridoma–monoclonal antibody technology resulted in the identification and dissection of a new class of antigen, and the Secher and Burke patent where special problems had been encountered in obtaining monoclonal antibody to interferon [45].

Following on from these precedents, monoclonal patent opportunities are likely to arise in three main circumstances:

1) The identification of subsets of monoclonal antibodies with special benefits over other monoclonal antibodies for the antigen in question; for instance, as the result of identification of a critical antigenic epitope.
2) The identification of an individual hybridoma cell line. Although specific cell lines have been successfully patented in this way, it is generally thought that the protection afforded by such a patent can be more effectively achieved by controlling the distribution of the cell line.
3) The development of new, useful technologies for preparation of monoclonal antibodies. An exciting opportunity would be a satisfactory solution to the problem of deriving human monoclonal antibodies.

7. Uses

Explosive growth in the annual number of new scientific publications relating to monoclonal antibodies (Fig. 6) reflects the enormous diversity of the uses of monoclonal antibodies. Extensive reviews can be found in [40], [46].

In this short section a few general statements will be made, followed by a highly selected but illustrative list of applications.

Research Applications. The opportunity that monoclonal antibodies offers to dissect complex antigenic mixtures (Chap. 2) is the key to the dramatic impact that the technology has had on biological research. Immunology, neurobiology, and embryology for example have all witnessed great leaps forward in the past decade as a result of the application of hybridoma technology to describe the complex structure of cell surfaces. Disciplines less familiar with immunology (e.g., botany and chemistry) were slower to realize the full potential of monoclonal antibodies, but are now catching up rapidly.

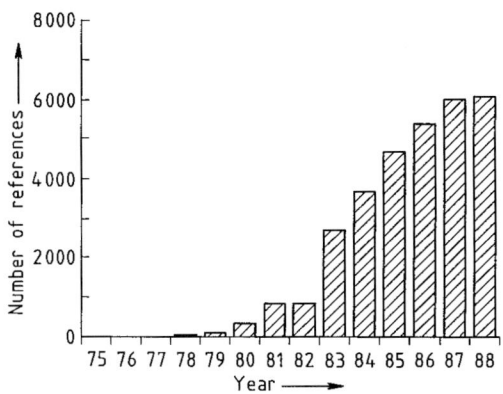

Figure 6. Number of scientific publications relating to monoclonal antibodies from 1975 to 1988 (Medline scientific references data bank)

In immunoassays, such as that used to measure thyroid hormones, monoclonal antibodies have almost completely replaced conventional antisera. Furthermore, because of their availability in large quantities and defined specificity and affinity, monoclonal antibodies have allowed the extension of immunoassay techniques into areas that were previously impractical. For example, assays using labeled antibodies (immunometric assay) are overtaking assays in which the antigen is labeled (radioimmunoassay).

The use of antibodies immobilized onto chromatographic matrices for immunopurification had previously been described using polyclonal antibodies but only with monoclonal antibodies has this technique become of general practical use, e.g., for the purification of interferon [19]. The exquisite specificity of monoclonal antibodies has also had an impact on the reliability and quality of immunohistochemistry.

Monoclonal antibodies are now also used as pharmacological tools to neutralize biologically active molecules or to interfere with specific ligand–receptor interactions.

Clinical Practice. The first important clinical use of monoclonal antibody was for the typing of human blood.

Diagnostics Monoclonal antibodies have allowed the development and commercial introduction of prepacked "kits" for the measurement of a variety of clinically relevant molecules. The introduction of diagnostic methods relying on enzyme rather than radioisotope labels has led to a rapid expansion of immunoassays in clinical practice.

Within clinical laboratories, the tendency towards full automation has been greatly accelerated by the introduction of monoclonal antibodies. Of equal or even greater importance is the development of simple kits that can be used in the doctor's office or even at home. The best example of this trend is the phenomenal growth in the market for home-based pregnancy detection and fertility testing. Another growth area is the development of tests for infectious disease agents. Here the specificity of monoclonal antibodies has been of paramount importance in developing tests of high diagnostic efficiency.

Rapid growth has also been seen in oncology, where tests for cancer markers, which are soluble tumor-related substances in the blood, can be used to monitor the progress of tumor therapy. Although not yet widely available, a different diagnostic approach for cancer is tumor imaging. This consists of injecting the patient with tumor-specific monoclonal antibody labeled with a γ-ray emitting radioisotope. Tumors can then be detected by γ-ray cameras after the labeled antibody has targeted to the cancer cells. This methodology may soon be extended to the detection of other abnormalities, such as atherosclerotic plaques, blood clots, and sites of infection.

Therapy. Compared to the success of diagnostic monoclonal antibodies the development of therapy based on monoclonals has been disappointing.

Proposed therapeutic uses of monoclonal antibodies range from the replacement of conventional antisera (e.g., in passive immunization against rabies) to more speculative novel applications (e.g., in the control of autoimmune disease; see also →Immunotherapy and Vaccines). In practice only a single monoclonal antibody for therapy has received a product licence in the United States (Orthoclon, Ortho Pharmaceutical Corporation). Orthoclon is an antibody to T-lymphocytes that is used to suppress patients' immune response following kidney transplantation, thus reducing the incidence of organ rejection. Antibodies are in clinical trials in three main areas:

1) The treatment of cancer using antibodies to target radioisotopes.
2) The use of monoclonal antibodies linked to cytotoxic drugs or biological toxins such as ricin for "purging" bone marrow before transplantation to remove leukemic cells or T-lymphocytes that are responsible for Graft versus Host Disease.
3) Neutralization of noxious substances such as rabies virus, bacterial endotoxins, or Tumor Necrosis Factor.

Four major development areas may bring about more widespread introduction of therapeutic monoclonal antibodies:

1) The problem of making human monoclonal antibodies will be solved by the approaches described in Chapter 3. This will allow repeated administration of monoclonal antibodies without the adverse reactions currently experienced.
2) The limited access of monoclonal antibodies to tumors and the prolonged residence time in the blood will be improved by genetically engineering antibody fragments.
3) These "cut-down" antibodies may be further modified by replacing the non-antigen-binding parts of the antibody molecule with other biologically active proteins or protein fragments.
4) Improvements in expression systems and large-scale production and purification of monoclonal antibodies will drastically reduce their manufacturing cost, particularly if the production of monoclonal antibody fragments in microorganisms can become generally applicable [47], [48].

8. References

[1] D. S. Secher, "Structure of immunoglobulins," in E. S. Lennox (ed.): *Defense and Recognition*, 2nd ed., University Park Press, Baltimore 1979, pp. 1–48.

[2] G. Galfre, G. W. Butcher in T. L. Wang (ed.): *Immunology in Plant Science*, University Press, Cambridge 1986, p. 5.

[3] G. Kohler, C. Milstein, "Continuous cultures of fused cells secreting antibody of predefined specificity," *Nature (London)* **256** (1975) 495–497.

[4] F. Sanchez-Madrid, P. Szklut, T. A. Springer, "Stable hamster – mouse hybridoma producing IgG and IgM hamster monoclonal antibodies of defined specificity," *J. Immunol.* **130** (1983) 309–312.

[5] T. J. G. Raybould, M. Takamashi, "Production of stable rabbit – mouse hybridomas that secrete rabbit monoclonal antibody of defined specificity," *Science (Washington D.C.)* **240** (1988) 1788–1790.

[6] D. M. Strickland, S. A. Saeed, M. L. Casey, M. D. Mitchell, "Bovine mouse hybridomas that secrete bovine immunoglobulin Gl.," *Science (Washington D.C.)* **220** (1982) 522–524.

[7] G. Galfre, C. Milstein, "Preparation of monoclonal antibodies: strategies and procedures," *Methods Enzymol.* **73** (1981), part B, "Immunochemical techniques".

[8] J. F. Kearney, A. Radbruch, B. Liesegang, K. Rajewsky, "A new mouse myeloma cell line that has lost immunoglobulin expression but permits the construction of antibody-secreting hybrid cell lines," *J. Immunol.* **123** (1979) 1548–1550.

[9] G. Köhler, S. C. Howe, C. Milstein, "Fusion between Ig-secreting and non-secreting myeloma cell lines," *Eur. J. Immunol.* **6** (1976) 292–295.

[10] G. Galfre, C. Milstein, "Preparation of monoclonal antibodies: strategies and procedures," *Methods Enzymol.* **73** (1981) part B, "Immunochemical techniques" 3–46.

[11] M. Shulman, C. D. Nilole, G. Köhler, "A better cell line for making hybridomas secreting specific antibodies," *Nature (London)* **276** (1978) 269–270.

[12] M. L. Gefter, D. H. Margulies, M. D. Scharff, "A simple method for polyethylene glycol-promoted hybridisation of mouse myeloma cells," *Somatic Cell Genet.* **3** (1977) 231–236.

[13] J. Sharon, S. L. Harrison, E. A. Kabat, "Formation of hybridoma clones in soft agarose: effect of pH and of medium," *Somatic Cell Genet.* **6** (1980) 435–441.

[14] I. S. Trowbridge, "Interspecies spleen – myeloma hybrid producing monoclonal antibodies against mouse lymphocyte surface glycoprotein, T 200," *J. Exp. Med.* **148** (1978) 313–323.

[15] H. Bazin, "Production of rat monoclonal antibodies with the lou rat non secreting IR 983F myeloma cell line," C. I. Hybridoma Technology, 615–618, reprinted from: Protides of the Biological Fluids 29th Colloqium 1981-Ed-H Peeters-Pergamon Press, Oxford-New York 1982.

[16] G. Galfre, C. Milstein, B. Wright, "Rat x rat hybrid myelomas and a monoclonal anti-Fd portion of mouse IgG," *Nature (London)* **277** (1979) 131–133.

[17] J. V. Kilmartin, B. Wright, C. Milstein, "Rat monoclonal antitubulin antibodies derived by using a new non-secreting rat cell line," *J. Cell. Biol.* **93** (1982) 576–582.

[18] V. Zimmermann, "Electrofusion of cells: principles and industrial potential," *Trends Biotechnol.* **1** (1983) 149–155.

[19] D. S. Secher, D. C. Burke, "A monoclonal antibody for large-scale purification of human leucocyte interferon," *Nature (London)* **285** (1980) 446–450.

[20] D. M. Weir: *Handbook of Experimental Immunology*, 4th ed., Blackwell Scientific Publications, Oxford 1986.

[21] C. Milstein, G. Galfre, D. S. Secher, T. Springer, "Monoclonal antibodies and cell surface antigens," *Hum. Genet.: Possibilities Realities,* Excerpta Medica, June 1979.

[22] W. Knapp et al., "Towards a better definition of human leucocyte surface molecules," *Immunol. Today* **10** (1989) no. 8.

[23] C. A. K. Borrebaeck, L. Danielsson, S. A. Moller, "Human monoclonal antibodies produced by primary in vitro immunisation of peripheral blood lymphocytes," *Proc. Nat. Acad. Sci. U.S.A.* **85** (1988) 3995–3999.

[24] J. M. McCune et al., "The SCID-hu mouse: murine model for the analysis of human hematolymphoid differentiation and function," *Science (Washington D.C.)* **241** (1988) 1632–1639. D. E. Mosier, R. J. Gulizia, S. M. Baird, D. B. Wilson, "Transfer of a functional human immune system to mice with severe combine immunodeficiency," *Nature (London)* **335** (1988) 256–259.

[25] U. Storb et al., "Expression of immunoglobulin genes in transgenic mice and transfected cells," *Ann. N.Y. Acad. Sci.* **(5 NM) 546** (1988) 51–56.

[26] K. James, G. T. Bell, "Human monoclonal antibody production. Current status and future prospects," *J. Immunol. Methods* **100** (1987) 5–40.

[27] D. Colcher et al., "Characterization and biodistribution of recombinant and recombinant/chimeric constructs of monoclonal antibody B 72.3," *Cancer Res.* **49** (1989) 1738–1745.

[28] L. Reichmann, M. Clark, H. Waldmann, G. Winter, "Reshaping human antibodies for therapy," *Nature (London)* **332** (1988) 323–327. M. Verhoeyen, L. Riechmann, "Engineering of antibodies," *BioEssays* **8** (1988) no. 2, 74–78.

[29] J. Feder, W. R. Talbert (eds.): *Large-Scale Mammalian Cell Culture,* Academic Press, New York 1985.

[30] K. Pullen et al., *Devel. Biol. Stand.* **70** (1988) 175–177.

[31] M. Rhodes, J. R. Birch, "Large-scale production of proteins from mammalian cells," *Bio/Technology* **6** (1988) 518.

[32] J. R. Birch et al., "The industrial production of monoclonal antibodies in cell culture," in J. M. Cardoso Duarte, L. J. Archeg, A. T. Bull, G. Holt (eds.): *Prospective in Biotechnology,* Plenum Publishing, New York 1987, 97–109.

[33] A. L. Van Wezel, "Growth of cell strains and primary cells on micro-carriers in homogeneous culture," *Nature (London)* **216** (1967) 64–65.

[34] C. R. Keese, I. Giaever, "Cell growth on liquid interfaces: Role of surface active compounds," *Proc. Nat. Acad. Sci. U.S.A.* **80** (1983) 5622–5626.

[35] F. Lim, A. Sun, "Microencapsulated islets as bioartificial endocrine pancreas," *Science (Washington D.C.)* **210** (1980) 908–910.

[36] R. G. Rupp, "Use of cellular microencapsulation in large-scale production of monoclonal antibodies," in J. Feder, W. R. Talbert (eds.): *Large-Scale Mammalian Cell Culture,* Academic Press, New York 1985.

[37] P. Calabresi et al., "Monoclonal antibody projection in artificial capillary cultures," *Proc. Am. Assoc. Cancer Res.* **22** (1981) 362–365.

[38] G. Klement, N. Scheiver, H. N. D. Katinger, "Construction of a large-scale membrane reactor system with different compartments for cells, medium and product," *Devel. Biol. Stand.* **66** (1987) 221–226.

[39] B. K. Lydersen et al., "The use of a ceramic matrix in a large scale cell culture system," in J. Feder, W. R. Talbert (eds.): *Large Scale Mammalian Cell Culture,* Academic Press, New York 1985, pp. 39–58.

[40] A. C. Kenney, "Large-scale production of monoclonal antibodies," *Monoclonal Antibodies: Production and Application,* Alan R. Liss Inc., New York 1989, pp. 143–160.
[41] H. Juarez-Salinas, G. S. Ott, US 4 704 366, 1987.
[42] H. Koprowski, C. M. Croce, US 4 172 124, 1979.
[43] G. S. Davis, H. E. Greene, UK 2 086 041 A, 1981; US 4 376 110, 1983.
[44] Ortho Diagnostic, US 4 361 549, US 4 361 550, 4 364 932, 4 364 933, 4 364 934, 4 364 935, 4 364 936, 1982 (P. C. Kung, G. Goldstein).
[45] D. S. Secher, US 4 423 147, 1983; US 4 514 507, 1985.
[46] J. W. Golding: *Monoclonal Antibodies: Principles and Practice. Production and Application of Monoclonal Antibodies in Cell Biology, Biochemistry and Immunology,* Academic Press, New York 1987.
[47] E. S. Ward et al., "Binding activities of a repertoire of single immunoglobulin variable domains secreted from Escherichia coli," *Nature (London)* **341** (1989) 544.
[48] W. D. Huse et al., "Generation of a large combinatorial library of the immunoglobulin repertoire in phage," *Science (Washington D.C.),* in press.

Ophthalmological Agents

Terence M. Dolak, Bausch & Lomb, Rochester, New York 14 692, United States
O. William Lever, Jr., Bausch & Lomb, Rochester, New York 14 692, United States
David Marsh, Bausch & Lomb, Rochester, New York 14 692, United States
Irene Moran, Bausch & Lomb, Rochester, New York 14 692, United States
Scott Sutton, Bausch & Lomb, Rochester, New York 14 692, United States

1.	Introduction	2002
2.	Anesthetics	2002
2.1.	Topical Anesthetics	2003
2.2.	Regional Anesthetics	2003
3.	Antimicrobial Agents	2004
3.1.	Antibacterial Agents	2004
3.1.1.	Sulfonamides	2004
3.1.2.	β-Lactam Antibiotics	2006
3.1.2.1.	Penicillins	2006
3.1.2.2.	Cephalosporins	2006
3.1.3.	Tetracycline Antibiotics	2007
3.1.4.	Aminoglycoside Antibiotics	2007
3.1.5.	Polypeptide Antibiotics	2009
3.1.6.	Miscellaneous Antimicrobial Agents	2011
3.2.	Antifungal Agents	2012
3.3.	Antiprotozoal Agents	2014
3.4.	Antiviral Agents	2014
4.	Anti-inflammatory Agents	2015
4.1.	Corticosteroids	2015
4.2.	Miscellaneous Anti-inflammatory Agents	2018
5.	Antiglaucomatous Agents	2019
5.1.	Sympathomimetic Drugs	2020
5.2.	β-Adrenergic Blocking Agents	2021
5.3.	Carbonic Anhydrase Inhibitors	2022
5.4.	Hyperosmotic Agents	2023
5.5.	Miotics (Parasympathomimetic Agents)	2023
5.5.1.	Cholinergic Agonists	2024
5.5.2.	Anticholinesterases	2025
6.	Mydriatics and Cycloplegics	2026
6.1.	Sympathomimetics	2026
6.2.	Parasympatholytics	2027
7.	Vasoactive (Adrenergic) Agents	2029
8.	Diagnostic Agents	2030
9.	Dry Eye Medications	2032
10.	Miscellaneous Ophthalmic and Contact Lens Preparations	2035
10.1.	Disinfectants and Preservatives	2037
10.1.1.	Biguanides	2037
10.1.2.	Mercurial Preservatives	2037
10.1.3.	Quaternary Ammonium Compounds	2038
10.1.4.	Miscellaneous Disinfectants and Preservatives	2038
10.2.	Contact Lens Cleaners and Rewetting Agents	2039
10.2.1.	Surfactant Cleaners	2039
10.2.2.	Enzymatic Cleaners	2040
11.	References	2040

1. Introduction

The eye is an external organ, and is susceptible to injury, infection, and irritation, as well as to organic and functional disorders. The products and preparations for ophthalmic use therefore comprise a broad category and include medications, diagnostic agents, and medical device formulations.

Over-the-counter and prescription medications are used to treat chronic ailments (e.g., glaucoma, blepharitis, and dry eye), as well as temporary conditions (e.g., infections, allergic reactions, and injuries).

Ocular diagnostic agents are a specialized group of drugs and dyes utilized primarily for ocular examination and the identification of potential disease conditions.

Medical device products are used for the hydration, wetting, cleaning, and disinfection of contact lenses.

2. Anesthetics

Topical anesthetics diminish sensory nerve impulse generation and conduction locally, and are used to anesthetize the ocular surface during short corneal or conjunctival procedures (e.g., tonometry, removal of sutures or foreign bodies) [1]–[4]. Proxymetacaine (0.5%), oxybuprocaine (0.4%), or tetracaine (0.25–1.0%) solutions are commonly used [1]; proxymetacaine is particularly useful [4]. These agents provide rapid onset of local anesthesia of relatively short duration (e.g., one or two drops of a 0.5% proxymetacaine solution take effect within 30 s with anesthesia lasting about 15 min). Cocaine exhibits significant corneal epithelial toxicity and is now rarely used; however, it may still be employed for debridement of the corneal epithelium, as in herpetic epithelial keratitis [3], [4]. Topical application of combinations of local anesthetics does not provide additive effects but increases the risk of side effects [2].

Regional anesthetics are employed in ophthalmic surgical procedures. They are administered by injection for infiltration anesthesia or for regional nerve block (facial or retrobulbar) anesthesia [1]–[3]. Onset of action is typically 3–11 min, and anesthesia may last from 30–45 min (procaine) up to 300–600 min (etidocaine) or longer (480–720 min for bupivacaine with epinephrine) [2]. Combination of injectable anesthetics with vasoconstrictors such as epinephrine decreases the rate of systemic absorption and prolongs the duration of anesthesia.

Most of the anesthetics listed in Sections 3.1 and 3.2 are described in detail elsewhere, see → Local Anesthetics.

2.1. Topical Anesthetics

Cocaine hydrochloride [*53-21-4*], $C_{17}H_{21}NO_4 \cdot HCl$, M_r 339.8. Cocaine is not commercially available as a solution for clinical use.

Proxymetacaine hydrochloride [*5875-06-9*], 2-diethylaminoethyl 3-amino-4-propoxybenzoate monohydrochloride, proparacaine hydrochloride, $C_{16}H_{26}N_2O_3 \cdot HCl$, M_r 330.9.

Trade Names: Ak-Taine (Akorn); Alcaine (Alcon); Kainair (Pharmafair); Keracaine, Kerakain (Chibret); Ocu-caine (Ocumed); Ophthaine (Squibb); Ophthetic (Allergan).

Tetracaine hydrochloride [*136-47-0*], 2-dimethylaminoethyl 4-butylaminobenzoate monohydrochloride, $C_{15}H_{24}N_2O_2 \cdot HCl$, M_r 300.8.

Trade Names: Minims Amethocaine Hydrochloride (Smith & Nephew), Pontocaine (Winthrop).

Oxybuprocaine hydrochloride [*5987-82-6*], 2-diethylaminoethyl 4-amino-3-butoxybenzoate monohydrochloride, benoxinate hydrochloride, $C_{17}H_{28}N_2O_3 \cdot HCl$, M_r 344.9.

Trade Names: Alcon Opulets Benoxinate (Alcon); Cebesine (Chauvin-Blanche); Conjuncain (Mann); Minims Benoxinate Hydrochloride, Minims Oxybuprocaine Hydrochloride (Smith & Nephew); Novesin (Dispersa); Novesina (Sandoz); Novesine (Chibret, Wander); Oftalmocaina, Oxibuprokain Minims (Smith & Nephew); Poen Caina, Prescaina (Llorens).
Oxybuprocaine hydrochloride (0.4%) is available in combination with sodium fluorescein (0.25%) under the trade name Fluress (Sola/Barnes – Hind) for applanation tonometry.

2.2. Regional Anesthetics

Bupivacaine hydrochloride [*18010-40-7*] (monohydrate [*14252-80-3*]), 1-butyl-*N*-(2,6-dimethylphenyl)-2-piperidinecarboxamide monohydrochloride monohydrate, $C_{18}H_{28}N_2O \cdot HCl \cdot H_2O$, M_r 342.9.

Etidocaine hydrochloride [*36637-19-1*], *N*-(2,6-dimethylphenyl)-2-(ethylpropylamino)butanamide monohydrochloride, $C_{17}H_{28}N_2O \cdot HCl$, M_r 312.9.

Hexylcaine hydrochloride [*532-76-3*], 2-cyclohexylamino-1-methylethyl benzoate monohydrochloride, $C_{16}H_{23}NO_2 \cdot HCl$, M_r 297.8.

Lidocaine hydrochloride [*73-78-9*] (anhydrous), monohydrate [*6108-05-0*], 2-diethyl-aminoaceto-*N*-(2,6-dimethylphenyl)acetamide monohydrochloride monohydrate, $C_{14}H_{22}N_2O \cdot HCl \cdot H_2O$, M_r 288.8.

Mepivacaine hydrochloride [1722-62-9], N-(2,6-dimethylphenyl)-1-methyl-2-piperidinecar-boxamide hydrochloride, $C_{15}H_{22}N_2O \cdot HCl$, M_r 282.8.

Prilocaine hydrochloride [1786-81-8], N-(2-methylphenyl)-2-(propylamino)propanamide hydrochloride, $C_{13}H_{20}N_2O \cdot HCl$, M_r 256.8.

Procaine hydrochloride [51-05-8], 2-diethylaminoethyl 4-aminobenzoate hydrochloride, $C_{13}H_{20}N_2O_2 \cdot HCl$, M_r 272.8.

Tetracaine hydrochloride (see Section 2.1).

3. Antimicrobial Agents

3.1. Antibacterial Agents

Successful therapy for bacterial infection with antibiotics relies upon five considerations:

1) Appropriate indications for use
2) Isolation and identification of the infective agent
3) Efficacy of the antibiotic for the infective agent
4) Adequate levels of the antibiotic at the site of infection
5) Low toxicity of the antibiotic for the host [5]

Antibiotics act upon bacteria through a variety of mechanisms. The penicillins, cephalosporins, vancomycin, and bacitracin inhibit development of the bacterial cell wall. Tetracycline, erythromycin, clindamycin, chloramphenicol, and the aminoglycosides inhibit protein synthesis.

3.1.1. Sulfonamides

See also → Synthetic Chemotherapeutic Agents.

Although the use of sulfonamides for the treatment of major bacterial infections has been largely supplanted by antibiotics, they are still prescribed for the topical treatment of minor ocular infections. They exhibit broad-spectrum bacteriostatic activity against many gram-positive and gram-negative organisms, and are useful against the protozoa *Toxoplasma gondii* and *Plasmodium falciparum*. Some strains of *Pseudomonas*, *Chlamydia*, and *Actinomyces* are also sensitive to these drugs. A major disadvantage is the frequent occurrence of cross-resistant microorganisms. Many side effects have been reported following the topical administration of sulfonamides. These drugs are contraindicated for the treatment of keratoconjunctivis sicca or penetrating corneal wounds.Sulfacetamide sodium is the most widely used sulfonamide at a concentration of 10%, or in combination with the steroid prednisolone acetate (0.2–0.5%).

The sulfonamides are antimetabolites. Their structural similarities to 4-aminobenzoic acid allow them to act as competitive inhibitors in the folic acid biosynthetic pathway.

Sulfacetamide sodium [*144-80-9*], N-[(4-aminophenyl)sulfonyl]acetamide, $C_8H_{11}N_2NaO_3S$, M_r 238, *mp* 182–184 °C, is incorrectly called sulfacetimide. For synthesis, see [6].

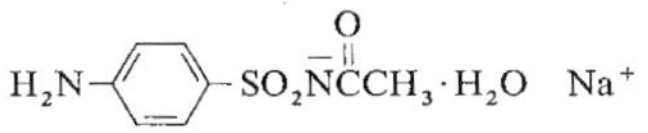

$$H_2N-\langle\rangle-SO_2\overset{-}{N}\overset{O}{\overset{\|}{C}}CH_3 \cdot H_2O \quad Na^+$$

Trade Names: Ak-Cide, Ak-Sulf (Akorn); Bleph-10, Blephamide Liquifilm/S.O.P. (Allergan); Cetamide Ointment, Cetrapred Ointment (Alcon); FML-S (Allergan); Isopto Cetmide Solution (Alcon); Metimyd Ophthalmic Solution/Ointment—Sterile (Schering); Ocu-Lone-C Sterile Ophthalmic Ointment/Suspension (Ocumed); Predsulfair Ointment/Suspension (Pharmafair); Sodium Sulamyd Ophthalmic Ointment 10%, Sodium Sulamyd Ophthalmic Suspension 10% (30%)—Sterile (Schering); Sulf-10 Dropperettes 10%/Ophthalmic Solution 10% (IOLAB); Sulfair (Pharmafair); Sulphrin (Bausch & Lomb); Sulpred Suspension (Pharmafair); Sulten-10 (Bausch & Lomb); Vasocidin Ointment/Solution, Vasosulf (IOLAB).

Sulfadiazine [*68-35-9*] (sodium [*547-32-0*], silver [*22199-08-2*]), 4-amino-N-2-pyrimidinylbenzenesulfonamide, $C_{10}H_{10}N_4O_2S$, M_r 250.3, *mp* 252–256 °C. For synthesis, see [7].

Sulfamerazine sodium [*127-58-2*], 4-amino-N-(4-methyl-2-pyrimidinyl)benzenesulfonamide sodium salt, $C_{11}H_{11}N_4NaO_2S$, M_r 286.3. For synthesis, see [7]. Sulfamerazine sodium is a short-acting sulfonamide with action similar to that of sulfamethoxazole. It is usually administered with other sulfonamides.

Sulfamethizole [*144-82-1*], 4-amino-N-(5-methyl-1,3,4-thiadiazol-2-yl)benzenesulfonamide, $C_9H_{10}N_4O_2S_2$, M_r 270.3, *mp* 208 °C. For synthesis, see [8].

Sulfisoxazole diolamine [*4299-60-9*], 4-amino-N-(3,4-dimethyl-5-isoxazolyl)benzenesulfonamide diethanolamine, M_r 372.4. For synthesis, see [9].

Trade Names: Ganstrin Ophthalmic Ointment/Solution, Suladrin, Sulfium (Roche).

3.1.2. β-Lactam Antibiotics

3.1.2.1. Penicillins

Penicillins are effective antimicrobials, although they have two drawbacks: (1) microbial resistance is relatively common, and (2) penicillins may be antigenic; anaphylactic shock is therefore a concern when they are administered systemically. See also → Antibiotics.

Ampicillin [69-53-4], (sodium derivative [69-52-3], trihydrate [7177-48-2]), $C_{16}H_{19}N_3O_4S$, M_r 349.42, mp 202 °C (decomp.). Ampicillin is a semisynthetic derivative of penicillin. For synthesis, see [10]. See also → Antibiotics.

Methicillin sodium [132-92-3], $C_{17}H_{19}N_2NaO_6S$, M_r 402.42, mp 196–197 °C. Methicillin sodium is a semisynthetic antibiotic related to penicillin. For synthesis, see [11]. See also → Antibiotics.

3.1.2.2. Cephalosporins

The cephalosporins originated from a mold of the genus *Cephalosporium*. Although cephalosporins do not usually cause an allergic reaction in patients sensitized to penicillin, allergy to cephalosporins can also develop. The cephalosporins are not inactivated by penicillin-resistant bacteria and have a broader spectrum of activity than penicillin. See also → Antibiotics.

Carbenicillin [4697-36-3] (disodium derivative [4800-94-6], indanyl sodium derivative [26605-69-6], phenyl sodium derivative [21649-57-0]), $C_{17}H_{18}N_2O_6S$, M_r 378.42. For synthesis, see [12]. See also → Antibiotics.

Cefazolin [25953-19-9] (sodium salt [27164-46-1]), $C_{14}H_{14}N_8O_4S_3$, M_r 454.50, mp 198–200 °C. For synthesis, see [13]. For structure, see → Antibiotics.

Cephalothin [153-61-7] (cephalothin sodium [58-71-9]), $C_{16}H_{15}N_2NaO_6S_2$, M_r 418, mp 160 °C. Cephalothin is a semisynthetic cephalosporin antibiotic. For synthesis, see [14]. For structure, see → Antibiotics.

3.1.3. Tetracycline Antibiotics

Tetracyclines reversibly block protein synthesis by specifically binding to the bacterial ribosome. Tetracycline and chloramphenicol were the first broad-spectrum antimicrobials discovered. The tetracycline antibiotics listed below are described in detail elsewhere; see → Antibiotics.

Chlortetracycline hydrochloride [64-72-2] (chlortetracycline [57-62-5]), $C_{22}H_{24}Cl_2N_2O_8$, M_r 493.9. Chlortetracycline is isolated from *Streptomyces aureofaciens* [15]. See also → Antibiotics.

Trade Name: Aureomycin Ophthalmic (IOLAB).

Oxytetracycline [79-57-2] (hydrochloride [2058-46-0]), $C_{22}H_{24}N_2O_9$, M_r 460.44, is an antibiotic isolated from the fermentation of *Streptomyces rimosus* [16]. See also → Antibiotics

Trade Name: Terramycin (5 mg/mL) with Polymyxin B Ointment (10 000 U/mL) (Roerig).

Tetracycline [60-54-8] (hydrochloride [64-75-5], phosphate complex [1336-20-5]), $C_{22}H_{24}N_2O_8$, M_r 444.43, is an antibiotic produced by *Streptomyces* spp. [17], [18]. See also → Antibiotics.

Trade Names: Achromycin and Achromycin Ophthalmic Ointment 1% (IOLAB).

3.1.4. Aminoglycoside Antibiotics

The aminoglycoside antibiotics exhibit broad-spectrum antimicrobial activity, particularly against gram-negative organisms (→ Antibiotics). They inhibit protein synthesis and decrease translational efficiency through interaction with the 30 S subunit of bacterial ribosomes. Cross resistance exists among the aminoglycosides.

Aminoglycosides are generally administered topically because systemic adsorption may produce oto- and nephrotoxicity. Aminoglycosides and β-lactam antibiotics are incompatible in the same formulation. Aminoglycosides are not effective under acid conditions, in high salt concentrations, or against intracellular bacteria.

Framycetin sulfate [4146-30-9], [28002-70-2], the sulfate of neomycin B, $C_{23}H_{46}N_6O_{13} \cdot 3\,H_2SO_4$, M_r 908.9. Neomycin B is isolated from the mixture of neomycins A, B, and C produced by *Streptomyces fradiae*; see also → Antibiotics [19]. Framycetin sulfate has broad-spectrum activity similar to that of neomycin but is more efficacious against *Pseudomonas* spp. Unlike most other aminoglycosides, framycetin

2007

sulfate can be administered by subconjunctival injection; however, because of its high degree of systemic absorption, extreme caution must be employed to avoid toxicity.

Trade Names: Framygen (Fisons), Soframycin (Roussel), with 0.5% hydrocortisone acetate in Framycort (Fisons), with 0.05% dexamethasome and 0.005% gramicidin in Sofradex (Roussel).

Gentamicin sulfate [*1405-41-0*] (gentamycin [*1403-66-3*]) is a complex mixture of the sulfates of gentamicin C_1, C_{1a}, and C_2. This complex is produced by fermentation of *Micromonospora purpurea* [20]; see also → Antibiotics. Gentamicin sulfate is used as a drop, ointment, or buffered preserved solution, generally at a concentration of 0.3%.

Trade Names: Alcomicin (Alcon); Cidomycin (Roussel); Garamycin (Kirby-Warrick); Garamycin Ophthalmic Ointment/Solution — Sterile (Schering); Genopyic (Allergan); Gentacidin (CooperVision); Gentafair (Pharmafair); Genticina (Antibioticos); Genoptic Liquifilm Sterile Ophthalmic Solution (Allergan); Genoptic S.O.P. Sterile Ophthalmic Ointment (Allergan); Gentafair Cream and Ointment/Solution (Pharmafair); Gentak Ointment and Solution (Akorn); Genticidin Ointment/Solution (IOLAB); Gentrasul Ointment/Solution (Bausch & Lomb); Minims Centamicin Sulfate (Smith & Nephew); Ocu-Mycin Sterile Ophthalmic Ointment/Solution (Ocumed); Pred-G Liquifilm Sterile Ophthalmic Suspension (Allergan); Pred-G S.O.P. Sterile Ophthalmic Ointment (Allergan).

Micronomicin sulfate (micronomicin [*52093-21-7*]), $(C_{20}H_{41}N_5O_7)_2 \cdot 5\,H_2SO_4$, M_r 1417.5, mp 260 °C. This gentamicin C complex antibiotic is produced by *Micromonospora sagamiensis* var. *nonreducans* [21].

Trade Names: Sagamicin (Kyowa), Santemycin (Santen).

Neomycin [*1404-04-2*] is an antibiotic complex made up of neomycins A, B, and C from *Streptomyces fradiae* [22]–[24]. See also → Antibiotics.

Trade Names: Ak-Spore (H.C.) Ointment/Solution, Ak-Trol Ointment & Suspension (Akorn); Cortisporin Ophthalmic Ointment/Suspension (Burroughs Wellcome); Dexacidin Ointment/Suspension (IOLAB); Dexasporin Ointment/Suspension (Pharmafair); Infectrol Ointment/Suspension (Bausch & Lomb); Maxitrol Ophthalmic Ointment/Suspension (Alcon); NeoDecadron Sterile Ophthalmic Ointment (Merck Sharp & Dohme); Neo Dexair Ophthalmic Solution (Pharmafair); Neosporin Ophthalmic Ointment/Solution (Burroughs Wellcome); Neotricin Ointment/Solution (Bausch & Lomb); Ocu-Cort Sterile Ophthalmic Ointment, Ocu-Spor-B Sterile Ophthalmic Ointment, Ocu-Spor-G Sterile Ophthalmic Solution (Ocumed); Ocutricin HC Ointment/Suspension, Ocutricin Ointment/Solution (Pharmafair); Ocu-Trol Sterile Ophthalmic Ointment/Suspension (Ocumed); Poly-Pred Liquifilm (Allergan).

Sisomicin [*32385-11-8*] (sulfate [*53179-09-2*]), $C_{19}H_{37}N_5O_7$, M_r 447.6, monohydrate from ethanol, mp 198–201 °C. Sisomicin is an antibiotic produced by *Micromonospora inyoesis* [25]. It is usually administered systemically and not commonly used as an ophthalmic antibiotic. See also → Antibiotics.

Streptomycin sulfate [*3810-74-0*] (streptomycin [*57-92-1*]), $C_{42}H_{84}N_{14}O_{36}S_3$, M_r 1457.40. Streptomycin is produced by *Streptomyces griseus* [26]; see also → Antibiotics. Corneal penetration of streptomycin is poor but can be greatly enhanced by iontophoresis.

Tobramycin [*32986-56-4*] (sulfate [*79645-27-5*]), $C_{18}H_{37}N_5O_9$, M_r 467.5. Tobramycin is produced by *Streptomyces tenebrarius* [27]. See also → Antibiotics. Tobramycin has a broad spectrum of activity similar to that of gentamicin. Cross resistance of tobramycin and gentamicin has been observed. Tobramycin is four times more effective against *Pseudomonas aeruginosa*, and less effective against *Serratia*, than gentamicin. Unlike other aminoglycosides, tobramycin is ineffective against mycobacteria.

Trade Names: TobraDex Ophthalmic Suspension and Ointment, TobraLex, Tobral, Tobramaxin, Tobrex (Alcon); Tobra-Gobens (Normon); Tobra-Laf (Andalucia).

3.1.5. Polypeptide Antibiotics

Bacitracin [*1405-87-4*], bacitracin A (major component): $C_{66}H_{103}N_{17}O_{16}S$, M_r 1422.71. Commercial bacitracin is a mixture of at least nine bacitracins produced by *Bacillus subtilis* and *B. licheniformis* [28], [29]; see → Antibiotics. It is available singly only as an ointment (500 mg/g) and in combination with other antimicrobials (see Table 1).

Trade Names: Ak-Poly-Bac Ointment, Ak-Spore (H.C.) Ointment (Akorn); Bacitracin Ophthalmic Ointment—Sterile (Pharmafair); Cortisporin Ophthalmic Ointment Sterile, Neosporin Ophthalmic Ointment Sterile (Burroughs Wellcome); Neotricin Ointment/Solution (Bausch & Lomb); Ocumycin (HC) Ointment (Pharmafair); Polysporin Ophthalmic Ointment Sterile (Burroughs Wellcome).

Colistimethate sodium [*8068-28-8*], $C_{58}H_{105}N_{16}Na_5O_{28}S_5$, M_r 1749.8. Colistimethate sodium is an injectable form of colistin, a cyclopolypeptide produced by *Bacillus colistinus* [30]. See also → Antibiotics.

Trade Name: Coly-Mycin M (Parke-Davis).

Polymyxin B sulfate [*1405-20-5*] is a mixture of polymyxins B_1 and B_2 produced by *Bacillus polymyxa* [31], [32]. See also → Antibiotics. Polymyxins are cyclic polypeptides that are similar to detergents in having basic groups with a fatty acid side chain. They exert their antimicrobial effect by solubilizing the cell membrane. Polymyxin B sulfate is available as a sterile powder for reconstitution or in combination with other antimicrobials (see Table 1).

Trade Names: Ak-Poly-Bac Ointment, Ak-Spore (H.C.) Ointment/Suspension, Ak-Trol Ointment & Suspension (Akorn); Cortisporin Ophthalmic Ointment/Suspension (Burroughs Wellcome); Dexaci-

Ophthalmological Agents

Table 1. Combination antibiotic solutions and ointments *

Product	Polymyxin B sulfate, units/g	Neomycin, mg/g	Bacitracin, units/g	Oxytetracycline, mg/g	Gramicidin, mg/mL	Company
Neotal Ophthalmic Ointment	5000	5	400			Hauck
Terramycin with Polymyxin B Ointment	10000			5		Roerig
Ak-Poly-Bac Ointment	10000		500			Akorn
Polysporin Ointment	10000		500			Burroughs Wellcome
Neomycin Sulfate–Polymycin B Sulfate–Gramicidin Solution	10000*	1.75**			0.025	Rugby
Ak-Spore Solution	10000*	1.75**			0.025	Akorn
Neocidin Solution	10000*	1.75**			0.025	Major
Neosporin Ophthalmic Solution	10000*	1.75**			0.025	Burroughs Wellcome
Neotricin Opthalmic Solution	10000*	1.75			0.025	Bausch & Lomb
Statrol Ointment	10000	3.5				Alcon
Ak-Spore Ointment	10000	3.5	400			Akorn
Neosporin Opathalmic Ointment	10000	3.5	400			Burroughs Wellcome
Neotricin Ointment	10000	3.5	400			Bausch & Lomb
Mycitracin Ophthalmic Ointment	10000	3.5	500			Upjohn
Statrol Solution	16250*	3.5**				Alcon

* Units/mL. ** mg/mL.

din Ointment/Suspension (IOLAB); Dexasporin Ointment/Suspension (Pharmafair); Infectrol Ointment/Suspension (Bausch & Lomb); Maxitrol Ophthalmic Ointment/Suspension (Alcon); Neosporin Opthalmic Ointment/Ophthalmic Solution Sterile (Burroughs Wellcome); Neotricin Ointment/Solution (Bausch & Lomb); Ocu-Cort Sterile Ophthalmic Ointment (Ocumed); Ocumycin Ointment (Pharmafair); Ocu-Spor-B Sterile Ophthalmic Ointment, Ocu-Spor-G Sterile Ophthalmic Solution (Ocumed); Ocutricin HC Ointment/Suspension, Ocutricin Ointment/Solution (Pharmafair); Ocu-Trol Sterile Ophthalmic Ointment/Suspension (Ocumed); Opthocort (Parke-Davis); Polymyxin B Sulfate Sterile Ophthalmic (Pfizer); Poly-Pred Liquifilm (Allergan); Polysporin Ophthalmic Ointment Sterile (Burroughs Wellcome).

3.1.6. Miscellaneous Antimicrobial Agents

Chloramphenicol [56-75-7], $C_{11}H_{12}Cl_2N_2O_5$, M_r 323.1, *mp* 150.5–151.5 °C, is a broadspectrum antibiotic produced by *Streptomyces venezuelae* [33]–[35]. Chloramphenicol reversibly blocks the bacterial ribosome, inhibiting protein synthesis. See also → Antibiotics.

Trade Names: Ak-Chlor Sterile Ophthalmic Ointment/Solution (Akorn); Chlorofair Ointment/Solution (Pharmafair); Chloromycetin Ophthalmic Ointment/Solution/Powder, Hydrocortisone Ophthalmic Solution (Parke-Davis); Chloroptic S.O.P., Chloroptic Sterile Ophthalmic Solution (Allergan); Ocu-Chlor Sterile Opthalmic Ointment/Solution (Ocumed); Opthochlor Ophthalmic Solution (Parke-Davis).

Clindamycin [18323-44-9] (palmitate hydrochloride [25507-04-4], hydrochloride monohydrate [58207-19-5]), $C_{18}H_{33}ClN_2O_5S$, M_r 424.98, is a semisynthetic antibiotic prepared from lincomycin [36]. See also → Antibiotics.

Trade Name: Cleocin Phosphate (Upjohn).

Erythromycin [114-07-8], $C_{37}H_{67}NO_{13}$, M_r 733.9, *mp* 135–140 °C, resolidifies with second *mp* 190–193 °C. Erythromycin is produced by *Streptomyces erythreus* [37] and is a mixture of erythromycin A, B, and C, with A predominant. It is the most widely used of the macrolide antibiotics and exerts its effect at the level of protein synthesis by reversibly inhibiting the bacterial ribosome. See also → Antibiotics.

Trade Names: Ak-Mycin Ointment (Akorn), Erythromycin Ophthalmic Ointment (Pharmafair), Ilotycin (Dista).

Lincomycin [154-21-2], $C_{18}H_{34}N_2O_6S$, M_r 406.6, is an antibiotic produced by *Streptomyces lincolnensis* var. *lincolnensis* [38], [39]. Lincomycin reversibly inhibits the bacterial ribosome. See also → Antibiotics.

Trade Name: Lincocin (Upjohn).

Ofloxacin [*82419-36-1*], $C_{18}H_{20}FN_3O_4$, M_r 361.4, *mp* 250–257 °C, is a broad-spectrum, fluorinated quinoline antibacterial. See also → Synthetic Chemotherapeutic Agents. For synthesis, see [40]. It is currently in clinical trials by Allergan.

Silver nitrate [*7761-88-8*], $AgNO_3$, M_r 169.9, *mp* 212 °C, is used to treat the eyes of newborns. Caution should be used with repeated applications because cauterization of the cornea and blindness may result. Silver nitrate is incompatible with sulfacetamide preparations.

Trade Name: Silver Nitrate (Lilly).

Vancomycin [*1404-90-6*] (hydrochloride [*1404-93-9*]), $C_{66}H_{75}Cl_2N_9O_{24}$, M_r 1449.2. Vancomycin is an amphoteric glycopeptide produced by *Streptomyces orientalis* [41]. It inhibits cell wall synthesis. See also → Antibiotics.

Trade Names: Vancocin (Lilly), Vancoled (Lederle), Vancor (Adria).

3.2. Antifungal Agents

For general information see → Antimycotics.

Ocular fungal infection is a rare disease but is increasing in occurrence [42]–[44]. Commensal fungi commonly present in the eye include *Aspergillus, Penicillium, Candida, Fusarium,* and *Rhodotorula*. These may become opportunistic pathogens if the eye's defenses are compromised by physical trauma or an underlying disease such as diabetes.

Although natamycin is the only antifungal agent currently available commercially in the United States as a topical ophthalmic formulation, other antifungal drugs (including polyenes and imidazoles) have been used topically in dilute solution or suspension [45].

The *polyene antimycotics* (amphotericin B, nystatin, natamycin) are insoluble in water and unstable in oxygen, light, water, heat, and at extreme pH. Their biological specificity for yeast relies on preferential binding to yeast membranes via ergosterol over their affinity for cholesterol, the primary sterol in mammalian cell membranes. Cell death is induced by increasing the permeability of the fungal membrane, allowing depletion of intracellular components.

The *imidazoles* (miconazole and ketoconazoles) have a broad antifungal and antimicrobial spectrum, with significantly less toxicity than the polyenes. Antifungal effects may result from the inhibition of estergol synthesis, leading to cell membrane permeability. Miconazole physically disrupts the membrane, leading to cell lysis. High concentrations of miconazole also increase intracellular concentrations of hydrogen peroxide, presumably through inhibition of cytochrome C peroxidase.

Flucytosine is a widely used *pyrimidine antimycotic*.

Amphotericin B [*1397-89-3*], $C_{47}H_{73}NO_{17}$, M_r 924.1, is a polyene antibiotic produced by *Streptomyces nodosus* [46]. Amphotericin can be administered parenterally for systemic mycosis. Topical administeration of 0.1 and 0.25% has been effective for the treatment of keratomycoses. See also → Antimycotics.

Trade Name: Fungizone (Squibb).

Nystatin [*1400-61-9*] is a polyene antifungal antibiotic complex containing three active components (A_1, A_2, and A_3) produced by several *Streptomyces* spp. [47]. See also → Antimycotics. Nystatin is effective in topical application for ocular infection.

Trade Names: Mycostatin (Squibb), Nilstat (Lederle), Nystex (Savage), O-V Statin (Squibb).

Natamycin [*7681-93-8*], $C_{33}H_{47}NO_{13}$, M_r 665.75, *mp* 280–300 °C (decomp.), is a polyene antibiotic produced by *Streptomyces natalensis* and *S. chattanoogensis* [48]. See also → Antimycotics. Natamycin is the least toxic, least irritating, and most stable polyene. It is only useful, however, for the treatment of superficial keratomycosis because it penetrates tissue poorly.

Trade Name: Natacyn (Alcon).

Miconazole [*22916-47-8*] (nitrate [*22832-87-7*]), $C_{18}H_{14}Cl_4N_2O$, M_r 416.12. See also → Antimycotics. For synthesis, see [49].

Trade Names: Micatin, Monistat-Derm (Ortho).

Ketoconazole [*65277-42-1*], $C_{26}H_{28}Cl_2N_4O_4$, M_r 531.44, *mp* 146 °C, is an orally active, broad-spectrum antimycotic. See also → Antimycotics. For synthesis, see [50].

Trade Name: Nizoral (Janssen).

Flucytosine [*2022-85-7*], 5-fluorocytosine, $C_4H_4FN_3O$, M_r 129.1, *mp* 295–297 °C (decomp.). For synthesis, see [51]. Flucytosine, a fluorinated pyrimidine, is metabolized to either 5-fluorouracil, an inhibitor of RNA synthesis, or 5-fluoro-2′-deoxyuridylic acid, a potent inhibitor of DNA synthesis. Flucytosine has a limited spectrum of activity but is effective in topical application against external infections. It exhibits poor ocular penetration but can be administered orally. See also → Antimycotics.

Trade Name: Ancobon (Hoffmann–La Roche).

3.3. Antiprotozoal Agents

Broline [496-00-4], 4,4′-(trimethylenedioxy)bis(3-bromobenzamidine), $C_{21}H_{30}Br_2N_4O_{10}S_2$, M_r 722.4, mp 226 °C, is used to treat *Acanthamoeba* keratitis infections [52]. For synthesis, see [53].

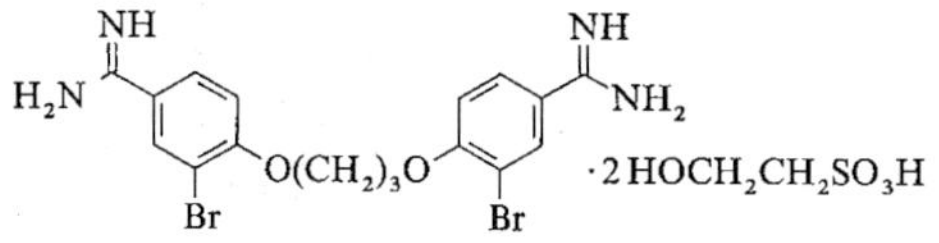

$$H_2N-C(=NH)-C_6H_3(Br)-O(CH_2)_3O-C_6H_3(Br)-C(=NH)NH_2 \cdot 2\,HOCH_2CH_2SO_3H$$

Trade names: Brolene Eye Drops (Bausch & Lomb); Broline Ointment, Brulidene (May & Baker).

Sulfadiazine [68-35-9] (silver salt [22199-08-2]), 4-amino-N-2-pyrimidinylbenzenesulfonamide, $C_{10}H_{10}N_4O_2S$, M_r 250.3, mp 252 – 256 °C. For synthesis, see [7]. See also → Synthetic Chemotherapeutic Agents. This compound is used to treat toxoplasmic retinochoroiditis in triple sulfonamide mixtures with sulfamerazine and sulfamethazine.

Trade Names: Coco-Diazine (Lilly), Eskadiazine (SKF).

Antibiotic – Corticosteroid Preparations. A variety of ophthalmic antibiotic – corticosteroid combination preparations are available for treating protozoal infections. Ophthalmic applications of corticosteroids and corticosteroid – antibiotic combination preparations are discussed in Section 4.1.

3.4. Antiviral Agents

Currently marketed antivirals are available only as topical applications for the treatment of infection with the herpes simplex virus (HSV). These agents inhibit viral replication at the level of DNA synthesis. The prophylactic use of antivirals may reduce the frequency of viral latency [54]. The use of high dosages does not rid the host of latent virus but may reduce the frequency of recurrences during therapy [55].

Acyclovir [59277-89-3], $C_8H_{11}N_5O_3$, M_r 225.2, mp 256.5 – 257 °C, is an orally active acyclic nucleoside used against HSV. It is also used as a 3% ointment in HSV infections of the epithelium. For synthesis, see [56]. See also → Synthetic Chemotherapeutic Agents.

Trade Name: Zovirax (Burroughs Wellcome).

Idoxuridine [54-42-2], 2′-deoxy-5-iodouridine, $C_9H_{11}IN_2O_5$, M_r 354.1. Idoxuridine is a functional analogue of the nucleoside thymidine; see → Synthetic Chemotherapeutic

Agents. Incorporation into DNA results in death of the viral particle through DNA base-pair mismatches and mutation. The toxicity of the substance is attributable to its incorporation into the host cell DNA. This agent is best used topically on a superficial lesion and is available as a 0.1% solution or a 0.5% ointment. For synthesis, see [57].

Trade Names: Herpex (Allergan); Stoxil Ointment, Stoxil Solution (SKF).

Trifluridine [*70-00-8*], 2′-deoxy-5-(trifluoromethyl)uridine, $C_{10}H_{11}F_3N_2O_5$, M_r 296.2, *mp* 186–189 °C. Trifluridine is a thymidine analogue but unlike idoxuridine is rapidly metabolized and thus less toxic. Trifluridine is effective against HSV [58] and, in addition, is compatible with corticosteroids [59]. See also → Synthetic Chemotherapeutic Agents. Trifluridine is generally available as a 1% solution or a 3% ointment. For synthesis, see [60].

Trade Names: TFT-Ophtiole (Mann), TFT Eye Ointment (Mann), Viroptic (Burroughs Wellcome).

Vidarabine (anhydrous [*5536-17-4*], monohydrate [*24356-66-9*]), $C_{10}H_{13}N_5O_4 \cdot H_2O$, M_r 285.3, *mp* 257–257.5 °C, is an adenosine analogue produced by *Streptomyces antibioticus*. Its mechanism of action is similar to that of idoxuridine. The number and severity of adverse reactions of the two drugs are not significantly different, although vidarabine has been used to treat cases unresponsive to idoxuridine [61]. Vidarabine is available as a 3% ointment. See also → Synthetic Chemotherapeutic Agents.

Trade Name: Vira-A (Parke-Davis).

4. Anti-inflammatory Agents

A variety of agents are available for the treatment of ocular inflammatory conditions, including corticosteroids, nonsteroidal anti-inflammatory agents that block mediator synthesis or release, and, more rarely, immunosuppressive agents. For general information, see → Anti-inflammatory – Antirheumatic Drugs.

4.1. Corticosteroids

For general information, see → Hormones.

Corticosteroids are widely used for the treatment of ocular inflammation [62]–[64]. Prednisolone, hydrocortisone, dexamethasone, medrysone, and fluorometholone are available in topical ocular preparations (suspension, ointment, or solution). In certain cases, topical application may be supplemented or replaced by systemic or periocular administration (e.g., severe anterior uveitis and inflammation of the posterior segment, optic nerve, or orbit) [62], [63].

Table 2. Steroid and antibiotic drops

Product	Steroid	Antibiotic	Company
Chlormycetin Hydrocortisone Powder	0.5% hydrocortisone acetate	0.25% chloramphenicol	Parke-Davis
Neo-Cortef Suspension	0.5% hydrocortisone acetate	neomycin sulfate equivalent to 0.35% base	Upjohn
Bacticort Suspension	1% hydrocortisone	neomycin sulfate equivalent to 0.35% base and 10 000 units polymyxin B sulfate/mL	Rugby
Triple-Gen Suspension	1% hydrocortisone	neomycin sulfate equivalent to 0.35% base and 10 000 units polymyxin B sulfate/mL	Goldline
Cortisporin Suspension	1% hydrocortisone	neomycin sulfate equivalent to 0.35% base and 10 000 units polymyxin B sulfate/mL	Burroughs Wellcome
Terra-Cortil Suspension	1.5% hydrocortisone acetate	0.5% oxytetracycline	Roerig
Ak-Neo-Cort Suspension	1.5% hydrocortisone acetate	neomycin sulfate equivalent to 0.35% base	Akorn
Cor-Oticin Suspension	1.5% hydrocortisone acetate	neomycin sulfate equivalent to 0.35% base	Americal
Ortho Drops Suspension	1.5% hydrocortisone acetate	neomycin sulfate equivalent to 0.35% base	Vortech
Poly-Pred Suspension	0.5% prednisolone acetate	neomycin sulfate equivalent to 0.35% base and 10 000 units polymyxin B sulfate/mL	Allergan
Pred-G Suspension	1% prednisolone acetate	gentamycin sulfate equivalent to 0.3% gentamycin base	Allergan
NeoDecadron Solution	0.1% dexamethasone phosphate	neomycin sulfate equivalent to 0.35% base	MSD
TobraDex Suspension	0.1% dexamethasone	0.3% tobramycin	Alcon
Dexasporin Suspension	0.1% dexamethasone	neomycin sulfate equivalent to 0.35% base and 10 000 units polymyxin B sulfate/mL	Pharmafair, Rugby, and others
Ak-Trol Suspension	0.1% dexamethasone	neomycin sulfate equivalent to 0.35% base and 10 000 units polymyxin B sulfate/mL	Akorn
Dexacindin Suspension	0.1% dexamethasone	neomycin sulfate equivalent to 0.35% base and 10 000 units polymyxin B sulfate/mL	IOLAB
Maxitrol Suspension	0.1% dexamethasone	neomycin sulfate equivalent to 0.35% base and 10 000 units polymyxin B sulfate/mL	Alcon

The use of corticosteroids is not without potential ocular and systemic side effects (e.g., elevated intraocular pressure, posterior subcapsular cataracts, and inhibition of epithelial wound healing [60], [62], [65]–[67]). Because corticosteroids have immunosuppressive properties, resistance to infection is lowered. Accordingly, a number of ophthalmic preparations containing combinations of corticosteroids and antimicrobial agents are available for anti-inflammatory/anti-infective therapy (Tables 2 and 3).

Hydrocortisone [*50-23-7*], 17-hydroxycorticosterone, cortisol, $C_{21}H_{30}O_5$, M_r 362.5. See also → Hormones.

Hydrocortisone (1%) is available in antimicrobial–anti-inflammatory ointment or suspension combination products (Tables 2 and 3).

Trade Names: Ak-Spore ointment (Akorn; hydrocortisone, neomycin sulfate and polymyxin B sulfate); Cortisporin ointment (Burroughs Wellcome: hydrocortisone, polymyxin B sulfate, neomycin sulfate, and bacitracin zinc); Cortisporin suspension (Burroughs Wellcome: hydrocortisone, neomycin sulfate, and polymyxin B sulfate); Ocu-Cort ointment (Ocumed: hydrocortisone, polymyxin B sulfate, neomycin sulfate, and bacitracin zinc); Ocutricin HC ointment (Pharmafair: hydrocortisone, bacitracin

Table 3. Steroid and antibiotic ointments

Product	Steroid	Antibiotic	Company
Orthocort	0.5% hydrocortisone acetate	1% chloramphenicol and 10000 units polymyxin B/g	Parke-Davis
Cortisporin	1% hydrocortisone	neomycin sulfate equivalent to 0.35% base, 400 units bacitracin zinc, and 10000 units polymyxin B sulfate/g	Burroughs Wellcome
Coracin	1% hydrocortisone acetate	0.5% neomycin sulfate, 400 units bacitracin zinc and 10000 units polymyxin B sulfate/g	Coracin
NeoDecadron	0.05% dexamethasone phosphate	neomycin sulfate equivalent to 0.35% neomycin base	MSD
AK-Trol	0.1% dexamethasone	neomycin sulfate equivalent to 0.35% base and 10000 units polymyxin B sulfate/g	Akorn
Dexacidin	0.1% dexamethasone	neomycin sulfate equivalent to 0.35% base and 10000 units polymyxin B sulfate/g	IOLAB
Dexasporin	0.1% dexamethasone	neomycin sulfate equivalent to 0.35% base and 10000 units polymyxin B sulfate/g	Pharmafair, Rugby, and others
Maxitrol	0.1% dexamethasone	neomycin sulfate equivalent to 0.35% base and 10000 units polymyxin B sulfate/g	Alcon

zinc, polymyxin B sulfate, and neomycin sulfate); Ocutricin HC suspension (Pharmafair: hydrocortisone, neomycin sulfate, and polymyxin B sulfate).

Hydrocortisone acetate [50-03-3], hydrocortisone 21-acetate, cortisol acetate, $C_{23}H_{32}O_6$, M_r 404.5.

Trade Names: Hydrocortone (Merck Sharp & Dohme), Optef (Upjohn); also see Tables 2 and 3.

Prednisolone [50-24-8], 1,2-dehydrohydrocortisone, $C_{21}H_{28}O_5$, M_r 360.5. See also → Hormones.

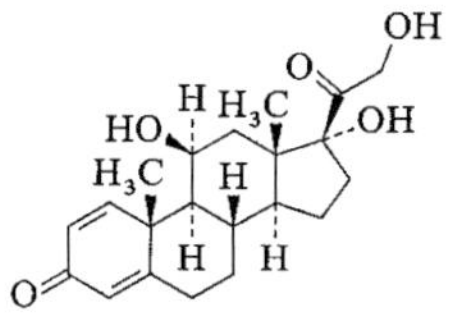

Prednisolone acetate [52-21-1], prednisolone 21-acetate, $C_{23}H_{30}O_6$, M_r 402.5.

Trade Names: Ak-Tate (Akorn); Econopred, Econopred Plus (Alcon); Ocu-pred A (Ocumed); Pred Forte, Pred Mild (Allergan); see also Tables 2 and 3.

Prednisolone sodium phosphate [125-02-0], prednisolone 21-(disodium orthophosphate), $C_{21}H_{27}Na_2O_8P$, M_r 484.4, is water soluble and is available as solutions.

Trade Names: Ak-Pred—0.125% and 1% (Akorn); I-Pred and I-Pred 1% (Americal); Inflamase Forte 1%, Inflamase Mild (IOLAB); Ocu-Pred, Ocu-Pred Forte (Ocumed); Predair, Predair Forte (Pharmafair); see also Tables 2 and 3.

Dexamethasone [50-02-2], 9α-fluoro-16α-methylprednisolone, $C_{22}H_{29}FO_5$, M_r 392.5. See also → Hormones.

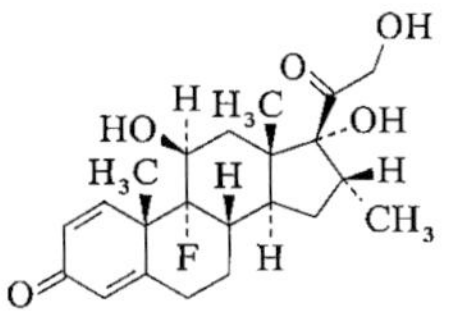

Trade Names: Ak-Dex, Ak-Trol (Akorn); Baldex (Bausch & Lomb); Dexacidin (IOLAB); Decadron Phosphate (Merck Sharp & Dohme); Dexair (Pharmafair); Dexasporin (Pharmafair); Infectrol (Bausch & Lomb); Maxidex, Maxitrol (Alcon); I-Methasone (Americal); Ocu-Trol (Ocumed); Tobradex (Alcon); see also Tables 2 and 3.

Medrysone [*2668-66-8*], 11β-hydroxy-6α-methylprogesterone, $C_{22}H_{32}O_3$, M_r 344.5.

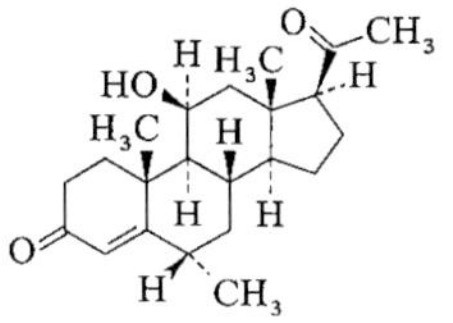

Trade Names: HMS Liquifilm, Ipoflogin (Allergan); Medrifar (Farmila); Medriusar (Difa); Ophtocortin (Winzer); Sedesterol (Poen); Spectramedryn (Allergan); Visudrisone (LOA).

Fluorometholone [*426-13-1*], $C_{22}H_{29}FO_4$, M_r 376.5, is used ophthalmically as a 0.1 or 0.25 % suspension or as a 0.1 % ointment. See also → Hormones.

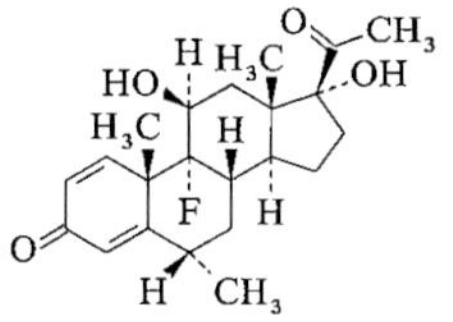

Trade Names: Flucon, Isopto-Flucon (Alcon); FML Liquifilm, FML Forte Liquifilm, FML S.O.P. (Allergan); Fluor-Op (IOLAB).

4.2. Miscellaneous Anti-inflammatory Agents

Disodium cromoglycate [*15826-37-6*], DSCG, cromolyn sodium, sodium cromoglycate, $C_{23}H_{14}Na_2O_{11}$, M_r 5123, appears to achieve its antiallergic and anti-inflammatory effects by stabilizing mast cells and inhibiting antigen-induced mast cell degranulation and the associated release of a variety of inflammatory mediators [62], [68]. See also → Antiallergics. Topical ophthalmic DSCG (Opticrom, Fisons) is available for prophylactic treatment of allergic disorders such as vernal or allergic keratoconjunctivitis. Maintaining DSCG therapy throughout the allergy season may be more advantageous than sporadic treatment of flare-ups [68]. Therapy with DSCG may permit the reduction or elimination of topical corticosteroid treatment [62].

Trade Names: Opticrom (Fisons), Vividrin (Mann).

Nonsteroidal anti-inflammatory drugs (NSAIDs) inhibit the enzyme cyclooxygenase, which is responsible for the first step in the synthesis of prostaglandins from arachidonic acid. The prostaglandins mediate certain steps of the inflammatory process. The role of arachidonic acid metabolites in ocular inflammation and pressure elevation continues to emerge [62]. Oral administration of NSAIDs has been approached as an alternative to corticosteroid anti-inflammatory therapy [62], [65]. Oral administration of classical cyclooxygenase inhibitors (e.g., acetylsalicyclic acid and indomethacin) has met with mixed success in the treatment of ocular inflammation, partially due to systemic intolerance (e.g., gastrointestinal side effects). A number of more recently developed NSAIDs such as diflusinal (Dolobid) and naproxen (Naprosyn) may be better tolerated orally.

Flurbiprofen [*5104-49-4*], 2-fluoro-α-methyl-(1,1′-biphenyl)-4-acetic acid, $C_{15}H_{13}FO_2$, M_r 244.3. The sodium salt dihydrate of flurbiprofen is available as an ophthalmic solution (0.03%) for topical application (Ocufen Liquifilm, Allergan) for inhibition of intraoperative miosis.

Tetracycline (see Section 3.1.3) is used to treat ocular rosacea [69].

Immunosuppressive agents, including alkylating agents such as cyclophosphamide, the folic acid antimetabolite methotrexate, and the antibiotic cyclosporin A, are systemically cytotoxic and dangerous drugs. Although their use in ocular therapy is rare, some of them have been used to treat intractable and progressively destructive ocular conditions [65].

5. Antiglaucomatous Agents

Glaucoma is a leading cause of blindness and may be defined as a loss of visual function and optic nerve damage associated with an elevated intraocular pressure (IOP). Ocular hypertension alone is not predictive of glaucoma or visual impairment [70]. In a healthy eye, the IOP (ca. 2 – 2.6 kPa, 15 – 20 mm Hg) [71] is maintained by a balance of aqueous humor formation and outflow. Glaucoma therapy often involves a reduction in IOP by increasing outflow or decreasing production. A variety of therapeutic agents have ocular hypotensive effects [68] – [74] and include adrenergic agents, miotics, carbonic anhydrase inhibitors, and hyperosmotic agents. The applications and adverse effects of ocular hypotensives are discussed in [70] – [80].

Glaucoma associated with congenital ocular abnormalities is treated surgically or with a combination of surgical and medical therapy. Glaucomas may be classified as primary (direct disturbance of the aqueous circulation) or secondary (arising from other disease states). They are also classified as open angle or narrow angle, depending on the anterior chamber angle (the anatomy of the structures within the angle of the eye as visualized by gonioscopy). These classifications are important in determining whether surgery or drug therapy should be employed and which ocular hypotensive drug therapies may be appropriate.

5.1. Sympathomimetic Drugs

Epinephrine [51-43-4], adrenaline, $C_9H_{13}NO_3$, M_r 183.2, is used widely in primary open-angle glaucoma but is contraindicated in narrow-angle glaucoma [70], [71] because of its mydriatic effects (dictation of the pupil). Topically applied epinephrine has mydriatic, vasoconstricting, and ocular hypotensive effects. Although its mechanism of action is not fully understood, it may lower the IOP by reducing aqueous humor production and also increasing outflow [70]–[74]. Epinephrine is often used in combination with other ocular hypotensives (e.g., pilocarpine).

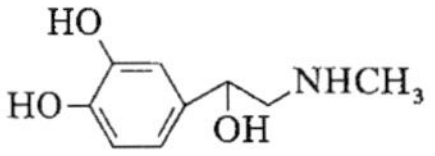

Trade Names: Simplene (Smith & Nephew), 1 and 2% solutions (Pharmafair); U.S.P. Ophthalmic Solution: a solution of epinephrine prepared with hydrochloric acid.

Epinephrine hydrochloride [55-31-2], Epifrin (0.25, 0.5, 1, or 2%; Allergan), Glaucon (1 or 2%; Alcon).
Epinephrine bitartrate [51-42-3], adrenaline bitartrate, $C_9H_{13}NO_3 \cdot C_4H_6O_6$, M_r 333.3, Epitrate (Wyeth-Ayerst). Epinephrine bitartrate is also available in combination with pilocarpine hydrochloride (E-Pilo-1 – 6, IOLAB)
Epinephryl borate [5579-16-8]: Eppy/N (0.5, 1.0, and 2%, Sola/Barnes Hind).

Dipivefrin hydrochloride [52365-63-6], dipivalylepinephrine hydrochloride, dipivalyladrenaline hydrochloride, $C_{19}H_{29}O_5N \cdot HCl$, M_r 387.9. For synthesis, see [81]. Dipivefrin hydrochloride is a dipivalyl prodrug that is converted to epinephrine by esterase action. It has enhanced corneal penetration and hypotensive potency relative to epinephrine [71], [73]. Dipivefrin 0.1% is about as potent as 2% epinephrine.

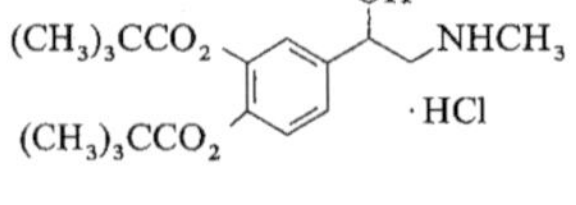

Trade Names: d Epifrin (Allergan), Glaucothil (Thilo), Glaudrops (Cusi), Propine (Allergan).

5.2. β-Adrenergic Blocking Agents

Although systemic blockers of the β-adrenoceptors are not approved for ocular hypotensive indications, some reduce IOP in addition to having cardiovascular activity against arrhythmias, angina, and hypertension. The β-blockers lower the IOP by decreasing aqueous humor production [72]–[74]. Although topical preparations are generally well tolerated, systemic (CNS, cardiovascular, and respiratory) side effects have been observed [72]–[74], [76]–[78].

Timolol maleate [26921-17-5] (timolol [26839-75-8]), (S)-1-[(1,1-dimethylethyl)amino]-3-[(4-morpholinyl-1,2,5-thiadiazol-3-yl)oxy]-2-propanol maleate, $C_{13}H_{24}N_4O_3S \cdot C_4H_4O_4$, M_r 432.5, is available as 0.25 and 0.5% solutions. For synthesis, see [82]. See also → Antihypertensive Agents. Timolol can provide an additive effect when used concurrently with other ocular hypotensive agents such as miotic agents or carbonic anhydrase inhibitors, although the effects with epinephrine are more controversial [72]–[74].

Trade Names: Timoptic, Timoptol (Merck Sharp & Dohme).

Levobunolol hydrochloride [27912-14-7] (levobunolol [47141-42-4]), (−)-bunolol hydrochloride, (−)-5-[3-(*tert*-butylamino)-2-hydroxypropoxy]-3,4-dihydro-1(2*H*)-naphthalenone hydrochloride, $C_{17}H_{25}NO_3 \cdot HCl$, M_r 327.9. For synthesis, see [83].

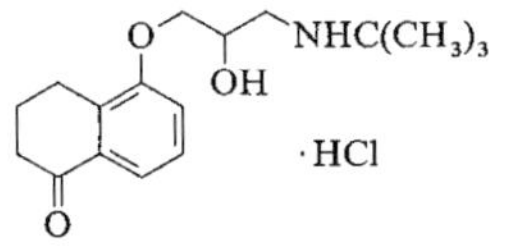

Trade Name: Betagan (0.5%, Allergan).

Betaxolol hydrochloride [63659-19-8] (betaxolol [63659-18-7]), 1-[4-[2-(cyclopropylmethoxy)ethyl]phenoxy]-3-isopropylamino-2-propanol hydrochloride, $C_{18}H_{29}NO_3 \cdot HCl$, M_r 343.9. For synthesis, see [84].

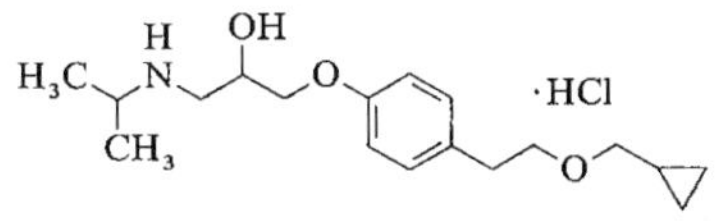

Trade Name: Betoptic (0.5% Alcon).

Metipranolol [22664-55-7], 4-(2-hydroxy-3-isopropylaminopropoxy)-2,3,6-trimethylphenyl acetate, $C_{17}H_{27}NO_4$, M_r 309.4. For synthesis, see [85], [86].

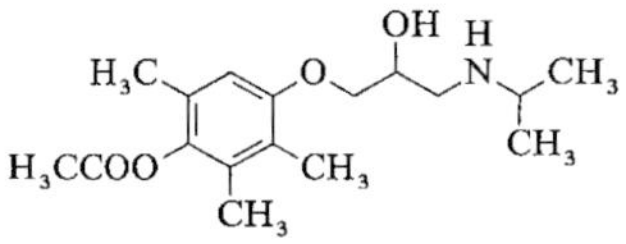

Trade Names: Betamann, Beta-Ophtiole (Mann); Glauline, Minims Metipranolol (Smith & Nephew); Optipranolol (Bausch & Lomb); Normoglaucon (Mann).

5.3. Carbonic Anhydrase Inhibitors

Carbonic anhydrase inhibitors decrease the IOP by blocking bicarbonate formation in the ciliary process required for aqueous humor production [74], [79]. These agents are administered systemically and are useful in treating glaucoma cases that do not respond to topical therapy [72], [74]. Virtually all (90–99%) of the carbonic anhydrase activity must be blocked before the IOP is lowered [72], [79], and oral doses are usually administered several times a day. The sulfonamide carbonic anhydrase inhibitors are diuretics that are readily absorbed from the gastrointestinal tract. Effective oral doses of these agents for glaucoma therapy depend on their pharmacokinetic properties [72], [74], [79]. Systemic side effects [72], [74], [76], [79] may limit the use of oral carbonic anhydrase inhibitors, especially in the elderly. Unfortunately, no topically effective preparations of these agents are available, possibly due to poor corneal penetration. Orally administered carbonic anhydrase inhibitors may provide additive effects when used in combination with topically applied antiglaucoma agents (e.g., pilocarpine and timolol) [74], [79].

Acetazolamide [*59-66-5*], 2-acetylamino-1,3,4-thiadiazole-5-sulfonamide, $C_4H_6N_4O_3S_2$, M_r 222.2. See also → Diuretics.

Trade Names: Acetamide (Nessa); Diamox, Diamox Sequels (Lederle); Diuriwas (IFI); Edemox (Wasserman); Glaupax (Erco, Dispersa); Theraplix (Cycanamide, Cyanamide-Novalis).

Dichlorphenamide [*120-97-8*], 4,5-dichloro-1,3-benzenedisulfonamide, $C_6H_6Cl_2N_2O_4S_2$, M_r 305.2. For synthesis, see [87].

Trade Names: Antidrasi (ISF, Merck Sharp & Dohme); Daranide (Merck Sharp & Dohme); Fenamide (Farmigea); Glauconide (Llorens); Glaumid (SIFI); Hipotensor Oftalmico, Oralcon, Oratrol (Alcon).

Methazolamide [*554-57-4*], N-[5-(aminosulfonyl)-3-methyl-1,3,4-thiadiazol-2(3H)-ylidene]-acetamide, $C_5H_8N_4O_3S_2$, M_r 236.3. For synthesis, see [88].

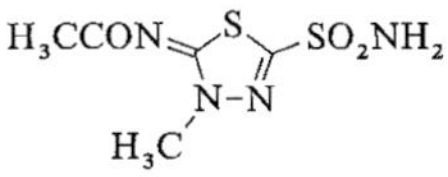

H₃CCON—S—SO₂NH₂
      N–N
  H₃C

Trade Names: Neptazane, Theraplix (Lederle).

5.4. Hyperosmotic Agents

Orally (glycerol, isosorbide, urea) or intravenously (mannitol) administered hyperosmotic agents produce a rapid reduction of IOP as a result of migration of water from the eye to ocular blood vessels [72], [75]. These drugs may have significant side effects [72], [75], [76]. They are used primarily to treat acute IOP elevation and in ocular surgical procedures [72].

Glycerol [*56-81-5*], propane-1,2,3-triol, $C_3H_8O_3$, M_r 92.1.

Trade Names: Ophthalgan (Wyeth-Ayerst), Osmoglyn (Alcon).

Isosorbide [*652-67-5*], 1,4:3,6-dianhydrosorbitol, $C_6H_{10}O_4$, M_r 146.1.

Trade Name: Ismotic (Alcon).

Mannitol [*69-65-8*], cordycepic acid, mannite, $C_6H_{14}O_6$, M_r 182.2.

Trade Name: Mannitol Injection (Astra).

Urea [*57-13-6*], carbamide, ureum, CH_4N_2O, M_r 60.1.

Trade Name: Ureaphil (Abbott).

5.5. Miotics (Parasympathomimetic Agents)

Miotics are agents that cause constriction of the pupil. Two classes of miotics, cholinergics and anticholinesterases (parasympathomimetic agents), have therapeutic roles in glaucoma. Acetylcholine itself is not generally useful because of its short biological half-life and topical ineffectiveness, although it can induce miosis when applied directly to the iris during surgery [73]. Cholinergic agonists (pilocarpine, carbachol) act directly on cholinergic effector cells to mimic the effects of acetylcholine; anticholinesterases (physostigmine, demecarium bromide, echothiophate iodide, isoflurophate) prolong the action of endogenous acetylcholine by blocking its hydrolysis by cholinesterases. In the eye, miotics appear to reduce IOP through activation of muscarinic acetylcholine receptors, which contract the ciliary muscle and enhance flow through the trabecular meshwork. Muscarinic activation also constricts the pupillary sphincter and produces miosis, which is sometimes an undesirable side effect of antiglaucoma therapy. Other ocular and systemic side effects may be associated with the use of miotics for antiglaucoma therapy [72], [73], [76], [77], [80].

5.5.1. Cholinergic Agonists

Carbachol [51-83-2], O-carbamoylcholine chloride, $C_6H_{15}ClN_2O_2$, M_r 182.7. Carbachol is inherently more potent and longer acting than pilocarpine but has poorer corneal penetration and more severe ocular side effects. Enhancement of corneal penetration may be achieved with 0.03% benzalkonium chloride (BAK) in the formulation, or with lower levels of BAK along with methyl cellulose or hydroxypropyl methyl cellulose to prolong contact with the ocular surface [73]. Carbachol is used when the response to pilocarpine is inadequate or when a pilocarpine allergy develops [73], [76].

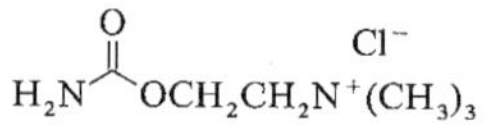

Trade Names: Isopto Carbachol (Alcon), also available as a 0.01% solution; Miostat (Alcon) for intraocular injection for miosis during surgery.

Pilocarpine [92-13-7], (3S,4R)-3-ethyldihydro-4-[(1-methyl-1H-imidazol-5-yl)methyl]-furan-2(3H)-one, $C_{11}H_{16}N_2O_2$, M_r 208.3. Pilocarpine is one of the most useful drugs for management of glaucomas, including primary open-angle and acute angle-closure glaucomas [72], [73]. Principal ocular side effects of pilocarpine solutions (0.25–10%) include miosis, ciliary spasm, and visual blurring [72], [73], [76]. A sustained-release drug delivery system (Ocusert) for pilocarpine base is also available [72], [73], [76].

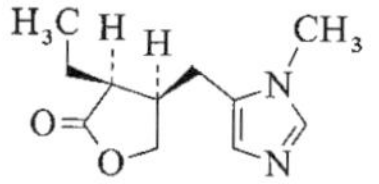

Trade Names: Ocusert Pilo-20 and Ocusert Pilo-40 Ocular Therapeutic System (20 or 40-µg pilocarpine delivery per hour for one week, Alza).

Pilocarpine hydrochloride [54-71-7], $C_{11}H_{16}N_2O_2 \cdot HCl$, M_r 244.7.

Trade Names: Akarpine (Akorn); I-Pilocarpine (Americal); Isopto Carpine (Alcon); Ocu-Carpine (Ocumed); Pilocar (IOLAB); Pilokair (Pharmafair); Pilopine (Alcon), and in various combinations with epinephrine bitartrate: E-Pilo-1 – 6 (IOLAB).

Pilocarpine nitrate [148-72-1], $C_{11}H_{16}N_2O_2 \cdot HNO_3$, M_r 271.3.

Trade Name: Pilagan Liquifilm (Allergan).

5.5.2. Anticholinesterases

Anticholinesterases may be divided into two classes. The short-acting "reversible" inhibitors (e.g., physostigmine and demecarium) transfer a carbamoyl group to the enzyme, which is removed within a few hours to regenerate the enzyme. The long-acting irreversible inhibitors (e.g., echothiophate and isoflurophate) phosphorylate the enzyme, and cholinesterase activity is restored only through synthesis of new enzyme.

The anticholinesterases are contraindicated for angle-closure glaucoma [72], and their ocular use may produce a variety of systemic and ocular side effects [72], [73], [76]. Anticholinesterases are used primarily when other antiglaucoma medications have failed and are the least commonly used antiglaucoma agents [72].

Physostigmine [57-47-6], eserine, $C_{15}H_{21}N_3O_2$, M_r 275.4.

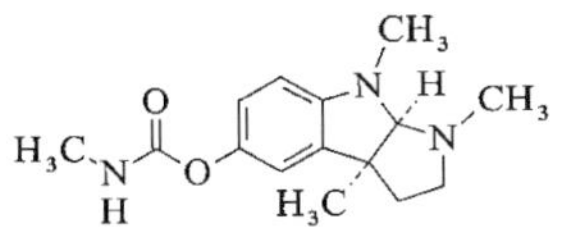

Trade Names: Eserine sulfate (0.25%) Sterile Ophthalmic Ointment (IOLAB).

Demecarium bromide [56-94-0], $C_{32}H_{52}Br_2N_4O_4$, M_r 716.6.

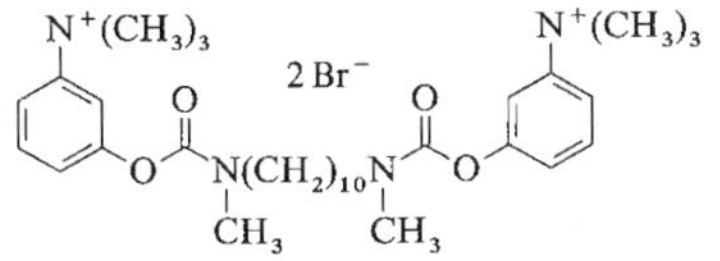

Trade Names: Humorsol (Merck Sharp & Dohme), Tosmilen (Chem Linz, Sinclair).

Echothiophate iodide [513-10-0] (echothiophate [6736-03-4]), mercaptoethyltrimethylammonium iodide *O,O*-diethyl phosphorothioate, $C_9H_{23}INO_3PS$, M_r 383.2.

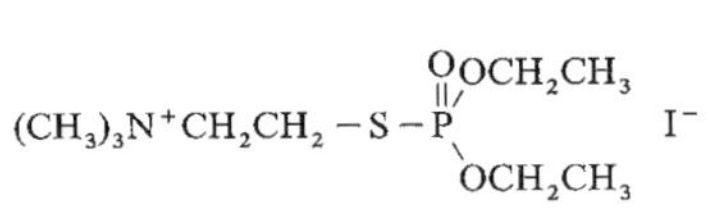

Trade Names: Echofilina, Iodeto de fosfolina, Phospholine (Sodip); Phospholine Iodide (Wyeth-Ayerst, Promedica, Chinoin); Phospholinjodid (Winzer).

Isoflurophate [55-91-4], diisopropyl flurophosphate, $C_6H_{14}FO_3P$, M_r 184.1.

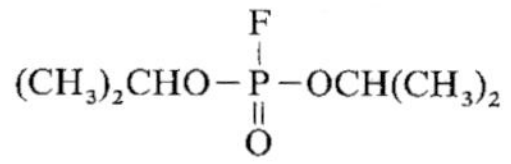

$$(CH_3)_2CHO-\underset{\underset{O}{\|}}{\overset{\overset{F}{|}}{P}}-OCH(CH_3)_2$$

Trade Names: Diflupyl (Labaz), DFP-Oel (Winzer), Floropryl (Merck Sharp & Dohme).

6. Mydriatics and Cycloplegics

Mydriatic and cycloplegic agents are used routinely in ophthalmic practice for dilating the pupil to facilitate examination of the retina [89]–[98]. In addition to dilation, cycloplegics cause paralysis of accommodation for near vision and are used primarily as an aid in refraction and in the treatment of uveitis [97]. They are also used as postoperative agents in cataract and retinal detachment surgery.

Mydriatics and cycloplegics are divided into sympathomimetics and parasympatholytics. They can be used alone or in combination. Sympathomimetic agents imitate (direct acting) or potentiate (indirect acting) the action of norepinephrine at sympathetic nerve terminals and produce mydriasis by stimulating the iris dilator muscle fibers. Parasympatholytic agents block the action of acetylcholine at the neuromuscular junction and produce pupil dilation with loss of accommodation by immobilizing the iris sphincter and ciliary muscle.

Although serious side effects from mydriatics and cycloplegics are rare, individual response to these drugs varies [98]. They should be used with caution in patients with closed-angle glaucoma or in patients with a narrow angle between the iris and the cornea because they may increase the IOP and precipitate an acute attack [98].

6.1. Sympathomimetics

Phenylephrine hydrochloride [61-76-7], (S)-1-(3-hydroxyphenyl)-2-methylaminoethanol hydrochloride, $C_9H_{13}NO_2 \cdot HCl$, M_r 203.7. Phenylephrine hydrochloride is used topically (2.5 and 10% solutions) to dilate the pupil both for ophthalmoscopy in the treatment of uveitis and for cataract surgery [97]. Maximum dilation occurs within 45–60 min and recovery in about 6 h. A 1% solution can be used in the diagnosis of Horner's syndrome [97].

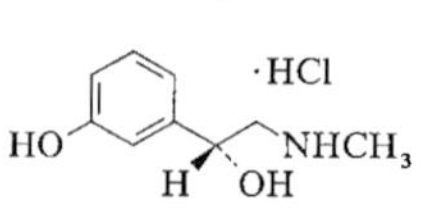

Trade Names: Ak-Dilate, Ak-Vernacon (Akorn); Blephamide (Allergan); I Phrine (Americal); Isonefrine (Allergan); Isopto Frin (Alcon); Minims Phenylephrine Hydrochloride (Smith & Nephew); Mydfrin

(Alcon); Neosynephrine (Winthrop); Opstin (Allergan); Phenylephrine Hydrochloride (IOLAB); Prefin (Allergan); Vasosulf (Cooper Vision); Vistosan (Allergan); Zincfrin (Alcon).

Hydroxyamphetamine hydrobromide [306-21-8], (±)-4-(2-aminopropyl)phenol hydrobromide, $C_9H_{13}NO \cdot HBr$, M_r 232.1. For synthesis, see [99]. Hydroxyamphetamine hydrobromide is an indirect-acting sympathomimetic agent. It is used in a 1% solution as a mydriatic (effect comparable to 2.5% phenylephrine) and in the diagnosis of Horner's syndrome [98]. Hydroxyamphetamine hydrobromide may be used in patients allergic to phenylephrine.

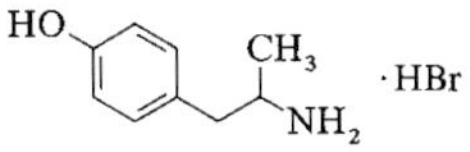

Trade Name: Paredrine (SKF).

Cocaine [53-21-4], $C_{17}H_{21}NO_4 \cdot HCl$, M_r 339.8, is an indirect-acting sympathomimetic agent. In addition to its mydriatic action, it is also a local anesthetic and vasoconstrictor. Cocaine is not available commercially as a solution for clinical use.

6.2. Parasympatholytics

Atropine sulfate [5908-99-6] (monohydrate [55-48-1]), $(C_{17}H_{23}NO_3)_2 \cdot H_2SO_4 \cdot H_2O$, M_r 694.8. See also → Antiasthmatics. Atropine is the most effective cycloplegic agent and has the longest duration of action. It is used to maintain a dilated pupil after intraocular surgery. Pupil dilation occurs within 30–40 min and lasts up to 2 weeks.

Trade Names: Alcon Opulets Atropine 1% (Alcon); Atropine-Care Ophthalmic (Akorn); Atropine Minims (Smith & Nephew); Atropine Sulfate Ophthalmic (Alcon, Allergan, Moore, Rugby, Schein, Steris, Balon, Fougera, Lilly); Atropine Sulfate S.O.P. (Allergan); Atropisol (IOLAB); Cicloplegyl (Winzer); Isopto Atropine (Alcon); I-Tropine (Americal); Minims Atropine Sulfate (Smith & Nephew); SMP Atropine (Cooper Vision).

Cyclopentolate hydrochloride [5870-29-1], $C_{17}H_{25}NO_3 \cdot HCl$, M_r 327.9. For synthesis, see [100]. Cyclopentolate hydrochloride has a rapid onset (30–60 min) but shorter duration of action (< 24 h) than atropine [97].

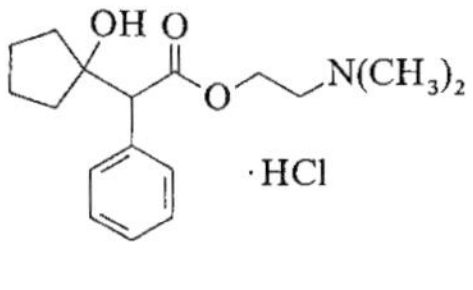

Trade Names: Alcon Opulets Cyclopentolate (Alcon); Ak-Pentolate (Akorn); Ciclolux (Allergan); Ciclopegic, Colircusi ciclopejico, Colirio Oculos Cicloplegic (Llorens), Cyclopen, Cyclopentol Colircusi,

Cyclomydril, Cyclogyl (Alcon); Cyclopentolat Minims, Cyplegin, Mydplegin, Minims Cyclopentalate Hydrochloride (Smith & Nephew); Mydrilate (Boehringer, Ingelheim); Skiacol (Pos); Zyklolat (Mann).

Homatropine hydrobromide [*51-56-9*], $C_{16}H_{21}NO \cdot HBr$, M_r 356.3. Homatropine hydrobromide is weaker and less toxic than atropine. The onset of action occurs in about 15 – 20 min and lasts ca. 3 h. Complete recovery time is about 36 – 48 h. Homatropine is used for refraction, in the treatment of inflammatory conditions of the uveal tract, and as a pre- and postoperative agent. It is also used as an optical aid for axial lens opacities [97].

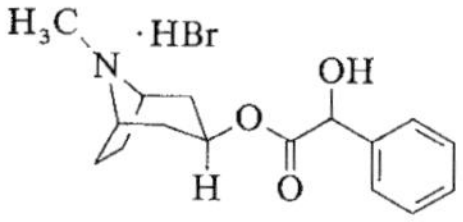

Trade Names: Ak-Homatropine (Akorn), Allergan Homatropine (Allergan), Ancatropine Infant Drops (Anca), Homatrine (Americal), Homatropine HBr Ophthalmic (IOLAB), Isopto Homatropine (Alcon), Minims Homatropine Hydrobromide (Smith & Nephew).

Scopolamine hydrobromide [*114-49-8*], (trihydrate [*6533-68-2*]), $C_{17}H_{21}NO_4 \cdot HBr \cdot 3\,H_2O$, M_r 438.3. Scopolamine hydrobromide is an effective cycloplegic and is used in the treatment of uveitis, in refraction of children, postoperatively, and in patients sensitive to atropine. Mydriasis and cycloplegia occur within 20 – 60 min and last 3 – 7 d. The duration of action is much shorter in eyes with inflammation.

Trade Names: Contac (Allergan), Isopto Hyoscine (Alcon), Minims Hyoscine Hydrobromide (Smith & Nephew), Murocoll-2 (Bausch & Lomb).

Tropicamide [*1508-75-4*], *N*-ethyl-*N*-(4-pyridylmethyl)tropamide, $C_{17}H_{20}N_2O_2$, M_r 284.4. For synthesis, see [101]. Tropicamide is used as a shorter-acting substitute for atropine or scopolamine when prolonged mydriasis and cycloplegia are not required. It is an effective mydriatic with weak cycloplegic activity and is, therefore, useful for ophthalmoscopy and some preoperative and postoperative states. Onset of mydriasis takes place within 15 – 20 min and lasts ca. 7 h [98]. Mydriasis may be counteracted by local application of pilocarpine. Maximum cycloplegia occurs within 20 – 25 min; the duration of this peak effect is 15 – 20 min and has passed in 6 – 8 h.

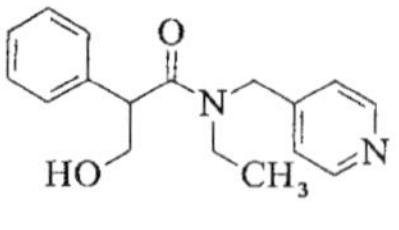

Trade Names: I-Picamide (Americal), Minims Tropicamide (Smith & Nephew), Mydriacyl (Alcon), Mydrian (Dispersa), Mydriaticum (Merck Sharp & Dohme), Phenyltrope (Akorn), Tropicacyl (Akorn), Tropikamid Minims (Smith & Nephew), Tropimil (Farmigea), Visumidriatic (Merck Sharp & Dohme).

7. Vasoactive (Adrenergic) Agents

In addition to their mydriatic effect, adrenergic agents constrict the vascular system of the conjunctiva within minutes (α-adrenergic effect) and are used to treat congestion and relieve minor allergy irritation and itching of the conjunctiva. Commercially available ophthalmic vasoconstrictive agents usually contain ephedrine, naphazoline, oxymetazoline, phenylephrine, or tetrahydrozoline. Side effects are not typically observed because of the relatively low concentrations used but rebound congestion can occur with extended use [97].

Naphazoline hydrochloride [550-99-2], 2-(1-naphthylmethyl)-2-imidazoline hydrochloride, $C_{14}H_{14}N_2 \cdot HCl$, M_r 246.7. For synthesis, see [102]. One or two drops of naphazoline solution (0.012–0.1%) are administered in the eye every 3–4 h.

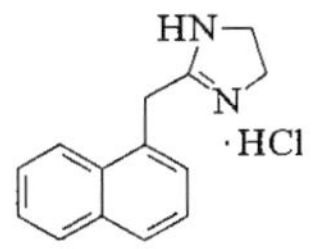

Trade Names: Ak-Con (Akorn); Albalon (Allergan); Allerest Eye Drops (Pharmacraft); Allergy Drops (Bausch & Lomb); Clear Eyes (Ross); Clera (Person and Covey); Comfort Eye Drops, Degest 2 (Barnes-Hind); Imidazyl (Allergan); Naphcon (Alcon); I-Naphline (Americal); Opcon (Bausch & Lomb); Optazine (Lederle); Otrivin, Privin, Privina, Privine (Ciba); Vasocon Regular (IOLAB).

Naphazoline formulations that also contain an antihistamine are also available: Ak-Con-A (Akorn); Opcon-A (Bausch & Lomb); Naphcon-A, Vasocon-A (IOLAB); Vasoclear (IOLAB); Vistalbalon, Albalon A Liquifilm (Allergan); Zincfrin-A also contains zinc sulfate (Alcon).

Oxymetazoline hydrochloride [2315-02-8], 2-(4-*tert*-butyl-3-hydroxy-2,6-dimethylbenzyl)-2-imidazoline hydrochloride, $C_{16}H_{24}N_2O \cdot HCl$, M_r 296.8.

Trade Names: Ocuclear Eye Drops (Schering), Visine L.R. Eye Drops (Pfizer).

Phenylephrine hydrochloride [61-76-7], (*S*)-1-(3-hydroxyphenyl)-2-methylaminoethanol hydrochloride, $C_9H_{13}NO_2 \cdot HCl$, M_r 203.7. Phenylephrine is used at a concentration of 0.12%. This relatively low concentration causes vasoconstriction with little or no pupillary dilation. One or two drops are instilled in the eye(s) three or four times daily to relieve burning and itching in mild cases of noninfectious conjunctivitis.

Trade Names: Ak-Dilate, Ak-Vernacon, Ak-Nephrin, (Akorn); Isonefrine (Allergan); Isopto Frin, Isopto Phenylephrine (Alcon); Metaoxedrin Minims, Minims Phenylephrine Hydrochloride (Smith & Nephew); Mydfrin (Alcon); Ocu-Phrin (Ocumed); Optistin, Prefrin, Prefrin Liquifilm, Prefrin-A, Prefrin-Z (Allergan); Vasosulf (CooperVision); Relief (Allergan); Zincfrin (Alcon).

Tetrahydrozoline hydrochloride [522-48-5], 2-(1,2,3,4-tetrahydro-1-naphthyl)-2-imidazoline hydrochloride, $C_{13}H_{16}N_2 \cdot HCl$, M_r 236.7. For synthesis, see [103]. Tetrahydrozoline hydrochloride solutions (0.05%) are used to decrease swelling of the mucosa in conjunctivitis. One or two drops are instilled in the eye up to four times daily. The onset of vasoconstriction becomes apparent within minutes.

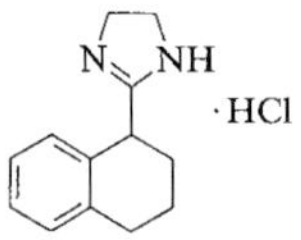

Trade Names: Berberin Ophtiole (Mann); Collyrium (Wyeth); Optigene 3 (Pfeiffer); Mallazine Drops (Hauck); Murine (Abbott); Murine Plus (Ross); Soothe Eye Drops (Alcon); Visine A.C., Visine Eye Drops (Leeming); Visine Plus (G.P. Laboratories).

8. Diagnostic Agents

Diagnostic agents are used to examine the eye for signs of systemic disease and to diagnose ocular abnormalities. Dyes used for this purpose include fluorescein, rose bengal, indocyanine green [3599-32-4], trypan blue [72-57-1], alcian blue [12040-44-7], methylene blue [61-73-4], and fluorexon. Of these, only fluorescein, rose bengal, and fluorexon are commercially available [104], [105]. Several drugs are also used as ophthalmic diagnostic agents, including edrophonium chloride, methacholine, and pilocarpine.

Fluorescein C.I. 45 350 [2321-07-5] (fluorescein sodium [518-47-8]), $C_{20}H_{12}O_5$, M_r 332.3, *mp* 314 – 316 °C (decomp.). The synthesis, properties, and histological uses of fluorescein have been reviewed [106], [107]. Fluorescein sodium is used typically as a 1 or 2% solution or as impregnated filter paper strips for detecting foreign bodies and examination of the corneal surface. During corneal surgery, topical administration of fluorescein can be used to detect leakage from the anterior chamber. In cataract surgery, leaks are detected as bright green rivulets [104], [105].

Fluorescein is used for the fitting and management of hard contact lenses. In the presence of fluorescein the tear layer fluoresces green under a cobalt blue light and contrasts with a blue fluorescence or the absence of fluorescence where the lens comes in contact with the cornea. Fluorescein is not useful for fitting soft contact lenses because it penetrates the lens matrix, providing insufficient contrast between the lens and the tissue [104], [105].

Fluorescein is also used intravenously in concentrations of 5 – 25% for examination of the ophthalmic vasculature and the presence of ocular lesions.

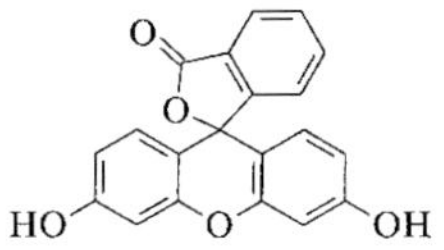

Trade Names: Ak-Fluor (Akorn), Fluorescite (Alcon), Fluorets (Akorn), Fluor-I-Strip (Wyeth-Ayerst), Ful-Glo (Barnes-Hind), Funduscein (CooperVision); with proparacaine hydrochloride: Fluoracaine (Akorn).

Rose bengal, C.I. 45 440 [*11121-48-5*], (disodium salt [*632-69-9*]), $C_{20}H_2Cl_4I_4Na_2O_5$, M_r 1017.6. The dipotassium or disodium salt of rose bengal dissolves in water, giving a bluish red color [106]. Unlike fluorescein, rose bengal stains degenerated corneal and conjunctival epithelial tissue red. The stain intensity correlates with the state of degeneration, with dead cells staining intensely [108]. The dye is useful in highlighting abnormal epithelial cells in "dry eye" conditions. It is also a useful adjunct in determining the area of epibulbar squamous neoplasms [109].

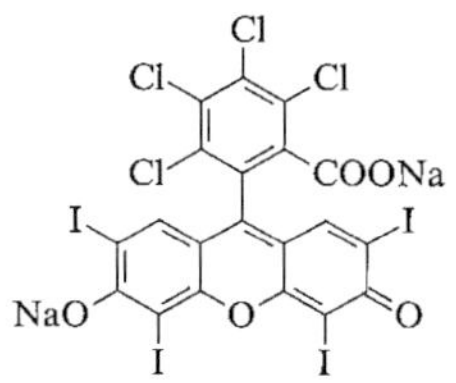

Trade Names: EV Rose Bengal (Eagle Vision), Rose Bengal (Akorn).

Fluorexon [*1461-15-0*], $C_{30}H_{26}N_2O_{13}$, M_r 618.54; tetrasodium salt $C_{30}H_{22}N_2Na_4O_{13}$, M_r 710.5. Unlike fluorescein, fluorexon is slow to penetrate the matrix of soft contact lens materials and may be used as an adjunct for fitting soft contact lenses. (However, it is not recommended for lenses with > 60% water content [110].)

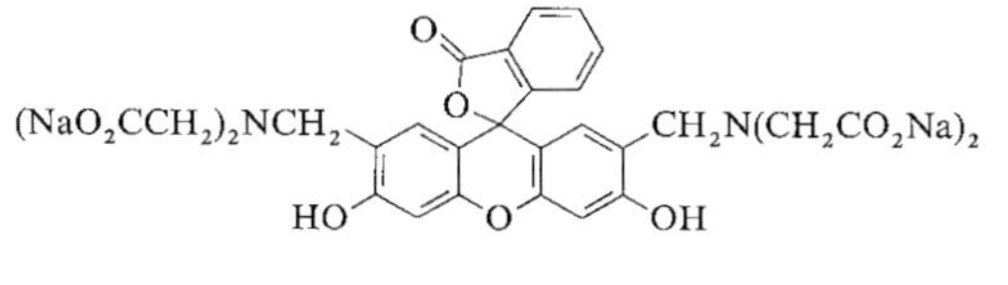

Trade Name: Fluoresoft (Holles).

Edrophonium chloride [*116-38-1*], ethyl-(2-hydoxyphenyl)dimethylammonium chloride, $C_{10}H_{16}ClNO$, M_r 201.7, mp 162–163 °C (decomp.). For synthesis, see [111]. Edrophonium chloride (an anticholinesterase) may be injected intravenously to determine whether the cause of ptosis (drooping of the upper eyelid) is myasthenia gravis (a neuromuscular disease).

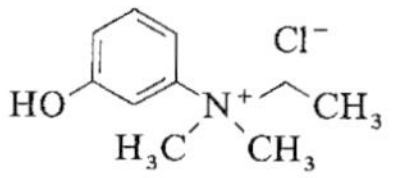

Trade Name: Enion (Anaquest), Tensilon (Roche).

Methacholine chloride [62-51-1], acetyl-β-methylcholine chloride, $C_8H_{18}ClNO_2$, M_r 195.7, mp 172–173 °C. A 2.5% methacholine (a cholinergic drug) solution is applied topically to the conjunctival sac to diagnose Adie's pupil (a disruption of cholinergic innervation to the iris). Normal pupils do not respond significantly to methacholine, whereas the Adie's pupil responds with intense miosis.

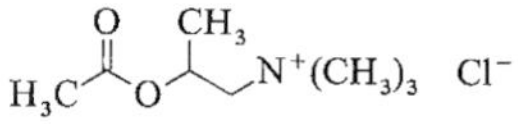

Trade Name: Provocholine (Hoffmann-La Roche).

Pilocarpine, see Section 5.5.1. A 0.1–0.125% solution of pilocarpine can be used to diagnose Adie's pupil. A 0.5–1% solution can be used in the differential diagnosis of the fixed, dilated pupil. The dilated pupil responds to the drug if the dilation is of neurologic origin, where it does not respond if the etiology is anticholinergic in nature [104], [112].

9. Dry Eye Medications

"Dry eye" denotes a number of conditions including insufficient lacrimation, mucin deficiency, lipid abnormality, impaired blinking, or primary ocular disorder [113]. The seriousness of dry eye ranges from mild "dry and scratchy" eyes to serious degeneration and keratinization of corneal tissue, which may lead to blindness. The latter condition is often found in developing countries in areas of malnutrition (deficiency of protein and vitamin A). However, in highly developed nations, a milder form of dry eye occurs, which is characterized by chronic eye irritation and decreased visual acuity [114]. Symptoms are normally subclinical but become symptomatic in a dry, windy, or dusty atmosphere, in air conditioning; or when contact lenses are worn.

If dietary deficiency is the underlying cause of the dry eye syndrome, a proper diet should be implemented. *trans*-Retinoic acid ointment may be applied topically to counter vitamin A deficiency [115]. In dry eye conditions induced by infection, treatment with an antimicrobial agent should be initiated.

The most common treatment of nonspecific, incipient dry eye is with tear substitutes [113]. Commercial dry eye preparations typically contain at least one *demulcent*, which lubricates and wets the eye, and a *preservative* to prevent microbial contamination of the product. Nonpreserved preparations are also available in single-dose containers. Some demulcents serve as thickening agents (e.g., substituted cellulose ethers) based on the concept that highly viscous tear substitutes should have a longer retention time in the eye. Because contact lens wearers generally experience improved comfort from the lubricant and wetting properties of demulcents, many lens preparations include one or more of these agents (see Section 10.1).

Other components of dry eye medications include *emollients* (such as petrolatum, mineral oil, and lanolin), which lubricate and protect the eye from drying, and *lipids*, which are intended to supplement a deficient superficial lipid layer of tear fluid. Both emollients and lipids may, however, interfere with visual acuity.

The most common demulcents are listed below.

Carboxymethyl cellulose sodium [*9004-32-4*], exists as white granules, with water solubility dependent on the degree of substitution. It is available in various viscosities.

Trade Names: Bro-Lac (Riker), Celluvisc (Allergan).

Dextran [*9004-54-0*], a polysaccharide with a backbone of D-glucose units is produced from bacteria cultured with a sucrose substrate.

Trade Names: With hydroxymethyl cellulose: Muro Tears TM (Muro), Tears Naturale R (Alcon); with hydroxypropyl methyl cellulose and dextran 70: Moisture Drops (Bausch & Lomb).

Gelatin is a mixture of water-soluble proteins obtained after collagen-containing animal tissues (e.g., skin, tendons, ligaments, bones) have been boiled with water.

Trade Names: Lacril R (Allergan).

Gycerol [*56-81-5*], propane-1,2,3-triol, $C_3H_8O_3$, M_r 92.1.

Trade Names: Dry Eye Therapy (Bausch & Lomb); with dextran 70 and hydroxypropyl methyl cellulose: Moisture Drops (Bausch & Lomb).

Ophthalmic Preparation: Glycerin Ophthalmic Solution, U.S.P.

Hydroxyethyl cellulose [*9004-62-0*], nearly odorless, yellowish-white, white, or grayish-white, hygroscopic granules or powder.

Trade Names: Adsorbo Tears (Alcon); Lyteers (Barnes-Hind); with polyvinyl alcohol: Comfort Tears, Neo Tears (Barnes-Hind); with lipid: TearGard (Bioproducts).
Contact lens products: Soft Mate ps (Barnes-Hind), Clerz (CooperVision)

Hydroxypropyl cellulose [*9004-64-2*] is an off-white powder that softens at 130 °C, produces aqueous solutions with a wide range of viscosities, and precipitates from solution at 40–45 °C.

Trade Name: Lacrisert (Merck Sharp & Dohme), a sterile (prescription only) ophthalmic insert that slowly releases hydroxypropyl cellulose.

Hydroxypropyl methyl cellulose [*9004-65-3*] is a powder that slowly dissolves in cold water (insoluble in hot water) to give solutions with a wide range of viscosities.

Trade Names: Isopto Alkaline, Isopto Plain, Isopto Tears (Alcon); Tearisol (CooperVision); Ultratears (Alcon); with gelatin and polysorbate 80: Lacril (Allergan); with dextran: Tears Naturale (Alcon); with dextran 70 and glycerol: Moisture Drops (Bausch & Lomb).
Lens care products: With poly(vinyl alcohol): Liquifilm.

Methyl cellulose [*9004-67-5*], consists of white granules that are soluble in cold water (insoluble in hot water), with solubility and viscosity dependent on the degree of substitution.

Trade Names: Methopto Forte (Bioproducts), Methulose (Warner-Lambert), Milroy Artificial Tears (Milroy), Murocel Ophthalmic Solution (Muro), Visculose (Warner-Lambert).

Polyethylene glycol [*25322-68-3*], is a clear liquid or white solid that dissolves in water to form transparent solutions of various viscosities.

Trade Names: With poly(vinyl alcohol): HypoTears (CooperVision).
Contact lens products: With polyoxyl 40 stearate: Blink-N-Clean (Allergan).

Poly(vinyl alcohol) [*9002-89-5*] is a white to cream colored powder that softens at ca. 200 °C; it is available in a variety of viscosities and solubilities in water.

Trade Names: Liquifilm, Pre-Sert (Allergan); with poly(ethylene glycol); HypoTears (CooperVision); with hydroxyethyl cellulose: NeoTears (Barnes-Hind); with polyvinylpyrrolidone: Refresh, Tears Plus (Allergan).
Contact lens products: Lens-Wet, Pre-Sert, Total, Wet-N-Soak (Allergan), Hy-Flow (Cooper-Vision); with hydroxypropyl methyl cellulose: Liquifil (Allergan), Barnes-Hind Wetting Solution, Visalens Wetting Solution; with povidone and hydroxyethyl cellulose: Barnes-Hind Wetting and Soaking Solution (Barnes-Hind); with poly-(ethylene glycol) and hydroxyethyl cellulose: Lensine 5 (CooperVision).

Povidone [*9003-39-8*], polyvinylpyrrolidone, a pale yellow solid, M_r 10 000 – 700 000, gives colloidal solutions in water.

Trade Names: With other demulcents: AdsorboTear (Alcon); Refresh, Tears Plus (Allergan).

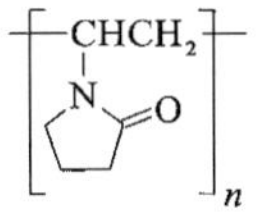

[structure: poly(N-vinylpyrrolidone) repeat unit —CHCH₂— with N attached to pyrrolidone ring]

Contact lens products: Soft Care (Barnes-Hind).

Propylene glycol [*57-55-6*], 1,2-propane-diol, $C_3H_8O_2$, M_r 76.09, *mp* –59 °C.

Contact lens products: Allergan Hydrocare (Allergan).

10. Miscellaneous Ophthalmic and Contact Lens Preparations

Contact lenses are broadly classified as "hard" or "soft". *Hard lenses* are typically composed of poly(methyl methacrylate) (PMMA); they are inflexible and hydrophobic, and retain their shape when removed from the eye. *Soft ("hydrophilic") lenses* are flexible, hydrophilic (typically with a water content of 30–79 %), and conform to the contour of the supporting surface. They are composed of any one of a wide variety of cross-linked polymers that form a hydrophilic gel network (e.g., polymacon, a copolymer of 2-hydroxyethyl methacrylate with 2-hydroxyethyl-2-methyl-2-propenoate) [116]–[118]. In addition, *"semirigid" or gas-permeable lenses* are available. These lenses are commonly made from PMMA and silicone, and are a hybrid of soft and hard lenses.

The advantages and disadvantages of each type of lens are reviewed elsewhere [116]–[118]. The wide range of polymers used for contact lens materials results in a broad spectrum of physicochemical interactions between the lenses, tear fluid, eye tissue, and contact lens products. For example, 20–30 μg of tear-fluid protein is typically deposited on a polymacon soft lens, whereas > 500 μg of protein may be deposited on an etafilcon A soft lens during the same period of wear [119]. (Etafilcon A is a copolymer of 2-hydroxyethyl methacrylate, sodium methacrylate, and the trimethacrylate ester of 2-ethyl-2-hydroxymethyl-1,3-propanediol). Moreover, the compositions of the tear-fluid proteins deposited on each lens type differ: the negatively charged etafilcon A lens attracts more positively charged lysozyme than the nonionic polymacon material.

The effectiveness and cytotoxicity of the agents used for contact lens disinfection, cleaning, lubrication, wetting, etc., vary markedly with the properties of the lens material. Many factors may affect product efficacy, safety, and utility. Heat disinfection is unsatisfactory for some hydrophilic lens materials because it reduces lens life and makes deposits more difficult to remove after thermal disinfection. Chemical *disinfectants and preservatives* are described in Section 10.1. Hydrogen peroxide (3 %) is a good antimicrobial agent, but its use requires a neutralization step and has patient compliance problems. Thimerosal is also a good antimicrobial agent, but a substantial proportion of the population has become sensitized to it. Benzyl alcohol is an effective preservative but tends to irritate the eye [120]. Chlorobutanol is only a fair preservative because it has a slow kill rate, is unstable in solutions above pH 6, and is adsorbed by containers. Ethylenediaminetetraacetic acid is not an effective antimicrobial agent when used alone; it disrupts the integrity of bacterial cell walls and is used in combination with other preservatives.

Boric acid (often used as a buffering agent in ophthalmic and contact lens care solutions) and sorbic acid are preservatives with relatively low cytotoxicity, but they are not useful as lens disinfectants. Quaternary amines such as benzalkonium chloride are good disinfectants and preservatives but accumulate on negatively charged hydrophilic lenses with a high water content, causing cytotoxic responses [121]–[126]. Polyquad is the only quaternary amine currently in use for soft lens disinfection.

Biguanides are both potent disinfectants and good preservatives. Unfortunately, chlorhexidine has been associated with sensitivity reactions. In contrast, poly(hexamethylenebiguanide) (PHMB) has a very low frequency of adverse patient response and is safe for use with soft lens materials. Lenses disinfected with PHMB solution (0.5 ppm) can be placed directly on the eye, without the saline rinse required by most other disinfectants.

Considerable variation exists among individuals in both tear chemistry and tear deposits on contact lenses [127]–[132].

Most *lens cleaning agents* are used daily or weekly (see Section 10.2). Daily cleaners (e.g., macrogol esters and ethers, sodium lauryl sulfate, and block copolymers) remove proteinaceous and lipoidal lens deposits by their surfactant action. However, daily cleaners do not completely remove protein deposits, and proteins tend to accumulate with time, serving as a possible site for microbial growth and causing eye irritation and lens clouding. Weekly cleaners are used to remove accumulated protein deposits. They are generally formulations of proteolytic enzymes (e.g., pancreatin, papain, or subtilisin-A), which cleave the protein into smaller fragments that are more readily rinsed off the lens surface.

In addition to disinfectants, preservatives, and cleaning agents, contact lens preparations may contain *lubricants, wetting agents,* or *viscosity adjusters* to improve lens comfort. Formulations containing such materials are listed in Chapter 9 (contact lens products). Contact lens solutions also contain *buffers* (e.g., borate, citrate, and phosphate) and *osmolality adjusting agents* (usually sodium chloride).

10.1. Disinfectants and Preservatives

10.1.1. Biguanides

Chlorhexidine gluconate [*18472-51-0*], chlorhexidine digluconate, $C_{22}H_{30}Cl_2N_{10} \cdot 2\, C_6H_{12}O_7$, M_r 897.8.

Contact lens products: Bausch & Lomb Sterile Disinfecting Solution (Bausch & Lomb); Flexcare, Flexsol (Alcon); Soft-Mate (Barnes-Hind).

Poly(hexamethylenebiguanide) hydrochloride [*27083-27-8*], polyaminopropyl biguanide hydrochloride $[C_8H_{17}N_5]_n \cdot$ HCl. For synthesis, see [120].

Contact lens products: ReNu Multi-Purpose Solution; ReNu Saline Solution (Bausch & Lomb).

10.1.2. Mercurial Preservatives

Phenylmercury(II) acetate [*62-38-4*], $C_8H_8HgO_2$, M_r 336.8, *mp* 149 °C.

Trade Name: Blinx (Barnes-Hind).

Phenylmercury(II) nitrate [*8003-05-2*], $C_{12}H_{11}Hg_2NO_4$, M_r 634.4, *mp* 187 – 190 °C (decomp.).

Contact lens preparation: Clean-N-Soak (Allergan).

Thimerosal [*54-64-8*], $C_9H_9HgNaO_2S$, M_r 404.84.

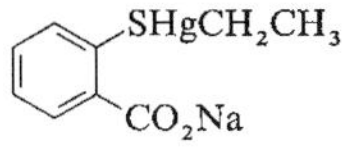

Trade Names: Absorbonac, Absorbotear (Alcon); Collyrium (Wyeth); Eye Cool, 20/20 Eye Drops (Milroy); Liquifilm Forte, Prefin Z Liquifilm (Allergan); M/Rinse (Milroy); Neo Tears (Barnes-Hind); Soothe (Alcon).

Contact lens products: Adapettes (Alcon); Cleaning & Disinfecting (Allergan); Clerz (Cooper Vision); Dual-Clean (Sherman); Flex-Care, Flexsol (Alcon); Gel-Clea (Barnes-Hind); LC-65, Lens-Wet (Allergan); Pena-Vel II (Serman Labs.); Soaclens (Alcon); Soft Care, Soft Mate (Barnes-Hind); Stay Brite (Sherman); Sterile Disinfecting Solution, Sterile Lens Lubricant (Bausch & Lomb).

10.1.3. Quaternary Ammonium Compounds

Benzalkonium chloride [*8001-54-5*] is a mixture of alkyldimethylbenzylammonium chlorides.

Trade Names and contact lens products are too numerous to list.

Benzethonium chloride [*121-54-0*], $C_{27}H_{42}ClNO_2$, M_r 448.1, mp 164–166 °C.

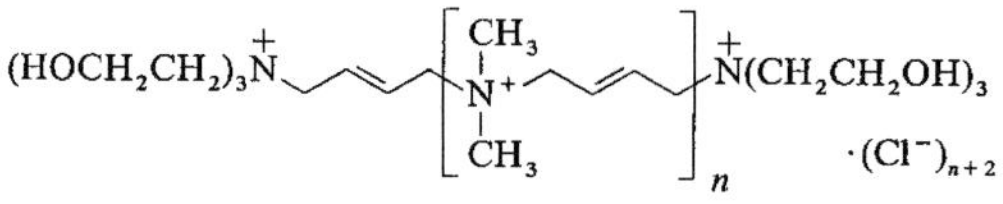

$$(HOCH_2CH_2)_3\overset{+}{N}\text{—}\left[\begin{array}{c}CH_3\\|\\\overset{+}{N}\\|\\CH_3\end{array}\right]_n\text{—}\overset{+}{N}(CH_2CH_2OH)_3 \cdot (Cl^-)_{n+2}$$

Cetylpyridinium chloride [*123-03-5*], 1-hexadecylpyridinium chloride monohydrate, $C_{21}H_{38}ClN \cdot H_2O$, M_r 358.0, mp 77–83 °C.

Cetyltrimethylammonium bromide [*505-86-2*], cetrimide, is chiefly trimethyltetradecylammonium bromide together with dodecyl- and hexadecyltrimethylammonium bromides.

Trade Names: Pilomann, Dexamytrex Ophtiole (Mann).

Polyquad [*75345-27-6*], poly[(dimethyliminio)-2-butene-1,4-diyl chloride], polyquarternium-1, polyquat.

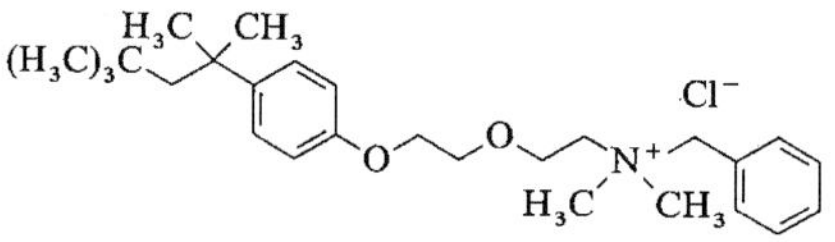

Contact lens products: Opti-Free (Alcon).

10.1.4. Miscellaneous Disinfectants and Preservatives

Benzyl alcohol [*100-51-6*], C_7H_8O, M_r 108.1, mp −15.2 °C.

Boric acid [*10043-35-3*], orthoboric acid, BH_3O_3, M_r 61.84, mp ca. 171 °C.

Commercial products: Boric acid is used in a variety of ophthalmic and contact lens care products.

Chlorobutanol [*57-15-8*], 1,1,1-trichloro-2-methyl-2-propanol, $C_4H_7Cl_3O$, M_r 177.5, mp 97 °C.

Trade Names: Gentamytrex Ophtiole (Mann); Lacril, Liquifilm Tears, Tears Plus (Allergan).

Ethylenediaminetetracetic acid [60-00-4], EDTA, $C_{10}H_{16}N_2O_8$, M_r 292.2, *mp* > 300 °C.

Commercial products: EDTA has widespread use in a variety of ophthalmic and contact lens care products.

Hydrogen peroxide [7722-84-1], H_2O_2, M_r 34.0.

Contact lens products: OxySept (Allergan); Consept (Barnes-Hind); Quik-Sept (Bausch & Lomb); AOSept (Ciba); MiraSept (CooperVision); Murine Pure Sept (Ross).

Sorbic acid [110-44-1], 2,4-hexadienoic acid, M_r 112.12, *mp* 134.5 °C.

Ophthalmic and contact lens products: Sorbic acid is used widely as a preservative in contact lens care products.

10.2. Contact Lens Cleaners and Rewetting Agents

Most lens cleaning agents are used daily or weekly (see Section 10.1). Daily cleaners remove proteinaceous and lipoidal deposits by their surfactant action. Weekly cleaners are used to remove accumulated protein deposits.

10.2.1. Surfactant Cleaners

Octylphenoxypolyethoxy ethanol [9002-93-1], poly(ethylene glycol) 4-isooctylphenyl ether, $C_{34}H_{62}O_{12}$, average M_r 647.

Contact lens products: Barnes-Hind Wetting & Soaking, Soft-Mate (Barnes-Hind).

Polyoxyl 40 stearate [9004-99-3], poly(ethylene glycol) monostearate.

Contact lens product: Blink-N-Clean (Allergan).

Poly(ethylene glycol) [25322-68-3], typical lens product M_r range 190–420; hygroscopic, viscous liquid.

Contact lens products: Blink-N-Clean (Allergan), ReNu Effervescent Enzymatic Contact Lens Cleaner (Bausch & Lomb).

Tyloxapol [*25301-02-4*], 4-(1,1,3,3-tetramethylbutyl)phenol polymer with formaldehyde and ethylene oxide, tyloxypal.

Trade Name: Enuclene (Alcon).

Sodium lauryl sulfate [*151-21-3*], sodium dodecyl sulfate, $C_{12}H_{25}NaO_4S$, M_r 288.4

Contact lens products: Lens Plus Daily Cleaner, Resolve/GP Daily Cleaner (Allergan).

Poloxamer 407 [*9003-11-6*], ethylene–propylene block copolymer, average M_r 4000 with 70 wt% polyethylene.

Contact lens products: Ciba Vision Lens Drops, MiraFlow Extra-Strength Cleaner (Ciba).

Poloxamine, [*110617-70-4*], Tetronic, copolymer of ethylene oxide, propylene oxide, and tetrakis ether of ethylenediamine with propanol.

Contact lens products: ReNu Multi-Purpose Solution (Bausch & Lomb).

10.2.2. Enzymatic Cleaners

Pancreatin [*8049-47-6*] is obtained from hog pancreas and has protease, amylase, and lipase activity.

Contact lens products: Optizyme (Alcon).

Papain [*9001-73-4*] is derived from papaya.

Contact lens products: Allergan Enzymatic, Extenzyme (Allergan).

Subtilisin A [*9014-01-1*] is extracted from *Bacillus licheniformis*.

Contact lens products: ReNu Effervescent Enzymatic Contact Lens Cleaner, ReNu Thermal Enzymatic Contact Lens Cleaner (Bausch & Lomb); Ultrazyme (Allergan).

11. References

[1] J. E. F. Reynolds (ed.): *Martindale, The Extra Pharmacoepia,* 29th ed., The Pharmaceutical Press, London 1989, pp. 1205–1227.
[2] F. W. Chang: "Local Anesthetics," in J. D. Bartlett, S. D. Jaanus (eds.): *Clinical Ocular Pharmacology,* chap. 4, Butterworths, Boston 1984.

[3] S. S. Gandhi: "Local Anesthetics," in D. W. Lamberts, D. E. Potter (eds.): *Clinical Ophthalmic Pharmacology*, Chap. 11, Little, Brown and Co., Boston 1987.
[4] J. D. Bartlett: "Topical Anesthesia," in J. D. Bartlett, S. D. Jaanus (eds.): *Clinical Ocular Pharmacology*, chap. 14, Butterworth, Boston 1984.
[5] D. B. Glasser, R. A. Hyndiuk in K. F. Tabbara, R. A. Hynduik (eds.): *Infections of the Eye*, Little, Brown and Co., Boston 1986, p. 211.
[6] M. L. Crossley, E. H. Northey, M. E. Hultquist, *J. Am. Chem. Soc.* **61** (1939) 2950.
[7] R. O. Roblin, Jr., J. H. Williams, P. S. Winnek, J. P. English, *J. Am. Chem. Soc.* **62** (1940) 2002.
[8] H. Lundbeck, US 2 447 702, 1948 (O. Huber).
[9] Hoffman-La Roche, US 2 430 094, 1947 (H. M. Wuest, H. Hoffer).
[10] F. P. Doyle, G. R. Fossker, J. H. C. Nayler, *J. Chem. Soc.* 1962, 1440.
[11] F. P. Doyle et al., *J. Chem. Soc.* 1962, 1453.
[12] E. G. Brain, J. H. C. Nayler, US 3 282 926, 1966.
[13] N. Koniwhi et al., US 3 516 997, 1970.
[14] R. R. Chauvette et al., *J. Am. Chem. Soc.* **84** (1962) 3401.
[15] B. M. Duggar, *Ann. NY Acad. Sci.* **51** (1948) 177.
[16] A. C. Finlay et al., *Science* **111** (1950) 85.
[17] J. H. Boothe et al., *J. Am. Chem. Soc.* **75** (1953) 4621.
[18] L. H. Conover et al., *J. Am. Chem. Soc.* **75** (1953) 4622.
[19] K. L. Rinehart, Jr. et al., *J. Am. Chem. Soc.* **82** (1960) 3938.
[20] M. J. Weinstein, G. M. Luedemann, E. M. Oden, G. H. Wagman, *Antimicrob. Agents Chemother.* 1963, 1.
[21] R. Okachi et al., *J. Antibiot.* **27** (1974) 793.
[22] S. A. Waksman, H. A. Lechevalier, *Science* **109** (1949) 305.
[23] S. A. Waksman, H. A. Lechevalier, D. A. Harris, *J. Clin. Invest.* **28** (1949) 934.
[24] E. A. Swart, S. A. Waksman, D. Hutchison, *J. Clin. Invest.* **28** (1949) 1045.
[25] M. J. Weinstein et al., *J. Antibiot.* **23** (1970) 551.
[26] A. Schatz, E. Bugie, S. A. Waksman, *Proc. Soc. Exp. Biol. Med.* **55** (1944) 66.
[27] W. M. Stark, N. G. Knox, R. M. Wilgus, *Folia Micobiol. (Prague)* **16** (1971) 205–217.
[28] B. A. Johnson, H. Anker, F. Meleney, *Science* **102** (1945) 376.
[29] H. S. Anker, B. A. Johnson, J. Goldberg, F. L. Meleney, *J. Bacteriol.* **55** (1948) 249.
[30] T. Suzuki, H. Inouye, K. Fujikawa, Y. Suketa, *J. Biochem. (Tokyo)* **54** (1963) 25.
[31] K. Vogler, R. O. Studer, *Experientia* **22** (1966) 345–354.
[32] H. Paulus: "Polymyxins," in D. Gottlieb, P. D. Shaw (eds.): *Antibiotics*, vol **2**, Springer-Verlag, New York 1967, 254–267.
[33] Q. R. Bartz, *J. Biol. Chem.* **172** (1948) 445.
[34] D. Gottlieb, P. K. Bhattacharyya, H. W. Anderson, H. E. Carter, *J. Bacteriol.* **55** (1948) 409.
[35] J. Ehrlich et al., *J. Bacteriol.* **56** (1948) 467.
[36] R. D. Birkenmeyer, F. Kagan, *J. Med. Chem.* **13** (1970) 616.
[37] J. M. McGuire et al., *Antibiot. Chemother.* **2** (1952) 281.
[38] D. J. Mason, A. Dietz, C. Deboer, *Antimicrob. Agents Chemother.* 1962, 555.
[39] R. R. Herr, M. E. Bergy, *Antimicrob. Agents Chemother.* 1962, 560.
[40] I. Hayakawa, T. Hiramitsu, US 4 382 892, 1983. H. Egawa, T. Miamoto, J. Matsumoto, *Chem. Pharm. Bull.* **34** (1986) 4098.
[41] M. H. McCormick et al., *Antibiotic Annu.* **1957–1958** (1958) 906.
[42] B. R. Jones, *Tr. Am. Acad. Ophthalmol. Otolaryngol.* **79** (1975) 15.
[43] T. H. Petit, R. J. Olson, R. Y. Foos, W. J. Martin, *Arch. Ophthalmol.* **98** (1980) 1025.

[44] J. Francois, M. Rijsselaere, *Ann. Ophthalmol.* **6** (1974) 207.
[45] D. P. Yolton: "Anti-Infective Drugs," in J. D. Bartlett, S. D. Jaanus (eds.): *Clinical Ocular Pharmacology*, 2nd ed., chap. 6, Butterworths, Boston 1989.
[46] W. Gold, H. A. Stout, J. F. Pagano, R. Donovick, *Antibiotic. Annu.* 1955–1956 (1956) 579.
[47] R. Brown, E. L. Hazen, *Trans NY Acad Sci* Ser. II **19** (1956–1957) 447–456.
[48] A. P. Struyk et al., *Antibiotic Annu.* 1957–1958 (1958) 878–885.
[49] E. F. Godefroi, J. Heeres, J. Van Cutsem, P. A. J. Janssen, *J. Med. Chem.* **12** (1969) 784.
[50] J. Heeres, L. J. J. Backx, J. H. Mostmanns, J. Van Custem, *J. Med. Chem.* **22** (1979) 1003.
[51] R. Duschinsky, E. Pleven, *J. Am. Chem. Soc.* **79** (1957) 4559.
[52] J. J. Wiens, W. B. Jackson, *Can. J. Ophthalmol.* **23** (1988) 107;
[53] S. S. Berg, G. Newbery, GB 598 911, 1948.S. S. Berg, G. Newbery, *J. Chem. Soc.* **1949**, 642.
[54] D. Pavan-Lanston, N. H. Park, J. H. Laas, *Arch. Ophthalmol.* **97** (1979) 1964.
[55] A. B. Nesburn, D. E. Willey, M. D. Trousdale, *Proc. Soc. Exp. Biol. Med.* **172** (1983) 316.
[56] H. Matsumoto et al., *Chem. Pharm. Bull.* **36** (1988) 1153.
[57] F. Prusof, *Biochim. Biophys. Acta* **32** (1959) 295.
[58] D. Pavan-Lanston, J. Lass, R. Campbell, *Arch. Ophthalmol.* **97** (1979) 1132.
[59] F. M. Polak, M. G. Burdette, L. A. Sidry, *Cornea* **1** (1982) 29.
[60] C. Heidelberger, D. Parsons, D. C. Remy, *J. Am. Chem. Soc.* **84** (1962) 3597.
[61] D. Pavan-Langston, R. A. Buchanan, *Trans. Am. Acad. Ophthalmol. Otolaryngol* **81** (1976) 813.
[62] S. D. Jaanus: "Anti-inflammatory Drugs," in J. D. Bartlett, S. D. Jaanus (eds.): *Clinical Ocular Pharmacology*, chap. 5, Butterworths, Boston 1989.
[63] B. J. Mondino, D. H. Aizuss, M. K. Farley: "Steroids," in D. W. Lamberts, D. E. Potter (eds): *Clinical Ophthalmic Pharmacology*, chap. 5, Little, Brown and Co., Boston 1987.
[64] J. E. F. Reynolds (ed.): *Martindale, The Extra Pharmacoepia*, 29th ed., The Pharmaceutical Press, London 1989, pp. 872–902.
[65] C. S. Foster: "Nonsteroidal Anti-Inflammatory and Immunosuppressive Agents," in D. W. Lamberts, D. E. Potter (eds.): *Clinical Ophthalmic Pharmacology*, chap. 6, Little, Brown and Co., Boston 1987.
[66] F. M. Wilson, II, *Survey Ophthalmology* **24** (1979) 57.
[67] F. T. Fraunfelder, S. M. Meyer, *Cornea* **5** (1986) 55.
[68] S. I. Butrus, J. H. Weston, M. B. Abelson: "Ocular Mast Cell Stabilizing Agents," in D. W. Lamberts, D. E. Potter (eds): *Clinical Ophthalmic Pharmacology*, chap. 18, Little, Brown and Co., Boston 1987.
[69] J. B. Walsh, A. Gold: *Physician's Desk Reference for Ophthalmology*, 17th ed., Medical Economics Company, Inc., Oradell, NJ 1989, p. 9.
[70] J. B. Eskridge, J. D. Bartlett: "The Glaucomas," in J. D. Bartlett, S. D. Jaanus (eds.): *Clinical Ocular Pharmacology*, chap. 29, Butterworths, Boston 1989.
[71] A. J. Flach, *J. Toxicol. Cutaneous Ocul. Toxicol.* **3** (1984) 31–51.
[72] Z. M. Shihab: "Antiglaucoma Therapy," in D. W. Lamberts, D. E. Potter (eds.): *Clinical Ophthalmic Pharmacology*, chap. 7, Little, Brown and Co., Boston 1987.
[73] S. D. Jaanus, V. T. Pagano, J. D. Bartlett: "Drugs Affecting the Autonomic Nervous System," in J. D. Bartlett, S. D. Jaanus (eds.): *Clinical Ocular Pharmacology*, chap. 3, Butterworths, Boston 1989.
[74] J. A. Hiett: "Inhibitors of Aqueous Formation," in J. D. Bartlett, S. D. Jaanus (eds.): *Clinical Ocular Pharmacology*, chap. 7, Butterworths, Boston 1989.
[75] S. D. Jaanus: "Hyperosmotic Drugs," in J. D. Bartlett, S. D. Jaanus (eds.): *Clinical Ocular Pharmacology*, chap. 8, Butterworths, Boston 1989.

[76] K. W. Benjamin: "Toxicity of Ocular Medications" in *Int. Ophthalmol. Clin.* **19** (1979) 199.
[77] B. L. Selvin, *South. Med. J.* **76** (1983) 349–358.
[78] F. T. Fraunfelder, S. M. Meyer, *Medical Toxicology* **2** (1987) 287–293; *Cornea* **5** (1986) 55–59.
[79] R. D. Newcomb, M. L. Priest: "Systemic Effects of Ocular Drugs" in J. O. Bartlett, S. D. Jaanus (eds.): *Clinical Ocular Pharmacology*, Butterworths, Boston 1989, chap. 31.
[80] B. L. Selvin: "Systemic Effects of Topical Ophthalmic Medications," in D. W. Lamberts, D. E. Potter (eds.): *Clinical Ophthalmic Pharmacology*, chap. 17, Little, Brown and Co., Boston 1987.
[81] A. Hussain, J. E. Truelove, US 3 809 714, 1974; 3 839 584, 1974. D. Henschler, J. Wagner, H. Hampel, US 4 085 270, 1978.
[82] L. M. Weinstock, R. J. Tull, M. D. Mulvey, US 3 619 370, 1971; B. K. Wasson, US 3 655 663, 1972; L. M. Weinstock, R. J. Tull, M. D. Mulvey, US 3 657 237, 1972; B. K. Wasson et al., *J. Med. Chem.* **15** (1972) 651.
[83] C. F. Schwender, J. Shavel, Jr., US 3 649 691, 1972.
[84] P. M. J. Manoury, I. A. G. Cavero, H. Mayer, D. P. R. L. Guidicelli, US 4 252 984, 1981.
[85] J. W. Clapp, R. O. Robin, Jr., US 2 554 816, 1951.
[86] P. Gianfranco, US 2 980 679, 1961.
[87] E. M. Schultz, US 2 835 702, 1958.
[88] R. W. Young, K. H. Wood, J. R. Vaughn, Jr., US 2 783 241, 1957; R. W. Young et al., *J. Am. Chem. Soc.* **78** (1956) 4649.
[89] A. D. Charap: "Ocular Pharmacology" in S. J. Rhode, S. P. Ginsberg (eds.): *Ophthalmic Technology*, Raven Press, New York 1987, pp. 469–478.
[90] N. Weiner in A. G. Gilman, L. S. Goodman, T. W. Rall, F. Murad (eds.): *The Pharmacological Basis of Therapeutics*, Macmillan Publishing Co., New York 1985, pp. 130–144.
[91] B. L. Selvin, *South. Med. J.* **76** (1983) 349.
[92] J. H. Calhoun, *1979 Annual Meeting of the Pennsylvania Academy of Ophthalmology and Otolaryngology*, Bedford, Pennsylvania, May 16–19, 1979, pp. 10–13.
[93] F. T. Fraunfelder, S. M. Meyer, *Medical Toxicology* **2** (1987) 287.
[94] G. A. Hopkins, W. M. Lyle, *J. Am. Optom. Assoc.* **48** (1977) 1241.
[95] T. O. Soine: "Autonomic Blocking Agents and Related Drugs," in C. O. Wilson, O. Gisvold, R. F. Doerge (eds.): *Textbook of Organic Medicinal and Pharmaceutical Chemistry*, 7th ed. J. P. Lippincott Co., Philadelphia, Toronto 1977, p. 481.
[96] P. P. Ellis in D. Vaughan, T. Asbury, K. F. Tabbara (eds.): *General Ophthalmology, Commonly Used Eye Medications*, Appleton and Lange, Norwalk, Connecticut; San Mateo, California 1989, pp. 399-407.
[97] J. D. Bartlett, S. D. Jaanus (eds.): *Ophthalmic Drug Facts*, J. B. Lippincott Company, St. Louis, Missouri 1989, p. 13–27.
[98] J. E. F. Reynolds (ed.): *Martindale, The Extra Pharmacoepia*, 29th ed., The Pharmaceutical Press, London 1989, pp. 522–545, pp. 1453–1486.
[99] L. L. Abell, W. R. Bruce, J. Seifter, US 2 590 079, 1952.
[100] G. R. Treves, US 2 554 511, 1951.
[101] G. Rey-Bellet, H. Spiegelberg, US 2 726 245, 1955.
[102] A. Sonn, US 2 161 938, 1939.
[103] M. E. Syderholm, L. H. Jules, M. Sahyun, US 2 731 471, 1956; J. F. Gardocki, D. E. Hutcheon, G. D. Lanbach, S. Y. P'an, US 2 842 478, 1958.
[104] S. D. Jaanus in J. D. Bartlett, S. D. Jaanus (eds.): *Clinical Ocular Pharmacology*, Butterworths, Boston 1984, pp. 311–326.

[105] W. A. Boothe in D. W. Lamberts, D. E. Potter (eds.): *Clinical Ophthalmic Pharmacology*, Little, Brown and Company, Boston 1987, pp. 361–415.
[106] S. Budavari (ed.): *The Merck Index*, 11th ed., Merck & Co., Inc., Rahway, N.J. 1989.
[107] R. F. Steiner, H. Edelhoch, *Chem. Rev.* **62** (1962) 457.
[108] M. S. Norn, *Acta Ophthalmol.* **48** (1970) 227.
[109] F. M. Wilson, *Ophthalmic Surg.* **7** (1976) 21.
[110] F. J. Holly, US 4 518 579, 1985.
[111] J. A. Aeschlimann, A. Stempel, US 2 647 924, 1953.
[112] J. B. Walsh, A. Gold, H. Charles in *Physicians' Desk Reference for Ophthalmology*, 18th ed., Medical Economics Company Inc., Oradell, N.J. 1990, pp. 15–16.
[113] F. J. Holly, M. A. Lemp, *Surv. Ophthalmol.* **22** (1977) 69.
[114] F. J. Holly in D. W. Lamberts, D. E. Potter (eds.): *Clinical Ophthalmic Pharmacology*, Little, Brown and Co., Boston 1987, pp. 497–518.
[115] S. C. G. Tseng, *Trans. Ophthalmol. Soc. U.K.* **104** (1985) 489.
[116] J. R. Boyd in *Handbook of Non-Prescription Drugs*, 8th ed., American Pharmaceutical Association, Washington D.C., pp. 453–476.
[117] J. B. Walsh, A. Gold, H. Charles in *Physicians' Desk Reference for Ophthalmology*, 18th ed., Medical Economics Company Inc., Oradell, N.J. 1990, pp. 209–256.
[118] *Contact Lens Forum*, August (1988) 11.
[119] W. Reindel et al., *International Contact Lens Clinic* **16** (1989) 232.
[120] G. L. Feldman, *Contact Lens Spectrum*, May (1989) 41.
[121] J. L. Sterling, A. S. Hecht, *Contact Lens Spectrum*, March (1988) 62.
[122] W. R. Baily, Jr., *Contact Lens Soc. Am. J.* **6** (1972) 33.
[123] S. S. Davis, M. A. Watson, *J. Pharm. Pharmacol.* **33** (1981) 109 P.
[124] S. Lerman, G. Sapp, *Can. J. Opthalmol.* **6** (1971) 1.
[125] A. R. Gasset, *Am. J. Ophthalmol.* **84**, Series 3 (1977) 169.
[126] A. R. Gasset, H. E. Kaufman, *Am. J. Ophthalmol.* **69** Series 3 (1970) 252.
[127] N. J. Van Haeringen, *Survey Ophthalmol.* **26** (1981) 84.
[128] F. J. Holly, *Contact Intraocul. Lens Med. J.* **4** (1978) no. 2, 14.
[129] F. J. Holly, *Contact Intraocul. Lens Med. J.* **4** (1978) no. 3, 52.
[130] W. G. Bachman, G. Wilson, *Investigative Ophthalmology Visual Sci.* **26** (1985) 1484.
[131] M. A. Lemp, *Clinical Ophthalmology* **4** (1987) 1.
[132] R. L. Farris, *Trans. Am. Ophthalmol.* **83** (1985) 501.

Prostaglandins

BERND BUCHMANN, Research Laboratories, Schering Aktiengesellschaft, Berlin, Federal Republic of Germany

HARTMUT REHWINKEL, Research Laboratories, Schering Aktiengesellschaft, Berlin, Federal Republic of Germany

1. History 2045
2. Structure, Nomenclature, and Physical Properties 2046
3. Occurrence and Biosynthesis 2050
4. Pharmacological Effects and Uses 2055
5. Syntheses. 2058
6. References 2071

1. History

Prostaglandins (PG) are a group of compounds formed in animals and humans that are derived from unsaturated C_{20} carboxylic acids. The physiological action of these compounds was described independently by U. S. VON EULER and M. W. GOLDBLATT as early as 1933/34 [33]. However, it was not until 1960 that S. BERGSTRÖM isolated pure $PGF_{1\alpha}$ and PGE_1 [34] and, a few years later, elucidated their structures [35]. Prostaglandins of the D series (PGD) were first characterized in 1966 [36]. The prostaglandin endoperoxides (PGG_2, PGH_2) [37] (1974, B. SAMUELSSON), thromboxane (TXA_2) [38] (1975, B. SAMUELSSON), and prostacyclin (PGI_2) [39] (1976, J. R. VANE) were presumably the last important derivatives of the cyclooxygenase pathway to be found.

In 1979, B. SAMUELSSON discovered another biosynthetic pathway that started from unsaturated C_{20} carboxylic acids, the lipoxygenase pathway [40]. The derivatives of the 5-lipoxygenase pathway are called leukotrienes (A–F). The physiological action of a mixture of some of these substances (then called SRS = slow reacting substances) was described in 1938 by W. FELDBERG and C. H. KELLAWAY [41]. Their absolute configurations were finally assigned by E. J. COREY and B. SAMUELSSON in 1980 [42].

In 1984, the structurally very closely related lipoxines were discovered by B. SAMUELSSON [43]. S. BERGSTRÖM, B. SAMUELSSON, and J. R. VANE received the Nobel Prize in 1982 for their extensive work.

2. Structure, Nomenclature, and Physical Properties

Prostaglandins are formed from *arachidonic acid*. All derivatives of arachidonic acid are called *eicosanoids*.

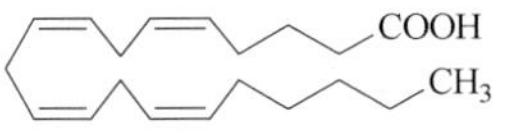

Arachidonic acid

The nomenclature of the *prostaglandins* (PG) is based on the hypothetical compound prostanoic acid [44]; the numbering of the carbon atoms is continuous and starts at the carboxyl group:

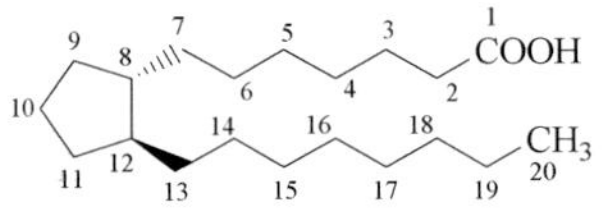

Prostanoic acid

The prostaglandins are classified (A–J) on the basis of the functional groups attached to the cyclopentane ring:

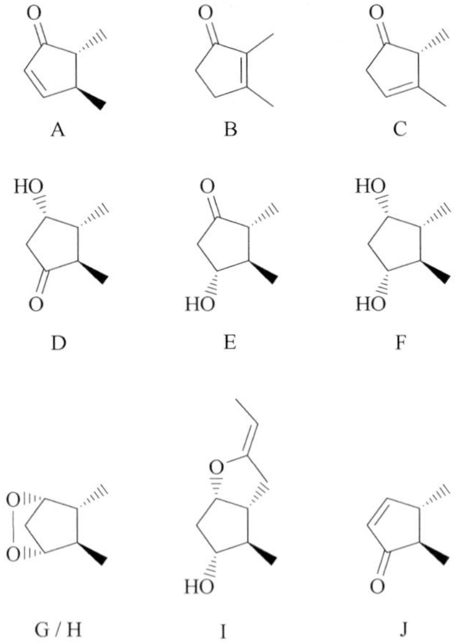

A B C

D E F

G/H I J

Additional numerals 1, 2, or 3 (subscripts) refer to the number, position, and stereochemistry of the double bonds in the side chains of the natural prostanes. Prostaglandins of the 1 series possess a double bond with an *E* configuration between C-13 and C-14. Those belonging to the 2 series have an additional double bond with a *Z* configuration between C-5 and C-6. Those of the 3 series possess a further double bond with a *Z* configuration between C-17 and C-18 (Table 1).

The subscript α denotes ring substituents on the same side as the carboxyl chain and β, on the opposite side. The *thromboxanes* (TX) have a tetrahydropyran ring instead of the cyclopentane ring of the prostaglandins.

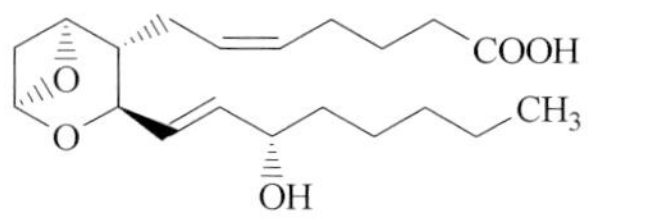

Thromboxane A$_2$ [57576-52-0] (TXA$_2$)

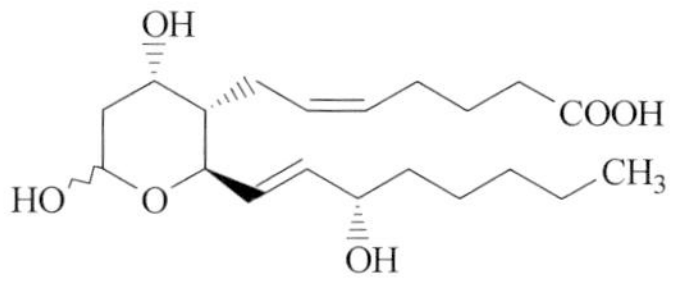

Thromboxane B$_2$ [54397-85-2] (TXB$_2$)

No uniform basic structure exists for the closely related *leukotrienesL* (LT) and *lipoxines* (LX). These compounds are all polyunsaturated, open-chain carboxylic acids. Similar to the prostanes, the numbering of the carbon atoms starts at the carboxyl group. The letter in the name of these compounds indicates the configuration and the position of the double bonds. The numerals refer to the number of double bonds. The 14, 15 double bond with the *Z* configuration is missing in the leukotrienes of the 3 series, and an additional double bond with a *Z* configuration in position 17 exists in compounds of the 5 series.

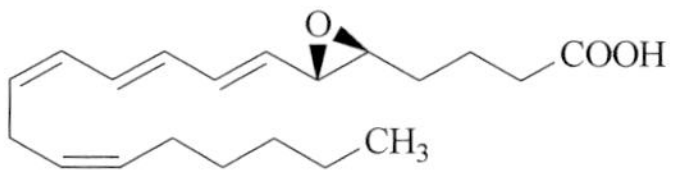

Leukotriene A$_4$ [72059-45-1] (LTA$_4$)

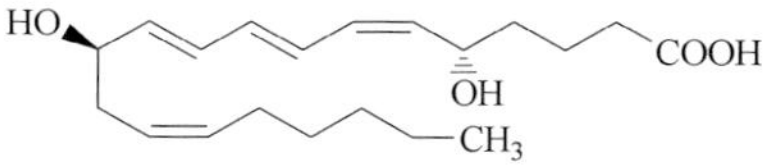

Leukotriene B$_4$ [71160-24-2] (LTB$_4$)

Table 1. Nomenclature and structure of some naturally occurring prostaglandins

Formula	Trivial name	Chemical name, CAS registry no.
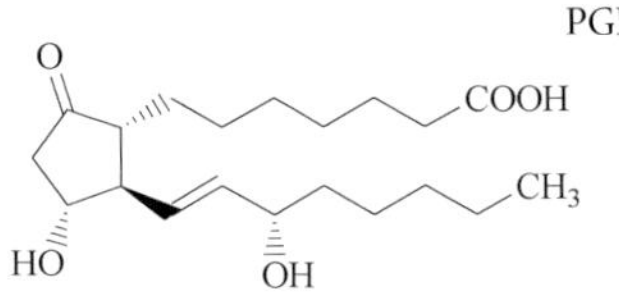	PGE$_1$	11α,15α-dihydroxy-9-oxo-13E-prostenoic acid [*745-65-3*]
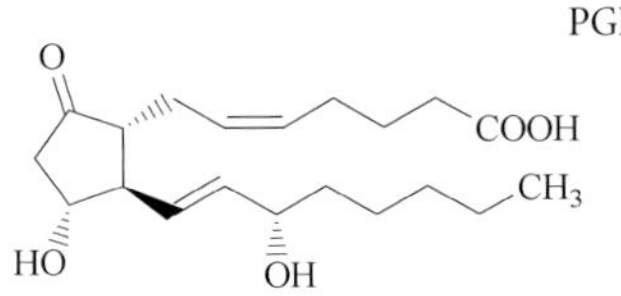	PGE$_2$	11α,15α-dihydroxy-9-oxo-5Z,13E-prostadienoic acid [*363-24-6*]
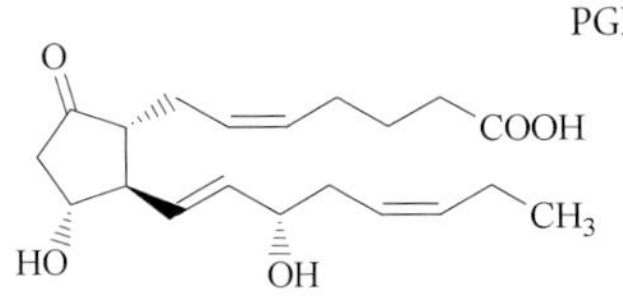	PGE$_3$	11α,15α-dihydroxy-9-oxo-5Z,13E,17Z-prostatrienoic acid [*802-31-3*]
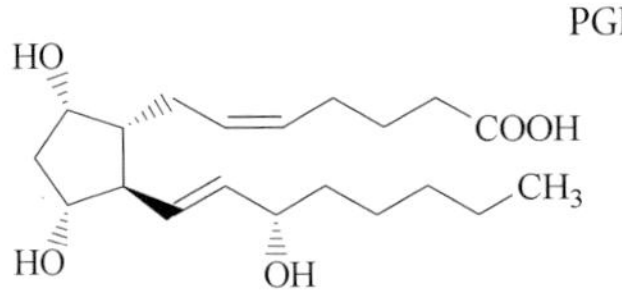	PGF$_{2α}$	9α,11α,15α-trihydroxy-5Z,13E-prostadienoic acid [*551-11-1*]
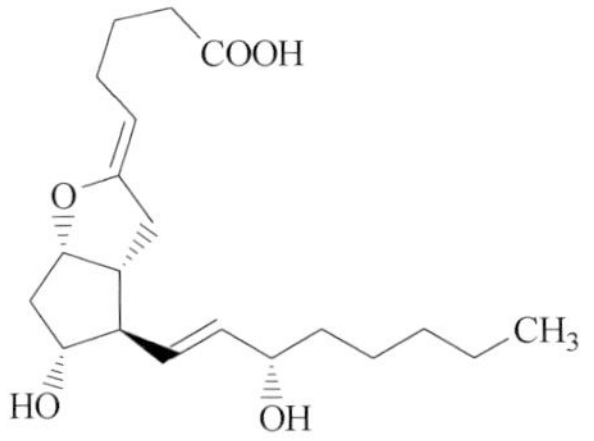	PGI$_2$	11α,15α-dihydroxy-6,9α-epoxy-5Z,13E-prostadienoic acid [*35121-78-9*]

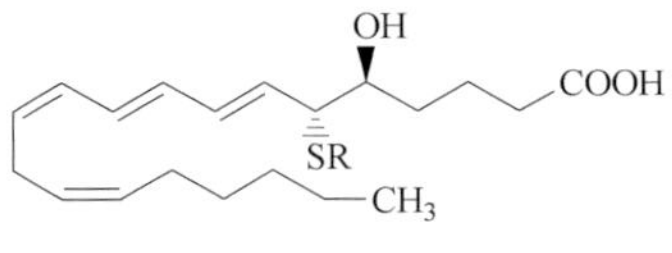

Leukotriene C$_4$–F$_4$
R = –Cys(γ-Glu)–Gly–OH, LTC$_4$ [*72025-60-6*];
 –Cys–Gly–OH, LTD$_4$ [*73836-78-9*];
 –Cys–OH, LTE$_4$ [*75715-89-8*];
 –Cys(γ-Glu)–OH, LTF$_4$ [*83851-42-7*]

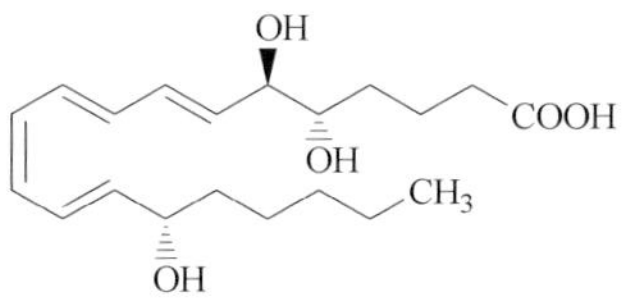

Lipoxine A₄ [89663-86-5] (LXA₄)

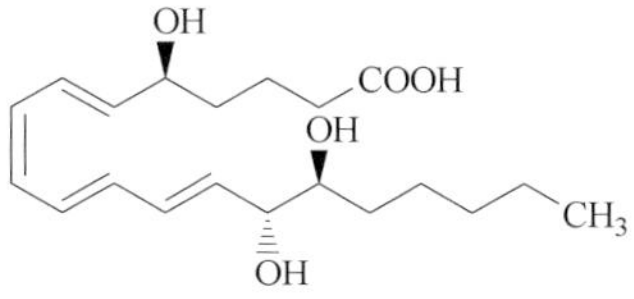

Lipoxine B₄ [98049-69-5] (LXB₄)

Physical Properties. Many of the natural prostaglandins, leukotrienes, and lipoxines are oils or solids with low melting points. Thus, they have to be characterized by spectroscopic methods (^{1}H- and ^{13}C-NMR, UV, and IR spectroscopy; mass spectrometry). Radioimmunoassays (RIA) based on antigen–antibody binding are most suitable for the determination of several prostanes and leukotrienes in biological media, because of their high sensitivity and selectivity. The stable metabolites of the prostanes are normally measured to obtain reproducible results.

PGE$_1$, C$_{20}$H$_{34}$O$_5$, M_r 354.49, fp 115–116 °C, $[\alpha]_D = -55°$ (tetrahydrofuran), UV spectrum after base treatment: λ_{max} (methanol) 278 nm, $\varepsilon = 25\,350$ L mol^{-1} cm^{-1}.

PGE$_2$, C$_{20}$H$_{32}$O$_5$, M_r 352.47, fp 64.5–65.5 °C, $[\alpha]_D = -61°$ (tetrahydrofuran).

PGF$_{1\alpha}$, C$_{20}$H$_{36}$O$_5$, M_r 356.51, fp 101–103 °C, $[\alpha]_D = +30°$ (ethanol).

PGF$_{2\alpha}$, C$_{20}$H$_{34}$O$_5$, M_r 354.49, fp 30–35 °C, $[\alpha]_D = +24°$ (tetrahydrofuran).

PGA$_2$, C$_{20}$H$_{30}$O$_4$, M_r 334.46, viscous oil, $[\alpha]_D = +140°$ (chloroform).

PGD$_2$, C$_{20}$H$_{32}$O$_5$, M_r 352.47, fp 58–59 °C, $[\alpha]_D = +9°$ (tetrahydrofuran).

PGI$_2$ sodium salt, C$_{20}$H$_{31}$O$_5$Na, M_r 374.45, fp 116–124 °C (capillary), fp 166–168 °C (block), $[\alpha]_D = -124°$ (95% ethanol), hygroscopic.

LTB$_4$, C$_{20}$H$_{32}$O$_4$, M_r 336.47, fp 25–28 °C, $[\alpha]_D = +13.1°$ (trideuterochloroform), UV spectrum: λ_{max} (methanol) 260, 269, 281 nm, $\varepsilon = 39\,000, 53\,000, 43\,000$ L mol^{-1} cm^{-1}.

LXA$_4$ methyl ester, C$_{21}$H$_{34}$O$_5$, M_r 366.50, $[\alpha]_D = +38.2°$ (methanol), UV spectrum: λ_{max} (methanol) 284, 296, 309 nm.

LXB$_4$ methyl ester C$_{21}$H$_{34}$O$_5$, M_r 366.50, colorless waxy solid, $[\alpha]_D = +19.2°$ (chloroform), UV spectrum: λ_{max} (methanol) 275, 287, 301, 315 nm.

Most eicosanoids have a very low chemical and metabolic stability. (Metabolic instability means that the biological degradation is high.) For instance, PGE$_2$ is only stable at pH 7; below pH 7 it is converted to PGA$_2$, above pH 7 it is converted to PGB$_2$. Pure PGI$_2$ has a half-life of a few minutes and is hydrolyzed to 6-oxo-PGF$_{1\alpha}$. The sodium salt of PGI$_2$ is, however, sufficiently stable for transportation at low temperatures.

While $PGF_{2\alpha}$ is stable enough, it is advisable to store and transport the leukotrienes and lipoxines in alcoholic solution, in a cool place, and as far as possible under exclusion of oxygen.

TXA_2 decomposes within a few minutes to give TXB_2. Hence, only stable TXA_2 analogues are commercially available.

3. Occurrence and Biosynthesis

Eicosanoids occur almost ubiquitously in animals and humans. The concentrations depend on the type of tissue and are generally very low (ca. 1 µg/g wet weight). Prostaglandin concentrations in human seminal fluid are especially high (300 µg/mL).

The occurrence of eicosanoids in corals is of interest. *Plexaura homomalla*, a marine soft coral found in the Carribean, contains the 1-methyl ester of $15R$-PGA_2-15-acetate in relatively large amounts (1 – 3 % of the dry weight) and the corresponding 15-epimer in smaller amounts [45]. Furthermore, new prostanoid structures called clavulones and punaglandins have been isolated from *Clavularia viridis* and *Telesto riisei*, respectively [46] (see Table 2). These compounds are of interest because they exhibit antitumor activity.

Prostaglandins also occur in plants [47]. The ubiquitous presence of these arachidonic acid derivatives in living organisms, their wide spectrum of activity, and high potency indicate that they play a central physiological role.

The enzyme phospholipase A_2 (PLA_2) releases arachidonic acid from reserves (mainly phosphatidyl choline) in the lipids of the cell membranes. The biosynthesis of prostaglandins and thromboxanes of the 2 series proceeds via PGH_2, which is formed from arachidonic acid by the enzyme cyclooxygenase (see Fig. 1).

The prostanes of the 1 and 3 series are synthesized from the corresponding unsaturated C_{20} carboxylic acids (dihomo-γ-linolenic acid or eicosapentaenoic acid).

Since 1990 a novel class of prostaglandin-like compounds, the isoprostanes, is known, the members of which may play a role as mediators of oxidant injury in vivo. These compounds are produced by a free radical process and in contrast to prostanes the stereochemical arrangement of the two side chains at the cyclopentane ring is cis [48].

The key compound in the biosynthesis of leukotrienes is LTA_4, which is formed from arachidonic acid by the enzyme 5-lipoxygenase. LTA_4 is then converted to LTB_4 by a hydrolase and to the leukotrienes C_4–E_4 by glutathione-S-transferase and further degradation (Fig. 2).

Arachidonic acid can be converted to lipoxines [49]–[51] by the combined action of 15- and 5-lipoxygenase (see Fig. 2).

The eicosanoids formed locally from the released arachidonic acid exert their specific, selective action via receptors [52], causing intracellular changes in the concentration of cyclic AMP and inositol triphosphate. Using the methods of molecular biology several

Table 2. Clavulones and punaglandins

Formula	Trivial name	CAS registry no.
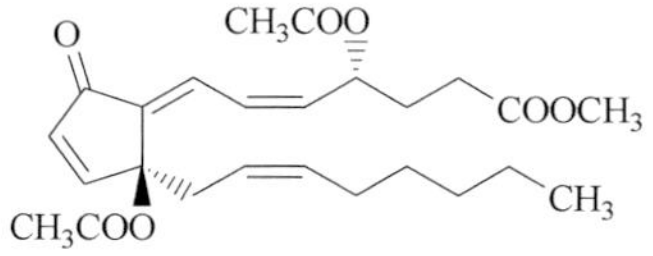	Clavulone I (claviridenone d)	[85700-42-1]
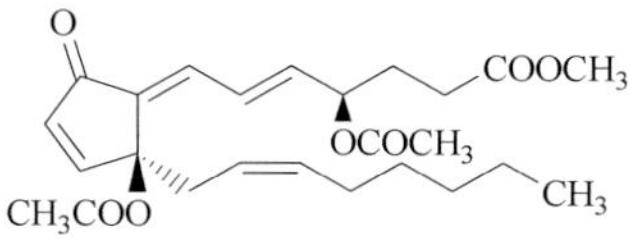	Clavulone II (claviridenone c)	[85700-43-2]
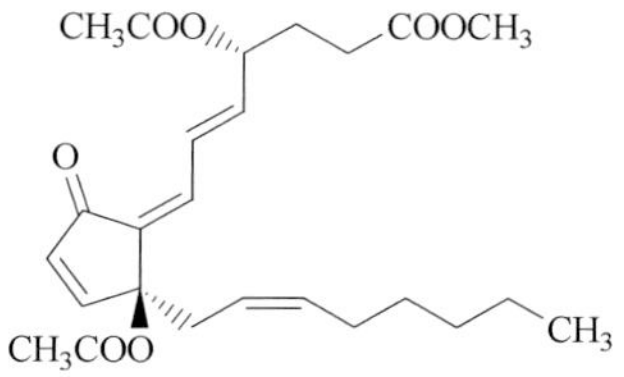	Clavulone III (claviridenone b)	[85700-44-3]
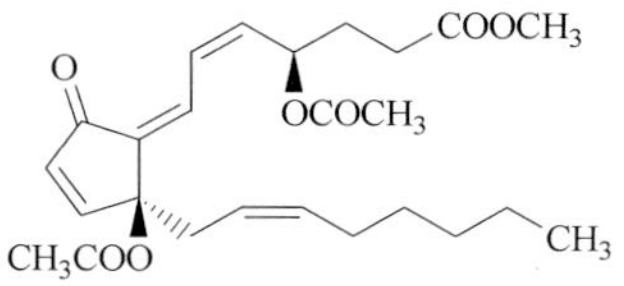	Clavulone IV (claviridenone a)	[85611-86-5]
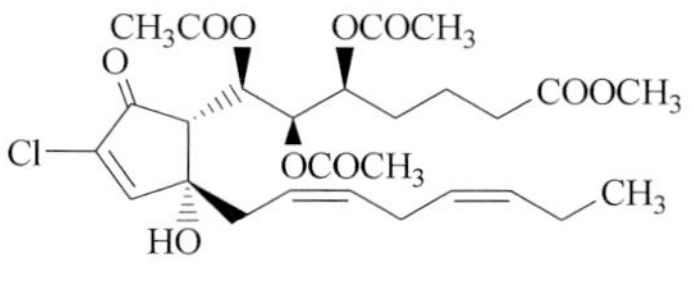	Punaglandin 1	[96055-63-9]
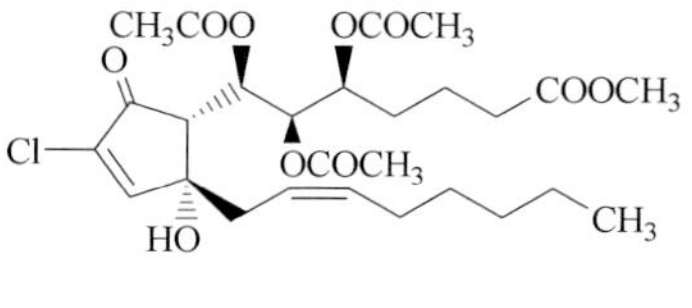	Punaglandin 2	[96055-64-0]

Table 2. (continued)

Formula	Trivial name	CAS registry no.
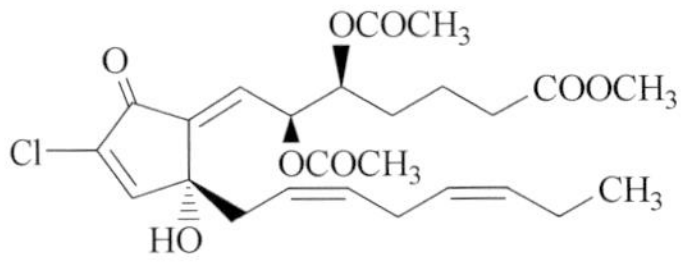	Punaglandin 3	[*96055-65-1*]
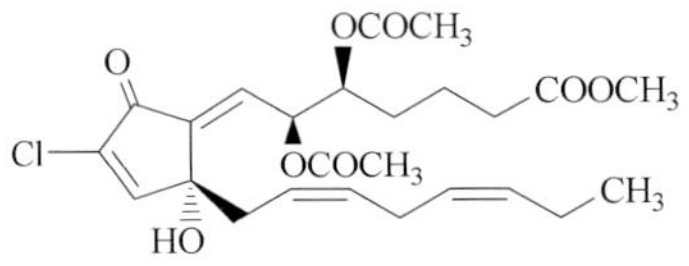	Punaglandin 4	[*96055-66-2*]

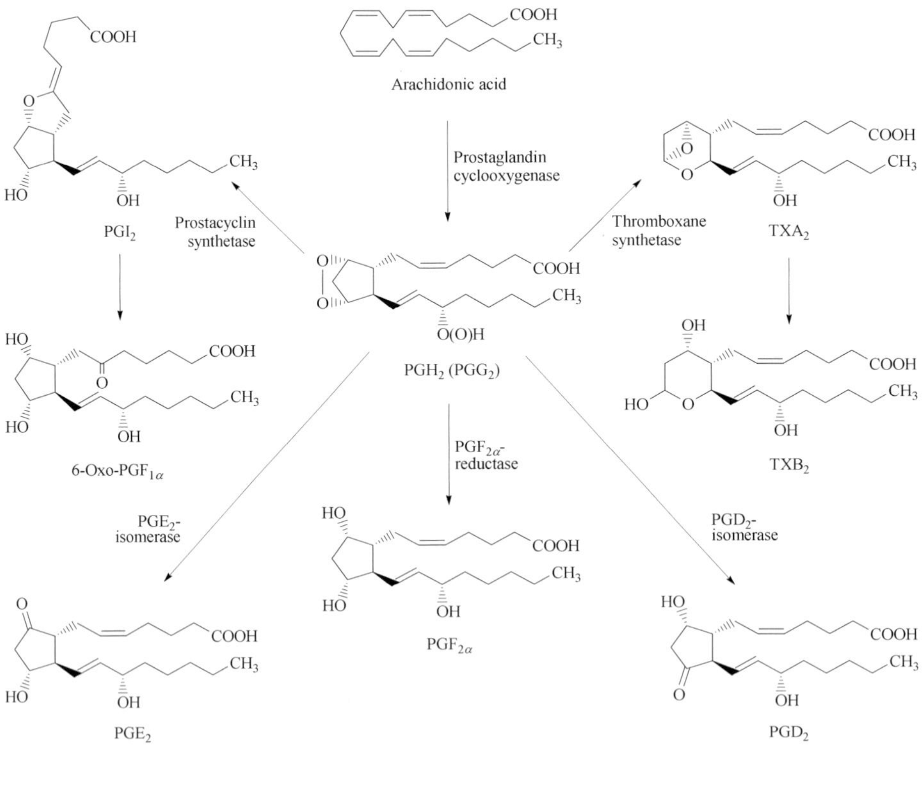

Figure 1. Biosynthesis of prostaglandins and thromboxanes from arachidonic acid

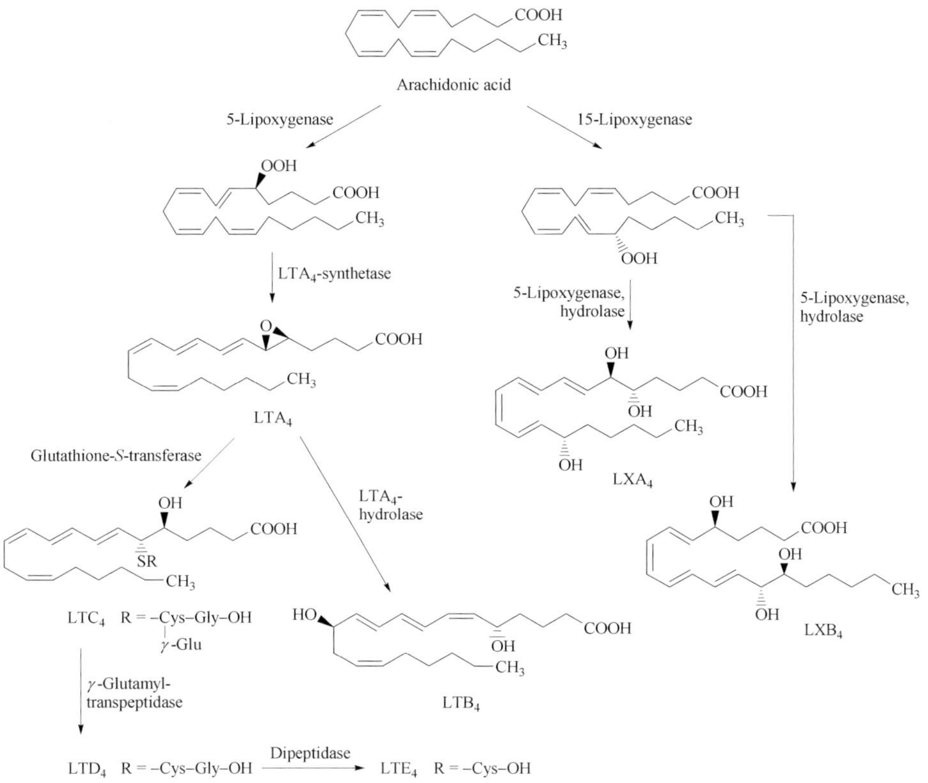

Figure 2. Biosynthesis of leukotrienes and lipoxines from arachidonic acid

eicosanoid receptors and their subtypes have been characterized and some of them have been cloned (see Table 3). The major importance of natural prostaglandins lies in the autocrine and paracrine regulation of physiological and pathological processes.

The biosynthesis of prostanes, leukotrienes, and lipoxines can be influenced by inhibitors of cyclooxygenase (nonsteroidal anti-inflammatory drugs, NSAIDs), and lipoxygenase [53], [54]. Aspirin, for example, causes irreversible inhibition of cyclooxygenase and ibuprofen reversible inhibition. This inhibition is responsible for the antiphlogistic activity of these two compounds. See also, → Anti-inflammatory – Antirheumatic Drugs.

Since 1991 it has been known that cyclooxygenase exists in two isoforms COX-1 and COX-2 [55] – [58]. Whereas COX-1 is responsible for the biosynthesis of prostaglandins under normal physiological conditions, COX-2 is expressed mainly in inflammatory tissue, causing the synthesis of prostaglandins which support this inflammation.

Under physiological conditions COX-2 is present for example in the brain and in the kidney but not in the gastric mucosa. Based on these findings the development of novel,

Table 3. Classification of prostanoid receptors

Receptor/subtype		Selective agonist, CAS registry no.	Selective antagonist, CAS registry no.
FP		cloprostenol [*40665-92-7*] fluprostenol [*40666-16-8*] prostalene [*54120-61-5*]	none
EP			
	EP_1	17-phenyl-18,19,20-trinor-PGE_2 [*38315-43-4*] sulprostone (also EP_3 receptor agonist) [*60325-46-4*] iloprost (also IP receptor agonist) [*78919-13-8*]	AH 6809 (also DP receptor antagonist) [*33458-93-4*] SC-19220 [*19395-87-0*]
	EP_2	butaprost [*69648-38-0*] AH 13205 [*148436-63-9*] misoprostol (also EP_3 receptor agonist) [*59122-46-2*]	none
	EP_3	enprostil [*73121-56-9*] GR 63799 [*183023-82-7*] sulprostone (also EP_1 receptor agonist) [*60325-46-4*] misoprostol (also EP_2 receptor agonist) [*59122-46-2*]	none
	EP_4	11-deoxy-PGE_1 (also EP_2 and EP_3 agonist) [*37786-00-8*]	AH 22921 (also TP receptor antagonist) [*744820-23-2*] AH 23848 (also TP receptor antagonist) [*81443-73-4*]
DP		BW 245C [*72814-32-5*] ZK 110841 [*105595-17-3*]	BW A868C [*118675-50-6*] AH 6809 (also EP_1 receptor antagonist) [*33458-93-4*]
IP		cicaprost [*94079-80-8*] iloprost [*78919-13-8*] octimibate [*89838-96-0*]	none
TP		U44069 [*56985-32-1*] U46619 [*56985-40-1*] SQ 26655 [*82337-14-2*] EP 011 [*75010-43-4*]	AH 23848 (also EP_4 receptor antagonist) [*81443-73-4*] GR 32191 [*87248-13-3*] EP 092 [*81806-67-9*] SQ 29548 [*98672-91-4*] Bay u3405 [*116649-85-5*] S 145 [*115266-92-7*]

more selective antirheumatic drugs with lower side effects concerning the stomach is possible.

Two of the most promising COX-2 inhibitors are celecoxib (SC-58635) [*169590-42-5*] and rofecoxib (MK-966) [*162011-90-7*].

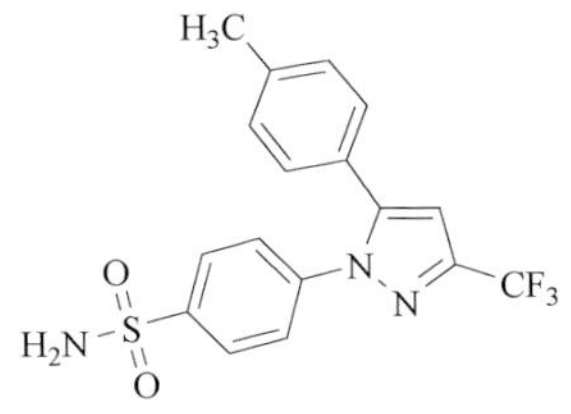

Celecoxib

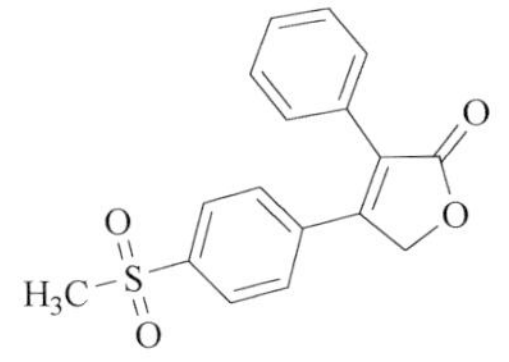

Rofecoxib

The 5-lipoxygenase inhibitor *zileuton* [*111406-87-2*] (trade name Zyflo) was introduced by Abbott as an anti-asthmatic and anti-inflammatory drug (→ Antiasthmatics).

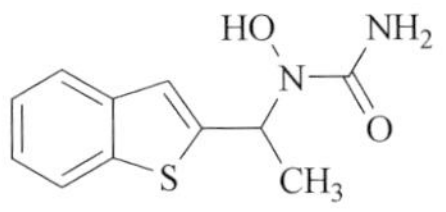

Zileuton

4. Pharmacological Effects and Uses

Eicosanoids exert different pharmacological effects on different organs. The high intrinsic activity at the site is accompanied by rapid metabolic inactivation (e.g., by 15-, β-, or ω-oxidation). The only effects of the structures in the cyclooxygenase and lipoxygenase pathways that are discussed here are those that are of possible therapeutic interest.

Reproduction. Prostaglandins of the E and F series contract uterine smooth muscle and are therefore suitable agents for the induction of parturition or abortion. As a result of their effects on other smooth muscle organs and consequent side effects (headache, nausea, vomiting, diarrhea), these compounds are seldom given systemically to induce labor. Even local application (intraamniotic, vaginal) is accompanied by side effects.

A combination of prostaglandin and antigestagen has proved useful for inducing abortions for medical reasons because it is tolerated better than an antigestagen alone [59].

Another economically important application of prostaglandins (mainly of the F series) is in veterinary medicine. Their luteolytic activity (the ability to induce the regression of the corpus luteum) is exploited for the induction of ovulation and synchronization of the menstrual cycle in cows, pigs, sheep, and horses.

Lungs. Prostaglandins of the E and I series administered as an aerosol dilate the smooth muscles of the bronchial tract and antagonize bronchoconstriction caused by histamine. In spite of these effects, the development of a prostaglandin agent for treating asthma has not been possible so far.

The healthy lung has almost no 5-lipoxygenase activity, whereas formation of leukotrienes and thromboxane A_2 is increased in the asthmatic lung. The bronchoconstricting effects of LTC_4 and LTD_4 are 1000 times that of histamine. The development of selective LTC_4/D_4 antagonists or inhibitors of the 5-lipoxygenase pathway [60] have therefore proven useful in asthma therapy ($\rightarrow$ Antiasthmatics). Antagonists of LTB_4 and TXA_2 have also been tested for this indication.

Stomach and Intestine. The outstanding significance of the prostanes (E, A, and I types) in the gastrointestinal tract is based on their inhibition of gastric acid secretion and their cytoprotective effect [61]. Cytoprotection here refers to the ability of prostaglandins, in very low concentrations, to protect the gastrointestinal tract against harmful noxious agents, e.g., nonsteroidal anti-inflammatory agents (aspirin etc.). This property is probably based on an increased formation of mucous membranes and improvement of circulation.

One of the main applications of prostaglandin derivatives is in ulcer therapy. Faster healing of peptic ulcers has been observed after the administration of PGE analogues [62]. In contrast to cytoprotection, relatively high doses are required to inhibit gastric acid production and thus achieve healing. Diarrhea is an undesirable side effect.

In patients suffering from ulcerative colitis, raised levels of LTB_4 and, to a lesser extent, PGE_2 and LTC_4 have been observed [63]. This indicates a possible therapeutic use of LTB_4 antagonists.

Heart and Circulation. (See also, $\rightarrow$ Cardioactive and VasoactiveDrugs) Prostaglandins of the E and A series lower blood pressure (hypotensive effect) in humans. In contrast, $PGF_{2\alpha}$ and TXA_2 increase blood pressure because they constrict blood vessels.

Prostacyclin, the natural antagonist of thromboxane A_2, also exerts a hypotensive effect, but its inhibition of platelet aggregation is more important.

The use of prostacyclin mimetics for thrombosis prophylaxis or thrombolysis [64] (e.g., peripheral circulatory disorders, Raynaud syndrome) is based on this effect.

Although less active than PGI_2, PGE_1 is also used for the same indication. Side effects include headache, nausea, flush, and diarrhea.

Antagonists of TXA_2 are also potential therapeutic agents against circulatory disorders [65].

Analogous to cytoprotection in the stomach, a cardioprotective effect of PGE_1 and prostacyclin has been proposed. This could lead to the application of these compounds for prophylaxis of heart attacks.

PGI_2 mimetics may prove useful in alleviating changes produced by arteriosclerosis because the level of prostacyclin synthetase in healthy vessel walls decreases with progressing arteriosclerosis [66].

Kidneys. PGI_2 and PGE_2 cause renal vasodilation and stimulate kidney circulation [67]. In addition, prostaglandins are involved in the excretion of salt and water. Their influence on other hormones (e.g., angiotensin II) is also of significance.

In the kidney, arachidonic acid is also metabolized by another enzyme system (epoxygenase pathway). The importance of the resulting metabolites in renal circulation has not yet been clarified [68].

Cerebral Effects. TXA_2 and $PGF_{2\alpha}$ contract brain vessels, whereas PGI_2 has a dilatory effect. The use of PGI_2 mimetics [69] and TXA_2 antagonists [70] in cerebral infarctions has therefore been discussed.

Much higher concentrations of PGD_2 are found in the brain than in other tissues. The equilibrium between PGD_2 and PGE_2 in brain tissue apparently plays an important role in the sleep–wake cycle [71].

Miscellaneous Effects. A higher capacity for prostaglandin synthesis is found in human tumors than in normal tissue. Prostanes are involved in tumor growth and metastasis. However, PGI_2 and mimetic agents exert an antimetastatic effect on melanoma cells, probably due to their strong antiaggregation effect [72]. Although the mechanisms of action are still unclear, there is hope that a therapeutic effect will be obtained by influencing the arachidonic acid cascade [73].

An interesting discovery is the fact that prostaglandins of the F and D series are capable of reducing intra-ocular pressure, and thus represent additional therapeutics for glaucoma [74].

In psoriasis, increased levels of LTB_4 have been found in inflamed areas of the skin [75]. LTB_4 seems to be involved in other inflammatory processes as well (e.g., rheumatic diseases) [76].

LTB_4 is assumed to be an important mediator of inflammation. It is a very effective chemotactic agent for neutrophilic leukocytes, i.e., the white blood cells move in the direction of increasing LTB_4 concentrations and release lysosomal enzymes which attack not only the inflamed, but also neighboring healthy tissue. LTB_4 antagonists may therefore be therapeutically useful as new antiphlogistic agents.

The natural eicosanoids are normally not considered for therapeutic applications because of their rapid metabolic degradation and their wide spectrum of activity. Modified derivatives are therefore synthesized to improve not only the selectivity, but also the metabolic stability of these compounds.

A selection of prostaglandin, prostacyclin, thromboxane, and leukotriene derivatives is presented in Tables 4–8, respectively. They include therapeutic agents that have already been introduced to the market and compounds that are still the subject of research and development.

Economic Aspects. The economically most important applications of eicosanoids are the treatment of cardiovascular, gastrointestinal, pulmonary, ophthalmic and gynaecological diseases.

For example, 5% of men over 50 suffer from peripheral circulatory disorders. Of these 10% develop a critical phase, i.e., 25% will lose a limb and 20% will die within one year. These patients strongly benefit from treatment with commercially available prostanes (e.g. Alprostadil, Epoprostenol, Iloprost, Beraprost).

Another example is the use of prostaglandins in patients with arthritis who must be treated with NSAIDs over a long period. About a quarter of them suffer from gastric ulcers which develop as a result of the daily intake of these drugs. Prostaglandins (e.g., Misoprostol or Enprostil) prevent these injuries effectively.

In the future additional eicosanoid derivatives may be suitable for treating certain cancers and inflammation diseases (e.g. psoriasis).

5. Syntheses

Many different syntheses exist for the compounds of the cyclooxygenase and lipoxygenase pathways (Chap. 2). A large number of these syntheses are described in numerous review articles and monographs [1]–[32]. Three basic strategies of synthesis that are exemplary for the prostaglandin field are presented here, followed by the first synthesis of the natural substances PGI_2, TXA_2, and LTB_4. All these syntheses yield not only the natural substances, but also allow the synthesis of analogues. Some of these methods are now used for industrial production.

Synthesis of $PGF_{2\alpha}$ According to E. J. Corey (Fig. 3). E. J. Corey was the first to develop a method for a stereocontrolled access to natural prostaglandins [77]. The cyclopentadiene derivative **1** is converted in a Diels–Alder reaction with α-chloroacrylic acid chloride to the bicyclic compound **2**. Curtius rearrangement with subsequent hydrolysis gives the ketone **3**, which is converted to the lactone **4** in a Baeyer–Villiger oxidation. After saponification of the lactone, the acid is subjected to a resolution of the enantiomers with an optically active amine (e.g., amphetamine). Iodine lactonization of the acid **5** and benzoylation give the benzoate **6**. Reductive

Table 4. Uses of prostaglandins in human and veterinary medicine

Structure	Generic name, CAS registry no.	Producer	Trade name or phase of development	(Potential) indication
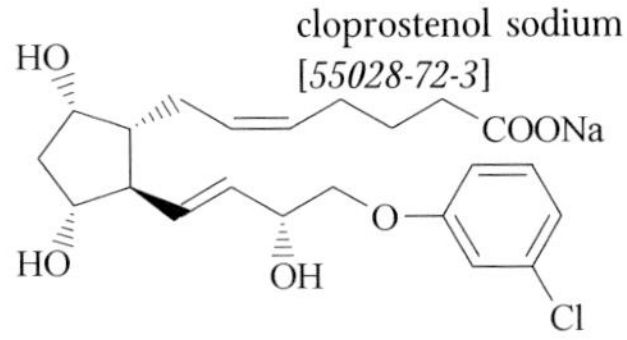	cloprostenol sodium [55028-72-3]	Zeneca	Estrumate	synchronization of menstrual cycle in cows
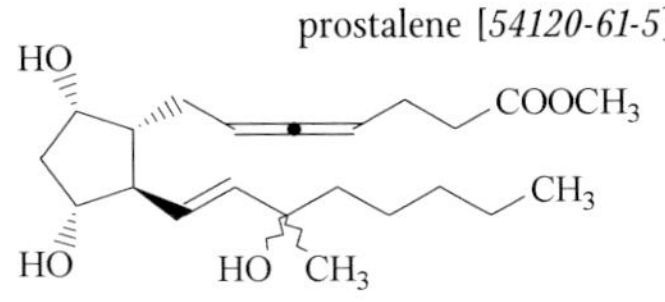	prostalene [54120-61-5]	Syntex	Synchrocept	synchronization of menstrual cycle in domestic animals
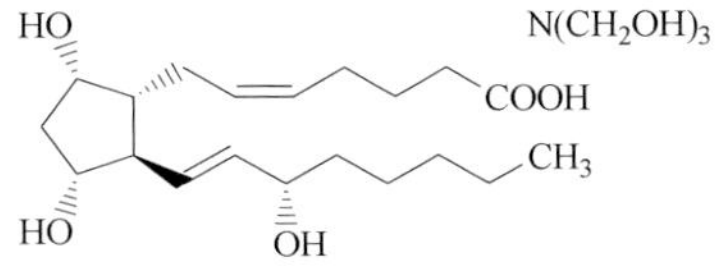	dinoprost [38562-01-5] N(CH$_2$OH)$_3$	Pharmacia & Upjohn	Minprostin F$_{2\alpha}$	termination of pregnancy
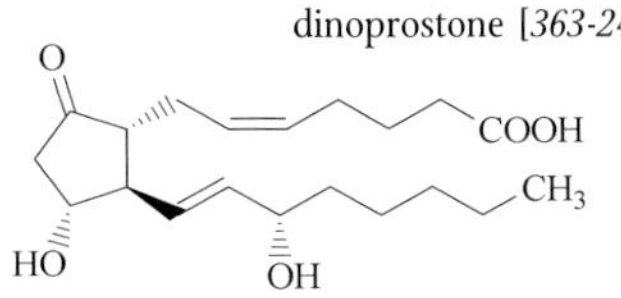	dinoprostone [363-24-6]	Pharmacia & Upjohn	Minprostin E$_2$	termination of pregnancy
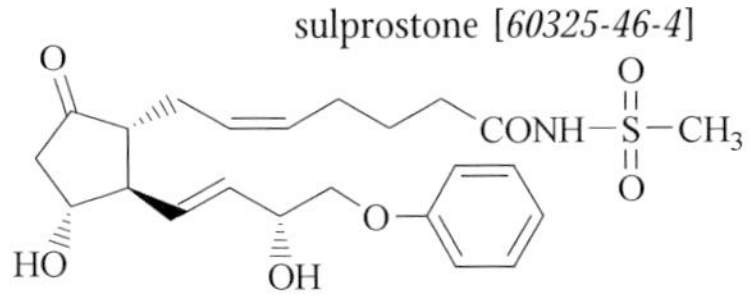	sulprostone [60325-46-4]	Schering AG, Pfizer	Nalador	termination of pregnancy
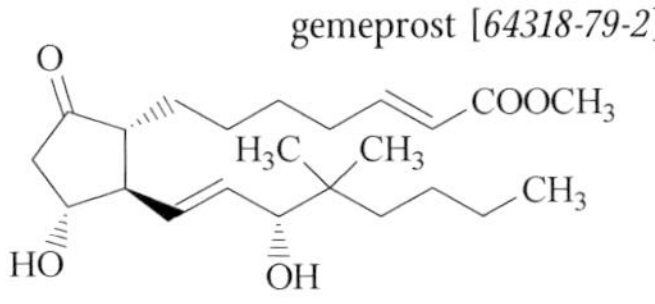	gemeprost [64318-79-2]	ONO	Gervagem	termination of pregnancy, initiation of menstruation

Table 4. (continued)

Structure	Generic name, CAS registry no.	Producer	Trade name or phase of development	(Potential) indication
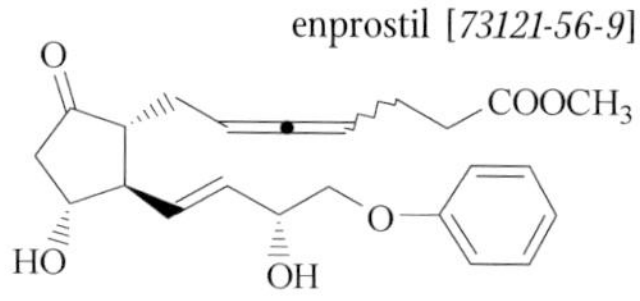	enprostil [73121-56-9]	Syntex	Gardrin	antiulcer
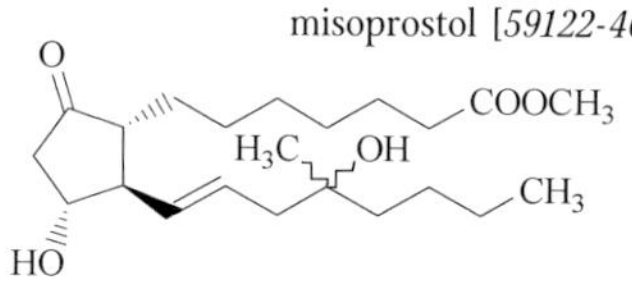	misoprostol [59122-46-2]	Searle	Cytotec	antiulcer
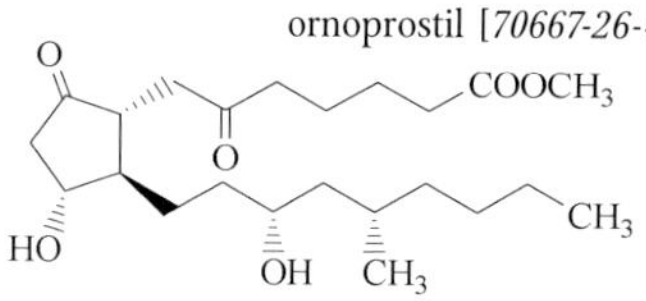	ornoprostil [70667-26-4]	ONO	Ronok	antiulcer
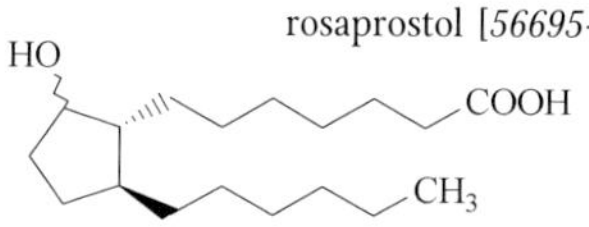	rosaprostol [56695-65-9]	IBI	Rosal	antiulcer
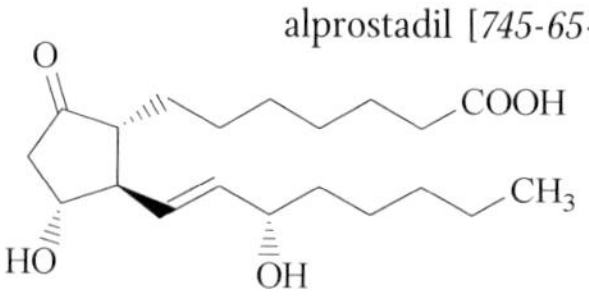	alprostadil [745-65-3]	Schwarz Pharma, ONO	Prostandin, Prostavasin	arterial circulatory disorders
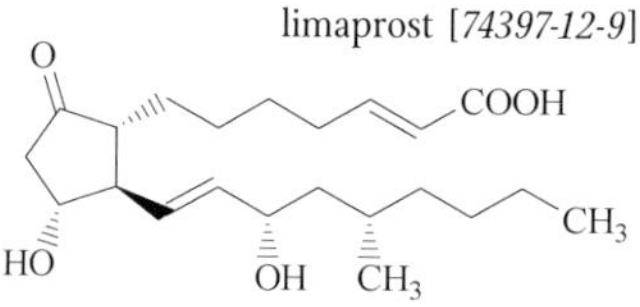	limaprost [74397-12-9]	ONO	Opalmon	peripheral circulatory disorders
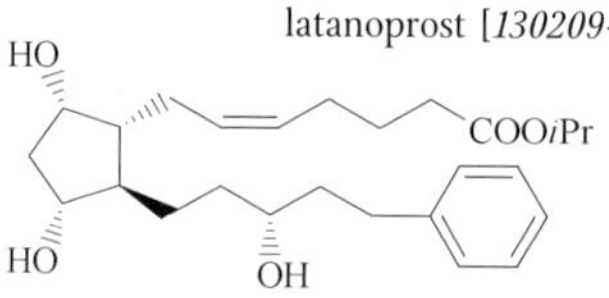	latanoprost [130209-82-4]	Pharmacia & Upjohn	Xalatan	glaucoma

Table 4. (continued)

Structure	Generic name, CAS registry no.	Producer	Trade name or phase of development	(Potential) indication
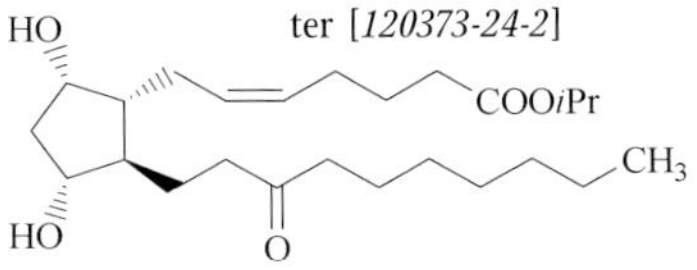 (structure)	unoprostone isopropylester [120373-24-2]	Ueno	Rescula	glaucoma

Table 5. Uses of prostacyclins in human medicine

Structure	Generic name, CAS registry no.	Producer	Trade name or phase of development	(Potential) indication
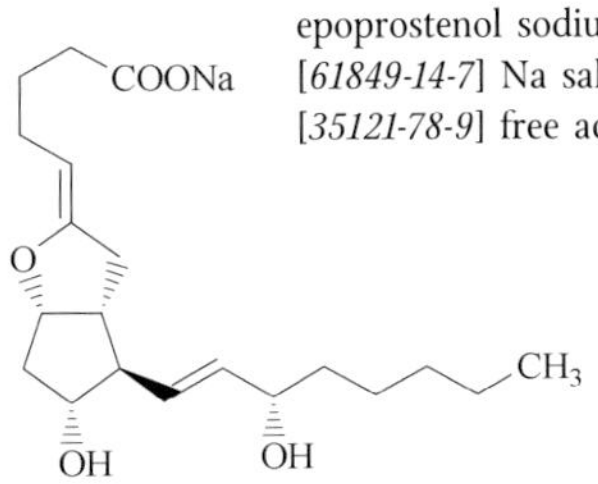 (structure)	epoprostenol sodium [61849-14-7] Na salt, [35121-78-9] free acid	Glaxo Wellcome, Pharmacia & Upjohn	Flolan, Cyclo-Prostin	circulatory disorders
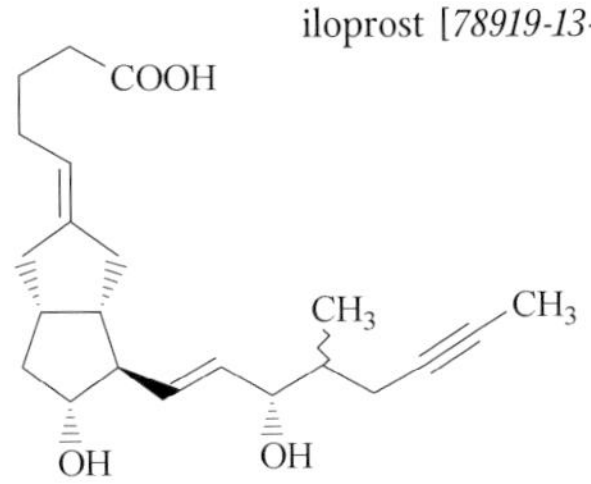 (structure)	iloprost [78919-13-8]	Schering AG	Ilomedin	peripheral circulatory disorders, Raynaud syndrom
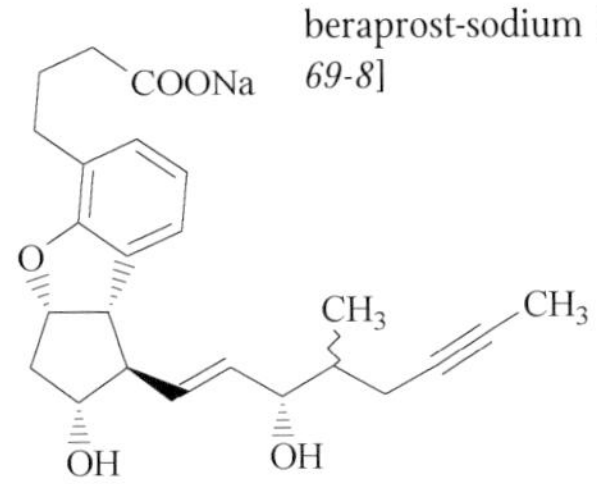 (structure)	beraprost-sodium [88475-69-8]	Toray-Kaken	Dorner, Procyclin	peripheral circulatory disorders

Table 5. (continued)

Structure	Generic name, CAS registry no.	Producer	Trade name or phase of development	(Potential) indication
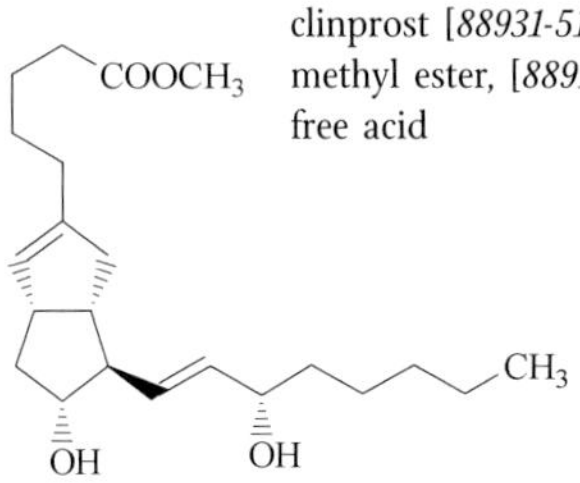	clinprost [88931-51-5] methyl ester, [88911-35-7] free acid	Teijin	clinical trials	peripheral circulatory disorders, cerebral thrombosis
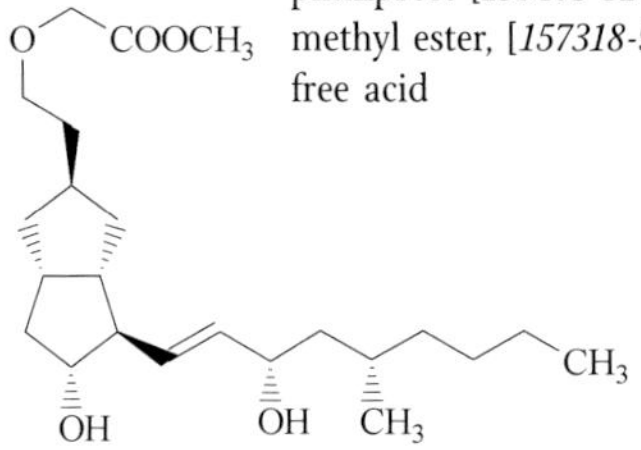	pimilprost [139403-31-9] methyl ester, [157318-92-8] free acid	Sumitomo	clinical trials	peripheral circulatory disorders

Table 6. Uses of thromboxane antagonists in human medicine

Structure	Generic name, CAS registry no.	Producer	Trade name or phase of development	(Potential) indication
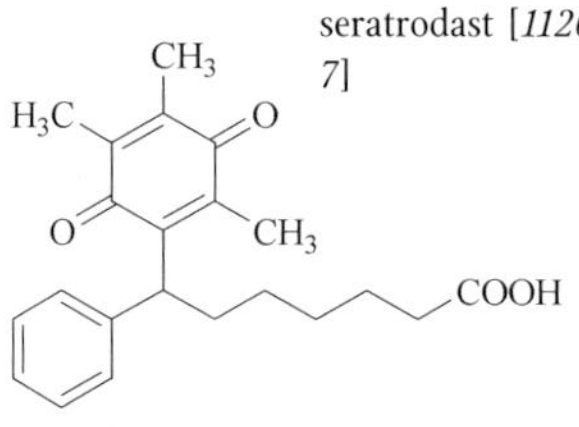	seratrodast [112665-43-7]	Takeda	Bronica	asthma, allergic rhinitis
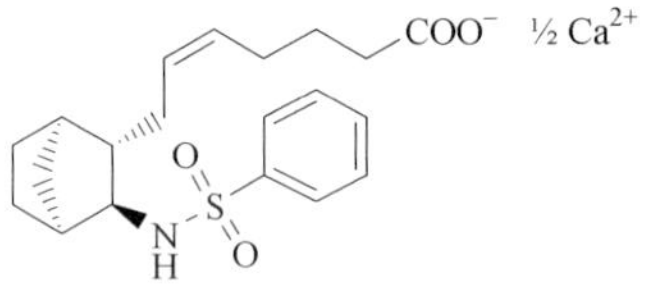	domitroban-calciumhydrate [132747-47-8]	Shionogi	Anboxan	asthma, antithrombotic

Table 6. (continued)

Structure	Generic name, CAS registry no.	Producer	Trade name or phase of development	(Potential) indication
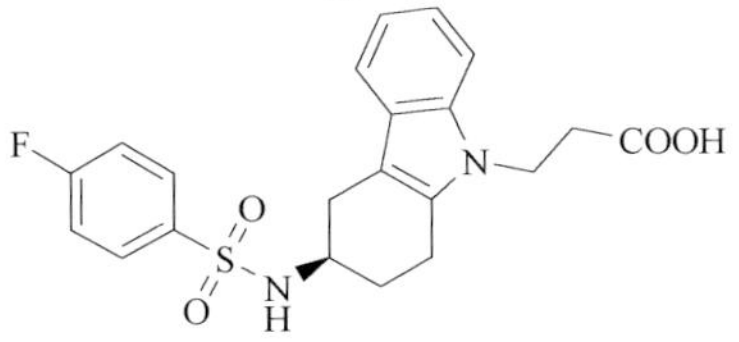	ramatroban [*116649-85-5*]	Bayer	clinical trials	asthma, allergic rhinits, antithrombotic angina
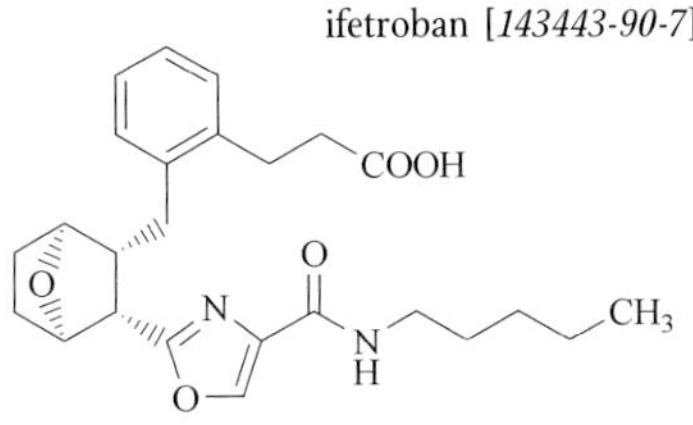	ifetroban [*143443-90-7*]	Bristol-Myers Squibb	clinical trials	antithrombotic, anti-ischaemic, topical for pressure-ulcers
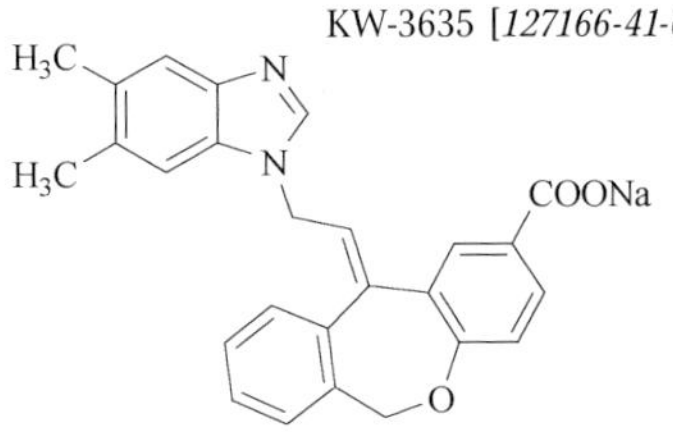	KW-3635 [*127166-41-0*]	Kyowa Hakko	clinical trials	asthma, antithrombotic
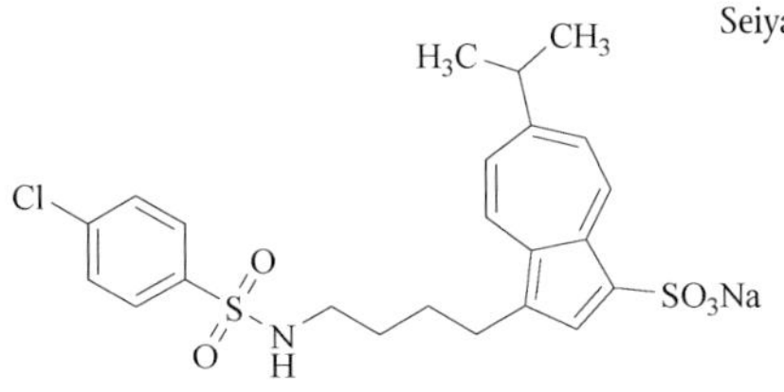	KT2-962 [*129648-96-0*]	Kotobuki Seiyaku	clinical trials	cardiovascular, renal diseases
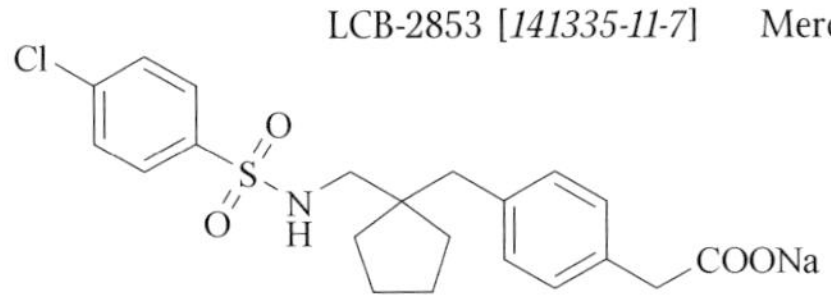	LCB-2853 [*141335-11-7*]	Merck KGaA	clinical trials	antithrombotic

Table 7. Uses of peptido leukotriene antagonists in human medicine

Structure	Generic name, CAS registry no.	Producer	Trade name or phase of development	(Potential) indication
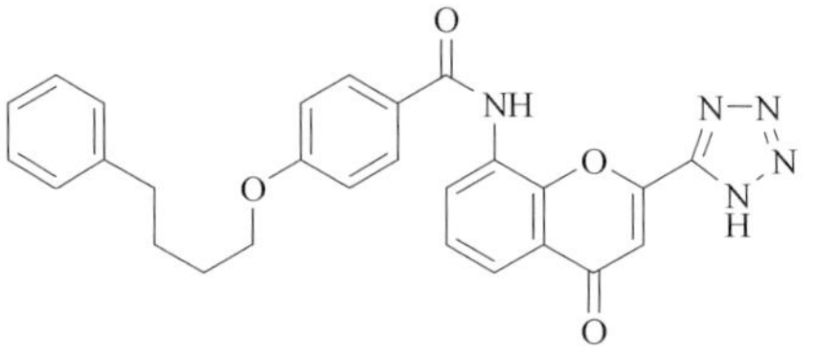	pranlukast [103177-37-3]	ONO	Onon, Ultair	asthma, allergy
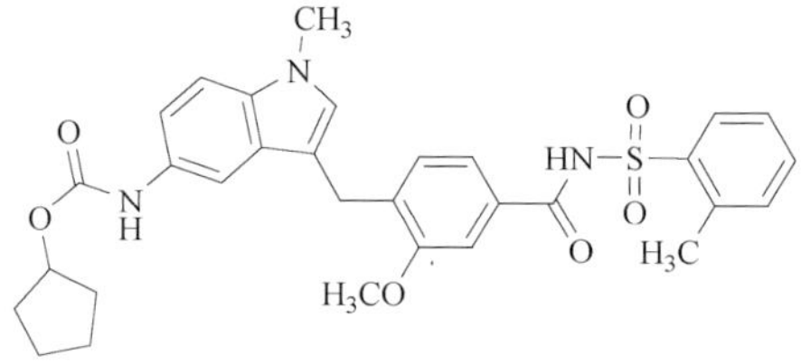	zafirlukast [107753-78-6]	Zeneca	Accolate, Vanticon	asthma, allergy
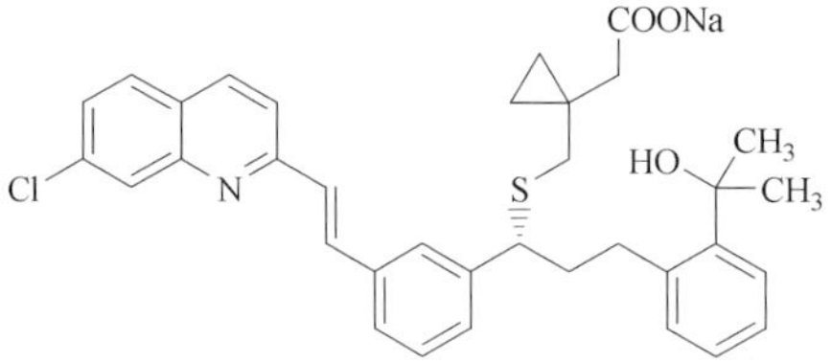	montelukast [151767-02-1]	Merck & Co.	Singulair	asthma
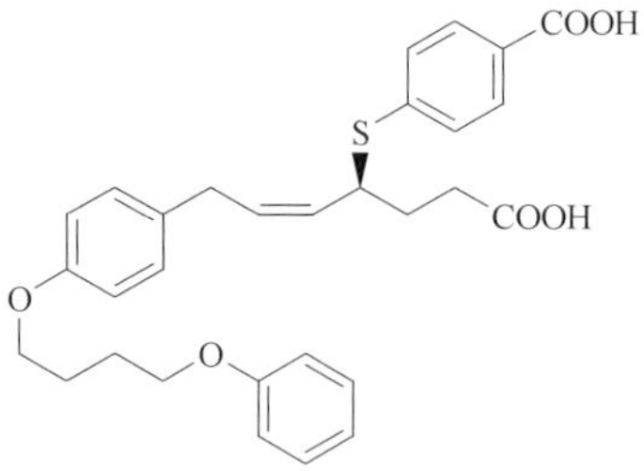	Bay x7195 [143538-27-6]	Bayer	clinical trials	asthma

Table 7. (continued)

Structure	Generic name, CAS registry no.	Producer	Trade name or phase of development	(Potential) indication
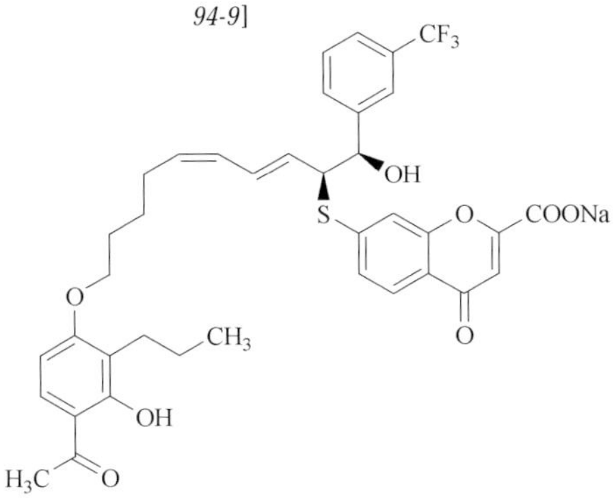	iralukast sodium [*125617-94-9*]	Novartis	clinical trials	asthma
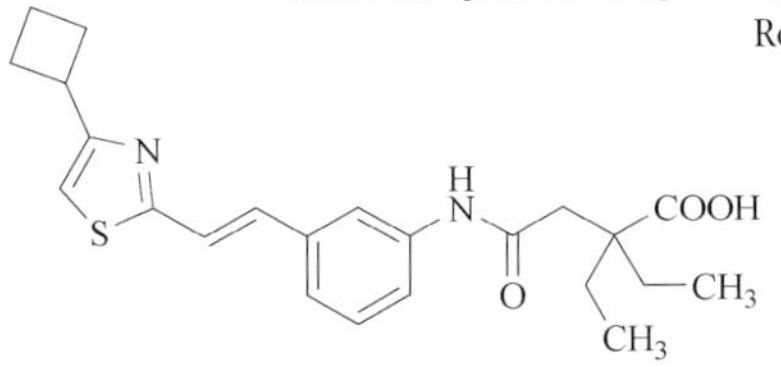	cinalukast [*128312-51-6*]	Hoffmann–La Roche	clinical trials	asthma

Table 8. Uses of leukotriene B_4 antagonists in human medicine

Structure	Generic name, CAS registry no.	Producer	Trade name or phase of development	(Potential) indication
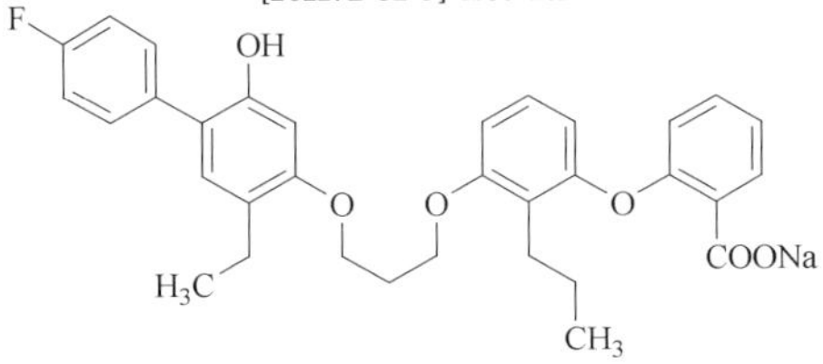	VML-295, LY 293111 [*152608-41-8*] Na salt, [*161172-51-6*] free acid	Eli Lilly	clinical trials	asthma

2065

Table 8. (continued)

Structure	Generic name, CAS registry no.	Producer	Trade name or phase of development	(Potential) indication
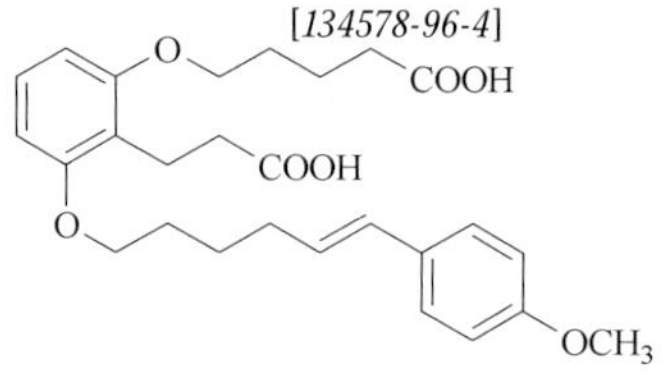	ONO LB-457, ONO 4057 [*134578-96-4*]	ONO	clinical trials	atopic dermatitis, ulcerative colitis, Behcet's disease
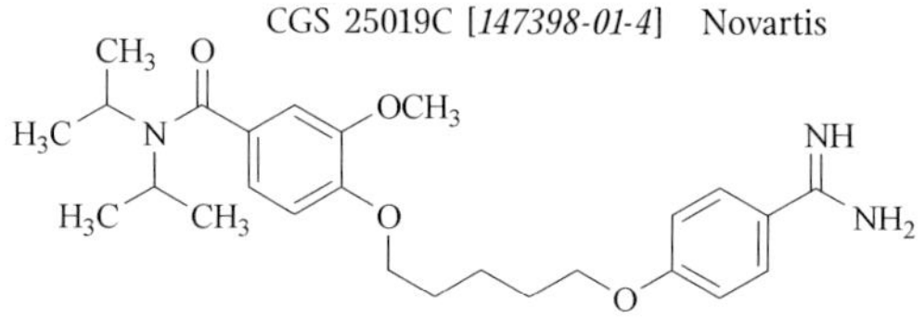	CGS 25019C [*147398-01-4*]	Novartis	clinical trials	psoriasis
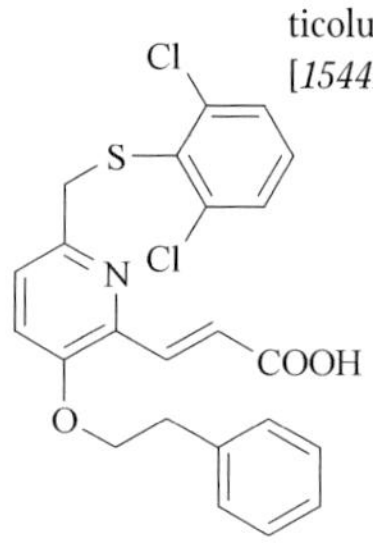	ticolubant, SB 209247 [*154413-16-3*]	SmithKline Beecham	clinical trials	eczema
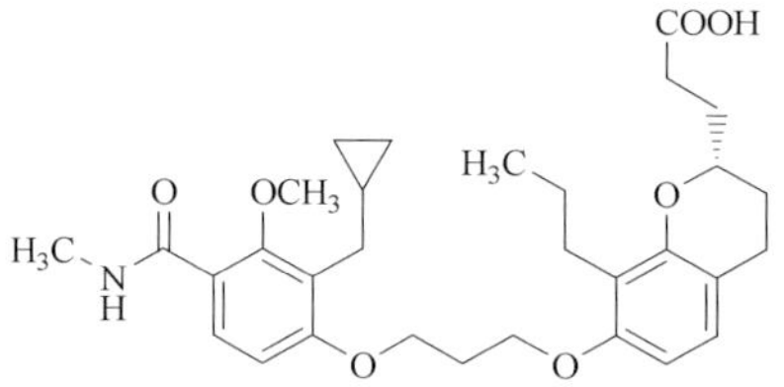	SC-53228 [*153633-01-3*]	Searle	clinical trials	psoriasis, inflammatory bowel disease
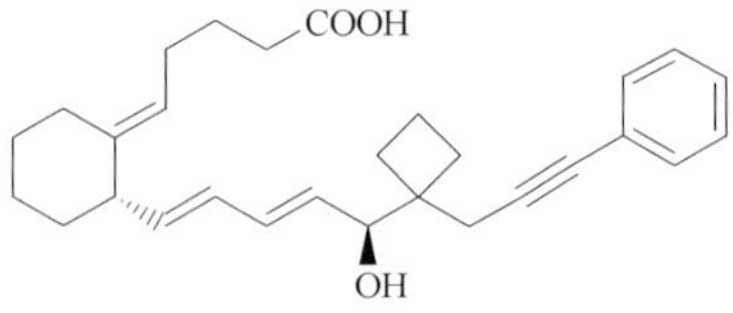	ZK 158252 [*162362-36-9*]	Schering AG	clinical trials	psoriasis

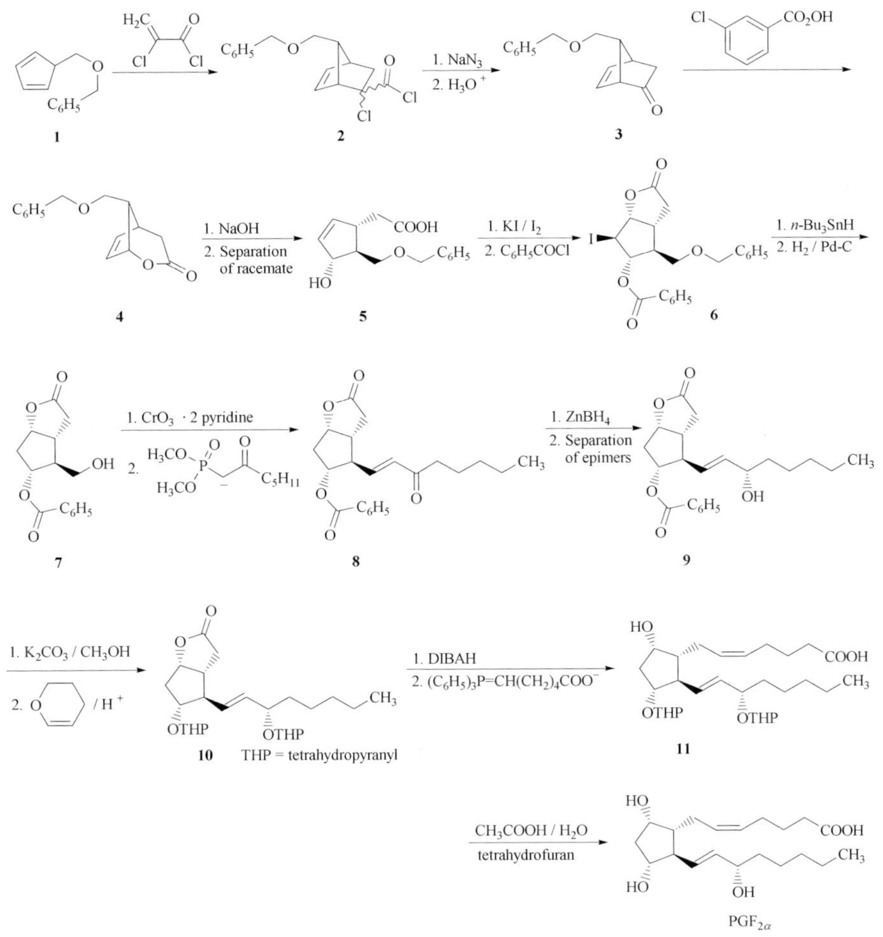

Figure 3. Synthesis of PGF$_{2\alpha}$ according to E. J. Corey

elimination of the iodine with tributyltin hydride and subsequent hydrogenolysis of the benzyl ether yield the Corey alcohol **7**. Oxidation to the aldehyde followed by a Horner–Emmons reaction gives the α,β-unsaturated ketone **8**. After reduction with zinc borohydride, the epimeric allyl alcohols are separated, giving the desired alcohol **9**. Cleavage of the benzoate and etherification of the two hydroxyl groups with dihydropyran yield the bis(tetrahydropyranyl) ether **10**. The reduction of the lactone to lactol with diisobutylaluminum hydride (DIBAH) and a subsequent Wittig reaction yield the acid **11**, which gives natural PGF$_{2\alpha}$ after cleavage of the protecting groups.

Cyclopentenone Method According to C. J. Sih (Fig. 4). As an example of this synthesis strategy, a method for the production of natural PGE$_2$ will be described [78]. Optical activity is introduced by a microbiological reaction. The triketone **12** can be

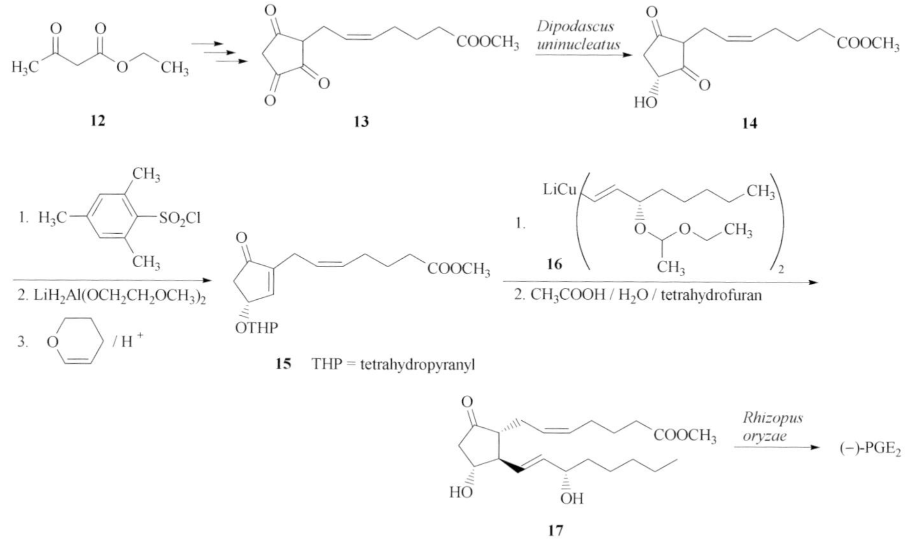

Figure 4. Cyclopentenone method according to C. J. Sih

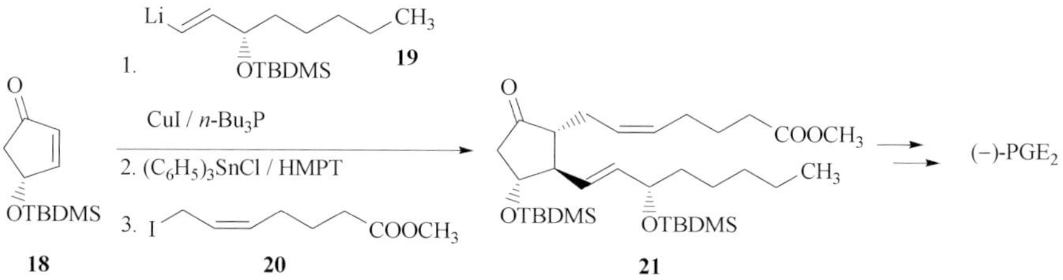

TBDMS = *tert*-Butyldimethylsilyl; HMPT = hexamethylphosphoric triamide

Figure 5. Three-component method according to R. Noyori

obtained in several steps from the acetoacetic ester **13**. The chiral alcohol **14** is made by a regio- and enantioselective, microbiological reduction of the racemic triketone **12**. Conversion to an enol sulfonate, reduction, and protection of the hydroxyl group yield the cyclopentenone **15**. The stereospecific 1,4-addition of the cuprate **16** to the α,β-unsaturated ketone **15** and cleavage of the protecting groups give the methyl ester of PGE_2 **17**, which yields natural PGE_2 by microbiological saponification. Saponification with a base is not possible because of the labile β-hydroxyketone moiety.

Three-Component Method According to R. Noyori [79] (Fig. 5). This is a "one-pot" process for the introduction of the α- and ω-chain. The optically active 3-hydroxycyclopentenone derivative **18** reacts with the organolithium compound **19** in a copper-catalyzed 1,4-addition. Then the resulting enolate reacts with the allyl iodide **20** via an organotin intermediate to give the PGE_2 derivative **21**. Cleavage of the protecting groups yields PGE_2.

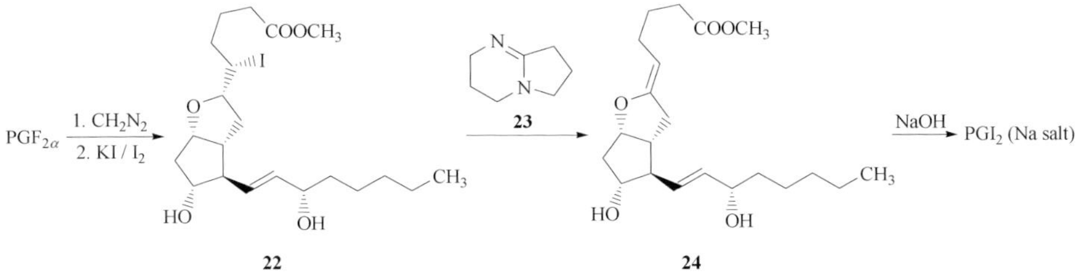

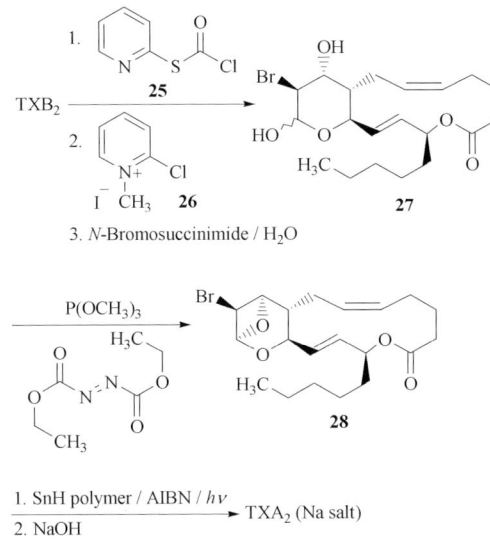

Figure 6. Synthesis of prostacyclin (PGI$_2$) from PGF$_{2\alpha}$

Figure 7. Synthesis of TXA$_2$ according to W. C. STILL

AIBN = azobisisobutyronitrile

Synthesis of PGI$_2$ from PGF$_{2\alpha}$ (Fig. 6). The synthesis of prostacyclin (PGI$_2$) from PGF$_{2\alpha}$ was described almost at the same time by five different teams independently of each other [80]. Esterification of natural PGF$_{2\alpha}$ with diazomethane and subsequent iodine lactonization yield the PGI$_1$ derivative **22**. Basic elimination with diazabicyclononene (**23**) gives the enol ether **24**, which, in turn, yields the sodium salt of PGI$_2$ after saponification. The free acid is normally not made because it is very unstable.

Synthesis of TXA$_2$ According to W. C. STILL [81] (Fig. 7). Starting with natural thromboxane B$_2$, a 1,15-macrolactone is synthesized as an intermediate with the help of the pyridine derivative **25**. Elimination of the 11-hydroxyl group with the Mukaiyama reagent **26** followed by a HOBr addition yields the bromohydrin **27**. An intramolecular Mitsunobu reaction gives the strained oxetane system **28**. Reductive elimination of the bromine with the help of polymer-bound tin hydride and subsequent saponification furnish the sodium salt of TXA$_2$. It has not yet been possible to synthesize free TXA$_2$ because of its great instability.

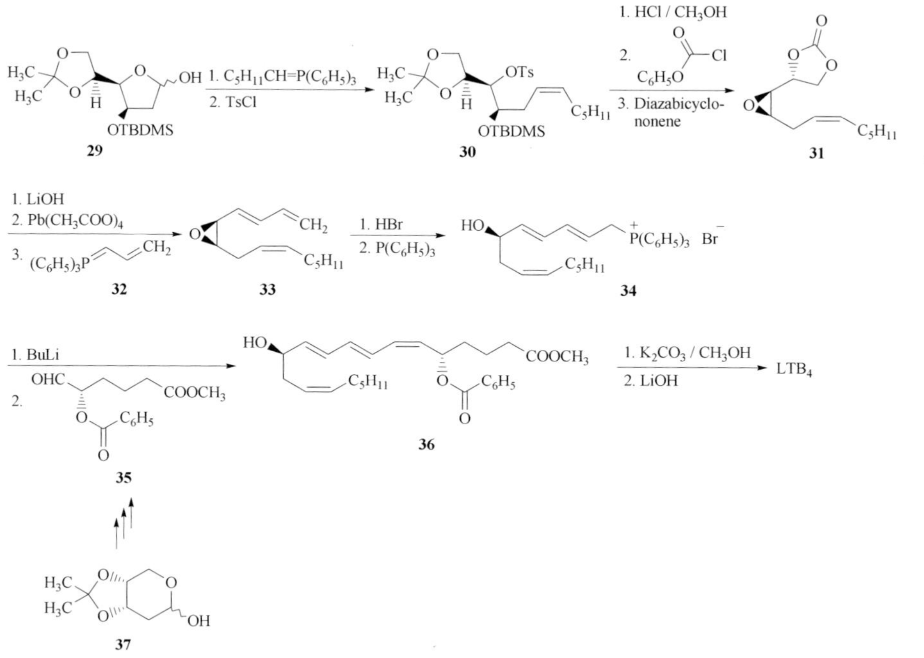

Figure 8. Synthesis of LTB$_4$ according to E. J. Corey

Synthesis of LTB$_4$ According to E. J. Corey [82] (Fig. 8). The lactol **29** can be made from D-(+)-mannose in a few steps and is then converted to the olefin **30** in a Wittig reaction with subsequent tosylation. Methanolic HCl simultaneously cleaves the acetonide and the silyl ether. The vicinal hydroxyl groups in the resulting triol are converted to the carbonate with phenyl chloroformate. Nucleophilic substitution of the tosylate by the remaining secondary hydroxyl group yields the epoxide **31**. The epoxy aldehyde obtained after saponification and glycol cleavage with lead tetraacetate is converted to the triene **32** in a Wittig reaction with the phosphorane **33**. The epoxide opening with HBr proceeds with simultaneous displacement of the diene system. The resulting primary bromide is converted to the Wittig salt **34** with triphenylphosphine. Deprotonation with butyllithium and reaction of the resulting ylide with the aldehyde **35** give the protected LTB$_4$ derivative **36**. Stepwise saponification produces natural LTB$_4$. The aldehyde **35** is synthesized in several steps from the acetonide of the deoxyribose **37**.

6. References

General References

[1]   J. S. Bindra, R. Bindra: *Prostaglandin Synthesis*, Academic Press, New York 1977.
[2]   A. Mitra: *The Synthesis of Prostaglandins*, Wiley-Interscience, New York 1977.
[3]   S. M. M. Karim (ed.): *The Prostaglandins*, Medical and Technical Publishing, Oxford 1972.
[4]   E. W. Horton: *Prostaglandins*, Springer Verlag, Berlin 1972.
[5]   P. W. Ramwell (ed.): *The Prostaglandins*, 2 vols., Plenum Press, New York 1974.
[6]   J. S. Bindra, R. Bindra, *Prog. Drug Res.* **17** (1973) 410.
[7]   T. O. Oesterling, W. Morozowich, T. J. Roseman, *J. Pharm. Sci.* **61** (1972) 1861.
[8]   K. C. Nicolaou, G. P. Gasic, W. E. Barnette, *Angew. Chem.* **90** (1978) 360; *Angew. Chem. Int. Ed. Engl.* **17** (1978) 293.
[9]   P. R. Marsham: *Aliphatic and Related Natural Product Chemistry (Specialist Periodical Report)*, vol. **1**, The Chemical Society, London 1979, p. 170.
[10] W. Bartmann, *Angew. Chem.* **87** (1975) 143; *Angew. Chem. Int. Ed. Engl.* **14** (1975) 337.
[11] W. Bartmann, G. Beck, *Angew. Chem* **94** (1982) 767; *Angew. Chem. Int. Ed. Engl.* **21** (1982) 751.
[12] B. Samuelsson, *Angew. Chem.* **94** (1982) 881; *Angew. Chem. Int. Ed. Engl.* **21** (1982) 902.
[13] B. Samuelsson, *Angew. Chem.* **95** (1983) 854; *Angew. Chem. Int. Ed. Engl.* **22** (1983) 805.
[14] S. Bergström, *Angew. Chem.* **95** (1983) 865; *Angew. Chem. Int. Ed. Engl.* **22** (1983) 858.
[15] J. R. Vane, *Angew. Chem.* **95** (1983) 782; *Angew. Chem. Int. Ed. Engl.* **22** (1983) 741.
[16] R. H. Green, P. F. Lambeth, *Tetrahedron* **39** (1983) 1687.
[17] K. Schrör: *Prostaglandine und verwandte Verbindungen*, Thieme Verlag, Stuttgart 1984.
[18] R. C. Nickolson, M. H. Town, H. Vorbrüggen, *Med. Res. Rev.* **5** (1985) 1.
[19] P. W. Collins, S. W. Djuric, *Chem. Rev.* **93** (1993) 1533.
[20] J. R. Vane, J. O'Grady (eds.): *Therapeutic Applications of Prostaglandins*, Edward Arnold, London – Boston – Melbourne – Auckland 1993.
[21] F. Scheinmann, J. Ackroyd: *Leukotriene Syntheses*, Raven Press, New York 1984.
[22] L. W. Chakrin, D. M. Bailey (eds.): *The Leukotrienes*, Academic Press, Orlando 1984.
[23] J. E. Pike, D. R. Morton (eds.): *Advances in Prostaglandin, Thromboxane, and Leukotriene Research*, vol. 14, Raven Press, New York 1985.
[24] G. Beck: *Arzneimittel, Fortschritte 1972 – 1985*, VCH Verlagsgesellschaft, Weinheim 1987, p. 835.
[25] A. L. Willis (ed.): *Handbook of Eicosanoids: Prostaglandins and Related Lipids*, 2 vols., CRC Press, Boca Raton, Florida 1987.
[26] T. W. Hart, *Nat. Prod. Rep.* **5** (1988) 1.
[27] J. Rokach (ed.): *Leukotrienes and Lipoxygenases*, Elsevier, Amsterdam 1989.
[28] W. T. Jackson, J. H. Fleisch, *Prog. Drug Res.* **46** (1996) 115.
[29] C. D. W. Brooks, J. B. Summers, *J. Med. Chem.* **39** (1996) 2629.
[30] S. L. Spector, *Annals of Allergy, Asthma, Immunol.* **75** (1995) 463.
[31] E. Aedelroth et al., *J. Allergy Clin. Immunol.* **99** (1997) 210.
[32] S. E. Dahlén et al. (eds.): *Leukotrienes as Mediators of Asthma and Inflammation: Basic and Applied Research*, Raven Press, New York, 1994.

Specific References

[33] U. S. von Euler, *Naunyn-Schmiedebergs Arch. Exp. Pathol. Pharmakol.* **175** (1934) 78. M. W. Goldblatt, *Chem. Ind. (London)* **52** (1933) 1056.

[34] S. Bergström, J. Sjövall, *Acta Chem. Scand.* **14** (1960) 1693. S. Bergström, J. Sjövall, *Acta Chem. Scand.* **11** (1957) 1086. S. Bergström, J. Sjövall, *Acta Chem. Scand.* **14** (1960) 1701.

[35] D. H. Nugteren et al., *Nature (London)* **212** (1966) 38. S. Bergström, R. Ryhage, B. Samuelsson, J. Sjövall, *Acta Chem. Scand.* **16** (1962) 501. S. Bergström, R. Ryhage, B. Samuelsson, J. Sjövall, *J. Biol. Chem.* **238** (1963) 3555.

[36] M. Hamberg, B. Samuelsson, *J. Am. Chem. Soc.* **88** (1966) 2349. D. H. Nugteren, R. K. Beerthuis, D. A. van Dorp, *Rec. Trav. Chim. Pays-Bas* **85** (1966) 405.

[37] M. Hamberg, B. Samuelsson, *Proc. Natl. Acad. Sci. USA* **71** (1974) 3400. M. Hamberg, J. Svensson, T. Wakabayashi, B. Samuelsson, *Proc. Natl. Acad. Sci. USA* **71** (1974) 345.

[38] M. Hamberg, J. Svensson, B. Samuelsson, *Proc. Natl. Acad. Sci. USA* **72** (1975) 2994.

[39] S. Bunting, R. Gryglewski, S. Moncada, J. R. Vane, *Prostaglandins* **12** (1976) 897.

[40] P. Borgeat, B. Samuelsson, *Proc. Natl. Acad. Sci. USA* **76** (1979) 2148.

[41] W. Feldberg, C. H. Kellaway, *J. Physiol. (London)* **94** (1938) 187.

[42] S. Hammarström et al., *Biochem. Biophys. Res. Commun.* **92** (1980) 946.

[43] C. N. Serhan, M. Hamberg, B. Samuelsson, *Biochem. Biophys. Res. Commun.* **118** (1984) 943. C. N. Serhan, M. Hamberg, B. Samuelsson, *Proc. Natl. Acad. Sci. USA* **81** (1984) 5335.

[44] N. A. Nelson, *J. Med. Chem.* **17** (1974) 911. R. A. Johnson, D. R. Morton, N. A. Nelson, *Prostaglandins* **15** (1978) 737.

[45] R. J. Light, B. Samuelsson, *Eur. J. Biochem.* **28** (1972) 232.

[46] M. Fukushima, T. Kato, *Adv. Prostaglandin Thromboxane Leukotriene Res.* **15** (1985) 415. A. N. Grechkin, *J. Lipid Mediat. Cell Signalling* **11** (1995) 205.

[47] E. D. Levin, V. E. Cherepanova, I. A. Zimovtseva, T. O. Sedlova, *Phytochemistry* **27** (1988) 3241.

[48] M. J. Müller, *Chemistry & Biology* **5** (1998) R323.

[49] K. C. Nicolaou, J. Y. Ramphal, N. A. Petasis, C. N. Serhan, *Angew. Chem.* **103** (1991) 1119; *Angew. Chem. Int. Ed. Engl.* **30** (1991) 1100.

[50] C. N. Serhan, *Biochimica et Biophysica Acta* **1212** (1994) 1.

[51] C. N. Serhan, *Prostaglandins* **53** (1997) 107.

[52] I. Kennedy, R. A. Coleman, P. P. A. Humphrey, P. Lumley, *Adv. Prostaglandin Thromboxane Leukotriene Res.* **11** (1983) 327. K.-H. Thierauch, H. Dinter, G. Stock, *J. Hypertens.* **11** (1993) 1315. K.-H. Thierauch, H. Dinter, G. Stock, *J. Hypertens.* **12** (1994) 1. M. Ushikubi, M. Hirata, S. Narumiya, *J. Lipid Mediators Cell Signalling* **12** (1995) 343. M. Hirata, F. Ushikubi, S. Narumiya, *J. Lipid Mediators Cell Signalling* **12** (1995) 393. R. R. A. Coleman, W. L. Smith, S. Narumiya, *Pharmacol. Rev.* **46** (1994) 205. T. Yokomizo et al., *Nature* **387** (1997) 620.

[53] G. Weissmann, *Sci. Am.* **264** (1991) 58.

[54] J. A. Salmon, *Adv. Drug Res.* **15** (1986) 111.

[55] M. G. Baumgärtner, *Deutsche Apotheker-Zeitung* **137** (1997) 2157.

[56] L. J. Marnett, A. S. Kalgutkar, *Curr. Opin. Chem. Biol.* **2** (1998) 482.

[57] D. O. Stichtenoth, H. Zeidler, J. C. Fröhlich, *Med. Klin.* **93** (1998) 407.

[58] J. R. Vane, R. M. Botting, *Inflamm. Res.* **47** (1998) Suppl. 2, S78.

[59] W. Elger et al., *J. Steroid Biochem.* **25** (1986) 835.

[60] G. Anderson, *Trends Pharmacol. Sci.* **11** (1990) 348. B. J. Lipworth, *Lancet* **353** (1999) 57.

[61] S. J. Konturek, *Scand. J. Gastroenterol.* **25** (1990) Suppl. 174, 15.

[62] A. Aly, *Scand. J. Gastroenterol.* **22** (1987) Suppl. 137, 43.

[63] W. F. Stenson, *Falk Symp.* **46** (1988) 143 (CA 109, 52 485 a).

[64] P. G. Adaikan, S. R. Kottegoda, *Drugs Future* **10** (1985) 765.

[65] S. E. Hall, *Med. Res. Rev.* **11** (1991) 503.

[66] A. L. Willis, D. L. Smith, *Eicosanoids* **2** (1989) 69.

[67] C. J. Lote, J. Haylor, *Prostaglandins Leukotrienes Essential Fatty Acids* **36** (1989) 203.
[68] F. A. Fitzpatrick, R. C. Murphy, *Pharmacol. Rev.* **40** (1988) 229.
[69] K. Hoshi, Y. Mizushima, *Prostaglandins* **40** (1990) 155.
[70] H. Yamazaki et al., *Adv. Prostaglandin Thromboxane Leukotriene Res.* **19** (1989) 289. H. Shirahase et al., *J. Cardiovasc. Pharmacol.* **10** (1987) 517.
[71] O. Hayaishi, *J. Biol. Chem.* **263** (1988) 14 593.
[72] K. V. Honn, B. Cicone, A. Skoff, *Science (Washington D.C.)* **212** (1981) 1270.
[73] A. M. Fulton, *Prostaglandins Leukotrienes Essential Fatty Acids* **34** (1988) 229. M. R. Schneider, M. Schirner, *Drugs Future* **18** (1993) 29. M. Schirner, *Wien. Klin. Wochenschr.* **107** (1995) 261. M. R. Schneider, D. G. Tang, M. Schirner, K. V. Honn, *Cancer Metastasis Rev.* **13** (1994) 349. *Drugs & Therapy Perspectives* **8** (1996) 1.
[74] A. Alm, *Curr. Opin. Ophthalmol.* **4** (1993) no. 2, 44.
[75] F. Meier, E. Gross, K.-M. Klotz, T. Ruzicka, *Skin Pharmacol.* **2** (1989) 61.
[76] E. Moilanen et al., *Agents Actions* **28** (1989) 290.
[77] E. J. Corey, J. Vlattas, K. Harding, *J. Am. Chem. Soc.* **91** (1969) 535. E. J. Corey et al., *J. Am. Chem. Soc.* **93** (1971) 1491.
[78] C. J. Sih et al., *J. Am. Chem. Soc.* **97** (1975) 865. J. B. Heather et al., *Tetrahedron Lett.* 1973, 2313.
[79] R. Noyori, M. Suzuki, *Chemtracts. Org. Chem.* **3** (1990) 173. M. Suzuki, A. Yanagisawa, R. Noyori, *J. Am. Chem. Soc.* **110** (1988) 4718.
[80] E. J. Corey, G. E. Keck, I. Székely, *J. Am. Chem. Soc.* **99** (1977) 2006. R. A. Johnson et al., *J. Am. Chem. Soc.* **99** (1977) 4182. N. Whittaker, *Tetrahedron Lett.* 1977, 2805. I. Tömösközi, G. Galambos, V. Simonidesz, G. Kovács, *Tetrahedron Lett.* 1977, 2627. K. C. Nicolaou et al., *J. Chem. Soc. Chem. Commun.* 1977, 630.
[81] S. S. Bhagwat, P. R. Hamann, W. C. Still, *J. Am. Chem. Soc.* **107** (1985) 6372.
[82] E. J. Corey, A. Marfat, G. Goto, F. Brion, *J. Am. Chem. Soc.* **102** (1980) 7984.

Related Technology

There are three subjects related to a review of the pharmaceutical market which are of interest to readers of this review. They include a discussion of drug testing, a review of pharmaceutical dosage forms and a specialized subclass of pharmaceutical products, veterinary drugs.

Pharmaceutical Dosage Forms

Paul Zanowiak, Temple University, Health Sciences Center, Philadelphia, Pennsylvania 19140, United States

1.	Dosage Forms as Drug Delivery Systems		2078
2.	Routes of Administration . .		2081
2.1.	Parenteral Administration . .		2081
2.2.	Oral Administration		2083
2.3.	Dermal Administration		2083
2.4.	Rectal Administration		2084
2.5.	Other Routes of Administration		2084
3.	Types of Dosage Forms . . .		2085
3.1.	Liquid Solution Dosage Forms		2086
3.1.1.	Solutions		2086
3.1.2.	Extractive Solutions		2086
3.1.3.	Parenteral Solutions (Injections)		2088
3.1.4.	Inhalations		2088
3.1.5.	Ophthalmic and Nasal Solutions (Drops)		2088
3.1.6.	Otic Solutions		2088
3.1.7.	Liniments		2088
3.1.8.	Enemas and Douches		2088
3.1.9.	Mouthwashes and Gargles . . .		2089
3.1.10.	Collodions.		2089
3.2.	Liquid Dispersions		2089
3.3.	Semisolid Dosage Forms . . .		2089
3.4.	Plastic Dosage Forms (Suppositories)		2091
3.5.	Solid Dosage Forms		2091
3.6.	Prolonged-Action Dosage Forms.		2100
3.7.	Pressurized Aerosol Dosage Forms.		2100
3.8.	Nonpressurized Gaseous Dosage Forms (Inhalants) . .		2101
3.9.	Radiopharmaceutical Dosage Forms.		2101
4.	Pharmaceutical Excipients . .		2101
5.	Quality Assurance		2105
6.	Pharmaceutical Containers .		2106
7.	Development of the Ideal Drug Delivery System		2109
7.1.	Controlled-Release Drug Delivery Systems		2111
7.2.	Sustained- or Prolonged-Release Technology		2114
7.2.1.	Oral Sustained-Release Dosage Forms		2114
7.2.2.	Other Sustained- and Prolonged-Release Drug Delivery Systems		2117
7.3.	Targeted Drug Delivery Systems		2121
8.	References		2122

I. Dosage Forms as Drug Delivery Systems [1]–[3]

Contemporary dosage forms are designed to achieve a safe and therapeutically effective response each time they are administered within appropriate regimens that reflect accurate doses and dosing intervals. Thus, drug products are drug delivery systems that accomplish the prescriber's intent and which complete the Physician – Patient – Pharmacist triad when used "as directed" in a compliant manner.

Each type of dosage form involves several concepts and components:

1) The active ingredient (drug) in correct and accurately measured amount (dose) and of appropriate purity
2) Nontherapeutic ingredients (adjuvants, excipients) needed for the safe and effective preparation and delivery of the drug moiety
3) Unit manufacturing processes (technologies) designed so as to ensure that the product elicits safe and effective responses each time (e.g., tablet-to-tablet) and for each batch
4) Container or packaging designed for patient convenience and to maintain stability and therapeutic performance
5) Comprehensive quality assurance to safeguard the drug during shelf-life, during appropriate storage and use

The *drug moiety* must be available in pure form with an appropriate particle size, crystalline structure, and physicochemical form and in an accurate amount.

Appropriate *excipients* must be selected to insure stability (e.g., antioxidants, antimicrobial preservatives, buffers), accurate dose delivery (diluents), and product performance. The latter excipients include disintegrating agents for compressed tablets and polymers for enteric coatings. Some excipients are processing aids, e.g., glidants to ensure effective flow of granules from hoppers and lubricants for tablet compression. Attention must also be given to excipients for patient acceptance and compliant use (flavors, colors). Excipients should be compatible with the drug moiety and should not form complexes with the drug that lessen therapeutic availability during production and storage. Similarly, excipients must not affect the stability of the drug product or any of its components.

Production technologies must guarantee reliable therapy, i.e., equally safe and effective responses each time a dosage form is administered. The technologies depend on the dosage form in question. They must not be deleterious to product stability, nor should they decrease the availability of the drug moiety upon administration.

The equipment used also varies in size from small laboratory units for research and development to large-scale manufacturing units. Each manufacturer utilizes a pilot plant or scale-up operation in which development pharmacists and engineers have to transform laboratory-scale batches to full production scale without altering product performance, accuracy, precision, or stability. The major operations utilized in the manufacture of drug products are listed below:

1) Dissolution
2) Filtration
3) Dispersion or wetting of solids
4) Homogenization
5) Lyophilization
6) Encapsulation of powders (hard gelatin shells and soft gelatin capsules)
7) Comminution of solids, including micronizing to ultrafine particle size
8) Blending of powders
9) Granulation of powders
10) Compression and tabletting
11) Coating of powders, granules or tablets (pan coating, compression coating, air suspension coating)
12) Microencapsulation (coacervation, etc.)
13) Heating and cooling of liquids and semisolids
14) Molding (suppositories)
15) Drying
16) Pressurized filling (of aerosols)
17) Sterilization (steam autoclave, ethylene oxide autoclave, membrane filtration, radiation, laminar flow hood)

The *design of the package*, including selection of appropriate container and sealing materials, also affects the performance of the drug product. The container must assure stability throughout the product storage and use. Its shape and size must insure ease of use and efficient delivery to foster patient compliance.

Throughout the preparation of the dosage form, *quality assurance* must be guaranteed by good manufacturing practice (e.g., cleanliness, record keeping, process control). Quality assurance also includes quality-control testing of raw materials and final products to comply with relevant specifications. In-process testing (e.g., of tablet hardness and weight) is performed at critical stages of production. Each process must be validated to show that it accomplishes its intended function(s) with the same results from batch-to-batch.

During the development of a new dosage form *stability testing* is performed to verify that degradation does not occur during processing and storage. A shelf life is determined for each product and expiration dates are obtained.

Product labeling insures product identity, provides appropriate storage and dating information, and cites necessary warnings.

Bioavailability [4]–[6] In order to achieve a therapeutic response after administration, the drug moiety must be made available to the body and efficiently released from the dosage form. For example, a compressed tablet designed for oral administration must first disintegrate in the gastrointestinal tract. The drug from the resulting particles must dissolve in the gastrointestinal fluids and be absorbed into the blood circulation for distribution throughout the body. The drug may, however, bind to plasma proteins

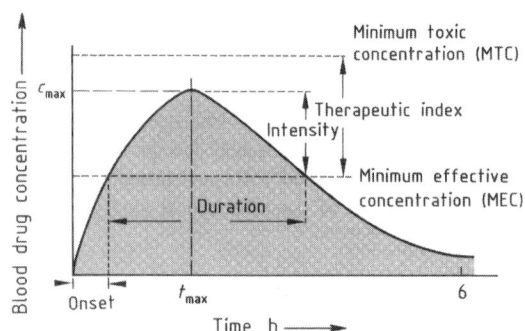

Figure 1. Blood drug concentration–time curve in a pharmacologic or therapeutic response
The shaded area is the "area under the curve" (AUC).

in the blood. The drug is metabolized in the liver but a certain proportion is excreted unchanged through the kidney. Absorption, distribution, metabolism, and excretion (ADME) occur simultaneously and at different rates that depend on the drug involved; the sex, race, age, and state of health of the patient; and other factors (e.g., presence of other drugs).

For a given drug dose to be effective, a sufficient amount must be absorbed (i.e., be available to the blood) at an appropriate rate. The kinetics of the subsequent distribution, metabolism, and excretion processes must ensure that the concentration of the drug at the site of action reaches the threshold level required to achieve the desired therapeutic response. This concentration must be maintained for a desired period. Subsequent administration of the drug must be timed appropriately to maintain this concentration at the target organ site. If too high a concentration is reached, safety may be compromised. If a subminimal concentration results, a period of non-therapy occurs.

A peak plasma concentration (c_{max}) occurs at a time t_{max} after each administration of a drug product (Fig. 1). This concentration should be somewhat greater than the minimum effective plasma concentration needed to achieve the desired pharmacologic effect, but not so high as to cause side effects and toxicity. The minimum toxic concentration (MTC) is the concentration of drug in the blood that initiates an untoward response (toxicity, side effect). The *therapeutic index* is the ratio of the minimum toxic concentration of drug (MTC) in the blood to the minimum effective concentration (MEC) in the blood (MTC/MEC); the greater the number the safer the drug is.

The parameter t_{max} is important in developing a rational regimen for dosage form administration. The point at which the threshold concentration (i.e., the MEC) is reached is termed the *onset of activity*. The time from the onset to the point when the concentration falls below the threshold level is termed the *duration of activity*.

Bioavailability is best defined as the amount of an administered drug that reaches the general circulation and the rate at which this occurs. Attention must be given to bioavailability during the design of a new drug product because it may be adversely affected by the selection of excipients and unit manufacturing operations. The area under the curve (AUC) is the total area enclosed by the blood drug concentration – time curve. It represents the extent of bioavailability of an administered dose of drug.

Excess compression pressure in tabletting may, for example, result in tablets that pass intact through the gastrointestinal tract. The relationship of formulation design and manufacture to bioavailability is generally termed biopharmaceutics. Biopharmaceutic concerns are important when a new (generic) dosage form is being designed to be pharmacologically and pharmacokinetically equivalent (i.e., bioequivalent) to an already marketed product that is no longer protected by patent.

2. Routes of Administration [7]

Drug products are administered to the human body by different routes and in various physical forms such as solutions, emulsions, suspensions, ointments, creams, suppositories, powders, granules, capsules, and compressed tablets. The most important administration route for achieving systemic therapeutic effects is the *enteral route*; the unit is swallowed, the drug dissolves in the gastrointestinal fluid and is absorbed through the mucous membrane of the gastrointestinal tract into the general circulation. Other major routes include *rectal administration* for local medication of the mucous membranes or for a systemic effect after absorption through the mucous membrane of the colon.

Parenteral injection through the skin, utilizing a syringe and needle, requires strict attention to sterility during administration and in the preparation and packaging of the dosage forms and equipment.

Dermal application is used for treatment of topical conditions (e.g., skin infections, psoriasis, acne, contact dermatitis). Drugs can, however, also be made available to the systemic circulation by percutaneous absorption.

2.1. Parenteral Administration [8], [9]

Dosage forms for parenteral administration are designed for injection through the skin with syringes and needles of appropriate size and diameter. When small doses are involved the drug is supplied in single-dose glass ampules or multi-dose vials. The vials have rubber stoppers through which the needle of a syringe is inserted to withdraw the desired dose.

For large-volume administration in hospitals or clinic settings, the drug is delivered intravenously from glass flasks or pliable plastic containers hung near the patient. Metered pumps are often used to insure accurate dosing (number of drops per minute) over long periods. Portable systems are also available in which the patient "wears" a self-contained unit consisting of a metering device and drug supply.

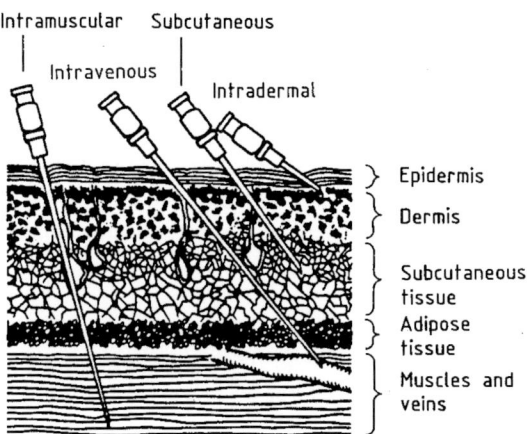

Figure 2. The most important types of parenteral administration [10]

Parenteral injections can deliver the drug directly to the general circulation intravenously (i.v.), subcutaneously (s.c.), or intramuscularly (i.m.), see Figure 2. The *intravenous route* can be used to administer a single dose (bolus) or for continuous administration over a long period (12 – 24 h). In the latter case, several drugs can be administerd in a sequential system ("piggy-backed"). Hyperalimentation solutions containing carbohydrates, amino acids, vitamins, and other nutrients are large-volume parenteral dosage forms that are administered intravenously to sustain debilitated patients for long periods of time. These nutrient solutions are administered continuously through indwelling catheters.

Parenteral dosage forms are manufactured as sterile units. They must also be free of pyrogens and foreign particles. Parenteral administration has the major disadvantage of being a potential source of infection, irritation, or pain at the site of injection. It is the route of choice when a quick therapeutic response is needed (e.g., in emergency situations or with comatose patients). It is also useful when oral drug administration cannot be tolerated (i.e., with nausea or after gastrointestinal surgery). Some drugs can only reach the general circulation by parenteral injection, for example, because of chemical instability in the gastric fluid or because they cannot be absorbed via the gastrointestinal tract (e.g., insulin).

Oily solutions or suspensions may be administered by *subcutaneous or intramuscular injection* to create "depot" doses. The drug slowly leaches from the injection site to extend the period of therapy. Small, sterile, rodlike pellets can also be implanted beneath the skin to achieve therapy periods of several months. Such implantation requires lancing and local anesthesia.

Specialized parenteral routes include intraarterial, intrathecal, intraspinal, intradermal, and intracisternal administration. Intra-arterial injection may require surgery to reach the artery and is used to administer radioopaque contrast media for viewing a specific organ. An intrathecal injection involves direct injection into the cerebrospinal fluid. Intraspinal injections are administered into the dural membrane surrounding the spinal cord. Intradermal administration occurs when the drug solution is injected into the dermis and epidermis of the skin (e.g., with solutions used to test for allergies). Intracisternal injection involves introduction of the drug solution into one of the subarachnoid cisternae of the brain.

Peritoneal dialysis is an example of a specialized large-volume parenteral dosage form. Drug solutions are injected intraperitoneally to treat the membranes of the abdominal cavity.

2.2. Oral Administration [11]

The oral route is popular because of its convenience. Oral dosage forms do not require specialized equipment, are noninvasive, and are generally more economical to prepare and administer. The main type of oral drug administration is the swallowing of solid (e.g., capsules, tablets) or liquid (e.g., syrups, solutions, emulsions) dosage forms. The drugs either medicate the gastrointestinal tract itself (e.g., antacids, anthelmintics) or are absorbed through the gastrointestinal mucous membranes into the systemic circulation to achieve a therapeutic effect at a distant target site or organ.

Orally administered dosage forms can also be held under the tongue (sublingual) or in the cheek pouch (buccal). The drug is absorbed through the oral mucous membrane into the blood stream. It enters the jugular vein and passes directly to the heart, avoiding the liver on its first pass through the body. This is in contrast to absorption through the gastrointestinal tract where the absorbed drug first passes through the portal vein to the liver before distribution to the rest of the body. Thus, drugs that are metabolized quickly by the liver (first-pass effect) can be administered more effectively in sublingual dosage forms. Also, drugs needed to treat the heart (e.g., nitroglycerin for angina) are delivered rapidly to the site of action from sublingually administered solid dosage forms that dissolve quickly.

Some dosage forms are designed for local treatment of the mucous membrane and gums of the mouth and throat (i.e., mouthwashes, gargles, lozenges).

2.3. Dermal Administration

Certain dosage forms are designed to treat the skin and its appendages (the sebaceous and sweat glands). Therapy is localized and nonsystemic; percutaneous absorption of the drug should be avoided. Dermal dosage forms can contain greasy base ingredients (pastes, ointments) or can be emulsified semisolids (cold creams, vanishing creams). Some topical applications are delivered from pressurized aerosol containers as solutions, suspensions, emulsions, foams, or powders. Solution dosage forms (soaks) are also used as astringent applications to weeping wounds.

Recent developments in drug delivery systems include the transdermal route of drug administration for prolonged periods of therapy. The drug is present either in a reservoir of drug and excipient ingredients or in a drug–excipient matrix that are attached to adhesive backings ("patches") for application to the skin. The drug is then released from the unit and transferred through the skin to the systemic circulation (see also Section 7.2.2).

2.4. Rectal Administration

Drugs can be administered rectally as enemas (solutions or suspensions) or suppositories to treat localized conditions of the rectum or colon (cathartics, hemmorhoid treatment). Such dosage forms can also be used for systemic therapy if the drug is absorbed through the mucous membrane of the colon. Treatment of comatose patients is possible by such rectal administration. This route is also useful in the treatment of infants, debilitated elderly patients, nauseated patients, and those having undergone recent gastrointestinal surgery. For example, debilitated cancer patients can gain relief from pain after administration of suppositories containing narcotic drugs.

2.5. Other Routes of Administration

Ophthalmic, Nasal, and Otic Routes. Various ophthalmic conditions (e.g., glaucoma) can be treated by local application of sterile, isotonic solutions or suspensions delivered from appropriate containers (glass dropper or plastic squeeze bottles). Sterile ophthalmic ointments are applied under the lower eye lid from specially designed tubes; any powdered drug present in an ophthalmic ointment or suspension must be of a very fine particle size (≤ 10 μm) to avoid irritation.

Nasal solutions, drops, or sprays from appropriately designed plastic containers are used to treat local conditions of the nasal mucosa (e.g., vasoconstrictor drugs to alleviate "stuffy nose" and anti-inflammatory agents to treat perennial rhinitis). Pressurized aerosols can also be used to treat nasal conditions.

Transnasal drug delivery via absorption through the nasal mucosa has gained strong interest for systematic therapy [12]. Absorption of polypeptides by this route is the subject of considerable research. It may be an alternative route of administration of insulin for diabetic patients. The assurance of accurate dosing is, however, still a problem.

Otic (aural) administration of drug products is used only for local treatment of the external auditory canal. Since this canal is lined with dermal tissue and not mucous membrane, infections are treated much like skin infections.

Intravaginal Administration. Dosage forms applied vaginally are intended for treatment of local infections. Solutions (douches), aerosol foams, creams, or suppositories are used. Spermicidal agents can be administered similarly. Intrauterine devices that release small amounts of contraceptive drugs for long periods of time are recent developments [13]. One such commercially available product (Progestasert, ALZA) acts as a progesterone reservoir that releases effective amounts of the drug for up to one year. Effects are local, thus minimizing the systemic side effects that accompany oral administration.

Oral Inhalation [14]. Pressurized aerosol drug products are available for inhalation and are used with adapters that are placed between the upper and lower teeth during actuation. Such solution or suspension products are intended for local pulmonary use (e.g., asthma attacks) or for systemic therapy. Rather complex technologies are required to ensure delivery of accurate doses (metered valves) and to control the particle size of the delivered products. Inhalation aerosols are also available for sublingual medication. Nitroglycerin, for example, can be delivered in metered doses directly under the tongue from pressurized aerosol containers for quick absorption and therapy of angina pain.

3. Types of Dosage Forms [15]

Dosage forms can be classed according to their use or physical form. In the former case the classes correspond to the routes of administration (Chap. 2): topical or external use (creams, ointments, pastes); oral use dosage forms (syrups, tablets, capsules); rectal suppositories; parenteral injections; transdermal patches; and localized drops (ocular, nasal, otic). In the present chapter, currently used dosage forms are classified according to their physical form as follows:

1) Liquid solutions
2) Liquid dispersions
3) Semisolid dispersions
4) Plastic dispersions
5) Solid dosage forms including coated solids
6) Prolonged action products
7) Pressurized aerosols
8) Nonpressurized gaseous dosage forms (inhalants)
9) Radiopharmaceuticals

3.1. Liquid Solution Dosage Forms [16]

Liquid solutions can be used for various pharmaceutical purposes:

1) As oral medication: sweetened, flavored, with aqueous or hydroalcoholic (water–ethanol) solvent system
2) For external application (soaks, baths) to skin wounds
3) Application to body cavities or apertures: enemas, douches, bladder irrigations
4) As inhalations, by mouth, from atomizers, nebulizers, and pressurized aerosols
5) As gargles or mouthwashes to treat the mucous membranes of the mouth and throat
6) As eye, ear, or nasal drops for local application
7) Parenteral injection

3.1.1. Solutions

Early compendia identified solutions as liquid preparations of one or more substances dissolved in liquid solvents, which by reason of their ingredients or method of preparation could not be classified in some other group of official dosage forms. They were termed "liquors" and were generally made by simple dissolution or by chemical reaction.

Official requirements for pharmaceuticals and their dosage forms are specified in national and international pharmacopeias. The Pharmacopea Internationalis and the European Pharmacopoeia are international works. National pharmacopeias include the British Pharmacopoeia, the Deutsches Arzneibuch (DAB 7, 1987), and the United States Pharmacopeia (U.S.P. XXII, 1990). In this article specifications are illustrated with examples from the U.S.P.

The United States Pharmacopeia and National Formulary (U.S.P./N.F.) currently identifies solutions according to their intended use (e.g., oral solutions, ophthalmic solutions, and inhalation solutions). Sterile parenteral solutions are termed injections.

3.1.2. Extractive Solutions

Several types of drug solution dosage forms that were originally developed years ago by extraction of plant (crude drug) materials with various solvents (menstrua) are still identified in compendia. Thus, various syrups, spirits, tinctures, and fluid extracts are extractive solutions that are listed in U.S.P. XXII and N.F. XVII. Some of these contain therapeutic ingredients (e.g., Ipecac Syrup, Senna Fluid Extract, Cascara Sagrada Fluid Extract, and Belladonna Tincture). Extracts that do not contain therapeutic ingredients were developed as flavored diluting vehicles for compounding admixture prescriptions (e.g., Peppermint Spirit, Vanilla Tincture, Tolu Balsam Tincture, and Tolu Balsam Syrup in N.F. XVII). Aromatic waters also belong to this type of solution. The last official waters in the U.S.P./N.F. are Peppermint Water, Rose Water, and Orange Flower Water.

Syrups. Aqueous solutions containing high concentrations of sucrose or other sugars are termed syrups. U.S.P. XXII lists the item "Syrup" as a near-saturated solution of sucrose in purified water. Certain polyols (e.g., glycerol, sorbitol) and flavoring agents can be added. Some syrups can be made by extraction; some are nonmedicated flavoring agents.

Aromatic waters are clear, saturated aqueous solutions of volatile oils or aromatic solids. They were used in the 1930s and 1940s as flavoring vehicles and had a minor carminative effect.

Elixirs. Sweetened, hydroalcoholic solutions for oral use are termed elixirs. Some are medicated; others were developed as official flavoring vehicles. Iso-alcoholic Elixir is an example. When this elixir was official it appeared in a high-alcoholic (75 vol %) and a low-alcoholic (10 vol %) form. By mixing proportions of each, a pharmacist could obtain varying percentages (10–75 vol %) of ethanol to insure solubility of prescribed drug entities in the extemporaneous compounding of prescriptions.

In commercial use the term "elixir" (unless followed by "U.S.P." or "N.F." in the United States) does not imply the official form. This is also true of terms such as "syrup" and "tincture".

Spirits and essences were alcoholic or hydroalcoholic solutions of volatile substances. Some were medicated, while others were flavoring vehicles to ensure a high ethanol content.

Tinctures were developed as alcoholic or hydroalcoholic solutions of plant drugs, synthetic drugs, or flavoring ingredients. Orginally, tinctures existed in three general types:

1) *Potent tinctures*: 100 mL contained the active ingredient content of 10 g of crude dried drugs (belladonna tincture)
2) *Nonpotent tinctures*: 100 mL contained the active ingredient content of 20 g of crude dried drug
3) *Flavoring tinctures*: 100 mL could contain the flavor principles of up to 50 g of crude, dried vegetable or fruit powder (sweet orange peel tincture, lemon tincture)

Fluid Extracts. Fluid extracts of plant materials were high in ethanol and could be viewed as standardized tinctures. They were designed so that each milliliter contained the extracted principle of 1 g of crude, dried drug.

3.1.3. Parenteral Solutions (Injections) [17]

Parenteral solutions are meant for injection through the skin and must be sterile, pyrogen-free, and isotonic. Their pH must be close to that of blood.

3.1.4. Inhalations

Inhalations are aqueous solutions for treatment of the lower respiratory tract. The smaller the droplet size delivered, the deeper into the respiratory tract the medication will reach. Administration necessitates use of inhalation devices, with nebulizers delivering smaller droplets than atomizers. Pressurized aerosols with mouthpiece adapters and actuators can deliver fine mists by means of specially designed valves. Such valves can deliver metered doses, this is an advantage over nebulizers and atomizers.

3.1.5. Ophthalmic and Nasal Solutions (Drops)

Ophthalmic and nasal solutions are used for local medication. Both should be isotonic and ophthalmic solutions must be sterile.

3.1.6. Otic Solutions

For application to the ear canal, otic solutions are best prepared with nonaqueous solvents. Anhydrous glycerol is a popular solvent.

3.1.7. Liniments

Liniments are liquid preparations meant for topical application with rubbing. They may be hydroalcoholic solutions, oily solutions, suspensions, or emulsions.

3.1.8. Enemas and Douches

Enemas and douches are solutions used to treat local rectal conditions or as vaginal medication, respectively.

3.1.9. Mouthwashes and Gargles

Mouthwashes and gargles generally are aqueous or hydroalcoholic solutions of aromatic principles and/or antiseptics. They are used to treat conditions of the mouth and throat or to refresh the buccal cavity. Antiseptic effects are, however, rather transient.

3.1.10. Collodions

Collodions are polymeric solutions (pyroxylin) that utilize diethyl ether and acetone as solvents. These viscous liquids are applied to the skin with applicator bottles. Evaporation of the solvent(s) leaves a protective film on the skin. Plasticizers can be added to provide a more flexible film. Drugs can also be added (e.g., salicylic acid as a keratolytic agent to treat corns and warts).

3.2. Liquid Dispersions [15], [16], [18], [19]

Liquid dispersions include suspensions of solids in liquids. Such suspensions can be formulated as oral products to reduce the bad taste of a drug or to control chemical degradation of the drug. When applied topically they generally are termed lotions.

This group also includes pourable oil-in-water (o/w) or water-in-oil (w/o) emulsions. Emulsification, like suspension, can be used to overcome bad taste or chemical degradation. Oral emulsions are o/w, whereas external-use emulsions (also called lotions) can be o/w (the vanishing type) or w/o (the cleansing or "cold cream" type). Suspensions and emulsions can be packaged for use as aerosol products.

3.3. Semisolid Dosage Forms [15], [18]

Dosage forms that are semisolids at room temperature include mucilages, gels (jellies), magmas, ointments, creams, and pastes.

Gels, Magmas, and Mucilages. This group of semisolids consists of colloidal particles dispersed in a liquid medium. They can be thixotropic systems that are semisolid upon standing but become liquid with shaking or stirring. Gels (jellies) are systems that contain either organic polymers (e.g., methyl cellulose, alginates) with entrapped liquid or inorganic particles (e.g., aluminum hydroxide) suspended in a liquid. The liquid phase of gels is generally water, but it may also be an oil. Mineral oil, for example, can form a gel structure with polyethylene for use as a topically applied base.

If the dispersed macromolecules are distributed throughout the liquid without specific boundaries or without a network between them and the liquid, the system is considered single phase. If the gel consists of larger aggregates of dispersed particles, it is considered a two-phase system and termed a magma (bentonite magma).

Single-phase gels can consist of synthetic organic molecules (carbamer or carboxypolymethylene, a carboxyvinyl polymer) or may be dispersions of natural gums (acacia, tragacanth). Such gum dispersions can be termed mucilages.

Gels can be applied topically as in tretonoin gel (U.S.P.), an acne treatment. Sodium fluoride and phosphoric acid gel (U.S.P.) is used as a dental application. Gels and magmas can be used orally; aluminum hydroxide gel and magnesia magma (U.S.P.) are antacid preparations. Some mucilages can be used as pharmaceutical adjuvants to increase the viscosity of preparations or to adjust mouthfeel.

Ointments, Creams, and Pastes. These types of dosage forms are designed for application to the skin. However, ophthalmic ointments are applied under the lower eyelids from specially designed tubes; they must be sterile and should not contain gritty powders.

Ointment bases can be classified in the following manner:

1) *Oleoginous, hydrocarbon bases* have a greasy consistency, cannot contain significant amount of admixed water, and are occlusive to the skin, i.e., they restrict water evaporation from the skin surface
2) *Absorptive bases* are either semisolid mixtures that form w/o emulsions upon sorption of water (Hydrophilic Petrolatum, U.S.P. XXII) or w/o emulsions that can take up more water (Cold Cream, U.S.P. XXI)
3) *Water-removable bases* are o/w emulsions (Hydrophilic Ointment, U.S.P. XXII; Vanishing Creams) that are water-washable and less occlusive than absorptive bases or oleoginous bases (water-washable denotes bases that are insoluble in water but are o/w emulsions and can therefore be washed off with water)
4) *Water-soluble bases* that are termed greaseless ointments are least occlusive, they are blends of poly(ethylene glycols) (Polyethylene Glycol Ointment, U.S.P. XXII)

The choice of an ointment base depends upon the condition being treated, the degree of desired occlusiveness, and the stability of the drug to be incorporated. Ophthalmic ointments are generally nonemulsified.

Pastes are semisolids that contain powders. Their base ingredients can be gels or oleoginous mixtures. Pastes can thus absorb more water than greasy ointments and can be used to absorb serous secretion in lesion oozing.

The term *cream* denotes an emulsified semisolid such as cold cream (w/o) and vanishing cream formulations (o/w). As dermal applications, they have more cosmetic appeal to users than greasy ointment bases. Creams can also be used to medicate body cavities (rectum, vagina). They can be delivered from pressurized aerosol containers.

3.4. Plastic Dosage Forms (Suppositories) [15], [20]

Suppositories are semirigid, plastic dosage forms designed for insertion into body cavities (rectum, vagina, or urethra). They can be used for systemic therapy (rectally) or for localized treatment.

Rectal suppositories are the route of choice for comatose patients or after gastrointestinal surgery. They also find use in pediatric and geriatric medication.

Depending upon the base used, suppositories either melt at body temperature upon insertion (cocoa butter) or dissolve in the cavity (poly(ethylene glycol) blends, glycerinated gelatin). Glycerinated gelatin base suppositories have a rubbery consistency (70% glycerol, 20% gelatin, 10% water plus drug).

Glycerin Suppositories (U.S.P. XXII) have a unique base: glycerin that is gelled or solidified with sodium stearate.

Suppositories based on cocoa butter can be prepared manually by pharmacists. However, heating mixtures of such base and drug, followed by pouring into calibrated molds and chilling, is the preferred method of preparation.

Poly(ethylene glycol) and glycerinated gelatin base suppositories must be made by fusion and molding. On a large industrial scale, all suppositories are made by heating and molding procedures.

3.5. Solid Dosage Forms [15], [21], [22]

Solid dosage forms have progressed from powder blends dispensed as bulk powders for administration by teaspoonful dosages in water or in measured amounts wrapped in folded paper squares (folded powders, chartulae) to modern compressed tablets. Pharmacists have also dispensed cachets, capsules, tablet triturates, granules, troches, and pills as solid dosage forms.

Bulk Powders and Effervescent Powders. Bulk powders are blends of drug and water-soluble diluents for dissolution in water. The resulting solution can be administered orally or applied externally. Dosing of such powder mixtures by teaspoonful measurement affords poor accuracy and precision. Small-dose, potent drugs should therefore be weighed accurately as a dose and enclosed in a folded paper. Alternately they can be dispensed in "zip-lock" plastic envelopes/bags or encapsulated in gelatin capsules.

Bulk powders can be formulated with ingredients that effervesce on addition of water (e.g., sodium bicarbonate plus blends of citric and tartaric acid). Such powders provide carbonated, aqueous solutions to help mask bad-tasting drugs. Flavoring and sweetening agents can also be added to the powder blend.

Dusting Powders. Dusting powders are applied externally to treat skin lesions or rashes; they should have an impalpable particle size (100–150 μm). They are sprinkled onto the affected skin from sifter-top containers. They generally contain absorbent substances (starch, kaolin) to absorb serous fluid and talc or metallic stearates as skin lubricants.

Insufflations are obsolete. They were powder blends of fine particle size that were propelled from atomizer-like devices (insufflators) into body cavities.

Lyophilized Powders. Several drug products exist as lyophilized (freeze-dried) powders for reconstitution into solutions or syrups by the pharmacist or patient. This dosage form is generally used when the drug would otherwise decompose in the presence of the solvent system during warehousing and distribution. The reconstituted product should be used in a relatively short period of time and should be refrigerated by the patient. An example of such a product is an antibiotic syrup for pediatric use. Parenteral dosage forms can also exist as lyophilized powders to which sterile solvents must be added under sterile conditions directly before use.

Capsules and Cachets. Cachets are mentioned here only for historical reasons. They were used before gelatin capsule shells became available. They consisted of two thin disks usually made of rice flour. Each disk had a raised edge creating a hollow area. Powder was placed in one of the disks, its edges were then carefully wetted, and the second disk was placed on top of it. On drying, a sealed edge was formed. Cachets, also called Konseals, were designed to hold ca. 0.5–1.0 g of powder. They were intended to be swallowed with water.

Gelatin capsules may be either hard or soft and are utilized to deliver small volumes of powder orally. The formulations of gelatin capsules do not require binders or disintegrating agents and are less complicated to formulate than compressed tablets. During the manufacture of the gelatin shells coloring agents, opaque pigments, and flavors can be added to the gelatin mixtures. Upon swallowing, the capsule's shell dissolves relatively quickly (< 10 min), making the enclosed powders available for gastrointestinal dissolution. Thus, bioavailability should be more complete and faster than that of compressed tablets.

Hard gelatin capsules are made from water and gelatin and can be produced in several sizes. They are of two-part construction (cap and base) and can be filled and dispensed by pharmacists extemporaneously or mass-produced. When the capsules are filled by high-speed machines, lubricants (metallic stearates, talc) can be added to the formulations to ensure uniform powder flow from the filling hoppers to each capsule shell. Such capsules range in size from number 000 to 5. The latter can hold ca. 65 mg of aspirin powder; number 00 can hold ca. 650 mg.

Soft gelatin capsules contain glycerol, as well as water and gelatin [23]. They cannot be filled extemporaneously because they have to be made by a heat-sealing process (Fig. 3). Two rolled sheets of gelatin are kept pliable with mild heat (37–40 °C). The

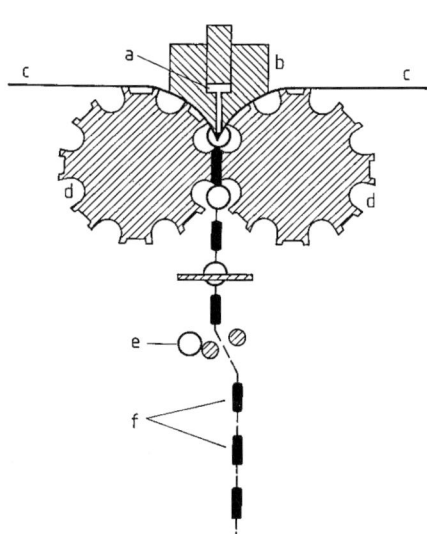

Figure 3. Rotary die process for production of soft gelatin (elastic) capsules (courtesy of R. P. Scherer Corp., Clearwater, Florida)
a) Fill material; b) Wedge; c) Gel ribbon; d) Die roll; e) Capsule; f) Waste gelatin

rolled sheets meet in an injection portion of the machine where the ingredient is injected into the resulting capsular form. This is sealed and ejected as a finished capsule. Excess gelatin is reprocessed. The encapsulated ingredients can be a powder, semisolid, or nonaqueous drug solution. Soft gelatin capsules can take any form from a globule (pearl) to the traditional capsular shape.

Pills. Although compressed tablet dosage forms are often called pills, this is a misnomer since pills were at one time a popular solid dosage form in themselves. Pills were an attempt to deliver a unit dose of a drug in a convenient, portable solid form. Their preparation entailed addition of adhesive liquids (e.g., syrup or gum mucilages) to medicated powder blends. Upon suitable manipulation the powders then became a pliable mass that was rolled into pencil-like cylinders called pipes. The pipes were each divided into an appropriate number of units on a gauged pill tile. Each unit was then rolled into a sphere, dried, and dusted with powder to minimize stickiness. Although attempts were made to add ingredients that would swell when exposed to gastrointestinal fluids to aid disintegration of such pills upon swallowing, bioavailability of the drug was poor. Solid dosage forms in pill shape are now made by granulation and compression (i.e., as a compressed tablet).

Tablet Triturates and Hypodermic Tablets. Tablet triturates were originally small, round tablets made in a small, hard, rubber mold that accomodated 50–100 tablets. They were intended to deliver a potent drug in an accurate amount that dissolved quickly and completely. They were composed of the drug, lactose, and/or sucrose. Binding of the ingredients was achieved by adding small amounts of 50 % ethanol to the drug–sugar blend before manual compaction into the mold.

Hypodermic tablets were similar products, intended for use by hypodermic injection after dissolution in sterile water. Today such products can be made with tablet compression equipment. One of the few such products used today is Nitroglycerin Tablets (U.S.P. XXII) for sublingual use to achieve quick relief of chest pain in angina patients. These tablets are very friable and crumble easily during handling and transport.

Troches and Lozenges are small solid disks or cubes that are designed to dissolve slowly in the mouth without disintegration to provide medication to the mucous membranes of the mouth and throat. Originally an adhesive kneadable mass was prepared, rolled into a sheet, cut into appropriate size and shape, and dried. Today such dosage forms are made by granulation (without disintegrating agents) and compression. Crystalline candy bases can also be formulated to which drug solutions can be added. Such syrup bases can then be molded into appropriate dosage units.

Granules are powder blends that have been either lyophilized or massed. Massing (wet granulation) involves addition of an adhesive binding solution (e.g., gelatin or polymer dispersion, gum, starch paste) to the powder blend. The mixture is then passed mechanically through appropriate sieves to obtain granules of the required size.

Granules can be used as such (in teaspoonful or tablespoonful doses) in water. Effervescent ingredients (see p. 2091) can also be included in granule formulations to add flavoring and carbonation to the medicated solution.

Granules prepared by massing can be used to prepare compressed tablets by the wet granulation method (see p. 2095). The granules are dried, resieved, and blended with glidant and ejection lubricating powders (talc, metallic stearates) prior to compression.

Compressed Tablets. The compressed tablet is the most popular dosage form. It offers convenience, ease of administration, stability, accurate and precise dosing, and good bioavailability. Once the formulation has been developed and proved clinically efficacious and safe, production validation and specifications can be developed to ensure batch-to-batch reliability. Millions of tablets can be produced daily, insuring economical preparation.

Appropriate quality control procedures during tablet manufacture and of the final product assure tablet-to-tablet consistency and reliability. Such tests include weight uniformity, content identity and uniformity, tablet hardness, disintegration time, and friability. Each manufacturer establishes an appropriate set of tests with specification limits. Bioavailability is determined during clinical testing. During preformulation studies, the properties of the drug are investigated to insure stability, safety, and therapeutic effectiveness.

Tableting Procedures and Equipment. Tableting may be performed by direct compression of powder blends or by preparation of suitable granulations followed by compression.

Direct compression involves the admixture of the necessary powders followed by compression with tablet compression presses. Powder blends of small particles must have acceptable flowability and compressibility properties. This is not always the case. Modern equipment is, however, more adaptable to direct compression than in the past because mechanical delivery to the dies has been improved. Excipient ingredients (diluents, binders, disintegrating agents) are now available in granular form to insure better flow and compressibility.

A granulation blend of 90–95% granules (ca. 10–40 mesh, 2 mm–425 µm) plus 5–10% fine particles generally provides better compressibility and flowability than fine powder blends.

When direct compression is not possible, wet or dry granulation (slugging) can be used. Granulation is required when the powdered drug with diluent etc. does not possess sufficient compressiblity calling for addition of a binding solution and granulation. Granulation is also advisable, if the powder blend does not flow well to ensure uniform content and weight of the tablets.

Wet granulation is widely used for tablets when the active ingredient is stable to water and heat. *Dry granulation (slugging)* is used when the drug is not stable to the conditions of wet granulation and the ingredients cannot be directly compressed. The ingredients are blended and compressed on a heavy-duty tablet press or slugger. Large hard tablets are produced, these "slugs" are 28–50 cm in diameter and weigh ca. 25–30 g. The slugs are then ground and sieved to appropriate mesh size and recompressed.

Various *excipient ingredients* are used in tablet production:

1) Diluents to insure appropriate size and convenient volume (e.g., lactose, calcium phosphate)
2) Disintegrating agents to aid tablet disintegration and dissolution of active ingredients (e.g., starch, microcrystalline cellulose)
3) Lubricating agents and glidants to ensure flowability of granulation, as well as antiadherents to ensure release from dies and punch faces (e.g., magnesium stearate, talc)
4) Binders to increase powder adhesion and ensure appropriate granule formation and compression (e.g., gelatin starch paste, methyl cellulose)

The major components of a *tablet press* are as follows (Fig. 4):

1) The *dies* are the units to which the granulation is delivered and whose shape determines the shape of the tablet
2) The *upper punch* fits snugly in the die and delivers the pressure
3) The *lower punch* also fits snugly in the die, it controls the volume of granulation delivered to the die and therefore tablet weight

The compression cycle (Fig. 5) begins with the delivery of the granulation from the hopper to the die by a feeder device that allows uniform filling. The lower punch is at its lowest position at this point. The upper punch is then lowered to deliver a defined pressure. It then moves upward together with the lower punch to eject the tablet. Single

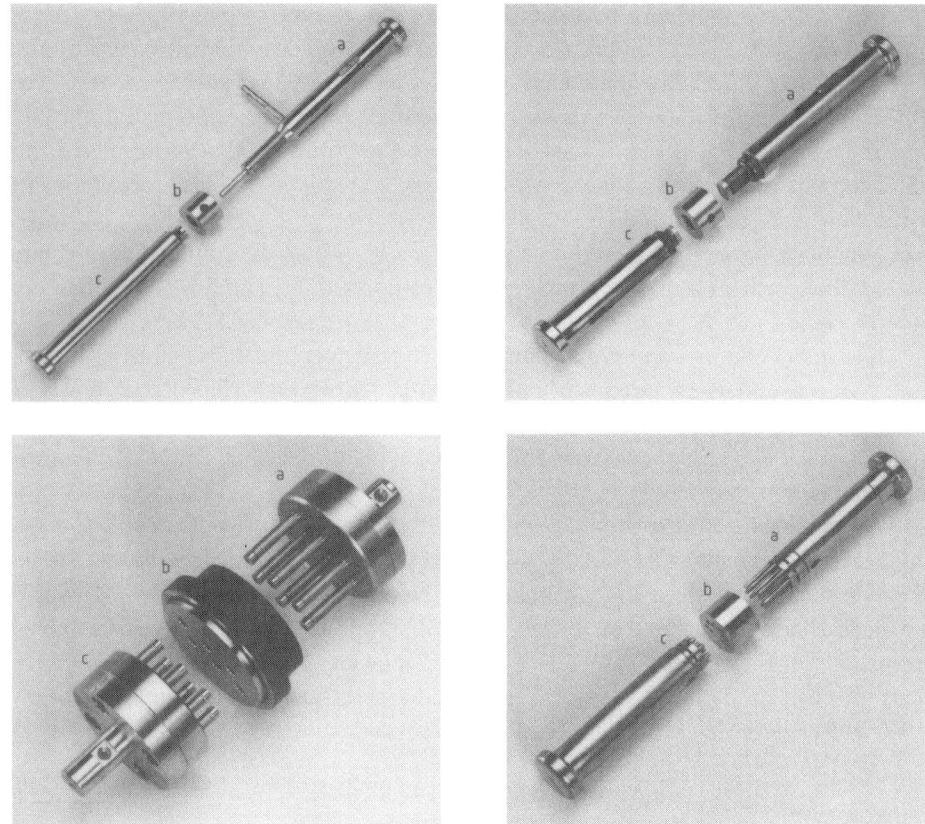

Figure 4. Tableting tools (courtesy of Fette GmbH, Hamburg, Germany)
a) Upper punch; b) Die; c) Lower punch

punch presses have only one die–punch unit and compression is fast. On a multistation rotary press (one station consists of one die, one upper punch, and one lower punch), more than 60 dies can be assembled in a rotating circular table (Fig. 6). The punches ride over and under the dies in well-tooled cam tracks or channels in the head and foot areas of the press to achieve the synchronized upward and downward stroking action.

Newer high-speed presses can produce > 10 000 tablets per minute. Such units are equipped to measure compression and ejection forces and monitor tablet weight. Weight variations can be adjusted automatically by computer program. Feeding of granulation in sufficient quantity and rate to the die can be a problem. Standard gravity feed from hopper to feeder frame is not sufficient. Rotary agitator mechanisms within the feeder-frame units are used to induce faster feeding to the dies.

Types of Compressed Tablets. Compressed tablets can be divided into three main groups (Fig. 7):

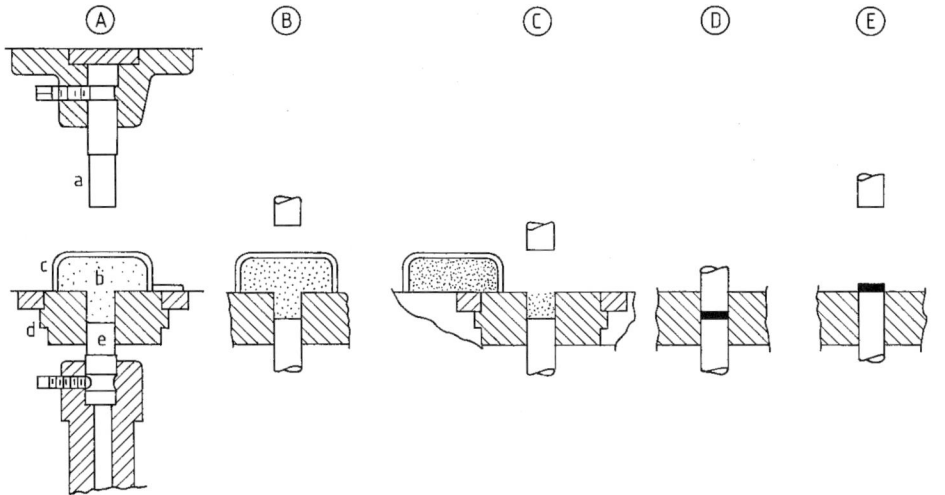

Figure 5. Compression on a single-punch tableting press (courtesy of Vector Corp., Marion, Iowa)
a) Upper punch; b) Granulation in die cavity; c) Feed shoe; d) Die; e) Lower punch
The feed shoe is filled with granulated materials from a hopper and is positioned over the die cavity (A) which then fills (B). The feed shoe retracts (C) and the upper punch lowers to compress the material (D). The upper punch retracts and the lower punch rises to eject the compact (E).

1) Single-layer, uncoated tablets
2) Layered tablets
3) Coated tablets

Single-layer, uncoated tablets may be classified as follows:

1) Tablets to be swallowed for systemic use or to medicate the gastrointestinal tract
2) Tablets that are to be chewed before swallowing; these are popular with young children (vitamins) and are generally flavored and sweetened
3) Buccal or sublingual tablets for quick dissolution and absorption and used for drugs that are not stable in gastrointestinal fluid, that are needed for quick cardiac therapy (angina), or have significant first-pass liver metabolism (see p. 2084)
4) Effervescent tablets
5) Lozenges or troches
6) Vaginal inserts
7) Pellets for parenteral implantation

Layered tablets may contain two or more layers and are prepared on special presses in which two or more different granulations are delivered from several hoppers. Layered tablets are used for two main reasons:

1) To separate ingredients that may be chemically or physically incompatible during processing and storage

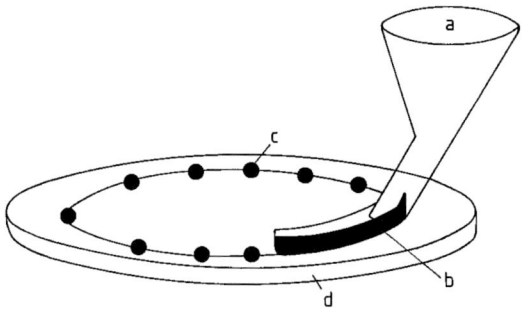
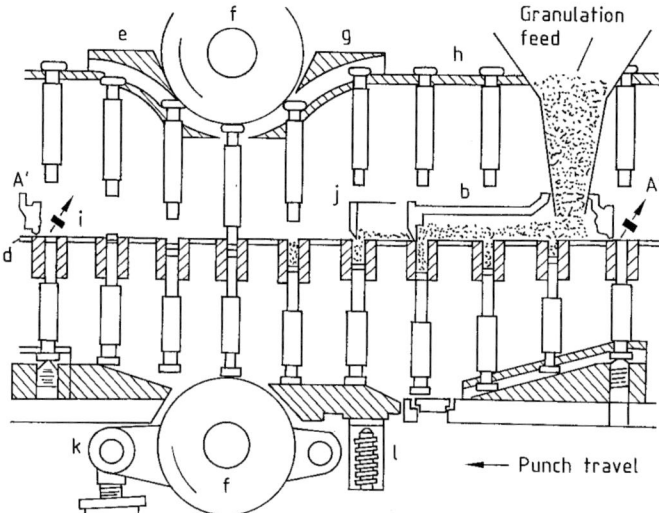

Figure 6. A tableting cycle (A′–A′) on a rotary press (courtesy of Stokes–Merril Inc., Bristol, Pensylvania)
a) Hopper; b) Feed frame; c) Die position; d) Die table; e) Raising cam; f) Pressure roll; g) Lowering cam; h) Dwell cam; i) Knock-off; j) Scrape-off; k) Roll carrier; l) Lower punch access plug

2) To provide an initial (loading) dose from one layer and a prolonged-release dose from another; a nonmedicated layer (e.g., lactose granulation) can be used to separate the drug-containing layers

In *coated tablets* the coating protects the ingredients from atmospheric oxygen, water, carbon dioxide, and/or light [24]. Tablets are also coated to mask bad taste and to control the site of disintegration and drug dissolution. In the case of enteric coatings, for example, the tablet remains intact in the stomach but disintegrates with drug dissolution in the duodenum. Powders, granules, and tablets are also coated to control the rate of drug release.

Sugar coatings are applied in rotating pans. The core tablets are somewhat harder and more spherical than tablets that are not to be coated so that they can withstand tumbling and thus facilitate application of the coating. Batches of core tablets are initially dusted to remove powder from their surface and are sealed with a thin coat of food-grade shellac. Alternate coatings of syrup and powdered sugar are then added with intermediate drying until a rounded geometry is obtained. Smoothing coats of syrup

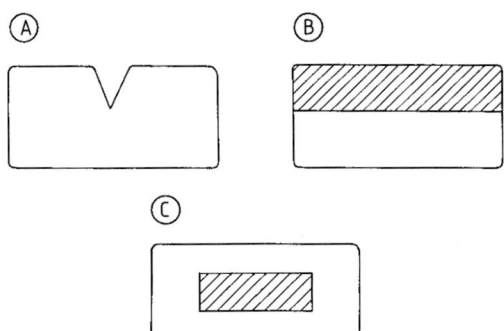

Figure 7. Compressed tablet types
A) Single-layer, uncoated, scored (bisected);
B) Bilayered, uncoated; C) Compression coated with core: different drugs can be located in the shell and the core can also be a delayed-release or sustained-release component

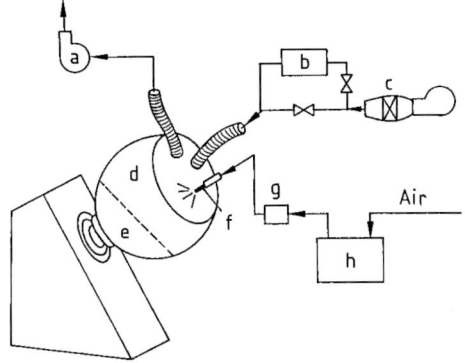

Figure 8. Hydraulic film coating system [25]
a) Exhaust; b) Heater; c) Cold/warm-air blower;
d) Dragee vessel; e) Position of the tablet cores;
f) Spraying gun; g) Filter; h) Pump

are applied with coloring agents, if called for. Finally, the tablet is polished with a thin coating of wax and imprinted with a logo.

Film coating is faster than sugar coating and allows tablet surface engravings to be seen. Unlike sugar coating it does not increase tablet weight significantly. Polymeric solutions (e.g., cellulose acetate phthalate for enteric coatings, sodium carboxymethyl cellulose for water-soluble coatings) are sprayed onto the tablets as they tumble in rotating pans (Fig. 8). Such solutions can contain plasticizers (e.g., poly(ethylene glycols), phthalates), coloring agents, and opaquing ingredients. Originally, organic solvents were used (e.g., acetone, chloroform). Modern technology now allows the use of aqueous polymeric dispersions. The polymers can be used to produce enteric film coatings, as well as coatings that dissolve in gastric fluids.

Film and sugar can be accomplished in equipment other than rotating pans. Air-suspension coating (Wurster apparatus) employs a cylindrical chamber in which the cores are suspended in a stream of air. Coating solutions are injected into this air stream with quick solvent evaporation. This technique can also be used to coat powders or beads. The coated beads or powders are then encapsulated or compressed into tablets to provide prolonged-release dosage forms (see p. 2119).

In *pressure coating*, coatings are compressed around core tablets. Two tablet presses are used in tandem: one to make the core tablet and the other to receive the core and compress the coating around it. The coating can be a protective one. However, it can include the initial-release portion of a prolonged-action tablet, with the core being a sustained-release unit. This situation also can be accomplished with sugar coating technology: the initially released drug portion added in the sugar coating and the sustained-release portion in the core.

3.6. Prolonged-Action Dosage Forms
[26] – [29]

The purpose of prolonged-action dosage forms is to extend periods of therapy from a single administration to two or three times those produced by conventional dosage forms. This can be accomplished by using orally administered dosage forms that release a portion of their drug content relatively quickly to induce onset of the therapeutic response and thereafter release the remaining drug content at a rate that approximates the rate of elimination of the drug from the body (Section 7.2.1). Prolonged therapy can also be achieved by transdermal drug systems [30], [31], ocular [32] and intrauterine [33] inserts, and parenteral dosage forms [34] (Section 7.2.2).

3.7. Pressurized Aerosol Dosage Forms
[19]

Aerosols are dispersions of liquid droplets or small solid particles in gases. Gaseous dispersions can be produced from atomizers, nebulizers, or insufflators. Pressurized containers utilizing various propellant gases for delivering drug products through appropriately designed valve systems and actuator devices have been available since the 1950s (see also Chap. 6). These propellants include liquefied gases or gas mixtures. Examples are the Freon series of fluorinated hydrocarbons (e.g., Freon 12, CCl_2F_2) and hydrocarbons (e.g., butane, propane). Nonliquefied compressed gases are also used (e.g., carbon dioxide, nitrogen monoxide, dinitrogen monoxide). Medications can be produced in the form of suspensions, emulsions, pastes, gels, solutions, and foams depending on the formulation, valve system, dip-tube, propellant, actuator, and container. They can be used in inhalation therapy, for external application, in dental care, for vaginal therapy, and for cosmetic and toiletry products (e.g., antiperspirants and shaving foams).

3.8. Nonpressurized Gaseous Dosage Forms (Inhalants) [35]

Some drugs have low vapor pressures at room temperature and atmospheric pressure (e.g., propylhexedrine and amyl nitrite). Inhalants are prepared by impregnating suitable inert (fibrous) materials with such drugs. Cylindrical rolls of the impregnated material are then fabricated into a size that can fit into a rigid plastic tube (inhaler), for use by gentle insertion of the inhaler tip into the nostrils and sniffing (e.g., Propylhexedrine Inhalant and Amyl Nitrite Inhalant, U.S.P. XXII). The gaseous drug then medicates the nasal membranes. Aromatic ingredients can be added.

3.9. Radiopharmaceutical Dosage Forms [36]

A radiopharmaceutical is a chemical or drug that contains a radioactive isotope for human use in diagnosing or treating a disease state. These specialized drugs are dispensed as traditional dosage forms. For example, cyanocobalamin ^{57}Co capsules are used as diagnostic aids for pernicious anemia; sodium ^{131}I iodohipparate injection is used for diagnosis of renal function conditions and sodium ^{32}P phosphate solution in diagnosis and localization of ocular tumors.

4. Pharmaceutical Excipients [37], [38]

Excipient ingredients, also called adjuvants or pharmaceutical necessities, are nontherapeutic ingredients that expedite the manufacture of dosage forms, make the delivery of the active ingredient possible, and/or ensure appropriate bioavailability (see Section p. 2079). For example, compressed tablets containing very small doses of drugs (<10 mg) also contain glidant/lubricant excipients (talc, starch) to aid the uniform delivery of the granulation from hopper to die for accurate dosing and tablet weight. Lactose, as a diluent, gives the needed bulk to such formulations. Gelatin can provide necessary adhesion as a binder agent, and a disintegrating agent (microcellulose) speeds up the rates of dissolution and gastrointestinal absorption of the drug.

Excipients should be pure, nontherapeutic, chemically stable, nonreactive with the drug, atmospheric oxygen, and/or water vapor, nonallergenic, odorless with an acceptable color and taste, and inexpensive. The physical nature of the excipient varies with the dosage forms and its physical state: liquid, semisolid, rigid but plastic (suppositories), and solid.

Solvents and Cosolvents. Purified Water (U.S.P./N.F.) is suitable for general use in oral and dermal drug products. Several U.S.P. types of water are used with parenteral dosage forms (Water for Injection, Bacteriostatic Water for Injection and Sterile Water for Injection, U.S.P./N.F).

Several water-miscible solvents are used in liquid dosage forms: ethanol (both pure and denatured), glycols, polyhydric alcohols (glycerol, sorbitol), and acetone.

A number of water-immiscible solvents are also employed: vegetable oils (peanut, soybean, cottonseed oils), mineral oil. Diethyl ether is used to solubilize pyroxylin for thickening in collodions.

Stabilizers. Several types of stabilizer are utilized to guarantee the required shelf-life: antioxidants (e.g., sodium sulfite), antimicrobial preservatives (e.g., parabens), and buffers (e.g., acetate – acetic acid, Sorensen phosphate solutions).

Excipients of Palatability [39]. Such ingredients assure product elegance and better patient compliance. They include colorants, flavors, sweeteners, and perfumes. Such ingredients must not harm product stability nor the patient (allergy, primary irritation).

Colorants include water-soluble and oil-soluble, coal-tar-derivative dyes. In the United States dyes are certified (or decertified) by the FDA. Colorants from natural sources include carmine from insects and caramel. Lakes are certified aluminum or calcium salts of coal-tar (aniline) dyes, that are water- and oil-insoluble. They are used in solid dosage forms. Pigment colorants [e.g., white titanium dioxide; red, yellow, and brown iron(III) oxide] find wide use in cosmetic formulas (e.g., face makeup).

Flavors can be complex mixtures of synthetic organic compounds formulated by specialty companies. They can also be of natural origin (e.g., peppermint, vanilla, glycyrrhiza, and anise). The U.S.P./N.F. still lists certain nonmedicated products (syrups, elixirs, fluid extracts, tinctures, aromatic waters, and spirits) as official entities, for use as flavored diluting solutions (e.g., Tola Balam Syrup, Aromatic Elixir, Peppermint Water, and Peppermint Spirit); see also Section 3.1.2.

Sweeteners may be sugars but several polyhydric alcohols also have a sweet taste (glycerol and sorbitol). Saccharin and aspartame are nonsugar sweetening agents that are used in preparations for diabetic patients.

Natural *perfume ingredients* and blends of synthetic aromatic compounds can be considered as excipients in cosmetic products. However, most dermal dosage forms omit such compounds to reduce the risk of irritation and allergic reactions.

Surfactants. Various surfactants (anionic, cationic, nonionic, and amphoteric) are used in several types of cosmetics and dosage forms [40]:

Anionic surfactants

1) Soaps: sodium oleate, potassium stearate, trolamine laurate
2) Fatty alcohol sulfates: sodium lauryl sulfate
3) Alkyl aryl sulfonates: sodium dioctyl sulfosuccinate

Cationic surfactants

> Quaternary ammonium salts: benzalkonium chloride, cetyltrimethylammonium bromide, cetylpyridinium chloride

Amphoteric surfactants

1) Natural: lecithin
2) Synthetic: dodecyl-β-alanine, N-dodecyl-N,N-dimethylglycine, N-alkyl-β-aminopropionate

Nonionic surfactants

1) Sorbitan esters: sorbitan trioleate
2) Glycerin esters: glycerylmonostearate
3) Poly(ethylene glycol) esters: polyoxyethylene 50 stearate
4) Polyoxyethylene ethers: polyoxyethylated lauryl ether, polyoxyethylated octyl phenol
5) Polyoxyethylene–polyoxypropylene block copolymers: polaxamers
6) Polyoxyethylated sorbitan ester: polysorbate 80

They serve as primary emulsifying agents (both o/w and w/o), wetting agents, and solubilizing agents (micellar solubilization). Thus, they are used in liquid (lotions) and semisolid (creams) emulsions, liniments, suspensions, mouthwashes, and solid dosage forms. Less than critical micellar concentrations aid in the dissolution of poorly water-soluble drugs from solid dosage forms in the gastrointestinal tract, thereby increasing the bioavailability of the drug. Cationic agents are seldom used as emulsifiers or wetting agents but are widely employed as antimicrobial preservatives. The synthetic amphoteric surfactants are generally expensive and are used mainly in cosmetic formulations.

Polymers can increase the viscosity of emulsions and suspensions, thereby increasing stability while delivering a more palatable or luxurious-feeling product.

Other polymers are used in solution to coat powders, granules, or compressed tablets; to alter the site of dissolution (in the small intestine rather than in the stomach); to slow the release of the drug for prolonged periods of therapy; or to increase product stability against environmental chemicals or other drugs that are present in the dosage form. An example of the latter would be the presence of two drugs in a tablet or capsule formulation which, when intimately mixed, would form a eutectic at room temperature. Polymeric coating of the powdered particles of one of the drugs prevents intimate contact with the other drug particles, thereby preventing eutectic formation and increasing product stability. More commonly used pharmaceutical polymers include sodium carboxymethyl cellulose, methyl cellulose, ethyl cellulose, hydroxypropylmethyl cellulose, cellulose acetate phthalate, carbamer, hydroxypropyl cellulose, hydroxyethyl cellulose, poly(vinyl alcohol), and polyvinylpyrrolidone.

Tonicity adjusters are nontherapeutic solutes that are added to aqueous drug solutions to increase their osmotic pressure to levels that are isotonic with body fluids (e.g., in parenteral, ophthalmic, or nasal solutions). The most common tonicity adjusters are dextrose (parenteral solutions), sodium chloride, and boric acid (ophthalmic solutions).

Emulsion excipients are ingredients that are added to the oil phase of emulsions to increase physical stability and obtain a more viscous or acceptable feel/taste. Such dosage forms can be used for dermal application (lotion, cream) or for oral medication. This type of excipient can also be added to greasy ointments bases. They include natural and mineral waxes (beeswax, paraffin); fatty acids (stearic, oleic), fatty alcohols (cetyl alcohol), and fatty acid esters (isopropyl palmitate); petrolatum; vegetable and mineral oils; and cocoa butter.

Polymeric excipients (protective colloids) are also used to increase the viscosity of the water phase of emulsions (e.g., sodium carboxymethyl cellulose, methyl cellulose, alginic acid derivatives from seaweed, and gelatin). Emulsified creams and lotions also include poly(ethylene glycols) and polyhydric alcohols that act as humectants for creams. Primary emulsifying agents for o/w or w/o emulsions are surfaceactive agents that can be anionic, cationic, amphoteric, or nonionic (see p. 2102). They can also be used as wetting agents in solid–liquid suspensions. Protective colloid dispersions are also used in suspension technology to affect stability and taste or "mouth feel".

Suppository Base Excipients. Cocoa butter (theobroma oil) is available as a natural product for oil-base suppositories. Some commercially available bases can be considered to be semisynthetic cocoa butter; they are more stable and yield a whiter product than cocoa butter itself. Examples include Cotomer, a partially hydrogenated cottonseed oil (Procter+Gamble); Suppositol, a hydrogenated coconut oil triglyceride (Fritz Wetz); Wecobees, a series of triglycerides derived from coconut and palm kernel oil (Drew Chemical Corporation); and Witepsol, a series of triglycerides (Riches-Nelson Inc.).

Water-dispersible suppository bases include glycerogelatin (ca. 20% gelatin, 70% glycerol, 10% water and drug) and blends of poly(ethylene glycols) of appropriate molecular mass to achieve appropriate consistency and form upon cooling.

Propellant Gases. Aerosol dosage forms for inhalation therapy or topical use utilize various propellant gases including hydrocarbons and fluorinated hydrocarbons (see section 3.7). Fluorinated hydrocarbons have been limited in use to medicated aerosols for metered-dose inhalers and contraceptive vaginal foams. Compressed nitrogen and carbon dioxide have also been used as aerosol propellants.

Solid Dosage Form Excipients. *Diluents*, such as lactose (water-soluble) and calcium phosphate (water-insoluble), are used to obtain convenient bulk and weight and as aids for homogeneous blending of solids.

Binders or granulating agents (e.g., gelatin, starch, and gum dispersions) add adhesiveness to powders to create granules for better flow and compression.

Disintegrating agents absorb water after the tablet or granule has been swallowed, causing swelling and eventual bursting or disintegration of the solid unit to powdered form. This increases the dissolution rate of the drug. Examples of such agents are corn starch, sodium alginate, and microcrystalline cellulose.

Glidant and lubricants ensure smooth and uniform flow of granulation blends from hoppers to dies (tableting) or capsule filling units. This is necessary for uniform dose and product weight. Anti-adherent (ejection) lubricants ensure the release of tablets from the metal surfaces of dies and compression punches. Examples of such excipients are talc, calcium, zinc, and magnesium stearate, stearic acid, corn starch, and colloidal silica.

Solid dosage forms (powders, granules, compressed tablets) that effervesce in water to create a carbonated solution before swallowing contain solid *effervescent excipients* that serve as sources of carbon dioxide and safe and nontherapeutic acid. Sodium bicarbonate is generally the carbon dioxide source; citric acid, tartaric acid, and/or sodium acid phosphate provide the acid source.

Colors, flavors, and surfactants (see p. 2102), can also be used.

5. Quality Assurance [41]–[43]

In the case of prescription drug products, the quality of a dosage form is measured by the degree to which it successfully delivers the prescriber's intent. In the case of over-the-counter or nonprescription drug products, quality is the functional result expected by the person selecting the product for self-medication. Quality is thus a gauge of the product's acceptability to the patient or user, its safety, and therapeutic effectiveness.

Several departments within a pharmaceutical company are involved in the design, formulation, and monitoring of quality parameters in dosage forms to assure satisfactory quality to patients, pharmacists, prescribers, and regulatory governmental agencies.

The *research and development units* establish the physical and chemical specifications for all of the product ingredients and packaging components. In conjunction with pilot-plant or scale-up personnel, they develop acceptable parameters for production processes and equipment. This includes their validation to ensure that they function as expected.

The major responsibility of the *production unit* is the preparation of the final product in a quality manner: measure and handling of ingredients, use of appropriate processes and equipment as documented in batch formulae, and packaging of the correct product in the correct package with correct labeling.

Quality control units are responsible for systems that measure the desired standards for a product. These include physical and chemical assays and measurements at three key points: raw materials, in-process, and final product. They also include documentation and audit of all pertinent records and usually the control of label distribution.

The *quality assurance unit* has the responsibility for the overall quality of its products. While this includes quality control's inspection, monitoring, and analytical responsibilities, it is more pervasive. It encompasses all personnel activities, processes, equipment, and facilities involved in the manufacture of drug products.

Regulatory concerns established by federal, state, and/or local agencies must also be satisfied. Thus, for example, national compendia (e.g., U.S.P./N.F.) establish specifications for various ingredients and processes. In the United States, the Food and Drug Administration (FDA) approves all new drugs and monitors the quality of prescription and nonprescription drug products marketed in the United States. Several other U.S. regulatory items that affect the assurance of quality of drug products are:

1) Establishment of good manufacturing practices (GMPs) to control all aspects of manufacture that must be attended to in order to deliver high-quality, safe, effective drug products (equipment; building/facilities; control of containers, closures, packaging and labeling; production and processes; personnel; warehousing and distribution; records and reports; and salvaging of returned products)
2) Good laboratory and good clinical procedures (GLPs and GCPs) to ensure these respective procedures
3) State health and licensing agencies with parallel requirements for the manufacture of drug products within their boundaries, which generally are very similar to federal laws and regulations
4) Local government regulations

Ecological environmental regulations affect pharmaceutical production at all levels of governance. A major concern is the quality of sewage a manufacturing unit produces and its effect on water pollution. Similarly, air quality must be monitored in terms of gaseous or aerosol effluent.

6. Pharmaceutical Containers [44], [45]

Dosage form containers, including their closures (seals), should not react with nor sorb any of the product ingredients. Similarly, package components should not leach into the drug product and the containers should be impervious to the environment. Plastic containers should not be permeable to atmospheric gases nor to product ingredients. The closure should also be impervious to gases entering or leaving the product.

In some cases the drug product must be protected against light by the use of opaque or amber glass. The product should not alter the physical state of the container: for example, plastics should not soften, become tacky, nor become disformed. Parenteral containers and closures must maintain sterility until used. All this must be assured for the entire designated shelf-life of the product, throughout storage and use.

Plastics used for the packaging of pharmaceutical dosage forms include: low-density polyethylene, high-density polyethylene, poly(vinyl chloride), poly(vinylidene chloride), polypropylene, polystyrene, and polyterephthalates. They differ in barrier and permeability properties, plasticity, and resistance to fracture.

Glass used for pharmaceutical containers, including parenteral dosage forms that may be exposed to autoclave temperatures, has been classified by the U.S.P./N.F. (U.S.P. XXII). The classes differ in chemical composition and resistance to hydrolysis:

1) Type I: borosilicate glass, highly resistant
2) Type II: soda-lime glass, internal surfaces are treated (dealkanized) to make them more resistant to acid pHs, than Type III
3) Type III: soda-lime glass
4) Type NP: soda-lime glass for general purpose use (nonparenteral glass)

Type I is intended for parenteral dosage forms at all pHs. Type II is used to package acidic or neutral pH parenteral products. Type III is used for nonparenteral drug products, but may be used for parenteral packaging if appropriate tests show it to be suitable. Type NP is used widely for oral or external-use dosage forms.

Every new container and closure material must be tested thoroughly with the drug product it is intended to contain before wide-scale manufacturing and marketing occur. Such tests should stress the product container over a suitable range of temperature, relative humidity, and light exposure to assure stability and, hence, safety and effective therapeutic activity.

The U.S.P./N.F. specifies various tests for establishing acceptable standards for container materials. In parenteral drug products, for example, materials are tested for light transmission (glass and plastic), chemical resistance of glass (the powdered glass test), container permeation, and foreign particulate matter.

Liquid and solid dosage forms for nonparenteral use are generally NP glass bottles or bottles made of a plastic that does not allow excessive gas permeation. The closure should similarly allow very little or no gas leakage into the container nor leakage of water vapor or other product components (e.g., perfume ingredients) out of the container. These containers are designed either for multiple use or as one-time, unit-dose containers. Amber glass is generally used to reduce the exposure of product to light. Clear glass is used when the drug product is light stable and its clarity improves product appearance. Plastic containers are lighter in weight and thus less expensive to ship. However, plastics are usually poorly biodegradable whereas glass can be recycled.

Parenteral and ophthalmic drug products utilize both glass and plastics as container components. Small-dose injections are delivered from glass ampules or vials whereas large-volume parenteral products are packaged in glass bottles and flasks or nonrigid plastic bags. Prefilled glass or rigid plastic syringes and syringe–vial combinations are sometimes used. Ophthalmic containers contain dropper units attached to the cap closure, or the dropper unit may be part of a squeezable plastic bottle. Such containers should not contribute particulate matter to the drug product; their form and chemical composition should also be stable to the sterilization procedure used.

External-use products are also packaged in glass or plastic containers. These can be bottles for the delivery of liquid solutions, suspensions, or emulsified lotions. They can also be squeezable plastic containers with specially designed plugs to aid product

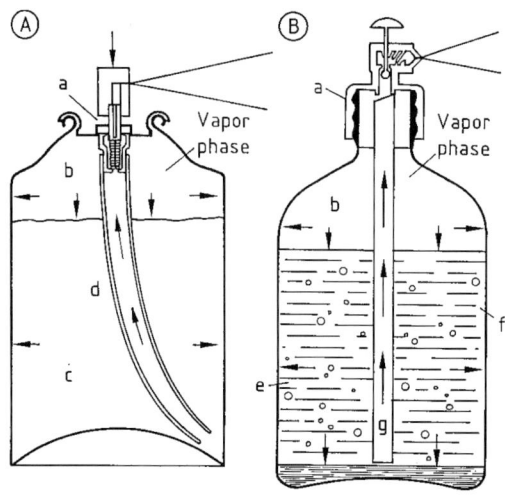

Figure 9. Two types of aerosol product containers (courtesy of Du Pont, Wilmington, Delaware) A) Two-phase system; B) Three-phase system a) Valve; b) Freon gas (fluorohydrocarbon) at 0.24 MPa, 294.26 K in container A and at 0.1 MPa, 294.26 K in container B; c) Solution of propellant and active ingredients; d) Standpipe; e) Aqueous solution of active ingredients; f) Liquid Freon; g) Dip tube

delivery. Many lotions are thixotropic and the design of the container should assist product flow.

Semisolid *ointments, creams, and pastes* can be delivered from glass or plastic jars of various diameters. The product is applied with the fingers, and widemouthed jars facilitate product access. Collapsible tubes (plastic, metal; coated, uncoated) are also used to deliver semisolid products. Less contact occurs between product and fingers, product contamination during use is thus lower. Viscous liquids and creams can be delivered from bottles with pump devices.

Suppositories can be packaged in cardboard boxes and may be wrapped individually in foil. Some suppositories are molded in rigid, plastic, torpedo-shaped molds, which then act as part of the final container.

Pressurized pharmaceutical aerosols are packaged in various containers. They must be able to "contain" their contents safely under the pressures that develop during production, storage, and use. Glass bottles coated with thermoplastic poly(vinyl chloride) are used for solution drug products. The external plastic coating is a protective measure to prevent shattering if the bottle is dropped. Suspensions and semisolids are generally enclosed in opaque containers. Aerosol containers may also be made of metal, with aluminum used for oral products and tinplated steel for topical drug production. To ensure stability, metal containers may be internally coated. These coatings must be acceptable in terms of safety and product compatibility.

The type of valve used depends upon the product use. It can be designed to deliver metered doses or to control the size of the delivered liquid droplets or solid particles. Various actuators or adaptor units can be attached for different purposes (e.g., to obtain different spray patterns). Oral inhalations to medicate an asthma episode need specially designed mouthpieces. Some aerosols include a flexible plastic dip tube or standpipe as an attachment to the valve unit (Fig. 9). Semisolid products (pastes, viscous creams) may utilize a rigid, plastic, domeshaped diaphragm that is pushed upward by the expanding propellant gases. Such products may also be enclosed in a flexible plastic pouch that is squeezed by expanding propellant. These adaptations ensure a more economical delivery of the product.

7. Development of the Ideal Drug Delivery System

Probably the most effective drug delivery system in use today consists of the intravenous administration of an appropriately formulated aqueous drug solution in a two-step procedure:

1) Intravenous administration of an initial or bolus dose of the solution to reach a therapeutically effective blood plasma concentration of the drug quickly
2) Subsequent intravenous administration of the drug solution at a delivery rate (drops per minute) that is adjusted to maintain the therapeutic blood plasma level for a long period of time (12–24 h)

Such a drug delivery system overcomes reduced bioavailability associated with the absorption of drug through membranes.

Delivery of a drug solution that is 100% bioavailable by this method avoids the "peak and trough" effect in blood concentration produced by conventional, orally administered tablets taken every 4–6 h. The difference between the maximum and minimum blood levels achieved by the intravenous drug can be minimized by control of the drug concentration in the delivered solution and its delivery rate. This effect reduces the potential for toxicity and results in a lessening of the "peak and trough" effect and a steadier therapeutic response throughout the duration of pharmacologic activity.

Several drawbacks exist in the utilization of the intravenous bolus–drip drug delivery system as a widely employed, patient-accepted system, especially for the ambulatory ill:

1) Patient inconvenience and discomfort
2) Need for special equipment and personnel
3) Potential source of infection and vein irritation
4) Expense

However, the temporal pattern of such administration (short onset with an extended duration of effect) has served as a model for noninvasive drug delivery systems that overcome such disadvantages by controlled release (Section 7.1).

The use of the oral route of administration to achieve a systemic effect is much more acceptable than the parenteral route, except in cases such as emergency medication, during surgery, and in comatose patients. An oral dosage form that mimics the intravenous bolus–drip by delivering a steady and controlled release of drug to the blood would have a great advantage (Section 7.2.1). Other routes of administration mimic this intravenous system (e.g., transdermal patches, subcutaneous implants, and medicated intravaginal inserts, Section 7.2.2).

All conventional oral dosage forms have a more delayed onset of activity than an intravenous bolus–drip administration since they must be transported across gastrointestinal membranes to reach the blood prior to distribution throughout the body. The sublingual route requires a somewhat shorter time to achieve the onset than the oral route. In order to mimic intravenous administration, release of drug from an oral dosage form and absorption (bioavailability) have to be designed so as to reduce the onset time.

In the drug delivery systems discussed above, the drug is distributed throughout the body before achieving its effective concentration at the desired site of pharmacologic activity. Tissues other than target tissues are therefore also exposed to various concentrations of the drug and pharmacologic responses may be elicited at other non-desired sites. This can result in unwanted side effects. For example, if the drug was to achieve a therapeutic effect at a particular organ innervated by the vagus nerve, undesirable responses could also occur at other organs controlled by vagal stimulation. This raises another concept in drug product development: targeted drug delivery in which the drug is released from the dosage form at a certain organ or tissue (site-specific release) or at a specific binding site within a tissue (receptor binding). Thus, systemically administered dosage forms would produce a highly localized effect (see Section 7.3). In the case of cancer chemotherapy, for example, side effects are many and serious, and such a drug delivery system would be extremely advantageous.

Ideal drug delivery systems should:

1) release a drug moiety to a target site, maintain therapeutic levels there for prolonged periods of time, and do so safely without side effects, by means of controlled-release technology;
2) have an onset of activity that approaches that of an intravenous bolus–drip;
3) be administered orally once a day, and be completely bioavailable to the target site;
4) be acceptable in terms of product elegance and utility to ensure compliance; and
5) be inexpensive.

While the development of drug delivery systems continues with the study of new technologies and materials, older concepts are being reinvestigated (e.g., medicated lollipops). The glycerogelatin base could perhaps be adapted for pediatric medication. Medicated, gummy "chews" could be used for administering vitamins, antipyretics, and cough and cold therapy to children too young to swallow or chew hard tablets or capsules and who refuse drops.

7.1. Controlled-Release Drug Delivery Systems

The ideal dosage form should deliver enough of its drug content to a specific pharmacodynamic site to achieve rapid initial onset of therapy. The therapeutic concentration should be maintained in a uniform, constant mode by slow release of the remaining portion of the drug over a period such that no more than two daily administrations (but preferably one) are needed. Furthermore, such a dosage form should be designed for oral administration and controlled release in the gastrointestinal tract to ensure good patient acceptance and compliance. In some instances where the biological half-life of the drug is < ca. 6 h, the "loading" portion of the dose may not be necessary since the slowrelease technology should reach effective therapeutic concentrations effectively without it.

In some instances, a longer period of therapy (> 24 h) may be desired from a single dose. This is not currently possible with orally administered dosage forms involving gastrointestinal release.

Transdermal controlled-release delivery from appropriate devices is a possible, noninvasive method for accomplishing such a therapy. Subcutaneous pellet implants, intravaginal insert devices, and portable or implanted infusion pump devices could be used as drug delivery systems for achieving therapy over several months. The infusion devices would need to have a built-in system of easily refillable drug reservoirs.

The design and manufacture of a controlled-release dosage form should produce a therapeutic plasma drug concentration at the site of activity followed by the maintenance of such a blood/target tissue level for an extended time period. Initial release of the drug to reach the therapeutic threshold concentration should be followed by constant, uniform release over a prolonged period. Once the initial threshold level is reached, the amount of drug released, absorbed, and distributed to the site of action should equal the rate of drug removal resulting from metabolism and excretion.

The most appropriate means to accomplish controlled release is to develop a technology in which the drug release rate is constant and independent of the amount of drug at the release site throughout the therapy, i.e., zero-order release. Most currently available oral dosage forms designed for extended periods of therapy are not true zero-order systems but slow, first-order release systems that depend on the concentration of available drug at the release site (e.g., release from core matrices and coated pellets). (The "osmotic pump" tablet discussed in Section 7.2.1 provides zero-order release throughout most of the extended duration of activity it is designed to deliver.)

Research and development on controlled-release, oral drug delivery products began in the 1950s. Since then, the associated terminology has developed into a plethora of marketing phrases without strict standardization and sufficient attention to specificity of definition. Many terms have been used synonymously and, thus, inappropriately: sustained release/action (SR/SA); prolonged release/action (PR/PA); repeat action

(RA); delayed action (DA); timed release (TR); and extended release/action (ER/EA). To avoid confusion, however, it is important to use these terms accurately within pharmaceutical and medical literature and especially in communication with health care personnel and patients [26], [27].

Initial differentiation must be made between conventional dosage forms and controlled-release products. The former are designed to release all of their active ingredients immediately for dissolution in body fluids; the latter release the drug in a controlled mode to extend the therapeutic response time and to deliver the drug to a desired target organ or tissue site. (Some controlled-release dosage forms may, however, release portions of the drug immediately to reach the onset of therapy more quickly.)

The term *controlled release* describes any drug delivery system that is used to adjust and control a therapeutic response, either temporally (extended duration) or spatially (site-specific, targeted delivery).

Sustained-release systems are designed to develop and maintain safe therapeutic responses at constant levels for extended periods that are some multiples of those achieved by conventional products (Fig. 10 A, B). With oral administration such periods can be 8–12 h, transdermal "patches" are effective for 24 h, and subcutaneous or intramuscular implant products can sustain therapy for 6 months or more. These systems are also termed controlled-release within the currently popular terminology if they involve some precise technology that achieves or approximates a zero-order rate of drug release (Fig. 10 C).

The term *prolonged release* also refers to products designed to achieve extended periods of therapeutic response. Sometimes this term is used to refer to systems that provide a lower degree of precision than a zero-order drug release rate, and cannot therefore be considered true controlled-release systems. However, they can be considered and identified as sustainedrelease products. Hence, within such terminology sustained and prolonged release are synonyms when the rate of release is not zero order (Figs. 10 B, C).

Zero-order release technologies are most accurately described as sustained-release and controlled-release systems. Slow, first-order release systems are, however, best referred to as prolonged-release products.

Repeat-action oral tablets can be viewed as prolonged-release products, since they provide longer periods of therapy than conventional tablets (Fig. 10 D). However, the blood concentration–time curves for such products have two (or three) peak levels of blood concentration with intermediate periods where the blood concentration falls below the median effective concentration (MEC). A concentration above the MEC is not sustained throughout the entire expected therapeutic period; hence the term sustained-release to describe repeat-action tablets could be questioned.

Delayed-release or *delayed-action* oral products (Fig. 10 E) are conventional dosage forms that have been designed to release all of the drug content at one time after a delay period. Thus, enteric coated tablets are examples of delayedaction/delayed-release oral

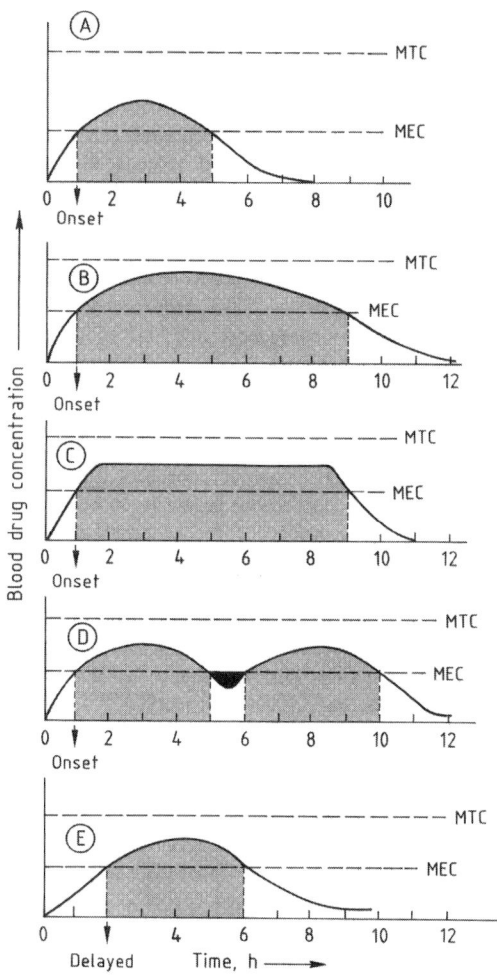

Figure 10. Blood drug concentration – time curves A) Conventional oral product; B) Sustained release/prolonged release; C) Sustained release/controlled release; D) Repeat action with period of nontherapeutic response; E) Delayed action (coated enteric product)
MTC = median toxic concentration;
MEC = median effective concentration

products because they begin releasing their drug content in the duodenum and not in the stomach.

Some researchers in the field feel that zero-order release has been overemphasized in the development of controlled-release products. Likewise, the overabundance of terms currently in use may confuse and distract attention from the objective of the controlled-release concept: the delivery of drugs by systems that provide excellent bioavailability to the systemic circulation for extended periods of time with targeted delivery to specific sites to eliminate both untoward side effects for the patient and undesirable effects of endogenous biochemicals upon the drugs [46].

Biotechnology has drawn attention to polyproteins, both endogenous and exogenous to the human body, as entities of therapy (e.g., insulin and growth hormone). They pose major problems of stability and effective delivery through biological membranes (i.e., bioavailability).

7.2. Sustained- or Prolonged-Release Technology

The advantages sustained- or prolongedrelease dosage forms include reduction of large fluctuations in blood concentration levels of the drug and longer duration of activity. This reduces side effects and increases patient compliance because fewer dosings are required each day.

7.2.1. Oral Sustained-Release Dosage Forms

The oral route of administration has received the greatest amount of developmental attention for sustained-release delivery systems. Oral sustained-release products mainly utilize diffusion-controlled or dissolution-controlled drug release technologies. Other systems include osmotic pump tablets, ion-exchange systems, and complexation.

The best candidates for inclusion in sustained- and prolonged-release systems are drugs that are uniformly well-absorbed throughout the gastrointestinal tract, that do not have long or very short biological half-lives (i.e., between 2 and 8 h), and that are meant for maintenance therapy of chronic disease states (rather than for acute therapy).

Dissolution-Controlled Oral Dosage Forms. In these dosage forms the dissolution rate of the drug is decreased to maintain a sustaining concentration in the blood and at the site of activity for extended durations. This is achieved in three ways:

1) *Formation of Small Drug Beads with Differing Coating Thicknesses.* The beads dissolve in the gastrointestinal fluids at varying rates depending upon the coating thickness. The coatings generally consist of lipoidal ingredients. The coated beads are encapsulated in gelatin shells or compressed into tablets.
2) Formation of Small, Layered Drug Beads. Drug layers are alternated with layers of rate-controlling cellulosic coatings (e.g., methyl cellulose). The outside layers can be the drug needed to reach the initial onset of activity; the remaining alternately coated layers produce the "pulsed intervals" of sustained release.
3) *Embedding of Drugs in Polymeric or Waxy Biodegradable Matrices.* Matrix dissolution products are obtained by compressing the drug with polymers (e.g., methyl cellulose, sodium carboxymethyl cellulose) or with hydrophobic waxy materials (e.g., glyceryl tristearate, cannuba wax) that are slowly biodegraded.

Diffusion-Controlled Oral Dosage Forms. Two major types of diffusion-controlled, sustained-release technologies have been described for oral administration: reservoir and matrix units.

Reservoir products utilize a central core (reservoir) of drug that is coated with a water-insoluble polymer (e.g., ethyl cellulose) through which the drug solution diffuses at a slow rate.

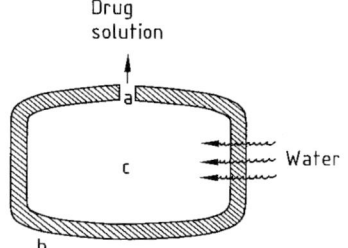

Figure 11. The OROS oral tablet system for prolonged and sustained therapy (courtesy of ALZA Corp., Palo Alto, California)
a) Delivery orifice; b) Semipermeable membrane; c) Osmotic core containing drug

In the *matrix system* the drug is dispersed throughout a water-insoluble polymer matrix. The drug located on the outermost layer of the matrix is the portion that is initially exposed to the gastrointestinal fluids and dissolves first. As the gastrointestinal fluid penetrates the matrix more drug dissolves and diffuses out. The matrix may be a water-insoluble cellulosic polymer or a plastic [e.g., polymethacrylate, polyethylene, or poly(vinyl acetate)]. Gradumet (Abbott) is a tablet using a plastic matrix. Both the spent matrix of the matrix diffusion system ("ghost") and the empty membrane "sac" of the reservoir diffusion system pass through the gastrointestinal tract undissolved.

Hydrocolloid Matrix [29]. In this system the drug is incorporated within the hydrocolloid matrix of a tablet. The powdered drug is dispersed in a solid hydrocolloid (e.g., cellulolose polymers). Upon exposure to gastrointestinal fluids the drug–matrix combination floats, having a bulk density <1. The outer layer of the matrix swells, slowing inward penetration of water. The outer layer gradually erodes and a second layer then swells and erodes. During this process the drug continually dissolves and diffuses into the gastrointestinal fluid. Such a design allows the drug unit to stay in the stomach for a longer period than with a conventional tablet and release continues in the small intestine. This technology is termed "Valrelease" and was developed by Roche Laboratories for diazepam.

Oral Sustained-Release Dosage Forms with Osmotic Control [29], [30], [47]. In this system osmotic pressure provides a driving force that produces zero-order controlled release. Tablets utilizing this technology contain a core of solid drug that is surrounded by a semipermeable membrane (e.g., ethylene–vinyl acetate copolymer or polypropylene) through which water can pass, but not the drug (Fig. 11). In the gastrointestinal tract, water flows through the membrane and dissolves the drug. The drug solution then exits at a constant rate through a laser-drilled hole in the tablet to the gastrointestinal lumen. This rate is controlled by the osmotic pressure developed by the drug solution.

The osmotic pressure concept for oral controlled-drug release was originally developed in the United States by the ALZA Corporation (OROS). In Elan's modification of this system (the multidirectional osmotic drug absorption system, MODAS), the enclosing membrane is permeable both to water and drug. Thus, water enters the unit to

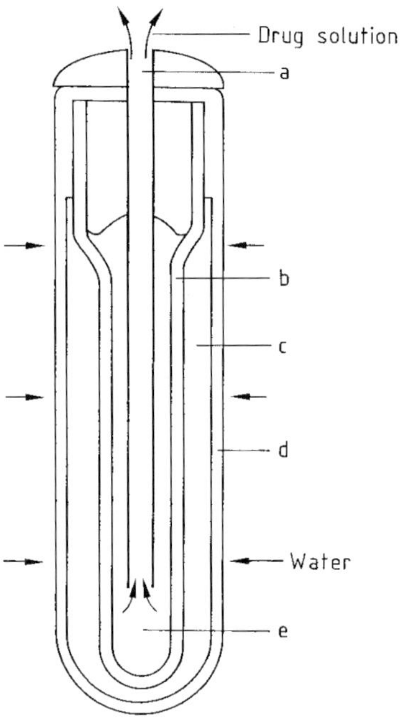

Figure 12. The Alzet osmotic pump (courtesy of ALZA Corp., Palo Alto, California)
a) Flow moderator; b) Flexible, impermeable reservoir wall; c) Saturated solution of osmotic agent; d) Rigid semipermeable membrane; e) Drug reservoir

dissolve the drug and the drug solution then passes out through the membrane in multiple directions, not through a single port.

In another variation of this system (Alzet osmotic pump, Fig. 12) a drug solution (e) is encased in an impermeable but flexible membrane (b) within a small capsular device. An electrolyte (c) surrounds the internal portion of the device. Water enters the device through a rigid semipermeable outer membrane (d) and dissolves the electrolyte. This creates a constant osmotic pressure that presses on the enclosed drug solution, causing it to exit through an orifice. This osmotic pump device has not yet been used in oral dosage forms, but as subcutaneous implants in experimental animals. The device has about the same size as a 000 capsule (diameter 6.5 mm, length 25 mm) and can deliver 2.5 µL/h for four weeks.

Ion-Exchange Dosage Forms for Oral Sustained Release [48]. Cationic or anionic drugs can be complexed to appropriate ion-exchange resins and then encapsulated in gelatin shells or compressed into tablets. The drug is slowly released from the resin in exchange for ions in the gastrointestinal fluid, the rate depending on pH and electrolyte concentration. An example of this technology is the encapsulation of such resin complexes of amphetamine and dextroamphetamine. Recently products have become available (Pennkinetic, PennWalt Co., now owned by Fisons Corp.) that utilize very small drug–resin beads coated with a polymer (ethyl cellulose) that is not soluble in gastrointestinal fluids but through which water, electrolytes, and drug can diffuse. The beads can be suspended in viscous liquid formulations to provide up to 12 h of therapy (e.g., cough syrups). The coating also masks the somewhat unpleasant taste of the resins (Delsym Syrup, Fisons Corp.).

Chemical Complexation Dosage Forms for Oral Sustained Release [29]. A few drugs have been complexed with organic compounds (tannic or polygalacturonic acid, for example) and formulated into oral sustained-release products. These complexes have a low solubility that usually depends on pH. Slow dissolution in the gastrointestinal fluid provides the prolonged release pattern.

Bioadhesives [49]. Recent research concerned with achieving prolonged duration of activity in oral dosage forms has included investigation of bioadhesive compounds for attachment to the mucous membranes. Various polymers have been tested [e.g., sodium carboxymethyl cellulose, carbomer, hydroxypropyl cellulose, cross-linked poly(meth)-acrylates]. Dosage forms utilizing such technology are envisaged as being attached to gastrointestinal or oral mucosal membranes and slowly releasing their drug content for systemic absorption. Thus, buccal tablets of various size and shape could be used, as well as dosage form units meant to be swallowed. If such units or portions of them could remain in place for long periods, the problem of gastric emptying time and overall gastrointestinal transit time as a limiting factor could be controlled.

7.2.2. Other Sustained- and Prolonged-Release Drug Delivery Systems [14]–[34]

There are several other sustained- and prolonged-release dosage forms that are not administered by the oral route:

1) Transdermal drug delivery
2) Vaginal inserts
3) Ophthalmic inserts
4) Parenteral implants and injections

Transdermal Drug Delivery Systems [30], [31], [50], [51]. Transdermal systems are dosage forms that are applied to intact, healthy skin to deliver their drug content percutaneously to the blood for a systemic, therapeutic effect with an extended duration of activity (e.g., 24 h). Until recently, transdermal systems were of two types: (1) drug release from a reservoir through a rate-controlling microporous semipermeable membrane made of ethylene–vinyl acetate copolymer or polypropylene (Fig. 13) and (2) direct drug release from a matrix directly to the epidermis (Fig. 14). These medicated reservoirs and matrices are attached to plastic "band-aid" type strips and can be used, for example, to administer therapeutic amounts of nitroglycerin to angina patients for periods >12 h. The drug diffuses from a matrix or through a membrane to the skin surface and then through the skin. As long as the concentration of the drug at the

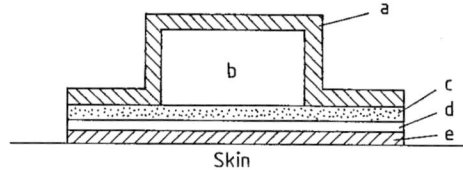

Figure 13. Transdermal patch using a semipermeable membrane to control drug release (Transderm Nitro, ALZA Corp. and Ciba)
a) Backing; b) Drug and excipient reservoir (fluid); c) Rate-release control membrane; d) Adhesive; e) Peel-away covering

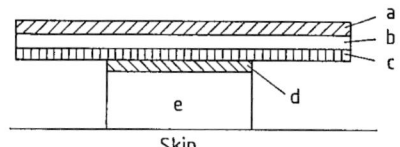

Figure 14. Transdermal patch using a matrix to control drug release (Nitrodisc, G. D. Searle; original form of Nitro-Dur, Schering–Key)
a) Occlusive outer backing; b) Absorbent pad; c) Adhesive; d) Foil base; e) Matrix (drug and excipient, semisolid formed)

skin – drug interface remains constant, the amount of drug in the unit does not affect the blood concentration. Such dosage forms are vastly different from the older, traditional plasters intended for local use (not systemic therapy). In such plasters the drug is embedded in an adhesive backing and no attempt is made to control the rate of drug release for prolonged action. For example, external analgesic drugs can be formulated so as to bring relief to some muscles.

The ALZA Corporation (1980) developed the first transdermal drug delivery system for scopolamine (Fig. 13). The product was designed to control nausea and motion sickness and was marketed by Ciba as Transderm Scop. With several excipients, the drug was contained in a fluid reservoir between an outer backing layer of aluminized polyester film and a microporous polypropylene membrane [52]. The membrane is covered externally with an adhesive for skin application. The adhesive contains sufficient impregnated drug to saturate the epidermis in contact with it and achieve a rapid onset of activity. Thereafter, the membrane controls the rate of drug diffusion to the skin to achieve prolonged therapy. This marketed unit has the appearance of a medium-sized "band aid" with an affixed pillow-shaped reservoir, resulting in the name "transdermal patch".

The transdermal system has since been used with several other drugs (e.g., nitroglycerin and estradiol). Other membrane materials have been used (e.g., an ethylene–vinyl acetate copolymer with the nitroglycerin product). Membrane polymers also include cellulose nitrate, cellulose acetate, polycarbonate, and polytetrafluoroethylene.

Another marketed version of a transdermal drug delivery product involves the use of a matrix and not a microporous membrane. One such product (Nitrodisc, G. D. Searle) utilizes a drug-impregnated, solid, silicone matrix that is bonded to an outer backing (Fig. 14). The drug is released by controlled diffusion to the skin surface. In a similar product developed by Key–Schering (Nitro-Dur) the matrix was a gellike structure made of polyvinylpyrrolidone poly(vinyl alcolhol), glycerol, water, lactose, and sodium citrate. This original Nitro-Dur form has been replaced by a new form in which the nitroglycerin is dispersed in acrylic-based polymer adhesives that are applied to an outer backing, and the product has the appearance of a piece of tan-colored adhesive

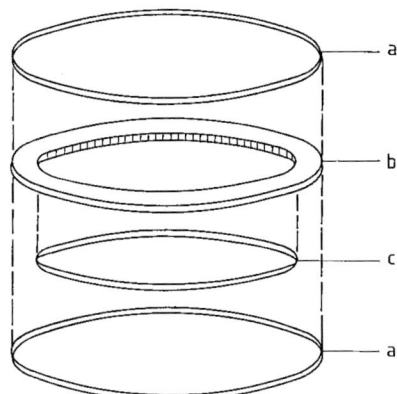

Figure 15. The Ocusert ocular therapy system for prolonged and sustained drug release after installation under the lower eyelid (courtesy of ALZA Corp., Palo Alto, California)
a) Transparent rate-controlling membrane; b) White ring of polymeric material; c) Drug reservoir film containing pilocarpine

tape. A resin cross-linking agent is also added to the adhesive to obtain the desired rate of controlled, diffusive release of nitroglycerin.

Intravaginal/Intrauterine System (Inserts) [33]. ALZA Corporation has developed the "Progestasert", which is designed to release progesterone by slow diffusion through a rate-controlling membrane for several months. The progesterone reservoir is built in a T-shaped intrauterine device.

Ophthalmic Controlled-Release System [32]. ALZA Corporation also has developed an ocular sustained-release device for the treatment of glaucoma. Pilocarpine diffuses from a drug reservoir through rate-controlling membranes composed of an ethylene–vinyl acetate copolymer. Marketed by Ciba-Geigy, these "Ocusert" disks are placed under the lower eyelid and intended to release medication for several days (Fig. 15).

Parenteral Sustained- and Prolonged-Release Systems[34]. Surgery or specialized injection equipment is required for *subcutaneous implants*. The implants are drug–polymer units designed for slow release of drug over prolonged periods (months). The polymers should be biodegradable; hydrogels and silicones have been used.

Parenteral administration of a drug as an intramuscular or subcutaneous *depot injection* can be used for extended periods of therapy. Such a formulation can be an oleoginous suspension or solution. The injected volume remains as a depot at the injection site and the drug slowly diffuses from it. The rate of drug release depends on the properties of the vehicles used (e.g., vegetable oils) and the drug.

Recent research concerning parenteral depot injections has involved use of colloidal dispersions for the depot unit. Depot injections of microencapsulated drug have been investigated; the polymeric coatings of the microcapsules act as membranes to control the dissolution and diffusion of drug.

Liposomes are of interest as components of site-specific or targeted drug delivery systems [27]. They are phospholipids that hydrate to form tiny sacs or vesicles (1 – 10 µm) with aqueous layers separating lipid layers. Liposomes can have a single internal aqueous compartment surrounded by a lipid membrane or may be multi-lamellar structures composed of concentric lipid and aqueous layers. A drug added during liposome formation can be located within the aqueous compartment (hydrophilic drug) or become part of the lipid layer (hydrophobic drug). Control of liposome vesicle size, as well as the chemical structure and properties of the phospholipid, can influence organ or tissue site disposition and drug release.

Microspheres (nanoparticles) are another area of current investigation as site-specific drug delivery systems [27], [53]. They have a diameter of 100 – 500 µm and are similar to liposomes in that they form colloidal dispersions which can be administered intravenously. They can be formed by a coacervation technique as employed in microencapsulation. Microspheres have been prepared from diverse matrix materials: lipoproteins, ion-exchange resins, ferromagnetic albumin, serum albumin, ethyl cellulose, gelatin, polystyrene, starch, polyacrylamide, and ferromagnetic alkylcyanoacrylate. The properties of microspheres (e.g., matrix material, size, density, cross-linking, molecular mass, drug concentration) affect the site, rate, and nature of drug release.

No liposome product is marketed currently. However, drug delivery is being investigated for parenteral (see Section 7.3), ocular, and inhalation administration. Similarly, no clinically proven drug product based on microsphere technology is commercially available yet. Proposed uses for microspheres and liposomes are similar, involving diseases of the reticuloendothelial systems (i.e., infections and cancers).

Circulating macrophages may also be able to assimilate such drug-rich particles for use within the vascular system. All such structures must be shown to exhibit matrix and lipid stability and purity, as well as drug stability within the structure. Particle size must also be strictly controlled.

Aqueous suspensions can be injected intramuscularly or subcutaneously to achieve prolonged therapy. For example, protamine, zinc, and insulin form a complex that slowly dissolves and acts for much longer periods than injections of simple insulin solutions. Control of the particle size of the suspended materials and their crystalline nature also contribute to the prolonged therapeutic pattern.

The parenteral use of a combination of an intravenous bolus injection with an intravenous drip (see section 7) achieves the goal of controlled, prolonged therapy as well.

7.3. Targeted Drug Delivery Systems
[53]–[58]

A drug delivered by a conventional dosage form interacts with various complex molecules and cellular components and traverses different membranes before it accumulates a threshold concentration at the desired site of therapeutic activity. This diminishes its potential effectiveness and may cause undesirable results in other organ tissues. Thus, an ideal drug delivery system must include the concept of drug delivery to a specific (target) site, and only that site. Targeted delivery could possibly be achieved by attaching the drug to a "carrier". The drug–carrier unit should remain intact and not undergo reaction with any endogenous metabolites or cellular structures until it reaches its target site: a particular organ, type of cell, or subcellular unit.

Site-specific (targeted) drug delivery is different from conventional dosage forms that are delivered to an anatomical site to achieve a localized effect (e.g., suppositories to relieve local symptoms of hemmorhoids, eye drops for local ocular therapy, and creams or lotions to treat skin rashes). Site-specific drug delivery involves systemic administration of a drug–carrier/dosage form with the anticipated therapeutic activity targeted at a site away from the point of administration.

The drug–carrier unit should be able to "recognize" its target unit and release the drug entity in a controlled manner to achieve prolonged duration of activity. The carrier should then be biodegraded and pass out of the body without any further biochemical or pharmacologic intervention. An "ideal" carrier for targeted drug delivery should therefore have the following properties:

1) biodegradability,
2) high "carrying" capacity,
3) selective drug delivery,
4) reproducible properties,
5) nontoxic,
6) nonimmunogenic,
7) controlled release of drug,
8) long shelf-life, and
9) low production costs.

Recent research on site-specific drug delivery has involved a variety of carrier substances including plasma proteins, glycoproteins, polymers, monoclonal antibodies, and microparticulate carriers. This last group includes liposomes, microspheres, nanoparticles, and erythrocytes. Site specificity can vary depending on the delivery level achieved.

The term *passive targeting* is used when the drug–carrier unit passes through the same natural distribution pattern in the body that the carrier substance would normally take. An example of passive targeting is the intravenous injection of microparticles of a drug–carrier for delivery to subcellular lysosomes in macrophages within the organs of the reticuloendothelial system (liver, spleen, bone marrow).

Active targeting refers to drug–carrier distribution to the target site that is in some way altered from the normal distribution in the body of the carrier substance. The use of magnetic, biodegradable colloidal particles as carriers might be viewed as an example of the latter. After intravenous injection, such a drug–carrier unit could be directed to the target site by an external magnetic device [59]–[61]. Once a suitable concentration has been achieved, the device then could release the drug at a controlled rate to achieve prolonged therapy.

8. References

[1] G. S. Banker: "The Drug Product As A Delivery System" in G. S. Banker, C. T. Rhodes (eds.): *Modern Pharmaceutics*, 2nd ed., Marcel Dekker, New York 1990, pp. 15–20.

[2] P. Zanowiak: "Dosage Forms: Nonparenterals" in J. Swarbrick, J. C. Boylan (eds.): *Encyclopedia of Pharmaceutical Technology*, vol. **4**, Marcel Dekker, New York 1990, in press.

[3] *Kirk-Othmer*, vol. **17**, pp. 272–290.

[4] H. C. Ansel, N. G. Popovich: "Dosage Form Design: Biopharmaceutic Considerations" in: *Pharmaceutical Dosage Forms and Drug Delivery Systems*, 5th ed., Lea & Febiger, Philadelphia 1990, pp. 51–91.

[5] D. W. Bourne, L. W. Dittert: "Bioavailability (Extent of Absorption)" in G. S. Banker, C. T. Rhodes (eds.): *Modern Pharmaceutics*, 2nd ed., Marcel Dekker, New York 1990, pp. 91–142.

[6] A. R. DiSanto: "Bioavailability and Bioequivalency Testing" in A. R. Gennaro (ed.): *Remington's Pharmaceutical Sciences*, 18th ed., Mack Publishing Co., Easton 1990, pp. 1451–1458.

[7] H. C. Ansel, N. G. Popovich: "Routes of Drug Administration" in: *Pharmaceutical Dosage Forms and Drug Delivery Systems*, 5th ed., Lea & Febiger, Philadelphia 1990, pp. 68–78.

[8] J. C. Boylan, A. L. Fites: "Routes of Parenteral Administration" in G. S. Banker, C. T. Rhodes (eds.): *Modern Pharmaceutics*, 2nd ed., Marcel Dekker, New York 1990, pp. 493–502.

[9] S. J. Turco, R. E. King: *Sterile Dosage Forms*, 3rd ed., Lea & Febiger, Philadelphia 1987, pp. 6, 7.

[10] P. H. List: *Arzneiformlehre*, Wissenschaftliche Verlagsgesellschaft, Stuttgart 1973.

[11] J. L. Colaizzi, W. H. Pitlick: "The Oral Route of Administration" in G. S. Banker, R. K. Chalmers (eds.): *Pharmaceutics and Pharmacy Practice*, J. B. Lippincott, Philadelphia 1982, pp. 184–190.

[12] V. H. L. Lee: "Changing Needs in Drug Delivery in the Era of Peptide and Protein Drugs" in V. H. L. Lee (ed.): *Peptide and Protein Drugs*, Marcel Dekker, New York 1990, pp. 1-56.

[13] G. M. Grass IV, J. R. Robinson: "Intravaginal and Intrauterine Systems", in G. S. Banker, C. T. Rhodes (eds.): *Modern Pharmaceutics*, 2nd ed., Marcel Dekker, New York 1990, pp. 666–668.

[14] J. J. Sciarra, A. J. Cutie: "Inhalation Aerosols" in [13] pp. 626–630.

[15] "Pharmaceutical Dosage Forms" in: *USP XXII, U. S. Pharmacopeial Convention*, Rockville 1990, pp. 1688–1697.

[16] J. G. Nairn: "Solution, Emulsions, Suspensions and Extractives" in A. R. Gennaro (ed.): *Remington's Pharmaceutical Sciences*, 18th ed., Mack Publishing, Easton 1990, pp. 1519–1544.

[17] S. J. Turco: "Dosage Forms: Parenterals" in J. Swarbrick, J. C. Boylan (eds.): *Encyclopedia of Pharmaceutical Technology*, Marcel Dekker, New York 1990, in press.

[18] H. C. Ansel, N. G. Popovich: "Oral Suspensions, Emulsions, Magmas and Gels" in: *Pharmaceutical Dosage Forms and Drug Delivery Systems*, 5th ed., Lea & Febiger, Philadelphia 1990, pp. 225–254.

[19] J. J. Sciarra, A. J. Cutie: "Aerosols" in A. R. Gennaro (ed.): *Remington's Pharmaceutical Sciences*, 18th ed., Mack Publishing, Easton 1990, pp. 1694–1714.

[20] H. C. Ansel, N. G. Popovich: "Suppositories, and Other Rectal, Vaginal and Urethral Preparations" in: *Pharmaceutical Dosage Forms and Drug Delivery Systems*, 5th ed., Lea & Febiger, Philadelphia 1990, pp. 373–390.

[21] R. E. O'Conner, E. G. Rippie, J. B. Schwartz: "Powders as a Dosage Form" in A. R. Gennaro (ed.): *Remington's Pharmaceutical Sciences*, 18th ed., Mack Publishing, Easton 1990, pp. 1629–1632.

[22] E. Rudnic, J. B. Schwartz: "Oral Solid Dosage Forms" in [20] pp. 1633–1665.

[23] H. Senger: "Soft Gelatin Capsules: A Solution to Many Tableting Problems", *Pharm. Technol.* **9** (1985) no. 9, 36.

[24] S. C. Porter: "Coating of Pharmaceutical Dosage Forms" in A. R. Gennaro (ed.): *Remington's Pharmaceutical Sciences*, 18th ed., Mack Publishing, Easton 1990, pp. 1666–1675.

[25] H. List, L. Hörhammer (eds.): *Hagers Handbuch der Pharmazeutischen Praxis*, vol. **VII**, "Arzneiformen und Hilfsstoffe" part A and B, Springer Verlag, Berlin 1977.

[26] M. A. Longer, J. R. Robinson: "Sustained-Release Drug Delivery System" in [25] pp. 1676–1691.

[27] G. M. Grass IV, J. R. Robinson: "Sustained- and Controlled-Release Drug Delivery Systems" in G. S. Banker, C. T. Rhodes (eds.): *Modern Pharmaceutics*, 2nd ed., Marcel Dekker, New York 1990, pp. 635–671.

[28] F. Theeuwes, W. Bayne: "Controlled-Release Dosage Form Design" in J. Urquhart (ed.): *Controlled-Release Pharmaceuticals*, American Pharmaceutical Association, Washington, D.C., 1981, pp. 61–93.

[29] H. C. Ansel, N. G. Popovich: "Rate-Controlled Dosage Forms and Drug Delivery Systems" in: *Pharmaceutical Dosage Forms and Drug Delivery Systems*, 5th ed., Lea & Febiger, Philadelphia 1990, pp. 183–191.

[30] C. D. Black: "Update: Programmed Drug Delivery Systems", *U. S. Pharmacist* **8,** November (1983) 49–60.

[31] Y. W. Chien: "Transdermal Controlled-Release Drug Administration" in: *Novel Drug Delivery Systems*, Marcel Dekker, New York 1982, pp. 149–218.

[32] "Ocular Controlled-Release Drug Administration" in [31] pp. 13–50.

[33] "Intravaginal/Intrauterine Controlled-Release Drug Administration" in [31] pp. 51–148.

[34] "Parenteral/Implantable Controlled-Release Drug Delivery Systems" in [31] pp. 219–412.

[35] H. C. Ansel, N. G. Popovich: "Aerosols, Inhalations, Inhalants and Sprays" in: *Pharmaceutical Dosage Forms and Drug Delivery Systems*, 5th ed., Lea & Febiger, Philadelphia 1990, pp. 390–405.

[36] G. D. Chase: "Radiopharmaceuticals Used in Medicine" in A. R. Gennaro (ed.): *Remington's Pharmaceutical Sciences*, 18th ed., Mack Publishing, Easton 1990, pp. 642–652.

[37] J. C. Boylan et al.: *Handbook of Pharmaceutical Excipients*, American Pharmaceutical Association, Washington, D.C., and the Pharmaceutical Society of Great Britain, London 1986, pp. 1–375.

[38] E. A. Swinyard, W. Lowenthal: "Pharmaceutical Necessities" in A. R. Gennaro (ed.): *Remington's Pharmaceutical Sciences*, 18th ed., Mack Publishing, Easton 1990, pp. 1286–1329.

[39] J. L. Colaizzi, W. H. Pitlick: "Palatability – A Special Requirement for Oral Medications" in G. S. Banker, R. K. Chalmers (eds.): *Pharmaceutics and Pharmacy Practice,* J. B. Lippincott, Philadelphia 1982, pp. 191–195.

[40] H. Schott, personal communications, Temple University, Philadelphia, Pennsylvania.

[41] A. M. Martin, R. V. Smith, V. A. Yanchick: "Product Quality Evaluation" in [39] pp. 33–40.

[42] S. L. Hem, R. G. Stoll: "Appraisal of Drug Product Quality and Performance" in G. S. Banker, C. T. Rhodes (eds.): *Modern Pharmaceutics,* 2nd ed., Marcel Dekker, New York 1990, pp. 741–802.

[43] C. R. Erskine, Jr.: "Quality Assurance and Control" in A. R. Gennaro (ed.): *Remington's Pharmaceutical Sciences,* 18th ed., Mack Publishing, Easton 1990, pp. 1513–1518.

[44] D. C. Liebe: "Packaging of Pharmaceutical Dosage Forms" in G. S. Banker, C. T. Rhodes (eds.): *Modern Pharmaceutics,* 2nd ed., Marcel Dekker, New York 1990, pp. 695–740.

[45] H. C. Ansel, N. G. Popovich: "Packaging, Labeling and Storage of Pharmaceuticals" in: *Pharmaceutical Dosage Forms and Drug Delivery Systems,* 5th ed., Lea and Febiger, Philadelphia 1990, pp. 116–133.

[46] J. Robinson, personal communications, University of Wisconsin, Madison, Wisconsin.

[47] F. Theeuwes: "Oral Dosage Form Design: Status and Goals of Oral Osmotic Systems Technology," *Pharm. Int.* **5** (1984) no. 12, 293–296.

[48] C. H. Nightingale: "The Technology of Controlled-Release Drug Delivery Systems" at a symposium: *Controlled-Release Drug Delivery and PennKinetic,* Omega Communications Inc., Springfield, New Jersey and PennWalt Corp., Rochester, New York 1983, pp. 26–31.

[49] K. Park, H. S. Ch'ng, J. R. Robinson: "Alternative Approaches to Oral Controlled Drug Delivery: Bioadhesives and In Situ Systems" in J. M. Anderson, S. W. Kim (eds.): *Recent Advances In Drug Delivery Systems,* Plenum Press, New York 1984, pp. 163–183.

[50] A. Karim: "Transdermal Delivery Systems" in: *Drug Delivery Systems,* Aster Publishing Corp., Springfield, Oreg., 1983, pp. 28–32.

[51] J. W. Fara: "Short- and Long-Term Transdermal Drug Delivery Systems" in [50] pp. 32–37.

[52] H. C. Ansch, N. G. Popovich: "Transdermal Drug Delivery Systems", *Pharmaceutical Dosage Forms and Drug Delivery Systems,* 5th ed., Lea and Febiger, Philadelphia 1990, pp. 310–320.

[53] E. Timlinson: "Site-Specific Drug Delivery" in G. S. Banker, C. T. Rhodes (eds.): *Modern Pharmaceutics,* 2nd ed., Marcel Dekker, New York 1990, pp. 673–694.

[54] S. N. Mills, S. S. Davis: "The Targeting of Drugs/Controlled Drug Delivery" in L. Illum, S. S. Davis (eds.): *Polymers in Controlled Drug Delivery,* IOP Publishing, Bristol 1987, pp. 4–6.

[55] P. Arthurson: "Site Specific Drug Delivery/The Fate of Microparticulate Drug Carriers after Intravenous Administration" in [54] pp. 15–24.

[56] R. L. Juliano, D. Layton: "Liposomes as a Drug Delivery System" in R. L. Juliano (ed.): *Drug Delivery Systems,* Oxford University Press, New York 1980, pp. 189–236.

[57] R. C. Oppenheim: "Nanoparticles" in [56] pp. 177–188.

[58] B. E. Ryman: "Liposomes: Possible Potential in Drug Delivery" in T. J. Roseman, S. Z. Mansdorf (eds.): *Controlled-Release Delivery Systems,* Marcel Dekker, New York 1983, pp. 27–42.

[59] D. S. T. Hsieh, R. Langer: "Zero-Order Drug Delivery Systems with Magnetic Control" in [58] pp. 121–132.

[60] R. M. Morris et al.: "Magnetic Microspheres in Drug Delivery" in J. M. Anderson, S. W. Kim (eds.): *Recent Advances In Drug Delivery Systems,* Plenum Press, New York 1984, pp. 221–228.

[61] R. Langer, L. Brown, E. Edelman: "Controlled Release and Magnetically Modulated Systems" in [60] pp. 249–258.

Drug Testing

ROLF KRETZSCHMAR, Knoll AG, Ludwigshafen, Federal Republic of Germany (Chap. 1)
KLAUS-JÜRGEN HAHN, Knoll AG, Ludwigshafen, Federal Republic of Germany (Chap. 2)
MANUEL ZAHN, Knoll AG, Ludwigshafen, Federal Republic of Germany (Chap. 3)

1.	Preclinical Testing	2126	3.4.	Validation of Analytical Procedures	2142
1.1.	Introduction	2126	3.5.	Process Validation and Qualification	2143
1.2.	Animal Experiments	2127	3.6.	Batch Release	2143
1.3.	Pharmacological Testing	2128	3.7.	Stability Testing	2144
1.4.	Toxicological Testing	2130	3.8.	Documentation	2145
2.	Clinical Trials	2134	3.9.	Variations — Changes	2148
2.1.	Phases of Clinical Drug Development	2134	3.10.	Change Control	2149
2.2.	Methods for Proving Effectiveness	2136	3.11.	Compliance	2149
2.3.	Adverse Drug Reactions	2138	3.12.	Changes Requiring New Applications — Line Extension	2149
2.4.	Good Clinical Practice	2138	3.13.	Inspections	2150
2.5.	Clinical Application Dossier	2140	3.14.	The Mutual Recognition Agreement (MRA)	2150
3.	Quality Control of Pharmaceutical Products	2140	3.15.	Pharmaceutical Inspection Convention (PIC)	2151
3.1.	Manufacturing of Drug Substances — Test of Starting Materials	2141	3.16.	Pharmaceutical Inspection Cooperation Scheme (PIC/S)	2152
3.2.	Control of Drug Substances: Impurities	2141	3.17.	Pharmacopeias	2152
3.3.	Control of Drug Products: Impurities	2142	3.18.	The EP Certificate of Suitability	2154
			4.	References	2155

1. Preclinical Testing

1.1. Introduction

Modern drug development can be divided into three major stages:

1) Finding the substance
2) Pharmacological – toxicological and pharmaceutical studies
3) Clinical studies

Rational methodology dominates from the very first phase of this process. The search for a new substance starts with the chemical synthesis of a compound or with its isolation from natural products or a fermentation broth of natural or genetically engineered microorganisms. In products produced by genetic engineering, the desired primary effect is usually known. This is, however, not the case with chemically synthesized products. Although a definite effect can often be predicted from structural data or from molecular biological concepts based on receptor models, this effect still has to be checked in pharmacological tests. These tests are normally carried out in a specific screening process that incorporates models for both the desired effect and possible side effects. The tests are performed on standardized models (animals, animal organs, cellular systems, microorganisms, or enzymes), and the most suitable substances are selected (screened) from a large number of analogous substances. In the 1950s, a 1:1000 chance of success for the discovery of a promising substance was assumed. Today, a chance of 1:7000 – 1:10 000 is expected for chemically synthesized substances because of the demand for increased selectivity of a desired effect.

The purpose of the preclinical evaluation of a new drug is to establish its complete spectrum of pharmacological and toxicological properties as well as its pharmaceutical quality. A detailed description of all experiments and results is required by the drug-approval authorities in all countries as evidence of drug efficacy and safety. Animal experiments play a central role in the elucidation of the pharmacology and toxicology of the active agent or mixtures in preclinical drug development. They are designed to identify all the effects of an active drug candidate which may justify its subsequent administration to humans or which are to be regarded as undesirable or harmful. A therapeutic dose range can be determined from a quantitative comparison of desirable and undesirable effects and the benefit – risk ratio can then be estimated. Chemical analysis of body fluids and organs should give insight into the chemical behavior of the substance in the animal. This includes its absorption, distribution, metabolism, and excretion. Pharmacokinetic and metabolic studies of this type characterize the animal species used in the pharmacological and toxicological experiments with regard to its suitability for predicting the corresponding characteristics in humans.

In toxicological tests, administration of often excessively high doses should produce local or systemic damage of the animal organism so that tolerance in humans can be predicted and possible target organs for acute or chronic poisoning can be determined.

Guidelines and Laws. Investigation of the pharmacological and toxicological properties of a new drug in accordance with the latest scientific knowledge is stipulated in all national and supranational (e.g., EC) drug approval procedures.

Most of the pharmacological and toxicological data have to be existent when clinical trials commence. In many countries (e.g., United States, Canada, United Kingdom, and Japan), clinical trials are officially permitted only after these results have been submitted. These tests should qualitatively and quantitatively establish the toxicity limits of the substance, as well as any unwanted or adverse effects, and the desired activity potential. Furthermore, as initiated by the FDA (USA), almost every country now stipulates that the safety testing of drugs be carried out under Good Laboratory Practice (GLP) guidelines (e.g., Council Directive 87/18/EEC). GLP requires that the reported results are comprehensible and that the quality of the data is secured and reviewable. Detailed recording of the planning and performance of experiments, precisely defined instructions (standard operating procedures, SOPs) for the staff, validation of methods, and independent quality control (quality assurance) are therefore necessary. Moreover, laws in many countries regulate the use of animals and the number and quality of animal experiments (e.g., Protection of Animals, Council Directive 86/609/EEC).

The necessity of animal experiments for establishing the pharmacological and toxicological profile of a drug is stipulated in national and supranational guidelines. For instance:

1) *EC:* Pharmacotoxicological Standards (Council Directive 75/318/EEC) and several Notes for Guidance concerning single tests
2) *United States*: Guidelines for the Format and Content of the Nonclinical/Pharmacology/Toxicology Section of the Application, US Department of Health and Human Services, Public Health Service, Food and Drug Administration, Washington, D.C./USA, 1987
3) *Japan*: General Toxicology Guidelines, Ministry of Health and Welfare, Tokyo, Japan, 1989

Several guidelines have been compiled by the WHO [1].

1.2. Animal Experiments

Animal experiments should allow prediction of the possible reaction of the human organism after single or repeated administration of a drug. The results of animal experiments are applicable to humans with some reservations. Neither the physiology nor the metabolism of commonly used experimental animals completely corresponds to that of humans. The information value can be improved by testing at least two different animal species (preferably one rodent and one nonrodent).

The physical reactions of humans do not generally differ from those of animals. Reactions vary between humans and animals to the same extent as between individual animal species. Thus, good or moderate agreement or even incompatibility is possible. It would, however, be incorrect to reject animal experiments on the basis of the rare cases where no relevance is observed, at least until other, more informative methods have been found.

Methodological difficulties in establishing drug activity or the emergence of unknown and, therefore, untested reactions (e.g., as in the thalidomide disaster of the

1960s) do not legitimate that subsequent drug-induced damage be blamed solely on the shortcomings of animal experiments. Numerous observations have shown that there is generally a good correlation between the reactions shown by animals (primarily vertebrates) and humans to a drug. Differences are frequently due to variations in the pharmacokinetics and metabolism of the active substance. These properties must therefore be carefully checked in all animal species. Animal experiments are certainly not a substitute for careful clinical trials and for the subsequent long-term observation of patients who have been treated with the new drug. However, they are still essential for the estimation of risk in the development period. Restriction to experiments involving important problems and based on validated methods should be an ethical obligation.

A special problem with animal experiments is that many human diseases do not occur spontaneously in animals. Many animal models have therefore been developed with the aim of "experimentally producing a comparable diseased condition" [2]. The applicability of results obtained from these systems is naturally less than that of results obtained in tests on animals with spontaneous pathological conditions. These limitations mainly apply to experiments for finding active substances that counteract specific diseases. For instance, it is easier to find an antihypertensive or antiarrhythmic agent using animal experiments than a psycho-pharmacologically active drug. Experiments for establishing drug safety are always carried out on healthy, purebred experimental animals.

The standardization of methods is another important factor for transferability of the results. Even minor changes in the test procedure (changing the breed of experimental animal, the time of testing, or temperature) can cause substantial deviations of results. All reports on pharmacological and toxicological testing must therefore be accompanied by a detailed description of the methodology used. Standard pharmacological methods are described in appropriate handbooks (e.g., [3], [4]). Comparative testing of a standard substance, if available, is of great importance. Ring experiments with standard substances involving several laboratories have proved useful for toxicological tests.

1.3. Pharmacological Testing

Studies of Special and General Pharmacodynamic Properties. Actual drug development begins when the signs of a therapeutically interesting effect found in the preliminary tests (in vitro screening tests or specific in vivo tests) can be reproduced and biometrically confirmed in comparison with standard substances. The aim of this phase of drug development is to analyze the special (i.e., therapeutically desired effect) and general pharmacodynamic (i.e., sum of all biological activities of a drug) profiles. The methods should be varied, for example by using different animal species. Tests include the determination of dose–response relationships, calculation of mean effective doses (ED_{50} = the dose that causes a defined effect in 50% of the animals or changes a quantifiable reaction in the animals by 50%), and the determination of time–response relationships. Different routes of administration should be used (generally oral and intravenous, but also subcutaneous, intraperitoneal, and intraduodenal). The duration and onset of the effect and the time of maximum response should be established.

Comparison of the mean effective doses for the special pharmacodynamic effect and the toxic doses determined in the toxicological experiments (e.g., LD_{50} = lethal dose in 50% of an animal group) furnishes important information on the "therapeutic range", i.e., whether a desired response can be achieved in animal experiments with doses that do not produce side effects or lethality.

The responses obtained in the acute experiments must be checked again after repeated dosage to determine whether the response decreases (tolerance) or increases (accumulation).

The general pharmacodynamic characterization of an active agent should show whether it also affects functions of the organism that are not necessarily connected with the primary, therapeutic effect. The functioning of all important organs (e.g., heart and circulation, lungs, nervous system, kidneys, endocrine glands, and gastrointestinal system) must therefore be tested in separate pharmacological experiments and changes in function must be quantitatively assessed by biometric methods. Experiments on isolated organs, isolated cell systems (e.g., receptor binding assays), and cell-free enzyme systems are of primary importance in this respect. If substance-dependent changes in functions are observed, the relevance of the results should be checked in the intact animal. If essential physiological reactions are inhibited and can no longer be evaluated as a result of anesthesia (e.g., central nervous reactions, central cardiovascular regulation), experiments must be performed on animals without anesthesia. Dose–response relationships must be determined if pharmacodynamic effects occur. The dosage level must be based on the dose required to give the therapeutically desired response. Furthermore, substance-dependent effects must be checked after repeated administration.

Some experiments are designed to find antagonistic drugs if poisoning by an overdose of the developing drug is the result of an exaggerated pharmacodynamic response. This antagonist should be used as an antidote for humans in the case of overreaction or poisoning. Studies on interaction with other drugs that are used frequently to treat the same indications should also be performed.

Pharmacokinetic Studies. Investigations which evaluate the "biological fate" of the new drug in the animal organism are an important part of its pharmacological characterization. The species used for these pharmacokinetic studies are the same as those employed for the pharmacodynamic and toxicological experiments. The path of the active substance is followed from the site of administration through the organism to the excretory organs. The elucidation of the pharmacokinetics of a drug, known as ADME, is divided into the following parts:

1) *Absorption or Application* (*A*) *Phase*. Determination of the degree and rate of absorption of the substance into the blood stream from the site of administration. The substance may be applied in a dissolved or suspended form, but also in a pharmaceutically prepared form.
2) *Distribution* (*D*) *Phase*. Determination of the concentration of the substance in body fluids and tissues after its distribution in the organism, including binding to special components (e.g., plasma proteins).
3) *Metabolism* (*M*) *Phase*. Determination of the metabolism of the substance, i.e., its biological degradation in the organism by the action of enzymes, primarily those in the liver, intestine, plasma, or kidneys.
4) *Elimination* (*E*) *Phase*. Determination of the routes and amounts by which the unchanged or metabolized substance is excreted.

Chemicoanalytical, immunological, and radiological methods are used to determine the substance in body fluids and organs. Radioactive substances (i.e., ^{3}H- or ^{14}C-labeled) or deuterium-labeled substances serve to identify metabolites.

Knowledge of the pharmacokinetics of a substance obtained in the animal species that are used for the pharmacodynamic and toxicological tests is essential for the interpretation of drug safety data. It is the most important criterion for estimating the relevance of the data for humans. It permits the association of a defined drug concentration in the plasma or at the site of action with a pharmacodynamic effect in animals and humans. Special features of the pharmacokinetic behavior of an active substance can be recognized in animal experiments and allow specific analysis in humans. Examples of such behavior include rapid metabolism during initial passage of the substance through the liver or the intestinal wall after enteral absorption (first pass effect), enterohepatic circulation (reabsorption of substances excreted into the intestine with the bile), and storage of the substance in certain tissues.

Conclusions about the dosage interval needed in long-term pharmacological and toxicological studies can be drawn from the retention time of a substance in the organism if, as in humans, a continuous plasma level of the drug and prevention of its accumulation are required.

Pharmacokinetic studies on animals are helpful for preparation of protocols for experiments on humans when the drug is first administered.

1.4. Toxicological Testing

Investigation of the toxicological properties of a potential drug is of considerable importance for its further development. Information about possible toxic effects and dose limitation must be obtained before performing the first trials on humans. In addition to acute toxicity tests, experiments are performed with repeated drug administration (varying duration) in different animal species. The duration of these animal tests depend on the planned duration of tests and subsequent administration in humans. Special toxicological tests are used to exclude local intolerance to parenterally administered drugs, reproductive damage, mutagenic, and, in the case of long-term human application for more than six months, carcinogenic effects. Tests for allergization and phototoxicity are required for dermatological drugs. As a rule, experiments must be conducted in such a manner that statistical analyses are possible. Toxicological tests must comply with GLP regulations and internationally standardized guidelines.

Single-Dose Acute Toxicity. Almost all countries require the testing of systemic tolerance after a single oral and parenteral administration of high doses to two animal species (mouse and rat), with a postobservation period of 14 days. Systemic toxic symptoms should be described separately for male and female animals, data on the dose dependency of symptoms and lethality (LD_{50}) should be recorded. The value of acute toxicity experiments is disputed. Their primary importance is the fact that appropriate animals deliver information for the experimental design of the more important toxicity tests involving repeated drug administration. Also, information is obtained on the poisoning profile of an acute overdosage in humans. Calculation of the therapeutic range from ED_{50} and LD_{50} values is discussed in Section 1.3.

It is generally sufficient if the approximate acute toxicity is determined on a small number of experimental animals (i.e., three per sex and species for each dose tested) without statistical evaluation (cf. EC Note for Guidance: single dose toxicity). This allows estimation of the mean, minimum, and maximum lethal doses.

In the case of new drug combinations, acute toxicity tests should establish whether the combination of active agents potentiates their action or leads to new toxic effects. If this is the case, extended tests are required to clarify the toxic properties in long-term application (full toxicological development program as with a single new drug).

Toxicity on Repeated Administration. Repeated administration of a drug (usually at daily intervals) should show which functional and morphological changes are produced by the active agent. The test must include a control group of animals that receives a preparation without an active ingredient (i.e., solvent alone). Doses range from therapeutic (i.e., in the range of the pharmacodynamic effective dose) to toxic. The highest dose should produce harmful effects and the lowest dose should be free of toxic effects (i.e., the results in the test group should not differ from those of the control group). As a rule, two series of experiments of varying duration are set up. In order to determine the subchronic toxicity (14 days to 3 months, daily administration) and chronic toxicity (3 – 12 months, daily administration) each series is performed on two animal species. A rodent and a nonrodent species should be chosen (usually rat and dog or hamster and monkey). In many countries, the appropriate animal species, the number of groups of experimental animals exposed to varying dosages, the number of animals in each group and of each sex, and the duration of treatment are stipulated by guidelines. In applications for drug approval, the required duration of drug administration to animals depends primarily on the expected maximum duration of the clinical treatment of patients. A survey of the requirements in some countries is given in Table 1.

Table 1. Toxicity study requirements in the EC, Japan, and USA

Duration of patient therapy (oral or parenteral administration)	Duration of toxicity studies		
	EC	Japan	USA*
≤ 1 day	2 weeks	28 days	14 days
2 – 7 days	4 weeks	90 days	90 days
8 – 30 days	3 months	26 weeks	26 weeks
> 30 days	6 months	52 weeks	26** or 52 weeks

* Depending on the stage of drug development, considerably shorter durations are sufficient for the performance of clinical studies before drug approval (IND phase).
** Duration of therapy up to three months.

The following investigations are used to evaluate tolerance or toxicity: functional tests (behavior, course of growth and food intake, acoustic and optic reactions, measurement of blood pressure, and electrocardiogram), clinical chemistry of blood and urine, status, autopsy including organ observation and weights, and light microscopy supplemented by electron microscopic investigations of all organs. Studies on the reversibility of the changes complete the toxicological tests.

Reproduction Studies. Reproduction tests have been an essential requirement since the thalidomide disaster in 1962. The effect of the test substance on fertility and mating behavior, as well as effects that lead to abortion of the fetus, fetal abnormalities, or damage of further progeny are studied. These tests are divided into three segments:

1) *Fertility Testing*. The influence of the substance on male and female fertility is generally tested in rats.
2) *Embryo Toxicity Testing*. These tests are carried out on at least two species, usually rat (or mouse) and rabbit. The test substance is administered to pregnant animals during the critical phase of organogenesis (day 6 – 15 of pregnancy in rats, day 6 – 18 in rabbits).
3) *Peri- and Postnatal Investigation*. The test substance is usually administered to pregnant rats in the last third of gestation and continued during the four-week lactation period. The influence on the size of the litter and breeding of progeny is recorded.

Mutagenicity tests are designed to detect any changes in genetic properties. As a rule, a series of at least four tests is conducted for each drug. Various types of mutation are investigated in different systems. Point mutations (i.e., gene mutations) are investigated in vitro in bacterial test systems (e.g., Ames test) and eucaryotic (e.g., mammalian) cells, predominantly lymphocytes, that are cultured in a suitable nutrient medium. The point mutation tests are also used as precarcinogenicity tests. Chromosomal changes are studied in vitro in mammalian bone marrow cells and in vivo in a rodent species. A series of other models can be used to supplement the investigations in individual cases [5].

Carcinogenicity tests are designed to establish whether a drug has cancer-producing (carcinogenic) effects. These tests are required if the drug is to be taken either regularly for more than six months or with interruptions, but frequently enough to produce the same total burden. They are also necessary if the chemical structure of the substance or previous chronic toxicity and mutagenic tests suggest a possible carcinogenic potential. The investigations should be performed on two rodent species that are treated with varying daily doses of the test substance for 18–24 months (mouse) and 24–30 months (rat).

Special Toxicity Tests. Pharmaceutical agents that are to be injected or applied to the skin or mucous membranes must be tested for local irritant effects. These tests are carried out on animals (dog, rabbit) by single and repeated application of different amounts of the final preparations at these sites. Special investigations should be conducted with dermatological drugs after combined systemic and topical application to reveal possible sensitization (Magnussen test), allergization, or phototoxicity. Systemic toxicity testing is also required for repeatedly administered dermatological drugs that penetrate the skin in relevant amounts (generally >1%). This test is most effectively carried out by subcutaneous application.

Safety Testing of Biotechnological Products. Human peptides, proteins, and monoclonal antibodies produced by biotechnological methods require preclinical evaluation of their safety for use in humans. This evaluation differs from that required for conventional, chemically synthesized substances. The development of antibodies to these products in test animals influences the test results (allergic reactions or neutralization) and limits the duration of administration. Exaggerated pharmacodynamic responses are usually responsible for the most serious toxicological problems caused by these proteins (e.g., recombinant insulin, erythropoietin, or tissue plasminogen activator). Sufficient information can therefore still be obtained under such restricted testing conditions provided the tests include as much pharmacological and biochemical data as possible (see Section 1.3). Toxic effects can, however, also occur after overdosage, i.e., when the concentrations of agents normally present in the body are greatly exceeded (e.g., with cytokines) during therapy or when mutants (homologues of human proteins) are tested.

If antibody development is detected in the animal species usually used for testing, studies on nonhuman primates are sometimes required to determine the most likely toxic profile of the products in humans. Another possibility would be to test a biotechnological product that has an identical amino acid sequence to the endogenous substance in the animal species used. This would, however, greatly increase development costs and would still only be a model.

2. Clinical Trials

The term clinical trials comprises the systematic, controlled and/or documented administration of potential drugs to a limited number of healthy or sick human subjects in order to provide proof of effectiveness (efficacy) and safety and to allow the subsequent, wide-scale, safe, application of the medicinal product in patient therapy. These studies must be performed in a group of sick subjects who are representative of the population for which the drug is intended. The animal studies described in Chapter 1 are an indispensable prerequisite for clinical studies: knowledge of the pharmacological and toxicological profile is needed to assess benefits and risks in future therapy, to estimate risks for initial administration in humans, and to design suitable clinical trials. However, the restricted applicability of data from animal experiments to humans must be taken into consideration (cf. Section 1.2). The effectiveness and tolerability of a substance in humans can only be determined in patients in clinical trials. The necessity for clinical studies is recognized throughout the world and is established in national laws. Before initiation of any clinical trial certain regulatory requirements have to be met, ranging from simple notification to the competent authority to official trial permission based on review of submitted data.

2.1. Phases of Clinical Drug Development

Clinical studies can be divided into four phases:

1) Human pharmacological testing (Phase I)
2) Tolerability and dose finding in patients (Phase II)
3) Proof of safety and effectiveness (Phase III)
4) Studies and follow-up checks after approval (Phase IV)

Although the phases are performed in consecutive order, wide overlap may occur. The definition of the individual phases is therefore not always standardized; Phases II and III are sometimes further subdivided.

Phase I. In human pharmacological testing, a new active substance (NAS) is administered to humans for the first time. The effect of the substance on the human body, namely pharmacodynamics (including effects and adverse drug reactions), is recorded with as much detail as possible. The effect of the body on the substance, namely pharmacokinetics (absorption, distribution, metabolism, and elimination: ADME), is also investigated; the four steps involved in the elucidation of the pharmacokinetic system are defined in Section 1.3. Sensitive analytical methods are used to determine the concentration of the substance in body fluids. For methodological and ethical reasons, Phase I investigations are usually carried out in a group of ca. 60 healthy volunteers. Studies in patients are only performed under special circumstances (e.g., in the case of cytostatic and immunologic agents).

The first application to subjects generally starts with the administration of a dose that is only a fraction of the dose effective in animals. This single dose is gradually increased until it exceeds the expected therapeutic range and/or adverse drug reactions occur. Statements made by the subject about his well-being are recorded. Physical and chemical data (e.g., vital functions, laboratory values, and the concentration of the substance in the blood and excretory products) are also measured. Important basic pharmacokinetic parameters (e.g., the distribution volume, half-life, and metabolism) are determined from the chemical data. The absolute bioavailability is calculated by comparing plasma levels after systemic (e.g., intravenous) and enteral (oral) administration. Since the metabolism of foreign substances is genetically determined and varies from person to person, the pharmacogenetics of the test persons must be defined.

The results of Phase I investigations provide important information for establishing the limiting dose and the dosing intervals for the first administration of the drug to patients in Phase II. Phase I investigations are completed by determining pharmacokinetics behavior on repeated administration, the bioavailability of the final dosage form, interactions with other drugs, and, finally, the influence of food, age, disease, and of impaired liver and kidney function on kinetics.

Phase II. In Phase II the potential drug is for the first time administered to patients suffering from the disease intended for treatment. The objectives are to obtain information on tolerability and to find the optimal dose and dosing scheme. Usually, several hundred patients are necessary. Since these investigations can only be carried out in patients, the accompanying conditions must be taken into account. As a result of the variety and variability of diseases and their symptoms, these studies demand high requirements and performance as regards planning, execution, diagnostic methodology, documentation, biometric evaluation, and clinical interpretation of the results. The choice of method depends on the indication and the test parameters; it may range from establishing influence of a drug on a clinically relevant, objectively measurable parameter (e.g., blood pressure) to the detection of effects that are only perceptible after long-term treatment and are based on subjective parameters such as patient's statements and physician's impressions (e.g., psychopharmacological drugs). The internationally standardized scales used for this purpose are validated in the patient's language (national language). Most of the relevant diagnostic procedures for a given disease are also suitable for demonstrating effects or efficacy of treatment; they are utilized selectively in clinical trials. Such efficacy parameters are used to determine the minimal and fully effective dosages. Suitable means of long-term recording methods provide information about the dosage intervals. The investigations should be performed on an adequate number of patients who are representative of the population to be treated later as regards the spectrum of the illness, age, sex, and race. The clinical situation comprises the common cold treated by the family doctor as well as the life-threatening arrhythmia of an acute myocardial infarction in the intensive care unit. Phase II investigations must comply with valid scientific standards which are defined in national and international guidelines for some groups of therapeutic agents. Furthermore, the value of the new therapy should be compared with established therapies, as regards recovery rates, incidence range of side effects, quality of life, secondary diseases, or rates of survival. Depending on the disease this information can be obtained from studies in as many as several hundred patients.

Phase III. The objectives of Phase III are to establish scientific proof of efficacy of the drug in question and to confirm the optimal dose in controlled randomized trials (see Section 2.2) in comparison to placebo and/or an established comparator drug. For these studies the considerations for clinical trials of Phase II also apply. In addition, special attention must be paid to drug safety and rarely occurring adverse drug reactions. Influences of accompanying therapy, interactions with other drugs and stimulants, effect on driving fitness and working ability should also be clarified. The investigations should extend over a period corresponding to the planned duration of application. In the case of drugs used for long-term treatment, the investigations should be performed for at least one year. The cross section of the diseased population should be as representative as possible and include widely differing forms of the disease. Since a large number of patients is required (several hundred to a thousand) the tests must be performed by a number of physicians according to a standard protocol. These multi-center studies are usually carried out on an international basis. A claim of prevention can usually be verified only after several thousand patients have been subjected to long-term therapy (several years) in multinational studies. Many of these intervention studies are therefore carried out after approval in Phase IV.

Phase IV. A new drug is granted marketing authorization by the relevant authorities on the basis of the results of pharmaceutical/quality data, preclinical testing, and clinical studies, particularly Phase I–III. Phase IV involves medicinal products that have already been approved. These studies are required to update information about the drug in keeping with scientific developments. Drug monitoring or pharmacovigilance (i.e., the collection and evaluation of adverse drug reactions) is often included in this phase. Phase IV begins with drug approval and ends with the discontinuation of sales.

2.2. Methods for Proving Effectiveness

As with all scientific experiments, clinical trials should provide generally valid evidence from a limited number of individual observations. The reproducibility of a result can be determined after repetition of the test and the probability of general validity can be calculated with statistical methods. Clinical tests must be therefore designed and evaluated on the basis of biostatistical criteria.

The basis for clinical trials for proof of efficacy is the therapeutic comparison in which the new drug is compared with an established drug (standard therapy) or placebo (dummy or sham therapy). Ideally, the two groups of subjects should differ only as regards the form of therapy they receive. In order to obtain a high degree of uniformity, factors affecting the groups (e.g., type, degree, and stage of illness; demographic characteristics) are defined by inclusion criteria. The structural uniformity can be further increased by stratification. Known interfering factors but also risk factors are eliminated by the definition of exclusion criteria. The patients must be randomly assigned to the treatment groups (randomization) to increase their comparability with respect to unknown factors and to lay the basis for the application of mathematical probability analysis.

In addition to structural uniformity, observational uniformity has to be established to eliminate subjective influences resulting from expectations or bias; this is achieved by performing blind studies. In a single-blind test only the patients are not informed about the treatment they receive. In a double-blind test the assignment of the patient to the test groups also remains unknown to the investigator. Coded test preparations with the same appearance and taste are randomly allocated to the test subjects. If a therapy is not to be compared with another treatment, a dummy or sham preparation (placebo) is used that looks exactly like the active preparation but does not contain an active agent. When comparing different galenic forms, blind testing is achieved by administering the two forms to both groups. Each form is administered as an active preparation and as a placebo (double dummy technique). For reasons of safety, provision for the decoding of the preparations must be made in case of an adverse event.

Noncontrolled, open studies do not generally permit scientifically valid conclusions to be made about the efficacy of a medication. A rare exception is if the drug has a striking effect on a previously incurable illness. These observation studies are, however, performed in other circumstances, e.g., if a drug is being administered for the first time, as a pilot experiment to check methodology, to establish dosages by dose titration, for long-term observation, and to document drug safety. Statistically valid results can only be obtained in comparative studies which are carried out in two main ways: as a comparison between different subjects in two groups (interindividual) or as a crossover experiment on a single subject (intraindividual).

Group (interindividual) comparison (parallel design) can be used to evaluate most medical problems, especially for dose confirmation and determination or comparison of drug effectiveness. However, this design is occasionally confronted with ethical limitations. If the treatment produces results that vary widely between different individuals or if small differences have to be verified, relatively large numbers of patients are required.

Variability and thus the number of observations can be significantly reduced if the study is conducted as an intraindividual comparison in the same patient. Each patient receives each of the test preparations in a random order for a sufficiently long period. This design can only be used in patients suffering from chronic illnesses with stable characteristics. Methodological problems sometimes restrict the use of this procedure, especially when the effect of one therapy continues into the subsequent treatment period (carry-over effect).

2.3. Adverse Drug Reactions

Very few therapeutic measures are completely free of risk. Adverse drug reactions occur unintentionally at the dosages normally used in diagnostics, therapy, or prophylaxis and are unpleasant or even harmful for the patient. Some adverse reactions can be expected from the pharmacological and toxicological properties of the drug and mainly occur after an overdose or as a result of hyperreactions in certain susceptible individuals (genetic origin). These reactions must be distinguished from immunological reactions which may result in a true drug allergy via sensitization or in a pseudoallergy via the release of mediator substances of the immune response. Since the value of a drug depends not only on its therapeutic benefit but also on its potential risks, proof of its tolerability must be substantiated with the same care as its efficacy. All adverse events are therefore recorded, irrespective of their causal relationship with drug administration because this relationship sometimes only becomes apparent after reoccurrence of these events. Each adverse event is recorded with respect to its symptomatology, onset, duration, and the degree of severity. Although the investigator evaluates the probability of a causal relationship, correct assignment can only be made later during retrospective overall evaluation. Standardized algorithms are available for determining the probability of a causal relationship. The entire range of adverse reactions of the test substance is compared with those of the control preparation, including the placebo. In addition to adverse drug reactions occurring during clinical development, physicians and pharmaceutical companies are obliged to report adverse drug reactions occurring during treatment with marketed drugs. The forms to be used, the time factor to be observed, and the authorities to be informed in the relevant notification system vary from country to country. The national centers report to the Collaborating Center of International Drug Monitoring of the WHO.

2.4. Good Clinical Practice

Quality standards and the assurance of their maintenance in drug production and in preclinical and clinical development are defined in guidelines such as Good Manufacturing Practice (GMP), Good Laboratory Practice (GLP), and Good Clinical Practice (GCP), see also Chapter 3. Compliance with these guidelines is subject to official control. The International Conference of Harmonization (ICH) has developed over the years binding guidelines on major aspects of drug development. The ICH guideline on good clinical practice (GCP) was adopted in 1997 by the regulatory authorities of the EU, Japan, and the United States. It represents the international ethical and scientific quality standard for clinical trials in humans. These guidelines have three main objectives:

1) The protection of the trial subjects
2) Definition of responsibilities
3) Data quality assurance

Written Standard Operating Procedures (SOP) ensure the adherence to these rules during the many activities involved in the planning, performance, monitoring, and evaluation of clinical studies.

Protection of Trial Subjects. The Declaration of Helsinki (last revision of Hong Kong 1989) of the World Medical Association gives guidance to physicians in biomedical research involving human subjects and is the accepted basis for clinical trial ethics.

Biomedical research must improve diagnostic, therapeutic, and prophylactic procedures and the understanding of the etiology and pathogenesis of disease. However, the right of the research subject to safeguard his or her integrity always has priority. Among the many principles to be regarded are careful benefit–risk assessment, adequate information and voluntary participation (informed consent), and insurance for the case of injury. A specially appointed committee independent of the investigator in conformity with the national laws (Ethics Committee, Institutional Review Board) should give its opinion as to whether these principles, legal requirements etc. are adequately taken in consideration.

Responsibilities of the persons involved in clinical trials (e.g., sponsor, monitor, investigator) are defined and must be laid down in written Standard Operating Procedures and in a trial protocol. All these measures serve the purpose of protecting the trial subjects/patients, complying with relevant national laws, and validating the data obtained from the study.

Data Quality Assurance in Clinical Trials. Since a new drug is granted approval by the authorities on the basis of the submitted data, considerable attention is paid in GCP guidelines to the careful collection, documentation, and validation of the data. The basic condition is that each result of a clinical report must be verifiable and must be retraceable to individual observations in individual patients. For pivotal studies, the Federal Code of Regulations of the United States specifies FDA inspections of the sponsor and of the testing location. The GCP guidelines of the EU also require audits at the site of the investigator and sponsor which are carried out either by an external institute or by an independent data quality assurance unit of the sponsor. The Japanese regulatory authority performs site and sponsor audits. Data audits are carried out on random samples and according to biometric criteria. The entire data handling must take into account the national data protection laws. The period of time for which the data must be kept at the individual locations is also regulated. Data must be kept for up to five years after termination of the marketing of the drug.

In Germany and some other countries, compliance with regulations and legal provisions that apply to the execution of clinical studies are subject to official surveillance.

2.5. Clinical Application Dossier

Although the formal requirements to be met by an application dossier vary in different countries, they are being harmonized. These requirements are contained in the "Notice to Applicants" in the EU and in the "Guideline for the Format and Content of the Clinical and Statistical Sections of a New Drug Application" in the United States. Japan has its own regulations. The official evaluation of a dossier and possible time limits are still regulated very differently at a national level. In addition to their own expertise, the authorities generally also make use of external professional commissions, advisory boards, or external expert opinions. After a drug has been granted approval and put on the market, it is still subject to official surveillance with respect to production quality, information of professional circles and patients, notification of adverse reactions, and promotional information.

3. Quality Control of Pharmaceutical Products

The manufacturers of pharmaceuticals products strive to achieve quality of their products both by controlling the quality of the outcome and also by manufacturing according to the strict rules and high standards laid down in the GMP guidelines.

Established quality standards are reached by agreement between the manufacturer and the health authorities prior to marketing pharmaceutical products. This agreement is similar to a legal contract in which the company producing and controlling the product is committed to meeting quality standards in future, i.e., prospectively, which are based on experience gained in the past while developing the product, i.e., retrospectively.

Quality standards are published by national regulatory authorities. Since the adoption of the first European Directive on pharmaceuticals (65/65/EEC) in 1965, the Member States of the European Union (EU) have been working towards harmonization of their requirements. These requirements have been published in "The Rules Governing Medicinal Products in the European Union" Volume III – Part 1 "Guidelines on the Quality, Safety, and Efficacy of Medicinal Products for Human Use" January 1996,

European Commission III/5380/96, to be found at http://dg3.eudra.org/eudralex/index.htm.

3.1. Manufacturing of Drug Substances—Test of Starting Materials

All starting materials have to be tested according to specifications applying suitable analytical procedures before manufacturing or synthesizing a drug substance. A chain of documents is generated as chemical synthesis proceeds. This documentation includes, amongst others, a written description of the process and appropriate production records, records of the raw materials used, records of batch numbers, records of the critical processing steps accomplished, and intermediate test results with meaningful standards (FDA Guide to Inspections of Bulk Pharmaceutical Chemicals, September 1991, reformatted in May 1994 with editorial changes.)

3.2. Control of Drug Substances: Impurities

Impurities, i.e. process and drug-related organic and inorganic impurities, may arise during the manufacturing process and/or storage of the drug substance. They may be identified or unidentified, volatile or nonvolatile, and include starting materials, by-products, intermediates, degradation products, reagents, and catalysts. Inorganic impurities are normally known and identified and include heavy metals, inorganic salts, and other materials, such as, filter aids and charcoal. The amount of impurities, identified or unidentified, must be limited to ensure the safety of patients. The procedure used to establish acceptable upper limits is described in detail in the "International Conference on Harmonisation of Technical Requirements for the Registration of Pharmaceuticals for Human Use" (ICH) *Harmonised Tripartite Guideline "Impurities in New Drug Substances"*, published in 1995 (to be found at http://www.ifpma.org/ich5q.html).

Solvents are used in the manufacture of drug substances and excipients or in the preparation of medicinal products. The levels of residual solvents in medicinal products should be no higher than can be supported by the safety data. Solvents that are known to cause unacceptable toxicity, such as, benzene, should be avoided in the production of drug substances, excipients, or medicinal products unless their use can be strongly justified in a risk-benefit assessment. Solvents associated with less severe toxicity, e.g., chloroform, methanol, or toluene, should be limited in order to protect patients from potential adverse effects. Ideally, less toxic solvents, such as acetone, ethanol, or ethyl ether, should be used where viable.

A method for classifying residual solvents by risk assessment is provided in the *ICH Guideline "Note for Guidance on Impurities: Residual Solvents" (CPMP/ICH/283/95), published 17 July 1997* (to be found at http://www.ifpma.org/ich5q.html).

3.3. Control of Drug Products: Impurities

As stated in an ICH guideline, "Impurities in drug products may be produced by degradation of the active ingredient or by reaction of the active ingredient with an excipient and/or with the immediate container/closure system." Consequently, the amount of degradation products allowed in a finished product should be such that the therapeutic efficacy of the product is not affected. The toxicity of degradation products must also be evaluated and assessed.

Levels of degradation products can be measured, for example, by comparison of an analytical response for a degradation product to that of an appropriate reference standard or to the response of the active ingredient itself. Methods for setting up limits for degradation products are described in detail in the *ICH Guideline "Note for Guidance on Impurities in New Medicinal Products" (CPMP/ICH/282/95)* of 6 November 1996 (to be found at http://www.ifpma.org/ich5q.html).

3.4. Validation of Analytical Procedures

The objective of analytical procedure validation is to demonstrate that the method in question is suitable for its intended purpose. The following analytical procedures have to be validated in particular:

- Identification tests intended to ensure the identity of an analyte in a sample by comparing a property of the sample to that of a reference standard
- Quantitative tests for the content of impurities
- Limit tests for the control of impurities
- Quantitative tests of the active moiety in the drug substance or drug product
- Dissolution testing for drug products
- Particle size determination for drug substances

Relevant information on validation of analytical procedures is summarized in *ICH Guideline "Validation of Analytical Procedures: Definitions and Terminology"* of November 1994 (to be found at http://www.ifpma.org/ich5q.html).

Guidance and recommendations on how to consider the various validation characteristics for each analytical procedure are provided in the *ICH Guideline "Note for*

Guidance on Validation of Analytical Procedures: Methodology" (CPMP/ICH/281/95) of 6 November 1996 (to be found at http://www.ifpma.org/ich5q.html).

3.5. Process Validation and Qualification

Since it is important to ensure that the defined quality standards of the pharmaceutical product are met, the manufacturers conduct "Process Validation and Qualification" on the product prior to marketing thereby establishing and documenting that:

- The facilities, equipment and processes have been designed in compliance with the requirements of the current GMP guidelines.
- The facilities and equipment have been built and installed as specified. This constitutes the Installation Qualification (IQ).
- The facilities and equipment are operated as specified. This constitutes the Operational Qualification (OQ).
- The facilities and equipment operate as specified and repeatedly and reliably produce a finished product of the requisite quality. This constitutes the Process Validation (PV).

Any aspect of the premises, facilities, equipment or processes, which may affect the quality of the product, either directly or indirectly, is subject to qualification and validation (*Convention for the Mutual Recognition of Inspections: "Principles of Qualification and Validation in Pharmaceutical Manufacture," Document PH 1/96*, January 1996).

3.6. Batch Release

Some types of pharmaceutical products, e.g., vaccines, sera, toxins, antitoxins, antigens, blood products, etc., have to be tested and released on a batch-by-batch basis by an independent state laboratory specializing in this particular analytical area. These products contain substances whose purity and potency cannot be measured by chemical methods alone. The batch release system is used in many countries, and the independent laboratory responsible for testing in Germany is the Paul Ehrlich Institut. In France, it is the Institut Pasteur.

3.7. Stability Testing

Stability studies are a major expense when developing new products. This applies in particular to drug products that are to be marketed in several strengths and package types. Multiple strengths and package types combined with multiple batches, various storage conditions, test parameters, and test intervals require a great number of samples to be tested at considerable cost. Further to this, the different requirements of the regulatory agencies must be taken into account. As a result, an enormous amount of stability testing, much of it redundant, was performed by multinational pharmaceutical companies seeking approvals in several countries. Therefore, the compilation of a common set of stability requirements for marketing authorizations was considered to be a top priority for the pharmaceutical industry when the International Conference on Harmonisation of Technical Requirements for the Registration of Pharmaceuticals for Human Use (ICH) was formed in 1991. Regulators and pharmaceutical industry representatives from the EU, Japan, and the United States with observers from the Canadian and Swiss Health Authorities and the WHO chose stability testing as one of the first issues to be discussed and harmonized. An ICH Guideline on Stability Testing was subsequently developed and published in October 1993, after which it was adopted throughout the ICH region [6]. This guideline describes the stability testing requirements for a registration application within the "ICH territory": EU, Japan, and the USA *(ICH Harmonized Tripartite Guideline "Stability Testing of New Drug Substances and Products")* of 27 October 1993, as amended (to be found at http://www.ifpma.org/ich5q.html).

As this guideline was limited to new active substances and products for human use sold in ICH countries in which climatic conditions are moderate, the WHO developed its own guideline to include established substances, and also hot or hot and humid climatic zones. The WHO guideline was published in 1996 (*"Guidelines for stability testing of pharmaceutical products containing well-established drug substances in conventional dosage forms" WHO Technical Report Series, No. 863, 1996*) [7]. Meanwhile, a variety of additional ICH stability guidelines have been developed, for example:

- *Photostability Testing* and a guideline concerning the *Stability of Biotechnology Products*, http://www.ifpma.org/ich5q.html
- *Reduced Stability Testing Plan—Bracketing and Matrixing (CPMP/QWP/157/96)*, dated 22 October 1997
- *Note for Guidance on Stability Testing of Existing Active Substances and Related Finished Products* (CPMP/QWP/556/96)

In the USA, the Code of Federal Regulations (CFR) provides guidance on stability testing, e.g., the GMP Guideline for Pharmaceutical Products, 21 CFR 211, contains a subpart 166 describing stability testing requirements, and 21 CFR 314.50(d)(1)(ii)(a) requires that the stability of a drug substance and product be studied after approval as per the stability protocols approved [8].

3.8. Documentation

Format of Registration Documentation. Regulatory authorities in various regions require applicants to submit marketing authorization applications according to their national guidelines and regulations. In *Europe*, the format for presenting quality related documents is outlined in detail in *The Rules governing Medicinal Products in the European Union, Volume 2B: Notice to Applicants*, January 1997, to be found at http://dg3.eudra.org/eudralex/index.htm. The format for applications in the *USA* is described in the *Code of Federal Regulations, Title 21, Chapter I, Subchapter D, Part 314, Subpart B* (http://www.fda.gov). In *Japan*, the Ministry of Health and Welfare regularly publishes updates of their Technical Requirements for New Drug Applications, updated versions are available from Yakuji Nippo, Ltd., Japan, editorial supervision by the Pharmaceuticals and Cosmetics Division, Pharmaceutical Affairs Bureau, Japanese Ministry of Health and Welfare.

The Common Technical Document — Quality. The ICH topic "Common Technical Document (CTD) Quality" proposes a common structure for the chemical/pharmaceutical section of dossiers for marketing authorization applications. The final ICH Guideline (expected to be released in November 2000) will be a major achievement and the result of a concerted effort on the part of the reviewers working in the health authorities and the quality control/regulatory affairs departments in the pharmaceutical companies. The aim of the Common Technical Document project is to harmonize both the structure (or Table of Contents) and data of regulatory dossiers submitted to the health authorities. It encompasses all documents except for administrative application forms and raw data. It includes Common Tabulated Summaries based on the structure of the documentation, as well as Common Written Summaries.

The following data have to be supplied to the regulatory authorities when submitting an application for marketing authorization:

Drug Substance

General information	
Nomenclature	Recommended INN[a], pharmacopoeial name, if relevant, chemical name(s), other name(s), company or laboratory code, regional name, national name, e.g., BAN[b], USAN[c], JAN[d], and CAS registry number.
Structure	NCE: structural formula, including relative and absolute stereochemistry, the molecular formula, and relative molecular mass.
	Biotech: schematic amino acid sequence indicating glycosylation sites where available and relative molecular mass.
General properties	Brief description of the physicochemical and other relevant properties of the drug substance.

Drug Substance

Manufacture	
Manufacturer(s)	Name, address, and responsibility of each manufacturer and each site or facility involved.
Manufacturing process description and process controls	Information on manufacturing route including, for example, a flow diagram and a description of the synthetic process. Also, a description of process parameters such as temperature, pressure, pH, and time. Tests at the appropriate control points. Additionally for Biotechnology: description of the manufacturing process including the cell culture, harvest, isolation, purification, concentration, filling, storage and shipping conditions.
Control of materials	Starting materials, solvents, reagents, catalysts, and any other materials used in the manufacture indicating where in the process each material is used. Acceptance criteria and testing. Additionally for biotech products produced from cell banks: description of genetic construct for recombinant cell substrates, development and stability, and cell bank system, and materials used for cell banking, with information on their origin.
Control of critical steps and intermediates	Critical steps: tests and acceptance criteria, with justification including experimental data, performed at critical steps of the manufacturing process to assure that the process is controlled. Intermediates: specifications and analytical procedures, if any, on intermediates including validation of analytical procedures, where appropriate.
Characterization	
Elucidation of structure and/or biological characterization	NCE: Confirmation of structure based on synthetic route and spectral analyses, for example. Information on the potential for isomerism and the identification of stereochemistry. Biotech: details on primary, secondary and higher order structure and information on biological activity, purity, and immunochemical properties (where relevant).
Impurities	
Control of Drug Substance Specification Analytical Procedures Validation Batch Analyses Justification	
Reference standards or materials	Information on the reference standards or reference materials used for testing of the drug substance and drug product.
Container closure system	description of the container closure system suitable for the storage and shipment of the drug substance, including components, composition, and specifications.

Drug Substance

Stability

Stability summary and conclusions	Summary of the types of studies conducted, protocols used, and the results obtained; conclusions with respect to storage conditions and retest date or expiry period.
Stability data	Results of the stability studies conducted in an appropriate format such as tabular, graphical, narrative; information on analytical procedures used and their validation.

[a]INN = International Non-Proprietary Names. [b]BAN = British Approved Names. [c]USAN = United States Adopted Names. JAN = Japanese Approved Names.

Drug product

Composition	Description of pharmaceutical form, list of all components, and their amounts on a per unit basis, their function and reference to their specifications, type of container and closure used.
Pharmaceutical development report	Description and, if necessary, justification of differences between clinical formulation(s) and the formualtion proposed for marketing.
	Pharmaceutical development studies on key parameters which might have an influence on the performance of the drug product.
	NCE: examples include physicochemical characteristics of the drug substance which affect drug product dissolution, e.g. polymorphic form, particle size.
	Biotech: an example is the selection of a stabilizing excipient.
	Identification and discussion of critical steps in the manufacturing process and justification of non-standard sterilization process.
	Justification of overages, if any.
	Microbiological attributes, suitability of container closure system.
Manufacture	
Manufacturer(s)	Name, address, and responsibility for each manufacturer and each site and facility involved.
Batch formula	Names and amounts of all components to be used in the manufacturing process.
Manufacturing process description and process controls	Information on manufacturing process including, for example, a flow diagram, a description of the process, and type of equipment; a description of process parameters and their control.
Control of intermediates	Critical steps: tests and acceptance criteria, with justification, including experimental data, performed at critical steps.
	Intermediates: specifications and analytical procedures, if any, for intermediates including validation of analytical procedures, where appropriate.
Process validation or evaluation	Process evaluation or validation protocol, based on experimental data on pilot or production scale batches, or validation data for production scale batches, where appropriate.

Drug product	
Control of excipients Specifications Analytical procedures Validation Justification	
Control of drug product Specifications Analytical procedures Validation Batch analyses Justification	
Container closure system	Description of the container closure system suitable for the storage and shipment of the drug product, including components, composition, and specifications.
Stability Stability summary and conclusions	A summary discussing the types of studies conducted, protocols used and the results obtained; conclusions with respect to storage conditions and expiry period, and, if applicable, in-use storage conditions and expiry period.
Stability data	Results of the stability studies conducted in an appropriate format such as tabular, graphical, narrative; information on analytical procedures used and their validation.

Drug Master Files. Drug Master Files (DMFs) are documents that are submitted to the U.S. FDA providing confidential detailed information on the facilities, processes, or articles used in the manufacturing, processing, packaging, and storage of medicinal products intended for human use. DMFs allow parties other than the holder of the DMF to reference information not disclosed to them (Center for Drug Evaluation and Research, FDA *Guideline for Drug Master Files,* September 1989, to be found at http://www.fda.gov).

In the EU, the use of Drug Master Files is restricted to active drug substances only (*European Drug Master File Procedure for Active Substances, The Rules governing medicinal products in the EU, Vol. III – Part 1,,* January 1996 to be found at http://dg3.eudra.org/eudralex/index.htm).

3.9. Variations — Changes

Drug manufacturers are required to amend and update manufacturing and control methods within the life-time of a medicinal product. In Europe, this is detailed in Council Directive 65/65/EEC, Article 11. Since the level of scientific development is constantly subject to change, modifications in a product's composition, manufacturing methods and control methods are unavoidable, and even desirable as they improve quality. However, every change that is made entails revision and amendments to the documentation. Sometimes approval has to be obtained from the health authorities before the proposed changes can be implemented (Commission Regulation (EC) No. 541/95, dated 10 March 1995, as amended by Commission Regulation (EC) 1146/98, EC Commission Guideline on Dossier Requirements for Type I Variations (III/5783/93), *Note for Guidance on Stability Testing for a Type II Variation to a Marketing Authorisation, CPMP/QWP/576/96*, dated 22 April 1998).

In the USA, regulation 21 CFR 314.70 provides a guideline for changes subject to FDA approval prior to implementation and for changes that can be reported to the FDA in an annual report. In addition, the FDA has developed various guidelines for testing requirements in the case of post-approval changes under the general titles of Scale-Up and Post-Approval Changes (SUPAC) and Bulk Active Chemicals Post-Approval Changes (BACPAC).

3.10. Change Control

To ensure that the systems concerned are continuously validated, the manufacturer must monitor all changes with respect to facilities, materials, equipment and processes used in the manufacture of drug substances and medicinal products. This commitment should be declared in the Validation Master Plan. As part of its Quality Management System, the manufacturer must establish a defined, formalized change control procedure (*Convention for the Mutual Recognition of Inspections: "Principles of Qualification and Validation in Pharmaceutical Manufacture," Document PH 1/96*, January 1996).

3.11. Compliance

Manufacturers must make sure that the documentation approved by the health authorities complies with the methods applied in the manufacturing and quality control of the drug substances and products.

3.12. Changes Requiring New Applications — Line Extension

Certain changes to a marketing authorization are considered to fundamentally change the terms of the authorization in question and therefore cannot be considered as a variation. For example:

– Adding or removing of one or more active ingredient(s)
– Changing the amount of active ingredient(s)

- Replacing the active ingredient(s) with a different salt/ester complex/derivative (with the same therapeutic moiety)
- Replacing an isomer, or using a different mixture of isomers, replacing a mixture with a single isomer (e.g., replacing a racemate with a single enantiomer)
- Replacing a biological substance or a biotechnology product with one of a different molecular structure; modification of the vector used to produce the antigen/source material, including a master cell bank from a different source
- Using a new ligand or coupling mechanism for a radiopharmaceutical

An abridged application for a new marketing authorization must be submitted for these changes and, in certain cases, the existing marketing authorization may have to be withdrawn. Otherwise, the new product is regarded as a "line extension" to the existing formulations and dosage forms.

3.13. Inspections

Manufacturers of both drug substances and drug products are inspected by the national authorities. The inspectors must establish whether the control methods are applied and the facilities adequate, the personnel properly trained and manufacturing procedures followed.The inspectors also have to check whether cross-contamination can be excluded, i.e. whether the substance or product is subject to contamination by material stored or used in the same building. Other factors to be considered are:

- The degree of exposure of the material to adverse environmental conditions
- Relative ease and thoroughness of clean-up
- Sterile vs. non-sterile operations

(*FDA Guide to Inspections of Bulk Pharmaceutical Chemicals*, September 1991, reformatted May 1994, with editorial changes). The inspections may take place during the clinical development of a new substance, prior to granting marketing authorization (pre-approval inspection), prior to the first launch of a drug product, or during the marketing phase.

3.14. The Mutual Recognition Agreement (MRA)

To cut down on the need for GMP inspections by regulatory authorities in foreign countries, agreements have been negotiated between the EU and the USA, the EU and Canada, the EU and Australia, the EU and New Zealand, and Switzerland and Japan. The aim is to achieve mutual recognition of GMP inspections conducted by local inspectors, with the results documented in an "Inspection Report".

The main advantage of an MRA would be the assurance that imported pharmaceutical products meet GMP requirements without resource expenditures on the part of the authorities of the importing country. Another important benefit of an MRA would be a reduction of costs for the pharmaceutical manufacturer, mainly as a result of inspection reports being accepted by more than one regulatory authority. This would eliminate the need for duplication of inspections and imply fewer inspections of the manufacturing sites. An MRA could also reduce regulatory review times as it would increase the efficiency of pre-marketing approval inspection activities. The current status of the MRAs between the EU and other countries is reported on the web site of the European Commission's DGIII (http://dg3.eudra.org/).

Agreement Between the USA and the EU. On June 20, 1997, the United States and the EU concluded negotiations on an agreement titled "Agreement on Mutual Recognition between the United States of America and the European Community". The MRA was signed on 18 May 1998 and came into force on 1 December 1998. The transition period will end on 30 November 2001. The complete text of this MRA is available on the Internet at the FDA's web site http://www.fda.gov. It has also been published in the US Federal Register April 10, 1998 (Volume 63, Number 69) Page 17744–17771.

3.15. Pharmaceutical Inspection Convention (PIC)

Another initiative to cut down on foreign GMP inspections is the "Convention for the Mutual Recognition of Inspections in respect of the Manufacture of Pharmaceutical Products" (shortened to Pharmaceutical Inspection Convention or PIC). This convention was signed in October 1970 by all EFTA countries at the time (i.e., Austria, Denmark, Finland, Iceland, Liechtenstein, Norway, Portugal, Sweden, Switzerland, and the United Kingdom). Since coming into force the convention has been extended to include Hungary, Ireland, Romania, Germany, Italy, Belgium, France, and Australia.

Similar to the MRAs mentioned above, the "PIC provides that an inspection of a pharmaceutical manufacturer, whose intention it is to export products in one of the Contracting States, is carried out by its national authority and shall be regarded and assessed by the health authority of the country of importation as if it had been carried out by its own inspectors."

A Committee of Officials supervises the operation of the PIC concerning the technical aspects of the manufacturing and control of pharmaceuticals. In the meantime, the convention's original scope has been extended to the training of laboratory staff, the mutual recognition of the quality standards of inspectorates, and also to cooperation concerning blood products, radiopharmaceuticals, medicinal gases, etc. Working groups meet regularly to develop guides, guidelines, and recommendations. Also seminars on technical subjects are organized, and the proceedings of these seminars and experts' meetings are published (http://www.efta.int/structure/EFTA/efta-sec.cfm).

3.16. Pharmaceutical Inspection Cooperation Scheme (PIC/S)

Some of the contracting states within the PIC are also Member States of the European Union with the result that certain rules and regulations were incompatible. Therefore, a new arrangement, which may eventually replace the PIC, called "Pharmaceutical Inspection Cooperation Scheme" (or PIC Scheme) has been set up. This new scheme is meant to be less formal and more flexible concerning cooperation between the inspectorates of the present PIC contracting states. Also the PIC Scheme allows for inspectorates of other countries to join. In addition to the recognition of foreign inspections, the PIC Scheme is in harmony with the main tasks of the PIC, namely, "the exchange of information, networking and confidence building between the national inspection authorities, the development of quality standards systems, the training of inspectors and other related experts and its work towards the global harmonisation of Good Manufacturing Practice (GMP)".

The contracting members of the new PIC Scheme are currently the following inspectorates: Australia, Canada (since 1999), the Czech Republic, Denmark, Finland, Hungary, Iceland, Ireland, Liechtenstein, the Netherlands, Norway, Romania, the Slovak Republic, Sweden, Switzerlan, and the United Kingdom. The remaining PIC inspectorates have confirmed their intention to join the PIC Scheme sooner or later. In addition, the following applications for membership are being considered: Estonia, Latvia, Singapore, South Africa, and Taiwan.

3.17. Pharmacopeias

National Pharmacopeias. In order to control drug quality, lists of drugs, their preparation and uses were developed and published in a "pharmacopeia". The earliest of these was probably the "New Compound Dispensatory" issued in 1498 by the "Florentine Guild of Physicians and Pharmacists". Various pharmacopeias were generated in the ensuing centuries: Barcelona (1535), Nürnberg (1546), Cologne (1565), Rome (1583), and London (1618). The first attempt to harmonize different standards and monographs was made by the Pharmacopeia Committees of London, Edinburgh and Dublin. As the result of their efforts, the first "British Pharmacopeia" documenting the agreement on harmonized international quality standards was published in 1864 [9].

International Pharmacopeia. The project of developing an "International Pharmacopeia" started in 1929 with an agreement signed by 19 countries. The next step was not made until 1948, when the first World Health Assembly established an expert committee to adapt and harmonize the existing monographs of major pharmacopeias, and also to develop new standards. As the committee also kept developing countries in mind, some of the drug substances included in the first edition of the International Pharmacopeia were for the treatment of tropical diseases. Analytical methods are adapted for the practice of quality control testing in countries that need inexpensive methods, such as on-the spot identification of counterfeit pharmaceuticals (WHO Drug Information Vol. 13, No. 1, 1999, pp. 13–14).

European Pharmacopeia. The need to unify the various national pharmacopeias in Europe arose as a result of the EU proposal to allow the free movement of medicinal products. Both public health and international trade require common manufacturing and quality control standards concerning drug substances for human and veterinary use. It also requires these standards be updated to stay in line with scientific progress. In 1964, the "European Pharmacopeia Convention" was signed by six Member States of the Council of Europe. It is open to European countries (members and, under certain conditions, non-members of the Council of Europe), as well as to an international European organization. Observer status is also possible.

Currently 27 parties are members of the Convention with 25 member states belonging to the Council of Europe (including the 15 EU member states): Austria, Belgium, Croatia, Cyprus, Czech Republic (since June 1998), Denmark, Finland, the former Yugoslav Republic of Macedonia, France, Germany, Greece, Iceland, Ireland, Italy, Luxembourg, the Netherlands, Norway, Portugal, the Slovak Republic, Slovenia, Spain, Sweden, Switzerland, Turkey and the United Kingdom; plus one state which is not a member of the Council of Europe, Bosnia-Herzegovina; and the Commission of the European Union.

A further 17 parties participate as observers: Albania, Bulgaria, Estonia, Hungary, Latvia (since 1998), Lithuania, Poland, Romania and the Ukraine, as well as the non-European countries Australia, Canada, China, Malaysia, Morocco, Syria, Tunisia, and the WHO. Some of these observers have implemented all or parts of the European Pharmacopoeia in their national pharmaceutical regulations.

In marketing authorization application dossiers submitted to regulatory authorities in the EU Member States, the monographs of the European Pharmacopoeia are obligatory. This requirement is based on Directive 75/318/EEC of 20 May 1975 of the Council of the Communities. Cooperation between the Council of Europe and the EU was further advanced on 26 May 1994, when the "European Pharmacopeia Secretariat" took over additional objectives, e.g., setting up a European network of laboratories involved in the quality control of pharmaceutical products. Consequently, the "European Pharmacopeia Secretariat" changed its name to the "European Department for the Quality of Medicines" (EDQM). This European network, called the Official Medicines Control Laboratories (OMCLs), is open not only to EU countries but also to members and observers of the European Pharmacopoeia Commission. Its main objectives are to achieve the mutual recognition of tests carried out at the national level for countries that belong to the EU, and the sharing of expertise, standardization and international collaboration for the other countries.

The European Pharmacopoeia is one of the cofounders of the Pharmacopoeial Discussion Group (PDG) that was set up in 1990 with Japan and the United States. Monographs and general methods of analysis proposed by national associations of manufacturers of pharmaceutical products are selected for harmonization in the three pharmacopoeias. The European Pharmacopeia Commission has set the goals of rapidly adding to the number of existing European monographs (1500 so far), reducing the time taken to compile them, and, if necessary, generating monographs on recent substances still protected by patent. The European Pharmacopeia (EP or Ph. Eur.) features more than 250 general analytical methods on which the specifications prescribed in the monographs are based. The methods range from physical, chemical and biological to microbiological, phytochemical, and so on. The EP also includes technical methods used to test the functional parameters of pharmaceutical forms such as the dissolution test for tablets.

The texts are updated at regular intervals to take into account scientific progress and any changes in the commercial products. The principles of monograph compilation are adapted to accommodate regulatory requirements in the public health area (licensing, control and inspection authorities), industrial constraints and recent advances in technology and science (http://www.pheur.org/).

The monographs and methods published in the European Pharmacopoeia are normally transferred to the national pharmacopoeias, e.g., the *Deutsches Arzneibuch* (DAB), the British Pharmacopoeia (BP), the Pharmacopeé Francaise, the Farmacopeia Ufficiale della Republica Italiana.

3.18. The EP Certificate of Suitability

Global expansion of international trade has given rise to a wide range of qualities of medicinal substances within the territory covered by the Convention on the Elaboration of a European Pharmacopoeia. The great variety of supply sources has made it necessary to take additional measures to protect public health so that the European Pharmacopoeia can continue to play its role as a reliable reference in defining the quality of medicines. It is essential to have a procedure that allows the drug manufacturer to prove that the purity of the substance is suitably controlled by the EP monograph. The procedure described in Resolution AP-CSP (98) 2 of the Council of Europe meets this need with a certificate of suitability of EP monographs issued by the EDQM.

As specified in Resolution AP-CSP (98) 2, the manufacturer has to submit a full dossier on the manufacturing method for the substance and its associated impurities for validation. The dossier is assessed according to a procedure that guarantees its confidentiality and, if the information received is pertinent, a certificate of suitability is issued. This certificate can be included in the dossier for marketing authorization application for all medicinal products containing the substance in question (http://www.pheur.org/).

4. References

[1] "Principles for Pre-clinical Testing of Drug Safety" *WHO Tech. Rep., Ser.* **341,** 1966. "Principles for Testing of Drugs for Teratogenicity" *WHO Tech. Rep. Ser.* **364,** 1967. "Principles for Testing of Drugs for Carcinogenicity" *WHO Tech. Rep. Ser.* **426,** 1969. "Evaluation and Testing of Drugs for Mutagenicity" *WHO Tech. Rep. Ser.* **482,** 1971.

[2] O. Eichler (ed.): "Erzeugung von Krankheitszuständen durch das Experiment," in *Handbook of Experimental Pharmacology,* vol. **XVI**, Parts 1–15, Springer Verlag, Berlin-Heidelberg-New York, 1962–1966.

[3] R. A. Turner: *Screening Methods in Pharmacology,* vol. **1**, vol. **2**, Academic Press, New York-London, 1965, 1971.

[4] L. Ther: *Grundlagen der experimentellen Arzneimittelforschung,* Wissenschaftliche Verlagsges. mbH, Stuttgart, 1965.

[5] J. Ashby: "The Prospects for a Simplified and Internationally Harmonized Approach to the Defection of Possible Human Carcinogens and Mutagens," *Mutagenesis* **1** (1986) 3–16.

[6] P. Jeffs: "The Importance of Stability Testing in the Registration of Pharmaceutical Products" in D. Mazzo (ed.): *International Stability Testing,* Interpharma Press, Inc., Buffalo Grove, IL 1999.

[7] S. Kopp-Kubel, M. Zahn: "The WHO Stability Guideline" in D. Mazzo (ed.): *International Stability Testing,* Interpharma Press, Inc., Buffalo Grove, IL 1999.

[8] D. Shah: "Postapproval FDA Stability Requirements" in D. Mazzo (ed.): *International Stability Testing,* Interpharma Press, Inc., Buffalo Grove, IL 1999.

[9] A. Cartwright, B. Matthews (eds.): *Pharmaceutical Product Licensing—Requirements for Europe,* Ellis Horwood Ltd., Chichester 1991.

Veterinary Drugs

RICHARD SAMS, The Ohio State University, Dept. of Veterinary Clinical Sciences, Columbus, Ohio 43210, United States

1.	Antibiotics and Antibacterial Drugs	2158	4.	Parasiticides	2173
1.1.	Sulfonamides	2158	4.1.	Avermectin Derivatives	2173
1.2.	β-Lactam Antibiotics	2160	4.2.	Benzimidazole Derivatives	2174
1.3.	Tetracyclines	2160	4.3.	Imidazothiazole Derivatives	2175
1.4.	Aminoglycosides/Aminocyclitols	2162	5.	Anti-inflammatory Drugs	2175
1.5.	Macrolide Antibiotics	2164	5.1.	Nonsteroidal Anti-inflammatory Drugs	2175
1.6.	Amphenicols	2167	5.2.	Steroidal Anti-inflammatory Drugs	2177
1.7.	Tiamulin	2168	6.	Respiratory Stimulants	2179
1.8.	Peptide Antibiotics	2169	7.	Anesthetics	2180
1.9.	Ionophore Antibiotics	2170	8.	Narcotic Agonists and Antagonists	2182
2.	Antimycotics	2171	9.	Antihistamines	2184
3.	Coccidiostats	2172	10.	References	2185

Drugs used in veterinary medicine include those originally developed for use in human medicine and later investigated for the treatment of animals. Such drugs are often used to treat horses and companion animals such as dogs and cats but are generally not used to treat foodproducing animals unless drug residues in food and other food safety issues have been investigated. Human drugs used in animals include various antibiotics (e.g., penicillins, cephalosporins, aminoglycosides, and tetracyclines), drugs affecting the cardiovascular system (e.g., digoxin, lidocaine, furosemide), anticonvulsant drugs (e.g., phenobarbital and phenytoin), anesthetics (e.g., halothane, pentobarbital), and analgesics (e.g., morphine, oxycodone, phenylbutazone, and aspirin).

Other drugs have been developed specifically for use in veterinary medicine. Some of these are used to prevent or treat parasitic infections unique to animals. Other veterinary drugs are used to increase weight gain in cattle, sheep, and pigs to decrease production costs and shorten the time until the animal is slaughtered for meat. Other veterinary drugs are close structural analogues of and are used similarly to drugs used in human medicine. For example, the animal tranquilizer acepromazine is a close structural analogue of chlorpromazine and other human tranquilizers.

The following chapters describe some of the veterinary drugs approved for use in the prevention or treatment of various diseases of animals. Human drugs that are not specifically approved by a regulatory agency for use in veterinary medicine are not included.

1. Antibiotics and Antibacterial Drugs

1.1. Sulfonamides

Sulfonamides are derivatives of 4-aminobenzenesulfonamide. These substances are weak acids with differing degrees of water solubility.

Sulfonamides are broad-spectrum bacteriostatic drugs, active against a wide range of gram-positive and gram-negative bacteria. Their antibacterial activity is increased by various diaminopyrimidines, in combination with which they are frequently marketed. Sulfaquinoxaline has both antibacterial and coccidiostat activity.

Sulfonamides are believed to act by inhibiting the conversion of p-aminobenzoic acid to dihydrofolic acid in sensitive bacteria. The diaminopyrimidines such as trimethoprim and ormetoprim inhibit conversion of dihydrofolic acid to tetrahydrofolic acid. Inhibition of both steps by administration of a combination of these two groups of drugs is common and leads to greater antibacterial efficacy than either drug alone. Some combinations are bactericidal to certain organisms.

Sulfonamides

Sulfadiazine

Sulfamethazine

Sulfaquinoxaline

Sulfadimethoxine

Sulfamerazine

Sulfapyridine

Sulfadoxine

Sulfathiazole

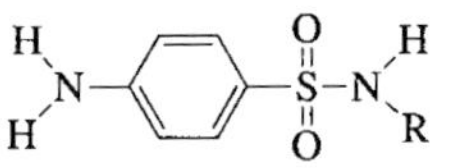

Sulfamethazine [*57-68-1*] (sulfadimidine) is used in the treatment of bacterial enteritis and to maintain weight in the presence of atrophic rhinitis in pigs. In combination with chlortetracycline it is used to maintain weight gains in beef cattle suffering from respiratory diseases such as shipping fever. In combination with chlortetracycline and penicillin it is used to promote growth and increase feed efficiency in swine weighing up to 34 kg. Sulfamethazine is frequently found as an illegal residue in meats in the United States.

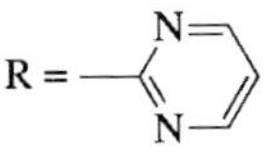

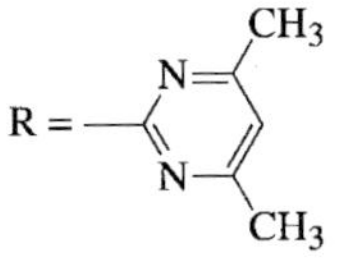

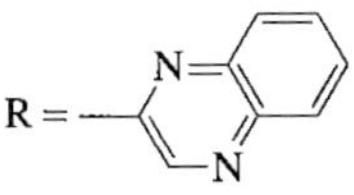

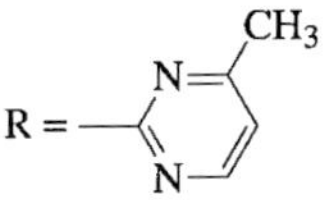

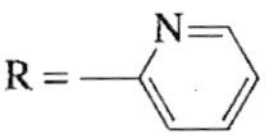

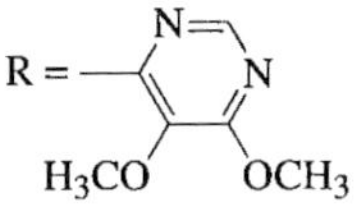

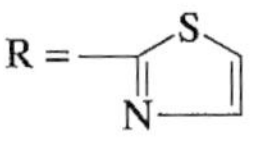

Sulfaquinoxaline [*59-40-5*] is used to treat or control outbreaks of coccidiosis caused by *Eimeria tenella, E. necatrix, E. acervulina, E. maxima,* or *E. brunetti* in chickens; by *E. meleagrimitis* or *E. adenoeides* in turkeys; and by *E. bovis* or *E. zurnii* in cattle. It is also used to treat or control fowl cholera caused by *Pasteurella multocida,* as well as fowl typhoid caused by sensitive organisms.

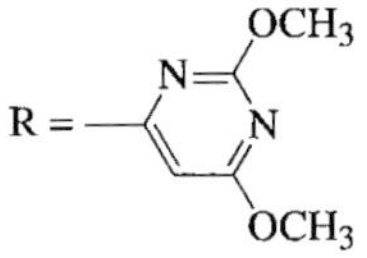

Sulfadimethoxine, sulfapyridine, sulfamerazine, sulfadoxine, and sulfadiazine are other sulfonamides used in veterinary medicine to treat infections caused by various bacteria including streptococci, staphylococci, escherichia, salmonella, klebsiella, proteus, or shigella organisms.

1.2. β-Lactam Antibiotics (see also → Antibiotics)

Penicillins are semisynthetic or synthetic derivatives of 6-aminopenicillanic acid produced by *Penicillium chrysogenum*, whereas cephalosporins are derivatives of 7-aminocephalosporanic acid, originally produced by cultures of *Cephalosporium acremonium*. Modifications of the side chains of both compounds have produced antibiotics with different antimicrobial spectra and pharmacokinetic properties.

β-Lactam antibiotics are bactericidal drugs with gram-positive activity in the older agents and broad-spectrum or gram-negative activity in some newer agents. Organisms producing β-lactamase hydrolyze the β-lactam ring of the penicillins and render them inactive. Clavulanic acid is a β-lactamase inhibitor that is marketed in combination with amoxicillin.

β-Lactam antibiotics inhibit cell wall synthesis in susceptible organisms. They inhibit transpeptidases, which prevents cell wall synthesis and eventually causes lysis of the bacterial cell.

The most commonly used penicillin in veterinary medicine is penicillin G. It is often formulated in combination with procaine or benzathine to produce a poorly soluble salt that dissolves slowly after intramuscular administration. Oral administration of penicillin results in relatively poor bioavailability due to hydrolysis in the acid environment of the stomach.

Semisynthetic penicillins such as ampicillin [*69-53-4*], amoxicillin [*26787-78-0*], hetacillin [*3511-16-8*], and ticarcillin are used in veterinary medicine to treat infections caused by susceptible organisms. Amoxicillin is a broad-spectrum antibiotic active against many gram-positive and gram-negative organisms, except those producing β-lactamase.

Cephalosporins approved for use in veterinary medicine include cefadroxil and ceftiofur. Cefadroxil [*66592-87-8*] is a first-generation cephalosporin with good activity against gram-positive bacteria and some activity agains gram-negative organisms. Ceftiofur [*80370-57-6*] is active against gram-positive and gram-negative bacteria, including those that produce β-lactamase.

1.3. Tetracyclines (see also → Antibiotics)

Chlortetracycline and oxytetracycline are produced by *Streptomyces aureofaciens* and *S. rimosus*, respectively. Tetracycline is produced from chlortetracycline, and doxycycline is a semisynthetic derivative. Many of the tetracyclines form chelates with cations such as calcium and magnesium, rendering them inactive. Chlortetracycline has a chlorine atom at C-7 and oxytetracycline bears a hydroxy group at C-5, compared to tetracycline. Doxycycline has a hydroxy group and a hydrogen atom at C-5 and a methyl group at C-6, compared to tetracycline. All are weak bases that form salts with strong acids such as hydrochloric acid.

The tetracyclines are broad-spectrum antibiotics that are bacteriostatic in vitro. Good activity against various rickettsiae such as those responsible for Potomac Horse Fever has been demonstrated. Bacterial resistance often develops after exposure of bacteria to the tetracyclines by plasmid-mediated interference with transport of the drug to the ribosome, increased rate of removal from the ribosome, or decreased effect on the ribosome.

The tetracyclines inhibit bacterial protein synthesis by reversible binding to the 30 S ribosomal subunit of sensitive bacteria.

Tetracycline [60-54-8] is marketed in oral formulations including tablets and a soluble powder for dissolution in drinking water and milk. It is used in the treatment of bacterial enteritis caused by *Escherichia coli* and bacterial pneumonia caused by various species of *Pasteurella, Hemophilus*, and *Klebsiella*.

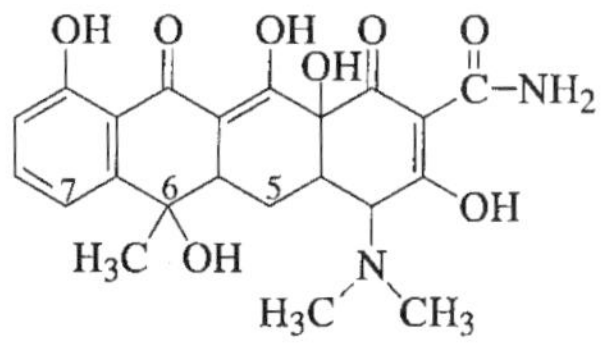

Tetracycline

Oxytetracycline [79-57-2] is marketed in a number of formulations including medicated feeds, parenteral products, and ointments. Terramycin Soluble Powder contains oxytetracycline hydrochloride and is used to treat bacterial enteritis in several species, shipping fever complex in cattle, liver abscesses in cattle, atrophic rhinitis in swine, leptospirosis in swine, enterotoxemia in sheep, as well as bluecomb, air-sac infection, sinusitis, fowl cholera, and hexamitiasis in poultry.

Chlortetracycline [57-62-5] is marketed in a number of products for veterinary use. As Aureomycin 2G Type C Medicated feed for swine, beef and dairy cattle, sheep, and horses it is used to promote growth and improve feed efficiency and as an aid in the prevention of bacterial swine enteritis. It is also employed in the treatment of bacterial enteritis and in reducing the spread of leptospirosis in swine. It is utilized in cattle to reduce losses due to shipping fever complex and for prevention of anaplasmosis, bacterial diarrhea, and bacterial foot rot. Aureomycin 70 Type A Medicated Article is used in the manufacture of various animal feeds. It is used to treat chronic air-sac infection and bluecomb and to prevent synovitis in chickens.

Doxycycline [17086-28-1] is used to treat infections caused by susceptible bacteria; in contrast to the other tetracyclines it is not usually administered as a feed additive.

1.4. Aminoglycosides/Aminocyclitols (see also → Antibiotics)

The aminoglycoside antibiotics are produced by species of *Streptomyces* ("mycins") and *Micromonospora* ("micins"). They consist of one or more aminosugars attached to an aminocyclitol ring through a glycosidic bond. The aminocyclitol is either streptidine (streptomycin and dihydrostreptomycin) or 2-deoxystreptamine (all other available aminoglycosides).

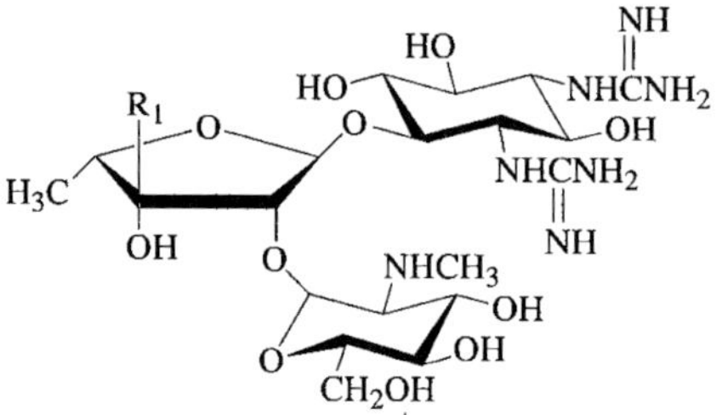

Streptomycin: $R_1 = CHO$
Dihydrostreptomycin: $R_1 = CH_2OH$

The aminoglycosides are highly polar drugs due to the large number of hydroxy and amino groups.

The aminoglycoside antibiotics are used primarily to treat infections caused by susceptible gram-negative aerobic bacteria. Bacterial resistance is plasmid mediated and results in bacterial modification of the aminoglycoside; the modified aminoglycosides are not clinically active. Strict anaerobes are resistant to aminoglycosides because they lack the mechanism necessary to transport the drug to the site of action. Due to their high polarity the aminoglycoside and aminocyclitol antibiotics are not well absorbed after oral administration. However, they are sometimes administered orally to reduce bacterial populations in the gastrointestinal tract. The aminoglycoside antibiotics most commonly used in food animal production are gentamicin, neomycin, streptomycin, and dihydrostreptomycin.

The aminoglycosides bind with the 30 S ribosomal subunit of susceptible bacteria and disrupt protein synthesis, resulting in a rapid bactericidal effect.

Aminoglycoside Antibiotics. *Streptomycin* [57-92-1] is produced by *Streptomyces griseus*; it is marketed as the sulfate salt in combination with bacitracin methylene disalicylate as Entromycin Powder for the treatment of bacterial enteritis and the associated diarrhea in dogs.

Dihydrostreptomycin [128-46-1] is a semisynthetic aminoglycoside produced by the reduction of streptomycin; it is no longer used in human medicine because of ototoxicity. It is marketed in combination with procaine penicillin G in an oil suspension as Quartermaster Suspension for intramammary use to reduce existing infection and to prevent new infection with *Staphylococcus aurea* in dry cows.

Kanamycin [8063-07-8] is an antibiotic complex produced by *Streptomyces kanamyceticus* and consists primarily of kanamycin A and two minor components, one of which is designated kanamycin B. Kanamycin is marketed in combination with bismuth subcarbonate and attapulgite as Amforol suspension and tablets for treatment of bacterial enteritis and the associated diarrhea in dogs.

Amikacin [37517-28-5] is a semisynthetic aminoglycoside derived from kanamycin A. It frequently has activity against organisms that have developed resistance to other aminoglycosides.

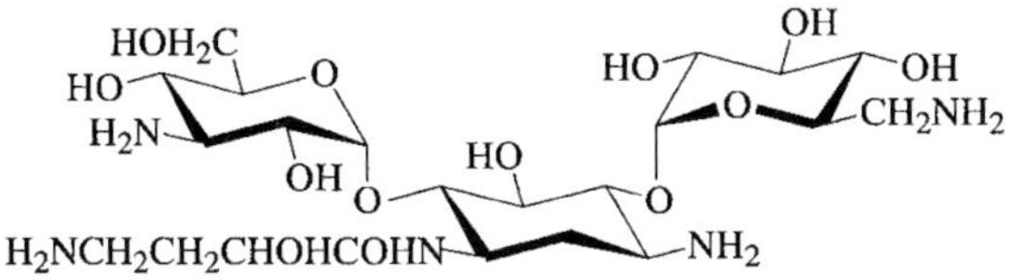

Amikacin

It is marketed as Amiglyde-V as the sulfate salt for injection and intrauterine infusion. The parenteral formulation is indicated for the treatment of skin and soft tissue infections caused by *Pseudomonas* species and *E. coli*, and urinary tract infections caused by *E. coli* and *Proteus* species. The intrauterine infusion is used for the treatment of uterine infections caused by susceptible organisms in the mare.

Gentamicin [1403-66-3] is an antibiotic complex produced by *Micromonospora purpurea* and *M. echinospora*; it consists of three substances designated gentamicins C_1, C_{1a}, and C_2.

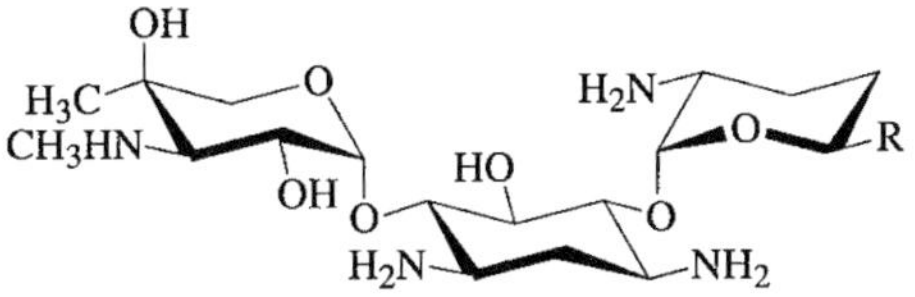

Gentamicin C_1:   R = $CHCH_3NHCH_3$
Gentamicin C_{1a}: R = CH_2NH_2
Gentamicin C_2:   R = $CHCH_3NH_2$

It is marketed as the sulfate salt as Garacin oral solution and in injectable form for treatment of colibacillosis in weanling swine caused by strains of *E. coli* sensitive to gentamicin, and swine dysentery associated with *Treponema hyodysenteriae*. Gentamicin is also available as Gentocin solution for parenteral administration to control bacterial infections of the uterus in horses and as an aid in improving conception in mares.

Aminocyclitol Antibiotics. *Spectinomycin* [1695-77-8] is an aminocyclitol antibiotic produced by *Streptomyces spectabilis*.

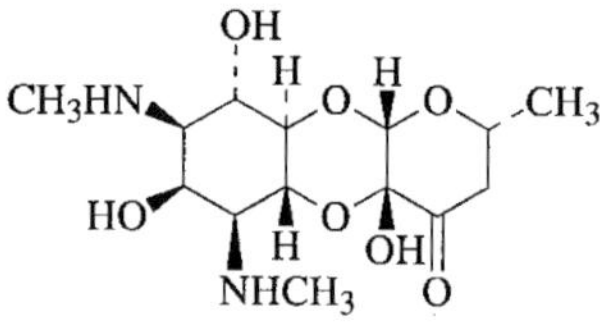

Spectinomycin

It is marketed as an oral solution as Spectam Scour-Halt for the treatment of infectious diarrhea in baby pigs caused by *E. coli*. It is also available as Spectam Injectable for treating airsaculitis and chronic respiratory disease associated with susceptible organisms in turkey poults.

Apramycin [37321-09-8] is an aminocyclitol antibiotic that is a component of the nebramycin complex and is produced by a strain of *Streptomyces tenebarius*. Apramycin is marketed as its sulfate salt as Apralan 75 Premix/Medicated, a type A medicated article for oral use in pigs, and Apralan Soluble Powder. Both products are used for the control of porcine colibacillosis caused by susceptible strains of *E. coli*.

1.5. Macrolide Antibiotics

Macrolide antibiotics (see also → Antibiotics) all possess a macrocyclic lactone ring attached to one or more amino sugars. The macrolide ring may contain 12, 14, 16, or 17 atoms.

Macrolide antibiotics have good antibacterial activity against many gram-positive organisms such as *Mycoplasma* and *Chlamydia* but have little or no activity against most gram-negative bacteria.

The macrolide antibiotics inhibit protein synthesis by binding to the 50 S ribosomal subunits of sensitive microorganisms. Gram-positive microorganisms accumulate much more macrolide antibiotic than do gram-negative microorganisms. Since bacterial cell walls are much more permeable to the unionized form, these drugs generally exhibit greater antimicrobial activity under alkaline conditions.

Erythromycin [114-07-8] is a mixture of three components, designated Erythromycin A, Erythromycin B, and Erythromycin C, produced by *Streptomyces erythreus*. Erythromycin A is the major component. All are macrolide antibiotics consisting of a fourteen-membered lactone ring attached to desosamine, an amino sugar, and cladinose or mycarose. Erythromycin, isolated in 1952, was the first macrolide antibiotic widely used in medicine. It is active against a wide variety of gram-positive bacteria and mycoplasmas and is widely used in veterinary medicine. Erythromycin is degraded by stomach acids and is therefore marketed as various esters (e.g., stearate and ethylsuccinate) or enteric coated products for oral administration. Erythromycin has a relatively large distribution volume and is extensively distributed to various tissues, particularly the lungs, kidneys, and liver, where it may attain tissue concentrations higher than the corresponding plasma concentration.

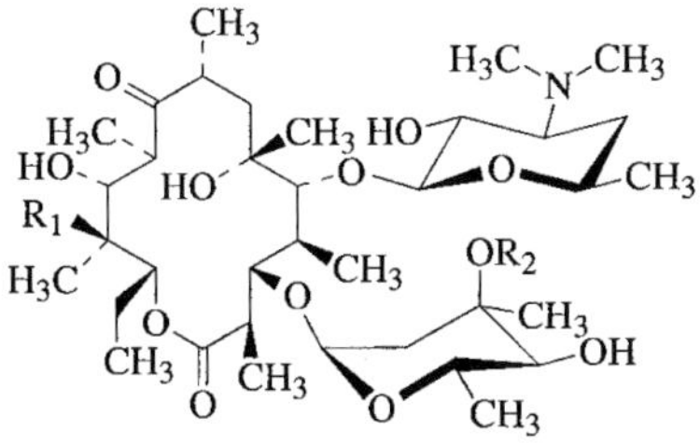

	R_1	R_2
Erythromycin A	OH	CH_3
Erythromycin B	H	CH_3
Erythromycin C	OH	H

Erytho-100 and Erythro-200 Injectable are erythromycin preparations in sterile solution containing 100 and 200 mg/L of erythromycin for intramuscular administration. In sheep it is indicated for the prevention of lamb dysentery and in the treatment of upper respiratory infections. In pigs it is used for treating respiratory syndrome and as an aid in the management of mastitis, metritis, and leptospirosis in sows at farrowing time and scours in young pigs. It is indicated in cattle for the treatment of pneumonia, shipping fever, mastitis, metritis, foot rot, and as an aid in reducing weight loss during shipping.

Gallamycin-Dry Cow and Gallimycin-36 are tubes containing erythromycin for administration into the udders of nonlactating and lactating cows, respectively, for the treatment of acute and chronic mastitis caused by *Staphylococcus aureus, Streptococcus agalactiae, S. dysgalactiae*, and *S. uberis*. Uddermate is indicated for mastitis treatment in lactating cows only.

Oleandomycin [3922-90-5] consists of a single component produced by *Streptomyces antibioticus*. It is a macrolide antibiotic consisting of a fourteen-membered lactone ring attached to desosamine and another sugar moiety. Oleandomycin has an antibacterial spectrum similar to that of erythromycin but is less potent. It is marketed as its phosphate salt and as its triacetyl derivative, troleandomycin.

Tylosin [1401-69-0] is a macrolide antibiotic containing a sixteen-membered lactone ring substituted with various sugar moieties. Tylosin is a mixture of four related compounds, designated tylosin A, B, C, and D, with tylosin A the major component.

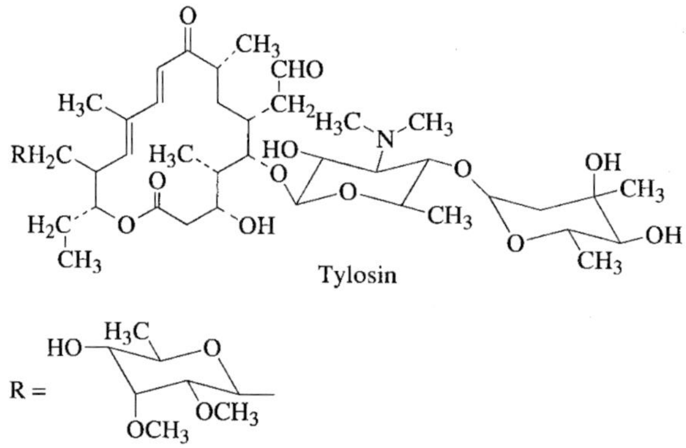

Tylosin is used as a feed additive for growth promotion and as an antibacterial drug in veterinary medicine. It is one of the most effective macrolide antibiotics against *Mycoplasma*. Tylosin was isolated from a strain of *Streptomyces fradiae* found in soil from Thailand.

Tylan 40 is a feed additive and antibiotic containing 40 g of tylosin as tylosin phosphate per pound. Tylan 40 is used in swine to maintain weight gain and feed efficiency in the presence of atrophic rhinitis. It is also used in swine to increase the rate of weight gain and to improve feed efficiency; it is employed to prevent, treat, and control swine dysentery. It is used to reduce the incidence of liver abscesses caused by *Spherophorus necrophorus* and *Corynebacterium pyogenes* in beef cattle. It is utilized in chickens to improve feed efficiency and to control chronic respiratory disease caused by *Mycoplasma gallisepticum*.

Tylosin Injection and Tylan Injection are formulated with 50 and 200 mg of the base per milliliter. The solution is administered intramuscularly to beef cattle and nonlactating dairy cattle to treat bovine respiratory complex associated with *Pasteurella multocida* and *Corynebacterium pyogenes*, foot rot and calf diphtheria caused by *Fusobacterium necrophorum*, and metritis caused by *Corynebacterium pyogenes*. Tylosin Injection is administered intramuscularly to swine to treat swine arthritis caused by *Mycoplasma hyosynoviae*, swine pneumonia caused by *Pasteurella* spp, swine erysipelas caused by *Erysipelothrix rhusiopathiae*, and acute swine dysentery associated with *Treponema hyodysenteriae*.

Spiramycin [*8025-81-8*] is a mixture of three macrolide antibiotics containing a sixteen-membered lactone ring, substituted with amino sugars and mycarosyl. The components are designated spiramycin I, II, and III, with spiramycin I being the most important. The antimicrobial spectrum of spiramycin is similar to that of other macrolide antibiotics, but its activity against *Treponema* is greater than that of erythromycin or oleandomycin.

1.6. Amphenicols

Chloramphenicol is produced by *Streptomyces venezuelae*; it is unique among natural compounds in having a nitrobenzene moiety and a dichloracetamide group. The active form of chloramphenicol has the *threo* configuration. Chloramphenicol is not ionizable at physiologic pH values and is relatively lipophilic. Thiamphenicol differs from chloramphenicol in that it has a methylsulfonyl group in place of the nitro group. Florfenicol differs from thiamphenicol by having a fluorine atom substituted for the hydroxy group at C-3. Thiamphenicol and florfenicol are totally synthetic.

The amphenicols are broad-spectrum antibiotics that are primarily bacteriostatic but may be bacteriocidal to certain bacteria. Gram-negative bacteria generally acquire resistance to chloramphenicol and thiamphenicol through plasmid-mediated transfer of acetyl transferase, which inactivates these drugs.

Florfenicol, lacking the hydroxy group that is acetylated in these other drugs, is not affected by the presence of acetyl transferases and therefore is effective against many chloramphenicol-resistant organisms.

Amphenicol antibiotics bind to the 50 S ribosomal subunits of susceptible bacteria, thereby inhibiting protein synthesis.

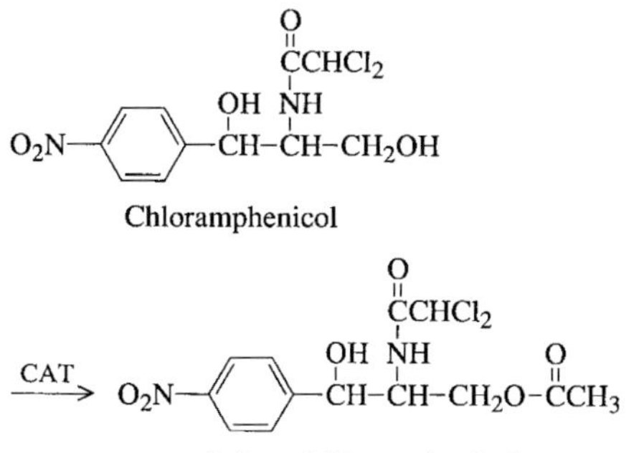

$$O_2N-\underset{\text{Chloramphenicol}}{\underset{|}{\bigcirc}}-\underset{|}{\overset{OH}{C}}H-\underset{|}{\overset{NH-\overset{O}{\overset{\|}{C}}CHCl_2}{C}}H-CH_2OH$$

$$\xrightarrow{CAT} O_2N-\underset{\text{3-Acetylchloramphenicol}}{\underset{|}{\bigcirc}}-\underset{|}{\overset{OH}{C}}H-\underset{|}{\overset{NH-\overset{O}{\overset{\|}{C}}CHCl_2}{C}}H-CH_2O-\overset{O}{\overset{\|}{C}}CH_3$$

Chloramphenicol [56-75-7] is a broad-spectrum antibiotic marketed as Bemacol ointment and Chlorbiotic ointment for topical treatment of bacterial conjunctivitis due to susceptible organisms in dogs and cats. It is also available as chloramphenicol tablets and capsules for treatment of bacterial infections in dogs. Chloramphenicol palmitate is an oral suspension for treatment of bacterial infections in dogs.

Chloramphenicol use in food-producing animals is not allowed due to the possibility of food contamination.

Thiamphenicol [15318-45-3] is a broad-spectrum antibiotic used in veterinary medicine. The D-form is used as an antimicrobial drug, and the DL-form in the control of fowl cholera.

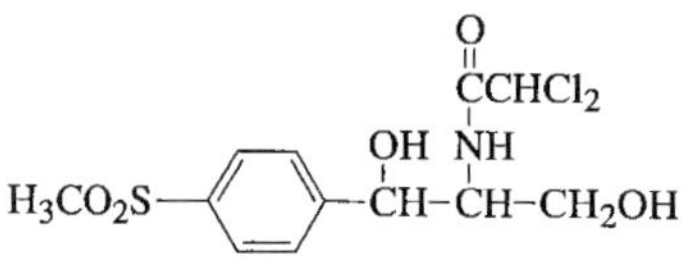

$$\text{H}_3\text{CO}_2\text{S}-\text{C}_6\text{H}_4-\overset{\overset{\text{OH}}{|}}{\text{CH}}-\overset{\overset{\text{NH}-\overset{\text{O}}{\overset{\|}{\text{C}}}\text{CHCl}_2}{|}}{\text{CH}}-\text{CH}_2\text{OH}$$

Thiamphenicol

Florfenicol [73231-24-2] is a broad-spectrum antibiotic used to treat bovine respiratory disease complex. It is also effective in the treatment of various fish diseases such as furunculosis in Atlantic salmon, pseudotuberculosis in yellowtail, and vibriosis in goldfish.

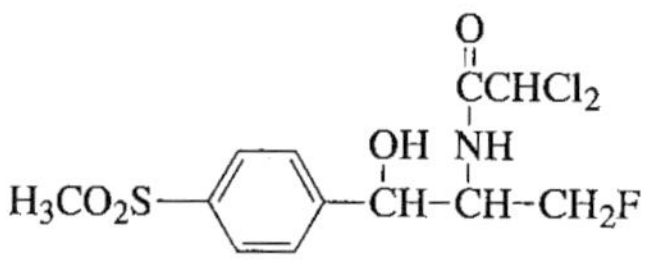

$$\text{H}_3\text{CO}_2\text{S}-\text{C}_6\text{H}_4-\overset{\overset{\text{OH}}{|}}{\text{CH}}-\overset{\overset{\text{NH}-\overset{\text{O}}{\overset{\|}{\text{C}}}\text{CHCl}_2}{|}}{\text{CH}}-\text{CH}_2\text{F}$$

Florfenicol

1.7. Tiamulin

Tiamulin [55297-95-5] is a diterpene antibiotic containing an eight-membered carbocyclic ring. It is a derivative of pleuromutilin, an antibiotic substance produced by *Pleurotus mutilis* and *P. passeckerianus*.

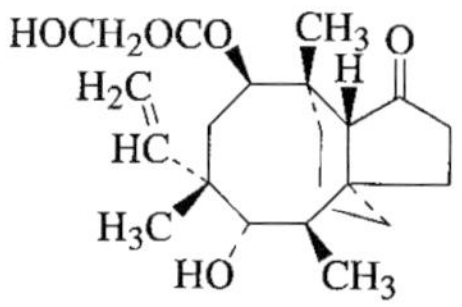

Tiamulin

Tiamulin is effective against gram-positive bacteria, *Mycoplasma*, and *Serpulina hyodysenteriae*. Tiamulin interferes with the metabolism and elimination of polyether ionophores (e.g., monensin, salinomycin, lasalocid, and narasin). This interaction is a potential cause of toxicity when used in combination with these antibiotics. For this reason the combination of tiamulin and ionophore antibiotics is generally contraindicated.

Denagard soluble antibiotic is a fine, white granular powder containing 45 wt% tiamulin designed for administration in the drinking water of swine to treat infections caused by *Treponema hyodysenteriae* and *Hemophilus pleuropneumonia*. Denagard 10 is a feed additive and antibiotic containing 10 g of tiamulin as the hydrogenfumarate salt per pound. This product is used to treat swine dysentery associated with *Treponema hyodysenteriae* and as a feed additive to increase weight gain. Swine being treated with Denagard 10 should not have access to feeds containing ionophore antibiotics.

1.8. Peptide Antibiotics

The peptide antibiotics (see also → Antibiotics) are a diverse group of compounds having peptides arranged in linear or ring structures. Many polypeptide antibiotics have been used in veterinary medicine as feed additives and drugs. Bacitracin, enramycin, and virginiamycin are widely used for growth promotion, improvement of feed efficiency, and treatment of disease.

The various polypeptide antibiotics have different antibacterial spectra and produce their antibacterial effects by a variety of mechanisms.

Actaplanins are a complex of antibiotics produced by *Actinoplanes missouriensis*. Six components designated A, B_1, B_2, B_3, C_1, and G have been isolated and characterized. The drugs have a central peptide core, with the amino sugar ristosamine and up to four neutral sugars attached. The actaplanins are used to promote growth and increase milk production in ruminants.

Avoparcin [37332-99-3] is a glycopeptide antibiotic complex produced by *Streptomyces candidus*. It consists of a mixture of α- and β-avoparacin (chlorine-substituted analogue) and is used as an antibacterial agent and growth enhancer.

Bacitracin [1405-87-4] is a polypeptide complex produced by *Bacillus subtilis* and *licheniformis*. The commercial product is a mixture of at least nine bacitracins. It is used as local antibiotic, growth promoter, and for enteric infections.

Virginiamycin is an antibiotic complex produced by a *Streptomyces* related to *S. virginiae*. It is a mixture of two principal components designated virginiamycin M_1 [21411-53-0] and virginiamycin S_1 [23152-29-6]. Virginiamycin is used as an antibiotic and as a feed additive.

Bambermycin [11015-37-5] is an antibiotic complex consisting of at least four components: moenomycins A, B_1, B_2, and C with moenomycin A the major component. It is obtained from cultures of *Streptomyces bambergiensis* and others. Bambermycin is used as an antibacterial agent and as a growth promoter in poultry, swine, and calves.

1.9. Ionophore Antibiotics

The ionophore antibiotics are classified as polyethers and contain an acidic functional group at one end of the molecule. The acid group forms a hydrogen bond with a hydroxyl group at the other end of the molecule, giving an inclusion complex with alkyl groups on the outer surface and polar groups on the inner surface. Cations bind to the polar groups on the inner surface.

Ionophore antibiotics were initially developed as poultry coccidiostats but are now also used to promote growth of cattle. They enhance propionic acid production in ruminants by altering rumen fermentation. They are usually added at a rate of 5–30 ppm of finished ration to improve feed efficiency of feedlot cattle.

Ionophore antibiotics facilitate transport of monovalent and divalent cations across biological membranes. They selectively transport sodium and potassium ions between extracellular and intracellular spaces and interfere with potassium ion transport across mitochondrial membranes.

Monensin [17090-79-8] is a major component of the antibiotic complex isolated from *Streptomyces cinnamonensis*. It has antibacterial, antifungal, and antiprotozoal activity and is a coccidiostat in chickens. Coban 45 and Coban 60 are feed additives and coccidiostats containing 45 and 60 g of monensin sodium salt per pound for use in poultry; Rumensin 60 is a feed efficiency enhancer for feedlot and pasture cattle containing 60 g of monensin sodium salt per pound.

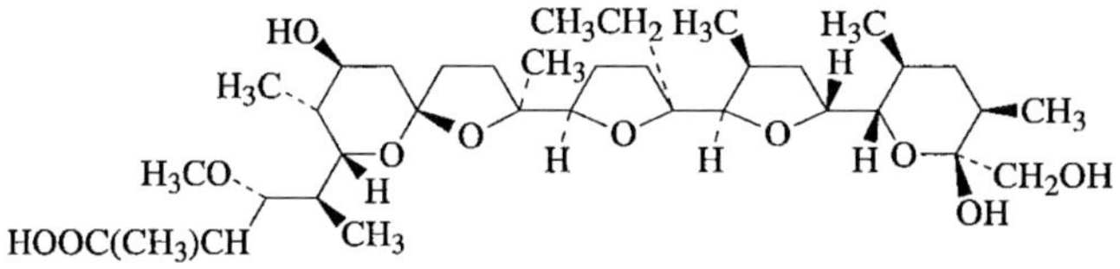

Monensin

Narasin [55134-13-9] is the main component of the polyether antibiotic complex produced by *Streptomyces aureofaciens*. It is used as a coccidiostat and as a growth stimulant. Monteban 45 is marketed as a feed additive and coccidiostat at a concentration of 45 g per pound. Maxiban 72 is a feed additive and coccidiostat containing both narasin and nicarbazine at a concentration of 36 g per pound for use in broiler chickens.

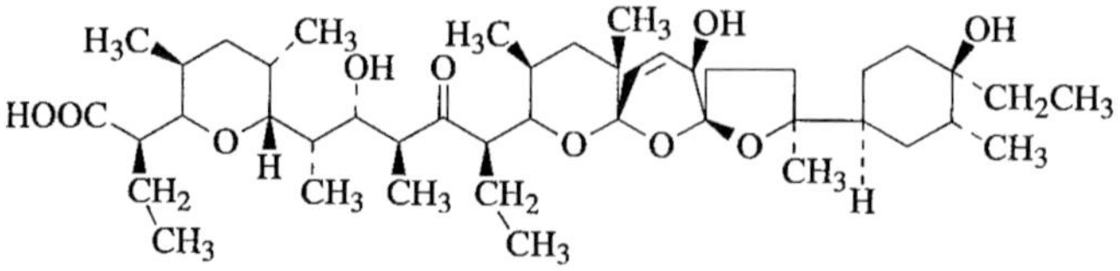

Narasin

Maduramicin [84878-61-5] is a polyether antibiotic isolated from a species of

Nocardia. It is used as a feed additive and coccidiostat. Cygro Type A Medicated Article is a feed additive and coccidiostat containing 4.54 g of maduramicin ammonium salt per pound for prevention of coccidiosis in broiler chickens caused by various species of *Eimeria*.

Lasalocid [25999-31-9] was originally isolated from an unknown species of *Streptomyces* from Hyde Park, Massachusetts. It is used as a feed additive and coccidiostat in chickens. Avatec is a medicated premix containing 20% of sodium lasalocid in a carrier suitable for use in feed for broiler chickens for prevention of coccidiosis. Bovatex is a medicated premix that contains 15% of sodium lasalocid in a carrier suitable for incorporation in feed to improve feed efficiency and increased rate of gain when used for cattle fed in confinement for slaughter.

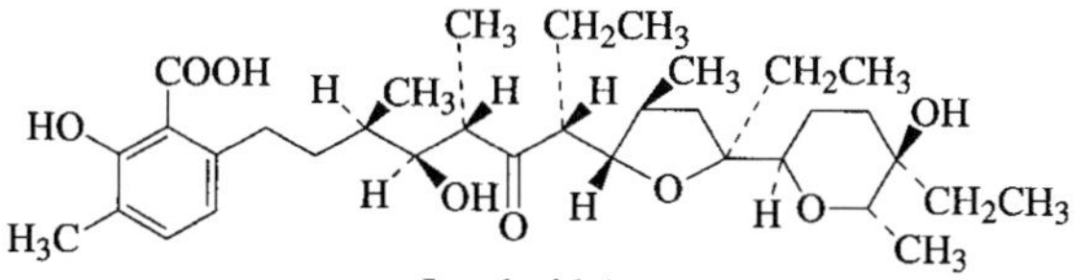

Lasalocid A

2. Antimycotics (see also → Antimycotics).

Miconazole and clotrimazole are substituted imidazoles containing one or more aromatic rings with chlorine substituents.

Miconazole [22916-47-8] is marketed as its nitrate salt in a number of topical products (e.g., Conofite Cream, Lotion, and Spray and Dermazole Shampoo) for the control of fungal infections caused by *Microsporum canis*, *M. gypseum*, and *Trichophyton mentagrophytes* in dogs and cats.

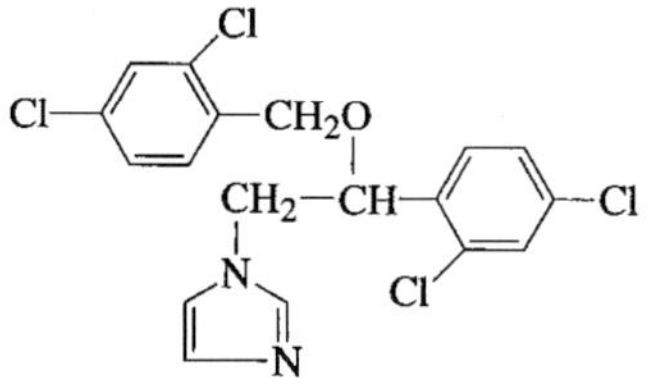

Miconazole

Clotrimazole [23593-75-1] is marketed as Veltrim Dermatologic Cream for treatment of fungal infections caused by *Microsporum canis* and *Trichophyton mentagrophytes* in dogs and cats.

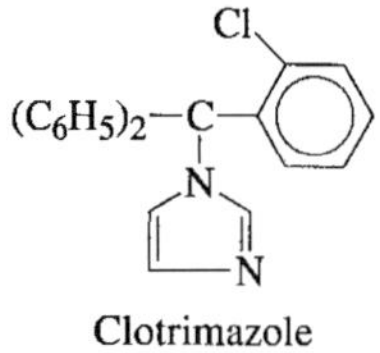

$(C_6H_5)_2-C$

Clotrimazole

3. Coccidiostats

Coccidiostats are a diverse group of drugs, including the ionophore antibiotics and other substances useful in the control or treatment of coccidiosis.

Coccidiostats affect various stages of the life cycle of species of *Eimeria* that infect the digestive tracts of birds, causing economic losses in commercial poultry operations. Resistance to coccidiostats may develop upon repeated use. Therefore, it has become common practice in the broiler industry to use two or more coccidiostats sequentially at various intervals.

Clopidol [2971-90-6] is mainly coccidiostatic and is active only against the sporozoite stage of *Eimeria*. It is effective only if administered on the day of exposure to the coccidial oocysts. Clopidol is marketed as a feed additive and is fed at a rate of ca. 125 ppm in the feed.

Amprolium [121-25-5] interferes with thiamine utilization in the parasite.

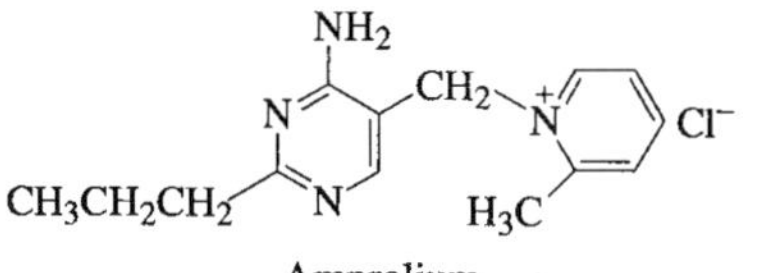

Amprolium

Amprolium is rarely used alone because it has relatively weak activity against *Eimeria maxima, E. mivati*, and other species. It is the only coccidiostat without a withdrawal time in the poultry industry. Amprolium is marketed as Aprol, Amprovine, and Corid in various soluble powders and oral solutions for treatment of coccidiosis in growing chickens, turkeys, and laying hens.

Decoquinate [18507-89-6] is a 4-hydroxyquinolone derivative that prevents development of the sporozoite stage by disruption of electron transport in the mitochondrial cytochrome system of the coccidia. Decoquinate is marketed as Deccox as a top dressing for addition to the feed for prevention of coccidiosis caused by *Eimeria bovis* and *Eimeria zurnii* in ruminating calves and cattle.

4. Parasiticides

4.1. Avermectin Derivatives

The avermectins are natural and semisynthetic derivatives of pentacyclic 16-membered lactones related to the milbemycins. The natural compounds are produced by *Streptomyces avermitilis*.

The avermectin antibiotics are very effective and highly potent anthelmintics that are effective by a number of administration routes. They are effective against a wide variety of nematodes as well as external parasites such as lice, flies, and bots. Ivermectin is also effective against the microfilarial and infective stages of canine heartworm disease caused by *Dirofilaria immitis*. Concerns have been expressed about their effects on the environment since they are excreted in the feces and may affect insects and other organisms responsible for the normal degradation of dung.

These drugs apparently inhibit motility of parasites by increasing the release of γ-aminobutyric acid (GABA) thereby inhibiting neurotransmission. Worms are thereby paralyzed and are expelled from the animal.

Ivermectin [70288-86-7], 22,23-dihydroabamectin, is a semisynthetic derivative of abamectin. Ivermectin contains at least 80% of 22,23-dihydroabamectin B_{1a} and not more than 20% of 22,23-dihydroabamectin B_{1b}.

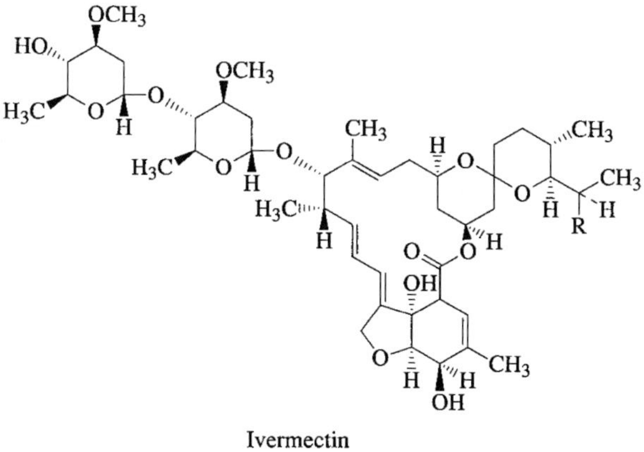

Ivermectin

Component B_{1a}: R = CH$_2$CH$_3$
Component B_{1b}: R = CH$_3$

Ivermectin is marketed as Eqvalan liquid and paste for use in horses for the treatment and control of large strongyles, small strongyles, hairworms, pinworms, ascarids, intestinal thread-worms, large-mouth stomach worms, bots, lungworms, summer sores, and cutaneous onchocerciasis. Ivermectin is also available as Heartgard-30 tablets to prevent canine heartworm disease by eliminating the tissue stage of heartworm larvae for a month after infection, and for the treatment and control of ascarids (*Toxacara canis* and *Toxascaris leonina*) and hookworms (*Ancylostoma caninum* and *Uncinario stenocephala*). Ivomec is a 1% ivermectin solution for treatment and control of internal and external cattle parasites and for treatment and control of gastrointestinal roundworms, lungworms, lice, and mange mites in swine. Ivomec sheep drench is a 0.08% solution of ivermectin for treatment of gastrointestinal roundworms, lungworms, and nasal bots in sheep.

Abamectin [*71751-41-2*], avermectin B_1, is produced by *Streptomyces avermitilis* and is structurally related to ivermectin. It is a mixture containing at least 80% of avermectin B_{1a} and not more than 20% of avermectin B_{1b}. Abamectin is marketed as Affirm, Avomec, and Vertimec for use as an anthelmintic drug.

4.2. Benzimidazole Derivatives

Benzimidazole anthelmintics are widely used in cattle, swine, and horses for controlling various gastrointestinal roundworms, tapeworms, liver flukes, and lungworms. They have a wide range of antiparasitic activity, a high degree of efficacy, and a good safety margin. Most are effective when administered orally. Thiabendazole is also used as a fungicide on many crops.

These drugs inhibit microtubules, inhibit glycogen uptake, and inhibit fumarate reductase in susceptible organisms.

Thiabendazole [*148-79-8*] was one of the first benzimidazole anthelmintic drugs approved for use in animals. It is marketed as Equizole A as a liquid horse wormer in combination with piperazine, and as Equizole suspension for control of gastrointestinal roundworms in cattle, horses, sheep, and goats. It is commercially available as TBZ 6 suspension for treatment and control of roundworm infection in cattle, and as Tresaderm in combination with dexamethasone and neomycin sulfate in solution for treatment of various bacterial, mycotic, and dermatoses and otitis externa in dogs and cats.

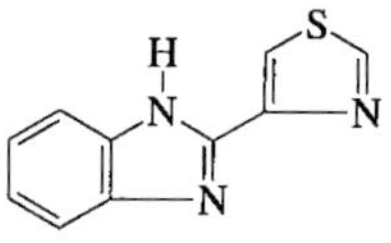

Thiabendazole

Other benzimidazole anthelmintics currently used in veterinary medicine include albendazole, fenbendazole, cambendazole, oxfendazole, oxibendazole, and mebendazole.

4.3. Imidazothiazole Derivatives

Imidazothiazole drugs are broad-spectrum anthelmintics used to treat and control a number of internal parasites affecting several animal species. They can be administered by various routes and are highly efficacious.

Butamisole [54400-59-8] is used to treat whipworm and hookworm infections in dogs. It is administered subcutaneously. It is marketed as Styquin, an aqueous solution of the hydrochloride. The product is not to be administered to heartworm-positive dogs or those treated with bunamidine because fatalities have been reported following its use in these animals.

Levamisole [16595-80-5] is an anthelmintic that causes paralysis of nematodes by ganglionic stimulation. It is available in several formulations including parenteral solutions, drenches, boluses, and oral gels. Tramisol injectable solution is a 13.65% solution of levamisole hydrochloride for subcutaneous injection in cattle, and Tramisol Type A Medicated Article is a premix for addition to the feed of cattle and swine.

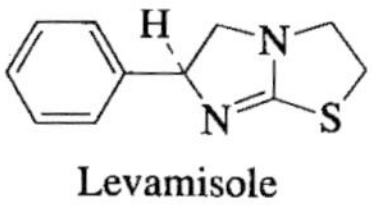

Levamisole

5. Anti-inflammatory Drugs

5.1. Nonsteroidal Anti-inflammatory Drugs

The nonsteroidal anti-inflammatory drugs are lipophilic acids of various structures. Ketoprofen and carprofen are members of the propionic acid class, which also includes ibuprofen, naproxen, and fenoprofen. Meclofenamic acid is a member of the fenamate class of nonsteroidal anti-inflammatory drugs.

The nonsteroidal anti-inflammatory drugs produce anti-inflammatory, antipyretic, and analgesic effects in animals.

They are believed to exert their pharmacologic effects at least in part by inhibiting cyclooxygenase activity, thereby reducing concentrations of prostaglandins, thromboxanes, and prostacyclins. They may also act centrally to reduce pain.

Flunixin [*38677-85-9*] is a nonsteroidal anti-inflammatory drug approved for use in horses. It is marketed as Banamine granules containing flunixin as its meglumine salt, equivalent to 250 or 500 mg of flunixin per 10 or 20 g packet for oral administration, for alleviation of pain and inflammation associated with musculoskeletal disorders of horses. Banamine paste contains 1500 mg of flunixin as its meglumine salt per 30 g syringe for oral administration to alleviate pain and inflammation associated with musculoskeletal disorders of horses. Banamine solution contains 50 mg of flunixin as its meglumine salt per milliliter for alleviation of pain and inflammation associated with musculoskeletal disorders and pain associated with colic in horses. Banamine solution may be administered intravenously or intramuscularly.

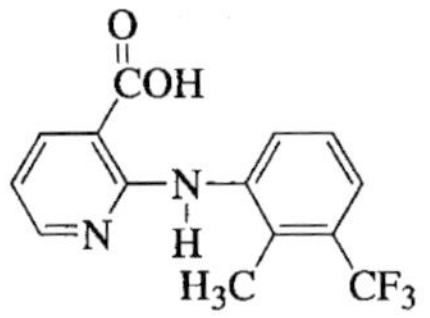

Flunixin

Meclofenamic acid [*644-62-2*] is marketed as Arquel granules containing 500 mg of meclofenamic acid per packet for oral administration to horses to alleviate pain and inflammation associated with musculoskeletal disorders.

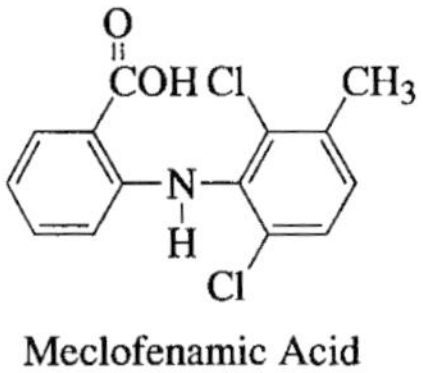

Meclofenamic Acid

Ketoprofen [*22071-15-4*] is a relatively new nonsteroidal anti-inflammatory drug. It is marketed as Ketofen solution containing 100 mg/mL of ketoprofen for intravenous administration to horses for alleviating pain and inflammation associated with musculoskeletal disorders.

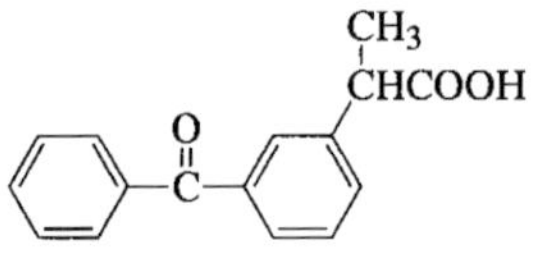

Ketoprofen

Phenylbutazone [50-33-9] is the oldest of the nonsteroidal anti-inflammatory drugs used in veterinary medicine. It is marketed in a number of parenteral and oral formulations. Phenylbutazone tablets containing 100 mg and 1000 mg of active ingredient are available for oral administration to dogs and horses, respectively, for treatment of inflammation associated with the musculoskeletal system. Phenylbutazone solution for intravenous injection contains 200 mg/mL of active ingredient and is indicated for the relief of inflammation associated with the musculoskeletal system of the horse.

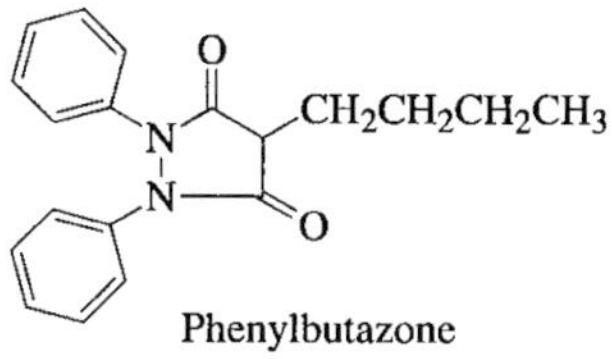

Phenylbutazone

Carprofen is the newest member of this group and is used to treat musculoskeletal injuries in horses.

5.2. Steroidal Anti-inflammatory Drugs

The steroidal anti-inflammatory drugs are fluorinated analogues of cortisol. Fluorine substitution has been shown to increase potency.

They are used locally or systemically to produce an anti-inflammatory effect. They have greater anti-inflammatory activity and less sodium-retaining activity than naturally occurring corticosteroids such as cortisol and cortisone.

The steroidal anti-inflammatory drugs inhibit release of arachidonic acid from cell membranes. They thereby interfere with the first step of the arachidonic acid cascade and prevent or reduce the biosynthesis of prostaglandins, prostacyclins, thromboxanes, and other cytokines involved in the inflammatory response.

Betamethasone [378-44-9] is marketed as Betasone, a mixture of betamethasone dipropionate and betamethasone sodium phosphate. The betamethasone dipropionate is in aqueous suspension and the betamethasone sodium phosphate is in solution in this product to provide immediate and sustained anti-inflammatory effects in the dog for relief of pruritis associated with dermatoses. Betamethasone is also available as its valerate ester in combination with gentamicin in a spray (Gentocin) and an ointment (Topagen) for topical use to treat superficial lesions caused by bacteria susceptible to gentamicin in the dog.

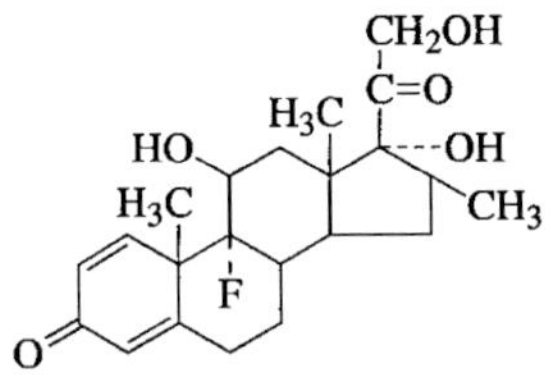

Betamethasone

Dexamethasone [*50-02-2*] is marketed as Azium powder for cattle and horses and as Azium solution for intramuscular and intravenous administration for treating primary bovine ketosis, and as an anti-inflammatory agent in cattle, horses, dogs, and cats. Dexamethasone is also marketed in combination with the diuretic trichlormethiazide as Naquasone for the treatment of physiological parturient udder edema in cattle.

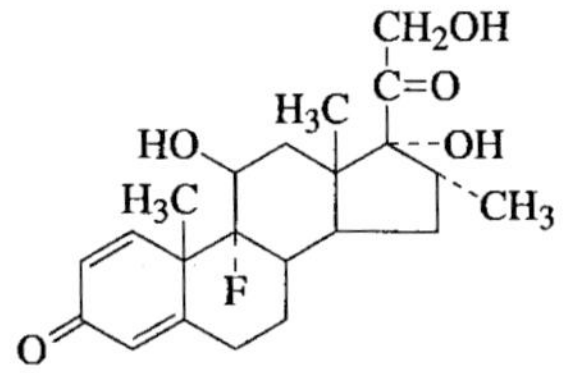

Dexamethasone

Flumethasosone [*2135-17-3*] is marketed in combination with neomycin sulfate and polymyxin B sulfate in Anaprime Ophthalmic Solution for the treatment of inflammation, edema, and secondary bacterial infections of the eye in dogs. It is also marketed as Flucort solution for intravenous, intramuscular, or subcutaneous administration to horses, dogs, and cats, and intra-articular administration to horses. Flucort tablets are available for oral therapy with flumethasone.

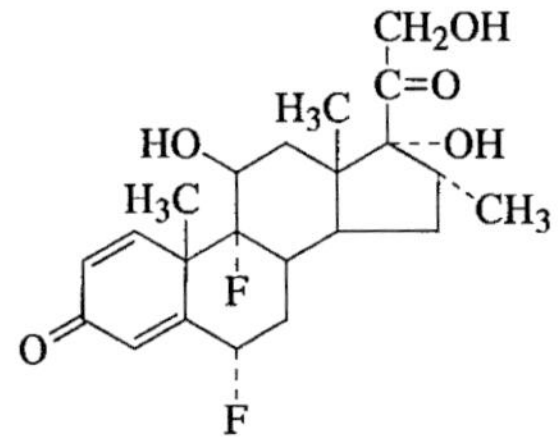

Flumethasone

Isoflupredone [*338-95-4*] is marketed as its acetate ester in combination with neomycin sulfate in Neo-Predef ointment for use in cattle, horses, dogs, and cats, and in combination with both neomycin sulfate and the local anesthetic tetracaine as Neo-Predef, as Tetracaine topical powder, and Tritop topical ointment for use in dogs, cats, and horses. Predef 2 X is a suspension of isoflupredone acetate for intramuscular or intra-synovial use in cattle, horses, and pigs.

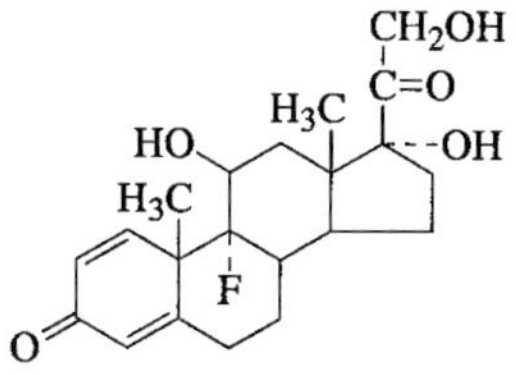

Isoflupredone

Triamcinolone [*124-94-7*] is marketed as its acetonide [*76-25-5*] in Vetalog cream, oral powder, tablets, and parenteral solution.

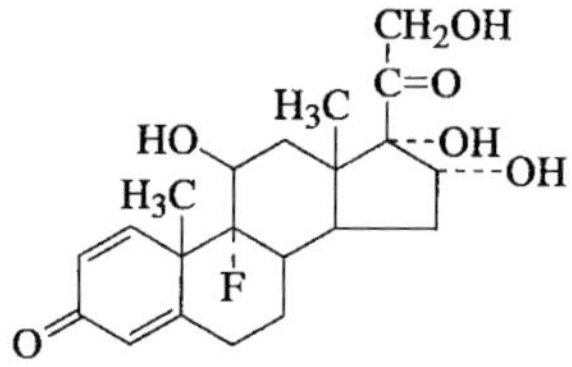

Triamcinolone

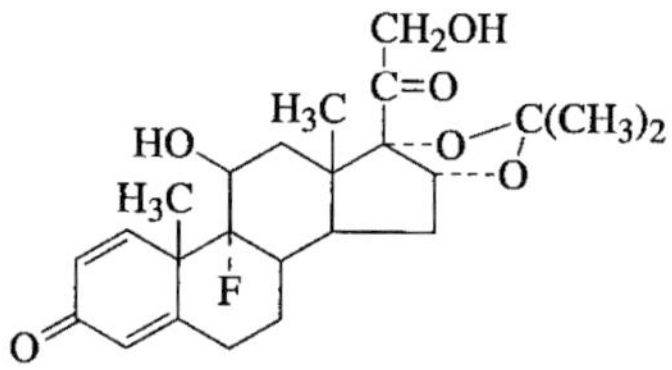

Triamcinolone acetonide

6. Respiratory Stimulants

Doxapram [*309-29-5*] is used as a respiratory stimulant during recovery from barbiturate or inhalation anesthesia in dogs, cats, and horses. It produces respiratory stimulation at doses considerably less than those producing CNS stimulation.

Doxapram stimulates chemoreceptors of the carotid and aortic regions and may stimulate the medullary respiratory center.

Doxapram is marketed as Dopram-V injectable for intravenous use in dogs, cats, and horses.

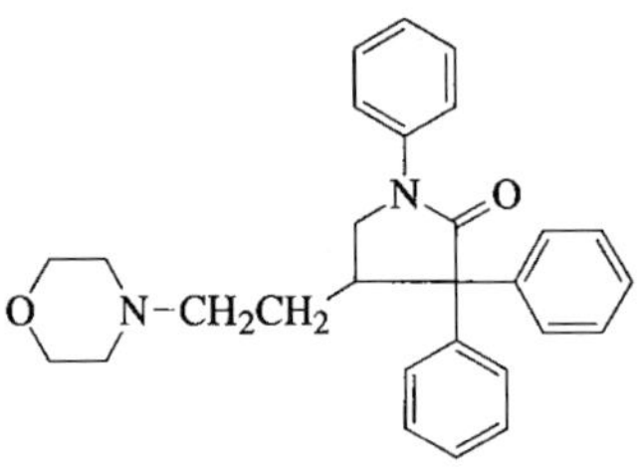

Doxapram

Amiphenazole [*490-55-1*] is a phenyl-substituted thiazolediamine that forms soluble salts with acids.

It is used as a barbiturate and morphine antagonist in animals, and is marketed as Daptizole and Fenamizol.

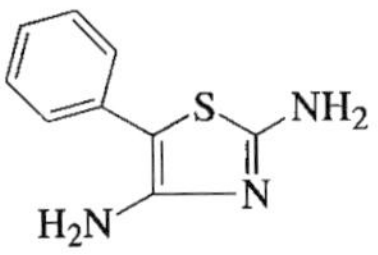

Amiphenazole

7. Anesthetics

Dissociative anesthetics are rapid-acting, nonnarcotic, nonbarbiturate drugs used to produce anesthesia in cats and dogs (tiletamine only) as well as restraint in nonhuman primates (ketamine only). These drugs produce profound analgesia, normal pharyngeal-laryngeal reflexes, as well as mild cardiac stimulation and respiratory depression.

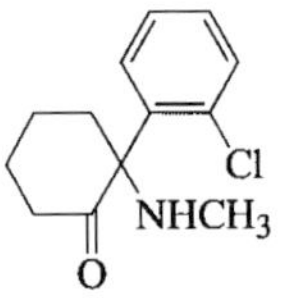

Ketamine 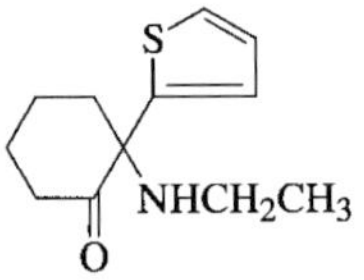   Tiletamine

Ketamine [*6740-88-1*] is marketed as Ketaset and Vetalar solutions for intramuscular administration to cats and subhuman primates.

Tiletamine is marketed as its hydrochloride salt in combination with zolazepam hydrochloride as Telazol solution for intramuscular administration to dogs and cats.

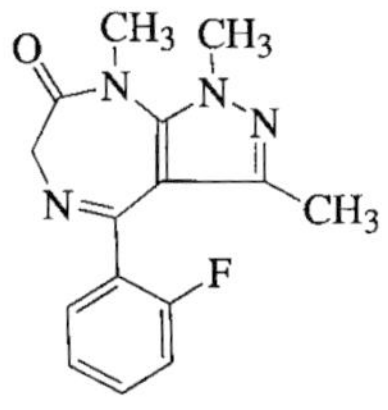

Zolazepam

α₂-Agonists. The α_2-agonists used in veterinary medicine consist of a substituted aromatic ring attached to a heterocyclic ring containing at least one nitrogen atom. Addition of a methyl group to detomidine produced medetomidine and created an asymmetric center.

The α_2-agonists are sedative and analgesic drugs used to facilitate minor surgical procedures.

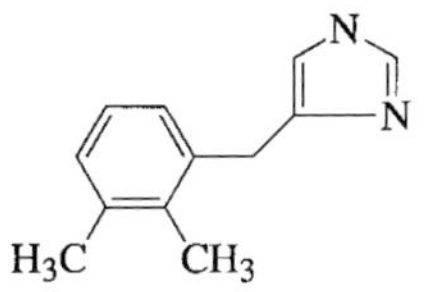

Detomidine

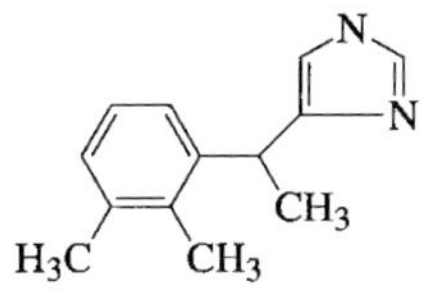

Medetomidine

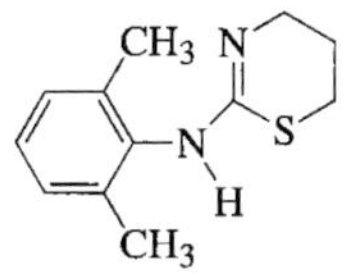

Xylazine

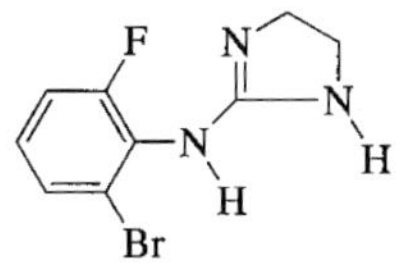

Romifidine

The α_2-adrenoreceptor agonist drugs produce sedation and analgesia via CNS depression. The action of the α_2-adrenoreceptor agonist drugs can be reversed by administration of α_2-adrenoreceptor antagonists such as yohimbine and atipamezole.

Xylazine [7361-61-7] is marketed as its hydrochloride salt as Rompun (100 mg/mL) and Anased (20 mg/mL) injectable solutions for intravenous administration to horses and dogs, respectively. The actions of xylazine may be reversed by the administration of yohimbine, an indolalkylamine alkaloid, that blocks those α_2-adrenoreceptors that are stimulated by xylazine.

Detomidine [76631-46-4] is marketed as its hydrochloride salt as Dormosedan solution for intramuscular or intravenous administration to horses.

Medetomidine [86347-14-0] is marketed as its hydrochloride salt as Domitor, and its dextrorotatory isomer, *dexmedetomidine*, are used to produce sedation and analgesia in many mammalian species but used mainly in the dog and domestic cat.

Romifidine is marketed as its hydrochloride salt.

Reversing Agents. *Yohimbine* [146-48-5] is an alkaloid found in *Corynanthe johimbe* and related trees as well as in *Raulwolfia serpentina*. It is marketed as its hydrochloride salt as Yobine to reverse the effects of the α_2-adrenoreceptor agonists.

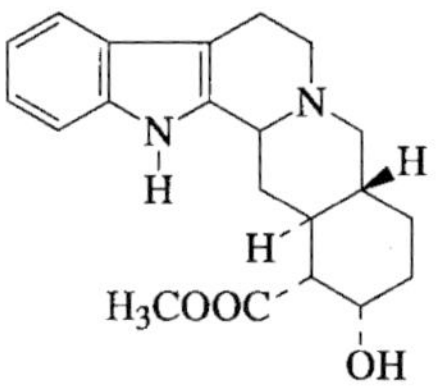

Atipamezole is a synthetic compound marketed as Antisedan to reverse the effects of the α_2-adrenoreceptor agonists.

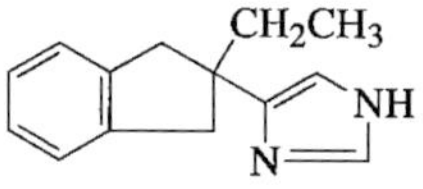

8. Narcotic Agonists and Antagonists

Narcotic analgesic drugs used in veterinary medicine are synthetic or semisynthetic derivatives of morphine (and related alkaloids) and derivatives of fentanyl.

Narcotic analgesics are used in veterinary medicine to produce tranquilization to assist in the capture of wild and zoo animals, as preanesthetic medications, and as analgesic drugs.

They are believed to exert their effects by acting as agonists at specific opiate receptors located in the brain and possibly other sites in the body. Stimulation of these receptors produces profound analgesia in animals. The actions of the narcotic analgesic drugs may be reversed by the administration of competitive antagonists such as diprenorphine and naloxone.

Numerous narcotic analgesic drugs such as morphine, codeine, oxymorphone (Numorphan), pentazocine (Talwin-V), and butorphanol (Torbugesic and Torbitrol) are used in veterinary medicine. Other narcotics with somewhat specialized application in veterinary medicine are discussed below.

Etorphine/diprenorphine is an agonist/antagonist combination used to produce tranquilization and reversal in large animals. Etorphine [14521-96-1] is marketed as its hydrochloride salt as M-99 for intramuscular administration. It is approximately 10 000 times more potent than morphine as an analgesic. It is usually dosed heavily to produce tranquilization and avoid hyperexcitability in wild animals, and then reversed with diprenorphine [14357-78-9] (M-5050). Etorphine is also marketed in combination with acepromazine, a phenothiazine tranquilizer, to produce neuroleptic analgesia.

Fentanyl [*437-38-7*] is marketed as its citrate salt in combination with droperidol, a butryphenone tranquilizer, as Innovar-Vet for intravenous or intramuscular use in the dog for various surgical procedures requiring analgesia and tranquilization.

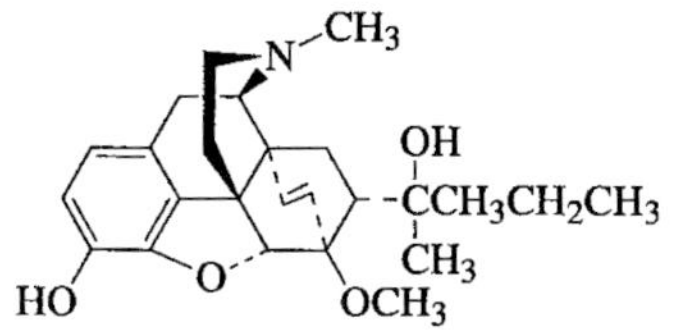

Etorphine

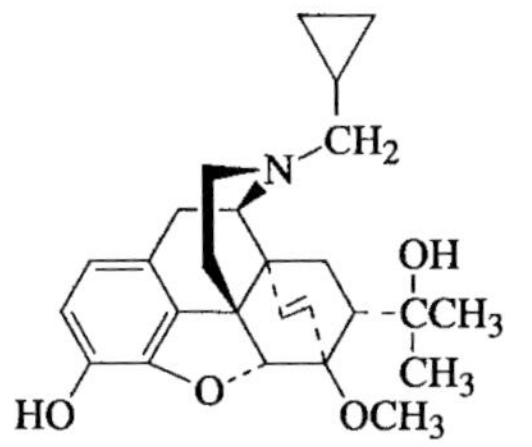

Diprenorphine

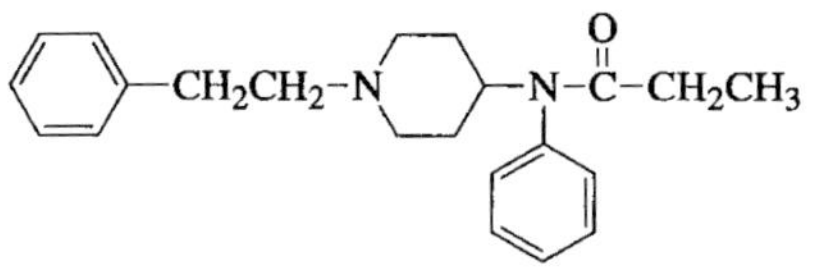

Fentanyl

Carfentanil is marketed as its citrate salt as Wildnil for immobilization of free-ranging or confined deer, elk, and moose. Carfentanil is ca. 10 000 times as potent as morphine. The effect of carfentanil is rapidly reversed by the administration of diprenorphine.

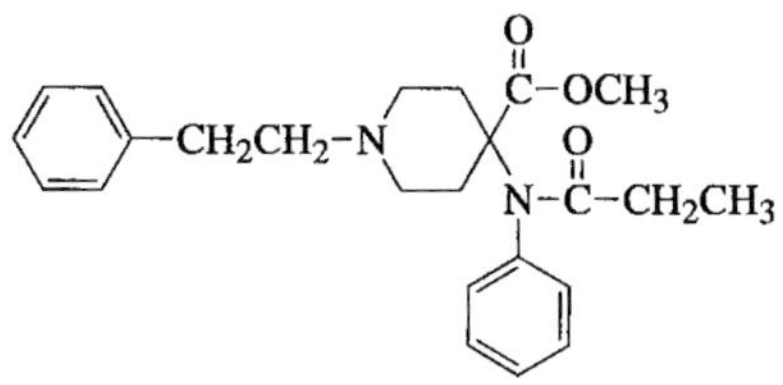

Carfentanil

9. Antihistamines (see also → Antiallergics)

The antihistamines used in veterinary medicine are classified as ethylenediamine derivatives (i.e., pyrilamine and tripelennamine), ethanolamine derivatives (i.e., doxylamine), and phenothiazine derivatives (i.e., trimeprazine).

Antihistamines antagonize many of the pharmacological effects of histamine by acting as competitive inhibitors of histamine H_1 receptors.

Pyrilamine [*91-84-9*] is marketed as Histavet-P containing 20 mg/mL of pyrilamine maleate and is indicated for use in the horse. The formulation may be administered intravenously, intramuscularly, or subcutaneously.

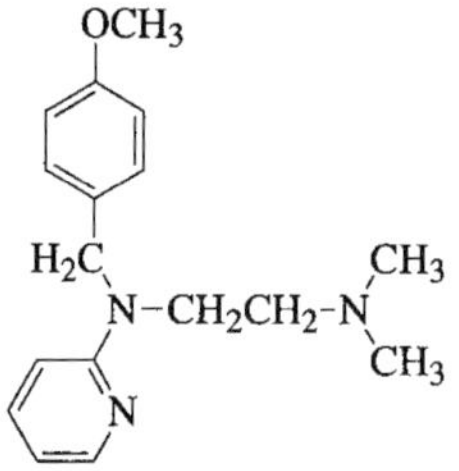

Pyrilamine

Doxylamine [*496-21-6*] is marketed as A-H Injection containing 11.36 mg/mL of doxylamine succinate for use as an antihistamine in horses, dogs, and cats. It may be administered intravenously to horses and intramuscularly or subcutaneously to dogs, cats, and horses. A-H Tablets contain 100 mg of doxylamine succinate for use as an antihistamine in horses.

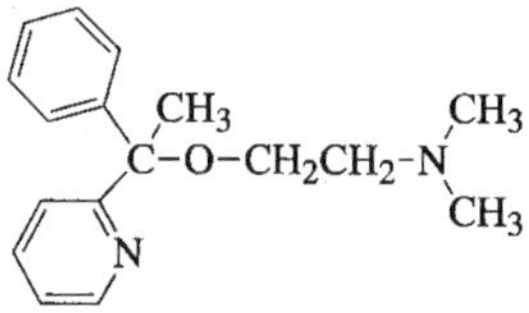

Doxylamine

Tripelennamine [*91-81-6*] is marketed as Re-Covr Injection containing 20 mg/mL of tripelennamine hydrochloride for use as an antihistamine in cattle, horses, dogs, and cats. It may be administered intravenously to cattle and intramuscularly to cattle, horses, dogs, and cats.

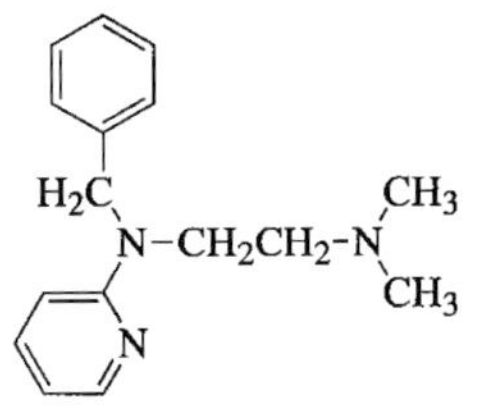

Tripelennamine

Trimeprazine [*84-96-8*] is a phenothiazine derivative marketed as Temaril-P tablets containing 5 mg of trimeprazine tartrate and 2 mg of prednisolone. The drug is used for the antipruritic and antitussive effects of trimeprazine and the anti-inflammatory effects of prednisolone in dogs.

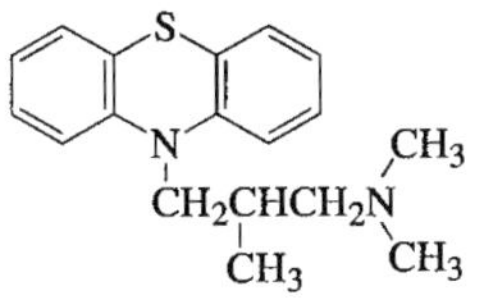

Trimeprazine

10. References

General References

[1] M. Sittig: *Veterinary Drug Manufacturing Encyclopedia*, Noyes Publications, Park Ridge, NJ, 1981.
[2] H. R. Adams: *Veterinary Pharmacology and Therapeutics*, 5th ed., The Iowa State University Press, Ames, IA, 1995.
[3] K. Bennett (ed.): *Compendium of Veterinary Products*, 1st ed., North American Compendiums, Inc., Port Huron, MI, 1991.
[4] R. B. Talbot (ed.): *Veterinary Pharmaceuticals and Biologicals*, 8th ed., Veterinary Medicine Publishing Company, Lenexa, KS, 1993/1994.

Author Index

Actor, Paul, Smith Kline & French Laboratories, Philadelphia, Pennsylvania 19101, United States, *Synthetic Chemotherapeutic Agents*, **3**

Akkermans, Louis M. A., Department of Surgery, University Hospital Utrecht, Utrecht, The Netherlands, *Gastroprokinetic Agents*, **2**

Albrecht, Hans P., BASF Aktiengesellschaft, Ludwigshafen, Federal Republic of Germany, *Cardiac Glycosides and Synthetic Cardiotonic Drugs*, **1**; *Cardioactive and Vasoactive Drugs*, **1**

Allegra, Carmen J., National Cancer Institute, Bethesda, Maryland 20205, United States, *Cancer Chemotherapy*, **4**

Andrews, Peter, Bayer AG, Wuppertal, Federal Republic of Germany, *Anthelmintics*, **3**

Beaulieu, Pierre L., Boehringer Ingelheim (Canada) Ltd., Bio-Méga Research Division, Laval (Québec), Canada, *HIV and AIDS Therapeutics*, **3**

Beckh, Sonja, Medizinische Klinik 3, Pneumologie, Klinikum Nürnberg Nord, Nürnberg, Federal Republic of Germany, *Antiallergics*, **2**

Binnig, Fritz, Knoll AG, Ludwigshafen, Federal Republic of Germany, *Antiarrhythmic Drugs*, **1**

Bischoff, Hilmar, Bayer AG, Wuppertal, Federal Republic of Germany, *Oral Antidiabetic Drugs*, **3**

Böhme, Georg Andrees, Rhône-Poulenc Rorer, Vitry-Sur-Seine, France, *Antipsychotics/Neuroleptics*, **2**

Bölcskei, Pal L., Semmelweis Medical University Medizinische Klinik 3, Pneumologie, Klinikum Nürnberg Nord, Budapest, Nürnberg, Hungary, Federal Republic of Germany, *Antiallergics*, **2**

Briejer, Michel R., Department of Gastrointestinal Pharmacology, Janssen Research Foundation, Beerse, Belgium, *Gastroprokinetic Agents*, **2**

Brune, Kay, Department of Experimental and Clinical Pharmacology and Toxicology, Friedrich Alexander University Erlangen-Nürnberg, Erlangen, Germany, *Anti-inflammatory – Antirheumatic Drugs*, **4**

Buchmann, Bernd, Research Laboratories, Schering Aktiengesellschaft, Berlin, Federal Republic of Germany, *Prostaglandins*, **4**

Bungardt, Edwin, Pharmakologisches Institut für Naturwissenschaftler der J. W. Goethe-Universität, Frankfurt/Main, Federal Republic of Germany, *Spasmolytics*, **2**

Buschmann, Helmut, Grünenthal GmbH, Center of Research, Aachen, Federal Republic of Germany, *Analgesics and Antipyretics*, **2**

Capet, Marc, Rhône-Poulenc Rorer, Vitry-Sur-Seine, France, *Antipsychotics/Neuroleptics*, **2**

Casals-Stenzel, Jorge, Gauting-Königswiesen, Federal Republic of Germany, *Antiasthmatics*, **2**

Cauwenbergh, Geert, Johnson & Johnson Consumer Products Worldwide, Skillman, New Jersey 08558, United States, *Drugs Used in Dermatology*, **4**

Chabner, Bruce A., National Cancer Institute, Bethesda, Maryland 20205, United States, *Cancer Chemotherapy*, **4**

Chao Yeh, Grace, National Cancer Institute, Bethesda, Maryland 20205, United States, *Cancer Chemotherapy*, **4**

Chow, Alfred W., Smith Kline & French Laboratories, Philadelphia, Pennsylvania 19101, United States, *Synthetic Chemotherapeutic Agents*, **3**

Christoph, Thomas, Grünenthal GmbH, Center of Research, Aachen, Federal Republic of Germany, *Analgesics and Antipyretics*, **2**

Crawley, Patrick E., Celltech Ltd., Slough, United Kingdom, *Monoclonal Antibodies*, **4**

Cross, Alan, Department of Bacterial Diseases, Walter Reed Army Institute of Research, Washington D.C. 20307-5100, United States, *Immunotherapy and Vaccines*, **4**

Cryz, Jr., Stanley J., Swiss Serum and Vaccine Institute Berne, Switzerland, *Immunotherapy and Vaccines*, **4**

Csomós, Géza, Hamburg, Federal Republic of Germany, *Gallbladder and Liver Therapeutics*, **2**

Curt, Gregory A., National Cancer Institute, Bethesda, Maryland 20205, United States, *Cancer Chemotherapy*, **4**

Dahlmanns, Simone M., Forschungszentrum Jülich GmbH, Jülich, Federal Republic of Germany, *Thyrotherapeutic Agents*, **3**

Dodt, Johannes, Paul-Ehrlich-Institute, Dept. of Haematology and Transfusion Medicine, Langen, Federal Republic of Germany, *Blood*, **1**

Dolak, Terence M., Bausch & Lomb, Rochester, New York 14 692, United States, *Ophthalmological Agents*, **4**

Dorn, Jr., Conrad P., Merck Sharp & Dohme Research Laboratories, Rahway, New Jersey 07065, United States, *Anti-inflammatory – Antirheumatic Drugs*, **4**

Dubroeucq, Marie-Christine, Rhône-Poulenc Rorer, Vitry-Sur-Seine, France, *Antipsychotics/Neuroleptics*, **2**

Dutko, Frank J., Sterling-Winthrop Research Institute, Rensselaer, New York 12144, United States, *Synthetic Chemotherapeutic Agents*, **3**

Engelhorn, Robert, Biberach, Federal Republic of Germany, *Laxatives*, **2**

Fine, Robert L., National Cancer Institute, Bethesda, Maryland 20205, United States, *Cancer Chemotherapy*, **4**
Finizio, Michael, New Drug Services, Longwood Corporate Center South, Kennett Square, Pennsylvania 19348–2412, United States, *Calcium Antagonists*, **1**
Franke, Albrecht, BASF Aktiengesellschaft, Ludwigshafen, Federal Republic of Germany, *Antiarrhythmic Drugs*, **1**
Friderichs, Elmar, Grünenthal GmbH, Center of Research, Aachen, Federal Republic of Germany, *Analgesics and Antipyretics*, **2**

Galfre, Giovanni L., Celltech Ltd., Slough, United Kingdom, *Monoclonal Antibodies*, **4**
Geiss, Karl-Heinz, BASF Aktiengesellschaft, Ludwigshafen, Federal Republic of Germany, *Cardiac Glycosides and Synthetic Cardiotonic Drugs*, **1**
Gendimenico, Gerard J., Johnson & Johnson Consumer Products Worldwide, Skillman, New Jersey 08558, United States, *Drugs Used in Dermatology*, **4**
Geschke, Frank-Ulrich, Bayer AG, Wuppertal, Federal Republic of Germany, *Antimycotics*, **3**
Gottstein, Bruno, Institut für Parasitologie, Universität Zürich, Zürich, Switzerland, *Immunotherapy and Vaccines*, **4**
Graham, Barrie R., Clinical Pharmacology, University College London, London WC1 E6JJ, United Kingdom, *Beta Blockers*, **1**
Granstrom, Marta, Departments of Clinical Microbiology and of Vaccine Production, National Bacteriological Laboratory, Karolinska Hospital, Stockholm, Sweden, *Immunotherapy and Vaccines*, **4**

Hahn, Do Won, R. W. Johnson Pharmaceutical Research Institute, Raritan, New Jersey 08869, United States, *Chemical Contraception*, **3**
Hahn, Klaus-Jürgen, Knoll AG, Ludwigshafen, Federal Republic of Germany, *Drug Testing*, **4**
Harder, Achim, Bayer AG, Wuppertal, Federal Republic of Germany, *Anthelmintics*, **3**
Haring, Michael, Henning Berlin GmbH, Berlin, Federal Republic of Germany, *Hormones*, **3**
Hepner, Leo, L. Hepner and Associates, Ltd., London, United Kingdom, *Antibiotics*, **3**
Hinz, Burkhard, Department of Experimental and Clinical Pharmacology and Toxicology, Friedrich Alexander University Erlangen-Nürnberg, Erlangen, Germany, *Anti-inflammatory – Antirheumatic Drugs*, **4**
Hoffmann, Hinrich, Hoechst Aktiengesellschaft, Frankfurt, Federal Republic of Germany, *Antibiotics*, **3**
Höfgen, Norbert, Arzneimittelwerk Dresden GmbH, Radebeul, Federal Republic of Germany, *Antiallergics*, **2**
Hofmeister, Alfred, Chemische Werke Minden GmbH, Minden/Westphalia, Federal Republic of Germany, *Analeptics*, **2**
Hollmann, Matthias, Mannheim, Federal Republic of Germany, *Cardioactive and Vasoactive Drugs*, **1**
Houlihan, William J., Sandoz Research Institute, East Hanover, New Jersey 07936, United States, *Appetite Suppressants*, **2**
Hropot, Max, Hoechst AG, Hoechst Marion Roussel, Frankfurt/Main, Federal Republic of Germany, *Diuretics*, **1**

Jennewein, Hans Michael, Boehringer Ingelheim Pharma KG, Ingelheim/Rhein, Federal Republic of Germany, *Antiasthmatics*, **2**

Kondo, Shinichi, Institute of Microbial Chemistry, Tokyo, Japan, *Antibiotics*, **3**
König, Herbert, Paul-Ehrlich-Institute, Dept. of Haematology and Transfusion Medicine, Langen, Federal Republic of Germany, *Blood*, **1**
König, Wolfgang, Hoechst Aktiengesellschaft, Frankfurt am Main, Federal Republic of Germany, *Peptides and Protein Hormones*, **3**
Kretzschmar, Rolf, Knoll AG, Ludwigshafen, Federal Republic of Germany, *Drug Testing*, **4**
Kromer, Wolfgang, Byk Gulden, Konstanz, Federal Republic of Germany, *Anti-ulcer Drugs*, **2**
Krüger, Uwe, Byk Gulden, Konstanz, Federal Republic of Germany, *Anti-ulcer Drugs*, **2**
Kutscher, Bernhard, ASTA Medica AG, Frankfurt am Main, Federal Republic of Germany, *Cancer Chemotherapy*, **4**

Lang, Hans-Jochen, Hoechst AG, Hoechst Marion Roussel, Frankfurt/Main, Federal Republic of Germany, *Diuretics*, **1**
Larrick, James, Genelabs Incorporated, Redwood City, California, United States, *Immunotherapy and Vaccines*, **4**
Lehmann, Hans Dieter, Knoll AG, Ludwigshafen, Federal Republic of Germany, *Blood Pressure Increasing Agents*, **1**; *Cardioactive and Vasoactive Drugs*, **1**

Leuschner, Ulrich, Johann Wolfgang Goethe Universität, Frankfurt, Federal Republic of Germany, *Gallbladder and Liver Therapeutics*, **2**
Lever,Jr., O. William, Bausch & Lomb, Rochester, New York 14 692, United States, *Ophthalmological Agents*, **4**
Lyssy, Ralf H., Institut für Pharmazeutische Chemie, J. W. Goethe-Universität, Frankfurt am Main, Federal Republic of Germany, *Parkinsonism Treatment*, **2**

Male, Celia, L. Hepner and Associates, Ltd., London, United Kingdom, *Antibiotics*, **3**
Manfré, Franco, Rhône-Poulenc Rorer, Vitry-Sur-Seine, France, *Antipsychotics/Neuroleptics*, **2**
Marsh, David, Bausch & Lomb, Rochester, New York 14 692, United States, *Ophthalmological Agents*, **4**
McGuire, John L., Johnson & Johnson, New Brunswick, New Jersey 08933, United States, *The Pharmaceutical Industry*, **1**; *Chemical Contraception*, **3**; *Drugs Used in Dermatology*, **4**
McKinlay, Mark A., Sterling-Winthrop Research Institute, Rensselaer, New York 12144, United States, *Synthetic Chemotherapeutic Agents*, **3**
Mezick, James A., Johnson & Johnson Consumer Products Worldwide, Skillman, New Jersey 08558, United States, *Drugs Used in Dermatology*, **4**
Moran, Irene, Bausch & Lomb, Rochester, New York 14 692, United States, *Ophthalmological Agents*, **4**
Mujagic, Hamza, National Cancer Institute, Bethesda, Maryland 20205, United States, *Cancer Chemotherapy*, **4**
Müller, Renate, Schering AG, Berlin, Federal Republic of Germany, *Steroids*, **3**
Müller-Gärtner, Hans-Wilhelm, Forschungszentrum Jülich GmbH, Jülich, Federal Republic of Germany, *Thyrotherapeutic Agents*, **3**
Mutschler, Ernst, Pharmakologisches Institut für Naturwissenschaftler der J. W. Goethe-Universität, Frankfurt/Main, Federal Republic of Germany, *Spasmolytics*, **2**

Nagabhushan, Tattanahalli L., Schering–Plough Research, Bloomfield, New Jersey 07003, United States, *Interferons*, **4**
Neef, Günter, Schering AG, Berlin, Federal Republic of Germany, *Hormones*, **3**

Öppinger, Heinz, Hoechst Aktiengesellschaft, Frankfurt, Federal Republic of Germany, *Antibiotics*, **3**
Oberdorf, Anton, Knoll AG, Ludwigshafen, Federal Republic of Germany, *Antiarrhythmic Drugs*, **1**
Ohno, Masaji, Faculty of Pharmaceutical Sciences, University of Tokyo, Tokyo, Japan, *Antibiotics*, **3**
Otsuka, Masami, Faculty of Pharmaceutical Sciences, University of Tokyo, Tokyo, Japan, *Antibiotics*, **3**
Owens, Christopher W. I., Clinical Pharmacology, University College London, London WC1 E6JJ, United Kingdom, *Beta Blockers*, **1**

Perrin, Luc, Division d'Hematologie, Hôpital Cantonal Universitaire, Genève, Switzerland, *Immunotherapy and Vaccines*, **4**
Postius, Stefan, Byk Gulden, Konstanz, Federal Republic of Germany, *Anti-ulcer Drugs*, **2**
Prezewowsky, Klaus, Schering AG, Berlin, Federal Republic of Germany, *Hormones*, **3**
Prichard, Brian N. C., Clinical Pharmacology, University College London, London WC1 E6JJ, United Kingdom, *Beta Blockers*, **1**
Proudfoot, John, Boehringer Ingelheim Pharmaceuticals Inc., Ridgefield, Connecticut 06877, United States, *HIV and AIDS Therapeutics*, **3**

Rackur, Gerhard, Hoechst Aktiengesellschaft, Frankfurt, Federal Republic of Germany, *Antiemetics*, **2**
Rehwinkel, Hartmut, Research Laboratories, Schering Aktiengesellschaft, Berlin, Federal Republic of Germany, *Prostaglandins*, **4**
Rippel, Robert, Hoechst Aktiengesellschaft (retired), Frankfurt, Federal Republic of Germany, *Local Anesthetics*, **2**
Rundfeldt, Chris, ASTA Medica Group, Corporate Research and Development, Dresden, Germany, *Antiepileptic Drugs*, **2**

Sams, Richard, The Ohio State University, Dept. of Veterinary Clinical Sciences, Columbus, Ohio 43210, United States, *Veterinary Drugs*, **4**
Sandow, Jürgen, Hoechst AG, Frankfurt, Federal Republic of Germany, *Hormones*, **3**
Schatton, Wolfgang, KliniPharm, Frankfurt am Main, Federal Republic of Germany, *Parkinsonism Treatment*, **2**
Scheiffele, Ekkehard, Henning Berlin GmbH, Berlin, Federal Republic of Germany, *Hormones*, **3**
Schmidt, Axel, Bayer AG, Wuppertal, Federal Republic of Germany, *Antimycotics*, **3**
Schromm, Kurt, Boehringer Ingelheim Pharma KG, Ingelheim/Rhein, Federal Republic of Germany, *Antiasthmatics*, **2**

Schuurkes, Jan. A. J., Department of Gastrointestinal Pharmacology, Janssen Research Foundation, Beerse, Belgium, *Gastroprokinetic Agents*, **2**

Scriabine, Alexander, Miles Laboratories, Inc., Institute for Preclinical Pharmacology, New Haven, Connecticut 06511, United States, *Antihypertensive Agents*, **1**

Secher, David S., Celltech Ltd., Slough, United Kingdom, *Monoclonal Antibodies*, **4**

Seeger, Ernst, Biberach, Federal Republic of Germany, *Laxatives*, **2**

Seitz, Rainer, Paul-Ehrlich-Institute, Dept. of Haematology and Transfusion Medicine, Langen, Federal Republic of Germany, *Blood*, **1**

Shen, Tsung Ying, Merck Sharp & Dohme Research Laboratories, Rahway, New Jersey 07065, United States, *Antiinflammatory–Antirheumatic Drugs*, **4**

Stache, Ulrich, Hoechst AG, Frankfurt, Federal Republic of Germany, *Hormones*, **3**

Stutzmann, Jean-Marie, Rhône-Poulenc Rorer, Vitry-Sur-Seine, France, *Antipsychotics/Neuroleptics*, **2**

Sukatsch, Dieter, Hoechst Aktiengesellschaft, Frankfurt, Federal Republic of Germany, *Antibiotics*, **3**

Sutton, Scott, Bausch & Lomb, Rochester, New York 14 692, United States, *Ophthalmological Agents*, **4**

Szelenyi, Istvan, Arzneimittelwerk Dresden GmbH, Radebeul, Federal Republic of Germany, *Antiallergics*, **2**

Taylor, David G., Miles Laboratories, Inc., Institute for Preclinical Pharmacology, New Haven, Connecticut 06511, United States, *Antihypertensive Agents*, **1**

Thyes, Marco, Knoll AG, Ludwigshafen, Federal Republic of Germany, *Blood Pressure Increasing Agents*, **1**; *Cardioactive and Vasoactive Drugs*, **1**

Tobia, Alfonso J., New Drug Services, Longwood Corporate Center South, Kennet Square, Pennsylvania 19348–2412, United States., *Calcium Antagonists*, **1**

Trotta, Paul P., Schering–Plough Research, Bloomfield, New Jersey 07003, United States, *Interferons*, **4**

Unverferth, Klaus, ASTA Medica Group, Corporate Research and Development, Dresden, Germany, *Antiepileptic Drugs*, **2**

von Philipsborn, Gerda, Knoll AG, Ludwigshafen, Federal Republic of Germany, *Antiarrhythmic Drugs*, **1**

Whittaker, Victor P., Arbeitsgruppe Neurochemie, Max-Planck-Institut für biophysikalische Chemie Johannes Gutenberg Universität, Göttingen, Mainz, Federal Republic of Germany, Federal Republic of Germany, *Neuropharmaceutical Agents*, **2**

Winbury, Martin M., Pharmaceutical Research Division, Warner-Lambert Company, Ann Arbor, Michigan, United States, *Drugs Affecting Circulation*, **1**

Wollweber, Hartmund, Bayer AG, Wuppertal, Federal Republic of Germany, *Anesthetics*, **2**; *Sedatives*, **2**

Yagisawa, Morimasa, Japan Antibiotics Research Association, Tokyo, Japan, *Antibiotics*, **3**

Zahn, Manuel, Knoll AG, Ludwigshafen, Federal Republic of Germany, *Drug Testing*, **4**

Zanowiak, Paul, Temple University, Health Sciences Center, Philadelphia, Pennsylvania 19140, United States, *Pharmaceutical Dosage Forms*, **4**

Zwaving, Jan H., University of Groningen, Groningen, The Netherlands, *Anthraquinone Laxatives*, **2**

CAS Registry Number Index

[29-74-8] **2**: 913
[50-0-0] **4**: 1765
[50-2-2] **3**: 1625; **4**: 2017, 2178
[50-3-3] **4**: 2017
[50-6-6] **2**: 479, 506
[50-7-7] **3**: 1070; **4**: 1896
[50-9-9] **2**: 444, 511
[50-10-2] **2**: 696
[50-12-4] **2**: 475
[50-13-5] **2**: 397
[50-14-6] **3**: 1637
[50-18-0] **4**: 1888
[50-19-1] **2**: 645
[50-21-5] **4**: 1723
[50-22-6] **3**: 1607
[50-23-7] **3**: 1607; **4**: 2016
[50-24-8] **3**: 1617; **4**: 2017
[50-27-1] **3**: 1330
[50-33-9] **2**: 374; **4**: 1694, 2177
[50-34-0] **2**: 697
[50-35-1] **2**: 511
[50-44-2] **4**: 1882
[50-47-5] **2**: 614
[50-48-6] **2**: 611
[50-49-7] **2**: 613
[50-52-2] **2**: 592
[50-53-3] **2**: 590
[50-55-5] **1**: 85
[50-56-6] **3**: 1398
[50-57-7] **3**: 1402
[50-58-8] **2**: 465
[50-59-9] **3**: 1029
[50-60-2] **2**: 706
[50-65-7] **3**: 1282
[50-67-9] **2**: 458
[50-70-4] **2**: 785
[50-76-0] **3**: 1062; **4**: 1904
[50-78-2] **1**: 35; **2**: 346; **4**: 1678, 1721
[50-91-9] **4**: 1873
[50-96-4] **2**: 866
[50-98-6] **2**: 463, 877
[51-5-8] **2**: 543; **4**: 2004
[51-6-9] **1**: 124
[51-12-7] **2**: 623
[51-21-8] **4**: 1871
[51-24-1] **3**: 1523, 1557
[51-34-3] **2**: 562, 741; **4**: 1715
[51-40-1] **1**: 60
[51-41-2] **1**: 60; **2**: 458; **3**: 1562
[51-42-3] **1**: 60; **2**: 864; **4**: 2020
[51-43-4] **1**: 60; **2**: 864; **4**: 2020
[51-45-6] **2**: 902; **4**: 1719
[51-48-9] **3**: 1523, 1558
[51-49-0] **1**: 22

[51-52-5] **3**: 1529
[51-55-8] **2**: 562, 690, 891
[51-56-9] **4**: 2028
[51-61-6] **1**: 60, 192
[51-64-9] **2**: 461
[51-68-3] **2**: 652
[51-71-8] **2**: 624
[51-75-2] **4**: 1886
[51-79-6] **2**: 499
[51-83-2] **4**: 2024
[52-1-7] **1**: 74, 298; **3**: 1616
[52-21-1] **4**: 2017
[52-24-4] **4**: 1889
[52-26-6] **2**: 389
[52-28-8] **2**: 390
[52-31-3] **2**: 507
[52-39-1] **3**: 1613, 1615
[52-43-7] **2**: 508
[52-49-3] **2**: 563
[52-53-9] **1**: 97, 131, 208
[52-67-5] **2**: 771; **4**: 1705
[52-68-6] **3**: 1281
[52-76-7] **3**: 1497
[52-86-8] **2**: 598
[52-88-0] **2**: 690, 891
[53-3-2] **3**: 1617; **4**: 1726
[53-6-5] **3**: 1607
[53-16-7] **3**: 1330, 1573
[53-21-4] **4**: 2003, 2027
[53-33-8] **3**: 1629
[53-34-9] **3**: 1629
[53-43-0] **3**: 1572
[53-45-9] **2**: 529
[53-46-3] **2**: 695
[53-73-6] **3**: 1391
[53-86-1] **2**: 355, 706; **4**: 1679, 1721
[54-5-7] **3**: 1161
[54-30-8] **2**: 702
[54-31-9] **1**: 72, 296
[54-42-2] **3**: 1178; **4**: 2014
[54-64-8] **3**: 1255; **4**: 2037
[54-71-7] **4**: 2024
[54-85-3] **3**: 1139
[54-92-2] **2**: 623
[55-3-8] **3**: 1523
[55-6-1] **3**: 1523
[55-31-2] **1**: 60; **4**: 2020
[55-48-1] **2**: 690, 891; **4**: 2027
[55-63-0] **1**: 223; **2**: 706; **4**: 1715
[55-65-2] **1**: 86
[55-86-7] **4**: 1728
[55-91-4] **4**: 2026
[56-4-2] **3**: 1529
[56-29-1] **2**: 444, 511

[56-53-1]	**3:** 1585	[60-56-0]	**3:** 1530
[56-54-2]	**1:** 124	[60-87-7]	**2:** 921
[56-75-7]	**3:** 1069; **4:** 2011, 2167	[60-91-3]	**2:** 565
[56-81-5]	**2:** 782; **4:** 2023, 2033	[60-99-1]	**2:** 591
[56-94-0]	**4:** 2025	[61-0-7]	**2:** 591
[57-13-6]	**4:** 1723, 1727, 2023	[61-12-1]	**2:** 549
[57-15-8]	**4:** 2038	[61-25-6]	**2:** 699
[57-22-7]	**4:** 1910	[61-56-3]	**2:** 486
[57-27-2]	**2:** 389	[61-57-4]	**3:** 1282
[57-29-4]	**2:** 418	[61-68-7]	**2:** 352; **4:** 1697
[57-30-7]	**2:** 506	[61-73-4]	**4:** 2030
[57-33-0]	**2:** 509	[61-75-6]	**1:** 130
[57-41-0]	**1:** 125; **2:** 474	[61-76-7]	**4:** 2026, 2029
[57-42-1]	**2:** 397, 404	[62-31-7]	**1:** 60, 192
[57-43-2]	**2:** 507	[62-38-4]	**4:** 2037
[57-44-3]	**2:** 506	[62-44-2]	**2:** 349
[57-47-6]	**2:** 653; **4:** 2025	[62-46-4]	**2:** 770
[57-53-4]	**2:** 644, 678	[62-51-1]	**4:** 2032
[57-53-6]	**3:** 1501	[62-67-9]	**2:** 418
[57-55-6]	**4:** 2035	[62-97-5]	**2:** 693
[57-62-5]	**4:** 2007, 2161	[63-42-3]	**2:** 785
[57-63-6]	**4:** 1726	[63-45-6]	**3:** 1163
[57-67-0]	**3:** 1137	[63-56-9]	**2:** 907
[57-68-1]	**4:** 2159	[63-74-1]	**1:** 71
[57-83-0]	**3:** 1497, 1571	[64-31-3]	**2:** 389
[57-88-5]	**3:** 1567	[64-43-7]	**2:** 507
[57-92-1]	**4:** 2009, 2162	[64-65-3]	**2:** 339
[57-94-3]	**2:** 672	[64-72-2]	**3:** 1036; **4:** 2007
[58-14-0]	**3:** 1130, 1164	[64-75-5]	**4:** 2007
[58-22-0]	**3:** 1330, 1574; **4:** 1726	[64-77-7]	**3:** 1658
[58-25-3]	**2:** 636	[64-85-7]	**3:** 1614
[58-32-2]	**1:** 231, 39	[64-86-8]	**2:** 771
[58-33-3]	**2:** 921	[65-29-2]	**2:** 670
[58-38-8]	**2:** 594, 744	[65-45-2]	**2:** 348
[58-39-9]	**2:** 594	[65-49-6]	**3:** 1130, 1141
[58-40-2]	**2:** 590	[65-85-0]	**4:** 1732
[58-54-8]	**1:** 76, 297	[65-86-1]	**2:** 769
[58-55-9]	**1:** 290; **2:** 887	[66-2-4]	**3:** 1558
[58-71-9]	**3:** 1028; **4:** 2006	[66-28-4]	**1:** 179
[58-73-1]	**2:** 566, 742, 909; **4:** 1720	[67-20-9]	**3:** 1145
[58-74-2]	**2:** 699	[67-45-8]	**3:** 1158
[58-82-2]	**3:** 1392	[67-73-2]	**3:** 1629
[58-93-5]	**1:** 71, 293	[67-92-5]	**2:** 702
[58-94-6]	**1:** 71, 292	[67-97-0]	**3:** 1636
[59-5-2]	**4:** 1728, 1868	[68-22-4]	**3:** 1497, 1575
[59-30-3]	**3:** 1129	[68-23-5]	**3:** 1497
[59-32-5]	**2:** 905	[68-35-9]	**3:** 1134; **4:** 2005, 2014
[59-33-6]	**2:** 906	[68-41-7]	**3:** 1069
[59-40-5]	**4:** 2159	[68-88-2]	**2:** 646, 915; **4:** 1720
[59-66-5]	**1:** 71, 291; **2:** 486; **4:** 2022	[68-89-3]	**2:** 369
[59-67-6]	**1:** 23	[68-91-7]	**1:** 85
[59-92-7]	**2:** 568	[69-9-0]	**2:** 681
[59-96-1]	**2:** 706	[69-23-8]	**2:** 595
[60-0-4]	**4:** 2039	[69-43-2]	**1:** 208
[60-26-4]	**1:** 84	[69-52-3]	**4:** 2006
[60-29-7]	**2:** 440	[69-53-4]	**3:** 1022; **4:** 2006, 2160
[60-54-8]	**3:** 1036; **4:** 1725, 2007, 2161	[69-65-8]	**2:** 785; **4:** 2023

[69-72-7]	**4:** 1724, 1727	[78-44-4]	**2:** 677
[70-0-8]	**3:** 1178; **4:** 2015	[79-1-6]	**2:** 439
[70-30-4]	**3:** 1254	[79-57-2]	**4:** 2007, 2161
[70-78-0]	**3:** 1558	[79-63-0]	**3:** 1326
[71-27-2]	**2:** 674	[80-8-0]	**3:** 1142; **4:** 1726
[71-58-9]	**3:** 1497	[80-49-9]	**2:** 691
[71-63-6]	**1:** 186, 188	[80-50-2]	**2:** 692
[71-68-1]	**2:** 394	[80-78-4]	**3:** 1333
[71-73-8]	**2:** 445	[81-23-2]	**2:** 749; **3:** 1329
[71-81-8]	**2:** 694	[81-24-3]	**2:** 749
[71-82-9]	**2:** 421	[81-25-4]	**2:** 749; **3:** 1329, 1611
[72-23-1]	**3:** 1607	[82-88-2]	**2:** 925
[72-33-3]	**3:** 1501	[82-92-8]	**2:** 742, 914
[72-44-6]	**2:** 513	[82-93-9]	**2:** 913
[72-57-1]	**4:** 2030	[82-95-1]	**2:** 913
[72-69-5]	**2:** 615	[82-99-5]	**2:** 705
[72-80-0]	**3:** 1254	[83-43-2]	**3:** 1623
[73-9-6]	**1:** 297	[83-44-3]	**2:** 749; **3:** 1329
[73-48-3]	**1:** 294	[83-46-5]	**1:** 22; **3:** 1569
[73-49-4]	**1:** 295	[83-48-7]	**3:** 1569
[73-78-9]	**1:** 125; **2:** 546; **4:** 2003	[83-73-8]	**3:** 1170; **4:** 1732
[74-55-5]	**3:** 1140	[83-89-6]	**3:** 1158
[75-19-4]	**2:** 439	[83-98-7]	**2:** 567, 680
[76-22-2]	**4:** 1723	[84-2-6]	**2:** 744
[76-25-5]	**4:** 1726, 2179	[84-4-8]	**2:** 592
[76-26-5]	**3:** 1620	[84-6-0]	**2:** 595
[76-38-0]	**2:** 442	[84-16-2]	**3:** 1585
[76-41-5]	**2:** 397	[84-17-3]	**3:** 1585
[76-42-6]	**2:** 393, 396	[84-96-8]	**2:** 591, 918; **4:** 2185
[76-45-9]	**3:** 1334	[84-97-9]	**2:** 593
[76-57-3]	**2:** 390	[85-73-4]	**3:** 1137
[76-58-4]	**2:** 391	[86-12-4]	**2:** 918
[76-68-6]	**2:** 510	[86-21-5]	**2:** 912
[76-74-4]	**2:** 509	[86-22-6]	**2:** 911
[76-75-5]	**2:** 445	[86-34-0]	**2:** 478
[76-76-6]	**2:** 507	[86-35-1]	**2:** 475
[76-90-4]	**2:** 695	[86-42-0]	**3:** 1162
[77-2-1]	**2:** 507	[86-54-4]	**1:** 99
[77-4-3]	**2:** 511	[87-8-1]	**3:** 1018
[77-7-6]	**2:** 394	[87-12-7]	**3:** 1254
[77-9-8]	**2:** 787	[87-33-2]	**1:** 223
[77-19-0]	**2:** 702	[90-33-5]	**2:** 700, 753
[77-21-4]	**2:** 512	[90-34-6]	**3:** 1163
[77-27-0]	**2:** 446	[90-39-1]	**1:** 126
[77-28-1]	**2:** 506	[90-49-3]	**2:** 501
[77-36-1]	**1:** 72, 295	[90-69-7]	**2:** 339
[77-37-2]	**2:** 564	[90-84-6]	**2:** 464
[77-38-3]	**2:** 742	[91-75-8]	**2:** 905
[77-41-8]	**2:** 479	[91-79-2]	**2:** 907
[77-46-3]	**3:** 1143	[91-80-5]	**2:** 906
[77-59-8]	**3:** 1333	[91-81-6]	**2:** 907; **4:** 1720, 2184
[77-65-6]	**2:** 501	[91-84-9]	**2:** 906; **4:** 1720, 2184
[77-66-7]	**2:** 501	[91-85-0]	**2:** 907
[77-67-8]	**2:** 478	[92-12-6]	**2:** 910
[77-75-8]	**2:** 497	[92-13-7]	**4:** 2024
[78-11-5]	**1:** 224	[93-54-9]	**2:** 750
[78-12-6]	**2:** 499	[94-9-7]	**2:** 542; **4:** 1723

[94-20-2]	**3**: 1655	[125-33-7]	**2**: 480
[94-25-7]	**2**: 542	[125-40-6]	**2**: 508
[94-36-0]	**4**: 1725	[125-42-8]	**2**: 509
[95-4-5]	**2**: 500	[125-58-6]	**2**: 404
[95-25-0]	**2**: 677	[125-60-0]	**2**: 693
[96-50-4]	**3**: 1529	[125-64-4]	**2**: 512
[97-18-7]	**3**: 1279	[125-84-8]	**4**: 1924
[97-23-4]	**3**: 1254	[125-88-2]	**2**: 507
[98-96-4]	**3**: 1140	[125-99-5]	**2**: 697
[99-26-3]	**2**: 731	[126-7-8]	**3**: 1069, 1241; **4**: 1731
[99-66-1]	**2**: 477	[126-27-2]	**2**: 549
[100-33-4]	**3**: 1149	[126-52-3]	**2**: 500
[100-51-6]	**2**: 552; **4**: 2038	[127-7-1]	**4**: 1728
[101-31-5]	**2**: 741	[127-31-1]	**1**: 66; **3**: 1619
[101-40-6]	**2**: 462	[127-33-3]	**3**: 1037
[102-76-1]	**4**: 1731	[127-48-0]	**2**: 479
[103-90-2]	**2**: 348	[127-58-2]	**4**: 2005
[104-6-3]	**3**: 1142	[127-69-5]	**3**: 1134
[108-1-0]	**2**: 652	[128-13-2]	**2**: 756; **3**: 1329
[108-46-3]	**4**: 1724	[128-46-1]	**3**: 1046; **4**: 2162
[108-95-2]	**4**: 1723	[129-3-3]	**2**: 924; **4**: 1720
[109-93-3]	**2**: 441	[129-16-8]	**3**: 1255
[110-44-1]	**4**: 2039	[129-18-0]	**2**: 374
[110-85-0]	**3**: 1284	[129-20-4]	**2**: 373; **4**: 1694
[112-38-9]	**3**: 1255; **4**: 1731	[129-46-4]	**3**: 1188, 1286
[113-18-8]	**2**: 497	[130-26-7]	**3**: 1170; **4**: 1732
[113-45-1]	**2**: 339	[132-17-2]	**2**: 562
[113-53-1]	**2**: 612	[132-18-3]	**2**: 917
[113-59-7]	**2**: 596	[132-20-7]	**2**: 912
[113-73-5]	**3**: 1057	[132-22-9]	**2**: 911; **4**: 1720
[113-92-8]	**2**: 911	[132-92-3]	**4**: 2006
[113-98-4]	**3**: 1017	[132-93-4]	**3**: 1018
[114-7-8]	**3**: 1050; **4**: 1725, 2011, 2164	[132-98-9]	**3**: 1018
[114-49-8]	**4**: 2028	[133-51-7]	**3**: 1154
[114-86-3]	**3**: 1660	[133-67-5]	**1**: 294
[115-38-8]	**2**: 480	[134-49-6]	**2**: 465
[115-44-6]	**2**: 510	[134-72-5]	**2**: 877
[115-63-9]	**2**: 694	[134-80-5]	**2**: 464
[116-38-1]	**4**: 2031	[135-7-9]	**1**: 294
[116-43-8]	**3**: 1137	[135-9-1]	**1**: 294
[117-89-5]	**2**: 594	[135-23-9]	**2**: 906
[118-23-0]	**2**: 908	[135-35-3]	**2**: 905
[118-42-3]	**3**: 1163	[136-47-0]	**2**: 545; **4**: 2003
[120-97-8]	**4**: 2022	[136-69-6]	**2**: 867
[121-25-5]	**4**: 2172	[136-70-9]	**2**: 867
[121-54-0]	**4**: 2038	[137-53-1]	**1**: 22
[122-9-8]	**2**: 462	[137-58-6]	**1**: 125
[123-3-5]	**4**: 2038	[138-56-7]	**2**: 910
[123-63-7]	**2**: 497	[139-12-8]	**4**: 1723
[123-99-9]	**4**: 1725	[140-64-7]	**3**: 1149
[124-72-1]	**2**: 440	[141-22-0]	**2**: 789
[124-90-3]	**2**: 393, 396	[143-62-4]	**1**: 179
[124-94-7]	**3**: 1620; **4**: 2179	[143-76-0]	**2**: 507
[124-99-2]	**1**: 183	[143-81-7]	**2**: 508
[125-2-0]	**4**: 2017	[143-82-8]	**2**: 507
[125-25-7]	**2**: 390	[144-2-5]	**2**: 506
[125-28-0]	**2**: 392	[144-12-7]	**2**: 697

[144-75-2]	**3:** 1143	[314-3-4]	**2:** 925
[144-76-3]	**3:** 1143	[315-72-0]	**2:** 619
[144-80-9]	**4:** 1725, 2005	[316-81-4]	**2:** 594
[144-82-1]	**3:** 1135; **4:** 2005	[317-34-0]	**2:** 888
[145-63-1]	**3:** 1188	[317-52-2]	**2:** 674
[146-22-5]	**2:** 481, 518	[318-98-9]	**1:** 129, 158, 226
[146-48-5]	**4:** 2182	[320-67-2]	**4:** 1875
[146-54-3]	**2:** 591, 744	[321-64-2]	**2:** 653
[147-20-6]	**2:** 917	[322-35-0]	**2:** 569
[147-24-0]	**2:** 496, 742, 909	[329-56-6]	**1:** 60
[147-94-4]	**4:** 1877	[329-65-7]	**3:** 1562
[148-64-1]	**2:** 905	[337-47-3]	**2:** 446
[148-65-2]	**2:** 905	[338-95-4]	**4:** 2178
[148-72-1]	**4:** 2025	[339-43-5]	**3:** 1655
[148-79-8]	**3:** 1287; **4:** 2174	[340-52-3]	**2:** 514
[148-82-3]	**4:** 1887	[340-56-7]	**2:** 513
[149-53-1]	**2:** 865	[342-10-9]	**3:** 1392
[149-64-4]	**2:** 691	[346-18-9]	**1:** 294
[150-13-0]	**4:** 1725	[350-12-9]	**3:** 1255
[151-21-3]	**4:** 2040	[357-7-3]	**2:** 397
[151-67-7]	**2:** 439	[357-8-4]	**2:** 422
[152-2-3]	**2:** 421	[357-56-2]	**2:** 406
[152-11-4]	**1:** 131, 208, 229	[358-52-1]	**2:** 500
[152-97-6]	**3:** 1633	[359-83-1]	**2:** 419
[153-61-7]	**4:** 2006	[362-29-8]	**2:** 921
[153-87-7]	**2:** 609	[363-24-6]	**4:** 2048, 2059
[154-21-2]	**3:** 1067; **4:** 2011	[364-62-5]	**2:** 729, 744
[154-23-4]	**2:** 765	[364-98-7]	**1:** 100
[154-41-6]	**2:** 463	[378-44-9]	**3:** 1625; **4:** 2177
[154-42-7]	**4:** 1728, 1882	[388-51-2]	**2:** 595
[154-69-8]	**2:** 907	[389-8-2]	**3:** 1117
[154-93-8]	**4:** 1891	[390-64-7]	**1:** 208, 229
[155-9-9]	**2:** 625	[396-1-0]	**1:** 299, 75
[156-8-1]	**2:** 461	[406-90-6]	**2:** 442
[287-76-7]	**3:** 1497	[426-13-1]	**3:** 1623; **4:** 2018
[298-46-4]	**2:** 423, 475	[427-51-0]	**3:** 1591; **4:** 1726, 1928
[298-55-5]	**2:** 914	[434-22-0]	**3:** 1575
[298-57-7]	**1:** 208; **2:** 914	[434-43-5]	**2:** 463
[298-81-7]	**4:** 1729	[437-38-7]	**2:** 399, 454; **4:** 2183
[299-39-8]	**1:** 126	[437-74-1]	**1:** 23
[299-42-3]	**2:** 463, 877	[438-60-8]	**2:** 615
[300-48-1]	**2:** 570	[439-14-5]	**2:** 481, 636, 678
[300-62-9]	**2:** 460	[443-48-1]	**3:** 1144
[302-17-0]	**2:** 498	[446-86-6]	**4:** 1728
[302-22-7]	**3:** 1497	[458-24-2]	**2:** 461
[302-23-8]	**3:** 1576	[465-15-6]	**1:** 179
[302-41-0]	**2:** 410	[465-16-7]	**1:** 179, 186
[302-79-4]	**4:** 1724, 1728	[465-21-4]	**1:** 179
[303-25-3]	**2:** 742, 914	[465-22-5]	**1:** 179
[303-45-7]	**3:** 1514	[465-65-6]	**2:** 422
[303-49-1]	**2:** 612	[465-90-7]	**1:** 179, 184
[305-3-3]	**4:** 1889	[466-6-8]	**1:** 183, 186, 191
[306-21-8]	**4:** 2027	[466-9-1]	**1:** 179
[306-52-5]	**2:** 498	[466-99-9]	**2:** 394
[309-29-5]	**4:** 2179	[467-36-7]	**2:** 446
[309-36-4]	**2:** 444	[467-51-69]	**3:** 1334
[309-43-3]	**2:** 510	[468-20-2]	**1:** 179

[468-31-5]	**3**: 1333	[525-66-6]	**1**: 129, 90
[468-65-5]	**2**: 446	[528-92-7]	**2**: 501
[469-21-6]	**2**: 909	[530-8-5]	**2**: 866
[469-62-5]	**2**: 407	[530-43-8]	**3**: 1070
[469-79-4]	**2**: 399	[530-78-9]	**2**: 351; **4**: 1697
[472-29-7]	**3**: 1326	[532-3-6]	**2**: 679
[474-25-9]	**2**: 756; **3**: 1329	[532-11-6]	**2**: 752
[474-45-3]	**3**: 1333	[532-76-3]	**4**: 2003
[475-31-0]	**2**: 749	[533-28-8]	**2**: 543
[477-27-0]	**4**: 1708	[536-21-0]	**1**: 61
[479-18-5]	**2**: 889	[536-33-4]	**3**: 1141
[479-81-2]	**2**: 702	[536-43-6]	**2**: 550
[479-92-5]	**2**: 371	[536-71-0]	**3**: 1150
[480-22-8]	**4**: 1728	[537-46-2]	**2**: 339, 461
[482-15-5]	**2**: 920	[541-22-0]	**2**: 673
[486-12-4]	**2**: 912	[541-79-7]	**2**: 499
[486-16-8]	**2**: 908	[545-26-6]	**1**: 179
[486-47-5]	**2**: 699	[545-74-4]	**2**: 507
[489-15-0]	**4**: 1890	[545-93-7]	**2**: 510
[490-55-1]	**2**: 339; **4**: 2180	[547-32-0]	**4**: 2005
[491-59-8]	**4**: 1728	[548-62-9]	**4**: 1732
[492-18-2]	**1**: 290	[548-66-3]	**2**: 703
[492-39-7]	**2**: 464	[548-68-5]	**2**: 705
[493-80-1]	**2**: 906	[548-73-2]	**2**: 454, 600
[494-79-1]	**3**: 1149	[550-70-9]	**2**: 912
[495-83-0]	**2**: 563	[550-99-2]	**4**: 2029
[496-0-4]	**4**: 2014	[551-11-1]	**3**: 1509; **4**: 2048
[496-21-6]	**4**: 2184	[551-27-9]	**3**: 1019
[496-67-3]	**2**: 501	[554-57-4]	**4**: 2022
[500-92-5]	**3**: 1164	[554-72-3]	**3**: 1150
[502-85-2]	**2**: 449	[554-92-7]	**2**: 910
[505-86-2]	**4**: 2038	[555-30-6]	**1**: 81
[508-52-1]	**1**: 179	[555-57-7]	**2**: 624
[508-75-8]	**1**: 182, 186	[560-53-2]	**1**: 182
[508-77-0]	**1**: 182, 186, 190	[560-54-3]	**1**: 179
[509-86-4]	**2**: 508	[561-27-2]	**2**: 391
[510-62-3]	**1**: 179	[561-77-3]	**2**: 703
[510-63-4]	**1**: 179	[562-9-4]	**2**: 742
[510-90-7]	**2**: 446	[562-10-7]	**2**: 496, 909
[511-12-6]	**1**: 65	[566-48-3]	**4**: 1924
[511-45-5]	**2**: 564	[569-59-5]	**2**: 925
[512-4-9]	**3**: 1569	[569-65-3]	**2**: 496, 743, 916
[512-16-3]	**2**: 752	[577-11-7]	**2**: 782
[512-48-1]	**2**: 502	[579-38-4]	**3**: 1171
[513-10-0]	**4**: 2025	[583-3-9]	**2**: 750
[514-36-3]	**1**: 66	[586-6-1]	**2**: 869
[514-39-6]	**1**: 179	[596-51-0]	**2**: 694
[514-65-8]	**2**: 563	[599-33-7]	**3**: 1628
[516-47-2]	**3**: 1394	[599-79-1]	**3**: 1136
[518-28-5]	**4**: 1912	[603-0-9]	**2**: 888
[518-47-8]	**4**: 2030	[603-50-9]	**2**: 787
[519-37-9]	**2**: 888	[604-75-1]	**2**: 522
[521-74-4]	**3**: 1170	[614-39-1]	**1**: 124
[522-0-9]	**2**: 565	[630-56-8]	**3**: 1504
[522-48-5]	**4**: 2030	[630-93-3]	**1**: 126; **2**: 475
[523-87-5]	**2**: 743	[631-72-1]	**3**: 1334
[524-81-2]	**2**: 925	[632-69-9]	**4**: 2031

[632-99-5]	**3**: 1255	[969-33-5]	**2**: 924
[633-47-6]	**2**: 339	[972-2-1]	**2**: 743
[634-3-7]	**2**: 465	[979-2-2]	**3**: 1570
[636-54-4]	**1**: 295	[980-71-2]	**2**: 911
[637-7-0]	**1**: 24	[982-24-1]	**2**: 597
[637-32-1]	**3**: 1164	[982-57-0]	**3**: 1070
[637-58-1]	**2**: 550	[985-13-7]	**2**: 699
[639-15-6]	**1**: 179	[987-78-0]	**2**: 652
[642-58-0]	**2**: 567	[990-73-8]	**2**: 399, 454
[642-83-1]	**3**: 1426	[1008-65-7]	**2**: 529
[643-22-1]	**3**: 1051	[1018-71-9]	**3**: 1070
[644-62-2]	**2**: 351; **4**: 1698, 2176	[1041-1-6]	**3**: 1524, 1558
[645-5-6]	**4**: 1896	[1041-90-3]	**2**: 418
[652-67-5]	**4**: 2023	[1069-66-1]	**2**: 477
[653-3-2]	**2**: 594	[1070-11-7]	**3**: 1140
[655-35-6]	**1**: 231	[1095-90-5]	**2**: 404
[657-24-9]	**3**: 1660	[1098-97-1]	**2**: 657; **4**: 1706
[664-95-9]	**3**: 1658	[1104-22-9]	**2**: 743
[665-66-7]	**3**: 1185	[1134-47-0]	**2**: 681
[671-16-9]	**4**: 1894	[1142-70-7]	**2**: 509
[673-31-4]	**2**: 645	[1155-49-3]	**2**: 550
[692-13-7]	**3**: 1660	[1156-19-0]	**3**: 1657
[709-55-7]	**1**: 63	[1159-93-9]	**2**: 923
[723-46-6]	**3**: 1132; **4**: 1725	[1163-37-7]	**2**: 699
[738-70-5]	**3**: 1130, 1133; **4**: 1725	[1172-18-5]	**2**: 520
[739-71-9]	**2**: 620	[1176-8-5]	**2**: 910
[742-20-1]	**1**: 294	[1179-69-7]	**2**: 744
[745-65-3]	**4**: 2048, 2060	[1182-87-2]	**1**: 186
[747-36-4]	**3**: 1163	[1225-60-1]	**2**: 920
[747-45-5]	**1**: 124	[1225-65-6]	**2**: 596
[749-2-0]	**2**: 599	[1229-35-2]	**2**: 920
[749-13-3]	**2**: 599	[1240-15-9]	**2**: 921
[751-97-3]	**3**: 1038	[1257-78-9]	**2**: 744
[768-94-5]	**2**: 573	[1263-89-4]	**3**: 1045
[777-11-7]	**4**: 1731	[1264-72-8]	**3**: 1058
[802-31-3]	**4**: 2048	[1304-85-4]	**2**: 731
[804-63-7]	**2**: 675	[1309-42-8]	**2**: 726
[846-49-1]	**2**: 522, 637	[1314-13-2]	**4**: 1723
[846-50-4]	**2**: 523	[1336-20-5]	**4**: 2007
[848-21-5]	**3**: 1580	[1392-21-8]	**3**: 1052
[848-75-9]	**2**: 523	[1393-25-5]	**3**: 1421
[853-34-9]	**2**: 372	[1394-2-1]	**3**: 1066
[865-21-4]	**4**: 1910	[1397-89-3]	**3**: 1065, 1232; **4**: 1732, 2013
[886-74-8]	**2**: 677	[1400-61-9]	**3**: 1065, 1237; **4**: 1732, 2013
[890-98-2]	**2**: 700	[1401-69-0]	**3**: 1054; **4**: 2165
[908-54-3]	**3**: 1150	[1403-66-3]	**4**: 2008, 2163
[911-45-5]	**3**: 1587	[1404-4-2]	**4**: 2008
[914-0-1]	**3**: 1037	[1404-26-8]	**3**: 1057
[915-30-0]	**2**: 408	[1404-90-6]	**3**: 1061; **4**: 2012
[938-73-8]	**2**: 347	[1404-93-9]	**4**: 2012
[943-17-9]	**1**: 63	[1405-10-3]	**3**: 1044
[947-8-0]	**2**: 446	[1405-20-5]	**4**: 2009
[949-36-0]	**2**: 865	[1405-37-4]	**3**: 1060
[958-93-0]	**2**: 907	[1405-41-0]	**3**: 1042; **4**: 2008
[959-24-0]	**1**: 130, 159, 226	[1405-87-4]	**3**: 1057; **4**: 2009, 2169
[961-71-7]	**2**: 907	[1420-55-9]	**2**: 744
[968-81-0]	**3**: 1655	[1421-14-3]	**2**: 447

CAS Registry Number Index

[1461-15-0]	**4**: 2031	[2315-2-8]	**4**: 2029
[1480-19-9]	**2**: 598	[2321-7-5]	**4**: 2030
[1490-4-6]	**4**: 1723	[2323-36-6]	**2**: 570
[1501-84-4]	**3**: 1185	[2338-21-8]	**2**: 922
[1508-65-2]	**2**: 704	[2398-96-1]	**3**: 1247; **4**: 1731
[1508-75-4]	**4**: 2028	[2430-49-1]	**2**: 508
[1508-76-5]	**2**: 564	[2438-72-4]	**2**: 354
[1620-21-9]	**2**: 913	[2447-57-6]	**3**: 1136
[1622-61-3]	**2**: 481	[2508-72-7]	**2**: 905
[1622-62-4]	**2**: 518	[2576-84-3]	**2**: 745
[1634-4-4]	**2**: 758	[2589-47-1]	**1**: 127
[1639-60-7]	**2**: 407	[2609-46-3]	**1**: 74, 299
[1642-54-2]	**3**: 1279	[2622-26-6]	**2**: 593
[1665-48-1]	**2**: 679	[2622-30-22]	**2**: 595
[1668-19-5]	**2**: 613; **4**: 1720	[2624-43-3]	**3**: 1587
[1672-46-6]	**1**: 179	[2668-66-8]	**4**: 2018
[1679-76-1]	**2**: 703	[2691-46-5]	**2**: 702
[1695-77-8]	**4**: 2163	[2709-56-0]	**2**: 597
[1707-14-8]	**2**: 465	[2751-9-9]	**3**: 1052
[1722-62-9]	**4**: 2004	[2751-68-0]	**2**: 595
[1744-22-5]	**2**: 575	[2753-45-9]	**2**: 701
[1778-88-7]	**1**: 179	[2773-92-4]	**2**: 551
[1786-81-8]	**2**: 546; **4**: 2004	[2870-71-5]	**2**: 690
[1808-12-4]	**2**: 908	[2897-83-8]	**2**: 512
[1812-30-2]	**2**: 637	[2922-44-3]	**2**: 406
[1830-32-6]	**2**: 753	[2955-38-6]	**2**: 639
[1841-19-6]	**2**: 600	[2971-90-6]	**4**: 2172
[1867-66-9]	**2**: 448	[3055-99-0]	**2**: 552
[1944-12-3]	**2**: 871	[3056-17-5]	**3**: 1298
[1951-25-3]	**1**: 130	[3092-17-9]	**1**: 63
[1977-10-2]	**2**: 605	[3116-76-5]	**3**: 1022
[1977-11-3]	**2**: 496	[3131-3-1]	**3**: 1059
[1982-37-2]	**2**: 920	[3215-70-1]	**2**: 868
[2011-67-8]	**2**: 522	[3254-89-5]	**2**: 743
[2013-58-3]	**4**: 1725	[3313-26-6]	**2**: 596
[2016-63-9]	**2**: 890	[3339-11-5]	**2**: 912
[2022-85-7]	**3**: 1238; **4**: 2013	[3362-45-6]	**2**: 618
[2030-63-9]	**3**: 1144	[3366-95-8]	**3**: 1156
[2043-38-1]	**1**: 294	[3385-3-3]	**2**: 894
[2045-52-5]	**2**: 907	[3397-23-7]	**3**: 1402
[2058-46-0]	**4**: 2007	[3416-26-0]	**1**: 229
[2058-52-8]	**2**: 604	[3459-20-9]	**3**: 1657
[2062-78-4]	**2**: 601	[3485-14-1]	**3**: 1024
[2062-84-2]	**2**: 599	[3485-62-9]	**2**: 692
[2078-54-8]	**2**: 450	[3486-86-0]	**2**: 509
[2079-0-7]	**3**: 1050	[3505-38-2]	**2**: 908
[2095-57-0]	**2**: 446	[3511-16-8]	**3**: 1024; **4**: 2160
[2135-17-3]	**3**: 1629; **4**: 2178	[3521-62-8]	**3**: 1051
[2152-34-3]	**2**: 340	[3544-35-2]	**2**: 622
[2181-4-6]	**3**: 1617	[3546-3-0]	**2**: 591
[2192-20-3]	**2**: 915	[3546-29-0]	**2**: 446
[2193-87-5]	**3**: 1627	[3546-41-6]	**3**: 1285
[2210-63-1]	**2**: 372	[3562-99-0]	**2**: 752
[2218-68-0]	**2**: 498	[3563-58-4]	**2**: 499
[2259-96-3]	**1**: 294	[3565-72-8]	**2**: 909
[2276-52-0]	**2**: 419	[3567-38-2]	**2**: 500
[2295-58-1]	**2**: 700	[3569-99-1]	**2**: 752

[3572-74-5]	**2**: 910	[5080-50-2]	**2**: 658
[3575-80-2]	**2**: 600	[5104-49-4]	**2**: 361; **4**: 1690, 2019
[3599-32-4]	**4**: 2030	[5119-48-2]	**3**: 1335
[3614-30-0]	**2**: 693	[5250-39-5]	**3**: 1022
[3614-69-5]	**2**: 912	[5355-48-6]	**1**: 186, 189
[3625-6-7]	**2**: 701	[5370-1-4]	**1**: 126
[3689-50-7]	**2**: 921	[5511-98-8]	**1**: 186, 189
[3689-76-7]	**3**: 1255	[5534-9-8]	**2**: 893
[3717-88-2]	**2**: 704	[5534-95-2]	**3**: 1419
[3736-81-0]	**3**: 1171	[5536-17-4]	**4**: 2015
[3737-9-5]	**1**: 125	[5579-13-5]	**2**: 502
[3778-73-2]	**4**: 1890	[5579-16-8]	**4**: 2020
[3810-74-0]	**3**: 1045; **4**: 2009	[5588-25-0]	**2**: 703
[3810-80-8]	**2**: 408	[5588-33-0]	**2**: 592
[3818-50-6]	**3**: 1278	[5593-20-4]	**4**: 1730
[3819-0-9]	**2**: 593	[5632-44-0]	**2**: 912
[3836-23-5]	**3**: 1504	[5633-20-5]	**2**: 704
[3847-29-8]	**3**: 1051	[5634-37-7]	**2**: 498
[3858-89-7]	**2**: 544	[5634-41-3]	**2**: 696
[3922-90-5]	**4**: 2165	[5636-83-9]	**2**: 912
[3930-20-9]	**1**: 130	[5786-21-0]	**2**: 604
[3946-23-4]	**2**: 514	[5817-39-0]	**3**: 1558
[3964-81-6]	**2**: 923	[5870-29-1]	**4**: 2027
[3978-86-7]	**2**: 923	[5875-6-9]	**2**: 544; **4**: 2003
[4146-30-9]	**4**: 2007	[5892-41-1]	**2**: 702
[4205-90-7]	**1**: 82; **2**: 424	[5897-19-8]	**2**: 742
[4205-91-8]	**2**: 424	[5908-99-6]	**2**: 690; **4**: 2027
[4268-36-4]	**2**: 645	[5942-95-0]	**2**: 603, 617
[4291-63-8]	**4**: 1880	[5956-46-7]	**2**: 907
[4299-60-9]	**4**: 2005	[5965-13-9]	**2**: 392
[4310-35-4]	**2**: 697	[5965-49-1]	**2**: 399
[4323-43-7]	**2**: 868	[5966-41-6]	**2**: 703
[4330-99-8]	**2**: 918	[5967-73-7]	**2**: 404
[4342-3-4]	**4**: 1895	[5985-38-6]	**2**: 395
[4360-12-7]	**1**: 127	[5987-82-6]	**2**: 544; **4**: 2003
[4378-36-3]	**2**: 465	[6027-28-7]	**2**: 548
[4394-0-7]	**2**: 352; **4**: 1698	[6028-35-9]	**3**: 1530
[4408-78-0]	**3**: 1183	[6036-95-9]	**2**: 906
[4419-39-0]	**2**: 893; **3**: 1627	[6059-47-8]	**2**: 390
[4428-95-9]	**2**: 763	[6078-56-4]	**2**: 865
[4498-32-2]	**2**: 618	[6108-5-0]	**4**: 2003
[4533-39-5]	**4**: 1903	[6108-5-7]	**2**: 546
[4562-36-1]	**1**: 186	[6130-64-9]	**3**: 1017
[4618-18-2]	**2**: 786	[6138-56-3]	**2**: 907
[4630-95-9]	**2**: 696	[6138-79-0]	**2**: 912
[4697-14-7]	**3**: 1020	[6151-30-0]	**3**: 1158
[4697-36-3]	**4**: 2006	[6153-33-9]	**2**: 925
[4759-48-2]	**4**: 1724	[6153-64-6]	**3**: 1037
[4779-94-6]	**1**: 61	[6168-76-9]	**2**: 339
[4784-40-1]	**2**: 921	[6170-42-9]	**2**: 905
[4800-94-6]	**3**: 1019; **4**: 2006	[6190-39-2]	**1**: 65
[4828-27-7]	**3**: 1634	[6202-23-9]	**2**: 678
[4945-47-5]	**2**: 917	[6211-15-0]	**2**: 389
[4985-25-5]	**2**: 374	[6385-2-0]	**2**: 351
[4985-46-0]	**3**: 1411	[6452-71-7]	**1**: 226
[5003-48-5]	**4**: 1678	[6452-73-9]	**1**: 157
[5051-62-7]	**1**: 84	[6474-85-7]	**2**: 465

CAS Registry Number	Reference
[6493-5-6]	**1:** 48
[6506-37-2]	**3:** 1156
[6533-0-2]	**3:** 1497, 1579
[6533-68-2]	**4:** 2028
[6556-11-2]	**1:** 23
[6577-41-9]	**2:** 696
[6621-47-2]	**1:** 229
[6673-35-4]	**1:** 158, 226
[6736-3-4]	**4:** 2025
[6740-88-1]	**2:** 448; **4:** 2180
[6746-59-4]	**2:** 391
[6785-34-8]	**3:** 1328
[6888-11-5]	**2:** 566
[6893-2-3]	**3:** 1523, 1558
[6980-18-3]	**3:** 1049
[6990-6-3]	**3:** 1071, 1327
[7008-26-6]	**3:** 1635
[7054-25-3]	**1:** 124
[7060-74-4]	**3:** 1051
[7081-38-1]	**2:** 373
[7081-44-9]	**3:** 1021
[7082-21-5]	**1:** 209; **2:** 705
[7104-40-7]	**2:** 869
[7177-48-2]	**4:** 2006
[7195-27-9]	**1:** 73, 295
[7240-38-2]	**3:** 1021
[7242-4-8]	**1:** 188
[7246-14-2]	**3:** 1021
[7246-20-0]	**2:** 498
[7297-25-8]	**1:** 224
[7308-48-2]	**1:** 295
[7361-61-7]	**4:** 2181
[7416-34-4]	**2:** 608
[7432-25-9]	**2:** 514
[7455-39-2]	**2:** 919
[7456-24-8]	**2:** 919
[7481-89-2]	**3:** 1297
[7487-88-9]	**2:** 785
[7488-56-4]	**4:** 1732
[7491-74-9]	**2:** 650
[7527-91-5]	**4:** 1732
[7601-55-0]	**2:** 671
[7647-63-4]	**2:** 919
[7654-3-7]	**2:** 622
[7681-93-8]	**3:** 1065, 1235; **4:** 2013
[7689-3-4]	**4:** 1913
[7704-34-9]	**4:** 1723
[7706-67-4]	**2:** 751
[7722-84-1]	**4:** 2039
[7757-82-6]	**2:** 785
[7761-88-8]	**4:** 2012
[7784-30-7]	**2:** 726
[7790-26-3]	**3:** 1528
[8001-54-5]	**3:** 1254; **4:** 2038
[8002-90-2]	**3:** 1171
[8003-5-2]	**4:** 2037
[8006-54-0]	**4:** 1723
[8009-3-8]	**4:** 1723
[8011-96-9]	**4:** 1723
[8012-95-1]	**4:** 1727
[8025-81-8]	**3:** 1053; **4:** 2166
[8049-47-6]	**4:** 2040
[8052-17-3]	**4:** 1732
[8063-7-8]	**4:** 2163
[8068-28-8]	**3:** 1058; **4:** 2009
[9000-30-0]	**3:** 1664
[9000-65-1]	**2:** 784
[9001-25-6]	**1:** 256
[9001-26-7]	**1:** 255
[9001-27-8]	**1:** 256
[9001-28-9]	**1:** 256
[9001-29-0]	**1:** 257
[9001-30-3]	**1:** 257
[9001-31-4]	**1:** 254
[9001-32-5]	**1:** 254
[9001-73-4]	**4:** 2040
[9001-91-6]	**1:** 258
[9002-18-0]	**2:** 783
[9002-60-2]	**3:** 1373
[9002-61-3]	**3:** 1355
[9002-62-4]	**3:** 1437
[9002-64-6]	**3:** 1363
[9002-67-9]	**3:** 1355
[9002-68-0]	**3:** 1355
[9002-69-1]	**3:** 1445
[9002-71-5]	**3:** 1357
[9002-72-6]	**3:** 1435
[9002-76-0]	**3:** 1418
[9002-89-5]	**4:** 2034
[9002-93-1]	**3:** 1495; **4:** 2039
[9003-11-6]	**4:** 2040
[9003-39-8]	**4:** 2034
[9004-10-8]	**3:** 1442
[9004-32-4]	**4:** 2033
[9004-54-0]	**4:** 2033
[9004-62-0]	**4:** 2033
[9004-64-2]	**4:** 2033
[9004-65-3]	**4:** 2034
[9004-67-5]	**4:** 2034
[9004-99-3]	**4:** 2039
[9007-12-9]	**3:** 1365
[9011-97-6]	**3:** 1414
[9013-55-2]	**1:** 257
[9013-56-3]	**1:** 257
[9014-1-1]	**4:** 2040
[9014-2-2]	**3:** 1062
[9015-71-8]	**3:** 1371
[9016-1-7]	**4:** 1708
[9034-38-2]	**3:** 1350
[9034-39-3]	**3:** 1432
[9034-40-6]	**3:** 1350
[9034-50-8]	**3:** 1399
[9035-54-5]	**3:** 1439
[9035-58-9]	**1:** 255

[9039-61-6]	**1**: 47	[14166-26-8]	**2**: 513
[9041-90-1]	**3**: 1389	[14252-80-3]	**4**: 2003
[9041-93-4]	**3**: 1055	[14357-78-9]	**4**: 2182
[9046-56-4]	**1**: 47	[14402-89-2]	**1**: 101
[9072-41-7]	**2**: 835	[14504-73-5]	**2**: 933
[9083-38-9]	**2**: 581; **3**: 1403	[14521-96-1]	**2**: 393; **4**: 2182
[10024-97-2]	**2**: 438	[14611-51-9]	**2**: 625
[10040-45-6]	**2**: 788	[14636-12-5]	**3**: 1402
[10043-35-3]	**4**: 2038	[14663-23-1]	**2**: 675
[10102-17-7]	**4**: 1732	[14698-29-4]	**3**: 1118
[10118-90-8]	**3**: 1038	[14759-6-9]	**2**: 592
[10238-21-8]	**3**: 1656	[14838-15-4]	**2**: 463
[10262-69-8]	**2**: 620	[14882-18-9]	**2**: 731
[10321-12-7]	**2**: 619	[14918-35-5]	**3**: 1046
[10405-2-4]	**2**: 692	[14929-11-4]	**1**: 25
[10457-90-6]	**2**: 598	[14976-57-9]	**2**: 909
[10539-19-2]	**2**: 699	[15180-3-7]	**2**: 669
[10540-29-1]	**4**: 1922	[15301-69-6]	**2**: 704
[11000-17-2]	**3**: 1398	[15302-12-2]	**2**: 919
[11006-76-1]	**3**: 1059	[15307-79-6]	**2**: 355; **4**: 1685
[11015-37-5]	**3**: 1068; **4**: 2169	[15307-86-5]	**2**: 355
[11018-89-6]	**1**: 181, 186, 190	[15318-45-3]	**4**: 2167
[11041-12-6]	**1**: 21	[15421-84-8]	**1**: 230
[11113-80-7]	**3**: 1050	[15500-66-0]	**2**: 671
[11115-82-5]	**3**: 1058	[15663-27-1]	**4**: 1918
[11121-48-5]	**4**: 2031	[15676-16-1]	**2**: 605
[11128-99-7]	**3**: 1389	[15686-51-8]	**2**: 909
[12040-44-7]	**4**: 2030	[15687-27-1]	**2**: 363; **4**: 1687, 1726
[12192-57-3]	**4**: 1702	[15793-40-5]	**1**: 209; **2**: 705
[12244-57-4]	**4**: 1703	[15825-70-4]	**1**: 224
[12284-76-3]	**2**: 730	[15826-37-6]	**2**: 895, 934; **4**: 1720, 2018
[12627-51-9]	**3**: 1397	[15879-93-3]	**2**: 499
[12687-51-3]	**3**: 1389	[16037-91-5]	**3**: 1154
[13010-47-4]	**4**: 1891	[16051-77-7]	**1**: 224
[13042-18-7]	**1**: 208, 229	[16110-51-3]	**2**: 895
[13055-82-8]	**2**: 873	[16378-21-5]	**2**: 566
[13093-88-4]	**2**: 496	[16449-54-0]	**2**: 351
[13292-46-1]	**3**: 1055	[16452-56-5]	**2**: 546
[13311-84-7]	**3**: 1591; **4**: 1928	[16590-41-3]	**2**: 422
[13355-0-5]	**3**: 1151	[16595-80-5]	**3**: 1280; **4**: 2175
[13392-18-2]	**2**: 706, 871	[16662-46-7]	**1**: 131, 208
[13392-28-4]	**3**: 1185	[16662-47-8]	**1**: 131, 208
[13463-41-7]	**4**: 1732	[16676-29-2]	**2**: 422
[13523-86-9]	**1**: 92, 157, 226	[16679-58-6]	**3**: 1402
[13539-59-8]	**4**: 1695	[16773-42-5]	**3**: 1156
[13636-18-5]	**1**: 208	[17086-28-1]	**3**: 1038; **4**: 2161
[13669-70-0]	**2**: 425	[17090-79-8]	**3**: 1064; **4**: 2170
[13707-88-5]	**1**: 226	[17140-81-7]	**3**: 1145
[13754-56-8]	**2**: 919	[17146-95-1]	**2**: 419
[13764-49-3]	**2**: 393	[17230-88-5]	**3**: 1588
[13838-16-9]	**2**: 441	[17479-19-5]	**2**: 655
[13900-14-6]	**3**: 1254	[17560-51-9]	**1**: 295
[13977-28-1]	**2**: 909	[17617-23-1]	**2**: 520
[13993-65-2]	**4**: 1684	[17650-98-5]	**3**: 1416
[14007-67-1]	**2**: 878	[17784-12-2]	**3**: 1135
[14028-44-5]	**2**: 611	[18010-40-7]	**2**: 547; **4**: 2003
[14107-37-0]	**2**: 453	[18016-80-3]	**2**: 576

CAS Registry Number Index

[18046-21-4] **4**: 1685
[18053-31-1] **2**: 339
[18296-44-1] **2**: 495
[18296-45-2] **2**: 495
[18323-44-9] **3**: 1067; **4**: 1725, 2011
[18378-89-7] **3**: 1066
[18464-39-6] **2**: 624
[18471-20-0] **4**: 1699
[18472-51-0] **4**: 2037
[18507-89-6] **4**: 2172
[18559-59-6] **2**: 868
[18559-94-9] **2**: 874
[18652-93-2] **2**: 444
[18833-13-1] **2**: 889
[18883-66-4] **4**: 1891
[19057-60-4] **3**: 1570
[19216-56-9] **1**: 87; **2**: 706
[19387-91-8] **3**: 1156
[19395-58-5] **2**: 754
[19395-87-0] **4**: 2054
[19396-3-3] **3**: 1050
[19396-6-6] **3**: 1050
[19504-77-9] **3**: 1069
[19562-30-2] **3**: 1120
[19774-82-4] **1**: 130
[19794-93-5] **2**: 632
[19982-8-2] **2**: 574
[20408-97-3] **3**: 1514
[20594-83-6] **2**: 417
[20684-6-4] **2**: 890
[20830-75-5] **1**: 186, 189
[20830-81-3] **4**: 1897
[21187-98-4] **3**: 1656
[21256-18-8] **2**: 381
[21362-8-3] **3**: 1058
[21363-18-8] **2**: 412
[21411-53-0] **4**: 2169
[21645-51-2] **2**: 725
[21649-57-0] **3**: 1019; **4**: 2006
[21715-46-8] **2**: 646
[21730-16-5] **2**: 615
[21738-42-1] **3**: 1283
[21829-25-4] **1**: 210, 229, 95
[21888-98-2] **2**: 565
[21898-19-1] **2**: 878
[22059-60-5] **1**: 125
[22071-15-4] **2**: 365; **4**: 1689, 2176
[22131-79-9] **4**: 1684
[22189-32-8] **3**: 1048
[22199-8-2] **4**: 2005, 2014
[22202-75-1] **3**: 1027
[22204-24-6] **3**: 1285
[22204-53-1] **2**: 366; **4**: 1688
[22232-54-8] **3**: 1530
[22232-71-9] **2**: 466
[22254-24-6] **2**: 891
[22263-79-2] **3**: 1335

[22316-47-8] **2**: 481, 644
[22345-47-7] **2**: 643
[22494-42-4] **2**: 346; **4**: 1678
[22601-59-8] **3**: 1057
[22609-73-0] **1**: 211
[22664-55-7] **4**: 2021
[22733-60-4] **3**: 1072
[22760-18-5] **4**: 1699
[22832-87-7] **3**: 1209; **4**: 1716, 2013
[22876-60-4] **2**: 463
[22888-70-6] **2**: 765
[22916-47-8] **3**: 1209; **4**: 1730, 2013, 2171
[22994-85-0] **3**: 1152
[23031-25-6] **2**: 871
[23031-32-5] **2**: 871
[23047-25-8] **2**: 614
[23092-17-3] **2**: 642
[23110-15-8] **4**: 1937
[23152-29-6] **4**: 2169
[23155-2-4] **3**: 1071
[23214-92-8] **3**: 1039; **4**: 1897
[23239-41-0] **3**: 1029
[23256-30-6] **3**: 1152
[23277-43-2] **2**: 417
[23288-49-5] **1**: 26
[23325-78-2] **3**: 1026
[23327-57-3] **2**: 425
[23465-76-1] **2**: 700
[23541-50-6] **3**: 1039
[23593-75-1] **3**: 1216; **4**: 1730, 2171
[23651-95-8] **2**: 569
[23784-10-3] **2**: 412
[23930-19-0] **2**: 452
[23964-57-0] **2**: 548
[24041-68-7] **3**: 1335
[24168-96-5] **3**: 1222
[24219-97-4] **2**: 621
[24280-93-1] **4**: 1728
[24305-27-9] **2**: 654; **3**: 1353
[24356-60-3] **3**: 1028
[24356-66-9] **2**: 764; **3**: 1179; **4**: 2015
[24356-94-3] **3**: 1504
[24358-65-4] **2**: 703
[24359-81-7] **2**: 905
[24527-27-3] **2**: 593
[24701-51-7] **2**: 618
[24729-96-2] **3**: 1068
[25046-79-1] **3**: 1657
[25122-46-7] **4**: 1727
[25126-32-3] **3**: 1416
[25161-41-5] **2**: 495
[25301-2-4] **4**: 2040
[25322-68-3] **4**: 2034, 2039
[25389-94-0] **3**: 1040
[25416-65-3] **3**: 1523, 1558
[25447-65-8] **2**: 655
[25447-66-9] **2**: 655

[25451-15-4]	**2**: 485	[29975-16-4]	**2**: 524
[25507-4-4]	**3**: 1068; **4**: 2011	[29984-33-6]	**2**: 764
[25546-65-0]	**3**: 1044	[30286-75-0]	**2**: 892
[25614-3-3]	**2**: 576	[30392-40-6]	**2**: 884
[25717-80-0]	**1**: 230	[30392-41-7]	**2**: 884
[25812-30-0]	**1**: 25	[30418-38-3]	**2**: 868
[25953-19-9]	**4**: 2006	[30484-77-6]	**1**: 209; **2**: 915
[25968-91-6]	**2**: 393	[30516-87-1]	**3**: 1188, 1295
[25999-31-9]	**3**: 1064; **4**: 2171	[30544-47-9]	**2**: 350
[26027-38-3]	**3**: 1495	[30578-37-1]	**1**: 64
[26095-59-0]	**2**: 695	[30685-43-9]	**1**: 186, 189
[26159-34-2]	**2**: 366	[30964-13-7]	**2**: 751
[26171-23-3]	**2**: 359; **4**: 1683	[31023-56-0]	**2**: 869
[26305-3-3]	**3**: 1063	[31282-4-9]	**3**: 1047
[26605-69-6]	**3**: 1020; **4**: 2006	[31314-38-2]	**2**: 564
[26629-87-8]	**2**: 629	[31362-50-2]	**3**: 1416
[26658-42-4]	**1**: 22	[31431-39-7]	**3**: 1281
[26675-46-7]	**2**: 442	[31477-60-8]	**3**: 1509
[26787-78-0]	**4**: 2160	[31637-97-5]	**1**: 26
[26807-65-8]	**1**: 295, 73	[31721-17-2]	**2**: 616
[26839-75-8]	**1**: 93; **4**: 2021	[31793-7-4]	**4**: 1693
[26844-12-2]	**1**: 88	[31828-71-4]	**1**: 126
[26864-56-2]	**2**: 601	[31842-61-2]	**2**: 867
[26921-17-5]	**1**: 160; **4**: 2021	[31879-5-7]	**4**: 1690
[26944-48-9]	**3**: 1655	[31883-5-3]	**1**: 133
[27035-30-9]	**4**: 1681	[31884-77-2]	**2**: 916
[27083-27-8]	**4**: 2037	[32222-6-3]	**3**: 1636
[27107-79-5]	**2**: 410	[32266-10-7]	**2**: 868
[27127-79-3]	**1**: 181	[32359-34-5]	**2**: 631
[27164-46-1]	**3**: 1031; **4**: 2006	[32385-11-8]	**3**: 1047; **4**: 2008
[27203-92-5]	**2**: 411	[32421-46-8]	**1**: 132
[27220-47-9]	**3**: 1219; **4**: 1730	[32780-64-6]	**1**: 155
[27223-35-4]	**2**: 641	[32808-9-6]	**2**: 699
[27367-90-4]	**2**: 916	[32838-28-1]	**2**: 502
[27523-40-6]	**3**: 1222	[32886-97-8]	**3**: 1026
[27848-84-6]	**2**: 655	[32953-89-2]	**2**: 867
[27912-14-7]	**4**: 2021	[32986-56-4]	**3**: 1041; **4**: 2009
[28002-70-2]	**4**: 2007	[32988-50-4]	**3**: 1061
[28289-54-5]	**2**: 559	[33005-95-7]	**2**: 368
[28395-3-1]	**1**: 297	[33032-12-1]	**2**: 906
[28657-80-9]	**3**: 1119	[33069-62-4]	**4**: 1915
[28721-7-5]	**2**: 476	[33086-27-0]	**3**: 1529
[28797-61-7]	**2**: 689, 729	[33103-22-9]	**3**: 1061
[28860-95-9]	**2**: 569	[33125-97-2]	**2**: 449
[28981-97-7]	**2**: 640	[33237-74-0]	**1**: 128
[29025-14-7]	**2**: 690	[33279-57-1]	**1**: 182, 186
[29094-61-9]	**3**: 1656	[33286-22-5]	**1**: 132, 213, 229
[29110-47-2]	**1**: 83	[33342-5-1]	**3**: 1657
[29122-68-7]	**1**: 152, 226	[33396-37-1]	**1**: 191
[29216-28-2]	**2**: 920	[33400-45-2]	**2**: 424
[29218-27-7]	**2**: 626	[33419-42-0]	**4**: 1912
[29342-5-0]	**3**: 1244	[33458-93-4]	**4**: 2054
[29400-42-8]	**3**: 1327	[33889-69-9]	**2**: 765
[29462-18-8]	**2**: 642	[34031-32-8]	**4**: 1703
[29767-20-2]	**4**: 1912	[34118-92-8]	**1**: 133
[29782-65-8]	**3**: 1325	[34140-59-5]	**2**: 701
[29782-68-1]	**2**: 765	[34148-1-1]	**4**: 1691

CAS Registry Number Index

[34183-23-8]	**1**: 129	[37932-96-0]	**2**: 931
[34195-34-1]	**2**: 393	[37933-66-7]	**1**: 181
[34273-10-4]	**3**: 1391	[38029-10-6]	**1**: 194; **2**: 876
[34348-60-2]	**4**: 1921	[38129-37-2]	**3**: 1072
[34368-4-2]	**1**: 192	[38194-50-2]	**2**: 358; **4**: 1682
[34433-66-4]	**2**: 405	[38304-91-5]	**1**: 101
[34444-1-4]	**3**: 1030	[38315-43-4]	**4**: 2054
[34493-98-6]	**3**: 1042	[38562-1-5]	**4**: 2059
[34552-83-5]	**2**: 409	[38677-81-5]	**1**: 194; **2**: 876
[34552-84-6]	**2**: 374	[38677-85-9]	**4**: 2176
[34580-13-7]	**2**: 895, 924	[38821-53-3]	**3**: 1027
[34580-14-8]	**2**: 924	[38916-34-6]	**3**: 1446
[34661-75-1]	**1**: 89	[39133-31-8]	**2**: 701
[34779-28-7]	**3**: 1020	[39562-70-4]	**1**: 96, 210, 229
[34786-70-4]	**3**: 1065	[39860-99-6]	**2**: 593
[34866-46-1]	**2**: 881	[39878-70-1]	**3**: 1025
[34866-47-2]	**2**: 881	[40034-42-2]	**3**: 1120
[34911-55-2]	**2**: 631	[40054-69-1]	**2**: 640
[35121-78-9]	**1**: 37; **4**: 2048, 2061	[40180-4-9]	**1**: 295
[35189-28-7]	**3**: 1497; **4**: 1726	[40665-92-7]	**4**: 2054
[35457-80-8]	**3**: 1055	[40666-16-8]	**4**: 2054
[35543-24-9]	**1**: 50	[40762-15-0]	**2**: 523
[35607-66-0]	**3**: 1033	[40819-93-0]	**1**: 133
[35711-34-3]	**2**: 359	[41194-16-5]	**3**: 1047
[35832-9-8]	**4**: 1896	[41340-25-4]	**2**: 382
[35891-93-1]	**1**: 126	[41342-53-4]	**3**: 1051
[35941-65-2]	**2**: 616	[41372-2-5]	**3**: 1017
[36104-80-0]	**2**: 639	[41372-20-7]	**2**: 741
[36167-63-2]	**3**: 1165	[41468-34-2]	**2**: 372
[36282-47-0]	**2**: 411	[41570-61-0]	**2**: 881
[36298-23-4]	**2**: 374	[41621-49-2]	**3**: 1244; **4**: 1731
[36322-90-4]	**2**: 377; **4**: 1696	[41708-72-9]	**1**: 126
[36330-85-5]	**4**: 1686	[41859-67-0]	**1**: 25
[36393-56-3]	**2**: 464	[42116-76-7]	**3**: 1157
[36505-84-7]	**2**: 647	[42200-33-9]	**1**: 92, 156, 226
[36637-19-1]	**2**: 547; **4**: 2003	[42399-41-7]	**1**: 98, 132, 213
[36735-22-5]	**2**: 523	[42408-82-2]	**2**: 415
[36791-4-5]	**2**: 764; **3**: 1181	[42794-76-3]	**1**: 63
[36894-69-6]	**1**: 94	[42835-25-6]	**3**: 1127
[37065-29-5]	**3**: 1121	[42924-53-8]	**2**: 360
[37115-32-5]	**2**: 633, 640	[42971-9-5]	**2**: 656
[37148-27-9]	**2**: 878	[43200-80-2]	**2**: 527
[37213-49-3]	**3**: 1376	[43229-80-7]	**2**: 882
[37221-79-7]	**3**: 1423	[46817-91-8]	**2**: 632
[37248-47-8]	**3**: 1049	[47141-42-4]	**4**: 2021
[37296-80-3]	**1**: 22	[47562-8-3]	**1**: 133
[37321-9-8]	**4**: 2164	[47739-98-0]	**2**: 603
[37332-99-3]	**4**: 2169	[49474-41-2]	**3**: 1448
[37350-58-6]	**1**: 226, 91	[49562-28-9]	**1**: 25
[37517-28-5]	**3**: 1043; **4**: 2163	[49564-56-9]	**2**: 670
[37517-30-9]	**1**: 152, 226, 93	[49745-95-1]	**1**: 192
[37526-80-0]	**3**: 1151	[49760-92-1]	**2**: 581
[37612-13-8]	**1**: 132	[50567-35-6]	**2**: 369
[37640-71-4]	**1**: 128	[50679-8-8]	**2**: 930
[37661-8-8]	**3**: 1025	[50700-72-6]	**2**: 672
[37686-84-3]	**2**: 576	[50838-36-3]	**3**: 1246
[37786-0-8]	**4**: 2054	[51012-32-9]	**2**: 606

[51022-70-9]	**2**: 874	[54504-70-0]	**1**: 26
[51037-30-0]	**1**: 24	[54527-84-3]	**1**: 210, 229
[51264-14-3]	**4**: 1903	[54739-18-3]	**2**: 628
[51322-75-9]	**2**: 681	[54749-90-5]	**4**: 1891
[51333-22-3]	**2**: 894	[54801-81-7]	**2**: 525
[51481-61-9]	**2**: 723; **4**: 1720	[54910-89-3]	**2**: 628
[51481-65-3]	**3**: 1023	[54965-21-8]	**3**: 1277
[51627-14-6]	**3**: 1030	[55028-72-3]	**4**: 2059
[51762-5-1]	**3**: 1028	[55079-83-9]	**4**: 1729
[51773-92-3]	**3**: 1165	[55134-13-9]	**4**: 2170
[51781-21-6]	**1**: 154	[55142-85-3]	**1**: 40
[51803-78-2]	**2**: 383	[55242-55-2]	**2**: 656
[51931-66-9]	**2**: 410	[55268-74-1]	**3**: 1284
[51940-44-4]	**3**: 1120	[55268-75-2]	**3**: 1032
[52093-21-7]	**3**: 1043; **4**: 2008	[55294-15-0]	**1**: 77, 297
[52152-93-9]	**3**: 1029	[55297-95-5]	**3**: 1072; **4**: 2168
[52212-2-9]	**2**: 672	[55726-47-1]	**4**: 1879
[52214-84-3]	**1**: 26	[55837-21-3]	**2**: 917
[52239-63-1]	**2**: 744	[55837-25-7]	**1**: 50
[52365-63-6]	**4**: 2020	[55837-29-1]	**2**: 701
[52443-21-7]	**4**: 1680	[55905-53-8]	**2**: 607
[52468-60-7]	**1**: 209; **2**: 915	[55985-32-5]	**1**: 210
[52479-85-3]	**2**: 657	[56030-54-7]	**2**: 403
[52485-79-7]	**2**: 414	[56180-94-0]	**3**: 1663
[52662-76-7]	**4**: 1891	[56219-57-9]	**3**: 1187
[52806-53-8]	**4**: 1929	[56281-36-8]	**4**: 1724
[53003-10-4]	**3**: 1064	[56341-8-3]	**2**: 880
[53152-21-9]	**2**: 414	[56391-57-2]	**3**: 1048
[53164-5-9]	**2**: 353; **4**: 1680	[56392-17-7]	**1**: 156
[53179-9-2]	**4**: 2008	[56420-45-2]	**4**: 1897
[53179-11-6]	**2**: 409	[56583-43-8]	**2**: 551
[53230-10-7]	**3**: 1165	[56689-45-3]	**3**: 1053
[53583-79-2]	**2**: 606	[56695-65-9]	**4**: 2060
[53643-48-4]	**4**: 1911	[56767-76-1]	**2**: 361
[53648-55-8]	**2**: 416	[56776-1-3]	**2**: 881
[53714-56-0]	**4**: 1932	[56796-20-4]	**3**: 1034
[53772-83-1]	**2**: 597	[56980-93-9]	**1**: 155
[53808-88-1]	**2**: 358	[56985-32-1]	**4**: 2054
[53813-83-5]	**2**: 649	[56985-40-1]	**4**: 2054
[53862-81-0]	**1**: 127	[56995-20-1]	**2**: 424
[53882-12-5]	**2**: 934	[57010-31-8]	**1**: 209
[53885-35-1]	**1**: 40	[57010-32-9]	**1**: 209
[54024-22-5]	**3**: 1497	[57076-71-8]	**2**: 659
[54063-52-4]	**2**: 704	[57109-90-7]	**2**: 638
[54063-53-5]	**1**: 129	[57236-36-9]	**2**: 416
[54063-54-6]	**2**: 872	[57296-63-6]	**1**: 77
[54120-61-5]	**4**: 2054, 2059	[57331-4-1]	**3**: 1325
[54143-55-4]	**1**: 128	[57469-78-0]	**2**: 365
[54143-56-5]	**1**: 128	[57495-14-4]	**2**: 365
[54143-57-6]	**2**: 745	[57526-81-5]	**1**: 193
[54182-58-0]	**2**: 730	[57574-9-1]	**2**: 617
[54188-38-4]	**2**: 625	[57576-44-0]	**3**: 1040
[54240-36-7]	**2**: 880	[57576-52-0]	**4**: 2047
[54340-58-8]	**2**: 417	[57648-21-2]	**2**: 599
[54350-48-0]	**4**: 1729	[57773-63-4]	**3**: 1352; **4**: 1932
[54397-85-2]	**4**: 2047	[57808-66-9]	**2**: 728
[54400-59-8]	**4**: 2175	[57821-32-6]	**3**: 1495

2205

[57982-77-1] **3**: 1352; **4**: 1932
[57982-78-2] **2**: 564
[58001-44-8] **3**: 1035
[58066-85-6] **4**: 1938
[58186-27-9] **2**: 654
[58207-19-5] **4**: 2011
[58581-89-8] **2**: 931
[58786-99-5] **2**: 415
[58934-46-6] **1**: 128
[58957-92-99] **4**: 1898
[58970-76-6] **3**: 1063
[58994-96-0] **4**: 1893
[59122-46-2] **2**: 727; **4**: 2054, 2060
[59128-97-1] **2**: 524
[59209-40-4] **4**: 1681
[59263-76-2] **2**: 417
[59277-89-3] **2**: 763; **3**: 1177; **4**: 2014
[59392-49-3] **3**: 1429
[59467-70-8] **2**: 450, 525
[59467-94-6] **2**: 450
[59703-84-3] **3**: 1023
[59729-31-6] **1**: 128
[59729-33-8] **2**: 627
[59763-91-6] **3**: 1408
[59767-13-4] **2**: 926
[59804-37-4] **2**: 377
[59859-58-4] **2**: 628
[59865-13-3] **4**: 1729
[60142-96-3] **2**: 483
[60282-87-3] **3**: 1497
[60325-46-4] **3**: 1509; **4**: 2054, 2059
[60560-33-0] **2**: 706
[60561-17-3] **2**: 403
[60569-19-9] **2**: 704
[60607-34-3] **2**: 916
[60617-12-1] **3**: 1380
[60628-96-8] **3**: 1212; **4**: 1730
[60643-86-9] **2**: 481
[60719-84-8] **1**: 195
[60762-57-4] **2**: 626
[61197-93-1] **2**: 526
[61260-5-7] **1**: 193
[61318-90-9] **3**: 1226; **4**: 1730
[61318-91-0] **3**: 1226
[61336-70-7] **3**: 1024
[61413-54-5] **2**: 633
[61732-85-2] **2**: 917
[61825-94-3] **4**: 1921
[61849-14-7] **1**: 37; **4**: 2061
[61869-8-7] **2**: 629
[62046-94-0] **3**: 1439
[62305-86-6] **2**: 581
[62305-91-3] **2**: 581
[62340-29-8] **3**: 1428
[62402-31-7] **4**: 1877
[62473-79-4] **2**: 659
[62568-57-4] **2**: 492

[62571-86-2] **1**: 78
[62613-82-5] **2**: 651
[62816-98-2] **4**: 1921
[62893-20-3] **3**: 1030
[62929-91-3] **2**: 884
[62959-43-7] **2**: 477
[63585-9-1] **2**: 763; **3**: 1184
[63610-9-3] **2**: 934
[63638-91-5] **2**: 633
[63659-18-7] **4**: 2021
[63659-19-8] **1**: 153; **4**: 2021
[63675-72-9] **1**: 211, 229
[63677-95-2] **4**: 1881
[63774-77-6] **3**: 1440
[64211-45-6] **3**: 1224
[64211-46-7] **3**: 1224; **4**: 1730
[64221-86-9] **3**: 1035
[64228-81-5] **2**: 669
[64294-95-7] **2**: 926
[64299-19-0] **1**: 193
[64318-79-2] **3**: 1509; **4**: 2059
[64335-73-5] **2**: 907
[64485-93-4] **3**: 1032
[64490-92-2] **2**: 359
[64706-54-3] **1**: 209
[64795-35-3] **2**: 577
[64872-76-0] **3**: 1214
[64872-77-1] **3**: 1214
[64953-12-4] **3**: 1034
[65043-22-3] **2**: 658
[65141-46-0] **1**: 230
[65277-42-1] **3**: 1206; **4**: 1730, 2013
[65472-88-0] **3**: 1249
[65473-14-5] **3**: 1249; **4**: 1731
[65652-44-0] **2**: 876
[65807-2-5] **3**: 1352; **4**: 1932
[65899-73-2] **3**: 1230; **4**: 1730
[66051-63-6] **3**: 1165
[66085-59-4] **1**: 210, 229; **2**: 659
[66104-22-1] **2**: 577
[66195-31-1] **1**: 192, 291
[66309-69-1] **3**: 1031
[66357-35-5] **2**: 724
[66532-85-2] **2**: 349
[66532-86-3] **2**: 349
[66575-29-9] **1**: 196
[66592-87-8] **3**: 1027; **4**: 2160
[66644-81-3] **2**: 608
[66722-44-9] **1**: 153
[66852-54-8] **4**: 1727
[66981-73-5] **2**: 617
[67467-83-8] **4**: 1731
[67763-96-6] **3**: 1439
[67915-31-5] **3**: 1228
[68247-85-8] **4**: 1906
[68291-97-4] **2**: 485
[68302-57-8] **2**: 934

[68401-82-1]	**3:** 1031	[73121-56-9]	**4:** 2054, 2060
[68497-62-1]	**2:** 659	[73151-29-8]	**3:** 1220
[68797-31-9]	**3:** 1219	[73231-24-2]	**4:** 2168
[68813-55-8]	**3:** 1283	[73573-87-2]	**2:** 882
[68844-77-9]	**2:** 927	[73590-58-6]	**2:** 720
[69049-73-6]	**2:** 896, 934	[73771-4-7]	**4:** 1722
[69049-74-7]	**2:** 896, 934	[73836-78-9]	**4:** 2047
[69124-5-6]	**3:** 1180	[74011-58-8]	**3:** 1124
[69388-84-7]	**3:** 1025	[74103-6-3]	**2:** 379
[69431-45-4]	**3:** 1412	[74103-7-4]	**2:** 379
[69552-46-1]	**1:** 38	[74252-25-8]	**2:** 355
[69630-19-9]	**1:** 193	[74381-53-6]	**3:** 1352
[69648-38-0]	**4:** 2054	[74397-12-9]	**4:** 2060
[69655-5-6]	**3:** 1296	[74512-12-2]	**3:** 1223
[69712-56-7]	**3:** 1034	[74578-69-1]	**3:** 1033
[69756-53-2]	**3:** 1165	[74764-40-2]	**1:** 131, 209, 229
[69979-46-0]	**3:** 1182	[74978-16-8]	**2:** 726
[70161-11-4]	**3:** 1054	[75010-43-4]	**4:** 2054
[70209-81-3]	**3:** 1054, 1279	[75330-75-5]	**1:** 27
[70288-86-7]	**3:** 1054; **4:** 2173	[75345-27-6]	**4:** 2038
[70356-3-5]	**3:** 1026	[75507-68-5]	**2:** 424
[70374-39-9]	**2:** 375	[75530-86-6]	**1:** 211
[70384-29-1]	**3:** 1056	[75558-90-6]	**2:** 609
[70458-92-3]	**3:** 1125	[75607-67-9]	**4:** 1880
[70458-96-7]	**3:** 1122	[75695-93-1]	**1:** 212
[70476-82-3]	**4:** 1902	[75696-2-5]	**2:** 523
[70667-26-4]	**4:** 2060	[75715-89-8]	**4:** 2047
[70704-3-9]	**2:** 660	[75738-58-8]	**3:** 1032
[70879-28-6]	**2:** 401	[75821-71-5]	**2:** 358
[71125-38-7]	**2:** 382	[75847-73-3]	**1:** 80
[71160-24-2]	**4:** 2047	[76554-66-0]	**3:** 1186
[71195-58-9]	**2:** 401	[76631-46-4]	**4:** 2181
[71320-77-9]	**2:** 625	[76824-35-6]	**2:** 723
[71439-68-4]	**4:** 1903	[76932-56-4]	**3:** 1352
[71486-22-1]	**4:** 1912	[76963-41-2]	**2:** 723
[71567-776-3]	**3:** 1428	[76990-56-2]	**2:** 609, 624
[71620-89-8]	**2:** 633	[77086-21-6]	**2:** 574
[71675-85-9]	**2:** 607	[77174-66-4]	**3:** 1218
[71751-41-2]	**4:** 2174	[77175-51-0]	**3:** 1218
[71752-56-0]	**2:** 578	[77326-96-6]	**2:** 894
[71771-90-9]	**1:** 193	[77528-1-9]	**2:** 655
[71827-3-7]	**3:** 1279	[77599-17-8]	**4:** 1924
[72025-60-6]	**4:** 2047	[77614-16-5]	**3:** 1385
[72059-45-1]	**4:** 2047	[77671-31-9]	**1:** 195
[72275-67-3]	**3:** 1049	[78110-38-0]	**3:** 1035
[72301-79-2]	**3:** 1185	[78273-80-0]	**2:** 724
[72332-33-3]	**2:** 884	[78415-72-2]	**1:** 195
[72432-3-2]	**3:** 1663	[78613-35-1]	**3:** 1252
[72432-10-1]	**2:** 650	[78613-38-4]	**3:** 1253
[72479-26-6]	**3:** 1220	[78628-80-4]	**3:** 1250
[72496-41-4]	**4:** 1898	[78628-80-5]	**4:** 1731
[72509-76-3]	**1:** 211	[78664-73-0]	**2:** 654
[72558-82-8]	**3:** 1033	[78919-13-8]	**1:** 38; **4:** 2054, 2061
[72803-2-2]	**1:** 212	[79483-69-5]	**4:** 1871
[72814-32-5]	**4:** 2054	[79516-68-0]	**2:** 930
[72956-9-3]	**1:** 154	[79517-1-4]	**3:** 1448
[73110-56-2]	**3:** 1179, 1186	[79547-78-7]	**2:** 930

[79617-96-2]	**2**: 629	[85650-56-2]	**2**: 609
[79645-27-5]	**4**: 2009	[85700-42-1]	**4**: 2051
[79660-53-0]	**3**: 1128	[85700-43-2]	**4**: 2051
[79660-72-3]	**3**: 1128	[85700-44-3]	**4**: 2051
[79794-75-5]	**2**: 932; **4**: 1720	[85721-33-1]	**3**: 1123
[79805-24-6]	**3**: 1386	[85798-8-9]	**2**: 579
[79902-63-9]	**1**: 28	[86030-63-9]	**3**: 1392
[80012-43-7]	**2**: 932	[86197-47-9]	**1**: 194
[80043-53-4]	**3**: 1416	[86304-28-1]	**3**: 1183
[80125-14-0]	**2**: 607	[86347-14-0]	**4**: 2181
[80370-57-6]	**4**: 2160	[86386-73-4]	**3**: 1201; **4**: 1730
[80471-63-2]	**3**: 1600	[86393-37-5]	**3**: 1126
[80474-14-2]	**2**: 895	[86780-90-7]	**1**: 213
[80497-65-0]	**3**: 1361	[86933-74-6]	**3**: 1394
[80879-63-6]	**3**: 1664	[86933-75-7]	**3**: 1394
[81093-37-0]	**1**: 28	[87051-43-2]	**2**: 633
[81098-60-4]	**2**: 728	[87233-61-2]	**2**: 932
[81131-70-6]	**1**: 28	[87248-13-3]	**4**: 2054
[81409-90-7]	**2**: 578	[87495-31-6]	**3**: 1187
[81443-73-4]	**4**: 2054	[87848-99-5]	**2**: 927
[81732-46-9]	**2**: 885	[88058-88-2]	**2**: 577
[81732-65-2]	**2**: 885	[88150-42-9]	**1**: 212
[81733-79-1]	**3**: 1385	[88475-69-8]	**4**: 2061
[81801-12-9]	**1**: 193	[88895-24-3]	**3**: 1382
[81806-67-9]	**4**: 2054	[88895-25-4]	**3**: 1384
[82182-51-2]	**3**: 1333	[88911-35-7]	**4**: 2062
[82337-14-2]	**4**: 2054	[88931-51-5]	**4**: 2062
[82410-32-0]	**3**: 1182	[89365-50-4]	**2**: 875
[82419-36-1]	**4**: 2012	[89663-86-5]	**4**: 2049
[82626-1-5]	**2**: 647	[89778-26-7]	**4**: 1922
[82626-48-0]	**2**: 529	[89838-96-0]	**4**: 2054
[82640-4-8]	**4**: 1924	[89899-81-0]	**3**: 1189
[82785-45-3]	**3**: 1409	[90182-92-6]	**2**: 609
[82952-645-4]	4 1871	[90293-1-9]	**2**: 656
[83314-1-6]	**4**: 1937	[90357-6-5]	**4**: 1930
[83366-66-9]	**2**: 633	[90729-43-4]	**2**: 929
[83380-47-6]	**3**: 1125	[90808-12-1]	**2**: 649
[83480-29-9]	**3**: 1663	[91161-71-6]	**3**: 1250
[83519-4-4]	**4**: 1938	[91188-0-0]	**3**: 1128
[83621-6-1]	**3**: 1223	[91374-21-9]	**2**: 578
[83647-29-4]	**4**: 1924	[91524-16-2]	**1**: 226
[83652-28-2]	**3**: 1367	[92118-27-9]	**4**: 1893
[83799-24-0]	**4**: 1720	[92623-85-3]	**2**: 633
[83851-42-7]	**4**: 2047	[93479-97-1]	**3**: 1656
[83881-51-0]	**2**: 927	[93957-54-1]	**1**: 28
[83881-52-1]	**2**: 927	[93957-55-2]	**1**: 29
[83910-44-5]	**2**: 649	[94079-80-8]	**4**: 2054
[83919-23-7]	**4**: 1722	[94470-67-4]	**2**: 706
[83928-76-1]	**2**: 633	[94749-8-3]	**2**: 875
[84057-84-1]	**2**: 482	[95058-81-4]	**4**: 1879
[84371-65-3]	**3**: 1510, 1600	[95729-65-0]	**2**: 581
[84379-13-5]	**2**: 648	[95734-82-0]	**4**: 1921
[84625-61-1]	**4**: 1730	[96055-63-9]	**4**: 2051
[84625-61-6]	**3**: 1203	[96055-64-0]	**4**: 2051
[84878-61-5]	**4**: 2170	[96055-65-1]	**4**: 2052
[85611-86-5]	**4**: 2051	[96055-66-2]	**4**: 2052
[85650-52-8]	**2**: 633	[97240-79-4]	**2**: 483

[97322-87-7]	**3**: 1666		[115956-13-3]	**2**: 609
[97682-44-5]	**4**: 1914		[116057-75-1]	**4**: 1924
[97979-65-2]	**2**: 514		[116243-73-3]	**3**: 1403
[98049-69-5]	**4**: 2049		[116649-85-5]	**4**: 2054, 2063
[98105-99-8]	**3**: 1127		[117148-67-1]	**3**: 1449
[98224-3-4]	**2**: 609		[117628-82-7]	**3**: 1359, 1362
[98672-91-4]	**4**: 2054		[118292-35-6]	**4**: 1907
[98726-64-8]	**3**: 1359		[118292-36-7]	**4**: 1907
[99294-93-6]	**2**: 529		[118292-40-3]	**4**: 1725
[99300-78-4]	**2**: 633		[118675-50-6]	**4**: 2054
[99592-32-2]	**3**: 1225		[118909-22-1]	**2**: 659
[99592-39-9]	**3**: 1225		[118929-49-0]	**2**: 609
[99593-25-6]	**2**: 526		[119386-96-8]	**2**: 571
[101827-46-7]	**4**: 1731		[119413-54-4]	**4**: 1914
[102625-70-7]	**2**: 720		[119418-4-1]	**3**: 1431
[102676-96-0]	**4**: 1926		[119813-10-4]	**4**: 1907
[103177-37-3]	**4**: 2064		[119975-64-3]	**3**: 1385
[103577-45-3]	**2**: 720		[120287-85-6]	**4**: 1934
[103733-2-4]	**4**: 1935		[120287-93-6]	**4**: 1934
[103878-84-8]	**2**: 571		[120373-24-2]	**4**: 2061
[104592-54-3]	**2**: 659		[120511-73-1]	**4**: 1926
[104625-48-1]	**3**: 1359		[121840-95-7]	**4**: 1926
[104632-26-0]	**2**: 577		[122111-3-9]	**4**: 1879
[105595-17-3]	**4**: 2054		[122320-73-4]	**3**: 1666
[105979-17-7]	**1**: 212		[124904-93-4]	**4**: 1934
[106266-6-2]	**2**: 609		[125572-93-2]	**2**: 579
[106685-40-9]	**4**: 1725		[125617-94-9]	**4**: 2065
[106819-53-8]	**2**: 670		[127166-41-0]	**4**: 2063
[107753-78-6]	**2**: 896; **4**: 1721, 2064		[127779-20-8]	**3**: 1305
[108612-45-9]	**2**: 930		[127932-90-5]	**4**: 1934
[110140-89-1]	**1**: 36		[128312-51-6]	**4**: 2065
[110314-48-2]	**4**: 1907		[129453-61-8]	**4**: 1924
[110588-56-2]	**2**: 930		[129618-40-2]	**3**: 1301
[110617-70-4]	**4**: 2040		[129648-96-0]	**4**: 2063
[110618-38-7]	**2**: 379		[129655-21-6]	**4**: 1907
[111205-55-1]	**2**: 649		[129731-10-8]	**4**: 1926
[111374-72-2]	**2**: 609		[130209-82-4]	**4**: 2060
[111406-87-2]	**2**: 897; **4**: 1721, 2055		[130929-57-6]	**2**: 572
[111745-44-9]	**3**: 1411		[132112-35-7]	**2**: 547
[111757-17-6]	**2**: 572		[132539-7-2]	**2**: 403
[111841-85-1]	**2**: 648		[132747-47-8]	**4**: 2062
[112192-4-8]	**2**: 609		[132875-61-7]	**2**: 403
[112529-15-4]	**3**: 1666		[132875-62-8]	**2**: 403
[112665-43-7]	**4**: 2062		[133454-47-4]	**2**: 609
[112809-51-5]	**4**: 1926		[134308-13-7]	**2**: 572
[112887-68-0]	**4**: 1871		[134457-28-6]	**4**: 1934
[112922-55-1]	**2**: 633		[134523-0-5]	**1**: 29
[112938-42-8]	**3**: 1369		[134523-3-8]	**1**: 29
[112965-21-6]	**4**: 1727		[134578-96-4]	**4**: 2066
[113440-58-7]	**4**: 1908		[134678-17-4]	**3**: 1299
[113665-84-2]	**1**: 41		[135046-48-9]	**1**: 41
[114797-28-3]	**4**: 1908		[135062-2-1]	**3**: 1658
[114949-23-4]	**3**: 1359		[135215-95-1]	**4**: 1935
[114977-28-5]	**4**: 1915		[135558-11-1]	**4**: 1921
[115103-54-3]	**2**: 484		[136236-51-6]	**2**: 571
[115266-92-7]	**4**: 2054		[136310-93-5]	**2**: 892
[115767-74-3]	**4**: 1924		[136470-78-5]	**3**: 1299

2209

[136817-59-9] **3**: 1301
[136989-30-5] **4**: 1934
[138995-18-3] **2**: 477
[139403-31-9] **4**: 2062
[139404-48-1] **2**: 892
[141335-11-7] **4**: 2063
[141807-96-7] **2**: 580
[143201-11-0] **1**: 29
[143443-90-7] **4**: 2063
[143538-27-6] **4**: 2064
[143653-53-6] **1**: 44
[145599-86-6] **1**: 29
[147221-93-0] **3**: 1301
[147398-1-4] **4**: 2066
[148436-63-9] **4**: 2054
[149845-6-7] **3**: 1305
[150378-17-9] **3**: 1312
[151277-78-5] **4**: 1934
[151767-2-1] **2**: 897; **4**: 2064
[152044-53-6] **4**: 1918
[152044-54-7] **4**: 1918
[152608-41-8] **4**: 2065

[153439-40-8] **2**: 929
[153633-1-3] **4**: 2066
[154413-16-3] **4**: 2066
[154598-52-4] **3**: 1302
[155213-67-5] **3**: 1310
[157318-92-8] **4**: 2062
[157716-52-4] **4**: 1938
[157810-81-6] **3**: 1312
[158364-59-1] **2**: 733
[158966-92-8] **2**: 897; **4**: 1721
[159989-64-7] **3**: 1307
[159989-65-8] **3**: 1307
[161172-51-6] **4**: 2065
[161814-49-9] **3**: 1308
[162011-90-7] **2**: 384; **4**: 1701, 1721, 2054
[162362-36-9] **4**: 2066
[169590-42-5] **2**: 381; **4**: 1700, 1721, 2054
[171655-91-7] **2**: 573
[183023-82-7] **4**: 2054
[188062-50-2] **3**: 1299
[190786-44-8] **2**: 929
[744820-23-2] **4**: 2054

Index

A–56620 **3**: 1126
A-75998 **4**: 1935
A-76154 **4**: 1934
A-86929 **2**: 578
1-(Aaminomethyl)cyclohexaneacetic acid **2**: 483
Abacavir **3**: 1299
Abamectin **4**: 2174
Aban **2**: 655
Abasin **2**: 501
Abbott 43 326 **1**: 154
Abciximab **1**: 43
Abecarnil **2**: 648
Abilit **2**: 606
Abortifacients **3**: 1509
Absence seizures **2**: 470
Absentol **2**: 479
Absolute arrhythmia **1**: 118
Absorbonac **4**: 2037
Absorbotear **4**: 2037
Absorption phase **4**: 2130
ABT-431 **2**: 578
Acanthamoeba **3**: 1169
Acarbose **3**: 1663
Accional **2**: 608
Accolate **2**: 897; **4**: 2064
Accumulation **4**: 2129
ACE **1**: 78
Acebutolol **1**: 93, 152, 227
Acecainide **1**: 133
Acecarbromal **2**: 501
Acedapsone **3**: 1143
ACE inhibitors **1**: 78
6-Acelaminopenem-3-carboxylic acid **3**: 972
Acemetacin **2**: 353; **4**: 1680
Acemix **4**: 1680
Acephylline piperazine **2**: 889
Acepifylline **2**: 889
Acepromazine **2**: 591
Acetamide **4**: 2022
1-(4-Acetamidophenoxy)-3-isopropylamino-2-propanol **1**: 158
Acetaminophen **2**: 348
Acetazolamide **1**: 71, 291; **2**: 486; **4**: 2022
Acetexa **2**: 615
Acethropan **3**: 1375
Acetohexamide **3**: 1655
Acetophenazine **2**: 594, 595
1,1′-[3,17-Bis(acetoxy)-androstane-2,16-diyl]bis[1-methylpiperidinium] dibromide **2**: 671
4,4′-[3,17-Bis(acetoxy)androstane-2,16-diyl]bis(1,1-dimethylpiperazinium) dibromide dihydrate **2**: 672
3β-Acetoxy-5-androsten-17-one synthesis **3**: 1572
2-acetoxybenzoic acid **2**: 346

17β-acetoxy-4-chloro-4-androsten-3-one **3**: 1593
17-Acetoxy-6-chloro-4,6-pregnadiene-3,20-dione **3**: 1596
3β-Acetoxydehydropregnenolone synthesis **3**: 1570
cis-(+)-3-Acetoxy-5-(2-dimethylaminoethyl)-2,3-dihydro-2-(4-methoxyphenyl)-1,5-benzothiazepin-4(5H)-one **1**: 132, 213
5-Acetoxy-3-ethenyldodecahydro-6,10,10b-trihydroxy-3,4a,7,7,10a-pentamethyl-[3R-(3α,4aβ,5β,6β,6aα,10α,10bα)]-1H-naphtho[2,1-b]pyran-1-one **1**: 196
17β-Acetoxy-1-methyl-5αandrost-1-en-3-one **3**: 1594
17-Acetoxy-6-methyl-4,6-pregnadiene-3,20-dione **3**: 1596
17-Acetoxy-6α-methyl-4-pregnene-3,20-dione **3**: 1596
17-Acetoxy-4-pregnene-3,20-dione **3**: 1596
4-Acetylamino-N-[2-(diethylamino)ethyl]benzamide **1**: 133
4-(Acetylamino)phenyl N,N-diethylglycinate **2**: 349
2-Acetylamino-1,3,4-thiadiazole-5-sulfonamide **1**: 291; **4**: 2022
8-acetyl-10-[(3-amino-2,3,6-trideoxy-α-L-lyxo-hexopyranosyl)oxy]-7,8,9,10-tetrahydro-6,8,11-trihydroxy-1-methoxy-5,12-naphthacenedione **4**: 1897
N-(4-Acetylbenzenesulfonyl)-N′-cyclohexylurea **3**: 1655
Acetylcarbromal **2**: 501
Acetylcarnitine **2**: 658
Acetylcholine **2**: 328, 688
  in synaptic transmission **2**: 327
  influence on sleep **2**: 491
cis-1-acetyl-4-[4-[[2-(2,4-dichlorophenyl)-2-(1H-imidazol-1-ylmethyl)-1,3-dioxolan-4-yl]methoxy]phenyl]piperazine **3**: 1206
β-Acetyldigoxin **1**: 189
3‴-Acetyldigoxin **1**: 189
4‴-Acetyldigoxin **1**: 189
2-Acetyl-10-(3-dimethylaminpropyl)-10H-phenothiazine **2**: 591
2-Acetyl-10-{3-[4-(2-hydroxyethyl)-1-piperazinyl]propyl}-10H-phenothiazine **2**: 595
2-Acetyl-10-[3-[4-(2-hydroxyethyl)piperidino]propyl]-10H-phenothiazine **2**: 593
N-(3-Acetyl-4-{2-hydroxy-3-[(1-methyl ethyl)amino]propoxy}-phenyl)butanamide **1**: 93
Acetylin **4**: 1678
l-α-Acetylmethadol **2**: 405
Acetyl-β-methylcholine chloride **4**: 2032
1-(2-Acetyl-4-n-butyramidophenoxy)-2-hydroxy-3-isopropylaminopropane **1**: 152
2-(Acetyloxy)benzoic acid **1**: 35
3-(Acetyloxy)-5-[2-(dimethylamino) ethyl]-2,3-dihydro-2-(4-methoxyphenyl)-1,5-benzothiazepin-4(5H)-one **1**: 98

1-[3,17-bis(acetyloxy)-2-(1-piperidinyl)androstan-16-yl]-
    1-methylpiperidinium bromide   **2**: 672
Acetylsalicylic acid   **1**: 15, 35; **2**: 344, 346; **4**: 1678
Acetylspiramycin   **3**: 1053
N-[3-Acetyl-4-(3′-*tert*-butylamino-2′-hydroxy)propoxy]-
    phenyl-N′-diethylurea   **1**: 155
Acevaltratum   **2**: 495
Acholeplasmataceae   **3**: 1111
Acid-fast bacilli   **3**: 1138
Acid pump antagonists   **2**: 732
Acid-related diseases
    of the gastrointestinal tract   **2**: 713
Acimetten   **4**: 1678
Acipimox   **1**: 24
Acitretin   **4**: 1729
Aclacinomycin   **3**: 955
Aclacinomycin A   **4**: 1901
Aclarubicin   **3**: 981
Acne vulgaris   **4**: 1724
Acovenosigenin   **1**: 179
ACP C-VET   **2**: 591
Acrisorcin   **4**: 1732
Acrivastine   **2**: 927
Actal   **2**: 726
Actamer   **3**: 1279
Actaplanins   **4**: 2169
Amoxycillin   **3**: 1024
Amperozide   **2**: 609
Amphenicols   **4**: 2167
Amphetamine   **2**: 460, 467
Ampho-Moronal   **3**: 1233
Amphomycin   **3**: 1008
Amphotericin B   **3**: 1065, 1232; **4**: 1732, 2013
Amphoteric surfactants   **4**: 2103
Ampicillin   **3**: 962, 1022; **4**: 2006, 2160
    staphylococcus aureus inhibition   **3**: 1091
Ampicillin ethoxycarbonyloxyethyl hydro-
    chloride   **3**: 1025
Ampicillin phthalidyl   **3**: 1025
Amplan   **2**: 593
Amprenavir   **3**: 1308
Amprolium   **4**: 2172
Amprovine   **4**: 2172
Amrinone   **1**: 195
m-AMSA   **4**: 1903
Amsacrine   **4**: 1903
Amsal   **2**: 507
Amsebarb   **2**: 507
Amsidine   **4**: 1904
Amudane   **3**: 1241
Amuno   **2**: 358; **4**: 1680
Amycal   **2**: 507
Amycor   **3**: 1213
Amydorm   **2**: 507
α-*Amylase*   **3**: 1662
Amylbarb Sodium   **2**: 507
Amylin   **3**: 1369

Amylobeta   **2**: 507
Amyloid precursor protein   **3**: 1397
Amyloid A4 Protein   **3**: 1397
Amytal   **2**: 507
Amytal Sodium   **2**: 507
Anabar   **2**: 348
Anabet   **1**: 156
Anaboleen   **3**: 1594
Anabolic Agents   **3**: 1588
Anabol-Tablinnen   **3**: 1594
Anadrol   **3**: 1590
Anaesthesie   **2**: 550
Anaesthesin   **2**: 542
Anafranil   **2**: 612
Anagregal   **1**: 41
Anahelp   **1**: 60
Ana-Kit   **1**: 60
Analeptics   **2**: 339
Analgesics   **2**: 341
Analgizer   **2**: 442
Analux   **2**: 652
Anamenth   **2**: 439
Anandron   **4**: 1929
Ananxyl   **2**: 647
Anaphylactic reactions   **2**: 902
Anaphylaxie-Besteck   **1**: 60
Anaplasmataceae   **3**: 1110
Anaprime   **4**: 2178
Anaprox   **4**: 1689
Anastrazole   **4**: 1926
Anatensol   **2**: 595
Anaus   **2**: 911
Anavar   **3**: 1594
Anboxan   **4**: 2062
Ancatropine Infant   **4**: 2028
Anco   **2**: 365
Ancobon   **3**: 1239; **4**: 2013
Ancotil   **3**: 1239
Ancrod   **1**: 47
Ancylostoma duodenale   **3**: 1273
Ancylostoma species   **3**: 1275
Andergin   **3**: 1210
Androgen antagonists   **4**: 1928
Androgens   **3**: 1564, 1588
Androlin   **3**: 1590
Andrometh   **3**: 1590
Androstadienedione
    production by fermentation   **3**: 1578
Androstane gestagens   **3**: 1598
Androstanolone   **3**: 1594
Androstanolone derivatives   **3**: 1594
Androstanolone propionate   **3**: 1594
5-Androstene-3β,17β-diol   **3**: 1565
Androstene-3,17-dione   **3**: 1565
Androstenolone synthesis   **3**: 1572
Androsterone   **3**: 1588
Anectine chloride   **2**: 674

Anesthesia
   stages   **2**: 435
Anesthetics   **4**: 2180
   general   **2**: 435
   local   **2**: 539
   ophthalmological   **4**: 2002
Anetamin   **2**: 450
Anethole trithione   **2**: 752
Aneural   **2**: 645, 679
ANF   **3**: 1405
Angass   **2**: 731
Angettes   **4**: 1678
Angilol   **1**: 159
Angina   **1**: 217
Anginal   **1**: 39
Angina pectoris   **1**: 217
Angionorm   **1**: 66
Angiotensin   **1**: 78
Angiotensin–Kinin System   **3**: 1388
Angiotensins   **3**: 1389, 1547
Angolamycin   **3**: 999
Angopril   **1**: 209
ACTH   **3**: 1543
Pro-Actidil   **2**: 913
Actifed   **2**: 913
Actinomycetaceae   **3**: 1110
Actinomycetes   **3**: 1110
Actinomycin   **3**: 953
Actinomycin D   **3**: 1062; **4**: 1904
Actinomycins   **3**: 1007
Actinospectacin   **3**: 984
Action potential amplitude   **1**: 119
Action potential duration (APD)   **1**: 119
Action potentials
   cardiac   **1**: 118
Actiphan   **2**: 645
Activated factor V   **1**: 256
Active immunization   **4**: 1762
Active targeting   **4**: 2122
Activins   **3**: 1358, 1359, 1513
Actol   **2**: 353; **4**: 1699
Actos   **3**: 1666
Acular   **2**: 380
Acumen   **2**: 652
Acupan   **2**: 426
Acute pain   **2**: 343
Acute toxicity   **4**: 2131
Acyclovir   **2**: 763; **3**: 1177; **4**: 2014
3-Acylamino-2-oxoazetidine-1-sulfonic acid   **3**: 976
Adalat   **1**: 95, 210
Adalin   **2**: 501
1-Adamantanamine hydrochloride   **3**: 1185
Adapalene   **4**: 1725
Adapettes   **4**: 2037
Adapin   **2**: 613
Adapress   **1**: 210
ADD   **4**: 1903

Addison's disease   **3**: 1543
Additives
   in fermentation of antibiotics   **3**: 1076
Adenosine A1 receptor antagonists   **1**: 300
Adenosine A_{2A} Antagonists   **2**: 580
Adenosine triphosphatase   **1**: 186
Adenoviridae   **3**: 1174
Adenylate cyclase stimulants   **1**: 196
Adepril   **2**: 611
Adiclair   **3**: 1238
Adinazolam   **2**: 633, 640
Adiparthrol   **2**: 461
Adipex-P   **2**: 467
Adiposetten   **2**: 464
Adityl   **2**: 501
Adjuvants   **4**: 2101
Adlone   **2**: 658
ADME   **4**: 2080, 2129
Administration of drugs   **4**: 2081
Admon   **1**: 211
Adofen   **2**: 628
Adolonta   **2**: 412
Adozelesin   **4**: 1907
Adprin   **4**: 1678
Adrane   **2**: 934
Adrenalin   **2**: 865
Adrenalin 1 : 10000   **1**: 60
Adrenalin 1 : 1000 JENA-PHARM   **1**: 60
Adrenalina ISM 1 : 1000   **1**: 60
Adrenaline   **1**: 60; **2**: 864; **3**: 1550, 1560, 1562; **4**: 2020
   in synaptic transmission   **2**: 327
Adrenaline bitartrate   **4**: 2020
Adrenal Medulla Hormones   **3**: 1560
Adrenal Steroid Hormones   **3**: 1601
Adrenal Steroids   **3**: 1552
Adrenam   **1**: 64
Adrenergic agents
   ophthalmological   **4**: 2029
β-Adrenergic blocking agents
   antiglaucomatons   **4**: 2021
Adrenergic Neuron Blocking Agents   **1**: 85
β-Adrenergie blocking agents
   Side effects   **1**: 150
Adrenoceptor Agonist bronchodilators   **2**: 864
α-Adrenoceptor Antagonists   **1**: 87, 89
β-adrenoceptor blocking agents
   treatment of ischemic heart disease   **1**: 225
β-Adrenoceptor blocking drugs   **1**: 137
   classification   **1**: 140
   Clinical uses   **1**: 144
   Pharmacodynamics   **1**: 139
   Pharmacokinetics   **1**: 143
Adrenoceptors
   Physiological effects   **1**: 138
Adriamycin   **4**: 1897
Adrostenolone   **3**: 1568

Adroyd **3**: 1590
AdsorboTear **4**: 2034
Adsorbo Tears **4**: 2033
Adumbran **2**: 522
Adverse drug reactions **4**: 2136, 2138
Advil **4**: 1688
Aerosol dosage forms **4**: 2100
Aerosol product containers **4**: 2108
Aether puriss **2**: 440
Aethoxysklerol **2**: 552
Affirm **4**: 2174
Aflodac **4**: 1683
Aflorix **3**: 1210
Afloxan **4**: 1682
Afluteston **3**: 1590
African Eye Worm **3**: 1277
African Trypanosomiasis **3**: 1148
Aftate **3**: 1248
AG 331 **4**: 1871
Agapurin **1**: 49
Agar **2**: 783
Agar Diffusion Test **3**: 1089
Agarol **2**: 781, 787
Agaroletten **2**: 782
Agedal **2**: 619
Agenerase **3**: 1309
Ageroplas **4**: 1700
Aggregation
 of blood platelets **1**: 30
Agicholit **2**: 749
Agilease **1**: 39
Agiolax **2**: 783
Agisten **3**: 1217
Agit depot sanol **1**: 66
Agit plus sanol **1**: 66
Agivilen **2**: 650
Agkistrodon rhodostoma venom proteinase **1**: 47
Agkistrodon serine proteinase **1**: 47
Aglycones **1**: 177
Agnosterol **3**: 1326
Agon **1**: 211
Agopton **2**: 720
Agradil **2**: 608
Agréal **2**: 608
Agriculture antibiotics **3**: 957
Agrypnal **2**: 480
Aguipiran **3**: 1285
Agupla **4**: 1921
DNA-gyrase **3**: 1117
AH 13205 **4**: 2054
AH 22921 **4**: 2054
AH 23848 **4**: 2054
AH 5158 **1**: 155
AH 6809 **4**: 2054
A-H injection **4**: 2184
AHR-11190 **2**: 609
AIDS **3**: 1291; **4**: 1832

AIDS Therapeutics **3**: 1291
Aiglonyl **2**: 606
Airlift fermenter **4**: 1990
Ajan **2**: 426
(17R,21R)-Ajmalan-17,21-diol **1**: 127
Ajmaline **1**: 14, 127
Ajmaline-17-chloroacetate **1**: 133
Akarpine **4**: 2024
Akatinol-Memantine **2**: 574
Ak-Chlor **4**: 2011
Ak-Cide **4**: 2005
Ak-Con **4**: 2029
Ak-Dex **4**: 2018
Ak-Dilate **4**: 2026, 2029
Ak-Fluor **4**: 2031
Ak-Homatropine **4**: 2028
Akinesia
 in Parkinsonism **2**: 556
Akineton **2**: 563
Aklavin **3**: 981
Ak-Mycin **4**: 2011
Akneclor **3**: 1217
Ak-Neo-Cort **4**: 2016
Ak-Nephrin **4**: 2029
Ak-Pentolate **4**: 2027
Ak-Poly-Bac **4**: 2009
Ak-Poly-Cac **4**: 2010
Ak-Pred **4**: 2017
Ak-Spore **4**: 2008, 2009, 2010, 2016
Ak-Sulf **4**: 2005
Ak-Taine **4**: 2003
Ak-Tate **4**: 2017
Aktil **4**: 1704
Aktren **4**: 1688
Ak-Trol **4**: 2008, 2009, 2016, 2017
Ak-Vernacon **4**: 2026, 2029
Alamon **2**: 646
Alapryl **2**: 643
Alaton **2**: 653
Albalon **4**: 2029
Albego **2**: 639
Albendazole **3**: 1272, 1274, 1277
Alber-T **3**: 1248
Albistat **3**: 1210
Albumin **1**: 252, 263
 therapeutic use **1**: 278
Albuterol **2**: 874
Alcaine **4**: 2003
Alclofenac **4**: 1684
Alclometasone dipropionate **3**: 1636
Alcoban **3**: 1239
Alcobon **3**: 1239
Alcomicin **4**: 2008
Alcon **4**: 2027
Alcon Opulets Benoxinate **4**: 2003
Alcon Opulets Cyclopentolate **4**: 2027
Alcopar(a) **3**: 1278

Alcuronium dichloride **2**: 669
Aldactone **3**: 1617
Aldipin **1**: 210
Aldosterone **3**: 1602, 1615
Aldosterone antagonists **1**: 298
Alemoxan **2**: 605
Alepam **2**: 522
Alesion **2**: 932
Aleudrin **2**: 866
Aleviatin **2**: 486
Alfabet **2**: 463
Alfadat **1**: 210
Alfenta **2**: 403
Alfentanil **2**: 401
Alfentanyl **2**: 453
Alfospas **2**: 701
Algifor **4**: 1688
Algocetil **4**: 1683
Algocor **1**: 208
Alimemazine **2**: 590, 591
Alimemazone **2**: 918
Alkaloids
   steroid **3**: 1333
Alka Seltzer **2**: 346; **4**: 1678
1-O-Alkyl-2-acetyl-sn-glycero-3-phosphocholine
   (platelet activating factor **4**: 1938
Alkylamines antihistamines **2**: 911
Alkylating agents
   cancer therapy **4**: 1885
6-Alkylpenem-3-carboxylic acid **3**: 972
Allegra **2**: 929
Allegron **2**: 615
Allerest Eye Drops **4**: 2029
Allergan **2**: 566
Allergan Enzymatic **4**: 2040
Allergan Homatropine **4**: 2028
Allergan Hydrocare **4**: 2035
Allergefon **2**: 908
Allergen **2**: 902
Allergen Preparations Desensitization **2**: 935
Allergens
   Properties **2**: 936
Allergen Standardization **2**: 936
Allergen Vaccines **2**: 940
Allergic asthma **2**: 861
Allergic contact dermatitis **4**: 1718
Allergic diseases **2**: 901
Allergodil **2**: 931
Allergy Drops **4**: 2029
Allobarbital **2**: 504, 508
Allobarbitone **2**: 508
All-*trans*-retinoic acid **4**: 1724
5-Allyl-5-(2-cyclopenten-1-yl(barbituric acid **2**: 510
1-[(6-Allylergolin-8(-yl)carbonyl]-1-[3-(dimethylamino)-
   propyl]-3-ethylurea **2**: 578
Allylestrenol **3**: 1598
17α-allyl-17β-hydroxy-4-estrene **3**: 1598

1-*N*-Allyl-3-hydroxymorphinane **2**: 421
2-allyl-2-isopropylacetylurea **2**: 501
5-Allyl-5-isopropylbarbituric acid **2**: 507
5-Allyl-5-(1-methylbutyl)-2-thiobarbituric acid **2**: 446
5-Allyl-5-(2-pentyl)barbituric acid **2**: 510
N-[(1-allyl-2-pyrrolidinyl)methyl]-5-sulfamoyl-5-veratra-
   mide **2**: 608
5-Allyl-5-*sec*-butylbarbituric acid **2**: 510
Almazine **2**: 638
Almizol **2**: 929
Alnert **2**: 656
Alodorm **2**: 518
Aloe-emodin **2**: 790
Aloe-emodinanthrone-10-C-glycoside **2**: 790
Aloes **2**: 790
Aloid **3**: 1210
Aloin **2**: 790
Alomide **2**: 934
Alonimid **2**: 512
Alonimide **2**: 513
Alopam **2**: 522
Alotec **2**: 870
Alphadolone **2**: 453
Alphadrol **3**: 1632
Alphadryl **2**: 910
Alphaxalone **2**: 452
Alpidem **2**: 647
Alprazolam **2**: 640
Alprenolol **1**: 227
Alprostadil **4**: 2060
Alrheumat **4**: 1690
Alrheumun **4**: 1690
Alsactide **3**: 1375
Alserin **1**: 86
Althesin **2**: 453
Altilev **2**: 615
Altinal **2**: 507
Altren **4**: 1680
Altretamine **4**: 1896
Aludrin **2**: 866
Aluminum acetate **4**: 1723
Aluminum hydroxide **2**: 725
Aluminum phosphate **2**: 726
Alupent **2**: 870
Alurate **2**: 508
Alzet osmotic pump **4**: 2116
AM – 715 **3**: 1123
AM – 833 **3**: 1128
Amantadine **2**: 573; **3**: 1184
Amaryl **3**: 1656
ProAmatine **1**: 63
Amazolon **2**: 574
Ambene **2**: 374; **4**: 1694
Ambilhar **3**: 1283
Ambilhar Ciba **3**: 1283
AmBisome **3**: 1233
Ambodryl **2**: 908

Amdinocillin pivoxil **3:** 1026
Amebas **3:** 1168
Amebiasis **3:** 1168
Amecrin **4:** 1904
Amenorone **3:** 1598
American trypanosomiasis **3:** 1151
amezinium **1:** 64
Ameziniummetilsulfate **1:** 64
Amfebutamone **2:** 631
Amfepranon **2:** 464
Amforol **4:** 2163
Amicel **3:** 1219
N-Amidino-3,5-diamino-6-chloropyrazinecarboxa-
    mide **1:** 74
N-Amidino-2-(2,6-dichlorophenyl) acetamide **1:** 83
Amidonal **1:** 128
Amifloxacin **3:** 1126
Amiglyde-V **4:** 2163
Amikacin **3:** 985, 987, 1043; **4:** 2163
Amiloride **1:** 74, 299
Amine Precursor Uptake and/or Decarboxylation
    cells **3:** 1538
Amineptine **2:** 617
5-[(2-aminoacetamido)methyl]-1-[4-chloro-2(2-chloro-
    benzoyl)-phenyl-N,N-dimethyl-1H-1,2,4-triazole-
    3-carboxamide **2:** 526
Amino acid analysis
    in interferon analysis **4:** 1973
Amino Acid Hormones **3:** 1555
Amino acids three letter symbols **3:** 1343
[1aR-(1aα,8β,8aα,8bα)]-6-amino-8-[(aminocarbonyl)ox-
    ymethyl]-1,1a,2,8,8a,8b-hexahydro-8a-methoxy-5-
    methylazirino [2',3':3,4]pyrrolo[1,2-a]indole-4,7-
    dione **4:** 1896
4-Aminobenzoic acid **3:** 1129
(E)-2-Amino-6-benzoyl-1-(isopropylsulfonyl)benz-
    imidazoleoxime **3:** 1185
Aminobenzyl penicillin **3:** 1022
4-Amino-N-(1-benzyl-4-piperidyl)-5- chloro-2-methoxy-
    benzamide **2:** 607
5-Amino-(3,4'-bipyridine)-6(1H)one **1:** 195
γ-Aminobutyric acid
    in Parkinsonism **2:** 557
3-(Aminocarbonyl)-O^4-deacetyl-3-de(meth-
    oxycarbonyl)vincaleukoblastine **4:** 1911
Aminocatechol bronchodilators **2:** 864
7-Aminocephamycinoic acid **3:** 968
4-Amino-5- chloro-N-[2-(diethylamino)ethyl]-2-methoxy-
    benzamide **2:** 607
4-Amino-5-chloro-N-[2-(diethylamino)-ethyl]-o-anisa-
    mide **2:** 729
cis-4-Amino-5-chloro-N-[1-[3-(p-fluorophenoxy)-propyl]-
    3-methoxy-4-piperidyl]-o-anisamide **2:** 728
γ-Amino-β-(4- chlorophenyl)butyric acid **2:** 681
1-(4-amino-3-chloro-5-trifluoromethylphenyl)-2-(tert-bu-
    tylamino)ethanol **2:** 880

Aminocyclitols
    veterinary **4:** 2162
(1S,4R)-4-[2-Amino-6-(cyclopropylamino)-9H-purin-9-
    yl]-2-cyclopentene-1-methanol **3:** 1299
4-Amino-1-β-D-arabinofuranosyl-2(1H)-pyrimi-
    done **4:** 1877
Aminodeoxykanamycin **3:** 986
[1-Amino-3-[[[2-[(diaminomethylene)-amino]-4-thiazo-
    lyl]methyl]thio]propylidene]-sulfamide **2:** 723
5-Amino-2-[1-(3,4-dichlorophenyl)-ethyl]-2,4-dihydro-
    3H-pyrazol-3-one **1:** 77
1-(4-Amino-3,5-dichlorophenyl)-2-(tert-butylamino)etha-
    nol **2:** 878
4-Amino-N-(2-diethylaminoethyl)benzamide **1:** 124
4-Amino-5,6-dihydro-1-β-D-ribofuranosyl-1,3,5-triazine-
    2(1H)one · monohydrochloride **4:** 1877
2-Amino-1,9-dihydro{[2-hydroxy-1-(hydroxymethy-
    l)ethoxy]methyl}-6H-purin-6-one **3:** 1182
2-Amino-1,7-dihydro-6H-purine-6-thione **4:** 1882
2-Amino-1- (3,4-dihydroxyphenyl)ethanol **3:** 1562
2-amino-N-[2-(2,5- dimethoxyphenyl)-2-
    hydroxyethyl]acetamide **1:** 62
1-(4-Amino-6,7-dimethoxy-2-quinazolinyl)-4-(2-furanyl-
    carbonyl)piperazine **1:** 87
4-Amino-N-(3,4-dimethyl-5-isoxazolyl)benzenesulfona-
    mide diethanolamine **4:** 2005
4-Amino-1-β-D-ribofuranosyl-1,3,5-triazine-2(1H)-
    one **4:** 1875
2-[(2-Aminoethoxy)methyl]-4-(2-chlorophenyl)-1,4-di-
    hydro-6-methyl-3,5-pyridinedicarboxylic acid 3-
    ethyl 5-methyl ester **1:** 212
4-(2-Aminoethyl)-1,2-benzenediol **1:** 60, 92
[R-(R*,R*)]-α-(1-Aminoethyl)benzenemethanol **2:** 464
N-(2-aminoethyl)-5-chloropyridine-2-carboxa-
    mide **2:** 571
4-Amino-N-[(1-ethyl-2-pyrrolidinyl)methyl]-5-(ethylsul-
    fonyl)-2-methoxybenzamide **2:** 607
4-Amino-5-fluoro-2(1H)-pyrimidinone **3:** 1238
Aminoglutethimide **4:** 1924
Aminogluthetimid **3:** 1554
Aminoglycoside
    economic aspects **3:** 1094
Aminoglycoside antibiotics **4:** 2007
Aminoglycosides **3:** 981
    veterinary **4:** 2162
4-Amino-5-hexenoic acid **2:** 481
1-N-[(S-)-4-Amino-2-hydroxy-butyryl]kanamy-
    cin **3:** 987
(2R,5S)-4-Amino-1-[2-(hydroxymethyl)-1,3-oxathiolan-5-
    yl]-2(1H)-pyrimidinone **3:** 1299
8-Amino-7-iodo-1,4-benzodioxan-5-carboxylic acid 1-
    butylpiperidin-4-yl-methyl ester **2:** 819
4-Amino-6-methoxy-1-phenylpyridazinium methyl sul-
    fate **1:** 64
α-(Aminomethyl)-3-hydroxybenzenemethanol **1:** 61
8-[(4-Amino-1-methylbutyl)amino]-6-methoxyquinoline
    phosphate **3:** 1163

2216

2-Amino-7-(1-methylethyl)-5-oxo-5H-[1]benzopyr-
  ano[2,3-b]pyridine-3-carboxylic acid **2:** 934
4-Amino-N-(4-methyl-2-pyrimidinyl)benzenesulfona-
  mide sodium salt **4:** 2005
4-Amino-N-(5-methyl-1,3,4-thiadiazol-2-yl)benzenesulfo-
  namide **4:** 2005
15-Amino-1-methyltricyclo[7.5.1.0^{127,255}]pentadeca-
  2,4,6-trien-4-ol **2:** 416
(±)–2-Amino-N-(2,6-xylyl)propionamide **1:** 126
(±)-2-(4-Aminophenyl)-2-ethylglutarimide **4:** 1924
N-[(4-Aminophenyl)sulfonyl]acetamide **4:** 2005
[3S-3R*(1R*,2S*)]-3-[[(4-aminophenyl)sulfonyl](2-
  methylpropyl)amino]-2-hydroxy-1-(phenylmethyl)prop **3:** 1308
Aminophylline **2:** 888
Aminopon **3:** 1448
(S)-(–)-2-Amino-6-(propylamino)-4,5,6,7-tetrahydroben-
  zothiazole **2:** 577
(±)-4-(2-Aminopropyl)phenol hydrobromide **4:** 2027
4-Amino-N-2-pyrimidinylbenzene-
  sulfonamide **4:** 2005, 2014
4-Aminosalicylic acid **3:** 1141
Aminosidin **3:** 984
3-(Aminosulfonyl)-4-chloro-N-(2,3-dihydro-2-methyl-
  1H-indol-1-yl)benzamide **1:** 73
5-(Aminosulfonyl)-4-chloro-2-[(2-furanylmethyl)ami-
  no]benzoic acid **1:** 72
N-[5-(Aminosulfonyl)-3-methyl-1,3,4-thiadiazol-2(3H)-
  ylidene]-acetamide **4:** 2022
N-[5-(Aminosulfonyl)-1,3,4-thiadiazol-2-yl]aceta-
  mide **2:** 486
9-Amino-1,2,3,4-tetrahydroacridine **2:** 653
Aminothiazole **3:** 1529
2-Aminothiazole **3:** 1529
(8S-cis)-10-[(3-Amino-2,3,6-trideoxy-β-L-arabino-hexo-
  pyranosyl)oxy]-7,8,9,10-tetrahydro-6,8,11-trihy-
  droxy-8-(hydroxyacetyl)-1-methoxy-5,12-naphtha-
  cenedione **4:** 1897
10-[(3-Amino-2,3,6-trideoxy-α-L-lyxo-hexopyranosy-
  l)oxy]-7,8,9,10-tetrahydro-6,8,11-trihydroxy-8-(hydroxyacetyl)-1-methoxy-5,12-naphthacene-
  dione **4:** 1897
(8S-cis)-10-[[3-Amino-2,3,6-trideoxy-4-O-(tetrahydro-
  2H-pyran-2-yl]-α-L-lyxo-hexopyranosyl]oxy]-
  7,8,9,10-tetrahydro-6,8,11-trihydroxy-8-(hydroxyacetyl)-1-methoxy-5,12-naphthacene-
  dione **4:** 1898
7-amino-4,5,6-triethoxy-3-(5,6,7,8-tetrahydro-4-meth-
  oxy-6-methyl-1,3-dioxolo[4,5-g]-isoquinolin-5-yl)-
  1(3H)isobenzofuranone **2:** 933
2-Amino-6-(trifluoromethoxy)benzothiazole **2:** 575
Amiodarone **1:** 130
Amiphenazole **4:** 2180
Amiphenazol hydrochloride **2:** 339
Amipress **1:** 155
Amisulpride **2:** 607
Amithiozone **3:** 1142
Amitid **2:** 611
Amitril **2:** 611
Amitriptyline **2:** 611
Amlexanox **2:** 934
Amlodipine **1:** 212
Ammonium 21-tungsto-9-antimonate **3:** 1189
Amniotin **3:** 1583
Amoban **2:** 529
Amobarbital **2:** 504, 507
Amodiaquine **3:** 1162
Amorolfine **3:** 1252; **4:** 1731
Amosene **2:** 645
Amovane **2:** 529
Amoxapine **2:** 611
Amoxicillin **3:** 962, 024; **4:** 2160
Angormin **1:** 208
Anhydron **1:** 294
Anhydrotetracyclines **3:** 978
Anifed **1:** 210
Animal experiments **4:** 2127
Anionic surfactants **4:** 2102
Aniracetam **2:** 650
Anisal **2:** 648
Ankylosing spondylitis **4:** 1674
Anotit **3:** 1210
Anovlar **3:**
ANP **3:** 1405, 547
Anquil **2:** 599
Ansaid **4:** 1691
Ansamycins antibiotics **3:** 1000
Ansatin **4:** 1698
Ansefal **4:** 1706
Ansepron **2:** 645
Anseren **2:** 641
Ansiced **2:** 648
Ansidyl **4:** 1904
Ansieten **2:** 641
Ansiolin **2:** 637
Ansopal **2:** 498
Ansudoral **2:** 690
Antacal **1:** 212
Antacids **2:** 715, 724
Antalon **2:** 602
Antalvic **2:** 408
Antarelix **4:** 1934
Antazoline **2:** 905
Antelepsin **2:** 481
Antepan **3:** 1354
Antepar 1953 **3:** 1284
Antergan **2:** 907
Anterior pituitary hormones **3:** 1539
Anthelmintics **3:** 1265
Anthen **2:** 564
Antheridiol **3:** 1335
9,10-Anthracenedicarboxaldehyde bis[(4,5-dihydro-1H-
  imidazol-2-yl] dihydrochloride **4:** 1903

2217

Anthracenes
   Intercalating **4**: 1902
Anthracycline **3**: 954
Anthracyclines **3**: 979
   Cancer therapy **4**: 1897
Anthralin **4**: 1728
Anthranilates
   as antiinflammatory drugs **4**: 1697
Anthraquinone Laxatives **2**: 789
Antiaging dermatological drugs **4**: 1732
Antiallergic asthma therapy **2**: 895
Antiallergics **2**: 901
Antiandrogens **3**: 1591; **4**: 1928
Antiangor **1**: 231
Antiarrhythmic agents
   Classification **1**: 120
Antiarrhythmic Drugs **1**: 115
   function **1**: 16
Antiasthmatics **2**: 861
Antiatherosclerotic Drugs **1**: 17
Antibacterial agents **3**: 1115; **4**: 2004
Antibacterial chemoprophylaxis **3**: 1114
Antibacterial drugs
   veterinary **4**: 2158
Antibiotic A 5283 **3**: 1236
Antibiotic disk method **3**: 1090
Antibiotics **3**: 951
   analyses **3**: 1089
   economic aspects **3**: 1093
   isotopically labeled **3**: 1092
   veterinary **4**: 2158
Antibiotic screening **3**: 1073
Antibiotics isolation **3**: 1084
Antibodies **4**: 1758, 981
Antibody production **4**: 1987
Anticellular agents **4**: 1961
Anticholinergic drugs
   in Parkinsonism treatment **2**: 559
   in the treatment of asthma **2**: 890
Anticholinesterases **4**: 2025
Anticholium **2**: 653
α_1 Antichymotrypsin **1**: 252
Anticore glycolipid vaccine **4**: 1778
Antidepressants **2**: 610, 657
Antidrasi **4**: 2022
Antiemetics **2**: 741
Antiepileptic drugs **2**: 469
Antiestrogens **3**: 1586; **4**: 1922
Antifibrinogenic drugs **2**: 771
Antiföhnon-N **2**: 348
Antifungal agents
   ophthamological **4**: 2012
Antifungals **4**: 1730
Antifungol **3**: 1217
Antigen **2**: 902; **4**: 1981
Antigens **4**: 1757
Antigestagens **3**: 1600

Antiglaucomatous agents **4**: 2019
Antigonadotropic Substances **3**: 1588
Antigreg **1**: 41
Antigrowth agents
   interferons **4**: 1961
Antihemophilic factor B **1**: 256
Antihemophilic globulin A **1**: 256
Antihistamines **2**: 495, 902
   as anti inflammatory dermatological drugs **4**: 1719
   veterinary **4**: 2184
Antihormones **3**: 1554; **4**: 1922
Antihypertensive Agents **1**: 69
Antihypertensive Drugs
   perspectives **1**: 102
Antihypertensives
   function **1**: 16
Antihypotensive agents **1**: 59
Antiinfectives **3**: 1067
Anti-inflammatory agents **4**: 2015
   in dermatology **4**: 1718
Anti-inflammatory drugs **4**: 1671
   for the treatment of asthma **2**: 893
   veterinary **4**: 2175
Antileukotrienes
   in asthma treatment **2**: 896
Antilirium **2**: 653
Antimalarial agents as antirheumatic drugs **4**: 1704
Antimalarial drugs **3**: 1160
Antimetabolites **4**: 1868
Antimicotico **3**: 1217
Antimicrobial agent selection **3**: 1113
Antimicrobial resistance **3**: 1111
Antimicrobial susceptibility **3**: 1113
Antiminth **3**: 1285
Antimuscarinic Drugs **2**: 729
M_1-antimuscarinics **2**: 714
Antimycotics **3**: 1199
   veterinary **4**: 2171
Antimyk Neu **3**: 1217
Anti-Naus **2**: 594
Antineoplastics **4**: 1938
Antiprotozoal agents
   ophthamological **4**: 2014
Antiprotozoan agents **3**: 1146
Antipruritics **4**: 1723
Antipsoriatics **4**: 1727
Antipsychotics/Neuroleptics **2**: 587
Antipyretics **2**: 341
Antirheumatic drugs **4**: 1671
Antisacer **2**: 475
Antispasmodic agents **2**: 689
Antistin **2**: 905
Antithrombin **1**: 252
Antithrombin III **1**: 260
Antitoxin preparations **4**: 1825
Antituberculosis agents **3**: 1139
Antitubulin agents **4**: 1909

Anti-ulcer Drugs 2: 713
Antivert 2: 743
Antiviral activity assessment 3: 1175
Antiviral agents 3: 1177
   interferons 4: 1960
   ophthamological 4: 2014
Antra 2: 720
Antrenyl 2: 696
Antrypol 3: 1149, 287
Anxiety 1: 149; 2: 634
Anxiolit 2: 522
Anxiolytics 2: 634
Anxon 2: 641
Aolan 2: 514
Aolept 2: 593
Aorta
   dissection 1: 146
AOSept 4: 2039
Apalcilling 3: 962
Apamin 2: 593
Apazone 4: 1695
Aphenylbarbit 2: 506
Apheresis techniques 1: 266
Aplactan 2: 914
Aplaket 1: 41
Aplexal 1: 209
Apocanda 3: 1217
Apocholic acid 3: 1328
Apodorm 2: 518
Aponal 2: 613
Apotomin 1: 209
Apozepam 2: 637
APP 3: 1397
Appetite suppressants 2: 457
Application phase 4: 2130
Apralan 75 Premix/Medicated 4: 2164
Apralan soluble powder 4: 2164
Apramycin 3: 1047; 4: 2164
Apranax 2: 368; 4: 1689
Apresoline-Esidrix 1: 99
Apresoline HCl 1: 99
Apresozide 1: 99
Aprical 1: 210
Aprindine 1: 128
Aprobarbital 2: 504, 507
Aprobit 2: 920
Aprol 4: 2172
Apronalide 2: 501
Apsolox 1: 157
APUD cells 3: 1538
Aquachloral 2: 498
Aquamox 1: 295
Aquaphor 1: 295
Aquaretics 1: 300
Aqueous 4: 1716
Aqueous suspensions 4: 2120
Ara-A 2: 764

9-β-D-arabinofuranosyl-adenine 2: 764; 3: 1179
Ara-C 4: 1877
Arachidonic acid 4: 2046
Arachidonic acid cascade 4: 1672
ara-CTP 4: 1877
Aralen hydrochloride 3: 1162
Aralen phosphate 3: 1162
Aranidipine 1: 205, 213
Arantil 2: 502
Arbid N 2: 917
Ardinex 4: 1688
Arduan 2: 672
Area under the curve 4: 2080
Arelix 1: 297
Arem 2: 518
Arenaviridae 3: 1174
Arensin 4: 1926
Argun 2: 358
Arildone 3: 1187
Arimedex 4: 1926
Aristeromycin 3: 994
Aristocort 3: 1622
Aristophan 3: 1622
Aristosol 3: 1622
Arlef 4: 1698
Arminol 2: 606
Aromatase Inhibitors 3: 1587; 4: 1924
Aromatic waters 4: 2087
Aropax 2: 629
Arpamyl 1: 208
Arpicoline 2: 564
Arquel 4: 2176
Arret 2: 409
Arrhythmias 1: 116, 46
Artamin 4: 1706
Artane 2: 563
Artate 1: 209
Arteolol 1: 154
Arteoptic 1: 154
Arterenol 1: 60
Arthrex 4: 1685
Arthrexin 4: 1680
Arthritis 4: 1674
Arthrocine 2: 359
Arthrotec 4: 1685
Arthus phenomenon 2: 902
Articaine hydrochloride 2: 548
Artificial blood 1: 273
Artolon 2: 645
Artosin 3: 1658
Artrocaptin 4: 1684
Arvynol 2: 497
Asaped 4: 1678
Asasantin 1: 39
Ascaridil 3: 1281
Ascaris lumbricoides 3: 1272
Ascending reticular activation system 2: 318

2219

Asendin **2:** 612
Asendis **2:** 612
Asexual blood stage malaria vaccine **4:** 1816
Aslapax **2:** 522
Asparenomycin **3:** 973
Aspartate **2:** 329
Aspirin **1:** 35, 271; **2:** 344, 346
Aspirine **2:** 346
Aspisol **2:** 346
Aspro **2:** 346; **4:** 1678
Assay of Androgenic and Anabolic Activity **3:** 1589
Astemizole **2:** 927
Asthmalitan **2:** 867
Asthma therapy **2:** 861
Asthmolysin **2:** 889
Astonin **3:** 1620
Astonin H **1:** 67
Astrocytes **2:** 311
Astromicin sulfate **3:** 1049
AT – 2266 **3:** 1125
Atamestan **3:** 1587
Atamestane **4:** 1924
Atamir **4:** 1706
Atarax **2:** 646, 915
Atarzine **2:** 590
AteHexal **1:** 152
Atemorin **2:** 497
Atempol **2:** 518
Atenase **3:** 1282
Atenolol **1:** 152, 227
Aterax **2:** 646
Athymil **2:** 621
AT III **1:** 260
Atilan **4:** 1686
Atipamezole **4:** 2182
Ativan **2:** 522, 638
Atock **2:** 883
Atonic seizures **2:** 471
Atopic dermatitis **4:** 1719
Atorvastatin **1:** 29
Atosil **2:** 921
ATPase **1:** 186
Atracurium besylate **2:** 669
Atrial fibrillation **1:** 117
Atrial flutter **1:** 117
Atrial Natriuretic Factor **3:** 1405
Atrial tachycardias **1:** 147
Atrionatriuretic peptide **3:** 1547
Atrioventricular block **1:** 118
Atrioventricular node (AV node) **1:** 116
Atromid **1:** 25
Atropen **2:** 690
Atropina **2:** 562
Atropine **2:** 562, 690, 891
Atropine-Care **4:** 2027
Atropine sulfate **4:** 2027
Atropinol **2:** 690

Atropisol **2:** 562, 690; **4:** 2027
Atropocil **2:** 690
Atrosol **2:** 690
Atrovent **2:** 891
Atumin **2:** 703
AUC **1:** 143; **4:** 2081
Aulin **2:** 384
Auranofin **4:** 1702, 703
Aurantex **2:** 453
Aureomycin **3:** 954; **4:** 2007
Aureomycin 70 Type A medicated article **4:** 2161
Aureomycin 2G Type C medicated feed **4:** 2161
Aureotan **4:** 1703
Aurolate **4:** 1703
Aurothioglucose **4:** 1702
Auroxix **2:** 626
Aut **3:** 1285
Automaticity **1:** 119
Autonomic nerves **2:** 312
Autonomic nervous system **2:** 686
Auxiloson **3:** 1627
Avacan **2:** 702
Avan **2:** 655
Avandia **3:** 1666
Avermectin B$_1$ **4:** 2174
Avermectins
  veterinary **4:** 2173
AVERT **1:** 27
Avil **2:** 912
Avlosulfon **3:** 1143
Avomec **4:** 2174
Avoparcin **4:** 2169
Avoxin **2:** 629
Axoaxonic synapses **2:** 319
Axodendritic synapses **2:** 319
Axon **2:** 310
Axonal Transport **2:** 337
Axosomatic synapses **2:** 319
AY 21 011 **1:** 158
AY 64 063 **1:** 159
Ayeramate **2:** 679
1-Azabicyclo[3.2.0]hept-2-ene **3:** 972
10-(1-Azabicyclo[2.2.2]oct-3-ylmethyl)-10H-phenothia-
  zine **2:** 920
5-Azacitidine **4:** 1875
5-Azacytidine **3:** 989; **4:** 1875
Azaline **4:** 1934
Azameno **3:** 1223
Azapropazone **4:** 1695
Azaron **2:** 907
Azatadine **2:** 923
Azathioprine **4:** 1707, 728
Azelaic acid **4:** 1725
Azelastine **2:** 931
Azepromazine **2:** 590
Azetirelin **2:** 581
3′-Azido-3′-deoxythymidine **3:** 1188, 1295

3′-Azidothymidine **3**: 1188
Azintamide **2**: 753
Azium **4**: 2178
1,1′-Azobis[3-methyl-2-phenylimidazo[1,2-α]pyridinium]dibromide **2**: 670
Azole antimicrobials **4**: 1730
Azole antimycotics **3**: 1201
Azotrex **3**: 1136
AZT **3**: 1188
Azthreonam **3**: 1035
Aztreonam **3**: 977, 035
Azugastan **2**: 606
Azupentat **1**: 49
Azuren **2**: 598
Azutrimazol **3**: 1217

Ba 39 089 **1**: 157
Babesia **3**: 1159, 166
Babylax **2**: 782
Bacampicillin **3**: 963
Bacampicillin hydrochloride **3**: 1025
Baccidal **3**: 1123
Bacillaceae **3**: 1110
Bacillary dysentery vaccines **4**: 1774
Bacitracin **3**: 1057; **4**: 2009, 2010, 2169
Bacitracins **3**: 1006
Baclofen **2**: 681
Baclon **2**: 681
Bacteria **3**: 955
  pathogenic **3**: 1109
Bacterial infections
  chemotherapy **3**: 1109
Bacterial resistance
  sulfonamides **3**: 1130
Bacterial resistance problems **3**: 1112
Bacterial vaccines **4**: 1756, 764
Bacteroidaceae **3**: 1110
Bacticort **4**: 2016
Bactrim **3**: 1133
Balantidium coli **3**: 1168
Baldex **4**: 2018
Bambec **2**: 886
Bambermycin **3**: 1068; **4**: 2169
Bambuterol **2**: 885
Bamifylline **2**: 890
Bamipine **2**: 917
Banamine **4**: 2176
Banflex **2**: 680
Banistyl **2**: 919
Banocide **3**: 1279
Banthine **2**: 695, 697
Baralgin **2**: 371
Baratol **1**: 88
Barbaloin **2**: 790
Barbamyl **2**: 507
Barbital **2**: 504, 506
Barbitone **2**: 506

Barbiturate addiction **2**: 504
Barbiturate habituation **2**: 504
Barbiturates **2**: 443
  comparison with benzodiazepines **2**: 516
Barbituric acids **2**: 443
  as sedatives **2**: 503
Barizin **1**: 210
Barnes-Hind Wetting & Soaking **4**: 2039
Barnes-Hind Wetting Solution **4**: 2034
Barnetil **2**: 606
Barnotil **2**: 606
Bartonellaceae **3**: 1110
Basal ganglia circuitry **2**: 558
Basen **3**: 1663
Basket cells **2**: 310
Basophil granulocytes **1**: 246
Basophils **1**: 239
Batch release **4**: 2143
Batrafen **3**: 1245
Batroxobin **1**: 47
Baumycin A2 **3**: 981
Baumycin B2 **3**: 981
Bausch & Lomb Sterile Disinfecting Solution **4**: 2037
Bay 09867 **3**: 1124
Bay 5097 **3**: 1216
BAY a 1040 **1**: 95
Baycaron **1**: 73, 295
Baycol **1**: 29
BAY D 7791 **3**: 1662
BAY E 4609 **3**: 1662
Bayer 205 **3**: 1149
Bayer 2502 **3**: 1152
Bay h 4502 **3**: 1212
BayK 8644 **1**: 196
Baymycard **1**: 211
BAY O 1248 **3**: 1664
Bayotensin **1**: 97, 210
Baypress **1**: 97, 210
Baytinal **2**: 446
Bay u3405 **4**: 2054
Bay x7195 **4**: 2064
BAY-Y-5959 **1**: 196
BB **3**: 1416
Bébégel **2**: 782
BB-K_8 **3**: 987
BBS **2**: 509
B-cells **1**: 248; **4**: 1760
Bec/Sapresta **1**: 213
Beclomethasone **2**: 893; **3**: 1627
Beclomethasone dipropionate **3**: 1636
Beef Tapeworm **3**: 1270
Befen **3**: 1278
Befenium **3**: 1278
Befeval **3**: 1278
N^4-Behenoyl-ara-C **4**: 1879
Bekanamycin **3**: 985, 986, 041
Belganyl **3**: 1149

Belladonna extract 2: 561
Bellafolin 2: 561
Beloc 1: 91, 56
Belseren 2: 639
Bemacol ointment 4: 2167
Bemegride 2: 339
Benadryl 2: 496, 566, 742, 909
Benadryl N 2: 909
Bendopa 2: 569
Bendroflumethiazide 1: 294
Benedorm 2: 512
Beneficat 2: 632
Benefluor 4: 1880
Benfofen 4: 1685
Benidipine 1: 212
Benmoxine 2: 622
Benoral 4: 1678
Benortan 4: 1678
Benorylate 4: 1678
Benoxinate hydrochloride 4: 2003
Benperidol 2: 599
Benserazide 2: 569
Bentazepam 2: 642
Bentex 2: 563
Bentnesol V 3: 1627
Bentyl 2: 703
Bentylol 2: 703
Benuride 2: 501
Benuron 2: 349
Benzalin 2: 518
Benzalkonium chloride 3: 1254; 4: 2038, 2103
Benzamide neuroleptics 2: 605
Benzamides
   Gastrointestinal prokinetic 2: 794
(±)-2-Benzamido-3-[4-(2-diethylaminoethoxy)phenyl]-
   N,N-dipropyl-propionamide 2: 701
Benzedrine 2: 460, 461
2-Benzenesulfonamido-5-(2-methoxyethoxy)-pyrimidine
   sodium 3: 1657
Benzethonium chloride 4: 2038
Benzhydryl antiemetics 2: 742
4-Benzhydrylidene-1,1-dimethylpiperidinium methylsulfate 2: 693
2-(Benzhydryloxy)-N,N-dimethylethylamine 2: 496
3-Benziloyloxy-1,1-dimethylpiperidinium bromide 2: 695
4-Benziloyloxy-1,1-dimethylpiperidinium bromide 2: 696
3-Benziloyloxy-1-methylquinuclidinium bromide 2: 692
3α-Benziloyloxyspiro[nortropane-8,1'-pyrrolidinium]
   chloride 2: 692
Benzimidazole anthelmintics
   veterinary 4: 2174
1,2-Benzisoxazole-3-methane-sulfonamide 2: 485
Benznidazole 3: 1152

(±)-N-(1-benzo[b]thien-2-ylethyl)-N-hydro-
   xyurea 2: 897
Benzocaine 2: 542
Benzodiazepine auxiolytics 2: 635
Benzodiazepines 2: 450, 481, 634
   as sedatives 2: 515
   comparison with barbiturates 2: 516
4-(4-Benzofurazanyl)-1,4-dihydro-2,6-dimethyl-3,5-pyri-
   dinedicarboxylic acid methyl 1-methylethyl ester 1: 212
Benzoic acid 2-(1-phenylethyl)hyrazide 2: 622
Benzothiadiazine
   as diuretics 1: 292
Benzothiazepines 1: 213
Benzotran 2: 522
5-Benzoyl-2,3-dihydro-1H-pyrrolizine-1-carboxylic
   acid 2: 379
3-Benzoyl-α-methylbenzeneacetic acid 2: 365; 4: 1689
3-benzoyloxy-1,3,5(10)-estratrien-17β-ol 3: 1584
Benzoyl peroxide 4: 1725
2-(3-Benzoylphenyl)propionic acid 2: 365
2-(5-benzoylthiophen-2-yl)propionic acid 2: 368
Benzozil 2: 522
Benzphetamine 2: 461, 467
Benztropine mesylate 2: 562
Benzyl alcohol 2: 552; 4: 2038
Benzyl Alcohol choleretics 2: 750
6-Benzyl-2,3-dihydro-2-thioxo-4(1H)-pyrimidi-
   none 3: 1529
1-Benzyl-3-dimethylamino-2-methyl-1-phenyl-propio-
   nate 2: 407
Benzyldimethyl(2-phenoxyethyl)ammonium 3-hydroxy-
   2-naphthoate 3: 1278
1-Benzyl-3-ethyl-6,7-dimethoxyisoquinoline 2: 699
Benzyl mandelate 2: 700
N-benzyl-2-nitroimidazole-1-acetamide 3: 1152
Benzylpenicillin 3: 960
Benzylpenicillin benzathine 3: 1017
Benzylpenicillin potassium 3: 1017
   biological activity 3: 958
Benzylpenicillin procaine 3: 1017, 958
Benzylpenicillin sodium 3: 958
(±)-Benzyl 2-phenylglycolate 2: 700
(+)-(S)-3-(1-Benzyl-4-piperidyl)-3-phenyl-2,6-piperidi-
   nedione 2: 565
Benzylthiouracil 3: 1525, 529
Bepadin 1: 209
Bephenium 3: 1269, 273, 278
Beprane 1: 159
Bepridil 1: 131, 209, 229
Berachin 2: 881
Beraprost-sodium 4: 2061
Berberin Ophtiole 4: 2030
Berenil 3: 1150
Berkatens 1: 208
Berotec 2: 872
Bespar 2: 648

Bestatin **3**: 1063
Beta-Blockers **1**: 137
Beta-Cardone **1**: 130, 159
Beta-Chlor **2**: 498
Betadren **1**: 158
Betagan **4**: 2021
Betaloc **1**: 91, 56
Betaloc SA **1**: 156
Betamac **2**: 606
Betamann **4**: 2022
Betamethasone **3**: 1625; **4**: 2177
   doses **3**: 1605
Betamethasone 17,21-dipropionae **3**: 1627
Betamethasone 17-valerate **3**: 1627
Beta-Ophtiole **4**: 2022
Betapace **1**: 159
Betapindolol **1**: 158
Betasone **4**: 2177
Beta-Tablinen **1**: 159
Betaxolol **1**: 153
Betaxolol hydrochloride **4**: 2021
Betim **1**: 93
Betnesol **3**: 1627
Betoptic **1**: 153; **4**: 2021
Betoptima **1**: 153
Betotastine Besilate **2**: 929
Bezafibrate **1**: 25
B-GF **3**: 1242
BHAC **4**: 1879
Biarison **4**: 1699
Biartac **4**: 1678
Bicalutamide **4**: 1930
   annual sales **4**: 1939
Bicor **1**: 209
Bicozamycin **3**: 1072
2-(Bicyclo[2,2,1]heptane-2-*endo*-3-*endo*-dicarboximido)-
   glutarimide **2**: 513
1-(Bicyclo[2.2.1]-hept-5-en-2-yl)-1-phenyl-3-piperidino-
   propanol **2**: 563
Bicyclomycin **3**: 1072
Bietamiverine **2**: 702
Bietanautine **2**: 566
Bifemelane **2**: 656
Bifiteral **2**: 786
Bifonazol **3**: 1213
Bifonazole **3**: 1212; **4**: 1730
Biglumide **2**: 513
Biguanides **4**: 2037
Biguanides antidiabetics **3**: 1660
Bilamid **2**: 752
Bilarcil **3**: 1282
Bile **2**: 748
Bile Acids **3**: 1328
Bile duct stones
   direct dissolution **2**: 759
Bile flow **2**: 748
Bilicanta **2**: 701

Bilicura **2**: 750
Biltricide **3**: 1285
Bilup **2**: 700
Bimanol **2**: 652
BIMU 1 **2**: 804
BIMU 8 **2**: 804
Binders **4**: 2104
Binding assays **4**: 1985
DNA–RNA Binding **4**: 1899
Bioadhesives **4**: 2117
Bioassays **2**: 326
Bioavailability **4**: 2079
Bioequivalent Allergy Unit **2**: 939
Biofanal **3**: 1238
Bioflutin-N **1**: 64
Biomioran **2**: 677
Bionicard **1**: 210
Biorphen **2**: 567
Bio-Tal **2**: 446
Biozolene **3**: 1202
Biperiden **2**: 563
1-([1,1'-Biphenyl]-4-ylphenylmethyl)-1*H*-imida-
   zole **3**: 1212
Biphetamine **2**: 461, 467
bis(2,2,2-Trichloroethyl)carbonate **2**: 498
Bisacodyl **2**: 787
Bisantrene hydrochloride **4**: 1903
Bismetin **1**: 208
Bismofalk **2**: 731
Bismuth aluminate **2**: 730
Bismuth antiulcer drugs **2**: 730
Bismuth citrate **2**: 730
Bismuth subgallate **2**: 731
Bismuth subnitrate **2**: 731
Bismuth subsalicylate **2**: 731
Bisobloc **1**: 153
Bisoprolol **1**: 153
Biston **2**: 476
Bithionol **3**: 1268, 279
Bitin **3**: 1279
Bitin S **3**: 1279
Bitolterol **2**: 884
Bitoscanate. **3**: 1273
Bizelesin **4**: 1907
BK **3**: 1174
Black pigment stones **2**: 755
Blasticidin S **3**: 995, 1050
Bleomycin **4**: 1905
Bleomycins **3**: 1002
Bleomycin sulfate **3**: 1055
Bleph-10 **4**: 2005
Blephamide **4**: 2005, 2026
Blink-N-Clean **4**: 2039
Blixophyllin **2**: 888
Blocadren **1**: 93, 60
Blocklin L **1**: 158

Blood **1**: 235
   eocnomic aspects **1**: 284
   quality and safety requirements and controls, legal aspects **1**: 278
   storage and transportation **1**: 283
Blood cells **1**: 238
   and blood viscosity **1**: 45
Blood coagulation **1**: 250
Blood coagulation factors **1**: 253
   therapeutic use **1**: 274
Blood donation **1**: 265
Blood drug concentration – time curve **4**: 2080
Blood Flow Affecting Drugs **1**: 44
Blood Flukes **3**: 1266
Blood iroducts in prophylaxis and therapy **4**: 1835
Blood Lipid Affecting Drugs **1**: 19
Blood plasma
   frozen **1**: 272
Blood plasma fractionation **1**: 267
Blood plasma proteinase inhibitors **1**: 259
Blood plasma proteins **1**: 251
Blood platelets **1**: 249
   therapeutic use **1**: 271
Blood pressure decreasing agents **1**: 69
Blood Pressure Increasing Agents **1**: 59
Blood Pressure Regulating Peptides **3**: 1388
Blood production **1**: 265
Blood Rheology Affecting Drugs **1**: 46
Blood Viscosity **1**: 45
B-Lymphocytes **1**: 248
BM 141 90 **1**: 154
BMS-181339 **4**: 1915
BNP **3**: 1405
Boforsin **1**: 196
Bolasterone **3**: 1593
Bolvidon **2**: 621
Bombesin **3**: 1416
Bonacid **1**: 210
Bonamine **2**: 743
Bone marrow **1**: 239
Bone Morphogenetic Proteins **3**: 1361
Bonifen **2**: 657; **4**: 1706
Bonine **2**: 916
Bon-Sonnil **2**: 514
Bonton **2**: 522
Bontril **2**: 465, 467
Bonyl **4**: 1689
Boric acid **4**: 2038, 2104
Borotropin **2**: 690
Bothrops venom proteinase **1**: 47
Botropase **1**: 48
Boxogetten **2**: 464
Bradalone **2**: 704
Bradyarrhythmias **1**: 120
Bradykinin **3**: 1392
Bradyphrenia
   in Parkinsonism **2**: 556

Brain natriuretic peptide **3**: 1405
Branigan **2**: 659
Brasofensine **2**: 573
Brassicasterol **3**: 1324
Breast-Feeding **3**: 1494
Bredinin **3**: 989
Brenal **2**: 652
Brenazol **3**: 1210
Brentan **3**: 1210
Bretazenil **2**: 648
Bretylate **1**: 130
Bretylium tosylate **1**: 130
Bretylol **1**: 130
Brevicon **3**: 1500
Brevinarcon **2**: 446
Brexidol **4**: 1697
Brexin **4**: 1697
Bricanyl **2**: 871
Bridal **2**: 907
Brinaldix **1**: 295
Britai **4**: 1693
Britane **3**: 1210
BRL-20627-A **2**: 797
BRL 43694 **2**: 800
BRL 46470 **2**: 801
Broad Tapeworm **3**: 1269
Brocadopa **2**: 569
Brocasipal **2**: 567
Brofaromine **2**: 633
Broflex **2**: 563
Bro-Lac **4**: 2033
Brolene **4**: 2014
Broline **4**: 2014
Bromadryl **2**: 910
Bromadyl **2**: 501
Bromazenil **2**: 637
Bromazepam **2**: 637
Bromazine **2**: 908
Bromdiphenhydramine **2**: 908
Bromidol **2**: 598
5-(2-Bromoallyl)-5-isopropylbarbituric acid **2**: 510
5-(2-Bromoallyl)-5-*sec*-butylbarbituric acid **2**: 509
2-Bromobenzyl(ethyl)dimethylammonium toluene-4-sulfonate **1**: 130
2-Bromo-4-(2-chlorophenyl)-9-methyl-6*H*-thieno[3,2-*f*]-[1,2,4]triazolo[4,3-*a*][1,4]diazepine **2**: 525
2-bromo-2-chloro1,1,1-trifluoroethane **2**: 439
Bromocriptine **2**: 576
N-Acetyl-*N'*-(2-bromo-2-diethylacetyl)urea **2**: 501
2-Bromo-2,2-diethylacetylurea **2**: 501
7-bromo-1,3-dihydro-5-(2-pyridinyl)-2*H*-1,4-benzodiazepin-2-one **2**: 637
(*S*)-3-bromo-2,6-dimethoxy-*N*-(1-ethyl-2-pyrrolidinyl)-methyl]-benzamide **2**: 607
(*E*)-5-(2-Bromoethenyl)-2'-deoxyuridine **3**: 1179

10-Bromo-11b-(2-fluorophenyl)-2,12;3,7- 11b- tetrahydro-oxazolo-[3,2-d][1,4]benzodiazepin-6(5<emphas **2**: 524
(5'α)-2-Bromo-12'-hydroxy-2'-(1-methylethyl)-5'-(2-methylpropyl)ergotaman-3',6',18-&SP **2**: 576
Bromoisoval **2**: 501
α-Bromoisovalerylurea **2**: 501
4-[4-(4-Bromophenyl)-4-hydroxypiperidino]-4'-fluoro-butyrophenone **2**: 598
2-[(4-Bromophenyl)-1-phenylethoxy]-N,N-dimethyletha-namine **2**: 909
2-[(4-Bromophenyl)phenylmethoxy]-N,N-dimethyletha-namine **2**: 908
3-(4-Bromophenyl)-3-(2-pyridinyl)-N,N-dimethylpropa-namine **2**: 911
Bromopride **2**: 795
5-Bromo-3-pyridinecarboxylate (8β)-1,6-dimethyl-10-methoxyergoline-8-methanol **2**: 655
2-bromo-1,1,1,2-tetrafluoroethane **2**: 440
Bromovaleryl **2**: 501
Bromovinyldeoxuridine **3**: 1179
Bromperidol **2**: 598
Brompheniramine **2**: 911
Bromuvan **2**: 501
Bromyl **2**: 501
Bronchial asthma **2**: 861
Bronchial hyperreactivity **2**: 862
Bronchodil **2**: 873
Bronchodilators **2**: 863
Broncholin **2**: 881
Bronchospasmin **2**: 873
Bronchovycrin **2**: 891
Broncodyl **2**: 888
Broncovaleas **2**: 875
Bronica **4**: 2062
Bronkosol **2**: 867
Bronsecur **2**: 882
Brontyl **2**: 889
Bros **2**: 658
Brotizolam **2**: 515, 525
Brotopon **2**: 598
Broxyquinoline **3**: 1170
Brufen **4**: 1688
Brugia malayi **3**: 1276
Brulidene **4**: 2014
Brumixol **3**: 1245
Bryostatin **4**: 1937
BTU **3**: 1525, 529
Bucastem **2**: 594
Buciclovir **3**: 1183
Buclifen **2**: 913
Buclizine **2**: 913
Budesonide **2**: 894; **3**: 1636
Budipine **2**: 564
Bufadienolides **1**: 178, 83
Bufalin **1**: 178, 79
Bufedil **1**: 50

Bufene **1**: 50
Bufexamac **2**: 354
Buflan **1**: 50
Buflo AbZ **1**: 50
Buflocit **1**: 50
Buflofar **1**: 50
Buflohexal **1**: 50
Buflomedil **1**: 49
BUFLO-PUREN **1**: 50
Bulking Agents **2**: 783
Bulk powders **4**: 2091
Bumetanide **1**: 297
Bunaftine **1**: 132
(−)-Bunolol hydrochloride **4**: 2021
Bunyamwera **3**: 1174
Bunyaviridae **3**: 1174
Bupivacaine hydrochloride **2**: 547; **4**: 2003
Buprenex **2**: 415
Buprenorphine **2**: 414
Burinex **1**: 297
Buronil **2**: 600
Buscapina **2**: 691
Buscopan **2**: 691
Buscotek **2**: 691
Buserelin **4**: 1932
Buspar **2**: 648
Buspirone **2**: 647
Buspisal **2**: 648
Buspon **2**: 648
Butabarbital **2**: 508
Butabarbital Sodium **2**: 509
Butabarpal **2**: 509
Butallyonal **2**: 504, 509
Butamben **2**: 542
Butamide **2**: 509
Butamisole **4**: 2175
Butanilicaine hydrochloride **2**: 548
Butaprost **4**: 2054
Butazolidin **2**: 374; **4**: 1694
Butazone **4**: 1694
Butenafine hydrochloride **4**: 1731
Butenil **2**: 506
Butesine **2**: 542
Buthalital sodium **2**: 446
Buthiazide thiabutazide **1**: 294
Buticaps **2**: 509
Butisol **2**: 509
Butix **2**: 920
Butobarbital **2**: 504, 506
Butobarbitone **2**: 506
Butoconazole **3**: 1214
Butoctamidesemisuccinate **2**: 502
Butorphanol **2**: 415
8-(4-Butoxybenzyl)-3α-[(S)-(−)-tropoyloxy]-1αH,5α H-tropanium bromide **2**: 690
2-butoxy-N-[2-(Diethylamino)ethyl]-4-quino-linecarboxamide monohydrochloride **2**: 549

4-[3-(4-Butoxyphenoxy)propyl]morpholine hydrochloride   **2**: 550
2-(4-Butoxyphenyl)-*N*-hydroxyacetamide   **2**: 354
1-(4- butoxyphenyl)-3-(1-piperidinyl)-1-propanone hydrochloride   **2**: 550
Butrex   **4**: 1694
Butriptyline   **2**: 616
Butropium bromide   **2**: 690
n-Butyl 4-aminobenzoate   **2**: 542
4-[2-(*tert*-butylamino)-1-hydroxyethyl]-1,2-phenylenebis(*p*-toluate)   **2**: 884
5-[3-(*tert*-Butylamino)-2-hydroxypropoxy]-3,4-dihydrocarbostyril   **1**: 154
(−)-5-[3-(*tert*-Butylamino-2-hydroxypropoxy]-3,4-dihydro-1(2*H*)-naphthalenone hydrochloride   **4**: 2021
S-(−)-3-(3-*tert*-Butylamino-2-hydroxypropoxy)-4-morpholino-1,2,5-thiadiazole   **1**: 160
1-(*tert*-butylamino)-3-[(5,6,7,8-tetrahydro-*cis*-6,7-Dihydroxy-1-naphthyl)oxy]-2-propanol   **1**: 156
(2-Butylbenzofuran-3-yl)-[4-(2-diethylaminoethoxy)-3,5-diiodophenyl]ketone   **1**: 130
N-Butylbiguanide   **3**: 1660
N-butyl-*N*-(2-Diethylaminoethyl)-1-naphthamide   **1**: 132
1-*n*-Butyl-*N*-(2,6-dimethylphenyl)-2-piperidinecarboxamide monohydrochloride   **2**: 547
1-Butyl-*N*-(2,6-dimethylphenyl)-2-piperidinecarboxamide monohydrochloride monohydrate   **4**: 2003
1-*tert*-Butyl-4,4-diphenylpiperidine   **2**: 564
4-Butyl-1,2-diphenylpyrazolidine-3,5-dione   **2**: 374; **4**: 1694
N-butyl-6β,7β-epoxy-3α-(−)-tropoyloxy-1αH,5αH-tropanium bromide   **2**: 691
5-Butyl-5-ethylbarbituric acid   **2**: 506
2-(4-*tert*-Butyl-3-hydroxy-2,6-dimethylbenzyl)-2-imidazoline hydrochloride   **4**: 2029
4-Butyl-1-(4-hydroxyphenyl)-2-phenylpyrazolidine-3,5-dione   **2**: 373; **4**: 1694
16α,17α-Butylidendioxy-11β-21dihydroxy-pregna-1,4-diene   **2**: 894
2-[(3-Butyl-1-isoquinolinyl)oxy]-*N*,*N*-dimethylethanamine monohydrochloride   **2**: 551
N-butyl-2-methyl-2-propyl-1,3-propyldicarbamate   **2**: 645
Butylone   **2**: 509
4-Butyl-1-phenylpyrazolidine-3,5-dione   **2**: 372
(1-Butyl-4-piperidinylmethyl)-8-amino-7-chloro-1,4-benzodioxan-5-carboxylate   **2**: 807
N-butylscopolaminium bromide   **2**: 691, 691
Butynoct   **2**: 506
2-butyryl-10-[3-(4-methyl-1-piperazinyl)propyl]-10*H*-phenothiazine   **2**: 594
BVDU   **3**: 1179
BW 245C   **4**: 2054
BW-683C   **3**: 1186
BW A868C   **4**: 2054

BW B759U   **3**: 1182
Bykonox   **2**: 508
Bylotensin   **1**: 210

C-2801 X   **3**: 967
C7E3 Fab   **1**: 44
Cabaseril   **2**: 579
Cabergoline   **2**: 578
Cabersase   **2**: 579
Cachets   **4**: 2092
Cafilon   **2**: 465, 466
Calamine   **4**: 1723
Calan   **1**: 97, 131, 208
Calcidrine   **2**: 878
Calciferols   **3**: 1553
Calcipotriene   **4**: 1727
Calcipotriol   **4**: 1727
Calcistin   **2**: 906
Calcitonin   **3**: 1362, 365, 546
Calcitonin Gene Related Peptide   **3**: 1367
Calcitriol   **3**: 1325, 637
Calcium antagonistic
   specific and nonspecific   **1**: 204
Calcium antagonists   **1**: 201
   as coronary dilators   **1**: 227
   classification   **1**: 203
Calcium bilirubinate   **2**: 759
Calcium channel blockers   **1**: 122, 31
Calcium channels   **1**: 202
Calcium Entry Blockers   **1**: 44
Calcium ions   **1**: 255
Calcium phosphate   **4**: 2104
Calcium-Regulating Hormones   **3**: 1546
Calcium valproate   **2**: 477
Calepsin   **2**: 476
Calicheamicin   **4**: 1908
Calicheamicins   **3**: 1013
Caliciviridae   **3**: 1174
Caliciviruses   **3**: 1174
Calmansial   **2**: 595
Calminal   **2**: 506
Calmivet   **2**: 591
Calmodulin   **1**: 203
Calmodulin antagonists   **1**: 204
Calmonal   **2**: 496
Calmotal   **2**: 590
Calodal   **2**: 592
Calsmin   **2**: 518
Calsynar   **3**: 1367
Caltridren   **1**: 154
Calvisken   **1**: 158
Calysterol   **3**: 1325
Camazepam   **2**: 639
Camont   **1**: 210
Camoquin hydrochloride   **3**: 1163
cAMP   **1**: 36
Campestane   **3**: 1323

Campesterol **3**: 1324
Camphor **4**: 1723
Campto **4**: 1914
Camptothecin **4**: 1913
Canarigenin **1**: 179
Cancer
 treatment with antibiotics **3**: 956
Cancer antimethabolites **4**: 1868
Cancer cells **3**: 956
Cancer chemotherapy **4**: 1865
Cancer drug market **4**: 1939
Cancer pain **2**: 343
Candex **3**: 1238
Candicidin D **3**: 1066
Candio-Hermal **3**: 1238
Candoral **3**: 1208
Canef **1**: 29
Canesten **3**: 1217
Canifug **3**: 1217
Cannogenin **1**: 179
Canofite **3**: 1210
Canrenone **1**: 298
Canstat **3**: 1238
Cantabilin **2**: 754
Cantabiline **2**: 701
Cantil **2**: 695
Cantril **2**: 695
Capoten **1**: 78
Capreomycin **3**: 1008
Capreomycin sulfate **3**: 1060
Caprodat **2**: 677
Capronor **3**: 1505
Capros **2**: 390
Capsules **4**: 2092
Captopril **1**: 15, 78
Capuride **2**: 502
Carbachol **4**: 2024
Carbacyclin **1**: 38
Carbagamma **2**: 476
Carbamates **2**: 644
Carbamazepine **2**: 423, 475
Carbamer **4**: 2103
Carbamide **4**: 2023
O-Carbamoylcholine chloride **4**: 2024
(3-Carbamoyl-3,3-diphenylpropyl)diisopropylmethyl-
 ammonium iodide **2**: 694
1-(3-cCarbamoyl-3,3-diphenylpropyl)-1-methylpiperidi-
 nium bromide **2**: 693
5-Carbamoyl-1H-imidazol-4-yl piperonylate **4**: 1885
4-Carbamoylimidazolium-5-olate **4**: 1885
1-p-Carbamoylmethylphenoxy-3-isopropylamino-2-pro-
 panolAtenolol **1**: 152
4-[(4-carbamoyl-4-piperidino)-4′-fluorobutyrophe-
 none **2**: 599
5-{3-[4-Carbamoyl-4-piperidino)piperidino]propyl}-
 10,11-dihydro-5H-dibenz[b, f]azepine **2**: 617

5-{[3-(4-Carbamoyl-4-piperidino]propyl}-3-chloro-10,11-
 dihydro-5H-dibenz[b, f]azepine **2**: 603
10-[3-(4-Carbamoylpiperidino)propyl]-2-chloro-10H-
 phenothiazine **2**: 592
Carbapenems **3**: 972
Carbased **2**: 501
(RS)-1-(9H-Carbazol-4-yloxy)-3-[[2-(2-
 methoxyphenoxy)ethyl] amino]-2-propa-
 nol **1**: 154
Carbelan **2**: 476
Carbenicillin **3**: 962; **4**: 2006
Carbenicillin disodium **3**: 1019
Carbenicillin indanyl sodium **3**: 1020
Carbenicillin phenyl sodium **3**: 1019
Carbidopa **2**: 569
Carbilazine **3**: 1279
Carbimazole **3**: 1525, 1530
 pharmacokinetic data **3**: 1527
Carbinoxamine **2**: 908
Carbocaine **2**: 547
Carbocalcitonin **3**: 1367
Carbocloral **2**: 499
Carbocyclic arabinofuranosyladenine **3**: 1182
Carbohydrate digestion inhibitors **3**: 1661
Carbomycin **3**: 999
Carbonic anhydrase blockers **1**: 291
Carbonic anhydrase inhibitors **4**: 2022
8,8′-{Carbonylbis[imino-3,1-phenylenecarbonyl-imino(4-
 methyl-3,1-phenylene)carbonylimino]}-bis(1,3,5-
 naphthalene)trisulfonic acid **3**: 1188
Carboplat **4**: 1920
Carboplatin **4**: 1920
Carbostesin **2**: 547
Carboxymethyl cellulose sodium **4**: 2033
2-Carboxy-5-methylpyrazine 4-oxide **1**: 24
(2R,3S)-N-Carboxy-3-phenylisoserine **4**: 1915
Carbromal **2**: 501
Carbutamide **3**: 1655
Carbuterol **2**: 881
Carcinil **4**: 1932
Carcinogenicity tests **4**: 2133
Cardanat **1**: 64
Cardene **1**: 210
Cardenolides **1**: 177, 178
Cardiac action potentials **1**: 118
Cardiac arrhythmia **1**: 146
Cardiac glycosides **1**: 176
**Cardiac glycosides and synthetic cardiotonic
 drugs** **1**: 173
Cardiac rhythm **1**: 115
Cardiagutt **1**: 208
Cardialgine **1**: 64
Cardibeltin **1**: 97, 131, 208
Cardilate **1**: 224
Cardioactive Drugs **1**: 13
Cardiomyopathy **1**: 148
Cardiorythmine **1**: 127

Cardioselectivity   **1**: 225
Cardiovascular diseases   **1**: 14
Cardiovascular Drugs   **1**: 11
Cardiovascular Hormones   **3**: 1547
Carditin-same   **1**: 208
Cardizem   **1**: 132, 213
Cardizem Herbesser   **1**: 98
Cardophylin   **2**: 888
CARE   **1**: 27
Carecin   **1**: 209
Carfecillin   **3**: 1019
Carfentanil   **4**: 2183
Carfimate   **2**: 500
Caricide   **3**: 1279
Caridorol   **1**: 159
Carindacillin   **3**: 1020
Carisoma   **2**: 677
Carisoprodol   **2**: 677
Carminomycin I   **3**: 981
Carmustine   **4**: 1891
Carnidazole   **3**: 1157
Caroverine   **2**: 700
Caroxazone   **2**: 624
Carpetimycin A   **3**: 974
Carpetimycin B   **3**: 974
Carphenazine   **2**: 595
Carpipramine   **2**: 603, 617
Carprofen   **4**: 2177
Carteol   **1**: 154
Carteolol   **1**: 154
Cartrol   **1**: 154
Carvedilol   **1**: 154
Carzelesin   **4**: 1907
Cascara   **2**: 790
Casodex   **4**: 1930
β-Casomorphin-7   **3**: 1386
Castellani's paint   **4**: 1732
Castor oil   **2**: 789
Catabon   **1**: 192
Catapres   **1**: 82; **2**: 424
Catapresan   **2**: 424
Catapressan   **1**: 82
Catarol   **2**: 632
(+)-Catechin   **2**: 765
Catecholamines hormones   **3**: 1550
Catechol-O-methyl-transferase inhibitors   **2**: 572
Catenulin   **3**: 984
Catergen   **2**: 765
Cateudyl   **2**: 514
Cathine   **2**: 464
Cationic surfactants   **4**: 2103
Cavinton   **2**: 656
Caytine   **2**: 867
CB 3717   **4**: 1871
CB-7432   **4**: 1924
CBDCA   **4**: 1920
CBZ   **3**: 1525, 530

CC-1065   **4**: 1907
CCK   **3**: 1413, 549
CCKRP   **3**: 1413
CCNU   **4**: 1891
2-CdA   **4**: 1880
CD-nomenclature   **1**: 247
Cebesine   **4**: 2003
Cebutid   **2**: 363; **4**: 1691
Cedur   **1**: 25
Cefacetrile   **3**: 1029
Cefaclor   **3**: 965, 026
Cefadroxil   **3**: 965, 027
Cefalexin   **3**: 1026
Cefaloglycin   **3**: 1027
Cefaloridine   **3**: 1029
Cefamandole   **3**: 965, 1030
Cefapirin   **3**: 1028
Cefatrizine   **3**: 965, 030
Cefazolin   **3**: 965; **4**: 2006
Cefazolin sodium   **3**: 1031
Cefmenoxime   **3**: 965
Cefmenoxime hydrochloride   **3**: 1032
Cefmetazole   **3**: 967, 034
Cefoperazone sodium   **3**: 1030
Cefotaxime   **3**: 965
Cefotaxime sodium   **3**: 1032
Cefotetan   **3**: 967, 034
Cefotiam   **3**: 965
Cefotiam hydrochloride   **3**: 1031
Cefoxitin   **3**: 967, 033
Cefradine   **3**: 1027
Cefroxadine   **3**: 965, 028
Cefsulodin   **3**: 965
Cefsulodin sodium   **3**: 1029
Ceftazidime   **3**: 1033
Ceftezole   **3**: 965
Ceftizoxime   **3**: 965
Ceftizoxime sodium   **3**: 1031
Ceftriaxone sodium   **3**: 1033
Cefuroxime   **3**: 965, 1032
Celebrex   **2**: 382; **4**: 1701
Celecoxib   **2**: 381; **4**: 1700, 2054
Celectol   **1**: 155
Celeport   **2**: 656
Celestan   **3**: 1627
Celestan V   **3**: 1627
Celiprolol   **1**: 155
Cell cultures
   industrial-scale   **4**: 1989
Cell cycle   **4**: 1867
Cell fusion   **4**: 1983
Cell-mediated immune response   **4**: 1761
Cellothyl   **2**: 784
Cellucon   **2**: 784
Cellulose acetate phthalate   **4**: 2103
Celluvisc   **4**: 2033
Celstone   **3**: 1627

Celtect   **2**: 916
Censpar   **2**: 648
Centchroman   **3**: 1509
Centractyl   **2**: 590
Central inhibition   **2**: 315
Centrally acting analgesics   **2**: 341, 386
Central nervous system   **2**: 312
   analeptics   **2**: 339
Centrax   **2**: 639
Centrelyse   **2**: 616
Centrophene   **2**: 919
Cepalexin   **3**: 965
Cephacetrile   **3**: 965
Cephacetrile sodium   **3**: 1029
Cephadol   **2**: 743
Cephalexin   **3**: 1026
Cephaloglycin   **3**: 965, 027
Cephaloridine   **3**: 965, 029
Cephalosporin C   **3**: 964
Cephalosporins   **3**: 953
   as ophthamological agents   **4**: 2006
   natural   **3**: 963
   semisynthetic   **3**: 964
Cephalothin   **3**: 965; **4**: 2006
Cephalothin sodium   **3**: 1028
Cephamycin A   **3**: 967
Cephamycin B   **3**: 967
Cephamycin C   **3**: 967
Cephamycins   **3**: 965
Cephapirin   **3**: 965
Cephapirin sodium   **3**: 1028
Cephradine   **3**: 965, 1027
Ceractin   **2**: 656
Cercine   **2**: 678
Cerebellum   **2**: 316
Cerebid   **2**: 699
Cerebolan   **1**: 209
Ceregulart   **2**: 637
Cerepar   **1**: 209
Cericlamine   **2**: 633
Cerivastatin   **1**: 29
Cerontin   **2**: 479
Certificate of suitability   **4**: 2154
Cerucal   **2**: 608, 745
Cerulein   **3**: 1414
Ceruletide   **3**: 1416
Ceruloplasmin   **1**: 252
Cerutil   **2**: 652
Cervical caps   **3**: 1495
Cervoxan   **2**: 652
Cesol   **3**: 1285
Cestocide   **3**: 1282
Cestodes   **3**: 1269
   vaccines   **4**: 1811
Cestox   **3**: 1285
Cetamide   **4**: 2005
Cetiprin   **2**: 693

Cetirizine   **2**: 927
Cetonax   **3**: 1208
Cetrane   **2**: 523
Cetrapred   **4**: 2005
Cetrimide   **4**: 2038
Cetrorelix   **4**: 1934
Cetylpyridinium chloride   **4**: 2038, 2103
Cetyltrimethylammonium bromide   **4**: 2038, 2103
CFU   **1**: 240
CG   **3**: 1355
CG-3509   **2**: 581
CG-3703   **2**: 581
CGP 2175   **1**: 156
CGP 28014   **2**: 572
CGP-32349   **4**: 1924
CGP-73547   **3**: 1314
CGRP see calcitonin gene related peptide   **3**: 1367
CGS-15873   **2**: 609
CGS 16949 A   **3**: 1587; **4**: 1926
CGS-20267   **4**: 1926
CGS-20625   **2**: 649
CGS 25019C   **4**: 2066
Chagas' disease   **3**: 1151
Chalcone Ro 09 – 0410   **3**: 1186
Change Control   **4**: 2149
Ca^{2+} Channel Antagonists   **1**: 95
Chemical Contraception   **3**: 1493
Chemionazolo   **3**: 1219
Chemoprophylaxis   **3**: 1114
Chemotherapeutic antibiotics   **3**: 956
Chemotherapeutics
   synthetic   **3**: 1107
Chemotherapy
   of bacterial infections   **3**: 1109, 1115
   of protozoan infections   **3**: 1146
   of viral infections   **3**: 1172
Chenodeoxycholic acid   **2**: 756; **3**: 1329
Chetopir   **2**: 372
Chicken Comb Test   **3**: 1589
Chinidin   **1**: 124
Chiniofon   **3**: 1171
Chinofungin   **3**: 1248
CHIP   **4**: 1921
Chlamydiaceae   **3**: 1111
Chlamydial
   treatment with monoclonal antibodies   **4**: 1842
Chlamydias   **3**: 1110
Chloral betaine   **2**: 498
Chloraldurat   **2**: 498
Chloral hydrate   **2**: 498
Chloralodol   **2**: 499
Chloralosan   **2**: 499
Chloralose   **2**: 499
Chlorambucil   **4**: 1889
Chloramphenicol   **3**: 954, 1009, 1069; **4**: 2011, 2167
   staphylococcus aureus inhibition   **3**: 1091
Chloramphenicol palmitate   **3**: 1070

2229

Chloramphenicol sodium succinate **3:** 1070
Chlorbiotic ointment **4:** 2167
Chlorcyclizine **2:** 913
Chlordiazepoxide **2:** 634, 636
Chlorethate **2:** 498
Chlorhexadol **2:** 499
Chlorhexidine digluconate **4:** 2037
Chlorhexidine gluconate **4:** 2037
Chlorisept **3:** 1248
Chlormadinone acetate **3:** 1497
Chlormadione acetate **3:** 1596
Chlormethiazole **2:** 529
N-4-[2-(Chlor-2-methoxybenzamido)-ethyl]-benzenesulfonyl)-N'-cyclohexylurea **3:** 1656
Chlormidazole **3:** 1255
Chlormycetin Hydrocortisone **4:** 2016
3-(p-Chloroanilino)-10-(p-chlorophenyl)-2,10-dihydro-2-(isopropylimino)phenazine **3:** 1144
11-Chloro-8,12b-dihydro-2,8-dimethyl-12b-phenyl-4H-[1,3]oxazino[3,2-d][1,4]benzodiazepine-4,7(6H)-dione **2:** 641
N-(4-Chlorobenzenesulfonyl)-N'-n-propylurea **3:** 1655
1-[4-Chlorobenzhydryl]-4-(3-tolyl)piperazine dihydrochloride **2:** 496
2-[4-[2-[(4-Chloro-benzoyl)amino]ethyl]phenoxy]-2-methylpropanoic acid **1:** 25
1-(4-Chlorobenzoyl)-N-hydroxy-5-methoxy-2-methyl-1H-indole-3-acetamide **4:** 1681
1-(4-Chlorobenzoyl)-5-methoxy-2-methyl-1H-indole-3-acetic acid **4:** 1679
[1-(4-Chlorobenzoyl)-5-methoxy-2-methyl-1H-indol-3-yl]-acetate **2:** 353
[1-(4-Chlorobenzoyl)-5-methoxy-2-methyl-1H-indol-3-yl]acetic acid **2:** 355
2-({[1-(4-Chlorobenzoyl)-5-methoxy-2-methyl-1H-indol-3-yl]acetyl}amino)–2-deoxy-D-glucose **4:** 1680
{[1-(4-Chlorobenzoyl)-5-methoxy-2-methylindol-3-yl]acetoxy}acetic acid **4:** 1680
N-{2-[1-(4-Chlorobenzoyl)–5-methoxy-2-methyl-3-indolylacetoxy]ethyl}-N'-[3-(N-benzoyl-N',N'-di-n-propyl-DL-isoglutaminyl)oxypropyl]piperazine (±)-dimaleate **4:** 1681
2-[4-(4-Chlorobenzoyl)phenoxy]-2-methylpropanoic acid 1-methylethyl ester **1:** 25
1-[1-(2-Chlorobenzyl)-1H-pyrrol-2-yl]-2-(di-sec-butylamino)ethanol **2:** 412
1-[1-[2-(3-Chlorobenzyloxy)phenyl]vinyl]imidazole **3:** 1218
Chlorobutanol **4:** 2038
6-Chloro-2-(4-chlorophenyl)-3,4-dihydro-4',6-di-chloroflavan **3:** 1186
7-Chloro-5-(2-chlorophenyl)-1,3-dihydro-3-hydroxy-2H-1,4-benzodiazepin-2-one **2:** 522, 637
7-Chloro-5-(2-chlorophenyl)-1,3-dihydro-3-hydroxy-1-methyl-2H-1,4-benzodiazepin-2-one **2:** 523
6-Chloro-2-(4-chlorophenyl)-N,N-dipropylimidazo[1,2-a]pyridine-3-acetamide **2:** 647

6α-Chlorocorticoids **3:** 1629
Chlorocortolone **3:** 1634
2,2'-[(2-Chloro-5-cyano-1,3-phenylene)diimino]bis[2-oxoacetic acid] **2:** 934
6-Chloro-5-cyclohexyl-2,3-dihydro-1H-indene-1-carboxylic acid **4:** 1691
(4S)-6-chloro-4-(cyclopropylethynyl)-1,4-dihydro-4-(trifluoromethyl)-2H-3,1-benzoxazin-2-one **3:** 1302
7-Chloro-1-(cyclopropylmethyl)-1,3-dihydro-5-phenyl-2H-1,4-benzodiazepin-2-one **2:** 639
6-Chlorodeoxyglucose **3:** 1514
7(S)-Chloro7-deoxylincomycin **3:** 1067
7-Chloro-1-(2-diethylaminoethyl)-5-(2-fluorophenyl)-1,3-dihydro-2H-1,4-benzodiazepin-2-one **2:** 520
6-chloro-9-{[4-(Diethylamino)-1-methylbutyl]-amino}-2-methoxyacridine dihydrochloride dihydrate **3:** 1158
7-Chloro-4-(4-diethylamino-1-methylbutylamino)quinoline **3:** 1161
1-Chloro-1-(difluoromethoxy)–2,2,2-trifluoroethane **2:** 442
2-Chloro-1-(difluoromethoxy)-1,1,2-trifluoroethane **2:** 441
6-Chloro-3,4-dihydro-2 H-1,2,4-benzothiadiazine-7-sulfonamide 1,1-dioxide **1:** 71
7-Chloro-2,3-dihydro-2,2-dihydroxy-5-phenyl-1H-1,4-benzodiazepine-3-carboxylic acid **2:** 638
3-chloro-10,11-dihydro-5-(3-dimethylaminopropyl)-5H-dibenz-[b,f]azepine **2:** 612
Endo-5-chloro-2,3-dihydro-2,2-dimethyl-N-(8-methyl-8-azabicyclo[3.2.1]oct-3-yl)-7-benzofuran carboxamide (Z)-& **2:** 802
8-Chloro-6,11-dihydro-11-(1-ethoxycarboxy-4-piperidinylidene)-5H-benzo[5,6]cyclohepta[1,2-b]-pyridine **2:** 932
7-Chloro-1,3-dihydro-3-hydroxy-1-methyl-5-phenyl-2H-1,4-benzodiazepin-2-one **2:** 523
7-Chloro-1,3-dihydro-3-hydroxy-5-phenyl-2H-1,4-benzodiazepin-2-one **2:** 522, 638
7-[(3-Chloro-6,11-dihydro-6-methyldibenzo[c,f][1,2]thiazepin-11-yl)amino]heptanoic acid S,S-d **2:** 617
7-Chloro-1,3-dihydro-1-methyl-5-phenyl-2H-1,4-benzodiazepin-2-one **2:** 481, 636, 678
7-Chloro-1,3-dihydro-3-(N,N-dimethylcarbamoyl)-1-methyl-5-phenyl-2H-1,4-benzodiazepinone **2:** 639
7-Chloro-1,3-dihydro-5-phenyl-1-(2,2,2-trifluoroethyl)-2H-1,4-benzodiazepin-2-one **2:** 642
3-Chloro-4-(2,5-dihydro-1H-pyrrol-1-yl)-α-methylbenzeneacetic acid **4:** 1693
6-Chloro-3,4-dihydro-7-sulfamoyl-2H-1,2,4-benzothiadiazine 1,1-dioxide **1:** 293
7-Chloro-10-(2-dimethylaminoethyl)-5,10-dihydro-11H-dibenzo[b,e][1,4]diazepin-11-one **2:** 923
8-Chloro-1-(dimethylamino)methyl-6-phenyl-4H-s-triazolo-[4,3-a][1,4]benzodiaze **2:** 640

2-Chloro-9-(3-dimethylaminopropylidene)-9H-thioxanthene **2**: 596

2-Chloro-10-(3-dimethylaminopropyl)-10H-phenothiazine **2**: 590

2-Chloro-10-(3-dimethylaminopropyl)phenothiazine hydrochloride **2**: 681

2-[p-[(Z)-4-Chloro-1,2-diphenyl-1-butenyl]phenoxy]-N,N-dimethylethylamine **4**: 1922

4-[bis(2-Chloroethyl)amino]benzenebutanoic acid **4**: 1889

4-[bis(2-Chloroethyl)amino]-L-phenylalanine **4**: 1887

6-Chloro-2-ethylamino-4-methyl-4-phenyl-4H-3,1-benzoxazine **2**: 646

N-(2-Chloroethyl)-N'-cyclohexyl-N-nitrosourea **4**: 1891

1-(2-Chloroethyl)-3-(4-methylcyclohexyl)-1-nitrosourea **4**: 1891

5-(2-Chloroethyl)-4-methylthiazole **2**: 529

2-({[(2-Chloroethyl)nitrosoamino]carbonyl}amino)-2-deoxy-D-glucose **4**: 1891

N,N'-bis(2-Chloroethyl)-N-nitrosourea **4**: 1891

N,N-bis(2-Chloroethyl)tetrahydro-2H-1,3,2-oxazaphosphorin-2-amine 2-oxide **4**: 1888, 890

Chlorofair **4**: 2011

8-Chloro-6-(2-fluorophenyl)-1-methyl-4H-imidazo-[1,5-a][1,4]benzodiazepine **2**: 525

7-Chloro-5-(2-fluorophenyl)-1, 3- dihydro-3-hydroxy-1-(2-hydroxyethyl)-2H-1,4-benzodiazepin-2-on **2**: 523

7-Chloro-5-(2-fluorophenyl)-2,3-dihydro-3-hydroxy-2-oxo-1H-1,4-benzodiazepine-1-propionitrile **2**: 523

8-Chloro-6-(2-fluorophenyl)–1-methyl-4H-imidazo[1,5-a]-[1,4]benzodiazepine maleate **2**: 450

7-Chloro-5-(2-fluorophenyl)-1-(trifluoro-2,2,2-ethyl)-1,3-dihydro-2H-1,4-benzodiazepine-2-thione **2**: 523

4-Chloro-N-furfuryl-5-sulfamoylanthranilic acid **1**: 72

4-Chloro-N-(2-furylmethyl)-5-sulfamoyl anthranilic acid **1**: 296

Chloroguanide **3**: 1164

Chloroguanide hydrochloride **3**: 1164

5-Chloro-2-hydroxy-benzoxazole **2**: 677

2-chloro-9-{3-[4-(2-hydroxyethyl)-1-piperazinyl]propylidene}-9H-thioxanthene **2**: 597

2-Chloro-10-{3-[4-(2-hydroxyethyl)-1-piperazinyl]propyl]-10H-phenothiazine **2**: 594

5-Chloro-8-hydroxy-7-iodoquinoline **3**: 1170

2-Chloro-5-(1-hydroxy-3-oxo-1-isoindolinyl)benzenesulfonamide **1**: 72

6-Chloro-4-hydroxy-2-methyl-1,1-dioxo-1,2-dihydro-1λ⁶-thieno[2,3-e][1,2]thiazine-3-carboxylic acid pyridin-2-yLamide **2**: 375

6-Chloro-4-hydroxy-2-methyl-N-(2-pyridyl)-2H-thieno[2,3-e]-1,2-thiazine-3-carboxamide 1,1-dioxide **2**: 375

5-chloro-4-(2-imidazolin-2-ylamino)-2,1,3-benzothiadiazole **2**: 681

4'-Chloro-N-(1-isopropyl-4-piperidyl)–2-phenylacetanilide **1**: 128

4-[6-(2-Chloro-4-methoxy)phenoxy]hexyl-3,5-heptanedione **3**: 1187

α-(Chloromethyl)-2-methyl-5-nitro-1H-imidazole-1-ethanol **3**: 1156

7-Chloro-2-methylamino-5-phenyl-3H-1,4-benzodiazepine-4-oxide **2**: 636

7-chloro-3-methyl-2H-1,2,4-benzothiadiazine-1,1-dioxide **1**: 100

4-Chloro-N-(2-methyl-1-indolinyl)-3-sulfamoylbenzamide **1**: 73

3'-Chloro-α-[methyl[(morpholinocarbonyl)-methyl]-amino]-o-benzotoluidide · HCl **2**: 339

7-Chloro-1-methyl-5-phenyl-1H-1,5-benzodiazepine-2,4(3H,5H)-dione **2**: 481, 644

N-(2-Chloro-6-methylphenyl)-2-(butylamino)acetamide hydrochloride **2**: 548

8- chloro-1-methyl-6-phenyl-4H-s-triazolo[4,3-a][1,4]benzodiazepine **2**: 640

2- chloro-11-(4-methyl-1-piperazinyl)-dibenz[b,f][1,4]oxazepine **2**: 605

8-Chloro-11-(4-methyl-1-piperazinyl)-5H-dibenzo[b,e][1,4]-diazepine **2**: 604

2-Chloro-11-(4-methyl-1-piperazinyl)-dibenzo[b,f][1,4]thiazepine **2**: 604

2-Chloro-10-[3-(4-methyl-1-piperazinyl)propyl]-10H-phenothiazine **2**: 594, 744

9α-Chloro-16β-methylprednisolone 17,21-dipropionate **3**: 1627

4-Chloro-N¹-methylN¹-[(tetrahydro-2-methyl-2-furanyl)-methyl]-1,3-benzenedisulfonamide **1**: 73

4-Chloro-N¹-methyl-N¹(tetrahydro-2-methylfurfuryl)-m-benzenedisulfonamide **1**: 73

4- chloro-N-(2-morpholinoethyl)benzamide **2**: 625

Chloromycetin **4**: 2011

Chloronase **3**: 1655

5-Chloro-1-[1-[3-(2-oxo-1-benzimidazolinyl)propyl]-4-piperidyl]-2-benzimidazolinone **2**: 728

8-[3-(2- Chloro-10H-phenothiazin-10-yl)propyl]-1-thia-4,8-diazaspirol[4,5]decan-3-one **2**: 593

(4-Chlorophenoxy)-acetic acid 2-isopropylhydrazide **2**: 622

2-(4-Chlorophenoxy)-2-methylpropanoic acid 1,3 propanediylester **1**: 26

2-(4-Chlorophenoxy)-2-methylpropanoic acid ethyl ester **1**: 24

2-(4-Chlorophenoxy)-2-methylpropanoic acid 1,3-propanediyl ester **1**: 25

3-(4-chlorophenoxy)-1,2-propanediol-1-carbamate **2**: 677

(±)-1-[4-(4-Chlorophenyl)-2-[(2,6-dichlorphenyl)thio]-butyl]-1H-imidazole **3**: 1214

5-(4-Chlorophenyl)-2,5-dihydro-3H-imidazo[2,1-a]isoindol-5-ol **2**: 466

6-(2-Chlorophenyl)-2,4-dihydro-2-[(4-methyl-1-piperazinyl)methylene]-8-nitro-1*H*-imidazo[1,2-*a*][1,4]benzodiazepin-1-one **2**: 526

5-(2-chlorophenyl)-1,3-dihydro-7-nitro-2*H*-1,4-benzodiazepin-2-one **2**: 481

1-[(2-Chlorophenyl)diphenylmethyl]-1*H*-imidazole **3**: 1216

5-(2-Chlorophenyl)-7-ethyl-1,3-dihydro-1-methyl-2*H*-thieno[2,3-*e*]-1,4-diazepin-2-one **2**: 641

[3-(4-chlorophenyl-1*H*-pyrazol-4-yl]acetic acid **2**: 358

4-(4-Chlorophenyl)-4-hydroxy-*N*,*N*-dimethyl-α,α-diphenyl-1-piperidinebutanamide **2**: 409

4-[4-(4-Chlorophenyl)-4-hydroxypiperidino]-4′-fluorobutyrophenone **2**: 598

4-[4-(4-Chlorophenyl)-4-hydroxypiperidin-1-yl]-*N*,*N*-dimethyl-2,2-diphenylbutyramide **2**: 409

2-(*o*-Chlorophenyl)-2-methylaminocyclohexan-1-one **2**: 448

N-[(4-chloro-phenyl)methyl]-*N*′,*N*′-dimethyl-*N*-2-pyridinyl-1,2-ethanediamine **2**: 905

4-[(4-chlorophenyl)methyl]-2-(hexahydro-1-methyl-1*H*-azepin-4-yl)-1(2*H*)-phthalazinone **2**: 931

5-[(2-chlorophenyl)methyl]-4,5,6,7-tetrahydrothieno[3,2-*c*]pyridine **1**: 40

1-[2-[[(4-Chlorophenyl)methyl]thio]-2-(2,4-dichlorophenyl)ethyl]-1*H*-imidazole **3**: 1226

2-[1-(4-Chlorophenyl)-1-phenylethoxy]-*N*,*N*-dimethylethyl-amine **2**: 742

1-[2[1-(4-Chlorophenyl)-1-phenylethoxy]ethyl]hexahydro-1*H*-azepine **2**: 926

1-[(4-Chlorophenyl)phenylmethyl]-4-[4-(1,1-dimethylethyl)phenyl]methyl]piperazine **2**: 913

1-[(4-chlorophenyl)phenylmethyl]–4-[(3-methylphenyl)methyl]-piperazine **2**: 743, 916

1-[(4-Chlorophenyl)phenylmethyl]-4-methylpiperazine **2**: 913

1-[(4-Chlorophenyl)phenylmethyl]-4-(3-phenyl-2-propenyl)piperazine **2**: 914

2-[4-[1-(4-Chlorophenyl)phenylmethyl]-1-piperazinyl]ethoxy]acetic acid **2**: 927

2-{2-{4-[(4-Chlorophenyl)phenylmethyl]-1-piperazinyl}ethoxy}-ethanol **2**: 646, 915

4-(4-Chlorophenyl)- 2-phenyl-5-thiazoleacetic acid **4**: 1685

2-{3-[4-(3-Chlorophenyl)-1-piperazinyl]propyl}-1,2,4-triazolo[4,3-*a*]pyridin-3(2*H*)-one **2**: 632

3-(4-Chlorophenyl)-3-(2-pyridinyl)-*N*,*N*-dimethylpropanamine **2**: 911

2-[(4-Chlorophenyl)-2-pyridinylmethoxy]-*N*,*N*-dimethylethanamine **2**: 908

(+)-(*S*)-4-[4-[1-(4-chlorophenyl)-1-(2-pyridyl)methoxy]-piperidin-1-yl]butanoic acid monobenzenesulfonate **2**: 929

8-Chloro-6-phenyl-4*H*-[1,2,4]triazolo -[4,3-*a*][1,4]benzodiazepine **2**: 524

(+)-6-[(4-Chlorophenyl)-1*H*-1,2,4-triazol-1-ylmethyl]-1-methyl-1*H*-benzotriazole **4**: 1926

2- chloro-11-(1-piperazinyc)dibenz[*b*, *f*][1,4]oxazepine **2**: 611

Chloroprocaine hydrochloride **2**: 544

3-Chloro-4-(2-propenyloxy)benzeneacetic acid **4**: 1684

Chloroptic **4**: 2011

Chloropyribenzamine **2**: 905

2-[(6-Chloro-3-pyridazinyl)thio]-*N*,*N*-diethylacetamide **2**: 753

6(5-chloro-2-pyridyl)-5-(4-methylpiperazin-1-yl)carbonyloxy-7-oxo- 6,7-dihydro-5*H*-pyrrolo[3,4-*b*]pyrazine **2**: 527

Chloroquinaldol **3**: 1254

Chloroquine **2**: 763; **3**: 1161; **4**: 1704

4-[(7-Chloro-4-quinolyl)amino]-α-(diethylamino)-*o*-cresol **3**: 1162

2-({4-[(7-Chloro-4-quinolyl)amino]pentyl}ethylamino)ethanolsulfate **3**: 1163

6-Chloro-7-sulfamoyl-2*H*-1,2,4-benzothiadiazine 1,1-dioxide **1**: 292

Chlorotestosterone acetate **3**: 1593

Chlorothen **2**: 905

Chlorothiazide **1**: 15, 71, 292

N-[(5-chloro-2-thienyl)methyl]-*N*′,*N*′-dimethyl-*N*-2-pyridinyl-1,2-ethanediamine **2**: 905

1-[2-[(2-Chloro-3-thienyl)methyloxy]-2-(2,4-dichlorophenyl)ethyl]-1*H*-imidazole **3**: 1230

9α-Chloro-11β,17α,21-trihydroxy-16β-methylpregna-1,4-diene-3,20-dione **2**: 893

7-Chloro-4,6-trimethoxy-6′-methylspiro[benzofuran-2(3*H*),1′-[2]cyclohexene]-3,4′-dione **3**: 1241

3-(β-Chlorovinyl)-3-hydroxy-pent-1-ine **2**: 497

Chlorozotocin **4**: 1891

Chlorphenesin carbamate **2**: 677

Chlorpheniramine **2**: 911; **4**: 1720

Chlorphenoxamine **2**: 742

1-[2-[(4-Chlorphenyl)methoxy]-2-(2,4-dichlorophenyl)-ethyl]-1*H*-imidazole **3**: 1219

Chlorpromazine **2**: 590, 763

Chlorpromazine hydrochloride **2**: 681

Chlorpropamide **3**: 1655

Chlorprothixene **2**: 596

Chlorpyramine **2**: 905

Chlortetracycline **3**: 954, 978; **4**: 2161

Chlortetracycline hydrochloride **3**: 1036; **4**: 2007

Chlorthalidone **1**: 71, 72, 295

Chlortritylimidazole **3**: 1216

Chlorylen **2**: 439

Chlorzoxazone **2**: 677

Chlotride **1**: 293

Choice-7 **3**: 1509

Cholagogue **2**: 748

Cholagogum capsules **2**: 750

Cholagutt-A **2**: 750

Cholane **3**: 1323

Chol-Arbuz **2**: 749

Cholecalciferol **3**: 1325, 636

Cholecalciferol synthesis **3:** 1639
Cholecystectomy **2:** 760
Cholecystokinin **3:** 1413, 549
Cholecystokinin releasing peptide **3:** 1413
Choleinic acid **3:** 1328
Cholekinetics **2:** 748
Cholenic acids **3:** 1328
Choleodoron **2:** 750
Cholera vaccine **4:** 1775
Choleretics **2:** 748
Cholesolvin **1:** 25
Cholestabil **1:** 22
Cholestane **3:** 1323
Cholesterol **1:** 20; **3:** 1322, 1564, 1567
Cholesterol stone **2:** 754
Cholic acid **2:** 749; **3:** 1328
Cholinergic agonists **4:** 2024
Cholinergic drugs **2:** 651
Cholinergic System
    action of gastrokinetic agens **2:** 798
Cholinesterase inhibitors **2:** 653
Cholipin **2:** 751
Cholonerton **2:** 701
Cholspasmin **2:** 701
Chondrillasterol **3:** 1325
Choragon **3:** 1357
Chorionic gonadotropin **3:** 1355, 546
Choriotropin **3:** 1546
Christmas factor **1:** 256
Chromonar **1:** 231
Chromopeptide antibiotics **3:** 1007
Chromosomal resistance **3:** 1112
Chronadalate **1:** 210
Chronic pain **2:** 343
Chrysophanol **2:** 790
Chrytemin **2:** 614
Chylomicrons **1:** 19
C.I. 45 350 **4:** 2030
C.I. 45 440 **4:** 2031
CI-825 **4:** 1881
CI–919 **3:** 1125
CI 988 **2:** 649
Ciatyl **2:** 597
Cibacalcin **3:** 1367
Ciba Vision Lens Drops **4:** 2040
Ciberon **2:** 908
Cicaprost **4:** 2054
Ciclacillin **3:** 1024
Ciclochem **3:** 1245
Ciclofalina **2:** 650
Ciclolux **4:** 2027
Ciclopegic **4:** 2027
Ciclopirox **3:** 1244; **4:** 1731
Ciclopirox ethanolamine **3:** 1244
Cicloplegyl **4:** 2027
Cidomycin **4:** 2008
Ciglitazone **3:** 1665

Cilergil **2:** 929
Ciliates **3:** 1168
Ciloprost **1:** 38
Cimetidine **2:** 723; **4:** 1720
Cinalukast **4:** 2065
Cinaperazine **1:** 209
Cinazyn **1:** 209
Cinchocaine hydrochloride **2:** 549
Cinerubin **3:** 981
C1-Inhibitor **1:** 252, 261
Cinitapride **2:** 795
    mode of action **2:** 798
Cinnacet **1:** 209
Cinnageron **1:** 209
Cinnamic Acid choleretics **2:** 751
Cinnamin **4:** 1695
Cinnarizine **1:** 208; **2:** 914
Cinolazepam **2:** 515, 523
    synthesis **2:** 520
Cinoxacin **3:** 1119
Cinubac **3:** 1119
Cipramil **2:** 627
Ciprofibrate **1:** 26
Ciprofloxacin **3:** 1123
Circanol **2:** 655
Circubid **2:** 699
Circulation
    coronary **1:** 218
Circupon RR **1:** 64
ECircuvit E **1:** 64
Cirvamycin A **3:** 999
Cisapride **2:** 728, 795
    mode of action **2:** 798
    Receptor binding affinities **2:** 808
Cisodinol **2:** 597
Cisplatin **4:** 1918
cis-Platinum **4:** 1918
Cisticid **3:** 1285
Citalopram **2:** 627, 657
Citanest **2:** 546
Citicoline **2:** 652
Citilat **1:** 210
Citopan **2:** 511
Citrate: Banflex **2:** 567
CL 12625 **3:** 1236
CL-216942 **4:** 1903
CL-232315 **4:** 1902
Cladribine **4:** 1880
Clanzol **2:** 607
Claritin **2:** 932
Clarmyl **2:** 644
Clauberg Test **3:** 1595
Claudicat **1:** 49
Clavam-2-carboxylic **3:** 971
Clavigrenin **1:** 66
Clavulanic acid **3:** 971, 035
Clavulanic acids **3:** 970

Clavulone **4**: 2051
Clazol **3**: 1217
Clédial **2**: 631
Cleaning & Disinfecting **4**: 2037
Clean-N-Soak **4**: 2037
Clean packing **3**: 1088
Clear Eyes **4**: 2029
Clebon **2**: 607
Clebopride **2**: 607, 795
   mode of action **2**: 798
   Receptor binding affinities **2**: 809
Cleboril **2**: 607
Clemastine **2**: 909
Clenbuterol **2**: 878
Cleocin Phosphate **4**: 2011
Clera **4**: 2029
Cleridium **1**: 39
Clerz **4**: 2033, 2037
Clianimon **2**: 599
Clidanac **4**: 1691
Clidinium bromide **2**: 692
Clift **1**: 191
Clincopal **4**: 1687
Clindamycin **3**: 1067; **4**: 1725, 2011
Clindamycin palmitate hydrochloride **3**: 1068
Clinical application dossier **4**: 2140
Clinical Trials **4**: 2134
Clinit **2**: 373
Clinoril **2**: 359; **4**: 1683
Clinovier **3**: 1596
Clinprost **4**: 2062
Clionasterol **3**: 1324
Clioquinol **3**: 1170; **4**: 1732
Clipoxide **2**: 693
Cliradon **2**: 399
Clistin **2**: 908
Clivoten **1**: 212
Clobazam **2**: 481, 644
Clobenzepam **2**: 923
Clobetasol propionate **4**: 1727
Clocapramine **2**: 603
Clocim **3**: 1217
Clocinizine **2**: 914
Clofazimine **3**: 1144
Clofekton **2**: 603
Clofibrate **1**: 15, 24
Clomiphene **3**: 1587
Clomipramine **2**: 612
Clonazepam **2**: 481
Clonic seizures **2**: 470
Clonidine **1**: 82; **2**: 424
Cloning **4**: 1986
Clonopin **2**: 481
Clonorchis sinensis **3**: 1268
Clopamide **1**: 295
Clopenthixol **2**: 597
Clopidogrel **1**: 41

Clopidol **4**: 2172
Clopixol **2**: 597
Cloprostenol **4**: 2054
Cloprostenol sodium **4**: 2059
Clorazepate **2**: 638
Clorets **2**: 498
Clot-basan **3**: 1217
Clothiapine **2**: 604
Clotiazepam **2**: 641
Clotrifug **3**: 1217
Clotrimaderm **3**: 1217
Clotrimazole **3**: 1216; **4**: 1730, 2171
Clotrimix **3**: 1217
clotri OPT **3**: 1217
Clotrizol **3**: 1217
Clotting cascade **1**: 254
Clotting system **4**: 1672
Clox **1**: 41
Cloxacillin **3**: 962
Cloxacillin sodium **3**: 1021
Cloxan **2**: 596
Clozan **2**: 641
Clozapine **2**: 604
Clozaril **2**: 605
Clozole **3**: 1217
Cluster of differentiation-nomenclature **1**: 247
β-CM-7 **3**: 1386
CMI response **4**: 1761
CMV **3**: 1174
CNP **3**: 1405
CNS **2**: 312
Coagulation factor I **1**: 254
Coagulation factor II **1**: 255
Coagulation factor III **1**: 255
Coagulation Factor IV **1**: 255
Coagulation factor IX **1**: 256
Coagulation factors **1**: 253
   therapeutic use **1**: 274
Coagulation Factor V **1**: 255
Coagulation factor VI **1**: 256
Coagulation factor VII **1**: 256
Coagulation factor VIII **1**: 256
Coagulation factor X **1**: 257
Coagulation factor XI **1**: 257
Coagulation factor XII **1**: 257
Coagulation factor XIII **1**: 257
Coated pits **4**: 1963
Coban 45 **4**: 2170
Coban 60 **4**: 2170
Cobantril **3**: 1285
Cocaine **2**: 540; **4**: 2027
Cocaine hydrochloride **4**: 2003
Cocci **3**: 1110
Coccidia **3**: 1159
Coccidiostats
   veterinary **4**: 2172
COCM **1**: 148

Cocoa butter **4:** 2104
Coco-Diazine **4:** 2014
Cocol **3:** 1239
Codeine **2:** 390
Co-dergocrine **2:** 655
Codethyline **2:** 391
Codicaps **2:** 391, 912
Codimal **2:** 391
Codipront **2:** 391, 910
Coforin **4:** 1881
Coformycin **3:** 989
Cofrim **3:** 1133
Cogard **1:** 156
Cogentin **2:** 562
Cogentinol **2:** 562
Cognex **2:** 653
Cohidrate **2:** 498
Cohn fractionation **1:** 267
Cohn – Oncley fractionation **4:** 1823
Colabid **4:** 1708
Colchicina **4:** 1708
Colchicine **2:** 771; **4:** 1708
Colectril **1:** 75, 299
Colestid **1:** 22
Colestipol **1:** 22
Colestyramine **1:** 21
Colfarit **4:** 1678
Colibantil **2:** 695
Coliopan **2:** 691
Colircusi ciclopejico **4:** 2027
Colirio Oculos Cicloplegic **4:** 2027
Colistimethate sodium **3:** 1058; **4:** 2009
Colistinmethanesulfonate sodium **3:** 1058
Colistin sulfate **3:** 1058
Colite **2:** 653
Collagenase **4:** 1673
Collins Elixir **2:** 391
Collodions **4:** 2089
Collox **2:** 742
Collyrium **4:** 2030, 2037
Colofac **2:** 701
Colony forming units **1:** 240
Colony stimulating factors **1:** 240
Colorants **4:** 2102
Colpro **3:** 1597
Colprone **3:** 1597
Colprosteron **3:** 1596
Colum **2:** 695
Coly-Mycin M **4:** 2009
Combantrin **3:** 1285
Combid **2:** 694
Combivir **3:** 1299
Comedolytic agents **4:** 1724
Comfort Eye Drops **4:** 2029
Comfort Tears **4:** 2033
Common tabulated summaries **4:** 2145
Common technical document — quality **4:** 2145

Common written summaries **4:** 2145
Compactin **1:** 15
Companies
 pharmaceutical **1:** 5
Compazine **2:** 594, 744
Compendium **2:** 637
Complamin **1:** 23
Complement system **4:** 1672
Complexation dosage forms **4:** 2117
Complex partial seizures **2:** 470
Compressed tablets **4:** 2094
Compression **4:** 2095
Comtan **2:** 572
Comtess **2:** 572
COMT inhibitors **2:** 572
Ca^{2+} concentration
 in heart muscle cells **1:** 175
Concentration – time curves **4:** 2113
Concor **1:** 153
Concordin **2:** 616
Condom **3:** 1495
Conduction velocity **1:** 119
Conessine **3:** 1333
Conestron **3:** 1583
Confidol **1:** 64
Conflictan **2:** 629
Confortid **4:** 1680
Congestive cardiomyopathy **1:** 148
Congestive heart failure **1:** 174
Coniel **1:** 213
Conjuncain **4:** 2003
Conoderm **3:** 1210
Conofite **3:** 1210; **4:** 2171
Consept **4:** 2039
Constan **2:** 641
Contac **4:** 2028
Contact lenses **4:** 2035
Containers **4:** 2106
Contalax **2:** 788
Contenton **2:** 574
Continuous fermentation
 in antibiotics production **3:** 1078
Contraception
 Barrier Methods **3:** 1495
 Chemical **3:** 1493
 Hormonal Methods **3:** 1497
 Intrauterine Devices **3:** 1506
 Natural Methods **3:** 1493
Contraceptives
 Long-Acting **3:** 1504
Contraceptive sponge **3:** 1495
Contraceptive Vaccines **3:** 1510
Contrafungin **3:** 1217
Contragestational Drugs :
Control **2:** 522
Controlled-release drug delivery systems **4:** 2111
Controlled release technology **4:** 2112

Convallatoxin **1:** 182
Convoluted Filaria **3:** 1277
Convulex **2:** 477
Convulsofin **2:** 477
Coomassie Blue R 250 **4:** 1973
Copharten **3:** 1282
Coracin **4:** 2017
Coracten **1:** 210
Corathiem **1:** 209
Cordalin **2:** 888
COrdan **1:** 208
Cordarex **1:** 130
Cordarone **1:** 130
Cordicant **1:** 210
Cordilan **1:** 210
Cordilox **1:** 97, 131, 208
Cordium **1:** 132, 209
Cordycepic acid **4:** 2023
Cordycepin **3:** 955, 990
Coredamin **1:** 208
Corenalin **2:** 653
Corgard **1:** 92
Corid **4:** 2172
Coritat **1:** 62
Corium **2:** 693
Coronarine **1:** 39
Coronary arteriolar vasodilators **1:** 231
Coronary artery dilators **1:** 230
Coronary circulation **1:** 218
Coronary steal **1:** 221
Coronaviridae **3:** 1174
Coronaviruses **3:** 1174
Corontin **1:** 208
Corosan **1:** 39
Cor-Oticin **4:** 2016
Corotoxigenin **1:** 179
Corotrend **1:** 210
Corotrope **1:** 195
Coroxin **1:** 39
Corrigast **2:** 697
Cortenil depot **3:** 1615
Cortexolone **3:** 1602
Cortexone **3:** 1602, 1614
Corticoid hormones **3:** 1601
Corticoid metabolism **3:** 1606
Corticoliberin **3:** 1369, 371
Corticosteroids
 ophthamological agents **4:** 2015
Corticosterone **3:** 1602, 1607
Corticotropin **3:** 1373, 1543
Corticotropin releasing hormone **3:** 1543
Cortiron **3:** 1615
Cortisol **4:** 2016
 doses **3:** 1605
Cortisol synthesis **3:** 1612
Cortisone **3:** 1602, 1605, 1607
Cortisone synthesis **3:** 1608

Cortisporin **4:** 2008, 2009, 2016, 2017
Corvaton **1:** 230
Corwin **1:** 193
Cosolvents **4:** 2102
Cospanon **2:** 700
Cotinazin **3:** 1140
Cotomer **4:** 2104
Coumarin **1:** 15
COX **2:** 344; **4:** 1672, 1721
COX-1 **2:** 344
COX-2 **2:** 345
COX inhibitors **4:** 1676
COX-1 inhibitors **4:** 2053
COX-2 Inhibitors **2:** 381; **4:** 2054
Coxsackievirus **3:** 1174
CP-15525 **2:** 609
CPE assay **4:** 1972
CPI **4:** 1907
CPT **4:** 1913
CPT-11 **4:** 1914
Cranoc **1:** 29
Cratodin **2:** 350
Creams **4:** 1715, 2090
Creosedin **2:** 637
Crepasin **1:** 208
CRH **3:** 1543
Crixivan **3:** 1313
CRL **3:** 1414
Croconazole **3:** 1218
Cromakalim **2:** 706
Cromolyn sodium **2:** 934; **4:** 1720, 2018
Cronil **2:** 501
Crotetamide **2:** 339
Crotonoside **3:** 990
Cryptosporidium **3:** 1159, 1167
Crystodigin **1:** 188
CSF **1:** 240
CT **2:** 325
CTD quality **4:** 2145
C-type natriuretic peptide **3:** 1405
Curantyl **1:** 39
Curare alkaloids **2:** 668
Curling factor **3:** 1241
Cutistad **3:** 1217
Cuvalit **2:** 576
CVT-124 **1:** 301
Cyabin **4:** 1903
Cyamemazine **2:** 590, 591
Cyanidanol **2:** 765
(+)3-cyanidanol **2:** 765
Cyanocobalamin ^{57}Co capsules **4:** 2101
(*E*)-2-Cyano-*N*,*N*-diethyl-3-(3,4-dihydroxy-5-nitrophe-
 nyl)propenamide **2:** 572
2-Cyano-1,4-dihydro-6-methyl-4-(3-nitrophenyl)-3,5-pyr-
 idinedicarboxylic acid 3-methyl 5-(1-methylethy-
 l)ester **1:** 211

2-Cyano-10-(3-dimethylamino-2-methylpropyl)-10*H*-phenothiazine   **2:** 591
1′-(3-Cyano-3,3-diphenylpropyl)-[1,4′]bipiperidinyl-4′-carboxamide   **2:** 410
1-[4-cyano-4-(4-Fluorophenyl)cyclohexyl]-3-methyl-4-phenyl-4-piperidinic acid   **2:** 930
2-Cyano-10-[3-(4-hydroxypiperidino)propyl]-10*H*-phenothiazine   **2:** 593
1-Cyano-2-methyl-3-[2-[[(5-methylimidazol-4-yl)methyl]thio]ethyl]guanidine   **2:** 723
Cybis   **3:** 1118
Cyclacillin   **3:** 962, 024
Cyclaradine   **3:** 1182
Cyclexedrine   **2:** 462
Cyclizine   **2:** 742, 914
Cycloartenol   **3:** 1326
Cyclobarbital   **2:** 504, 507
Cyclobarbitone   **2:** 507
Cyclobenzaprine hydrochloride   **2:** 678
(5α6α)-17-(Cyclobutylmethyl)-4,5-epoxymorphinan-3,6,14-triol   **2:** 417
11-Cyclobutylmethyl-1,2,3,4,9,10-hexahydro-4α,10-propanophenanthrene-6,10α-diol   **2:** 415
Cyclobutyrol   **2:** 752
Cycloeucalenol   **3:** 1326
Cyclofenil   **3:** 1587
Cyclogyl   **4:** 2028
[Sp-4-2-(1R-*trans*)](1,2-Cyclohexanediamine-*N,N*′)[ethanedioxato(2⁻⁻)-*O,O*′]-platinum   **4:** 1921
5-(Cyclohexen-1-yl)-1,5-dimethylbarbituric acid   **2:** 444
5-(1-Cyclohexen-1-yl)-5-ethylbarbituric acid   **2:** 507
Cycloheximide   **3:** 1008
2-Cyclohexylamino-1-methylethyl benzoate monohydrochloride   **4:** 2003
2-Cyclohexylcarbonyl-1,2,3,6,7,11*b*-hexahydro-4*H*-pyrazino[2,1-*a*]isoquinolin-4-one   **3:** 1284
3′-Cyclohexyl-3′-dephenylpaclitaxel   **4:** 1917
6-cyclohexyl-1-hydroxy-4-methyl-2(1*H*)-pyridinone   **3:** 1244
4-(β-cyclohexyl-β-hydroxyphenethyl)-1,1-dimethylpiperazinium methylsulfate   **2:** 694
(±)-*N*-(3-cyclohexyl-3-hydroxy-3-phenylpropyl)-*N*-,*N*-,*N*-triethylammonium chlor   **2:** 697
(±)-1-[(2-Cyclohexyl-2-phenyl-1,3-dioxolan-4-yl)methyl]-1-methylpiperidinium iodide   **2:** 696
α-Cyclohexyl-α-phenyl-1-piperidinepropanol hydrochloride   **2:** 563
(*RS*)-1-Cyclohexyl-1-phenyl-3-piperidinopropanol hydrochloride   **2:** 563
(*RS*)-1-Cyclohexyl-1-phenyl-3-(1-pyrrolidonyl)-1-propanol   **2:** 564
Cyclolaudenol   **3:** 1326
Cyclomat   **3:** 1352
Cyclomydril   **4:** 2028
Cyclonal Sodium   **2:** 445

Cycloneosamandione   **3:** 1334
Cyclonium iodide   **2:** 696
Cyclooxygenase   **4:** 1672
Cyclooxygenase Inhibitors   **1:** 35; **4:** 1676, 2053
Cyclooygenases   **4:** 1721
Cyclopal   **2:** 510
Cyclopan   **2:** 511
Cyclopen   **4:** 2027
Cyclopental   **2:** 510
Cyclopenta[α]-phenanthrene   **3:** 1321
Cyclopenthiazide   **1:** 71, 294
Cyclopentobarbital   **2:** 504, 510
Cyclopentolate hydrochloride   **4:** 2027
Cyclopentol Colircusi   **4:** 2027
(±)-3-(α-Cyclopentylmandeloyloxy)-1,1-dimethylpyrrolidinium bromide   **2:** 694
N-[4-[5-cyclopentyloxycarbonylamino)-1-methylindol-3-ylmethyl]-3-methoxybenzoyl]-2-methylbenzenesulfonamide   **2:** 896
17β-Cyclopentylpropionyloxy-4-androsten-3-one   **3:** 1590
17β-Cyclopentylpropionyloxy-4-estren-3-one   **3:** 1592
17β-cyclopentylpropionyloxy-4-hydroxy-4-estren-3-one   **3:** 1592
Cyclophosphamide   **4:** 1707, 888
Cycloplegics   **4:** 2026
Cyclo-Progynova   **3:** 1584, 598
Cyclopropane   **2:** 439
11-Cyclopropyl-5,11-dihydro-4-methyl-6*H*-dipyrido[3,2-b:2′,3′-e][1,4]diazepin-6-one   **3:** 1301
1-Cyclopropyl-6-fluoro-1,4-dihydro-4-oxo-7-(1-piperazinyl)-3-quinolinecarboxylic acid   **3:** 1123
1-[4-[2-(Cyclopropylmethoxy)ethyl]phenoxy]-3-isopropylamino-2-propanol hydrochloride   **4:** 2021
(5α)-17-(Cyclopropylmethyl)-4,5-epoxy-3,14-dihydroxymorphinan-6-one   **2:** 422
Cyclopropylpyrroloindoles   **4:** 1907
Cyclo-Prostin   **4:** 2061
Cycloserine   **3:** 1008, 069
Cyclosporin A   **4:** 1707
Cyclosporine   **4:** 1729
Cyclothiazide   **1:** 294
Cygro Type A medicated article   **4:** 2171
Cylinder-Plate Method   **3:** 1089
Cymarin   **1:** 190
Cymerin   **4:** 1893
Cynarin   **2:** 751
Cynarix   **2:** 751
Cynomel   **3:** 1561
Cyperon   **2:** 601
Cyplegin   **4:** 2028
Cyproheptadine   **2:** 924; **4:** 1720
Cyproterone acetate   **3:** 1591
Cyren A   **3:** 1586
Cyrpon   **2:** 645
Cysticercosis vaccines   **4:** 1811
Cystrin   **2:** 704

Cytarabine   **4**: 1877
Cytellin   **1**: 22
Cytidine 5′-sodium diphosphate choline ester   **2**: 652
Cytochrome   **2**: 717
Cytokines   **4**: 1673
Cytokins   **1**: 243, 245
   therapeutic use   **1**: 273
Cytolytic reactions   **2**: 902
Cytomegalovirus   **4**: 1830
Cytopathic effect inhibition assay   **4**: 1972
Cytosine arabinoside   **4**: 1877
Cytostatic agents
   interferons   **4**: 1961
Cytotec   **2**: 728; **4**: 2060

D-18506   **4**: 1938
D-19466   **4**: 1921
D-21266   **4**: 1938
Dacarbazine   **4**: 1895
Dacarel   **1**: 210
Dactinomycin   **3**: 1062; **4**: 1904
DAD   **4**: 1902
Dafnegin   **3**: 1245
Dagrabromyl   **2**: 501
Daktar   **3**: 1210
Daktarin   **3**: 1210
Dalargin   **3**: 1385
Dalgan   **2**: 416
Dalgol   **2**: 497
Dalmadorm   **2**: 522
Dalmane   **2**: 522
Dalmate   **2**: 522
Dalmene   **2**: 522
Dalzic Eraldin   **1**: 158
Danazol   **3**: 1509, 588
Dannetène   **2**: 497
Dantamicrin   **2**: 675
Danten   **2**: 475
Dantrirem   **3**: 1146
Dantrium   **2**: 675
Dantrolene sodium   **2**: 675
Daonil   **3**: 1656
Dapotum   **2**: 595
Dapsone   **3**: 1142; **4**: 1704
Daptizole   **4**: 2180
Daranide   **4**: 2022
Daraprim   **3**: 1130, 164
Darbid   **2**: 694
Daren   **2**: 932
Darodipine   **1**: 212
Dartal   **2**: 595
Dartalan   **2**: 595
Darvon   **2**: 408
Data quality assurance   **4**: 2139
DAU 6285   **2**: 804, 812
Daunomycin   **4**: 1897
Daunorubicin   **3**: 954, 981; **4**: 1897

Daunorubicin hydrochloride   **3**: 1039
Daxauten   **1**: 208
Daxolin   **2**: 605
Day-Barb   **2**: 509
Daypro   **2**: 381
Dazopride   **2**: 795
   mode of action   **2**: 798
2′-dCF   **4**: 1881
DCI   **1**: 139
DCNU   **4**: 1891
β-D-Cymarose   **1**: 180
β-D-Digitoxose   **1**: 180
Deacetoxyephalosporin C   **3**: 964
Deacetylephalosporin C   **3**: 964
Deacetylvinblastineamide   **4**: 1911
Deaner   **2**: 652
Deanol   **2**: 652
Deapril-ST   **2**: 655
Debephenium   **3**: 1278
Debridat   **2**: 702
Decacindin   **4**: 2016
Decadron   **3**: 1627
Decadron Phosphate   **4**: 2018
Deca-Durabolin   **3**: 1592
Decamethonium bromide   **2**: 673
Decamethylenebis(trimethylammonium) dibro-
   mide   **2**: 673
Decamethylenediguanidine   **3**: 1660
17β-decanoyloxy-4-estren-3-one   **3**: 1592
Decapeptyl   **4**: 1932
Decapryn   **2**: 496
Decaris   **3**: 1281
Deccox   **4**: 2172
Decetan   **2**: 595
Decholin   **2**: 749
Decomyc   **3**: 1210
Decoquinate   **4**: 2172
Decortilen   **3**: 1628
Decortin   **3**: 1618
Decortin-H   **3**: 1618
Decreten   **1**: 158
Deep-tank fermenters   **4**: 1990
Defekton   **2**: 603
Defenyl   **2**: 612
Defibrinogenation
   of blood plasma   **1**: 47
Deflam   **2**: 381
Defluina   **1**: 50
Defoperazone   **3**: 965
Deftan   **2**: 614
Degest 2   **4**: 2029
Dehydrobenperidol   **2**: 600
Dehydrobenzperidol   **2**: 454
24-Dehydrocholesterol   **3**: 1325
Dehydrocholic acid   **2**: 749
1,2-Dehydrocorticosteroids   **3**: 1617
11-Dehydrocorticosterone   **3**: 1602, 1607

8, 14-Dehydrodeoxycholic acid **3:** 1328
Dehydroepiandrosterone **3:** 1565
1,2-Dehydrohydrocortisone **4:** 2017
10,11-Dehydro-5-methyl-5*H*-dibenzo[*a,d*]cyclohepten-5,10-imine **2:** 574
Dehydropregnenolone **3:** 1568
Dehydropregnenolone acetate synthesis **3:** 1570
Deiten **1:** 210
Delacuratine **2:** 672
Deladroxat **3:** 1584, 597
Deladroxone **3:** 1597
Delanol **3:** 1636
Delatestryl **3:** 1590
Delatulin **3:** 1596
Delavirdine **3:** 1301
Delayed-hypersensitivity reactions **2:** 902
Delayed-release or delayed-action oral products **4:** 2112
Delestrogen **3:** 1584
Deliton **2:** 604
Delmeson **3:** 1624
Delmofulvina **3:** 1242
Delphicort **3:** 1622
Delpral **2:** 606
Delpregnin **3:** 1596
Deltacortril **3:** 1618
Deltamycin **3:** 999
Delta-sleep-inducing peptide **2:** 492; **3:** 1408, 1412
Deltorphins **3:** 1385
Delvinal **2:** 510
Demecarium bromide **4:** 2025
Demeclocycline **3:** 1037
Demerol **2:** 399
Demethylbleomycin **3:** 1002
Demethylchlortetracycline **3:** 978, 1037
5-*O*-Demethyl-22,23-dihydroavermectin A$_{1a}$ **3:** 1279
Demetrin **2:** 639
Demex **2:** 371
Demexiptiline **2:** 618
Demolox **2:** 612
Demulen **3:** 1500
Denagard 10 **4:** 2168
Denan **1:** 28
Denapol **1:** 209
Denbufylline **2:** 659
Dendrites **2:** 310
Dengue **3:** 1174
Denopamine **1:** 193
Denoral **2:** 914
Dentigoa **2:** 502
Dentinal fluid transport stimulating peptide **3:** 1421
Deoxycholic acid **2:** 749
2-Deoxycoformycin **3:** 990; **4:** 1881
17-Deoxycorticosteroids **3:** 1632
Deoxycorticosterone **3:** 1614
Deoxycytidine **4:** 1879
2'-Deoxy-2',2'-difluorocytidine **4:** 1879

21-Deoxy-9α-fluoro-6α-methylprednisolone **3:** 1623
2'-Deoxy-5-fluorouridine **4:** 1873
5'-Deoxy-5-fluorouridine **4:** 1875
2'-nor-2'-Deoxyguanosine (2'-NDG) **3:** 1182
15-Deoxy-16-hydroxy-16-methyl-PG-E$_1$ **2:** 727
2'-Deoxy-5-iodouridine **3:** 1178; **4:** 2014
3β-[(6-deoxy-α-L-mannopyranosyl)oxy]-1β,5,11α,14,19-pentahydroxy-5β,14β-card-20(22)-enolide octahydrate **1:** 190
1-Deoxy-1-(methylamino)-D-glucitol trioxoantimonate **3:** 1154
2-Deoxy-2{[(methylnitrosoamino)carbonyl]amino}-D-glucopyranose **4:** 1891
11-Deoxy-PGE$_1$ **4:** 2054
2'-Deoxy-5-(trifluoromethyl)uridine **4:** 2015
Depakene **2:** 477
Depakine **2:** 477
Depakote **2:** 477
Deparkin **2:** 565
Depas **2:** 640
Depasan **1:** 126
d Epifrin **4:** 2020
Depilorphin **3:** 1387
Depixol **2:** 597
Depo-Medrol **3:** 1624
Deponit **1:** 223
Depo Provera **3:** 1504
Depostat **3:** 1597
Depot **3:** 1594
Depo-Testadiol **3:** 1590
Depot-H-Insulin **3:** 1444
Depot injection **4:** 2119
Depot-Nortestonat **3:** 1592
Depot Ústronom **3:** 1586
Depovirin **3:** 1590
Deprenil **2:** 619
Deprenyl **2:** 570
Depressin **2:** 620
Depression **2:** 610
Deprimil **2:** 614
Deprinol **2:** 614
Deproceptin **3:** 1387
Deprolorphin **3:** 1387
Depyrel **2:** 632
Deracyn **2:** 640
Deralbine **3:** 1210
Dermacure **3:** 1210
Dermal administration **4:** 2083
Dermatological drugs **4:** 1713
Dermazol **3:** 1219
Dermazole **3:** 1219
Dermis **4:** 1714
Dermofix **3:** 1226
Dermonistat **3:** 1210
Dermorphin **3:** 1385
Dermoseptic **3:** 1226
Dermo-Trosyd **3:** 1231

Deronga **3**: 1236
Desconex **2**: 605
Desipramine **2**: 614
Desmopressin **3**: 1402
Desogen **3**: 1500
Desogestrel **3**: 1497, 1501, 1598
17-Desoximetasone **3**: 1633
Desoxyephedrine **2**: 461
Desoxyn **2**: 461, 467
Destomycin A **3**: 1046
Desyrel **2**: 632
Detajmium bitartrate **1**: 127
Detensiel **1**: 153
DET MS **1**: 66
Detomidine **4**: 2181
Detrulisin **2**: 693
Deturid **2**: 501
Develin **2**: 408
Dexacidin **4**: 2008, 2011, 2017, 2018
Dexa-Cortidelt **3**: 1627
Dexair **4**: 2018
Dexamethasone **3**: 1625; **4**: 2017, 2178
  doses **3**: 1605
Dexamytrex Ophtiole **4**: 2038
Dexa-Scheroson **3**: 1627
Dexasporin **4**: 2008, 2011, 2016, 2017, 2018
Dexa Tavegil **2**: 909
Dexedrine **2**: 461, 467
Dexetimide **2**: 565
Dexmedetomidine **4**: 2181
Dexonil **3**: 1627
Dextran **4**: 2033
Dextroamphetamine **2**: 461, 467
Dextromoramide **2**: 406
Dextropropoxyphene **2**: 407
Dextrose **4**: 2104
Dextrothyroxine **1**: 22
Dezocine **2**: 416
dFdC **4**: 1879
DFP-Oel **4**: 2026
DFT **3**: 1421
D-glucitol **2**: 785
β-D-Glucose **1**: 180
DHC Continus **2**: 393
D.H.E. 45 **1**: 66
DHE-PUREN **1**: 66
DHE-ratiopharm **1**: 66
DHFR **4**: 1870
D-HPG **3**: 1011, 182
Diabenese **3**: 1655
Diabetes mellitus **3**: 1442
  type 2 noninsulin dependent **3**: 1653
Diabetoral **3**: 1655
Diaboral **3**: 1658
9,10-Diacetoxy-2-propyl-4,5,5a(*R*),6,7,11b,(*S*)-hexahy-
  drobenzo[*f*]thieno[2,3-*c*]quinoline **2**: 578
Diacetylmorphine **2**: 391

Diacid **2**: 501
Diacylglycerol **2**: 688
Diadin **2**: 373
Diadol **2**: 508
Diagesil **2**: 392
Diagnostic agents **4**: 2030
Diagnostics
  monoclonal antibodies **4**: 1995
Dial **2**: 508
5,5-Diallylbarbituric acid **2**: 508
Diamicron **3**: 1656
cis-Diamine[1,1-cyclobutanedicarboxylato-(2)-O,O']pla-
  tinum(II) **4**: 1920
cis-Diamine(glycolato-O^1,O^2)platinum **4**: 1921
3,5-Diamino-*N*(aminoiminomethyl)-6-chloropyrazine-
  carboxamide **1**: 74
2,4-Diamino-5-(4-chlorophenyl)-6-ethylpyrimi-
  dine **3**: 1164
3,5-Diamino-6-(2,3-dichlorophenyl)-1,2,4-tria-
  zine **2**: 482
N^1-(diaminomethylene)-sulfanilamide **3**: 1137
2-[4-(4,6-Diamino-1,3,5-triazin-2-yl-amino)phenyl]-
  1,3,2-dithiarsolane-4-methanol **3**: 1149
2,4-Diamino-5-(3,4,5-trimethoxybenzyl)pyrimi-
  dine **3**: 1133
cis-Diamminedichloroplatinum **4**: 1918
Diamorphine **2**: 391
Diamox **1**: 292; **2**: 486; **4**: 2022
Diamox Sequels **4**: 2022
1,4:3,6-Dianhydro-D-glucitol dinitrate **1**: 223
1,4:3,6-Dianhydro-D-glucitol-5-nitrate **1**: 224
1,4:3,6-Dianhydrosorbitol **4**: 2023
Diaphragms **3**: 1495
Diarrhea **4**: 1782
Diasectral **1**: 152
Diaserd **2**: 408
Diasone sodium enterab **3**: 1143
Diastabol **3**: 1663
Diastatin **3**: 1238
Diazepam **2**: 481, 636, 678
Diazoxide **1**: 100
Dibekacin **3**: 985, 986, 042
Dibenzazepines neurleptics **2**: 603
5*H*-Dibenz[*b,f*]azepine-5-carboxamide **2**: 475
4-[3-(5*H*-dibenz[*b, f*]-azepin-5-yl)propyl]-1-(2-hydro-
  xyethyl)piperazine **2**: 619
Dibenzepin **2**: 618
5*H*-dibenzo[*a,d*]cyclohepten-5-one **2**: 618
4-(5*H*-dibenzo[*a,d*]cyclohepten-5-ylidene)-1-methyl-pi-
  peridine **2**: 924
Dibenzo[*b,f*]azepine-5-carboxamide **2**: 423
Dibenzthione **3**: 1255
Dibondrin **2**: 566
Dibro-Be mono **2**: 486
5,7-Dibromo-8-hydroxyquinoline **3**: 1170
Dibromosalane **3**: 1254
Diceplon **2**: 593

Dichlorisone **3**: 1635
4-[(Dichloroacetyl)methylamino]phenyl-2-furancarboxylate **3**: 1171
2-(2,6-Dichloroanilino)-2-imidazoline **1**: 82
4,5-Dichloro-1,3-benzenedisulfonamide **4**: 2022
2-[4-(2,2-Dichlorocyclopropyl) phenoxy]-2-methyl-propanoic acid **1**: 26
1,3-Dichloro-α-[2-(dibutylamino)-ethyl]-6-(trifluoromethyl)-9-phenanthrenemethanol hydrochloride **3**: 1165
2,2-Dichloro-1,1- difluoro-1-methoxyethane **2**: 442
[(6,7-Dichloro-2,3-dihydro-2-methyl-1-oxo-2-phenyl-1H-inden-5-yl)oxy]acetic acid **1**: 77
Dichlorodihydroxy-bis-(2-propanamine)platinum **4**: 1921
4′,6-Dichloroflavan **3**: 1186
2,2-Dichloro-4′-hydroxy-N-methylacetanilide **3**: 1171
Dichloroisoprenaline **1**: 139
[2,3-Dichloro-4-(2-methylene-1-oxybutyl)phenoxy]acetic acid **1**: 76
2-[(2,6-Dichloro-3-methylphenyl)amino]benzoic acid **2**: 351; **4**: 1698
2′,5-Dichloro-4′-nitrosalicylanilide **3**: 1282
Dichlorophene **3**: 1254
2-[(2,6-Dichlorophenyl)amino]benzeneacetic acid monosodium salt **4**: 1685
[2-(2,6-Dichlorophenylamino)phenyl]acetic acid **2**: 355
1-[2-(2,4-Dichlorophenyl)-2-[(2,4-dichlorophenyl)methoxy]ethyl]-1H-imidazole **3**: 1209
1-[2-(2,4-Dichlorophenyl)-2-[(2,6-dichlorophenyl)methoxy]ethyl]-1H-imidazole **3**: 1222
4-(2,3-Dichlorophenyl)-1,4-dihydro-2,6-dimethyl-3,5-pyridinedicarboxylic acid ethyl methyl ester **1**: 211
(2,6-Dichlorophenyl)-(4,5-dihydro-1H-imidazol-2-yl)amine **2**: 424
2-[(2,6-Dichlorophenyl)methylene]hydrazinecarboximidamide **1**: 84
(1R,2R,3S,5S)-3-(3,4-dichlorophenyl)-8-methyl-8-[3.2.1]-octan-2-carbaldehyde-(E)-O-methyloxime **2**: 573
(1S,4S)-4-(3,4-dichlorophenyl)-1,2,3,4-tetrahydro-N-methyl-1-naphthylamine **2**: 629
cis-1-[4-[[2-(2,4-Dichlorophenyl)-2-(1H-1,2,4-triazol-1-ylmethyl)-1,3-dioxolan-4-yl]methoxy] **3**: 1228
4-[4-[4-[[2,4-Dichlorophenyl)-2-(1H-1,2,4-triazol-1-ylmethyl)-1,3-dioxolan-4-yl]-methoxy]phenyl] **3**: 1203
Dichlorphenamide **4**: 2022
Di-Chlotride **1**: 293
Diclac **4**: 1685
Diclobene **4**: 1685
Diclofenac **2**: 355
Diclofenac sodium **4**: 1685
Diclophlogont **4**: 1685
Diclosifar **4**: 1685
Dicloxacillin **3**: 1022

Dicloxacillin **3**: 962
Dicodid **2**: 394
Dicodin **2**: 393
Dicyclomine **2**: 702
Didanosine **3**: 1296
3′,4′-Didehydro-4′-deoxy-C′-norvincaleukoblastine **4**: 1912
2′,3′-Didehydro-3′-deoxythymidine **3**: 1298
(5α,6α)-7,8-Didehydro-4,5-epoxy-3-ethoxy-17-methyl-morphinan-6-ol **2**: 391
(5α,6α)-7,8-Didehydro-4,5-epoxy-17-methylmorphinan-3,6-diol **2**: 389
(5α,6α)-7,8-Didehydro-4,5-epoxy-17-(2-propenyl)morphinan-3,6-diol **2**: 418
N′-[(8α)-9,10-didehydro-6-methylergolin-8-yl]-N,N-diethylurea **2**: 576
4,4′-Didemethyl-4,4′-di-2-propenyltoxiferinium dichloride **2**: 669
2′,3′-Dideoxycytidine **3**: 1297
1,5-Dideoxy-1,5-[(2-hydroxyethyl)imino]-D-glucitol **3**: 1663
3,4-Dideoxy-4-[[2-hydroxy-1-(hydroxymethyl)ethyl]amino]-2-C-(hydroxy-methyl)-D-epiinositol **3**: 1663
2′,3′-Dideoxyinosine **3**: 1296
3′,4′-Dideoxykanamycin B, DKB **3**: 986
Didrex **2**: 461, 467
Dienestrol **3**: 1585
Dientamoeba fragilis **3**: 1169
Diethazine **2**: 565
6,7-Diethoxy-1-(3,4-diethoxybenzyl)isoquinoline **2**: 699
2-Diethyl-aminoaceto-N-(2,6-dimethylphenyl)acetamide monohydrochloride monohydrate **4**: 2003
(±)-4-Diethylamino-2-butinyl α- cyclohexyl-α-phenylglycolate **2**: 704
2-Diethyl-amino-N-(2,6-dimethylphenyl)acetamide hydrochloride **2**: 546
N-[2-(diethylamino)-ethyl]-2-methoxy-5-(methylsulfonyl)benzamide **2**: 606
2-Diethylaminoethyl 4-aminobenzoate hydrochloride **4**: 2004
2-Diethyl-aminoethyl 4-aminobenzoate monohydrochloride **2**: 543
2-diethylaminoethyl 4-amino-3-butoxybenzoate monohydrochloride **2**: 544; **4**: 2003
2-Diethylaminoethyl 4-amino-2-chlorobenzoate monohydrochloride **2**: 544
2-Diethylaminoethyl 1- cyclohexyl-1- cyclohexanecarboxylate **2**: 702
2-Diethylaminoethyl 2- cyclohexyl-2-phenylacetate **2**: 703
S-2-diethylaminoethyl diphenylthioacetate **2**: 705
1-[2-(Diethylamino) ethyl]-3-(4-methoxybenzyl)-2(1H)-quinoxalinone **2**: 700

[[3-[2-(Diethylamino)-ethyl]-4-methyl-2-oxo-2H-1-benzopyran-7-yl]-oxy] acetic acid ethyl ester, coumadin **1**: 231
10-(2-diethylaminoethyl)phenothiazine **2**: 565
2-Diethylaminoethyl (2-phenyl-2-piperidine)acetate **2**: 702
α-(Diethylamino)-phenyl-1-propanone **2**: 464
N-(3-Diethylaminopropyl)-N-indan-2-ylaniline **1**: 128
(RS)-10-(2-Diethylaminopropyl)phenothiazine **2**: 565
5,5-Diethylbarbituric acid **2**: 506
N,N-Diethyl-(2-benzhydryloxy)ethylamine **2**: 567
Diethyl 4-(2,1,3-benzoxadiazol-4-yl)-1,4-dihydro-2,6-dimethylpyridine-3,5-dicarboxylate **1**: 212
Diethylcarbamazine **3**: 1275, 276, 277, 279
(±)-Diethyl[1-[3-(2-chloroethyl)-3-nitrosoureido]ethyl]-phosphonate **4**: 1893
N,N-diethyl-N-[2-(α-cyclohexylmandeloyloxy)ethyl]-N-methylammonium bromide **2**: 696
3,3-Diethyl-2,4-dioxo-1,2,3,4-tetrahydropyridine **2**: 511
Diethyl ether **2**: 440
N,N-Diethyl-N′-[(8α)-6-methylergolin-8-yl]urea **2**: 576
N,N-diethyl-4-methyl-1-piperazinecarboxamide citrate **3**: 1279
3,3,Diethyl-5-methyl-2,4-piperidinedione **2**: 512
N,N-Diethyl-5-methyl-[1,2,4]triazolo[1,5-a]pyrimidin-7-amine **1**: 230
N,N-diethyl-N-methyl-2-(9-xanthenylcarbonyloxy) ethylammonium bromide **2**: 695
N,N-diethyl-N-methyl-N-{2-[4-(2-octyloxybenzamido)-benzoyloxy]ethyl}ammonium bromide **2**: 695
2,2-Diethyl-4-pentenamide **2**: 502
Diethylpropion **2**: 464, 467
Diethylpropion HCl **2**: 464
Diethylstilbestrol **3**: 1585, 586
Diethylstilbestrol dipropionate **3**: 1586
Diethylstilbestrol phosphate **3**: 1586
(1,8-Diethyl-1,3,4,9-tetrahydro-pyrano[3,4-b]indol-1-yl)-acetic acid **2**: 382
Diffusion-controlled oral dosage forms **4**: 2114
Difhydan **2**: 474
Diflamil **4**: 1695
Diflonid **4**: 1678
Difloxacin **3**: 1127
Diflucan **3**: 1202
Diflunisal **2**: 346; **4**: 1678
6α,9α-Difluorocorticoids **3**: 1629
6,8-Difluoro-1-(2-fluoroethyl)-1,4-dihydro-7-(4-methyl-1-piperazinyl)-4-oxo-3-quinolinecarboxylic acid **3**: 1128
2′4′-Difluoro-4-hydroxybiphenyl-3-carboxylic acid **2**: 346
5-(Difluoromethoxy)-2-[[(3,4-dimethoxy-2-pyridyl)-methyl]sulfinyl]benzimidazole **2**: 720
α-(2,4-Difluorophenyl)-α-(1H-1,2,4-triazol-1-ylmethyl)-1H-1,2,4-triazole **3**: 1201
Diflupyl **4**: 2026

Diflurex **1**: 295
Diflusal **4**: 1678
Diforène **2**: 652
Digacin **1**: 189
Digimerck **1**: 188
Digitalis glycosides **1**: 178
Digitoxigenin **1**: 177, 79
Digitoxin **1**: 181, 188
Dignodolin **2**: 351
Dignotrimazol **3**: 1217
Dignover **1**: 208
Digoxigenin **1**: 179
Digoxin **1**: 189
9α,11β-Dihalocorticosteroids **3**: 1635
Dihexyverine **2**: 703
Dihycon **2**: 474
Di-Hydan **2**: 474
Dihydergot **1**: 66
Dihydergot plus **1**: 66
22,23-Dihydroabamectin **3**: 1279; **4**: 2173
22,23-Dihydroavermectin B$_1$ **3**: 1279
Dihydro-5-azacytidine **4**: 1877
3,12-Dihydrocholanic acid **2**: 749
Dihydrocodeine **2**: 392
7-[(10,11-Dihydro-5H-dibenzo[a,d]cyclohepten-5-yl)amino]heptanoic acid **2**: 617
3-(10,11-Dihydro-5H-dibenzo[a,d]cyclohepten-5-ylidene)-1-ethyl-2-methylpyrrolidine **2**: 566
9,13b-Dihydro-1H-dibenzo[c,f]imidazo[1,5-a]azepin-3-amine **2**: 932
10,11-Dihydro-10-(2-dimethylaminoethyl)-5-methyl-11-oxodibenzo[b,e][1,4]diazepine **2**: 618
10,11-Dihydro-5-(3-dimethylamino-2-methylpropyl)-5H-dibenzo-[a,d]cycloheptene **2**: 616
1,3-Dihydro-1-[3-(dimethylamino)propyl]-1-(4-fluorophenyl)-5-iso-benzofurancarbonitrile **2**: 627
10,11-Dihydro-5-(3-dimethylaminopropyl)-5H-dibenz[b,f]-azepine **2**: 613
6,11-Dihydro-11-(3-dimethylaminopropylidene)dibenz[b,e]oxepin **2**: 613
6,11-Dihydro-11-(3-dimethylaminopropylidene)dibenzo-[b,e]thiepin **2**: 612
10,11-Dihydro-5-(3-dimethylaminopropylidene)-5H-dibenzo[a,d]cycloheptene **2**: 611
6,11-dihydro-6-[2-(dimethylamino)propyl-5H-pyrido[2,3-b][1,5]-benzodiazepin-5-one **2**: 619
3,7-Dihydro-3,7-dimethyl-1-(5-oxohexyl)-1H-purine-2,6-dione **1**: 48
3,7-Dihydro-1,3-dimethyl-1H-purine-2,6-dione **2**: 887
1,2-Dihydro-1,5-dimethyl-4-(1-methylethyl)-2-phenyl-3H-pyrazol-3-one **2**: 371
1,4-Dihydro-2,6-dimethyl-4-(2-nitrophenyl)-3,5-pyridinecarboxylic acid dimethyl ester **1**: 95
1,4-Dihydro-2,6-dimethyl-4-(2-nitrophenyl)-3,5-pyridinecarboxylic acid methyl 2-oxopropyl ester **1**: 213

1,4-Dihydro-2,6-dimethyl-4-(2-nitrophenyl)-3,5-pyridinedicarboxylic acid dimethyl ester   **1**: 210

1,4-Dihydro-2,6-dimethyl-4-(3-nitrophenyl)-3,5-pyridinedicarboxylic acid ethyl methyl ester]   **1**: 96, 210

1,4-Dihydro-2,6-dimethyl-4-(3-nitrophenyl)-3,5-pyridinedicarboxylic acid 2-methoxyethyl 1-methylethyl ester   **1**: 210

1,4-Dihydro-2,6-dimethyl-4-(2-nitrophenyl)-3,5-pyridinedicarboxylic acid methyl 2-[methyl-(phenylmethyl)amino]ethyl ester   **1**: 210

1,4-Dihydro-2,6-dimethyl-4-(2-nitrophenyl)-3,5-pyridinedicarboxylic acid methyl 2-methylpropyl ester   **1**: 211

Dihydroergocristine   **2**: 655

Dihydro-α-ergocryptine   **2**: 655, 655

Dihydroergotamine   **1**: 65

9,10-dihydroergotamine   **1**: 65

Dihydroergotoxine   **2**: 655

Dihydrofolate reductase   **4**: 1868, 870

10,11-Dihydro-5H-dibenzo[a,d]cyclohepten-5-one   **2**: 618

Dihydrohydroxycodeinone   **2**: 396

3,7-dihydro-7-(2-hydroxyethyl)-1,3-dimethyl-1H-purine-2,6-dione   **2**: 888

3,7-Dihydro-7-[2-hydroxy-3-[(2-hydroxyethyl)-methylamino]propyl]-1,3-dimethyl-1-H-purine-2,6-dione   **1**: 23

(5′α,10α)-9,10-dihydro-12′-hydroxy-2′-methyl-5′-(phenyl-methyl)ergotaman-3′,6′,18-tri   **1**: 65

Dihydrohydroxymorphinone   **2**: 397

Dihydro-14-hydroxymorphinone   **2**: 397

3,7-Dihydro-7-(2-hydroxypropyl)-1,3-dimethyl-1H-purine-2,6-dione   **2**: 888

5-[[4-[(3,4-Dihydro-6-hydroxy-2,5,7,8-tetramethyl-2H-1-benzopyran-2-yl) methoxy]phenyl]methyl]-2,4-thiazolidinedione   **3**: 1666

N-{4-[2-(3,4-Dihydro-7-methoxy-4,4-dimethyl-1,3-dioxa-2(1H)-isoquinolyl)-ethyl]-benzenesulfonyl}-N-cyclohexy Lurea   **3**: 1657

5,8-Dihydro-5-methoxy-8-oxo-1,3-dioxolo[4,5-g]quinoline-7-carboxylic acid   **3**: 1121

10,11-Dihydro-5-methyl-10-(methylamino)-5H-dibenz[b,f]azepine   **2**: 615

10,11-Dihydro-5-[3-(methylaminopropyl)-5H-dibenz[b,f]-azepine   **2**: 614

10,11-Dihydro-5-(3-methylaminopropylidene)-5H-dibenzo-[a,d]cycloheptene   **2**: 615

1,6-Dihydro-2-methyl-6-oxo(3,4′-bipyridine)-5-carbonitrile   **1**: 195

3,7-Dihydro-3-methyl-1-(5-oxohexyl)-7-propyl-1H-purine-2,6-dione   **2**: 656

1,3-Dihydro-1-methyl-7-nitro-5-phenyl-2H-1,4-benzodiazepin-2-one   **2**: 522

5,11-Dihydro-11-[(4-methyl-1-piperazinyl)acetyl]-6H-pyrido[2,3-b][1,4]benzodiazepin-6-one   **2**: 729

6,11-Dihydro-11-(1-methyl-4-piperidylidene)-5H-benzo[5,6]cyclohepta[1,2-b]pyridine   **2**: 923

4,9-dihydro-4-(1-Methyl-4-piperidylidene)-10H-benzo[4,5]cyclohepta[1,2-b]thiophene-10(9H)-one   **2**: 924

2,3-Dihydro-6-methyl-2-thioxo-4(1H)-pyrimidinone   **3**: 1529

2,3-Dihydro-1-(4-morpholinylacetyl)-3-phenyl-4(1H)-quinazolinone   **2**: 754

Dihydrone   **2**: 396

3,7-Dihydro-7-[N-ethyl-2-(hydroxyethylamino)ethyl]-8-benzyl-1,3-dimethyl-1H-purine-2,6-dione   **2**: 890

1,3-Dihydro-7-nitro-5-phenyl-2H-1,4-benzodiazepin-2-one   **2**: 481, 518

5-{7-[4-(4,5-Dihydro-2-oxazolyl)phenoxy]heptyl}-3-methylisoxazole   **3**: 1187

10,11-Dihydro-10-oxo-5H-dibenz[b,f]azepin-5-carboxamide   **2**: 476

Dihydropenicillin F   **3**: 960

4,5-Dihydro-N-phenyl-N-(phenylmethyl)-1H-imidazole-2-methanamine   **2**: 905

Dihydrophylline   **2**: 889

2,3-Dihydro-6-propyl-2-thioxo-4(1H)-pyrimidone   **3**: 1529

1,7-Dihydro-6H-purine-6-thione   **4**: 1882

1,4-Dihydropyridines   **1**: 210

22,23-Dihydrostigmasterol   **1**: 22

Dihydrostreptomycin   **3**: 984, 1046; **4**: 2162

5α-Dihydrotestosterone   **3**: 1565

1,4-Dihydroxy-5,8-bis[(2-(2-hydroxyethyl)amino)ethyl]amino-9,10-anthracenedione   **4**: 1902

3α-7α-Dihydroxy-5β-cholestanic acid   **3**: 1328

1,4-Bis(3,4-dihydroxycinnamoyloxy)-3,5-dihydroxycyclohexanecarboxylic acid   **2**: 751

7′,12′-Dihydroxy-6,6′-dimethoxy-2,2′,2′-trimethyltubocuraranium chloride hydrochloride   **2**: 672

[±]-9-[2α,3β-Dihydroxy-4α-(hydroxymethyl)cyclopent-1α-yl]adenine   **2**: 1182

4,17β-dihydroxy-17α-methyl-4-androsten-3-one   **3**: 1593

3,4-Dihydroxy-4'-methyl-5-nitrobenzophenone   **2**: 572

11α,15α-Dihydroxy-9-oxo-13E-prostenoic acid   **4**: 2048

(2S)-3-(3,4-Dihydroxyphenyl)-2-hydrazino-2-methylpropionic acid monohydrate   **2**: 569

7-[-3-[[2-(3,5-Dihydroxyphenyl)-2-hydroxyethyl]amino]propyl]-3,7-dihydro-1,3-dimethyl-1H-purine-2,6-dione   **2**: 872

N,N′-bis[2-(3,4-dihydroxyphenyl)-2-hydroxyethyl]hexamethylenediamine   **2**: 868

1-(3,5-Dihydroxyphenyl)-2-(4-hydroxy-α-methylphenethylamino)ethanol   **2**: 871

1-(3,4-Dihydroxyphenyl)-2-(isopropylamino)butanol   **2**: 866

1-(3,4-Dihydroxyphenyl)-2-(isopropylamino)ethanol **2:** 865
1-(3,5-Dihydroxyphenyl)-2-(isopropylamino)ethanol **2:** 869
(−)-3-(3,4-Dihydroxyphenyl)-L-alanine **2:** 568
1-(3,4-Dihydroxyphenyl)-2-(methylamino)ethanol **2:** 864; **3:** 1562
1-(3,4-Dihydroxyphenyl)-2-(α-methyl-3,4-methylenedioxyphenethylamino)ethanol **2:** 867
1-(3,4-Dihydroxyphenyl)-1-(2-piperidinyl)methanol **2:** 867
1-(3,5-Dihydroxyphenyl)-2-(*tert*-butylamino)ethanol **2:** 871
3α,21-dihydroxy-5α-pregnane-11,20-dione-21-acetate **2:** 453
Dihydroxyprogesterone acetophenide **3:** 1504
Dihydroxyprogesterone acetophenonide **3:** 1597
7-(2,3-Dihydroxypropyl)-3,7-dihydro-1,3-dimethyl-1*H*-purine-2,6-dione **2:** 889
11α,15α-Dihydroxy-9-oxo-5*Z*,13*E*-prostadienoic acid **4:** 2048
11α,15α-Dihydroxy-9-oxo-5*Z*,13*E*,17*Z*-prostatrienoic acid **4:** 2048
1α,25-Dihydroxyvitamin D_3 **3:** 1637
Dihytamin **1:** 66
Diidergot **1:** 66
Diiodohydroxyquinoline **3:** 1170; **4:** 1732
5,7-Diiodo-8-quinolinol **3:** 1170
L-3,5-Diiodothyronine **3:** 1524, 1558
L-3,5-Diiodotyrosine **3:** 1558
Diisethionate salt **3:** 1149
4-Diisobutylphenoxypolyethoxy-ethanol **3:** 1495
Diisopromine **2:** 703
(±)–4-diisopropylamino-2-phenyl-2-(2-pyridyl)butyramide **1:** 125
N,N-Diisopropyl-3,3-diphenylpropylamine **2:** 703
Diisopropyl flurophosphate **4:** 2026
N,N-diisopropyl-*N*-methyl-*N*-[2-(9-xanthenecarbonyloxy)-ethyl]ammonium bromide **2:** 697
2,6-Diisopropylphenol **2:** 450
Dilantin **1:** 126; **2:** 474
Dilatrend **1:** 154
Dilaudid **2:** 394
Dilcoran **1:** 224
Diloderm **3:** 1635
Dilosyn **2:** 921
Diloxanide **3:** 1171
Diloxanide furoate **3:** 1171
Diltiazem **1:** 44, 98, 32, 206, 213
Diluents **4:** 2104
Dilzem **1:** 98, 32, 213
Dimecrotic acid **2:** 751
Dimegan **1:** 911
Dimelazine **2:** 919
Dimelor **3:** 1655
Dimenformon **3:** 1583
Dimenhydrinate **2:** 743

Dimethanesulfonate salt **3:** 1149
Dimethisterone **3:** 1598
Dimethothiazine **2:** 919
Dimethoxydiethylstilbene **3:** 1586
6,7-Dimethoxy-1-(3,4-dimethoxybenzyl)isoquinoline **2:** 699
7,8-Dimethoxy-1-(3,4-dimethoxyphenyl)-5-ethyl-4-methyl-5*H*-2,3-benzodiazepine **2:** 643
2,3-Dimethoxy-6-(10-hydroxydecyl)-5-methyl-1,4-benzoquinone **2:** 654
2,4-Dimethoxy-β-methylcinnamic acid **2:** 751
5,6-Dimethoxy-2-methyl-3-[2-(4-phenyl-1-piperazinyl)ethyl]indole **2:** 609
N-(3,4-Dimethoxyphenethyl)-2-(3,4-dimethoxyphenyl)-*N*-methyl-*m*-dithiane-2-propylamin-1,1,3,3- **1:** 209
(±)-5-[*N*-(3,4-Dimethoxyphenethyl)-*N*-methylamino]-2-(3,4-dimethoxyphenyl)-2-isopropylvaleronitrile **1:** 131
(±)-5-[*N*-(3,4-Dimethoxyphenethyl)-*N*-methylamino]-2-(3,4,5-trimethoxyphenyl)-2-isopropylvaleronitrile **1:** 131
α-(3-{[2-(3,4-Dimethoxyphenyl)ethyl]-methylamino}propyl)-3,4-dimethoxy-α-(1-methylethyl)benzeneacetonitrile **1:** 97, 208
α[3-[[2-(3,4-Dimethoxyphenyl)ethyl] methylamino]propyl]-3,4,5-trimethoxy-α-(1-methylethyl)1-benzeneacetonitrile **1:** 208
6-[(2,5-Dimethoxyphenyl)methyl]-5-methylpyrido[2,3-d]pyrimidine-2,4-diamine mono-2-hydroxyethanesulfonate **4:** 1871
N'-(5,6-Dimethoxy-4-pyrimidinyl)sulfanilamide **3:** 1136
(*E*)-8-(3,4-Dimethoxystyryl)-1,3-dipropyl-7-methylxanthine **2:** 580
11,17-Dimethoxy-18-[(3,4,5-trimethoxybenzoyl)oxy]yohimban-16-carboxylic acid methyl ester **1:** 85
3,5-Dimethyl-1-adamantanamine **2:** 574
6-(Dimethylamino)-2-[2-(2,5-dimethyl-1-phenyl-1*H*-pyrrol-3-yl)ethenyl]–1-methylquinolinium 4,4'-methylene-bis **3:** 1285
2-(Dimethylamino)ethanol **2:** 652
2-Dimethylaminoethoxyphenylmethyl-2-picoline **2:** 496
N-[[4-[2-(Dimethylamino)ethoxy]phenyl]methyl]-3,4,5-trimethoxybenzamide **2:** 910
2-Dimethylaminoethyl 4-butylaminobenzoate monohydrochloride **2:** 545; **4:** 2003
Dimethylaminoethyl 4-chlorophenoxyacetate **2:** 652
4-Dimethylamino-1-ethyl-2,2-diphenylpentyl acetate **2:** 405
O-[2-(dimethylamino)ethyl]oxime **2:** 618
4-(3-Dimethylamino-2-hydroxypropyl)ajmaline hydrogen tartrate **1:** 127
5-Dimethylamino-1-{4-(2-methylbenzoylamino)benzoyl}-2,3,4,5-tetrahydro-1*H*-benzazepine **1:** 300

10-(2-Dimethylamino-2-methylethyl)-10*H*-pyrido[3,2-*b*][1,4]-benzothiazine   **2**: 920

N-[2-[[5-[(dimethylamino)-methyl]furfuryl]-thio]ethyl]-*N'*-methyl-2-nitro-1,1-ethenediamine   **2**: 724

(S)-10-Dimethylaminomethyl-9-hydroxycamptothecin   **4**: 1914

cis-2-Dimethylaminomethyl-1-(3-methoxyphenyl)cyclohexanol   **2**: 411

10-[3-(Dimethylamino)-2-methylpropyl]phenothiazine   **2**: 918

10-(3-Dimethylamino-2-methylpropyl)-10H-phenothiazine   **2**: 591

10-(3-Dimethylamino-2-methylpropyl)phenothiazine-5,5-dioxide   **2**: 921

5-(Dimethylamino)-9-methyl-2-propyl-1*H*-pyrazolo[1,2*a*][1,2,4]benzotriazine-1,3-(2*H*)-dione   **4**: 1695

N-[2-[[[2-[(dimethylamino)methyl]-4-thiazolyl]methyl]-thio]ethyl]-*N'*-methyl-2-nitro-1,1-ethenediamine   **2**: 723

(±)-2-Dimethylamino-2-phenylbutyl 3,4,5-trimethoxybenzoate   **2**: 701

10-[2-(Dimethylamino)-propyl]-*N,N*-dimethyl-10*H*-phenothiazine-2-sulfonamide   **2**: 919

5-(3-Dimethylaminopropylidene)-dibenzo[*a,e*]cycloheptatriene hydrochloride   **2**: 678

10-(2-Dimethylamino-propyl)phenothiazine   **2**: 921

10-(3-Dimethylaminopropyl)-10*H*-phenothiazine   **2**: 590

10-[2-(Dimethylamino)propyl]-10H-phenothiazine-5,5-dioxide   **2**: 919

1-[2-(Dimethylamino)propyl]-10*H*-phenothiazin-2-yl]-1-propanone   **2**: 921

10-(3-Dimethylaminopropyl)-10*H*-pyrido[3,2-*b*][1,4]benzothiazine   **2**: 596

10-(3-Dimethylaminopropyl)-2-trifluoromethyl-10*H*-phenothiazine   **2**: 591

S-(+)-*N*, α-Dimethylbenzeneethanamine   **2**: 461, 462

Dimethylbiguanide   **3**: 1660

N-(1-Dimethylcarbamoylpropyl)-*N*-ethylcrotonamide   **2**: 339

N,αDimethylcyclohexaneethanamine   **2**: 462

1,5-Dimethyl-5-(cyclohexen-1-yl)barbituric acid   **2**: 511

N,N-dimethyl-2,2-diphenoxyethylamine   **2**: 631

N'-[(8α)-1,6-Dimethylergolin-8-yl]*N,N*-dimethylsulfamide   **2**: 577

5-[3-[(1,1-Dimethylethyl)amino]-2-hydroxypropoxyl]-1,2,3,4-tetrahydro-2,3-naphthalenediol   **1**: 92

1-[(1,1-dimethylethyl)amino]-3-{[4-morpholinyl-1,2,5-thiadiazol-3-yl]oxy}-2-propanol   **1**: 93

(*S*)-1-[(1,1-Dimethylethyl)amino]-3-[(4-morpholinyl-1,2,5-thiadiazol-3-yl)oxy]-2-propanol maleate   **4**: 2021

4-(1,1-dimethylethyl)-1-[1-hydroxy-4-[4-(1,1-diphenylhydroxymethyl)-1-piperidinyl]butyl]benzene   **2**: 930

N-(1,1-Dimethylethyl)-α-methyl-γ-phenylbenzenepropanamine   **1**: 209

1-[4-(1,1-Dimethylethyl)phenyl]-4-[4-(diphenylmethoxy)-1-piperidinyl]-1-butanone   **2**: 929

(*E*)-*N*-(6,6-Dimethyl-2-hepten-4-ynyl)-*N*-methyl-1-naphthalenemethanamine   **3**: 1250

7α,17α-Dimethyl-17β-hydroxy-4-androsten-3-one   **3**: 1593

3,4-Dimethyl-2-imino-5-phenylthiazolidine   **2**: 878

N^1-(3,4-dimethyl-5-isoxazolyl)sulfanilamide   **3**: 1134

DimethylÎstrogen-Holzinger   **3**: 1586

6,11-Dimethyl-3-(3-methylbut-2-enyl)-1,2,3,4,5,6-hexahydro-2,6-methanobenzo[*d*]azocin-8-ol   **2**: 419

N,N-dimethyl-2-(2-methyl-α-phenyloxy)ethylamine   **2**: 680

(±)-*cis*-2,6-Dimethyl-4-[2-methyl-3(*p-tert*-pentylphenyl)propyl]morpholine   **3**: 1252

α-2-Dimethyl-5-nitro-1*H*-imidazole-1-ethanol   **3**: 1156

(+)-*N*,α-Dimethyl-*N*-(phenylmethyl)benzeneethanamine   **2**: 461

5-(2,5-dimethylphenoxy)-2,2-dimethylpentanoic acid   **1**: 25

5-(3,5-Dimethylphenoxymethyl)-2-oxazolidinone   **2**: 679

2-[(2,3-Dimethylphenyl)amino]benzoic acid   **2**: 352; **4**: 1697

N-(2,6-Dimethylphenyl)-2-(ethylpropylamino)butanamide hydrochloride   **2**: 547

N-(2,6-Dimethylphenyl)-2-(ethylpropylamino)butanamide monohydrochloride   **4**: 2003

N,N-dimethyl-2-[2-(phenylmethyl)phenoxy]ethanamine   **2**: 910

N-(2,6-Dimethylphenyl)-1-methyl-2-piperidinecarboxamide hydrochloride   **4**: 2004

N-(2,6-Dimethylphenyl)-1-methyl-2-piperidinecarboxamide monohydrochloride   **2**: 546

N,N-dimethyl-*N'*-(phenylmethyl)-*N'*-2-pyridinyl-1,2-ethanediamine   **2**: 907

L-(+)−3*R-trans*-3,4-Dimethyl-2-phenylmorpholine   **2**: 465

N,N-dimethyl-*N'*-phenyl-*N'*-(phenylmethyl)-1,2-ethanediamine   **2**: 907

N,N-dimethyl-γ-phenyl-2-pyridinepropanamine   **2**: 912

N,N-dimethyl-2-[1-phenyl-1-(2-pyridinyl)ethoxy]ethanamine   **2**: 909

1,3-Dimethyl-3-phenyl-2,5-pyrrolidinedione   **2**: 479

6,17-Dimethyl-4,6-pregnadiene-3,20-dione   **3**: 1597

(*R*)-(−)-*N*-α-Dimethyl-*N*-2-propynylphenethylamine   **2**: 570, 625

N,N-dimethyl-3[1-(2-pyridinyl)ethyl]-1*H*-indene-2-ethanamine   **2**: 912

N,N-dimethyl-*N'*-2-pyridinyl-*N'*-(2-thienylmethyl)-1,2-ethanediamine   **2**: 906

2245

10-[(1,3-Dimethyl-3-pyrrolidinyl)methyl]-10H-phenothiazine **2**: 919
(RS)-N,N-Dimethyl-2-(α-2-tolylbenzyloxy)ethylamine **2**: 567
5-(3,3-Dimethyl-1-triazenyl)-1H-imidazole-4-carboxamide **4**: 1895
O,O-dimethyl-2,2,2-trichloro-1-hydroxyethylphosphonate **3**: 1281
3,5-Dimethyltricyclo[3.3.1.1^{3,7}]decylamine **2**: 574
Dimethyltubocurarine iodide **2**: 671
Dimetindene **2**: 912
Dimetylcystein **4**: 1706
Diminal **2**: 510
Diminazene aceturate **3**: 1150
Dimitron **1**: 209
Rho(D) immune globulin **4**: 1828
Dimycon **3**: 1202
Dinabol **3**: 1593
Dinacrin **3**: 1140
Dinaplex **1**: 209
Dinerfene **2**: 657
Dinoprost **3**: 1509; **4**: 2059
Dinoprostone **4**: 2059
Dinsidon **2**: 619
Dioctyl sodium sulfosuccinate **2**: 782
Diogyn **3**: 1583
Diogyn E **3**: 1584
Dioscin **3**: 1570
Diosgenin **3**: 1567, 569
2,2′-[(1,4-Dioxo-1,4-butanediyl)bis(oxy)]bis[N,N,N-trimethylethanaminium] dichloride **2**: 674
Dioxopromethazone **2**: 919
Di-Paralene **2**: 914
Dipavlon **2**: 649
Dipcosone **3**: 1627
Diphantoin **2**: 475
Diphemanil methylsulfate **2**: 693
Diphenhydramine **2**: 566, 742, 909
Diphenhydramine hydrochloride **2**: 496
Diphenidole **2**: 743
Diphenin **2**: 475
Diphenol Laxatives **2**: 786
Diphenoxylate **2**: 408
Diphenyhydramine **4**: 1720
2-(1,1-Diphenylethoxy)-N,N-dimethylethanamine **2**: 910
5,5-diphenylhydantoin **1**: 125
5,5-Diphenyl-2,4-imidazolidinedione **2**: 474
2-Diphenylmethoxy-N,N-diethylethylamine **2**: 567
2-(Diphenylmethoxy)-N,N-dimethylethanamine **2**: 909
2-Diphenylmethoxy-N,N-dimethylethylamine **2**: 566, 742
3-Diphenylmethoxy-8-methyl-8-azabicyclo[3.2.1]-octanemethanesulfonate **2**: 562
4-(Diphenylmethoxy)-1-methylpiperidine **2**: 917
3-(diphenylmethylene)-1,1-diethyl-2-methylpyrrolidinium bromide **2**: 696
2-[2-[2-[4-(Diphenylmethylene)-piperidinyl]ethoxy]ethoxy]ethanol **2**: 917
1-(Diphenylmethyl)–4-methylpiperazine **2**: 742, 914
1-(Diphenylmethyl)-4-(3-phenyl-2-propenyl)piperazine **1**: 208; **2**: 914
1-[3-[4-(Diphenylmethyl)-1-piperazinyl]propyl]–1,3-dihydro-2H-benzimidazol-2-one **2**: 916
3-(4,5-Diphenyl-oxazol-2-yl)propionic acid **2**: 381
1,1-Diphenyl-4-piperidinobutanol **2**: 743
1,1-Diphenyl-3-piperidino-1-propanol **2**: 564
Diphenylpyraline **2**: 917
2,2′-[(4,5-Diphenyl-2- oxazolyl)imino]diethanol **4**: 1699
Diphtheria
  treatment with monoclonal antibodies **4**: 1841
Diphtheria prophylaxis **4**: 1828
Diphtheria vaccine **4**: 1764
Diphyllobothrium latum **3**: 1269
Dipidolor **2**: 410
Dipiperal **2**: 599
2,2′,2″,2‴-(4,8-Dipiperidinopyrimido[5,4-d]pyrimidine-2,6-diyldinitrilo)tetraethanol **1**: 231
Dipiperon **2**: 599
Dipivalyladrenaline hydrochloride **4**: 2020
Dipivefrin hydrochloride **4**: 2020
Dipotassium 2{4-[(4,6-diamino-1,3,5-triazin-2-yl)amino]phenyl}-1,3,2-dithiarsolane-4,5-dicarboxylate **3**: 1151
Diprophylline **2**: 889
4-[2-(Dipropylamino)ethyl]-2-indolinone **2**: 578
6-Dipropylaminomethylidenimino)pyrid-2(1H)-one **2**: 572
(1,3-Dipropyl-8-[2-(5,6-epoxy)norbornyl]-xanthine **1**: 301
Dipyridamole **1**: 38, 231
Dipyrone[(1,5-dimethyl-3-oxo-2-phenyl-2,3-dihydro-1H-pyrazol-4-yl)methylamino]-methanesulfonic acid **2**: 369
Directim **2**: 617
Disinfectants
  for contact lenses **4**: 2037
Disintegrating agents **4**: 2105
Disipal **2**: 567
Dismenol **4**: 1688
Disoderm **3**: 1635
Disodium cromoglycate **2**: 895, 933, 934; **4**: 2018
Disodium 4,4′-dioxo-5,5′-(2-hydroxytrimethylenedioxy)-di-14H-chromene-2-carboxylate **2**: 895
Disodium 9-ethyl-6,9-dihydro-4,6-dioxo-10-propyl-4H-pyrano[3,2-g]quinoline-2,8-dicarboxylate **2**: 896
Disodium moxalactam **3**: 1034
Disodium sulfonylbis(p-phenyleneimino)dimethanesulfinate **3**: 1143
Disoprofol **2**: 450
Disopyramide **1**: 125

Disorders of cardiac rhythm   **1:** 116
Disoxaril   **3:** 1187
Dispersions   **4:** 2089
Dispert   **4:** 1708
Dissection of the aorta   **1:** 146
Dissolution-controlled oral dosage forms   **4:** 2114
Distamycin   **3:** 1008
Distensan   **2:** 641
Distesol   **2:** 501
Distraneurin   **2:** 529
Distribution phase   **4:** 2130
DIT   **3:** 1558
Ditazole   **4:** 1699
3,3′-(Dithiodimethylene)bis(5-hydroxy-6-methyl-4-pyri-
  dinemethanol)   **4:** 1706
5-(1,2-Dithiolan-3-yl)valeric acid   **2:** 770
Dithranol   **4:** 1728
Ditiazem   **1:** 229
Ditropan   **2:** 704
Diupres   **1:** 293
Diuretics   **1:** 70, 289
  K$^+$-Retaining   **1:** 74
Diuril   **1:** 293
Diuriwas   **4:** 2022
Divalvon-D   **2:** 657
Dividol   **2:** 413
Divinyl ether   **2:** 441
Dixibon   **2:** 606
Dizocilpine   **2:** 574
DL–8280   **3:** 1126
DM   **3:** 1385
DMDR   **4:** 1898
DMP-450   **3:** 1314
Doblexan   **4:** 1697
Dobren   **2:** 606
Dobutamine   **1:** 192
Dobutrex   **1:** 192
Docetaxel   **4:** 1915
Dociton   **1:** 129
Docitor   **1:** 90
Doctucid   **4:** 1680
Doctynol   **2:** 782
Documentation   **4:** 2145
Docusate sodium   **2:** 782
Dodecahydro-7,14-methano-2H,6H-dipyrido-[1,2-
  a : 1′,2′-e]-[1,5]diazocine   **1:** 126
Dodecyl-β-alanine   **4:** 2103
N-Dodecyl-N,N-dimethylglycine   **4:** 2103
Dogmatil   **2:** 606
Dog Tapeworm   **3:** 1271
Dolantin   **2:** 399
Dolexalan   **3:** 1217
Dolgit   **4:** 1685, 688
Dolmatil   **2:** 606
Dolobid   **2:** 347; **4:** 1678
Dolobis   **2:** 347; **4:** 1678
Dolophine   **2:** 405

Dolo-Puren   **4:** 1688
Dolormin   **4:** 1688
Dolosal   **2:** 399
Doloxene   **2:** 408
Dominal   **2:** 596
Domiphen bromide   **3:** 1254
Domistan   **2:** 906
Domitor   **4:** 2181
Domitroban-calciumhydrate   **4:** 2062
Domnamid   **2:** 524
Domperidone   **2:** 728
Donalgin   **4:** 1699
Donix   **2:** 522
Donorest   **4:** 1686
DONSS   **2:** 782
L-Dopa   **2:** 568
Dopacard   **1:** 194
Dopaflex   **2:** 569
Dopamin   **1:** 192
Dopamine   **1:** 60, 192; **2:** 329; **3:** 1562
  biosynthesis and metabolism   **2:** 560
  in Parkinsonism   **2:** 557
Dopamine Lucien   **1:** 61
Dopamine Nativelle   **1:** 61
Dopamine pathways   **2:** 337
Dopamine Pierre Fabre   **1:** 61
Dopamine receptor agonists   **2:** 575
Dopaminergic theory   **2:** 794
Dopamine uptake inhibitors   **2:** 617
Dopamin Fresenius   **1:** 61
Dopamin-ratiopharm   **1:** 61
Dopamin Solvay   **1:** 61
L-Dopa Monotherapy
  of Parkinsonism   **2:** 568
Dopergin   **2:** 576
Dopexamine   **1:** 194
Doppel-Spalt N   **2:** 700
Dopram-V   **4:** 2179
Dorico Solubile   **2:** 445
Doriden   **2:** 512
Dorindac   **4:** 1683
Dorison   **2:** 497
Dormador   **2:** 522
Dormalin   **2:** 523
Dormicum   **2:** 451, 525
Dormid   **2:** 501
Dormigene   **2:** 501
Dormileno   **2:** 506
Dorminal   **2:** 507, 509
Dormir   **2:** 514
Dormison   **2:** 497
Dormodor   **2:** 522
Dormogen   **2:** 514
Dormona   **2:** 511
Dormonil   **2:** 525
Dormonoct   **2:** 526
Dormo-Puren   **2:** 518

2247

Dormosedan **4**: 2181
Dormutil **2**: 514
Dorner **4**: 2061
Dosage forms **4**: 2077, 2085
Dose finding **4**: 2135
Dostinex **2**: 579
Dosulepin **2**: 612
Dothiepin **2**: 612
Double dummy technique **4**: 2137
Douches **4**: 2088
Doxacurium chloride **2**: 670
Doxans **2**: 523
Doxapram **4**: 2179
Doxefazepam **2**: 515, 523
  synthesis **2**: 520
Doxepin **2**: 613; **4**: 1720
Doxergan **2**: 921
Doxorubicin **3**: 954, 981, 039; **4**: 1897
Doxycycline **3**: 978, 038; **4**: 2161
Doxylamine **2**: 909; **4**: 2184
Doxylamine succinate **2**: 496
DQ 2466 **1**: 154
DQ-2805 **4**: 1914
Dracunculus medinensis **3**: 1276
Draganon **2**: 651
Dramamine **2**: 743
Drenusil **1**: 294
Dridase **2**: 704
Drofenine **2**: 703
Drogenil **4**: 1928
Drolasetron **2**: 609
Drolban **3**: 1594
Droleptan **2**: 600
Droloxifene **4**: 1922, 924
Droperidol **2**: 453, 454, 600
Drosteakard **1**: 208
Droxidopa **2**: 569
Droxone **3**: 1597
Drug administration **4**: 2081
Drug–carrier **4**: 2121
Drug delivery systems **4**: 2077, 2109
Drug monitoring **4**: 2136
Drugs affecting circulation **1**: 217
Drug testing **4**: 2125
Drupal **2**: 522
Dry eye medications **4**: 2032
Dry Eye Therapy **4**: 2033
DSIP **2**: 492; **3**: 1408, 412
DT **3**: 1385
β-D-Thevetose **1**: 180
D-thyroxine **1**: 22
DTIC **4**: 1895
DU-28853 **2**: 609
Duafen **4**: 1688
Dual-Clean **4**: 2037
Dubopride **2**: 795
Dulcolax **2**: 788

Dumicoat **3**: 1210
Dumirox **2**: 629
Dumolid **2**: 518
Dumopranol **1**: 159
Duocarmycin A **4**: 1907
Duolip **1**: 26
Duoluton **3**: 1598
DuP 753 **2**: 649
Duphaston **3**: 1597
Durabolin **3**: 1592
Durabolin O **3**: 1592
Duraboral **3**: 1592
Duracebrol **2**: 655
Duraclamid **2**: 608
Duraclon **2**: 424
Duradermal **2**: 354
Durafungol **3**: 1217
Duragesic **2**: 401
Dur-Anest **2**: 548, 548
Duranifin **1**: 95, 210
Durapental **1**: 49
Duraperidol **2**: 598
Durapindol **1**: 158
Durapirox **4**: 1697
Durapitrop **2**: 650
Duraquin **1**: 124
Duration of activity **4**: 2080
Durazanil **2**: 637
Durazepam **2**: 522
Durogesic **2**: 401
Duspatal **2**: 701
Duspatalin **2**: 701
Dusting powders **4**: 2092
Duvium **4**: 1678
Dwarf Tapeworm **3**: 1271
Dyazide **1**: 75
Dyclone **2**: 550
Dyclonine hydrochloride **2**: 550
Dydrogesterone **3**: 1597
Dynacirc **1**: 212
Dynacrine **1**: 212
Dynamectin **4**: 1680
Dynemicin **4**: 1908
Dynemicin A **3**: 1013
Dyneric **3**: 1587
Dynorphin **2**: 386
Dynothel **1**: 22
Dyphylline **2**: 889
Dyrenium **1**: 75, 299
Dyspnea **1**: 173
Dyspne-Inhal **1**: 60
Dytide H **1**: 293, 299

7E3 **1**: 44
E3810 **2**: 719
E-89/001 **4**: 1914
Eatan N **2**: 518

Ebastel **2:** 929
Ebastine **2:** 929
Ebola **3:** 1174
Ebramine **2:** 909
Ebrantil **1:** 89
Eburicoic acid **3:** 1326
EBV **3:** 1174
Ecalin **3:** 1219
Ecatril **2:** 618
Ecdysone **3:** 1331
Echinococcus species **3:** 1271
Echinostoma species **3:** 1269
Echofilina **4:** 2025
Echothiophate **4:** 2025
Echothiophate iodide **4:** 2025
Echovirus **3:** 1174
Ecodergin **3:** 1219
Ecodipin **1:** 210
Eco Mi **3:** 1219
Econazole **3:** 1219; **4:** 1730
Econopred **4:** 2017
Econopred Plus **4:** 2017
Ecorex **3:** 1219
Ecostatin **3:** 1219
Ecotam **3:** 1219
Ecrylène **2:** 439
Ectyl **2:** 501
Ectylurea **2:** 500
Eczematous dermatitis **4:** 1718
Edecrin **1:** 297
Edema **2:** 861
Edemox **4:** 2022
Edrophonium chloride **4:** 2031
Edrul **1:** 77, 297
EDTA **4:** 2039
EEE **3:** 1174
Efavirenz **3:** 1302
Eferox **3:** 1561
Effectin **2:** 885
Effectiveness proving **4:** 2136
Effective refractory period (ERP) **1:** 119
Effector cells **1:** 248
α-effects **1:** 138, 38
ReNu Effervescent Enzymatic Contact Lens Cleaner **4:** 2039, 2040
Effervescent excipients **4:** 2105
Effervescent powders **4:** 2091
Efficacy testing **4:** 2136
Effortil **1:** 64
Effortil plus **1:** 66
Efuranol **2:** 614
Eggobasin **2:** 462
Eglen **1:** 209
Ehtisterone **3:** 1598
Eicosanoids **2:** 344; **4:** 2046
   occurrence **4:** 2050
Ekko **2:** 474

Elantan **1:** 224
Elastase **4:** 1673
Elavil **2:** 611
Elazor **3:** 1202
Elbrol **1:** 159
Elcatonin **3:** 1367
Eldeprine **2:** 571, 625
Eldopar **2:** 569
Electrocardiogram (ECG) **1:** 116
Electrophoretic mobility **1:** 253
Elen **2:** 658
Elieten **2:** 608
Elimination phase **4:** 2130
Eliranol **2:** 590
Elixirs **4:** 2087
Elkapin **1:** 297
Elmarine **2:** 590
Elrodorm **2:** 512
Eltoprazine **2:** 609
Eltroxin **3:** 1561
Elzym **2:** 699
Embryo toxicity testing **4:** 2132
Embutal **2:** 509
Emconcor **1:** 153
Emcor **1:** 153
EMD 33.512 **1:** 153
EMD-49980 **2:** 609
Emdabol **3:** 1593
Emdalen **2:** 614
Emedastine **2:** 932
Emepronium bromide **2:** 693
Emergil **2:** 597
Emetics **2:** 741
Emex **2:** 608
Emflex **4:** 1680
Emiglitate **3:** 1664
Emodin **2:** 790
Emoflux **1:** 50
Emollients **2:** 781
Emotion **2:** 522
Emotival **2:** 522, 638
Empecid **3:** 1217
Emulsion excipients **4:** 2104
Emulsions **4:** 2089
Enalapril **1:** 80
Enatone **4:** 1932
Enavid **3:**
Enbol **2:** 657; **4:** 1706
Encainide **1:** 132
Encefabol **4:** 1706
Encefavol **2:** 657
Encephabol **4:** 1706
Encephalitis
   japanese **4:** 1803
   tick-borne **4:** 1802
Endak **1:** 154
Endep **2:** 611

Endocrine gland **3**: 1536
Endocrine peptide **3**: 1340
Endocrine secretion **3**: 1536
Endocringe Drugs **3**: 1319
Endocytosis **2**: 319
Endoprost **1**: 38
β-Endorphin **2**: 386; **3**: 1380
Endorphins **3**: 1544
Endothelin **3**: 1403
Enduracidin **3**: 1058
Enduracidins **3**: 1008
Enduron **1**: 294
Enemas **4**: 2088
Energona **1**: 62
Enflurane **2**: 441
Enidrel **2**: 522
Enion **4**: 2032
Enkade **1**: 132
Enkephalin **2**: 386
Enkephalins **3**: 1382, 544
Ennades **4**: 1912
Enocitabine **4**: 1879
Enovid **3**:
Enoxacin **3**: 1124
Enoximone **1**: 195, 196
Enprostil **4**: 2054, 2060
Enramycin **3**: 1058
Ensidon **2**: 619
Ensign **2**: 653
Entacapone **2**: 572
Entamoeba histolytica **3**: 1168, 1169
Enterobacteriaceae **3**: 1110
Enterobius vermicularis **3**: 1273
Enteroglucagon **3**: 1428
Entromycin powder **4**: 2162
Entumine **2**: 604
Enuclene **4**: 2040
Enviomycin **3**: 1061
Enviroxime **3**: 1185
Enzyme inhibitors
 cancer therapy **4**: 1937
Enzyme inhibitors antibiotics **3**: 956
Eosinophil granulocytes **1**: 246
Eosinophilic bronchial infiltration **2**: 863
Eosinophils **1**: 239
β-EP **3**: 1380
EP 011 **4**: 2054
EP 092 **4**: 2054
Epanutin **1**: 126; **2**: 475
Ephedrin **2**: 878
Ephedrine **2**: 463, 877
Ephetonin **2**: 878
4'-Epiadriamycin **4**: 1897
Epidermis **4**: 1714
Epidione **2**: 479
EPIFRIN **1**: 60; **4**: 2020
Epileo Petitmal **2**: 478

Epilepsy **2**: 469
Epilim **2**: 477
E-Pilo **4**: 2024
Epi-Monistat **3**: 1210
Epinastine **2**: 932
Epinephrine **1**: 60; **2**: 864; **3**: 1560, 1562; **4**: 2020
Epinephrine bitartrate **4**: 2020
Epinephrine hydrochloride **4**: 2020
Epinephryl borate **4**: 2020
EpiPen **1**: 60
Epi-Pevaryl **3**: 1219
Epirubicin **4**: 1897
Epithienamycin A **3**: 974
Epithienamycin B **3**: 974
Epithienamycin C **3**: 974
Epithienamycin D **3**: 974
Epithienamycin E **3**: 974
Epithienamycin F **3**: 974
Epitol **2**: 476
Epitopes **4**: 1757
Epitrate **4**: 2020
Epivir **3**: 1299
EPL **2**: 768
Epontol **2**: 447
Epoprostenol sodium **4**: 2061
Epostane **3**:
Epothilone **4**: 1918
(5α)-4,5-Epoxy-3,14-dihydroxy-17-(2-propenyl)morphinan-6-one **2**: 422
(5Z,9α,11α,13E,15S)-6,9-epoxy-11,15-dihydroxyprosta-5,13-dien-1-oic acid **1**: 37
5β,20-Epoxy-1,2α-4,7β,10β,13α-hexahydroxytax-11-en-9-one **4**: 1915
(5α)-4,5-Epoxy-14-hydroxy-3-methoxy-17-methylmorphinan-6-one **2**: 396
(5α)-4,5-Epoxy-3,14-hydroxy-17-methylmorphinan-6-one **2**: 397
(5α)-4,5-Epoxy-3-hydroxy-17-methylmorphinan-6-one **2**: 394
(5α,6α)-4,5-Epoxy-3-methoxy-17-methylmorphinan-6-ol **2**: 392
(5α)-4,5-Epoxy-3-methoxy-17-methylmorphinan-6-one **2**: 393
Epsom salts **2**: 785
Epsylone **2**: 506
Equalan **4**: 2173
Equanil **2**: 645, 679
Equatrate **2**: 679
Equilenin **3**: 1583
Equilid **2**: 606
Equilin **3**: 1583
Equilium **2**: 500
Equlum **2**: 606
Equipertine **2**: 609
Equipose **2**: 646
Equizole **4**: 2174
Ercoril **2**: 697

Ergenyl **2:** 477
Ergobel **2:** 655
Ergocalciferol **3:** 1325, 637
Ergolefrin **1:** 66
Ergoloid mesylates **2:** 655
Ergomimet **1:** 66
Ergomimet plus **1:** 66
Ergont **1:** 66
Ergostane **3:** 1323
Ergosterol **3:** 1201
Ergot Alkaloids **1:** 65
ergotam von ct **1:** 66
Erimin **2:** 522
Erythritol tetranitrate **1:** 224
Erythrityl tetranitrate **1:** 224
Erythro-100 **4:** 2165
Erythro-200 **4:** 2165
(DL-*Erythro*-α-2-piperidyl-2,8-bis(trifluoromethyl)-4-quinolinemethanol **3:** 1165
Erythrocyte count **1:** 239
Erythrocytes **1:** 242
   therapeutic use **1:** 271
Erythrol tetranitrate **1:** 224
Erythromycin **3:** 954, 050; **4:** 1725, 2011, 2164
Erythromycin estolate **3:** 1051
Erythromycin 2′-propionate dodecyl sulfate **3:** 1051
Erythromycins **3:** 996
Erythropoiesis **1:** 242
Erythropoietin **1:** 240, 243; **3:** 1550
Escherichia coli vaccine **4:** 1778
Escherichia coli vaccines **4:** 1782
Escre **2:** 498
Eserine **4:** 2025
Esidrix **1:** 72, 293
Esilgan **2:** 524
Eskabarb **2:** 480
Eskabarb Span **2:** 506
Eskadiazine **4:** 2014
Eskadole **3:** 1278
Eskazine **2:** 594
Esmarin **1:** 294
Esmind **2:** 590
Esperamicin **4:** 1908
Esperamicins **3:** 1013
Esperan **2:** 696
Espril **2:** 623
Esradin **1:** 212
Essences **4:** 2087
Essentiale **2:** 769
"Essential" Phospholipids **2:** 768
Estazolam **2:** 515, 524
   synthesis **2:** 521
Estracyt **4:** 1891
Estradiol **3:** 1501, 565, 583
Estradiol benzoate **3:** 1584
Estradiol valerate **3:** 1584
1,3,5(10)-Estraiene-3,16α,17β-triol **3:** 1583

Estramustine **4:** 1890
1,3,5(10)-Estratriene-3,17β-diol **3:** 1583
Estrimex **4:** 1923
Estriol **3:** 1583
Estrogen antagonists **4:** 1922
Estrogens **3:** 1551, 1564, 582
Estromedins **3:** 1549
Estrone **3:** 1565, 1568, 1583, 1573
Estrovis **3:** 1584
Estrumate **4:** 2059
Estulic **1:** 83
Estuline **1:** 83
ESWL **2:** 758
ET **3:** 1403
Etacrynic acid **1:** 297
Etafilcon A **4:** 2035
Eta-Lent **2:** 699
Etalontin **3:**
Etaperazin **2:** 595
Etaphydel **2:** 890
Etaqualone **2:** 514
Ethacrynic Acid **1:** 76
Ethambutol hydrochloride **3:** 1140
Ethanolamine antihistamines **2:** 908
Ethaquin **2:** 699
Ethatab **2:** 699
Ethaverine **2:** 699
Ethaverol **2:** 699
Ethchlorvynol **2:** 497
Ethenzamide **2:** 347
Ether **2:** 440
Ethical pharmaceuticals **1:** 6
Ethinamate **2:** 500
Ethinyl estradiol **3:** 1584; **4:** 1726
Ethionamide **3:** 1141
Ethmozine **1:** 133
Ethopropazine **2:** 565
Ethosuximide **2:** 473, 478
Ethotoin **2:** 475
2-ethoxybenzamide **2:** 347
1-[*N*-[1-(Ethoxycarbonyl)-3-phenylpropyl]-L-alanyl]-L-proline **1:** 80
1-(2-Ethoxyethyl)-2-(hexahydro-4-methyl-1*H*-1,4-diazepin-1-yl)-1*H*-benzimidazole **2:** 932
4′-Ethoxy-2′-hydroxy-4,6′-dimethoxychalcone **3:** 1186
1-(4-Ethoxy-2-hydroxy-6-methoxyphenyl)-3-(4-methoxyphenyl)-2-propen-1-one **3:** 1186
(*S*)-2-Ethoxy-4-[2-[[3-methyl-1-[2-(1-piperidinyl)phenyl]-butyl]amino]-2-oxoethyl]benzoic acid **3:** 1658
2-[(2-Ethoxyphenoxy)methyl]tetrahydro-1,4-oxazine **2:** 632
N-(4-Ethoxyphenyl)acetamide **2:** 349
Ethrane **2:** 442
Ethydan **2:** 441
Ethyl 4-aminobenzoate **2:** 542
Ethyl[2-amino-6-(4-fluorobenzylamino)pyridin-3-yl]carbamate **2:** 424

α-[(Ethylamino)methyl]-3-hydroxybenzenemethanol **1:** 63
Ethylbenzhydramine **2:** 567
5-Ethyl-5-(2-butyl)barbituric acid **2:** 508
Ethyl cellulose **4:** 2103
cis-(2-ethylcrotonoyl) urea **2:** 500
Ethyl 1-(3-cyano-3,3-diphenylpropyl)-4-phenylpiperidine-4-carboxylate **2:** 408
5-Ethyl-5-(1-cycloheptenyl)barbituric acid **2:** 508
1-Ethyl-1,4-dihydro-4-oxo[1,3]dioxolo[4,5-g]cinnoline-3-carboxylic acid **3:** 1119
9-Ethyl-6,9-dihydro-4,6-dioxo-10-propyl-4H-pyrano[3,2-g]quinoline-2,8-dicarboxylic acid **2:** 934
(3S,4R)-3-Ethyldihydro-4-[(1-methyl-1H-imidazol-5-yl)methyl]furan-2(3H)-one **4:** 2024
1-Ethyl-1,4-dihydro-7-methyl-4-oxo-1,8-naphthyridine-3-carboxylic acid **3:** 1117
5-Ethyl-5,8-dihydro-8-oxo-1,3-dioxolo[4.5-g]quinoline-7-carboxylic acid **3:** 1118
8-Ethyl-5,8-dihydro-5-oxo-2-(1-piperazinyl)pyrido[2,3-d]pyrimidine-6-carboxylic acid **3:** 1120
8-Ethyl-5,8-dihydro-5-oxo-2-(1-pyrrolidinyl)pyrido[2,3-d]pyrimidine-6-carboxylic acid **3:** 1120
5-Ethyldihydro-5-phenyl-4,6(1H,5H)-pyrimidinedione **2:** 480
1-Ethyl-1,4-dihydro-4-oxo-7-(4-pyridyl)-3-quinolinecarboxylic acid **3:** 1120
N^1-(1-Ethyl-1,2-dihydro-2-oxo-4-pyrimidinyl)sulfanilamide **3:** 1135
N-[1-[2-(4-Ethyl-4,5-dihydro-5-oxo-1H-tetrazol-1-yl)ethyl]-4-(methoxymethyl)-4-piperidinyl]N-phenylpropanamide hydrochloride **2:** 453
N-{1-[2-(4-Ethyl-5-oxo-4,5-dihydrotetrazol-1-yl)ethyl]-4-methoxymethylpiperidin-4-yl}-N-phenylpropionamide **2:** 401
Ethylenediamine antihistamines **2:** 904
Ethylenediaminetetracetic acid **4:** 2039
(+)-2,2′-(Ethylenediimino)-di-1-butanol dihydrochloride **3:** 1140
Ethylestrenol **3:** 1592
17α-Ethyl-4-estren-17β-ol **3:** 1592
1-Ethyl-7-{3-(ethylamino)methyl-1-pyrrolidinyl}-6,8-difluoro-1,4-dihydro-4-oxo-3-quinolinecarboxylic acid **3:** 1128
1-ethyl-6-fluoro-1,4-dihydro-4-oxo-7-(1-piperazinyl)-1,8-naphthyridine-3-carboxylic acid **3:** 1124
1-Ethyl-6-fluoro-1,4-dihydro-4-oxo-7-(1-piperazinyl)-3-quinolinecarboxylic acid **3:** 1122
1-ethyl-6-fluoro-7-(4-methyl-1-piperazinyl)-4-oxo-1,4-dihydroquinoline-3-carboxylic acid **3:** 1125
3-(3-ethylhexahydro-1-methyl-1H-azepin-3-yl)phenol **2:** 417
N-(2-Ethylhexyl)-3-hydroxybutyramide hydrogen succinate **2:** 502
Ethyl-(2-hydroxyphenyl)dimethylammonium chloride **4:** 2031
α-ethyl-1-Hydroxycyclohexaneacetic acid **2:** 752

5-Ethyl-5-isopentylbarbituric acid **2:** 507
5-Ethyl-5-isopropylbarbituric acid **2:** 507
3-(3-ethyl-1-methylazepan-3-yl)phenol **2:** 417
5-ethyl-5-(1-methyl-1-butenyl)barbituric acid **2:** 509
5-ethyl-5-(1-methylbutyl)-barbituric acid **2:** 509
(±)-5-Ethyl-6-methyl-7-(4-methoxyphenyl)-5-azaheptyl 3,4-dimethoxy-benzoate **2:** 701
3-Ethyl-2-methyl-5-(morpholinomethyl)-4,5,6,7-tetrahydro-1H-indol-4-one **2:** 608
5-Ethyl-3-methyl-5-phenyl-2,4-imidazolidinedione **2:** 475
α-Ethyl-2-(3-methyl-2-phenyl-4-morpholinyl) ethyl ester benzeneacetic acid **2:** 465
Ethyl 1-methyl-4-phenylpiperidine-4-carboxylate **2:** 397
5-Ethyl-1-methyl-5-phenyl-2,4,6(1H,3H,5H)-pyrimidinetrione **2:** 480
4-Ethyl-4-methyl-2,6-piperidinedione **2:** 339
3-Ethyl-3-methyl-2,5-pyrrolidinedione **2:** 478
Ethyl-3-methyl-2-thioxo-4-imidazoline-1-carboxylate **3:** 1530
2-Ethyl-3-methylvalerylurea **2:** 502
Ethylmorphine **2:** 391
Ethyl[10-(3-morpholinopropionyl)phenothiazin-2-yl]carbamate **1:** 133
Ethylpapaverine **2:** 699
Ethylphenacemide **2:** 501
5-Ethyl-5-phenylbarbituric acid **2:** 506
3-Ethyl-5-phenyl-2,4-imidazolidindione **2:** 475
3-(2-Ethylphenyl)-2-methyl-4(3H)-quinazolinone **2:** 514
3-Ethyl-3-phenylpiperidine-2,6-dione **2:** 512
5-Ethyl-5-phenyl-2,4,6(1H,3H,5H)-pyrimidinetrione **2:** 479
(+)-7-Ethyl-10-[4-(1-piperidyl)-1-piperidyl]carbonyloxycamptothecin **4:** 1914
4-[Ethyl (p-methoxy-α-methylphenethyl)-amino]butyl veratrate **2:** 701
(±)-5-[[4-[2-(5-Ethyl-2-pyridinyl)ethoxy]phenyl]methyl]-2,4-thiazolidinedione, hydrochloride **3:** 1666
(±)-3-Ethyl-3(4-pyridinyl)-2,6-piperidinedione **4:** 1926
N-[(1-ethyl-2-pyrro-lidinyl)methyl]-5-(ethylsulfonyl)-2-methoxybenz-amide **2:** 606
N-[(1-Ethyl-2-pyrrolidinyl)methyl]-2-methoxy-5-sulfamoylbenzamide **2:** 605
Ethyl (R)-(+)-1(α-methylbenzyl)-imidazole-5-carboxylate **2:** 449
1-[2-(Ethylsulfonyl)ethyl]-2-methyl-5-nitroimidazole **3:** 1156
2-Ethylthioisonicotinamide **3:** 1141
2-Ethylthio-10-[3-(4-methyl-1-piperazinyl)propyl]-10H-phenothiazine **2:** 744
Ethyl trans-2-dimethylamino-1-phenylcyclohex-3-enecarboxylate **2:** 410
17α-Ethynil-17β-hydroxy-4-androsten-3-one **3:** 1598
Ethynodiol diacetate **3:** 1497, 1500

17α-Ethynyl-17β-acetoxy-3-cyclopentyloxy-3,5-estradiene   **3**: 1598
1-Ethynylcyclohexylcarbamate   **2**: 500
17α-Ethynyl-3-cyclopentyloxy-1,3,5(10)-estratrien-17β-ol   **3**: 1584
17α-Ethynyl-3β,17β-diacetoxy-4-estrene   **3**:
17α-Ethynyl-1,3,5(10)-estratriene-3,17β-diol   **3**: 1584
17α-Ethynyl-13-ethyl-17β-hydroxy-4-gonen-3-one   **3**: 1598
17α-Ethynyl-17β-hydroxy-4,9,11-estrien-3-one   **3**: 1598
17α-Ethynyl-3-methoxy-1,3,5(10)-estratrien-17β-ol   **3**: 1584
Eticylin   **3**: 1584
Etidocaine hydrochloride   **2**: 547; **4**: 2003
Etifoxine   **2**: 646
etil 5 von ct   **1**: 64
Etilefrine   **1**: 63
ETI-PUREN   **1**: 64
Etizolam   **2**: 640
Etodolac   **2**: 382
Etofenamate   **2**: 350
Etofibrate   **1**: 26
Etofyllin clofibrate   **1**: 26
Etofylline   **2**: 888
Etomidate   **2**: 449
Etomine   **2**: 604
Etoposide   **4**: 1912
Etorphine   **2**: 393
Etorphine/diprenorphine   **4**: 2182
Etoscol   **2**: 869
Etoval   **2**: 506
Etozolin   **1**: 297
Etramon   **3**: 1219
Etretinate   **4**: 1729
Etumina   **2**: 604
Etumine   **2**: 604
Eu–5306   **3**: 1125
Eubine   **2**: 397
Eucil   **2**: 745
Eudatine   **2**: 624
Euglucon   **3**: 1656
Eugynon   **3**: 1598
Euhypnos   **2**: 523
Eukodal   **2**: 397
Eukraton   **2**: 339
Eulexin   **4**: 1928
Eumetabol   **2**: 653
Euminex   **4**: 1681
Eunal   **2**: 576
Eunerpan   **2**: 600
Eupaverin   **2**: 699, 700
Eupaverina   **2**: 700
Euphol   **3**: 1326
Euphyllin   **2**: 888
Euradel   **1**: 153
Eurodin   **2**: 524
European pharmacopeia   **4**: 2153

Eurosan   **2**: 637; **3**: 1217
Eusenium   **2**: 656
Euthyrox   **3**: 1561
Eutimox   **2**: 595
Eutonyl   **2**: 624
Evadene   **2**: 616
Evadyne   **2**: 616
Evasidol   **2**: 616
Eventin   **2**: 462
Evipal Sodium   **2**: 445
Evipan-Natrium   **2**: 445
Exceglan   **2**: 486
Excegram   **2**: 486
Excipient ingredients   **4**: 2101
Excipients   **4**: 2078
Excitability   **1**: 119
Excitatory ionotropic action   **2**: 333
Exelderm   **3**: 1227
Exemestane   **4**: 1924
Exifone   **2**: 657
Exirec   **2**: 877
Exirel   **1**: 194
Exocytosis   **2**: 319
Exoderil   **3**: 1249
Exorphins   **3**: 1386
Expidet   **2**: 522
Extenzyme   **4**: 2040
Extracorporeal Shock-Wave Lithotripsy   **2**: 758
Extractive solutions   **4**: 2086
Extrapyramidal tract   **2**: 315
Extrasystoles   **1**: 117
Eye Cool   **4**: 2037
20/20 Eye Drops   **4**: 2037
Ezon-T   **3**: 1248

F 1571   **2**: 904
F 929   **2**: 903
Fabadorm   **2**: 507
Fabontal   **2**: 447
Factor I   **1**: 254
Factor II   **1**: 255
Factor III   **1**: 255
Factor inhibitors   **4**: 1707
Factor IV   **1**: 255
Factor IX   **1**: 252, 256
Factor V   **1**: 252, 255
Factor VII   **1**: 252, 256
Factor VIII   **1**: 252, 256
Factor X   **1**: 252, 257
Factor XI   **1**: 257
Factor XII   **1**: 257
Factor XIII   **1**: 252, 257
Factor XIII deficiency   **1**: 276
Fadrozole   **4**: 1926
Faecosterol   **3**: 1324
FAHRAEUS-LINDQVIST effect   **1**: 45
Fallot's tetralogy   **1**: 148

Faltium **2**: 608
Falvin **3**: 1221
Famotidine **2**: 723
Fanasil **3**: 1136
2-F-Ara-AMP **4**: 1880
Fareston **4**: 1923
Farmorubicin **4**: 1898
Farmotal **2**: 445
Fasciolopsis buski **3**: 1268
Fasigyn **3**: 1156
Faslodex **4**: 1924
Fastin **2**: 467
Fastjekt **1**: 60
Faverin **2**: 629
Fazadinium bromide **2**: 670
Fazodon **2**: 670
Fazol **3**: 1222
FB b 5097 **3**: 1216
Fc-1157a **4**: 1922
5-FC **3**: 1238
5-FdUMP **4**: 1873
Fedal Telmin **3**: 1282
Fedal Uncin **3**: 1278
Feinalmin **2**: 614
Felbamate **2**: 485
Felbatol **2**: 485
Felden **2**: 377; **4**: 1697
Feldene **2**: 377
Feldéne **2**: 377
Felicur **2**: 750
Felison **2**: 522
Feloday **1**: 211
Felodipine **1**: 203, 211
Felunamin **4**: 1698
Femarfarmamide **3**: 1387
Femcare **3**: 1217
Femcosyn **3**: 1215
Femeron **3**: 1210
Femovan **3**:
Femoxetine **2**: 628
Femstat **3**: 1215
Fenadiazole **2**: 529
Fenamide **4**: 2022
Fenamizol **4**: 2180
Fenbufen **4**: 1686
Fendazol **4**: 1700
Fendilar **1**: 208
Fendiline **1**: 208, 229
Fenemal **2**: 506
Fenfluramine **2**: 461, 467
Fenileal **2**: 506
Fenipentol **2**: 750
Fenistil **2**: 912
Fenizolan **3**: 1221
Fenofibrate **1**: 25
Fenoprofen **4**: 1690
Fenopron **4**: 1690

Fenoterol **2**: 706, 871
Fenpiverinium bromide **2**: 693
Fentanyl **2**: 399, 453, 454; **4**: 2183
Fentanyl Janssen **2**: 401
Fentazin **2**: 595
Fentiazac **4**: 1685
Fenticonazole **3**: 1220
Fentiderm **3**: 1221
Fentigyn **3**: 1221
Fenylhist **2**: 909
Fepron **4**: 1690
Fermentation
   in antibiotics production **3**: 1073
Fermentation residues
   in penicillin production **3**: 1083
Fermentation technology **3**: 1079
Fermenters
   for monoclonal antibody production **4**: 1990
   in antibiotics production **3**: 1075
Fertile period **3**: 1493
Fertility testing **4**: 2132
Fertodur **3**: 1587
Fevarin **2**: 629
Fever
   treatment **2**: 341
Fexofenadine **2**: 929; **4**: 1720
FG-5606 **2**: 609
FG-7516 **2**: 649
FI-7045 **3**: 1225
FI-7056 **3**: 1225
FIAC **3**: 1180; **4**: 1879
Fibocil **1**: 128
Fiboran **1**: 128
Fibrillation **1**: 117
Fibrin degradation **1**: 258
Fibrinogen **1**: 252, 254
Fibrinolysis system **1**: 258
Fibrin stabilizing factor **1**: 257
Ficortril **3**: 1607
Filariasis vaccines **4**: 1810
Film coating **4**: 2099
Filoviridae **3**: 1174
Finlepsin **2**: 476
Fiostin **2**: 653
Fish Tapeworm **3**: 1269
Fitonal **3**: 1208
Fitonax **3**: 1219
FIX high-purity concentrates **1**: 269
FK-453 **1**: 301
Flagellates **3**: 1146
Flagyl **3**: 1145
Flavizol **3**: 1202
Flavophospholipol **3**: 1068
Flavoring tinctures **4**: 2087
Flavors **4**: 2102
Flavoxate **2**: 704
Flaxedil **2**: 671

Flecainide **1**: 128
Flexartal **2**: 677
Flex-Care **4**: 2037, 2037
Flexeril **2**: 678
Flexidin **4**: 1680
Flexsol **4**: 2037
Flodyl **1**: 211
Flogar **4**: 1681
Flogene **4**: 1686
Flolan **1**: 38; **4**: 2061
Flomed **1**: 50
Flopropione **2**: 700
Florfenicol **4**: 2168
Florid **3**: 1210
Florinef **1**: 67; **3**: 1620
Florinef Acetate **1**: 67
Florisan **2**: 782
Flormidal **2**: 525
Floropryl **4**: 2026
Floxacillin **3**: 1022
Floxuridine **4**: 1873
Floxyfral **2**: 629
Floxytral **2**: 629
Fluandrenolone **3**: 1629
Fluanisone **2**: 598
Fluanxol **2**: 597
Flucloxacillin **3**: 962, 022
Flucon **4**: 2018
Fluconazole **3**: 1201; **4**: 1730
Flucort **4**: 2178
Flucytosine **3**: 1238; **4**: 2013
Fludara **4**: 1880
Fludarabine phosphate **4**: 1880
Fludex **1**: 73
Fludrocortisone **1**: 67
Fludrocortisone **1**: 66
  doses **3**: 1605
FLUENT **1**: 27
Flufenamate **2**: 351
Flufenamic acid **2**: 351; **4**: 1697
Flugeral **1**: 209
Fluid Extracts **4**: 2087
Fluidol **2**: 601
Fluilast **1**: 41
Flukes **3**: 1266
Flumark **3**: 1125
Flumequine **3**: 1126
Flumethasone **3**: 1629
Flumethasosone **4**: 2178
Flunagen **1**: 209
Flunarizine **1**: 209
FlunarizineCyclimorph **2**: 915
Flunarl **1**: 209
Fluniget **4**: 1678
Flunipam **2**: 520
Flunisolide **2**: 894

Flunitrazepam **2**: 515, 518
  indications for use **2**: 519
  synthesis **2**: 520
Flunixin **4**: 2176
Fluocinolone acetonide **3**: 1629
Fluocortin butyl **3**: 1636
Fluocortolone **3**: 1633
  doses **3**: 1605
Fluoracaine **4**: 2031
Fluorescein **4**: 2030
Fluorescite **4**: 2031
Fluoresoft **4**: 2031
Fluorets **4**: 2031
Fluorexon **4**: 2031
Fluor-I-Strip **4**: 2031
1-[3-(4-Fluorobenzoyl)-propyl]-4-(2-oxo-1-benzimidazo-
  linyl)-1,2,3,6-tetrahydropyridine **2**: 454
8-[3-(4-Fluorobenzoylpropyl]-4-oxo-1-phenyl-1,3,8-tria-
  zaspiro[4,5]-decane **2**: 599
1-{1-[3-(4-Fluorobenzoyl)propyl]-4-piperidyl}-2-benzimi-
  dazolinone **2**: 599
1-{1-[3-(4-Fluorobenzoyl)propyl]-4-piperidyl}-2-mercap-
  tobenzimidazole **2**: 599
1-{1-[3-(4-Fluorobenzoyl)propyl]-1,2,3,6-tetrahydro-4-
  pyridyl}-2-benz-imidazolinone **2**: 600
1-(4-Fluorobenzyl)-2-[4-[N-(3,4-dihydro-4-oxopyrimi-
  din-2-yl)-N-methylamino]piperidin-1-yl]-
  ben **2**: 930
2-(3′-Fluorobiphenyl-4-yl)propionic acid **2**: 361
6α-Fluorocorticoids **3**: 1629
9α-Fluorocortisol **1**: 66; **3**: 1619
Fluorocortisone **3**: 1619
Fluorocytosine **3**: 1238
5-Fluorocytosine **4**: 2013
1-(2′-Fluoro-2′-deoxy-β′-D-arabinofuranosyl)-5-iodocyto-
  sine hydrochloride **3**: 1180
6-Fluoro-1,4-dihydro-1-(methylamino)-7-(4-methyl-1-pi-
  perazinyl)-4-oxo-3-quinolinecarboxylic
  acid **3**: 1126
6α-fluoro-11β,21-dihydroxy-16α,17α-isopropyldioxy-
  pregna-1,4-diene-3,20-dione **2**: 894
9-Fluoro-11β,17β-dihydroxy-17α-methyl-5-androsten-
  3β,17β-one **3**: 1590
9-Fluoro-6,7-dihydroxy-5-methyl-1-oxo-1H,5H-benzo[i,j]-
  quinolizine-2-carboxylic acid **3**: 1127
p-Fluorohexahydrosiladifenidol **2**: 706
9α-Fluorohydrocortisone **1**: 66
6α-Fluoro-16α-hydroxycortisol **3**: 1629
4′-Fluoro-4-[4-hydroxy-4-(4-
  methylphenyl)piperidino]butyrophenone **2**: 598
4′-Fluoro-4-[4-hydroxy-4-(3-trifluoromethylphenyl)pi-
  peridino]-butyrophenone **2**: 599
2′-Fluoro-5-iodo-1-β-D-arabinofuranosylcyto-
  sine **4**: 1879
Fluoromar **2**: 443

Fluorometholone **4**: 2018
4′-Fluoro-4-[4-(2-methoxyphenyl)-1-piperazinyl]butyrophenone **2**: 598
2-Fluoro-α-methyl[1,1′-biphenyl]-4-acetic acid **2**: 361; **4**: 1690, 2019
S-fluoromethyl-6α,9α-difluoro-11β-hydroxy-16α-methyl **2**: 895
(±)9-Fluoro-3-methyl-10-(4-methyl-1-piperazinyl)-7-oxo-2,3-dihydro-7H-pyrido[1,2,3-de] **3**: 1125
4′-Fluoro-4-(4-methylpiperidino)butyrophenone **2**: 600
9α-Fluoro-16α-methylprednisolone **3**: 1625, 625; **4**: 2017
2-Fluoro-9-(5-O-phosphono-β-D-arabino-furanosyl)-9H-purin-6-amine **4**: 1880
Fluor-Op **4**: 2018
(E)-2-(4-fluorophenetyl)-3-fluoroallylamine **2**: 571
1-[4,4-bis(4-Fluorophenyl)butyl]-4-(4-chloro-3-trifluoromethylphenyl)-4-hydroxypiperidine **2**: 601
8-[4,4-Bis(4-fluorophenyl)butyl]-1-phenyl-1,3,8-triazaspiro[4.5]-decan-4-one **2**: 600
1-{1-[4,4-Bis(4-fluorophenyl)butyl]-4-piperidinyl}-2-benzimidazolinone **2**: 601
5-(2-Fluorophenyl)-1,3-dihydro-1-methyl-7-nitro-2H-1,4-benzodiazepin-2-one **2**: 518
1-(p-fluorophenyl)-6-fluoro-1,4-dihydro-4-oxo-7-(1-piperazinyl)-quinoline-3-carboxylic acid hydrochloride **3**: 1127
(3R,5S,6E)-7-[4-(4-fluorophenyl)-5-methoxymethyl]-2,6-bis(1-methylethyl)-3-pyridinyl]-3,5-dihydroxy-6-heptenoic acid **1**: 29
(3R,5S,6E)-7-[3-(4-fluorophenyl)-1-(1-methylethyl)-1H-indol-2-yl]-3,5-dihydroxy-6-heptenoic acid **1**: 28
1-(4-fluorophenyl)-methyl-N-[1-[2-(4-methoxyphenyl)ethyl]-4-piperidinyl]-1H-benzimidazol-2-amine **2**: 927
(E)-1-[bis(4-Fluorophenyl)methyl]-4-(3-phenyl-2-propenyl)piperazine **1**: 209; **2**: 915
N-[3-[4-(4-fluorophenyl)-1-piperazinyl]-1-methylpropyl]-3-pyridinecarboxamide **2**: 916
(±)-4-[3-(4-Fluorophenylsulfonyl)-2-hydroxy-2-methylpropionamido]-2-(trifluoromethyl)benzonitrile **4**: 1930
6α-Fluoroprednisolone **3**: 1629
5-Fluoro-2,4-(1H, 3H)-pyrimidinedione **4**: 1871
Fluoropyrimidines cancer therapeutics **4**: 1871
Fluoroquinolones antibacterial agents **3**: 1121
(11β)-9-fluoro-11,17,21-trihydroxypregn-4-ene-3,20-dione **1**: 66
5-Fluorouracil **4**: 1871
Fluothane **2**: 440
Fluoxeren **2**: 628
Fluoxetine **2**: 628
Fluoxymestrone **3**: 1590
Flupentixol **2**: 597
Fluphenazine **2**: 594, 595
Flupirtine **2**: 424

Fluprednisolone **3**: 1629
Fluprednylidene **3**: 1628
Fluprostenol **4**: 2054
Flurazepam **2**: 515, 520
 synthesis **2**: 520
Flurbiprofen **2**: 361; **4**: 1690, 2019
Fluren **2**: 544
Fluress **4**: 2003
5-Flurocytosine **3**: 1238
Fluroxene **2**: 442
Fluspirilene **2**: 600
Flutamide **3**: 1591; **4**: 1928
 annual sales **4**: 1939
Fluticasone propionate **2**: 895
Flutide **2**: 895
Flutter **1**: 117
Fluvastatin **1**: 28
Fluvoxamine **2**: 628
Fluxarten **1**: 209; **2**: 915
FML Forte Liquifilm **4**: 2018
FML Liquifilm **4**: 2018
FML-S **4**: 2005
FML S.O.P. **4**: 2018
F-Mon **2**: 595
FMRF-amide **3**: 1387
Focusan **3**: 1248
Fokalepsin **2**: 476
Folcodal **1**: 209
Folic acid analogs **4**: 1868
Follistatin **3**: 1359, 362
Follitropin **3**: 1348, 355, 545
Fominoben hydrochloride **2**: 339
Fomocaine hydrochloride **2**: 551
Fongamil **3**: 1223
Fongarex **3**: 1223
Fongorex **3**: 1223
Fontex **2**: 628
Fonzylane **1**: 50
Footwork **3**: 1248
Foradil **2**: 883
Forane **2**: 442
Forhistal **2**: 912
Forit **2**: 609
Formestane **4**: 1924
N-formimidoylthienamycin **3**: 1035
Formoterol **2**: 882
Formycin **3**: 993
4′-Formylacetanilide thiosemicarbazone **3**: 1142
2-Formyloxymethyl-clavam **3**: 971
3-Formylrifamycin S **3**: 1000
Foromacidins **3**: 998
Fortecortin **3**: 1627
Fortimicin **3**: 987
Fortimicin A **3**: 988, 049
Fortimicin B **3**: 988
Fortovase **3**: 1307
Fortral **2**: 421

Foscarnet **3**: 1184
Foscarnet sodium **3**: 1184
Fosfomycin **3**: 1011, 071
Fotemustine **4**: 1893
Fotimin **3**: 1249
Fourneau 309 **3**: 1149
Fradiomycin **3**: 984, 1044
Framycetin sulfate **4**: 2007
Framycort **4**: 2008
Framygen **4**: 2008
Frangula bark **2**: 791
Frenactil **2**: 599
Frenolon **2**: 595
Frisin **2**: 644
Frisium **2**: 481, 644
Froben **2**: 363; **4**: 1691
Frontal **2**: 641
Frusemide **1**: 296
FSH **3**: 1545
FTA **4**: 1928
Ftorafur **4**: 1875
5-FU **4**: 1871
Fuchsin **3**: 1255
5-FUDR **4**: 1873
Fulcin **3**: 1242
Ful-Glo **4**: 2031
Fulvicin **3**: 1242
Fulvicina **3**: 1242
Fulvicin P/G **3**: 1242
Fumagillin **3**: 1010; **4**: 1937
Funduscein **4**: 2031
Fungal infections **4**: 1730
Fungal mycelium
 disposal **3**: 1082
Funganiline **3**: 1233
Fungarest **3**: 1208
Fungata **3**: 1202
Fungi
 pathogenic **3**: 1200
Fungibacid **3**: 1231
Fungicidin **3**: 1238
Fungicil **3**: 1208
Fungiderm **3**: 1210, 245
Fungifos **3**: 1246
Fungiframan **3**: 1217
Fungilin **3**: 1233
Fungi-med **3**: 1217
Funginazol **3**: 1210
Fungireduct **3**: 1238
Fungisdin **3**: 1210
Fungistat **3**: 1230
Fungivin **3**: 1242
Fungizid-ratiopharm **3**: 1217
Fungizone **3**: 1233; **4**: 2013
Fungo-Hubber **3**: 1208
Fungoral **3**: 1208
Fungosten **3**: 1217

Fungotox **3**: 1217
Fungowas **3**: 1245
Fungucit **3**: 1210
Fungur **3**: 1210
Funtumine **3**: 1333
Furadantin **3**: 1146
Furazolidone **3**: 1158
Furosemide **1**: 71, 72, 296
Furox **3**: 1159
Furoxone **3**: 1158, 159
Fusidic acid **3**: 1011, 1071, 1327
Fusion
 of cells **4**: 1983
Fydalex **2**: 501

GABA **2**: 329
 in Parkinsonism **2**: 557
Gabalon **2**: 681
Gabapentin **2**: 483
Gabitril **2**: 485
Gabrilen **4**: 1690
Gag gene **3**: 1292
GAL **3**: 1431
4-O-β-D-galactopyranosyl-D-fructose **2**: 786
4-O-β-D-galactopyranosyl-α-D-glucopyranose monohydrate **2**: 785
Galanine **3**: 1431
Gallamine triethiodide **2**: 670
Gallamycin-Dry **4**: 2165
Gallbladder **2**: 747
Galleb **2**: 750
Gallimycin-36 **4**: 2165
Gallo-Merz Spasmo Hymecromon **2**: 701
Gallopamil **1**: 131, 208
Gallo Sanol **2**: 749
Gallstone dissolution **2**: 754
Gallstones **2**: 754
Gamanil **2**: 614
Gamaquil **2**: 645
Gamma globulin preparations **4**: 1822
 adverse effects **4**: 1836
Gamma-OH **2**: 450
Gamonil **2**: 614
ganglia blockers **2**: 685
Ganirelix **4**: 1934
Ganstrin **4**: 2005
Gantrisin **3**: 1134
Garacin **4**: 2163
Garamycin **4**: 2008
Gardenal **2**: 506
Gardenale **2**: 506
Gardrin **4**: 2060
Gargles **4**: 2089
Gaseous dosage forms **4**: 2101
Gastrese LA **2**: 608
Gastric acid **2**: 714
Gastrin **3**: 1418, 549

Gastrin Inhibiting Peptide   3: 1429
Gastrin Releasing Peptide   3: 1416
Gastrodiagnost   3: 1420
Gastrodiscoides hominis   3: 1269
Gastrointestinal Drugs   2: 711
Gastrointestinal Hormones   3: 1549
Gastrointestinal nematode parasites
    vaccines   4: 1808
Gastromax   2: 608
Gastronerton   2: 745
Gastroprokinetic Agents   2: 793
Gastroprotection   2: 724
Gastrosil   2: 729
Gastrozepin   2: 729
Gc   1: 252
GCP   4: 2138
Geangin   1: 208
Gefulvine   3: 1242
Gelatin   4: 2033
Gelatin capsules   4: 2092
Gel-Clea   4: 2037
Gels   4: 1716, 2089
Gemcitabine   4: 1879
Gemeprost   3: 1510; 4: 2059
Gemfibrozil   1: 25
Gemzar   4: 1879
Genabol   3: 1592
Generalized seizures   2: 470
Generic pharmaceutical industry   1: 8
Gene therapy
    in Parkinsonism treatment   2: 580
Genetic engineering
    in vaccine development   4: 1764
Genin   1: 177
Genoptic   4: 2008
Genopyic   4: 2008
Genotropin   3: 1437
Gentacidin   4: 2008
Gentafair   4: 2008
Gentak   4: 2008
Gentamiccin C18   3: 985
Gentamicin   3: 986, 1042; 4: 2163
Gentamicin C Complex   3: 986
Gentamicin sulfate   3: 1042; 4: 2008
Gentamycin   4: 2008
Gentamytrex Ophtiole   4: 2039
Gentian violet   4: 1732
Genticidin   4: 2008
Genticina   4: 2008
Gentocin   4: 2163, 2177
Gentrasul   4: 2008
Genurin   2: 704
Gepirone   2: 633
Gerdaxyl   2: 631
Geriatric Drugs   2: 649
Germanin   3: 1149, 1287
German measles   4: 1790

Gervagem   4: 2059
Gesinal   3: 1596
Gestaforin   3: 1596
Gestagens   3: 1564, 593
Gestanin   3: 1598
Gestanon   3: 1598
Gestanyn   3: 1598
Gestodene   3: 1497, 1501
Geston   3: 1596
Gestonorone caproate   3: 1597
Gestovis   3: 1596
Gestovister   3: 1596
Gestratron   3: 1597
Gevilon   1: 25
GG   3: 1426
GH   3: 1434, 1435, 1545
GHRH   3: 1430, 1432, 1549
Giant Intestinal Fluke   3: 1268
Giardia lamblia   3: 1157
Giardiasis   3: 1157
Giganten   1: 209
Gilt   3: 1217
Gilurytmal   1: 127
Gingicain   2: 545
Gino-Canesten   3: 1217
Gino-Clotrimix   3: 1217
Gino-Lotramina   3: 1217
Gino-Tralen   3: 1231
Gioron   2: 450
GIP   3: 1429
Gitoxigenin   1: 179
Gitoxin   1: 188
Glandubolin   3: 1583
Glanil   1: 209
Glass containers   4: 2107
Glauber's salt   2: 785
Glaucoma   1: 149
Glaucoma therapy   4: 2019
Glaucon   4: 2020
Glauconide   4: 2022
Glaucothil   4: 2020
Glauco-Viskin   1: 158
Glaudrops   4: 2020
Glauline   4: 2022
Glaumid   4: 2022
Glaupax   1: 292; 4: 2022
Glia   2: 308
Glial cells   2: 310
Glibenclamide   1: 15
Glibenese   3: 1656
Glibornuride   3: 1655
Glicentin   3: 1428
Gliclazide   3: 1656
Glidant   4: 2105
Glimepiride   3: 1656
Glimide   2: 512
Glipizide   3: 1656

Gliquidone **3:** 1657
Glisoxepide **3:** 1657
Glitazones **3:** 1665
Globulins **1:** 252
Glomerulus **1:** 289
GLP **4:** 2138
GLP-1-(7–36)amide **3:** 1428
Gluborid **3:** 1655
Glucagon **2:** 772; **3:** 1426, 1548
Glucagon-Like Peptide-1-(7–36)amide **3:** 1428
Glucametacin **4:** 1680
Glucantime **3:** 1154
Glucobay **3:** 1663
Glucocorticoid hormones **3:** 1601
   therapeutic uses **3:** 1604
Glucocorticoids **3:** 1552
   as antiinflammatory dermatological drugs **4:** 1721
   as antiinflammatory drugs **4:** 1675
Glucocorticosteroids
   for the treatment of asthma **2:** 893
Glucor **3:** 1663
Glucose-dependent insulinotropic peptide **3:** 1429
Glucose-Regulating Hormones **3:** 1548
α-Glucosidase inhibitors **3:** 1661
Gludorm **2:** 512
Glurenorm **3:** 1657
Glutamate antagonists **2:** 574
Glutamine **2:** 329
Glutethimide **2:** 512
Glutisal **2:** 348
Glutril **3:** 1655
Glyburide **3:** 1656
Glycerogelatin **4:** 2104
Glycerol **2:** 782; **4:** 2023
Glycerol trinitrate **2:** 706
Glycerylmonostearate **4:** 2103
Glyceryl triacetate **4:** 1731
Glyciclamide **3:** 1658
Glycilax **2:** 782
Glycine **2:** 329
Glycopeptide antibiotics
   economic aspects **3:** 1094
α_2 HS Glycoprotein **1:** 252
8S α_3 Glycoprotein **1:** 252
9.5S α_1 Glycoprotein **1:** 252
Glycoprotein Hormones **3:** 1544
β_2 Glycoprotein I **1:** 252
β_2 Glycoprotein III **1:** 252
Glycopyrronium bromide **2:** 694
Glycylpressin **3:** 1402
Glymidine **3:** 1657
Glyphenarsine **3:** 1150
Glypressin **3:** 1402
Glyset **3:** 1663
Glysolax **2:** 782
Glyzac **2:** 624
Glyzan **2:** 624

GMP **1:** 266; **4:** 2138
GnRH **3:** 1544
Goeckerman regimen **4:** 1728
Gold
   as antirheumatic agents **4:** 1702
Golgi-method **2:** 310
Gonadal Steroids **3:** 1551
Gonadoliberin **3:** 1348
Gonadoliberin analogs **4:** 1931
Gonadorelin **3:** 1352
Gonadorelin analogs **4:** 1931
Gonadotropin releasing hormone **3:** 1544
Gonadotropins **3:** 1348, 1355, 1544
Gonadotropin secretion **3:** 1541
Gonads hormones **3:** 1350
Gonan **3:** 1321
Gonorrhea **4:** 1784
Good clinical practice **4:** 2138
Good clinical procedures **4:** 2106
Good laboratory practice **4:** 2138
Good laboratory procedures **4:** 2106
Good manufacturing practice **1:** 266; **4:** 2138
Good manufacturing practices **4:** 2106
Gorgostane **3:** 1323
Gorgosterol **3:** 1325
Goserelin **4:** 1932, 1933
   annual sales **4:** 1939
Gossypol **3:** 1514
Gout **4:** 1674
Govil **2:** 700
GP Daily Cleaner **4:** 2040
GPI-1046 **2:** 580
G-protein **2:** 688
GR 113808 **2:** 812
GR 125487 **2:** 807
GR 32191 **4:** 2054
GR 38032F **2:** 800
GR 63799 **4:** 2054
Gradient **1:** 209
Gram's stain **3:** 955
Gramalil **2:** 606
Gramicidin **4:** 2010
Gramicidins **3:** 1004, 057
Gram-negative bacterial cell envelope **4:** 1757
Grandaxin **2:** 644
Granisetron **2:** 800
Granulation **4:** 2095
Granules **4:** 2094
Granulocyte-colony stimulating factor **1:** 243
Granulocyte macrophage-colony stimulating factor **1:** 243
Granulocytes **1:** 245
Gratusminal **2:** 506
Greosin **3:** 1242
Gricin **3:** 1242
Grifulin **3:** 1242
Grifulvin V **3:** 1242

Gripponyl **2:** 350
Grisactin **3:** 1242
Griseo **3:** 1242
Griseoderm **3:** 1242
Griseofulvin **3:** 1009, 1069, 1241; **4:** 1731
Griseomed **3:** 1242
Griseostatin **3:** 1242
Griseo von ct **3:** 1242
Griséfuline **3:** 1242
Grisfulvin **3:** 1242
Grisol **3:** 1242
Grisovin **3:** 1242
Grisovina Fp **3:** 1242
Grisovin-FP **3:** 1242
Grivate **3:** 1242
Gromazol **3:** 1217
Grorm **3:** 1437
Growth Factors **3:** 1549
Growth hormone **3:** 1434, 435, 545
Growth hormone cascade **3:** 1430
Growth Hormone Releasing **3:** 1430
Growth Hormone Releasing Hormone **3:** 1432, 549
GRP **3:** 1416
Grysio **3:** 1242
g-Strophanthin **1:** 190
GT **3:** 1418
Guanabenz **1:** 84
Guanethidine **1:** 86
Guanfacine **1:** 83
Guar gum **3:** 1664
Guethine **1:** 86
Guinea Worm **3:** 1276
Gutron **1:** 63
Gycerol **4:** 2033
Gyki-13504 **4:** 1924
Gynaekosid **3:** 1598
Gynäsan **3:** 1583
Gyne-Lotremin **3:** 1217
Gyno-Daktar **3:** 1210
Gyno-Daktarin **3:** 1210
Gynoestryl **3:** 1583
Gyno-Monistat **3:** 1210
Gyno-Myfungar **3:** 1225
Gynomyk **3:** 1215
Gyno-Pevaryl **3:** 1219
Gynorest **3:** 1597
Gyno-Terazol **3:** 1230
Gyno-Travogen **3:** 1222
Gyno-Trosyd **3:** 1231
Gyonett **3:** 1584

H.93/26 **1:** 156
Hachimycin **3:** 1066
Hageman factor **4:** 1672
Hagemann factor **1:** 257
Halan **2:** 440
Halazepam **2:** 642

Halbmond **2:** 496
Halcion **2:** 525
Haldol **2:** 598
Haldrone **3:** 1632
Halobetasol propionate **4:** 1727
9α-Halocorticosteroids **3:** 1619
Halofantrine **3:** 1165
Halofantrine-β-glycerophosphate **3:** 1166
Halojust **2:** 598
Haloperidol **2:** 453, 598
Halopidol **2:** 598
Haloprogin **4:** 1731
Halosten **2:** 598
Halotestin **3:** 1590
Halothane **2:** 439
Halovis **2:** 440
Haloxazolam **2:** 515, 524
Haltran **4:** 1688
Haptoglobin **1:** 252
HATCHER method **1:** 185
Havlane **2:** 526
HCG **3:** 1546
HDL **1:** 19
Heart failure **1:** 173
Heartgard-30 **4:** 2173
Heart rhythm **1:** 115
Heavy-metal complexes
    cancer therapy **4:** 1918
Hebe **3:** 1278
Hebucol **2:** 752
Helfergin **2:** 652
Helicobacter pylori **2:** 733
Hellebrigenin **1:** 179, 184
Helmazine **3:** 1284
Helmex **3:** 1285
Helminth infections **3:** 1265
Helminth vaccines **4:** 1805
Helper cells **1:** 247
Helvolic acid **3:** 1327
Hematocrit **1:** 2391
    and blood viscosity **1:** 45
Hematopoiesis **1:** 238
Hematopoietic stem cells
    therapeutic use **1:** 272
Hemodilution **1:** 46
Hemoflagellates **3:** 1146
Hemoglobin **1:** 239, 242
Hemoglobin-derived oxygen carriers **1:** 273
Hemopexin **1:** 252
Hemophilias **1:** 274
Hemophilus influenzae type b vaccines **4:** 1785
Hemorrhage
    subarachvoid **1:** 149
Hemosporidians **3:** 1159
Hepadial **2:** 751
Hepadnaviridae **3:** 1174
Hepalande **2:** 752

Heparaxal **2:** 750
Hepata **2:** 750
Hepatitis A **3:** 1174
Hepatitis B vaccine **4:** 1794
Hepatitis B virus **3:** 1174
Hepatitis prophylaxis **4:** 1826, 1828
Hepatofalk **2:** 770
Heptabarb **2:** 504, 508
Heptabarbital **2:** 508
Heptabarbitone **2:** 508
17β-Heptanoyeoxy-4-androsten-3-one **3:** 1590
Herbesser **1:** 132, 213
SD-Hermal **3:** 1217
Herniocid **3:** 1238
Heroin **2:** 391
Herpes
  treatment with monoclonal antibodies **4:** 1843
Herpesviridae **3:** 1174
Herpes zoster **4:** 1799
Herpex **4:** 2015
Hetacillin **3:** 1024
Hetacilling **3:** 962
Heterohybridomas **4:** 1982
Heterophyes heterophyes **3:** 1269
Heterotopic disorders **1:** 117
Hetrazan **3:** 1279
Hexachlorophene **3:** 1254
2-[[(Hexadecyloxy)-hydroxyphosphinyl]oxy]-N,N,N-trimethylethanaminium hydroxide **4:** 1938
Hexadecylphosphocholine **4:** 1938
1-Hexadecylpyridinium chloride monohydrate **4:** 2038
Hexadecyltrimethylammonium bromides **4:** 2038
2,4-Hexadienoic acid **4:** 2039
Hexadilat **1:** 210
Hexafluorenium bromide **2:** 674
(2α,6α,11R*)-1,2,3,4,5,6-Hexahydro-6,11-dimethyl-3-(3-methyl-2-butenyl)-2,6-methano-3-benzazocin-8-ol **2:** 419
[2-(Hexahydro-1(2H)-azocinyl)ethyl]guanidine **1:** 86
10-{3-[Hexahydro-4-(2-hydroxyethyl)-1,4-diazepine-1-yl]propyl}-2-trifluoromethyl-10H-phenothiazine **2:** 595
1,3,6,7,8,9-Hexahydro-5-phenyl-2H-[1]benzothieno[2,3-e]-1,4-diazepin-2-one **2:** 642
(3aS-cis)-1,2,3,3a,8,8a-Hexahydro-1,3a,8-trimethylpyrrolo[2,3-b]-indol-5-ol methyl carbamate **2:** 653
2,3,3′,4,4′,5′-Hexahydroxybenzophenone **2:** 657
Hexamethonium **1:** 84
Hexamethylenebis(9-fluorenyldimethylammonium bromide) **2:** 674
Hexamethylmelamine **4:** 1896
N,N,N',N',N'',N''-Hexamethyl-1,3,5- triazine-2,4,6-triamine **4:** 1896
Hexanastab **2:** 445
17-Hexanoyloxy-19-nor-4-pregnene-3,20-dione **3:** 1597

17-Hexanoyloxy-4-pregnene-3,20-dione **3:** 1596
Hexapropymate **2:** 500
Hexa-3-pyridinecarboxylate **1:** 23
Hexasodium 8,8′-[carbonylbis[imino-3,1-phenylenecarbonylimino(4-methyl-3,1-phenylene)carbonylimino]]bis-1,3,5-naphthalenetrisulfonate **3:** 1286
Hexestrol **3:** 1585
Hexobarbital **2:** 444, 504, 511
Hexocyclium methylsulfate **2:** 694
Hexoprenaline **2:** 868
Hextol **2:** 657
Hexylcaine hydrochloride **4:** 2003
HGH **3:** 1545
HGPRT **4:** 1883
Hi-Alarzin **3:** 1248
Hiberna **2:** 921
High-ceiling diuretics **1:** 296
High density lipoproteins **1:** 19
FVIII High purity concentrates **1:** 269
Hilactan **1:** 209
Himecol **2:** 701
Hindbrain **2:** 316
1-(1H-Indol-4-yloxy)-3-[(1-methylethyl)amino]-2-propanol **1:** 92
H-Insulin **3:** 1444
Hipnosedon **2:** 520
Hipotensor Oftalmico **4:** 2022
Hipsal **2:** 518
His bundle electrography **1:** 116
Histamine **2:** 329, 902; **4:** 1719
Histamine receptors **2:** 903
Histamine Synthesis Inhibitoris **2:** 933
Histantin **2:** 914
Histapyrrodine **2:** 906
Histavet-P **4:** 2184
Histmanol **2:** 929
Histrelin **4:** 1932
HIV **3:** 1291; **4:** 1832
HIV-1 **3:** 1292
HIV-2 **3:** 1292
Hivid **3:** 1297
HIV Protease Inhibitors **3:** 1304
HIV Therapeutics **3:** 1291
HMG-CoA Reductase Inhibitors **1:** 27
HMS Liquifilm **4:** 2018
HN2 **4:** 1886
HOE 013 **4:** 1934
Hoe–280 **3:** 1126
Hoe 296 **3:** 1244
HOE 296b **3:** 1244
Hoe 777 **3:** 1636
Hoggar N **2:** 909
Hokunalin **2:** 881
Holarrhidine **3:** 1333
Holarrhimine **3:** 1333
Hollow-fiber reactors **4:** 1991
Holothurigenins **3:** 1335

2261

Holoxan **4**: 1890
Homatrine **4**: 2028
Homatropine HBr **4**: 2028
Homatropine hydrobromide **4**: 2028
Homatropine methylbromide **2**: 691
Homofenazine **2**: 595
Homogeneous suspension culture **4**: 1990
Hookworm disease vaccines **4**: 1809
Hookworms **3**: 1273
Horizon **2**: 678
Hormomed **3**: 1583
Hormone Antagonists **3**: 1554
Hormone cancer therapy **4**: 1922
Hormone-Producing Systems **3**: 1537
Hormones **3**: 1533
   influence on sleep **2**: 491
   steroid **3**: 1330
Hormone systems
   hierarchy **3**: 1539
Hospital-acquired bacterial infections **4**: 1776
Hostacain **2**: 548
Hostacortin **3**: 1618
Hostacortin-H **3**: 1618
Hostaginan **1**: 208
HP-029 **2**: 659
HP 873 **2**: 609
HPA 23 **3**: 1189
HSV **3**: 1174
Human adenoviruses **3**: 1174
Human growth hormone **3**: 1545
Human monoclonal antibodies **4**: 1987
Human pharmacological testing **4**: 2134
Human T-lymphotrophic viruses **3**: 1174
Humatrop **3**: 1437
Humegon **3**: 1357
Humorsol **4**: 2025
Humoryl **2**: 626
Hurricaine **2**: 542
Hybridoma clones **4**: 1987
Hybridomas **4**: 1982
Hycanthone **3**: 1267
Hydac **1**: 211
Hydantin **2**: 474
Hydantol **2**: 474
Hydatidosis vaccines **4**: 1811
Hydergine **2**: 655
Hydiphen **2**: 612
Hydralazine **1**: 15, 99
Hydralazine HCl **1**: 99
Hydrap-ES **1**: 99
α-Hydrazino-3,4-dihydroxy-α-methylbenzhydrocinnamic acid monohydrate **2**: 569
Hydrochloride **2**: 564
Hydrochlorothiazide **1**: 71, 293
Hydrocodone **2**: 393
Hydrocolloid Matrix **4**: 2115

Hydrocort **3**: 1607
Hydrocortisone **4**: 2016
Hydrocortone **4**: 2017
HydroDiuril **1**: 72, 293
Hydroflumethiazide **1**: 294
Hydrogen peroxide **4**: 2039
Hydromedin **1**: 297
Hydromorphone **2**: 394
Hydro-Saluric **1**: 293
Hydroxaprogesterone acetate **3**: 1596
17α-Hydroxaprogesterone caproate **3**: 1596
Hydroxidione **2**: 452
Hydroxyamphetamine hydrobromide **4**: 2027
17β-Hydroxy-5α-androstan-3-one **3**: 1594
4-Hydroxyandrost-4-ene-3,17-dione **4**: 1924
17β-Hydroxy-4-androsten-3-one **3**: 1590
2-hydroxybenzamide **2**: 348
Hydroxychloroquine **4**: 1704
Hydroxychloroquine sulfate **3**: 1163
17-Hydroxycorticosterone **4**: 2016
17α-Hydroxy-3-cyclopentyl-oxy-3,5-pregnadien-20-one **3**: 1596
3-Hydroxydiazepam **2**: 523
14-Hydroxydihydrocodeinone **2**: 396
14-Hydroxydihydromorphinone **2**: 397
L-3-[4-(4-Hydroxy-3,5-diiodophenoxy)-3,5-diiodophenyl]alanine **3**: 1523
L-3-[4-(4-Hydroxy-3,5-diiodophenoxy)-3,5-diiodophenyl]alanine sodium **3**: 1523
26-Hydroxy-β-ecdysone **3**: 1332
3-Hydroxy-1,3,5(10)-estratrien-17-one **3**: 1583
17β-hydroxy-4-estren-3-one **3**: 1592
2-(2-hydroxyethoxy)ethyl 2-(3-trifluoromethylphenylamino) benzoate **2**: 350
9-(2-Hydroxyethoxymethyl)guanine **2**: 763; **3**: 1177
Hydroxyethylamine Isosteres HIV protease inhibitors **3**: 1305
Hydroxyethyl cellulose **4**: 2033, 2103
(2-Hydroxyethyl)dimethyl-(1-methyl-2-phenothiazi-10-ylethyl) ammonium chloride **2**: 919
Hydroxyethylene Isosteres HIV protease inhibitors **3**: 1310
2,2′-[(2-Hydroxyethyl)-imino]bis[N-(1,1-dimethyl-2-phenylethyl)-N-methylacetamide] **2**: 549
N-(2-Hydroxyethyl)nicotinamide nitrate (ester) **1**: 230
2-hydroxyethyl nicotinate 2-(4-chlorophenoxy)-2-methylpropionate (ester) **1**: 26
9-{3-[4-(2-Hydroxyethyl)-1-piperazinyl]propylidene}-2-trifluoromethyl-9H-thioxanthene **2**: 597
10-[3-[4-(2-Hydroxyethyl)-1-piperazinyl]propyl]-2-propionyl-10H-phenothiazine **2**: 595
10-{3-[4-(2-Hydroxyethyl)-1-piperazinyl]propyl}-2-trifluoromethyl-10H-phenothiazine **2**: 595
10-{3-[4-(2-Hydroxyethyl)piperidino]propyl}-N,N-dimethyl-10H-phenothiazine-2-sulfonamide **2**: 593

Hydroxyethylpromethazine **2:** 919
8-Hydroxy-5-[1-hydroxy-2-(isopropylamino)butyl]-2(1H)-quinolinone **2:** 884
(RR,SS)-(±)-N-[2-hydroxy-5-[1-hydroxy-2-[[2-(4-methoxyphenyl)-1-methylethyl]amino]ethyl]phenyl]formamide **2:** 882
2-Hydroxy-5-{1-hydroxy-2-[(1-methyl-3-phenylpropyl)amino]- ethyl}benzamide **1:** 94
(±)-N-(2-[2-Hydroxy-3-(4-hydroxyphenoxy)propyl]amino-ethyl)-4-morpholinecarboxamide **1:** 193
6-{[(Hydroxyimino)phenyl]methyl}-1-[(1-methylethyl)sulfonyl]-1H-benzimidazol-2-amine **2:** 1185
4-(4-hydroxy-3-Iodophenoxy)-3,5-diiodophenylacetic acid **3:** 1523
L-3-[4-(4-Hydroxy-3-iodophenoxy)-3,5-diiodophenyl]alanine **3:** 1523
8-Hydroxy-7-iodo-5-quinolinesulfonic acid **3:** 1171
(±)-4′-(1-Hydroxy-2-isopropylaminoethyl)methanesulfoanilide **1:** 130
4′-[1-hydroxy-2-(isopropylamino)ethyl]methanesulfonamide **1:** 159
4-[2-Hydroxy-3-(isopropylamino)-propoxy]indole **1:** 157
4-(2-Hydroxy-3-isopropylaminopropoxy)-2,3,6-trimethylphenyl acetate **4:** 2021
Hydroxymagnesium aluminate **2:** 726
17β-Hydroxy-17α-methyl-1,4-androstadien-3-one **3:** 1593
17β-Hydroxy-1α-methyl-5α-androstan-3-one **3:** 1590
17β-Hydroxy-17α-methyl-4-androsten-3-one **3:** 1590
[7(S)-(1α,2β,4β,5α,7β)]-α-(Hydroxymethyl)benzeneacetic acid **2:** 562
1-3-Bis(hydroxymethyl)-2-benzimidazolinethione **3:** 1530
17β-Hydroxy-17α-methyl-1α,7α-bis(thioacetyl)-4-androsten-3-one **3:** 1593
2-Hydroxymethyl-clavam **3:** 971
7-Hydroxy-4-methylcoumarin **2:** 700, 753
17β-Hydroxy-17α-methyl-4-estren-3-one **3:** 1598
(−)-(S)-4-[2-Hydroxy-3-[(1-methylethyl)amino]propoxy]-phenol **1:** 193
17β-Hydroxy-17α-methyl-2-hydroxymethylen-5α-androstan-3-one **3:** 1590
bis(4-Hydroxymethyl-5-hydroxy-6-methyl-3-pyridylmethyl)disulfide **2:** 657
2-Hydroxymethyl-6-(1-hydroxy-2-tert-butylaminoethyl)-pyridine-3-ol **1:** 194
3-Hydroxy-α-methyl-L-tyrosine **1:** 81
[5S-(5R*,8R*,10R*,11R*)]-10-hydroxy-2-methyl-5-(1-methylethyl) **3:** 1310
4-Hydroxy-2-methyl-N-(5-methyl-3-isoxazolyl)-2H-1,2-benzothiazine-3-carboxamide 1,1-dioxide **2:** 374
4-Hydroxy-2-methyl-N-(5-methyl-2-thiazoyl)-2H-1,2-benzothiazine-3-carboxamide 1,1-dioxide **2:** 382
N-(Hydroxymethyl)nicotinamide **2:** 752
17β-Hydroxy-17α-methyl-2-oxa-5α-androstan-3-one **3:** 1594

Endo-(±)-α-(hydroxymethyl)phenylacetic acid 8-methyl-8-azabicyclo[3.2.1]oct-3-yl ester **2:** 891
5-[1-Hydroxy-2-[(1-methyl-3-phenylpropyl)amino]ethyl]salicylamide hydrochloride **1:** 155
11β-Hydroxy-6α-methylprogesterone **4:** 2018
4-Hydroxy-2-methyl-N-2-pyridinyl-2H-1,2-benzothiazine-3-carboxamide 1,1-dioxide **2:** 377; **4:** 1696
4-hydroxy-2-methyl-N-2-pyridinyl-2H-thieno[2,3-e]-thiazine-3-carboxamide 1,1-dioxide **2:** 377
Hydroxymycin **3:** 984
1-Hydroxy-2-naphthoic acid salt **2:** 875
(−)-3-Hydroxy-N-methylmorphinan **2:** 394
Hydroxyphenamate **2:** 645
L-3-[4-(4-Hydroxyphenoxy)-3,5-diiodophenyl]alanine **3:** 1524
N-(4-Hydroxyphenyl)acetamide **2:** 348
(±)-4-Hydroxy α^1-[[[6-(4-phenylbutoxy)hexyl]amino]-methyl]-1,3-benzenedimethanol **2:** 875
2-Hydroxy-2-phenylbutyl carbamate **2:** 645
D-(p-Hydroxyphenyl)glycine **3:** 1011
3,3-Bis-(4-hydroxyphenyl)-1-(3H)-isobenzofuranon **2:** 787
(S)-1-(3-Hydroxyphenyl)-2-methylaminoethanol hydrochloride **4:** 2026, 2029
N-(3-hydroxyphenyl)-N-(1-methylpiperidin-4-yl)propionamide **2:** 399
(±)-4-[2-[3-(4-Hydroxyphenyl)-1-methylpropyl]aminoethyl]-1,2-benzenediol **1:** 192
2-(2-Hydroxyphenyl)-1,3,4-oxadiazole **2:** 529
(±)-4-[3-Hydroxy-3-phenyl-3-(2-thienyl)propyl]-4-methylmorpholinium iodide **2:** 697
16α-Hydroxypregnolone **3:** 1565
3α-hydroxy-5α-pregnane-11,20-dione **2:** 452
17α-Hydroxyprogesterone **3:** 1565
17α-Hydroxyprogesterone caproate **3:** 1504
5,5′-[(2-Hydroxy-1,3-propanediyl)bis(oxy)]bis[4-oxo-4H-1-benzopyran-2-carboxylic acid] disodium salt **2:** 934
17β-Hydroxy-6α-(1-propinyl)-4androsten-3-one **3:** 1598
β-Hydroxypropionyl **3:** 971
(±)-2′-(2-Hydroxy-3-propylaminopropoxy)-3-phenylpropiophenone **1:** 129
Hydroxypropyl cellulose **4:** 2033, 2103
Hydroxypropyl methyl cellulose **4:** 2034, 2103
4-Hydroxy-2-oxo-1-pyrrolidineacetamide **2:** 651
Hydroxyquinine **3:** 1163
8-Hydroxyquinolines **3:** 1169
3α-Hydroxyspiro[1αH,5αH-nortropane-8,1′-pyrrolidinium] chloride benzilate **2:** 692
Hydroxyurea **4:** 1728
Hydroxyzine **2:** 646, 915; **4:** 1720
Hy-Flow **4:** 2034
Hygromycin B **3:** 1047
Hygroton **1:** 72, 295
Hymecromone **2:** 700, 753
Hymenolepis nana **3:** 1271

hyoscine butylbromide **2:** 691
(±)-Hyoscyamine **2:** 690, 741, 891
Hyosin **2:** 691
Hyperimmune globulins **4:** 1763
Hyperimmune globulins preparations **4:** 1825
Hyperlipidemia **1:** 19
Hyperlipoproteinemias **1:** 19
Hyperosmotic agents
  antiglaumatons **4:** 2023
Hyperstat I.V. **1:** 100
Hypertensin **3:** 1391
Hypertension **1:** 70, 45
Hyperthyreosis **3:** 1520
Hyperthyroidism **3:** 1557
Hyperysin **2:** 501
Hypnodine **2:** 496
Hypnodorm **2:** 520
Hypnol **2:** 509
Hypnolone **2:** 506
Hypnomidate **2:** 449
Hypnostan **2:** 445
Hypnotics **2:** 489
  classification **2:** 494
  influence on sleep **2:** 492
Hypnovel **2:** 451, 525
Hypnox **2:** 506
Hypodermic tablets **4:** 2093
Hypogammaglobulinemia **4:** 1826
HypoTears **4:** 2034
Hypotension **1:** 59
  orthostatic **1:** 146
Hypothalamic centers **3:** 1539
Hypothalamic Hormones **3:** 1542
Hypothyreosis **3:** 1520
Hypothyroidism **3:** 1557
Hypovase **1:** 87
Hypoxanthine guanine phosphoribosyltransferase **4:** 1883
Hyprenan **1:** 193
Hyptor **2:** 514

Ibidomide hydrochloride **1:** 155
Ibinolo **1:** 152
Ibopamine **1:** 192
Ibufen **4:** 1688
Ibunet **4:** 1688
Ibuprofen **2:** 363; **4:** 1687
Icaden **3:** 1222
ICI-176334 **4:** 1930
ICI-182780 **4:** 1924
ICI 204,636 **2:** 609
ICI 45 520 **1:** 159
ICI 46474 **4:** 1922
I.C.I. 50 172 **1:** 158
ICI 66 082 **1:** 152
ICI-D1033 **4:** 1926
Icramin **2:** 703

ICS 205-930 **2:** 800
Icteryl **2:** 752
Idamycin **4:** 1898
Idarubicin **4:** 1898
Ideal drug delivery systems **4:** 2110
Idebenone **2:** 654
IDL **1:** 19
Idom **2:** 613
Idoxifene **4:** 1924
Idoxuidine **3:** 1178
Idoxuridine **4:** 2014
Idrogesten **3:** 1596
Idulian **2:** 923
Ifetroban **4:** 2063
Ifex **4:** 1890
IFNα-2 a
  clinical studies **4:** 1977
IFNα-2 b amino acid **4:** 1965
IFNs see interferons **4:** 1957
Ifosfamide **4:** 1890
IgA **1:** 252; **4:** 1759
IgD **1:** 252; **4:** 1760
IgE **1:** 252; **4:** 1760
IGF-I **3:** 1439
IgG **1:** 252, 264
IgM **1:** 252; **4:** 1759
IK 20 349 **3:** 1230
Ikaran **1:** 66
IL-1 **4:** 1673
Ilmofosine **4:** 1938
Ilomedin **1:** 38; **4:** 2061
Ilomedine **1:** 38
Iloprost **1:** 38; **4:** 2054, 2061
Ilvin **2:** 911
Imap **2:** 601
Imavate **2:** 614
Imbun **2:** 365
Imeson **2:** 518
IMI-28 **4:** 1897
IMI-30 **4:** 1898
Imidazole antimyotics **3:** 1201
Imidazothiazoles
  veterinary **4:** 2175
Imidazyl **4:** 2029
Imipenem **3:** 1035
Imipramine **2:** 613
Immenoctal **2:** 511
Immobilized cell cultures **4:** 1990
Immobilon **2:** 393
Immune reaction **1:** 248
Immune Response **4:** 1760
Immune serum globulin **4:** 1829
Immune serum globulin preparation **4:** 1823
Immune stimulation
  interferons **4:** 1962
Immunization **4:** 1755

Immunoassays
  monoclonal antibodies **4:** 1995
Immunogen **4:** 1757
Immunogenicity
  of blood components **1:** 281
Immunoglobulin A **4:** 1759
Immunoglobulin D **4:** 1760
Immunoglobulin E **4:** 1760
Immunoglobulin G **1:** 44; **4:** 1759
Immunoglobulin G subclasses **4:** 1758
Immunoglobulin M **4:** 1759
Immunoglobulins **1:** 262; **4:** 1756
  therapeutic use **1:** 278
Immunoglobulins preparations **4:** 1822
Immunophilins
  in Parkinsonism treatment **2:** 579
Immunoregulants **4:** 1707
Immunosorbent chromatography **4:** 1970
Immunosuppressive agents
  ophthamological **4:** 2019
Immunotherapy **2:** 935; **4:** 1821
Immunotherapy and vaccines **4:** 1753
Imodium **2:** 409
Imovance **2:** 529
Imovane **2:** 529
Imperan **2:** 745
Impromen **2:** 598
Impurities **4:** 2141
Impurities solvents **4:** 2141
Inacid **4:** 1680
Inaktin **2:** 446
Inapetyl **2:** 461
I-Naphline **4:** 2029
Inapsine **2:** 600
Incazan **2:** 625
Incoran **1:** 208
Indacrinone **1:** 77
Indanal **4:** 1693
Indapamide **1:** 71, 73, 295
Indeloxazine **2:** 658
2-[(Inden-7-yloxy)methyl]morpholine hydrochloride **2:** 658
Inderal **1:** 90, 29
Inderex **1:** 90, 59
Indinavir **3:** 1312
Indobene **4:** 1680
Indocid **2:** 358
Indocin **2:** 358; **4:** 1680
Indolamines hormones **3:** 1550
N-[1-[2-(1H-Indol-3-yl)ethyl]-4-piperidinyl]benzamide **1:** 88
Indomethacin **2:** 355, 706; **4:** 1679
Indoramin **1:** 88
Indorm **2:** 922
Industry
  pharmaceutical **1:** 3
Infantile paralysis **4:** 1792

Infectokrupp Inhal **1:** 60
Infectrol **4:** 2008, 2011, 2018
Infiltrative anesthetics **2:** 539
Inflamase **4:** 2017
Inflammation
  treatment **2:** 341
Inflammatory response to injury **4:** 1671
Influenza
  treatment with monoclonal antibodies **4:** 1841, 1843
Influenza A **3:** 1174
Influenza vaccine **4:** 1798
Infumorph **2:** 390
INH **3:** 1140
Inhacort **2:** 894
Inhalants **4:** 2101
Inhalation **4:** 2085
Inhalation anesthetics **2:** 436, 438
Inhalations **4:** 2088
Inhibin **3:** 1348, 513
Inhibins **3:** 1358
Inhibitor deficiencies **1:** 277
Inhibitory ionotropic action **2:** 334
Inhibostamin **2:** 933
Injection **4:** 2081
Innovar **2:** 454
Innovar-Vet **4:** 2183
Inocore **1:** 195
Inoculum **3:** 1080
Inofal **2:** 592
Inokosterone **3:** 1331
Inolin **2:** 868
Inopamil **1:** 193
Inositol-1,4,5-triphosphate **2:** 688
Inositol nicotinate **1:** 23
Inotropic agents **1:** 175
Inotropic drugs **1:** 191
Inovan **1:** 192
Insidon **2:** 619
Insomin **2:** 518
Insomnia **2:** 502
Insom Rapido **2:** 509
Inspections **4:** 2150
Installation qualification **4:** 2143
Insufflations **4:** 2092
Insulin **2:** 772; **3:** 1442, 548
Insulin-like growth factor **3:** 1439
Insulin secretion stimulators **3:** 1654
Insulton **2:** 475
Insumin **2:** 522
Intal **2:** 896, 934
Integrin **2:** 609
Intenkordin **1:** 231
Intensain **1:** 231
Interferon **1:** 274; **2:** 763
  clinical studies **4:** 1975
Interferon classification **4:** 1959

Interferon purification **4**: 1968
Interferon quality specifications **4**: 1972
Interferon receptors **4**: 1963
Interferons **1**: 243; **4**: 1957
Interferon solutions **4**: 1968
Interferon structure **4**: 1964
Interleukin **4**: 1673
Interleukin-1 **1**: 243
Intermediate density lipoproteins **1**: 19
International pharmacopeia **4**: 2153
Interneurons **2**: 309
Inter-α-trypsin inhibitor **1**: 252
Intestinal flagellates **3**: 1155
Intocostrin-T **2**: 672
Intra-arterial injection **4**: 2082
Intra-atrial disturbances **1**: 118
Intracardiac pressure **1**: 185
Intracisternal injection **4**: 2083
Intradermal administration **4**: 2083
Intralgin **2**: 348
Intramuscular injection **4**: 2082
Intraspinal injections **4**: 2082
Intrathecal injection **4**: 2082
Intrauterine devices
  for contraception **3**: 1506
Intravaginal administration **4**: 2085
Intraval **2**: 445
Intravenous anesthetics **2**: 437, 443
Intravenous Immunoglobulin in prophylaxis and
  therapy **4**: 1830
Intravenous immunoglobulin preparation **4**: 1824
Intravenous injection **4**: 2082
Intraventricular disturbances **1**: 118
Intrinsic sympathomimetic activity **1**: 139, 42, 226
Introcar **1**: 210
Intron A **4**: 1958
  clinical studies **4**: 1975
Intron A amino acid **4**: 1965
Intropin **1**: 192
investigations **4**: 2136
Invirase **3**: 1307
Iodamoeba butschlii **3**: 1169
Iodeto de fosfolina **4**: 2025
Iodination Inhibitors **3**: 1525
Iodine supply **3**: 1520
Iodization Inhibitors **3**: 1525
Iodochlorhydroxyquin **3**: 1170
Iodochlorhydroxyquinoline **4**: 1732
Iodogorgoic acid **3**: 1558
Iodoquinol **3**: 1170; **4**: 1732
L-3-Iodotyrosine **3**: 1558
Ionamin **2**: 467
Ion channels **2**: 321
Ion-exchange dosage forms **4**: 2116
Ion-Exchange Resins **1**: 21
Ionophore antibiotics
  veterinary **4**: 2170

Ionotropic action **2**: 333
IP_3 **2**: 688
Ipamox **1**: 73
Ipec **3**: 1210
Ipersed **2**: 518
Ipoflogin **4**: 2018
Ipolab **1**: 155
Iporal **1**: 86
Ipral Calcium **2**: 507
Ipral Sodium **2**: 507
Ipratropium bromide **2**: 891
I-Pred **4**: 2017
Iproclozide **2**: 622
Iproniazid **2**: 623
Ipronid **2**: 623
Iproplatin **4**: 1921
Iralukast sodium **4**: 2065
Irehdiamine **3**: 1333
Irinotecan **4**: 1914
Irritren **2**: 358
Irrodan **1**: 50
Irrorin **1**: 208
ISA **1**: 139, 42, 226
ISA drugs **1**: 141
Ischemic heart disease **1**: 144
Iscover **1**: 43
Islet amyloid polypeptide **3**: 1369
Ismelin **1**: 86
ISMO **1**: 224
Ismotic **4**: 2023
Isoamytal **2**: 509
Isobarb **2**: 509
Isobarin **1**: 86
Isobec **2**: 507
Isobromyl **2**: 501
N-(3-Isobutoxy-2-pyrrolidin-1-ylpropyl)-N-phenylbenzy-
  lamine **1**: 131
2-(4-Isobutylphenyl)propionic acid **2**: 363
Isocarboxazid **2**: 623
Isochinol **2**: 551
Isoconazole **3**: 1222
Isodin **2**: 522
Isoelectric focusing **4**: 1971
Isoetharine **2**: 866
Isoflupredone **4**: 2178
Isoflurane **2**: 442
Isoflurophate **4**: 2026
Isofosphamide **4**: 1890
Isogyn Ginecologico **3**: 1222
Isoket **1**: 224
Isoleucine-gramicidines **3**: 1005
Iso Mack **1**: 224
Isomenyl **2**: 866
Isonal **2**: 507
Isonefrine **4**: 2026, 2029
Isoniazid **3**: 1139
Isonicotinic 2-isopropylhydrazide **2**: 623

Isonicotinic acid 2-(2-benzylcarbamoylethyl)hydrazide **2:** 623
Isonicotinic acid hydrazide **3:** 1139
Isopenicillin **3:** 960
Isoprenaline **2:** 865
Isopropamide iodide **2:** 694
(±) -1-[[α-(2-Isopropoxyethoxy)-*p*-tolyl]oxy]-3-(isopropylamino)-2-propanol **1:** 153
1-(Isopropylamino)-2-hydroxy-3-[*o*-(allyloxy)phenoxy]-propane **1:** 157
1-(Isopropylamino)-3-(1-naphthyloxy)-2-propanol **1:** 90, 29, 58
(±)-1-(Isopropylamino)-3-[*p*-(cyclopropylmethoxyethyl)-phenoxy]-2-propanol **1:** 153
(±)-1-(Isopropylamino)-3-[*p*-(β-methoxyethyl)phenoxy]-2-propanol **1:** 156
1-isopropyl-4,4-diphenylpiperidine **2:** 564
Isopropyl 2-methoxyethyl 1,4-dihydro-2,6-dimethyl-4-(3-nitrophenyl)-3,5-pyridinedicarboxylate **2:** 659
5-Isopropyl-2-methyl-phenoxyethyl)-diethylamine **2:** 903
1-(2-isopropyl-4-pentenoyl)-urea **2:** 501
4-Isopropyl-2-[3-(tri-fluoromethyl)-phenyl]morpholine **2:** 629
Isoproterenol **2:** 865
Isoptin **1:** 97, 31, 208
Isopto Alkaline **4:** 2034
Isopto-Atropine **2:** 690; **4:** 2027
Isopto Carbachol **4:** 2024
Isopto Carpine **4:** 2024
Isopto Cetmide Solution **4:** 2005
Isopto-Flucon **4:** 2018
Isopto Frin **4:** 2026, 2029
Isopto Homatropine **4:** 2028
Isopto Hyoscine **4:** 2028
Isopto Phenylephrine **4:** 2029
Isopto Plain **4:** 2034
Isopto Tears **4:** 2034
Isordil **1:** 224
Isosorbide **4:** 2023
Isosorbide-5-mononitrate **1:** 224
Isosorbide dinitrate **1:** 223
Isosporiasis **3:** 1166
Isosulfazecin **3:** 976
Isoten **1:** 153
Isothipendyl **2:** 920
Isoval **2:** 501
Isovex **2:** 699
Isoxicam **2:** 374
Isradipine **1:** 212
Issium **1:** 209
Istin **1:** 212
Isuprel **2:** 866
Itraconazole **3:** 1203; **4:** 1730
Itrizole **3:** 1204
I-Tropine **4:** 2027

Iturate **2:** 509
IUD **3:** 1506
Ivermectin **3:** 1054, 1272, 1274, 1277, 1279; **4:** 2173
IVIG **4:** 1830
IVIG preparations **4:** 1824
Ivomec **3:** 1280; **4:** 2174
Ixertol **1:** 209

Janimine **2:** 614
Japanese encephalitis vaccine **4:** 1803
Jatroneural **2:** 594
Jatropur **1:** 299
Jatrox **2:** 731
JC **3:** 1174
Jellin **3:** 1632
Jenamazol **3:** 1217
Jenest **3:** 1502
Jervine **3:** 1334
Jexin **2:** 672
JM 216 **4:** 1921
Jodthyrox **3:** 1561
Josamycin **3:** 998, 1053
Jumex **2:** 571, 625
Jumexal **2:** 571, 625
Junin **3:** 1174
Juvenimicin A$_2$ **3:** 999
Juvenimicin A$_4$ **3:** 999

Kaban **3:** 1635
KadeFungin **3:** 1217
Kainair **4:** 2003
Kalgut **1:** 194
Kallidin **3:** 1392
Kanamycin **3:** 985; **4:** 2163
Kanamycin A **3:** 986
Kanamycin B **3:** 986, 1041
Kanamycin sulfate **3:** 1040
Kapanol **2:** 390
Karidium **2:** 644
Karil **3:** 1367
Kasugamycin **3:** 1049
Katadolon **2:** 425
Katoseran **1:** 209
KC 9147 **3:** 1246
K^+ concentrations in neurophysiology **2:** 321
Kebuzone **2:** 372
Kedarcidin **3:** 1013
Kemadrin **2:** 564
Kemi **1:** 159
Kemithal **2:** 447
Kenacort **3:** 1622
Keracaine **4:** 2003
Kerakain **4:** 2003
Keratinocytes **4:** 1714
Kerlone **1:** 153
Keselan **2:** 598
Ketaject **2:** 448

Ketalar **2:** 448
Ketamine **2:** 448; **4:** 2180
Ketanest **2:** 448
Ketaset **2:** 448; **4:** 2180
Ketazol **3:** 1208
Ketazolam **2:** 641
Ketazon **2:** 372
Keteocort **3:** 1618
Ketobemidone **2:** 399
Ketoconazole **3:** 1206; **4:** 1730, 2013
Ketoderm **3:** 1208
Ketogan **2:** 399
Ketoisdin **3:** 1208
Ketonan **3:** 1208
Ketoprofen **2:** 365; **4:** 1689, 2176
Ketoral **3:** 1208
Ketorolac **2:** 379
Ketotifen **2:** 924
Ketrax **3:** 1281
Kevopril **2:** 616
KF17837 **2:** 580
Kidney function **1:** 289
Killer cells **1:** 248
Kilmicen **3:** 1246
Kinidin **1:** 124
Kinins **3:** 1388, 1392, 1547
Kinupril **2:** 616
Kitasamycin **3:** 1052
Klebsiella vaccine **4:** 1777
Klimanosid **3:** 1584
Klimax **3:** 1586
Klodin **1:** 41
Klysma-Sorbit **2:** 785
Klyxenema **2:** 782
KNAFFL-LENZ method **1:** 185
KNI-272 **3:** 1314
Kollateral **2:** 700
Kolpolyn **3:** 1584
Kolpon **3:** 1583
Kolton **2:** 348
Komb-H-Insulin **3:** 1444
Konogen **3:** 1583
Kordafen **1:** 210
Korostatin **3:** 1238
Kredex **1:** 154
Kreislauf Katovit **1:** 64
Kryptocur **3:** 1352
Kryptosterol **3:** 1326
KT2-962 **4:** 2063
Kuilil **2:** 645
KW-2307 **4:** 1912
KW-3285 **4:** 2063
KW-3902 **1:** 301

L 365,260 **2:** 649
LAAM **2:** 405
Labelol **1:** 155

Labetalol **1:** 94, 55
Labrocol **1:** 155
Labrodax **2:** 700
Labyrin **1:** 209
Lacril **4:** 2034, 2039
Lacril R **4:** 2033
Lacrisert **4:** 2034
β-Lactam antibiotics **3:** 953
   as ophthamological agents **4:** 2006
   veterinary **4:** 2160
Lactamin **1:** 208
β-Lactams antibiotics **3:** 959
   economic aspects **3:** 1094
Lactic acid **4:** 1723
Lactic dehydrogenase-x **3:** 1512
Lactose **2:** 785; **4:** 2104
Lactulose **2:** 786
Ladormin **2:** 526
Laevilac **2:** 786
Lagazepam **2:** 518
Lamictal **2:** 483
L-amino acids
   one-letter symbols **3:** 1342
Lamisil **3:** 1251
Lamivudine **3:** 1299
Lamoryl **3:** 1242
Lamotrigine **2:** 482
Lampit **3:** 1152
Lampren **3:** 1144
Lanchloral **2:** 498
Langerhans cells **4:** 1714
Lanicor **1:** 189
Lanitop **1:** 190
Lanosterol **3:** 1326
Lanoxin **1:** 189
Lansoprazole **2:** 714, 719, 720
Lantanon **2:** 621
Laracor **1:** 157
Largactil **2:** 590, 681
Larocord **1:** 210
Larodopa **2:** 569
Laroxyl **2:** 611
Larpose **2:** 522
Lasalocid **3:** 1064; **4:** 2171
Lasilix **1:** 296
Lasinavir **3:** 1314
Lasix **1:** 72, 296
Lassa **3:** 1174
Latamoxef **3:** 970, 034
Latanoprost **4:** 2060
Laubeel **2:** 522
Laughing gas **2:** 438
Laxariston-Granulat **2:** 784
Laxatives **2:** 779
Laxiplant **2:** 783
Laxoberal **2:** 788
Layered tablets **4:** 2097

Lazabemide **2:** 571
LB 46 **1:** 158
LC-65 **4:** 2037
LCB-2853 **4:** 2063
LCMV **3:** 1174
LDH-x **3:** 1512
LDL **1:** 19
LDL receptor **1:** 19
Lecibis **3:** 1278
Lecibral **1:** 210
Lecithin **4:** 2103
Lectopam **2:** 637
Lederfen **4:** 1687
Lederlind **3:** 1238
Leflunomide **4:** 1707
Legalon 140 **2:** 767
Legalon 70 **2:** 767
Legalon Liguidum **2:** 767
Legalon Sil Ampullen **2:** 767
Legatrin **2:** 676
Legionellaceae **3:** 1110
Lehydan **2:** 475
Leishmania **3:** 1151, 153
Leishmaniasis **3:** 1153
Lembrol **2:** 523
Lendorm **2:** 526
Lendormin **2:** 526
Lensine 5 **4:** 2034
Lens Plus Daily Cleaner **4:** 2040
Lens-Wet **4:** 2034, 2037
Lentaron **4:** 1924
Lentizol **2:** 611
Leo **2:** 645
Leopental **2:** 445
Lepinal **2:** 480
Lepinaletten **2:** 506
Leponex **2:** 605
Lepra
    treatment with monoclonal antibodies **4:** 1841
Leprosy **3:** 1138
Leptilan **2:** 477
Leptryl **2:** 496
Lergobine **2:** 917
Lescodil **1:** 210
Lescol **1:** 29
Lesser Intestinal Flukes **3:** 1269
Lethidrone **2:** 419
Lethobarb **2:** 509
Letrozole **4:** 1926
Leucomycin **3:** 1052
Leucomycins **3:** 998
Leukemia **4:** 1843
Leukocytes **1:** 245
    therapeutic use **1:** 272
Leukotriene A_4 **4:** 2047
Leukotriene B_4 **4:** 2047
Leukotriene $C_4 - F_4$ **4:** 2047

Leukotrienes **4:** 2047
    biosynthesis **4:** 2053
    occurrence **4:** 2050
    physical properties **4:** 2049
Leuprorelin **4:** 1932
    annual sales **4:** 1939
Leuprorelin acetate **4:** 1933
Leustatin **4:** 1880
Levallorphan **2:** 421
Levamisole **3:** 1272, 1273, 1280; **4:** 2175
Levanil **2:** 501
Levanxene **2:** 523
Levarterenol **3:** 1562
Levium **2:** 637
Levobunolol hydrochloride **4:** 2021
Levocabastine **2:** 930
Levocon **2:** 930
Levodopa **2:** 568
Levo-Dromoran **2:** 396
Levomepromazine **2:** 590, 591
Levomeprome **2:** 591
Levomethadone **2:** 404
Levomethadyl acetate **2:** 405
Levomezine **2:** 591
Levonorgestrel **3:** 1497, 500, 502
Levonormal **2:** 591
Levophed **1:** 60
Levophed Bitartrate **1:** 60
Levophed special **1:** 60
Levopraid **2:** 606
Levorin A_2 **3:** 1066
Levoroxine **3:** 1561
Levorphanol **2:** 394
Levothroid **3:** 1561
Levothyrox **3:** 1561
Levothyroxine **3:** 1523, 1558
    pharmacology **3:** 1521
Levothyroxine sodium **3:** 1523, 561
Levothyroxine sodium x-water **3:** 1523
Lexomil **2:** 637
Lexotan **2:** 637
Lexotanil **2:** 637
LH **3:** 1545
LHRH **3:** 1512, 544
LHRH analogs **4:** 1931
Librax **2:** 693
Libraxin **2:** 693
Librelease **2:** 636
Libritabs **2:** 636
Librium **2:** 636
Librizan **2:** 636
Lidanil **2:** 592
Lidex **3:** 1632
Lidocaine **1:** 125
Lidocaine hydrochloride **2:** 546; **4:** 2003
Lidoflaxine **1:** 229
Lidone **2:** 608

Lifopristone **3**: 1510
Lifril **2**: 626
Lignocaine **1**: 125
Likuden **3**: 1242
Limaprost **4**: 2060
Endolimax nana **3**: 1169
Limbial **2**: 638
Limbic system **2**: 317
Limpidon **2**: 639
Lincocin **4**: 2011
Lincomycin **3**: 1067; **4**: 2011
    staphylococcus aureus inhibition **3**: 1091
Line extension **4**: 2149
Liniments **4**: 2088
Lini Semen **2**: 783
Linseed **2**: 783
Linton **2**: 598
Lintopride **2**: 796
Lioresal **2**: 681
Liothyronine **3**: 1523, 558
    pharmacology **3**: 1521
Liothyronine sodium **3**: 1523
Liotropina **2**: 690
Lipanor **1**: 26
Lipantyl **1**: 25
Lipid Hypothesis **1**: 17
Lipid metabolism
    in activated platelets **1**: 32
Lipitor **1**: 29
Lipobay **1**: 29
Lipo-Merz **1**: 26
β Lipoprotein **1**: 252
$α_1$ Lipoprotein **1**: 252
Lipoprotein Catabolism **1**: 21
Lipoproteins **1**: 19
Liposomes **4**: 2120
Lipotropins **3**: 1544
Lipoxine A_4 **4**: 2049
Lipoxine B_4 **4**: 2049
Lipoxines **4**: 2047
    biosynthesis **4**: 2053
    occurrence **4**: 2053
    physical properties **4**: 2049
5-Lipoxygenase **4**: 1672
Lipoxygenases **4**: 1721
Liquid dispersions **4**: 2089
Liquid solution dosage forms **4**: 2086
Liquifil **4**: 2034
Liquifilm **4**: 2034
Liquifilm Forte **4**: 2037
Liquifilm Tears **4**: 2039
Lisino **2**: 932
Liskantin **2**: 480
Listica **2**: 645
Lisuride **2**: 576
Litalgin **2**: 704
Lithium neuroleptics **2**: 630

Lithium salts **3**: 1528
Litholysis **2**: 754
Livalfa **2**: 933
Liver Flukes **3**: 1268
Liver Therapeutics **2**: 747, 761
Lividomycin A **3**: 985
Livocab **2**: 930
Livornex **4**: 1699
LNG-20 **3**: 1507
Loa loa **3**: 1277; **4**: 1810
Lobaplatin **4**: 1921
Lobelin **2**: 339
Locacorten **3**: 1632
Local anesthetics **2**: 539
Local corticoid therapy **3**: 1604
Localicid **3**: 1217
Loceryl **3**: 1253
Lochol **1**: 29
Locéryl **3**: 1253
Lodales **1**: 28
Loderix **2**: 926
Lodine **2**: 382
Lodoxamide **2**: 934
Loestrin **3**: 1500
Lofepramine **2**: 614
Loftram **2**: 641
Loftyl **1**: 50
Logical **2**: 478
Lombpiareu **3**: 1285
Lomexin **3**: 1221
Lomide **2**: 934
Lomidine **3**: 1149
Lomir **1**: 212
Lomotil **2**: 408
Lomustine **4**: 1891
Lonazolac **2**: 358
Longalgic **4**: 1678
Longanoct **2**: 506
Longoran **2**: 601
Loniten **1**: 101
Loop saluretics **1**: 296
Lo-Ovral **3**: 1500
Loperamide **2**: 409
Lopid **1**: 25
Lopinavir **3**: 1314
Loporox **3**: 1245
Loprazolam **2**: 515, 526
Lopresor **1**: 91, 56
Lopresor S.R. **1**: 156
Lopressor **1**: 91, 56
Lora **2**: 499
Lorajmine **1**: 133
Loram **2**: 522
Loramet **2**: 523
Lorans **2**: 522
Lorasolid **2**: 638
Loratadine **2**: 932; **4**: 1720

Lorax **2:** 522, 638
Lorazepam **2:** 515, 522, 637
  synthesis **2:** 520
Lorcainide **1:** 128
Lorelco **1:** 27
Lorinon **2:** 678
Lorivan **2:** 522
Lormetazepam **2:** 515, 523
  indications for use **2:** 519
  synthesis **2:** 520
Lormin **3:** 1596
Lornoxicam **2:** 375
Losec **2:** 720
Lotions **4:** 1716
Lotramina **3:** 1217
Lotremin **3:** 1217
Lotrimin **3:** 1217
Lotusate **2:** 510
Lovastatin **1:** 27
Low density lipoproteins **1:** 19
LOX **4:** 1721
Loxapac **2:** 605
Loxapine **2:** 605
Loxen **1:** 210
Loxitane **2:** 605
Lozenges **4:** 2094
LTA$_4$ **4:** 2047
LTB$_4$ **4:** 2047, 2049
LTB$_4$ synthesis **4:** 2070
Lubricants **2:** 781; **4:** 2105
Ludiomil **2:** 621
Lumcalcio **2:** 506
Luminal **2:** 480, 506
Lumirelax **2:** 680
Lung Flukes **3:** 1267
Lupron **4:** 1932, 933
Lupus erythematosus **4:** 1674
Lurselle **1:** 27
Lustral **2:** 630
Luteinizing Hormone Releasing Hormone **3:** 1512, 544
Luteinizing hormone-releasing hormone analogs **4:** 1931
Luteogan **3:** 1596
Luteosid **3:** 1596
Lutocylin **3:** 1598
Lutoral **3:** 1598
Lutropin **3:** 1348, 545
Luvatren **2:** 598
Luvatrena **2:** 598
Luvatrene **2:** 598
Luvistin **2:** 906
Luxoben **2:** 606
LXA$_4$ **4:** 2049
LXA$_4$ methyl ester **4:** 2049
LXB$_4$ **4:** 2049
LY-139481 **4:** 1924
LY-188011 **4:** 1879
LY 277359 **2:** 802
LY 293111 **4:** 2065
Lymphatic Filariae **3:** 1276
Lymphocytes **1:** 239, 247
Lyophilized powders **4:** 2092
Lyorodin **2:** 595
Lysalgo **4:** 1697
Lysanxia **2:** 639
Lyseen **2:** 564
Lysenyl **2:** 576
Lysozyme **1:** 252
Lysthenon **2:** 674
Lystin **3:** 1238
Lyteers **4:** 2033

M-141 **3:** 984
M-5050 **4:** 2182
M-99 **4:** 2182
Mabertin **2:** 523
Mabuterol **2:** 880
Machupo **3:** 1174
Macitrol **4:** 2016
Macrodantin **3:** 1146
α_2 Macroglobulin **1:** 252
Macrolide Antibiotics
  veterinary **4:** 2164
Macrolides antibiotics **3:** 954, 995
Macrophages **1:** 246
Madopar **2:** 569
Maduramicin **4:** 2170
Maduropeptin **3:** 1013
Magaldrate **2:** 726
Magmas **4:** 2089
Magnesium hydroxide **2:** 726
Magnesium sulfate **2:** 785
Magnesium valproate **2:** 477
Magnetic resonance imaging **2:** 325
Magnussen test **4:** 2133
Maintate **1:** 153
Maiorad **2:** 701
Majeptil **2:** 594
Makisterone **3:** 1331
Malaria **3:** 1159
  treatment with monoclonal antibodies **4:** 1843
Malaria vaccine **4:** 1812
Male Fertility Control **3:** 1514
Malestrone **3:** 1590
Malexil **2:** 628
Mallazine Drops **4:** 2030
Mallorol **2:** 592
Maltyl **2:** 745
Malysol **2:** 339
Maneon **2:** 617
Manexin **1:** 224
Mannite **4:** 2023
Mannitol **2:** 785; **4:** 2023

Mannitol hexanitrate   **1:** 224
Mannitol nitrate   **1:** 224
Mansil   **3:** 1283
Mantadine   **2:** 574
Mantadix   **2:** 574
m antibiotics
   economic aspects   **3:** 1094
MAO   **2:** 570
MAO-A Inhibitors   **2:** 625
MAO-B   **2:** 559
MAOI   **2:** 622
Maolate   **2:** 677
Maprotiline   **2:** 620
Marburg   **3:** 1174
Marcaine   **2:** 547
Marcen   **2:** 641
Marezine   **2:** 742
Marflex   **2:** 567
Market
   pharmaceutical   **1:** 5
Maronil   **2:** 612
Marplan   **2:** 623
Marplon   **2:** 623
MARS   **1:** 27
Marsilid   **2:** 623
Marucotol   **2:** 652
Marvelon   **3:** 1500, 1598
Marziné   **2:** 915
Masked fluoropyrimidines   **4:** 1875
Mast cells   **4:** 1714
Mast Cell Stabilizing Agents   **2:** 933
Masterid   **3:** 1594
Matrix system   **4:** 2115
Maxeran   **2:** 608, 745
Maxiban 72   **4:** 2170
Maxibolin   **3:** 1592
Maxidex   **4:** 2018
Maximed   **2:** 616
Maximum upstroke velocity   **1:** 119
Maxitrol   **4:** 2008, 2011, 2017, 2018
Maxolon   **2:** 608, 745
Mayeptil   **2:** 594
Mazanor   **2:** 466, 467
Mazindol   **2:** 466, 467
MBC   **3:** 1114
MC696-SY2-A   **3:** 974
MCH   **1:** 239
MCHC   **1:** 239
MCNU   **4:** 1893
MCV   **1:** 239
MDL-73147   **2:** 609
Mean corpuscular hemoglobin   **1:** 239
Mean corpuscular hemoglobin concentration   **1:** 239
Mean corpuscular volume   **1:** 239
Measles   **3:** 1174
   treatment with monoclonal antibodies   **4:** 1843
Measles – mumps – rubella vaccine   **4:** 1791

Measles prophylaxix   **4:** 1826
Measles vaccine   **4:** 1787
Mebaral   **2:** 480
Mebendazole   **3:** 1270, 1271, 1272, 1273, 1274, 1281
mebendazolealbendazole   **3:** 1275
Mebeverine   **2:** 701
Mebhydroline   **2:** 925
MEC   **4:** 2080
Mechloral   **2:** 499
Mechlorethamine hydrochloride   **4:** 1728
Mechloroethamine   **4:** 1886
Meclastine   **2:** 909
Meclizine   **2:** 496
Meclocycline   **4:** 1725
Meclodol   **4:** 1698
Meclofenamate   **2:** 351
Meclofenamic acid   **2:** 351; **4:** 1698, 2176
Meclofenoxate   **2:** 652
Meclomen   **2:** 352; **4:** 1698
Meclopran   **2:** 608
Mecloral   **2:** 499
Mecloxate   **2:** 652
Meclozine   **2:** 496, 743, 916
Mectizan   **3:** 1280
Medapan   **2:** 508
Medetomidine   **4:** 2181
Medifoxamine   **2:** 631
Medigoxin   **1:** 189
Medihaler Epi   **2:** 865
Medil   **1:** 50
Mediphen   **2:** 506
Medomin   **2:** 508
Medrate   **3:** 1624
Medrifar   **4:** 2018
Medriusar   **4:** 2018
Medrogestone   **3:** 1597
Medrol   **3:** 1624
Medroxyprogesterone acetate   **3:** 1596
Medroxy-progesterone acetate   **3:** 1497, 504
Medrysone   **4:** 2018
Mefenamic acid   **2:** 352; **4:** 1697
Mefloquine   **3:** 1165
Mefloquine Quinate   **3:** 1165
Mefruside   **1:** 71, 73, 295
Megabyl   **2:** 703
Megadon   **2:** 518
Megage   **3:** 1596
Megakaryocytes   **1:** 242
Megaphen   **2:** 590
Megestrol acetate   **3:** 1596
Megimide   **2:** 339
Meglumine antimonate   **3:** 1154
Mekos   **2:** 500, 645
Melanocytes   **4:** 1714
Melanostatin   **2:** 581; **3:** 1403
Melanotropin   **3:** 1376

Melarsenoxide potassium dimercaptosuccinate **3:** 1151
Melarsonyl potassium **3:** 1151
Melarsoprol **3:** 1149
Melatonin **3:** 1550
   influence on sleep **2:** 491
Melfiat **2:** 465, 467
Melipramin **2:** 614
Mellaril **2:** 592
Melleril **2:** 592
Melodal **2:** 412
Melosan **2:** 501
Meloxicam **2:** 382
Melperone **2:** 600
Melphalan **4:** 1887
Melsed **2:** 514
Melsedin **2:** 514
Melur **3:** 1223
Mel W **3:** 1151
Memantine **2:** 574
Membrane stabilizing activity **1:** 139
Memoq **2:** 655
M & B 17 803 A **1:** 152
Mendon **2:** 639
Menfegol **3:** 1495
Meningitis
   treatment with monoclonal antibodies **4:** 1841
Meningitis vaccine **4:** 1779
Meningococcal meningitis vaccine **4:** 1779
Menstrogen **3:** 1598
4-Menthanylphenylpolyoxyethylene(8,8)ether **3:** 1495
Menthol **4:** 1723
5-MeOT **2:** 804
Mepenzolate bromide **2:** 695
Meperidine **2:** 397
Mephenhydramine **2:** 910
Mepivacaine hydrochloride **2:** 546; **4:** 2004
Meprobamate **2:** 634, 644, 678
Meproscillarin **1:** 191
Meprospan **2:** 679
Meprotabs **2:** 679
Meptazinol **2:** 417
Meptid **2:** 417
Meptin **2:** 884
Mepyramine **2:** 906
Mequelon **2:** 514
Mequin **2:** 514
Mequitazine **2:** 920
Merbentyl **2:** 703
Merbromin **3:** 1255
Mercaptobutanedioic acid monogold(I) sodium salt **4:** 1703
3-Mercapto-D-valine **4:** 1705
Mercaptoethyltrimethylammonium iodide O,O-diethyl phosphorothioate **4:** 2025
1-(3-Mercapto-2-methyl-1-oxopropyl)-L-proline **1:** 78
6-Mercaptopurine **4:** 1882

Mercurial preservatives
   for contact lenses **4:** 2037
Mercurochrome **3:** 1255
Meregon **1:** 133
Mereprine **2:** 909
Merinax **2:** 500
Merisyl **2:** 509
Merlit **2:** 522
Merozoite surface antigens **4:** 1817
Mersalyl **1:** 290
Mervan **4:** 1684
Mesoridazine **2:** 592
Mesotin **2:** 888
Mesterolone **3:** 1590
Mestoran **3:** 1590
Mestranol **3:** 1501, 584
Mesulergine **2:** 577
Metabolic Drugs **3:** 1319
Metabolism cholecalciferol **3:** 1638
Metabolism phase **4:** 2130
Metabolotropic action **2:** 335
Metacycline **3:** 1037
Metagonimus yokogawai **3:** 1269
Metalcaptase **2:** 772; **4:** 1706
Metamin **2:** 597
Metamizol **2:** 369
Metandienone **3:** 1593
Metandiol **3:** 1590
Metandren **3:** 1590
Metaoxedrin Minims **4:** 2029
Metaplexan **2:** 920
Metapramine **2:** 615
Metaproterenol **2:** 869
Metaxalone **2:** 679
Meteprine **2:** 496
Metformin **3:** 1660
Methacholine chloride **4:** 2032
Methacycline **3:** 978, 037
Methadone **2:** 404
Methalutin **3:** 1598
Methamphetamine **2:** 461, 467
Methamphetamine hydrochloride **2:** 339
Methandriol **3:** 1590
Methanthelinium bromide **2:** 695
Methapyrilen **2:** 906
Methaqualone **2:** 513
Methasedil **2:** 514
I-Methasone **4:** 2018
Methazolamide **4:** 2022
Methdilazine **2:** 920
Methenolone acetate **3:** 1594
Methenolone enanthate **3:** 1594
Methicillin **3:** 962
   staphylococcus aureus inhibition **3:** 1091
Methicillin sodium **3:** 1021; **4:** 2006
Methimazole **3:** 1525, 530
   pharmacokinetic data **3:** 1527

Methocarbamol **2**: 679
Methocel **2**: 784
6/9 method **4**: 1823
Methofane **2**: 442
Methohexital sodium **2**: 444
Methopto Forte **4**: 2034
Methostan **3**: 1590
Methotrexate **4**: 1707, 728, 868
2-Methoxy-4-amino-5-chloro-benzoic acid 2-(diethylamino)ethyl ester **2**: 804
1-(4-Methoxybenzoyl)-2-pyrrolidone **2**: 650
7 α-Methoxy-cephalosporin C **3**: 967
6′-Methoxycinchonan-9-ol sulfate **2**: 675
2-Methoxy-10-(3-dimethylamino-2-methylpropyl)-10H-phenothiazine **2**: 591
1-[4-(2-Methoxyethyl)phenoxy]-3-[(1-methylethyl)amino]-2-propanol **1**: 91
Methoxyflurane **2**: 442
3-Methoxy-L-tyrosine **2**: 570
5-Methoxy-2-[[(4-methoxy-3,5-dimethyl-2-pyridyl)methyl]sufinyl]benzimidazole **2**: 720
(+)-6-Methoxy-αmethyl-2-naphthaleneacetic acid **4**: 1688
N-[4-(Methoxymethyl)-1-[2-(2-thienyl)ethyl]-4-piperidinyl]-N-phenylpropanamide **2**: 403
N-[4-Methoxymethyl-1-(2-thiophen-2-yl-ethyl)piperidin-4-yl]-N-phenylpropionamide **2**: 403
4-(6-Methoxynaphthalen-2-yl)butan-2-one **2**: 360
(+)-(S)-2-(6-Methoxynaphthalen-2-yl)propionic acid **2**: 366
β-(1-Methoxy-4-naphthoyl)propionic acid **2**: 752
3-(2′-Methoxy-4′-nitrophenyl)-2-methyl-4(3H-)-quinazolinone **2**: 514
3-Methoxy-19-norpregna-1,3,5(10)-trien-20-yn-17-ol **3**: 1501
1-[3-(2-Methoxyphenothiazin-10-yl)-2-methylpropyl]-4-piperidinol **2**: 496
(3R-trans)-3-[4-(methoxy)phenoxymethyl]-1-methyl-4-phenylpiperidine **2**: 628
3-(2-Methoxyphenoxy)-1,2-propanediol 1-carbamate **2**: 679
5-(4-Methoxyphenyl)-3H-1,2-dithiole-3-thione **2**: 752
N-[(4-methoxyphenyl)methyl]-N′,N′-dimethyl-N-2-pyridinyl-1,2-ethanediamine **2**: 906
N-[(4-methoxyphenyl)methyl]-N′,N′-dimethyl-N-2-pyrimidinyl-1,2-ethanediamine **2**: 907
1-(2-Methoxyphenyl)-4-[4-(2-phtalimidobtyl]-piperazine **2**: 814
6-({3-[4-(2-Methoxyphenyl)-1-piperazinyl]propyl}amino)-1,3-dimethyl-2,4(1H,3H)-pyrimidinedione **1**: 89
8-Methoxypsoralen **4**: 1729
(+)-(αS)-α-(6-methoxy-4-quinolyl)-α-[(2R,4S,5R)-(vinylquinuclidin-2-yl)]methanol **1**: 124
5-Methoxy-4′-(tri-fluoromethyl)valerophenone(E)-O-(2-aminoethyl)-oxime **2**: 628
5-Methoxytryptamine **2**: 804

Methsuximide **2**: 479
Methulose **4**: 2034
2-(4-Methylaminobutoxy)diphenylmethane **2**: 656
4-[2-(Methyl-amino)ethyl]benzene-1,2-diol diisobutyrate **1**: 192
[R-(R*,S*)]-α-[1-(Methylamino)ethyl]benzenemethanol **2**: 463
O-[2-(methylamino)-ethyl]oxime **2**: 618
2-Methylamino-1-phenylpropanol **2**: 877
5-(3-Methylaminopropyl)-5H-dibenzo[a,d]cycloheptene **2**: 615
1-(3-Methylaminopropyl)dibenzo[b, e]bicyclo[2.2.2]octadiene **2**: 620
17α-Methyl-5α-androstano-[3,2-c]pyrazol-17β-ol **3**: 1594
17α-Methyl-5-androstene-3β,17β-diol **3**: 1590
Endo-N-(8-methyl-8-azabicyclo[3.2.1]oct-3-yl)-2,3-dihydro-3-ethyl-2-oxo-1H-benzimidazole-1-carboxamide **2**: 804
Endo-n-(8-methyl-8-azabi-cyclo[3.2.1]oct-3-yl) 3,2-dihydro-3,3-dimethylindole-1-carboxamide **2**: 801
Endo-8-methyl-8-azabicyclo[3.2.1]oct-3-yl-2,3-dihydro-6-methoxy-2-oxo-1H-benzimidazole-1-carboxylate **2**: 804
Endo-N-(8-methyl-8-azabicyclo[3.2.1]oct-3-yl)-2,3-dihydro-3-(1-methyl) ethyl-2-oxo-1H-benzimidazole-1-carboxamide **2**: 804
(±)-αMethylbenzeneethanamine **2**: 460, 461
Methyl (5-benzoyl-1H-benzimidazol-2-yl)carbamate **3**: 1281
2-Methyl-1,2,3,4,10,14b-hexahydro-2H-pyrazino[1,2-f]morphanthridine **2**: 621
2-Methyl-2-butenoic acid [1α,3α(E),5α]-8-methyl-8-azabicyclo[3.2.1]-oct-3-yl ester **2**: 563
5-(1-Methylbutyl)-5-vinylbarbituric acid **2**: 508
Methyl-CCNU **4**: 1891
Methylcellulose **2**: 784; **4**: 2034, 2103
Methylcellulosum **2**: 784
N-methyl-N-(4- chlorobenzoylmethyl)-3-(10,11-dihydro-5H-dibenzo[b, f]azepin-5-yl)propylamine **2**: 614
Methyl (+)-(S)-α-(2-chlorophenyl)-6,7-dihydrothieno[3,2-c]pyridine-5(4H)-acetate **1**: 41
Methylclothiazide **1**: 71, 294
16-Methylcorticosteroids **3**: 1624
6α-Methylcorticosteroids **3**: 1622
Methylcyclohexylchloroethylnitrosourea **4**: 1891
β-Methyldigoxin **1**: 189
4′′′-Methyldigoxin **1**: 189
(+)-5-Methyl-10,11-dihydro-5H-dibenzo[a,d]cyclohepten-5,10-imine **2**: 574
Methyl 2-[[(2,3-dimethylimidazo[1,2-a]pyridin-8-yl)amino]methyl]-3-methylcarbanilate **2**: 733
α-Methyldopa **1**: 15, 81
3-O-Methyldopa **2**: 570
24-Methylenecholesterol **3**: 1324
16-Methylenecorticoids **3**: 1627
16-Methyleneprednisolone **3**: 1628

Methylestradiol 3: 1584
17α-Methyl-1,3,5(10)-estratriene-3,17β-diol 3: 1584
Methylestrenolone 3: 1598
1-[3-[(1-Methylethyl)amino]-2-pyridinyl]-4-[[5-[(methylsulfonyl)amino]-1H-indol-2-yl]carbonyl]piperazine 3: 1301
4,4'-[(1-Methylethylidene)bis(thio)]bis[2,6-bis(1,1-dimethylethyl)phenol] 1: 26
2,3:4,5-bis-O-(1-Methylethylidene)-β-D-fructopyranose sulfamate 2: 483
N-(1-Methylethyl)-4-[(2-methylhydrazino)methyl]-benzamide 4: 1894
3-[(5-Methyl-2-furanyl)methyl]-N-4-piperidinyl-3H-imidazo[4,5-b]pyridin-2-amine 2: 930
1-Methyl-4-(9H-thioxanthen-9-ylidene)piperidine 2: 925
2-Methyl-2-[4-[1-hydroxy-4-[4-(1,1-diphenylhydroxymethyl)-1-piperidinyl]butyl]phenyl]propionic acid 2: 929
1-Methyl-2-imidazole thiol 3: 1530
5-Methyl-3-isoxazolecarboxylic acid 2-benzylhydrazide 2: 623
4-{4-[β-(5-Methylisoxazolyl-3-carboxamido)-ethyl]-benzenesulfonyl}-1,1-hexamethylenesemicarbazide 3: 1657
N^1-(5-Methyl-3-isoxazolyl)sulfanilamide 3: 1132
N-Methyllorazepam 2: 523
2-Methylmercapto-10-[2-(1-methyl-2-piperidyl)ethyl]-10H-phenothiazine 2: 592
2-Methylmercapto-10-[2-(1-methyl-2-piperidyl)ethyl]-10H-phenothiazine-2-oxide 2: 592
Methyl (3-methlphenyl)carbamothioic acid O-(1,2,3,4-tetrahydro-1,4-methanonaphtalen-6-yl) ester 3: 1246
Methyl 1-(2-methoxycarbonylethyl)-4-(phenylpropionylamino)piperidine-4-carboxylate 2: 403
Methyl 4-(methoxycarbonyl)-4-[(1-oxopropyl)phenylamino]-1-piperidinepropanoate 2: 403
3-methyl-8-methoxy-3H-1,2,5,6-tetrahydropyrazino-[1,2,3ab]-β-carboline 2: 625
1-Methyl-5-(4-methylbenzoyl)-1H-pyrrole-2-acetic acid 4: 1683
[1-Methyl-5-(4-methylbenzoyl)-1H-pyrrol-2-yl]acetic acid 2: 359
7-Methyl-1-(1-methylethyl)-4-phenyl-2(1H)-quinazolinone 4: 1699
17α-Methyl-18-methyl-17β-hydroxy-4-estren-3-one 3: 1592
O-methyl [2-(2-methyl-5-nitro-1H-imidazol-1-yl)ethyl]thiocarbamate 3: 1157
Methyl (6S,7S)-3-methyl-8-oxo-7-phtalimido-5-oxa-1-azabicyclo[4,2,0]-oct-2-ene-2-carboxylate 3: 970
4-Methyl-3-(4-methylthiobenzoyl)imidazolin-2-one 1: 195
17-Methylmorphinan-3-ol 2: 394

(+)-(S)-3-Methyl-4-morpholin-4-yl-2,2-diphenyl-1-pyrrolidin-1-yl-butan-1-one 2: 406
1-Methylnitrazepam 2: 522
2-Methyl-5-nitroimidazole-1-ethanol 3: 1144
2-Methyl-N-[4-nitro-3-(trifluormethyl)phenyl]-propanamide 4: 1928
4-Methyl-3-[2-(n-propylamino)propionamido]-2-thiophenecarboxylic acid methyl ester hydrochloride 2: 548
2-Methyl-3-o-tolyl-4-(3H)quinazolinone 2: 513
N-methyloxazepam 2: 523
Methylpentynol 2: 497
3-Methyl-1-pentyn-3-ol 2: 497
Methylphenidate hydrochloride 2: 339
Methylphenobarbital 2: 480
10-Methylphenothiazine-2-acetic acid 4: 1684
α-Methyl-3-phenoxybenzeneacetic acid 4: 1690
(3-Methylphenyl)carbamothioic acid O-2-naphthalenylmethylester 3: 1247
N-(1-Methyl-2-phenylethyl)-γ-phenylbenzenepropanamine 1: 208
(R)-(−)-N-Methyl-N-phenylisopropyl-2-propynylamine 2: 570
16α,17-(1'-Methyl-1'-phenylmethylenedioxy)-4-pregene-3,20-dione 3: 1597
Methyl 1-(phenylmethyl)-3-piperidinyl ester 1: 212
1-Methyl-N-phenyl-N-(phenylmethyl)-4-piperidinamine 2: 917
(E)-N-Methyl-N-(3-phenyl-2-propenyl)-1-naphthalenemethenamine 3: 1249
N-(2-Methylphenyl)-2-(propylamino)propanamide hydrochloride 4: 2004
N-(2-Methylphenyl)-2-(propylamino)propanamidemonohydrochloride 2: 546
1-Methyl-3-phenyl-2,5-pyrrolidinedione 2: 478
(E)-2-[1-(4-methylphenyl)-3-(1-pyrrolidinyl)-1-propenyl]pyridine 2: 912
5-Methyl-1-phenyl-3,4,5,6-tetrahydro-1H-benzo[f][1,4]oxazocine 2: 425
1-Methyl-4-phenyl-1,2,3,6-tetrahydropyridine 2: 559
1-Methyl-N-phenyl-N-(2-thienylmethyl)-4-piperidinamine 2: 918
4-[5-(4-Methylphenyl)-3-(trifluoromethylpyrazol-1-yl]benzenesulfonamide 4: 1700
2-(4-Methylphenyl)-6-N,N-trimethylimidazo-[1,2-a]pyridine-3-acetamide-L-(+)-tartrate 2: 529
6-(4-methyl-1-piperazinyl)-11 H-dibenz [b,e]azepine 2: 496
10-[3-(4-Methyl-1-piperazinyl)propyl]-N,N-dimethyl-10H-phenothiazine-2-sulfonamide 2: 594
10-[3-(4-Methyl-1-piperazinyl)propyl]-10H-phenothiazine 2: 593
Cis-9-[3-(4-methyl-piperazinyl)propylidene]-N,N-dimethyl-9H-thioxanthene-2-sulfonam 2: 596
10-[3-(4-Methyl-1-piperazinyl)-propyl]-2-trifluoromethyl-10H-phenothiazine 2: 744

Methyl 2-[4-(2-piperidinoethoxy)benzoyl]benzoate   **2:** 704
3-(2-Methylpiperidino)propyl benzoate hydrochloride   **2:** 543
1-Methyl-4-piperidyl 2,2-diphenyl-2-propoxyacetate   **2:** 704
(±)-2′-[2-(1-Methyl-2-piperidyl)ethyl]-*p*-anisanilide   **1:** 132
6α-Methylprednisolone   **3:** 1623, 636
  doses   **3:** 1605
2α-Methyl-17β-propionyloxy-5α-androstan-3-one   **3:** 1594
β-[(2–Methylpropoxy)methyl]-*N*-phenyl-*N*-(phenylmethyl)-1-pyrrolidineethanamine   **1:** 209
α-[[*bis*(1-Methylpropyl)amino]methyl]-1-[(2-chlorophenyl)methyl]-1*H*-pyrrole-2-methanol   **2:** 412
2-Methyl-2-propyl-1,3-propanediol dicarbamate   **2:** 678
2-methyl-2-propyl-1,3-propyldicarbamate   **2:** 644
Methyl(5-(propylthio)-2-benzimidazol)carbamate   **3:** 1277
N-methyl-*O*-(2-propylvaleryl)tropinium bromide   **2:** 692
N-methyl-*N*-2-propynylbenzylamine   **2:** 624
N-{4-[5-Methylpyrazine-2-carboxamido)-ethyl]-benzenesulfonyl}-*N*′-cyclohexylurea   **3:** 1656
5-(4-[2-(Methylpyridin-2-ylamino)ethoxy]benzyl)thiazolidine-2,4-dione   **3:** 1666
10-[(1-Methyl-3-pyrrolidinyl)methyl]-10*H*-phenothiazine   **2:** 920
Methyl Sterols   **3:** 1326
2-Methylsulfonyl-10-[2-(1-methyl-2-piperidyl)ethyl]-10*H*-phenothiazine   **2:** 592
4-[4-(methylsulfonyl)phenyl]-3-phenylfuran-2(5*H*)-one   **2:** 384; **4:** 1701
[1-[2-Methylsulphonyl)amino]ethyl]-4-piperidinyl]methyl-2-methoxy-5-fluoro-1*H*-indole-3-carboxylate   **2:** 807
[1-[2-methylsulphonyl)amino]ethyl]-4-piperidinyl]methyl-1-methyl-1*H*-indole-3-carboxylate   **2:** 812
Methyltestosterone   **3:** 1590
8-Methyl-1,2,3,4-tetrahydro-1,10-trimethylenepyrazino[1,2-a]indole   **2:** 626
*N*¹-(5-Methyl-1,3,4-thiadiazol-2-yl)sulfanilamide   **3:** 1135
8β-(Methylthiomethyl)-6-propylergoline   **2:** 577
N-[(6-Methyl-5-oxo-3-thiomorpholinyl)carbonyl]-L-histidine-L-prolinamide   **2:** 581
Methylthiouracil   **3:** 1525, 529
  pharmacokinetic data   **3:** 1527
2-Methyl-(2′,2′,2′-trichloro-1′-hydroxy-4-ethoxy)-pentan-2-ol   **2:** 499
α-Methyl-tricyclo[3.3.1.1^{3,7}]decane-1-methanamine   **1:** 1185
2-[[[3-Methyl-4-(2,2,2-trifluoroethoxy)-2-pyridyl]methyl]sulfinyl]benzimidazole   **2:** 720

N-methyl-3-[4-trifluoromethylphenoxy-3-phenylpropylamine   **2:** 628
(±)-1-Methyl-2-(2,6-xylyloxy)ethylamine   **1:** 126
Methymycin   **3:** 996
Methyprylon   **2:** 512
Metiazinic acid   **4:** 1684
Meticillin   **3:** 1021
Metildigoxin   **1:** 189
Metimyd   **4:** 2005
Metipranolol   **4:** 2021
Metligine   **1:** 63
Metoclol   **2:** 745
Metoclopramide   **2:** 607, 729, 744, 795, 801
  mode of action   **2:** 798
  Receptor binding affinities   **2:** 808
Metocurine iodide   **2:** 671
Metofenazate   **2:** 595
Metolazone   **1:** 295
Metoprolol   **1:** 91, 56, 227
Metoros   **1:** 156
Metralindole   **2:** 625
Metramid   **2:** 608
Metrifonate   **3:** 1267, 281
Metrodin   **3:** 1357
Metro I.V.   **3:** 1145
Metronid   **3:** 1145
SK-Metronidazol   **3:** 1145
Metronidazole   **3:** 1144, 156, 158, 276
Metryl   **3:** 1145
Metubine iodide   **2:** 671
Metycaine   **2:** 543
Mevacor   **1:** 28
Mevalotin   **1:** 28
Mevinacor   **1:** 28
Mevinolin   **1:** 15
Mexiletine   **1:** 126
Mexitil   **1:** 126
Mezlizine   **2:** 916
Mezlocillin   **3:** 962, 1023
Mianserin   **2:** 620, 621
MIC   **3:** 956, 114
Micatin   **3:** 1210; **4:** 2013
Micefal   **2:** 601
Miclast   **3:** 1245
Micocert   **3:** 1219
Micoderm   **3:** 1210
Micoespec   **3:** 1219
Micofim   **3:** 1210
Micofun   **3:** 1213
Micofungal   **3:** 1219
Micogel   **3:** 1210
Micogin   **3:** 1219
Micogyn   **3:** 1219
Micomicen   **3:** 1245
Micomisan   **3:** 1217
Miconal   **3:** 1210
Miconazole   **3:** 1209; **4:** 1730, 2013, 2171

Micoral   **3:** 1208
Micoren   **2:** 339
Micos   **3:** 1219
Micosten   **3:** 1219
Micostyl   **3:** 1219
Micotar   **3:** 1210
Micotef   **3:** 1210
Micotek   **3:** 1208
Micoter   **3:** 1217
Micoticum   **3:** 1208
Micoxolamin   **3:** 1245
Microbead culture   **4:** 1991
Microbial pathogens   **3:** 955
Microbiological analysis
  of antibiotic samples   **3:** 1089
Microbiological conversion fermentation of steroids   **3:** 1577
Microcarriers   **4:** 1991
Micrococcaceae   **3:** 1110
Microcytes   **2:** 311
Microencapsulated Steroids   **3:** 1505
Microgynon   **3:** 1500, 1598
Microklist   **2:** 782
Microlut   **3:** 1598
Micronett   **3:**
Micronomicin   **3:** 1043; **4:** 2008
Micronomicin sulfate   **4:** 2008
Microorganism stains
  maintenance in antibiotics fermentation   **3:** 1079
Microspheres   **4:** 2120
Mictonetten   **2:** 705
Mictonorm   **2:** 705
Mictrol   **1:** 209; **2:** 705
Micturin   **1:** 209; **2:** 705
Micturol   **1:** 209
Midamor   **1:** 75, 299
Midarine   **2:** 674
Midazolam   **2:** 515, 525, 634
  indications for use   **2:** 519
  synthesis   **2:** 521
Midazolam maleate   **2:** 450
Midecamycin   **3:** 1055
Midodrine   **1:** 62
Midronal   **1:** 209
MIF   **2:** 581; **3:** 1403
Mifepristone   **3:** 1510
Miglitol   **3:** 1663
Migraine   **1:** 149
Migralave N   **2:** 913
Migranal   **1:** 66
Migristene   **2:** 919
Mikamycin   **3:** 1008, 1059
Mikamycin B   **3:** 1059
Mikedimide   **2:** 339
Mikelan   **1:** 154
Mikoderm   **3:** 1248
Mikostatin   **3:** 1238

Milacemide   **2:** 609, 624
Millicorten   **3:** 1627
Milnacipran   **2:** 633
Miloxacin   **3:** 1121
Milrinone   **1:** 195
Milroy Artificial Tears   **4:** 2034
Miltaun   **2:** 645
Miltefosine   **4:** 1938
Miltex   **4:** 1938
Miltown   **2:** 679
Minaxolone   **2:** 452
Mineralocorticoid hormones   **3:** 1601
Mineralocorticoids   **1:** 66; **3:** 1553
Mineral oil   **2:** 781; **4:** 1727
Minetoin   **2:** 475
Minias   **2:** 523
Mini Diab   **3:** 1656
MIN-I-JET Atropinsulfat   **2:** 690
Minimal bactericidal concentration   **3:** 1114
Minimal inhibitory concentration   **3:** 1114
Minims amethocaine hydrochloride   **4:** 2003
Minims Atropine Sulfate   **4:** 2027
Minims Benoxinate Hydrochloride   **4:** 2003
Minims Centamicin Sulfate   **4:** 2008
Minims Cyclopentalate   **4:** 2028
Minims Homatropine   **4:** 2028
Minims Hyoscine   **4:** 2028
Minims Metipranolol   **4:** 2022
Minims Oxybuprocaine Hydrochloride   **4:** 2003
Minims Phenylephrine   **4:** 2026, 2029
Minims Tropicamide   **4:** 2028
Minimum effective concentration   **4:** 2080
Minimum inhibitory concentration   **3:** 956
Minimum toxic concentration   **4:** 2080
Mini-pill   **3:** 1503
Minipress   **1:** 87
Minirin   **3:** 1402
Minocycline   **3:** 978, 038
Minoxidil   **1:** 101
Minozinan   **2:** 591
Minprostin   **4:** 2059
Mintezol   **3:** 1287
Minzolum   **3:** 1287
Mioblock   **2:** 672
Miolaxene   **2:** 680
Mioril   **2:** 677
Miostat   **4:** 2024
Miotics
  antiglaumatons   **4:** 2023
Miproxifene phosphate   **4:** 1924
Miradol   **2:** 606
MiraFlow Extra-Strength Cleaner   **4:** 2040
Miranax   **4:** 1689
Mirapex   **2:** 577
MiraSept   **4:** 2039
Mirbanil   **2:** 606
Mirontin   **2:** 478

Mirtazapine **2**: 633
MIS **3**: 1550
Miscellaneous Drugs **4**: 1669
Misoprostol **2**: 727; **4**: 2054, 2060
MIT **3**: 1558
Mitanoline **2**: 564
MIT-C **4**: 1896
Mitekol **3**: 1219
Mithramycin **3**: 1066
Mitidin **2**: 518
Mitomycin C **3**: 1070; **4**: 1896
Mitomycins **3**: 1010
Mitotan **3**: 1554
Mitoxantrone **4**: 1902
Mitral valve prolapse **1**: 146
Mitronal **1**: 209; **2**: 914
Mizolastine **2**: 930
Mizollen **2**: 930
MJ-1999 **1**: 159
MK–0366 **3**: 1123
MK-950 **1**: 93, 60
MLCu-250 **3**: 1507
MLCu-375 **3**: 1507
M-Long **2**: 390
MMI **3**: 1525
MMR vaccine **4**: 1791
Moban **2**: 608
Mobec **2**: 383
Mobiflex **2**: 378
Moclobemide **2**: 625
Modalina **2**: 594
MODAS **4**: 2115
Moditen **2**: 595
Modrasone **3**: 1636
Moduretic **1**: 293, 299
Moenomycin **3**: 1068
Mofebutazone **2**: 372
Mofegiline **2**: 571
Mofesal **2**: 373
Mogadan **2**: 518
Mogadon **2**: 481, 518
Moisture Drops **4**: 2033
Molevac **3**: 1286
Molindone **2**: 608
Molipaxin **2**: 632
Molsidolat **1**: 230
Molsidomine **1**: 230
Mometasone furoate **4**: 1722
Mondus **1**: 209
Monensin **3**: 1064; **4**: 2170
Monensins **3**: 1010
Monistat **3**: 1210
Monistat-Derm **4**: 2013
Monoamine oxidase-B **2**: 559
Monoamine oxidase inhibitors **2**: 570, 622
Monoamine reuptake inhibitors **2**: 573
Monobactams **3**: 976

Mono-Baycuten **3**: 1217
Monoclonal antibodies **4**: 1759, 1981
  immunotherapeutic uses **4**: 1839
Monoclonal antibody patents **4**: 1993
Monocor **1**: 153
Monocortin **3**: 1632
Monocytes **1**: 239, 246
Monoiodotyrosine **3**: 1558
MonoMack **1**: 224
Monophylline **2**: 889
Monopina **1**: 212
Monosodium N-(carbamoylmethyl)arsanilate **3**: 1150
Monostat 7 **3**: 1210
Monovalent vaccines **4**: 1762
Monteban 45 **4**: 2170
Montelukast **4**: 1721, 2064
Montelukast sodium **2**: 897
Montirelin **2**: 581
Moperone **2**: 598
Moquizone **2**: 754
Moracizine **1**: 133
Moranyl **3**: 1149, 287
Morial **1**: 230
Moriperan **2**: 608, 745
Moronal **3**: 1238
Morphine **2**: 389
Moryspan **2**: 691
Mosapride **2**: 797
Moscontin **2**: 390
Motazomin **1**: 230
Motilin **2**: 835; **3**: 1549
Motility disorders **2**: 794
Motilium **2**: 729
Motor neurons **2**: 309
Motrin **2**: 365
Mouthwashes **4**: 2089
Movergan **2**: 571, 625
Moxadil **2**: 612
Moxaverine **2**: 699
Mozambin **2**: 514
MPTP **2**: 559
MRI **2**: 325
MRIH **2**: 581
M/Rinse **4**: 2037
MSA **1**: 139
MSH **3**: 1544
MSH-release inhibiting hormone **2**: 581
MST continuous **2**: 390
MTC **4**: 2080
MTU **3**: 1525, 529
Mucilages **4**: 2089
Mucinol **2**: 753
Mucopolysaccharides **2**: 783
Muellerian Inhibiting Substance **3**: 1361, 1550
Multergan **2**: 922
Multibarrelled electrode **2**: 322

Multidirectional osmotic drug absorption system **4**: 2115
Multifuge **3**: 1284
Multilind **3**: 1238
Multiload-250 **3**: 1507
Multiload-375 **3**: 1507
Multipax **2**: 646
Multiphasic oral contraceptives **3**: 1500
ReNu Multi-Purpose Solution **4**: 2037, 2040
Multivalent vaccines **4**: 1762
Mumps **3**: 1174
Mumps prophylaxis **4**: 1828
Mumps vaccine **4**: 1788
Munobal **1**: 211
Mupaten **3**: 1222
Murelax **2**: 522
Muricalm **2**: 926
Murine **4**: 2030
Murine monoclonal antibodies **4**: 1983
Murine Pure Sept **4**: 2039
Murocel **4**: 2034
Muro Tears TM **4**: 2033
Muscarine
 in synaptic transmission **2**: 327
Muscarinic Receptors **2**: 687
Muscle relaxants **1**: 95
 skeletal **2**: 667
Muscle relaxants skeletal **2**: 667
Muscle Spasms
 acute **2**: 676
Musculotropic spasmolytics **2**: 698
Mutagenicity tests **4**: 2132
Mutual recognition agreement **4**: 2150
Mutum **3**: 1202
Muzolimine **1**: 77, 297
Myagen **3**: 1593
Myambutol **3**: 1140
Mycelex **3**: 1217
Mycelex-7 **3**: 1217
Mycelex-G **3**: 1217
Mycinamicins **3**: 999
Mycitracin **4**: 2010
Myclo **3**: 1217
Mycobacteriaceae **3**: 1110
Mycobacterial infections **3**: 1138
Mycobacterium bovis **3**: 1138
Mycobacterium leprae **3**: 1138
Mycobacterium tuberculosis bovis **3**: 1138
Mycodermil **3**: 1221
Mycofug **3**: 1217
Myco-Hermal **3**: 1217
Mycophenolic acid **3**: 953; **4**: 1728
Mycophyt **3**: 1236
Mycoplasmas **3**: 1111
Mycoplasmataceae **3**: 1111
MycoPosterine N **3**: 1238
Mycoril **3**: 1217

Mycoses **3**: 1200
Mycospor **3**: 1213
Mycosporin **3**: 1213, 1217
Mycostatin **3**: 1238; **4**: 2013
Mycoster **3**: 1245
Mycotrim **3**: 1217
Mydfrin **4**: 2026, 2029
Mydplegin **4**: 2028
Mydriacyl **4**: 2028
Mydrian **4**: 2028
Mydriatics **4**: 2026
Mydriaticum **4**: 2028
Mydrilate **4**: 2028
Myelin **2**: 311
Myeloma cell lines **4**: 1985
Myelopoiesis **1**: 242
Myfungar **3**: 1225
MYK **3**: 1227
MYK-1 **3**: 1227
Mykinac **3**: 1238
Myko Crdes **3**: 1217
Mykofungin **3**: 1217
Mykohaug C **3**: 1217
Mykopevaryl **3**: 1219
Myko Posterine **2**: 551
Mykotral **3**: 1210
Mykundex **3**: 1238
Mylaxen **2**: 674
Mylepsin **2**: 480
Mylepsinum **2**: 480
Mylproin **2**: 477
Myobid **2**: 699
Myocard **1**: 152
Myocardial infarction **1**: 144
Myoclonic seizures **2**: 470
Myocrisin **4**: 1703
Myo-inositol **1**: 23
Myospan **2**: 681
Myprozine **3**: 1236
Myroxim **2**: 629
Mysalfon **2**: 577
Mysoline **2**: 480

N-0923 **2**: 579
Nabumetone **2**: 360
Naclex **1**: 294
Nacom **2**: 569
Na^+ concentrations
 in neurophysiologie **2**: 321
Nadisan invenol **3**: 1655
Nadolol **1**: 92, 56, 226
Nadostine **3**: 1238
Naegleria **3**: 1169
Nafarelin **4**: 1932
Nafcillin **3**: 962
Nafteryl **3**: 1249
Naftifine **3**: 1249

Naftifine hydrochloride **4**: 1731
Naftin **3**: 1249
Naganol **3**: 1149
Nagel Batrafen **3**: 1244
Nalador **4**: 2059
Nalbuphine **2**: 417
Nalfon **4**: 1690
Nalgesic **4**: 1690
Nal-Glu **4**: 1935
Nalidixic acid **3**: 1116, 1117
N-Alkyl-β-aminopropionate **4**: 2103
Nalline **2**: 419
Nalorex **2**: 423
Nalorphine **2**: 418
Nalorphine Serb **2**: 419
Naloxone **2**: 422
Naltrexone **2**: 422
NAN-190 **2**: 814
Nandrolone **3**: 1592
Nanoparticles **4**: 2120
Nansidol **2**: 593
Napa **1**: 133
Napanol **4**: 1687
Napental **2**: 509
Naphazoline hydrochloride **4**: 2029
Naphcon **4**: 2029
Naphthiomate-T **3**: 1247
2-(1-Naphthylmethyl)-2-imidazoline hydrochloride **4**: 2029
Naphuride **3**: 1149
Naprosyn **2**: 368
Naproxen **2**: 366; **4**: 1688
Naquasone **4**: 2178
Narasin **4**: 2170
Narbomycin **3**: 997
Narcan **2**: 422
Narcanti **2**: 422
Narcoren **2**: 509
Narcotan **2**: 440
Narcotic analgesic drugs
 veterinary **4**: 2182
Narcozep **2**: 520
Nardelzine **2**: 624
Nardil **2**: 624
Narkosid **2**: 439
Narkothion **2**: 446
Naropin **2**: 547
Nasal administration **4**: 2084
Nasal solutions **4**: 2088
Nasemo **3**: 1282
Nastenon **3**: 1590
Natacyn **3**: 1236; **4**: 2013
Natafucin **3**: 1236
Natam **2**: 522
Natamycin **3**: 1065, 235; **4**: 2013
Nateglimide **3**: 1658
National pharmacopeias **4**: 2152

Natrii sulfas decahydricus **2**: 785
Natrilix **1**: 73, 295
Natrium sulfuricum **2**: 785
Naturacil **2**: 783
Naturetin **1**: 294
Nausea **2**: 741
Nautamine **2**: 567
Navane **2**: 597
Navaron **2**: 597
Navelbine **4**: 1912
Navidrex **1**: 294
Naxagolide **2**: 577
Naxogin **3**: 1156
Nebularine **3**: 990
Necator americanus **3**: 1273
Nedaplatin **4**: 1921
Nedocromil **2**: 934
Nedocromil sodium **2**: 896
H. S. Need **2**: 498
Nefazodone **2**: 633
Nefopam **2**: 425
Negative inotropic potency **1**: 205
NegGram **3**: 1118
Neisseriaceae **3**: 1110
Neisseria gonorrhoeae vaccine **4**: 1784
Nelaxan **2**: 645
Nelbon **2**: 518
Nelfinavir **3**: 1307
Nemactil **2**: 593
Nematodes **3**: 1272
Nematode vaccine **4**: 1808
Nembutal **2**: 509
Nemexin **2**: 423
Neo-Antergan **2**: 906
Neo-Barb **2**: 509
Neo-Bridal **2**: 906
Neocarzinostatin **3**: 1008, 1013, 1062
Neocidin **4**: 2010
Neo-Cortef **4**: 2016
NeoDecadron **4**: 2008, 2016, 2017
Neo Dexair **4**: 2008
Neodit **2**: 618
Neodorm **2**: 509
Neoendorphin **2**: 386
Neo-Gilurytmal **1**: 127
Neogynon **3**: 1598
Neo Hombreol M **3**: 1590
Neolior **2**: 617
Neolutin forte **3**: 1596
Neomethymycin **3**: 996
Neomycin **3**: 984; **4**: 2008
Neomycin B **3**: 985
Neomycin C **3**: 984
Neomycin sulfate **3**: 1044
Neomycin Sulfate–Polymycin B Sulfate–Gramicidin
 Solution **4**: 2010
Neophyllin **2**: 888

Neo-Predef   4: 2178
Neosporin   4: 2008, 2009, 2010, 2011
Neosteron   3: 1590
Neosynephrine   4: 2027
Neotal   4: 2010
Neo Tears   4: 2033, 2034, 2037
Neothylline   2: 889
Neotricin   4: 2008, 2009, 2010, 2011
Neox   3: 1284
Nephron   1: 289
Neplanocin A   3: 994
Neptall   1: 93, 52
Neptazane   4: 2023
Nerdipina   1: 210
Nerusil   2: 622
Nerve block anesthetics   2: 539
Nerve cells   2: 308
Nervifene   2: 498
Nervistop L   2: 522
Nesacaine   2: 544
Nesdomal   2: 445
NET EN   3: 1504
(±)-N-ethyl-N,N-dimethyl-(4,4-diphenyl-2-butyl) ammonium bromide   2: 693
N-ethyl-α-methyl-3-(trifluoromethyl)benzeneethanamine   2: 461
Netilmicin sulfate   3: 1048
Neuchlonic   2: 518
Neulactil   2: 593
Neuleptil   2: 593
Neulin   2: 888
Neupan   2: 651
Neuractiv   2: 651
Neuralex   2: 622
Neur-Amyl   2: 507
Neuroanatomy   2: 318
Neurobiochemistry   2: 323
Neurochemistry   2: 323
Neurocil   2: 591
Neuroendocrine secretion   3: 1536
Neurogard   2: 575
Neuroimmophilins
   in Parkinsonism treatment   2: 580
Neurokin   2: 501
Neurokinins   3: 1394
Neuroleptanalgesics   2: 437, 453
Neuroleptics   2: 588
Neurolithium   2: 631
Neurological Spasticity
   Chronic   2: 680
Neuromedin U   3: 1408, 411
Neuromet   2: 651
Neuromuscular blocking agents   2: 667
Neuromuscular junction   2: 668
Neuromyfar   2: 606
Neuronal integration   2: 314
Neurons   2: 308

Neurontin   2: 483
Neuropeptide K   3: 1395
Neuropeptides   2: 330
Neuropeptide Y   3: 1408, 1409
Neuropharmaceuticals   2: 305
Neuropharmacology   2: 325
Neurophysiology   2: 320
Neuroplegil   2: 590
Neuroprocin   2: 501
Neurotranq   2: 591
Neurotransmitters   2: 310, 328
   in control of appetite   2: 458
Neurotrophic growth factors
   in Parkinsonism treatment   2: 579
Neurotrophic immunophilin ligands   2: 579
Neurotropic – Musculotropic spasmolytics   2: 698
Neurotropic Spasmolytics   2: 687
Neuroxin   2: 657; 4: 1706
Neutraphylline   2: 889
Neutrophil granulocytes   1: 245
Neutrophils   1: 239
Nevirapine   3: 1301
Nevripan   2: 495
New chemical entities   1: 8
Nexen   2: 384
Niacin   1: 23
Nialamide   2: 623
Niamide   2: 623
Niaprazine   2: 916
Nibal   3: 1594
Nicant   1: 210
Nicapress   1: 210
Nicardal   1: 210
Nicardipine   1: 203, 210, 229
Nicarpin   1: 210
Nicergoline   2: 655
Nicergolyn   2: 655
NicerHexal   2: 655
Nicetile   2: 659
Nicholin   2: 653
Niclocide   3: 1282
Niclosamide   3: 1269, 1270, 1271, 1282
Nicobid   1: 23
Nicodel   1: 210
Nicolar   1: 23
Nicolin   2: 653
Nicolsint   2: 653
Nicorandil   1: 230
Nicotine
   in synaptic transmission   2: 327
Nicotinic Acid   1: 23
Nicotinic receptor agonists   2: 580
NIDDM   3: 1653
Nidrel   1: 210
Nifaltide   3: 1384
Nifedicor   1: 210
Nifedin   1: 210

Nifedipine   **1**: 44, 95, 203, 210, 229
Nifelan   **1**: 210
Nifelat   **1**: 210
Nifensar XL   **1**: 210
Niflam   **4**: 1699
Niflugel   **2**: 353
Niflumic acid   **2**: 352; **4**: 1698
Niflurid   **2**: 353
Nifluril   **4**: 1699
Nifurtimox   **3**: 1152
Niloric   **2**: 655
Nilstat   **3**: 1238; **4**: 2013
Niludipine   **1**: 203, 211
Nilutamide   **4**: 1929
Nilvadipine   **1**: 211
Nimesulide   **2**: 383
Nimetazepam   **2**: 522
  synthesis   **2**: 520
Nimicor   **1**: 210
Nimodipine   **1**: 203, 210, 229; **2**: 659
Nimorazole   **3**: 1156, 158
Nimotop   **1**: 211; **2**: 659
Nipent   **4**: 1881
Nipodal   **2**: 594
Nipride   **1**: 101
Niridazole   **3**: 1267, 276, 282
Nirvetil   **2**: 500
Nirvotin   **2**: 500
Nisidana   **2**: 619
Nisoldipine   **1**: 203, 211, 229
N-isopropyl-2-methyl-2-propyl-1,3-propanediol dicarbamate   **2**: 677
Nitepam   **2**: 518
Nitracin   **4**: 1903
Nitrados   **2**: 518
Nitrates
  treatment of angina pectoris   **1**: 222
Nitrazepam   **2**: 515, 518
  indications for use   **2**: 519
  synthesis   **2**: 520
Nitrempax   **2**: 518
Nitrendipine   **1**: 96, 203, 210, 229
Nitro-bid   **1**: 223
Nitro compounds   **1**: 14
Nitrodisc   **4**: 2118
Nitrodisk   **1**: 223
Nitrodur   **1**: 223
Nitrofurantoin   **3**: 1145
1-[(5-Nitrofurfurylidene)amino]hydantoin   **3**: 1145
4-[(5-Nitrofurfurylidene)amino]-3-methyl-4-thiomorpholine-1,1-dioxide   **3**: 1152
3-[(5-nitrofurfurylidene)amino]-2-oxazolidinone   **3**: 1158
Nitrogen monoxide   **2**: 438
Nitrogen mustard   **4**: 1886
Nitroglycerin   **1**: 222, 223
4-[2-(5-Nitroimidazolyl-1-yl)ethyl]morpholine   **3**: 1156

Nitrolingual   **1**: 223
Nitromannite   **1**: 224
Nitromannitol   **1**: 224
Nitromethaqualone   **2**: 514
N-[(2-Nitrooxy)-ethyl]-3-pyridinecarboxamide   **1**: 230
2,2-Bis[(nitrooxy)methyl]-1,3-propanediol dinitrate   **1**: 224
N-(4-Nitro-2-phenoxyphenyl)methanesulfonamide   **2**: 383
1-[[5-(4-nitrophenyl)furfurylidene]amino]hydantoin   **2**: 675
Nitropress   **1**: 101
Nitroprusside   **1**: 101
Nitrosoureas   **4**: 1891
Nitrostat   **1**: 223
1-(5-Nitro-2-thiazolyl)-2-imidazolidinone   **3**: 1282
Nitrous oxide   **2**: 438
Nivadil   **1**: 211
Nivaquine   **3**: 1162
Nizatidine   **2**: 723
Nizax   **2**: 724
Nizoral   **3**: 1208; **4**: 2013
Nizshampoo   **3**: 1208
NK-601   **4**: 1928
NK-622   **4**: 1922
NKA   **3**: 1394
NKB   **3**: 1394
NKS-01   **4**: 1924
NMR   **2**: 325
NMU   **3**: 1411
Nobadorm   **2**: 514
Noberastine   **2**: 930
Nocardiaceae   **3**: 1110
Nocardicin A   **3**: 976
Nocardicin B   **3**: 976
Nocardicin C   **3**: 976
Nocardicin D   **3**: 976
Nocardicin E   **3**: 976
Nocardicin F   **3**: 976
Nocardicin G   **3**: 976
Nocardicins   **3**: 975
Nociceptin   **2**: 386
Noctal   **2**: 510, 524
Noctamid   **2**: 523
Noctazepam   **2**: 522, 638
Nocton   **2**: 523
Noctosom   **2**: 522
Noemin   **2**: 730
Nogedal   **2**: 619
Noiafren   : 644
Noin   **2**: 658
Noleptan   **2**: 339
Noludar   **2**: 512
Nolurate   **2**: 512
Nolvadex   **4**: 1922
Nometine   **2**: 593
Nomotopic disorders   **1**: 117

Nonallergic asthma **2**: 862
Nonallergic contact dermatitis **4**: 1718
Noncardioselective agents **1**: 226
Nonglycosidic inotropic agents **1**: 191
Noninsulin dependent diabetes mellitus **3**: 1653
Nonionic surfactants **4**: 2103
non-ISA drugs **1**: 141
Nonoxynol-9 **3**: 1495
Nonpotent tinctures **4**: 2087
Nonpressin **2**: 564
Nonrapid eye movement sleep **2**: 490
Nonsteroidal anti-inflammatory drugs **2**: 341, 344; **4**: 1676
 in dermatology **4**: 1721
 ophthamological **4**: 2019
 veterinary **4**: 2175
7-Nonylphenoxypolyethoxyethanol **3**: 1495
Noostan **2**: 650
Nootrop **2**: 650
Nootropics **2**: 650
Nopar **2**: 691
Nopron **2**: 916
[8-(noradamantan-3-yl)-1,3-dipropylxanthine] **1**: 301
Noradrenalin 1 : 1000 JENA-PHARM **1**: 60
Noradrenalina tartrato **1**: 60
Noradrenaline **1**: 60; **2**: 328; **3**: 1550, 1560, 1562
 in synaptic transmission **2**: 327
 influence on sleep **2**: 491
Noradrenaline uptake inhibitors **2**: 611
Noral **2**: 648
Norandrol **3**: 1592
Norandrostane derivatives **3**: 1592
19-Norandrostane gestagens **3**: 1598
Norcuron **2**: 673
Nordette **3**: 1500
Nordic Biological Unit **2**: 939
Norditropin **3**: 1437
Norephedrine **2**: 463
Norepinephrine **1**: 60; **2**: 458; **3**: 1560, 1562
norethindrone **3**: 1497, 1500, 1502
Norethindrone acetate **3**: 1500
Norethindrone enanthate **3**: 1504
Norethisterone **3**: 1497
Norethisterone synthesis **3**: 1575
Norethynodrel **3**: 1497
Norethynordrel **3**: 1497
Norfenefrine **1**: 61
Norfenefrin retard forte-ratiopharm **1**: 62
Norfenefrin "Ziethen" **1**: 62
Norflex **2**: 567, 680
Norfloxacin **3**: 1122
Norgestimat **3**:
Norgestimate **3**: 1497, 1500, 1502; **4**: 1726
Norgestrel **3**: 1598
D-Norgestrel synthesis **3**: 1579
Norgestrienone **3**: 1598
Norgestrienone synthesis **3**: 1580

D-Norgestrol **3**: 1598
Norglycin **3**: 1657
Norlutin **3**: 1657
Norlutin A **3**: 1657
Normacol **2**: 784
Normadate **1**: 155
Normalmin **2**: 594
Norma-sterin **3**: 1598
Normison **2**: 523
Normoc **2**: 637
Normodyne **1**: 155
Normoglaucon **4**: 2022
Normonal **1**: 295
Normorest **2**: 514
Norofulvin **3**: 1242
Noroxin **3**: 1123
Norpace **1**: 125
Norplant **3**: 1505
Norpolake **2**: 614
Norpramin **2**: 614
19-Nor-17-pregna-1,3,5(10)-trien-20-yne-3,17-diol **3**: 1501
19-Norpregnane gestagens **3**: 1596
Norpseudoephedrine **2**: 464
Norstenol **3**: 1592
Nortestonat **3**: 1592
Nortestosterone **3**: 1592
Nortestosterone cyclopentylpropionate **3**: 1592
Nortestosterone decanoate **3**: 1592
Nortestosterone phenylpropionate **3**: 1592
19-Nortestosterone synthesis **3**: 1575
Nortimil **2**: 614
Nortrilen **2**: 615
Nortriptyline **2**: 615
Norvasc **1**: 211, 212
Norvedan **4**: 1686
Norvir **3**: 1311
Norwalk **3**: 1174
Noscal **3**: 1666
Nosocomial infections **4**: 1776
Notezine **3**: 1279
Nourilax **2**: 788
Novadral **1**: 62
Novaldex **3**: 1587
Novalgin **2**: 371
Novamobarb **2**: 507
Novanox **2**: 518
Novanthrone **4**: 1902
Nova-Rectal **2**: 509
Nova T **3**: 1507
Novatropina **2**: 691
Novesin **4**: 2003
Novesina **4**: 2003
Novesine **4**: 2003
Novesine Wander **2**: 544
Novidorm **2**: 525

Novobiocin **3:** 1009
   staphylococcus aureus inhibition **3:** 1091
Novocain **2:** 544
Novocamid **1:** 125
Novocetam **2:** 650
Novochlorhydrate **2:** 498
Novodigal **1:** 189
Novodil **1:** 39
Novodorm **2:** 525
Novoflupam **2:** 522
Novolurazine **2:** 594
Novonal **2:** 502
Novo-Naprox **4:** 1689
Novonorm **3:** 1658
Novopentobarb **2:** 509
Novoperidol **2:** 598
Novoridazine **2:** 592
Novosecobarb **2:** 511
Novothyral **3:** 1561
Novoxapin **2:** 613
Noxair **2:** 877
Noxiptilin **2:** 618
Noxybel **2:** 514
Nozinan **2:** 591
NP-27 **3:** 1248
N-Phenyl-N-[1-(2-phenylethyl)-4-piperidinyl]-propana-
   mide **2:** 454
N-phenyl-N-(phenylmethyl)-1-pyrrolidineethana-
   mine **2:** 906
NPY **3:** 1408, 409
NREM sleep non REM **2:** 490
NS-2214 **2:** 581
N(α)-([(S)-4-Oxo-2-azetidinyl]carbonyl)-L-histidine-L-
   prolinamide **2:** 581
NSAID **2:** 341
   in dermatology **4:** 1721
NSAIDs **2:** 344
   ophthamological **4:** 2019
NSC-102816 **4:** 1875
NSC-109724 **4:** 1890
NSC-118742 **4:** 1896
NSC-119875 **4:** 1918
NSC-122819 **4:** 1912
NSC-123127 **4:** 1897
NSC-125066 **4:** 1905
NSC-125973 **4:** 1915
NSC-13875 **4:** 1896
(NSC-141540 **4:** 1912
NSC-156303 **4:** 1903
NSC-178248 **4:** 1891
NSC-208734 **4:** 1901
NSC-218321 **4:** 1881
NSC-239336 **4:** 1879
NSC-241240 **4:** 1920
NSC-245467 **4:** 1911
NSC-249992 **4:** 1903
NSC-256439 **4:** 1898

NSC-26271 **4:** 1888
NSC-264880 **4:** 1877
NSC-270561 **4:** 1893
NSC-27640 **4:** 1873
NSC-301739 **4:** 1902
NSC-3053 **4:** 1904
NSC-3088 **4:** 1889
NSC-312887 **4:** 1880
NSC-337766 **4:** 1903
NSC-363812 **4:** 1921
NSC-375101D **4:** 1921
NSC-409962 **4:** 1891
NSC-45388 **4:** 1895
NSC-49842 **4:** 1910
NSC-609699 **4:** 1914
NSC-616348 **4:** 1914
NSC-628503 **4:** 1915
NSC-63878 **4:** 1877, 882
NSC-67574 **4:** 1910
(NSC-740) [*59-05-2*], N-(4-{(2,4-Diamino-6-
   pteridinyl)methyl}methylamino}benzoyl)-L-gluta-
   mic acid **4:** 1868
NSC-755 **4:** 1882
NSC-762 **4:** 1886
NSC-77213 **4:** 1894
NSC-79037 **4:** 1891
NSC-82151 **4:** 1897
NSC-85998 **4:** 1891
NSC-8806 **4:** 1887
NSC-91 523 **1:** 159
NSC-94941 **4:** 1891
Nubain **2:** 418
Nuclear magnetic resonance **2:** 325
Nucleic acid **3:** 1172
Nucleoside Analogs reverse transcriptase inhibi-
   tors **3:** 1295
Nucleoside analogues **3:** 1177
Nucleosides antibiotics **3:** 988
Nuctane **2:** 525
Nuetalon **2:** 524
Numal **2:** 508
Numbon **2:** 518
Numorphan **2:** 397
Numotac **2:** 867
Nupercainal **2:** 549
Nuran **2:** 924
Nuredal **2:** 623
Nuromax **2:** 670
NVB **4:** 1912
Nyaderm **3:** 1238
Nydrazid **3:** 1140
Nysert **3:** 1238
Nystacid **3:** 1238
Nystaderm **3:** 1238
Nysta-Dome **3:** 1238
Nystan **3:** 1238
Nystatin **3:** 1065, 1234; **4:** 1732, 2013

Nystatin A1 **3**: 1237
Nystat-Rx **3**: 1238
Nystavescent **3**: 1238
Nystex **3**: 1238; **4**: 2013

Obedrin **2**: 461
Obetrol-10 **2**: 461
Obetrol-20 **2**: 461
Obstinol **2**: 787
Obstinol Mild **2**: 781
Obstipation **2**: 780
Obytin **3**: 1245
OC-340 **2**: 660
Occluding platelet aggregates **1**: 32
Oceral **3**: 1225
(–)-(4a*R*,8a*R*)-4,4a,5,6,7,8a,9-Octahydro-5-propyl-1*H*-pyrazolo[3,4-g]quinoline **2**: 579
Octatropine methylbromide **2**: 692
Octimibate **4**: 2054
Octoxynol **3**: 1495
Octreotide **3**: 1448
Octylphenoxypolyethoxy ethanol **4**: 2039
Ocu-caine **4**: 2003
Ocu-Carpine **4**: 2024
Ocu-Chlor **4**: 2011
Ocuclear Eye Drops **4**: 2029
Ocu-Cort **4**: 2008, 2011, 2016
Ocufen **2**: 363
Ocular fungal infection **4**: 2012
Ocu-Lone-C **4**: 2005
Ocu-Mycin **4**: 2008, 2009, 2011
Ocu-Phrin **4**: 2029
Ocu-Pred **4**: 2017
Ocu-pred A **4**: 2017
Ocupress **1**: 154
Ocusert **4**: 2024, 2119
Ocu-Spor **4**: 2008, 2011
Ocutricin HC **4**: 2008, 2011, 2016
Ocu-Trol **4**: 2008, 2011, 2018
Oekolb **3**: 1586
Ofloxacin **3**: 1125; **4**: 2012
Oftalmocaina **4**: 2003
4-OHA **4**: 1924
OHP **4**: 1921
Ointments **4**: 1715, 2090
Olamin **1**: 209
Olbemox **1**: 24
Olbetam **1**: 24
Oleandomycin **3**: 997; **4**: 2165
Oleandomycin phosphate **3**: 1051
Oleandrigenin **1**: 179
α-L-Oleandrose **1**: 180
Oleptan **2**: 339
2′,5′-Oligoadenylate synthetase **4**: 1963
Oligodendrocytes **2**: 311
Olivanic acid MM 4550 **3**: 974
Olren **2**: 699

Omca **2**: 595
Omeprazole **2**: 714, 717, 720
Omeril **2**: 925
Omnipress **2**: 612
Omoconazole **3**: 1223
Onapristone **3**: 1510
Onchocerca volvulus **3**: 1277
Oncovin **4**: 1910
Ondansetron **2**: 800
Onkotrone **4**: 1902
ONO 4057 **4**: 2066
ONO LB-457 **4**: 2066
Onon **4**: 2064
Onset of activity **4**: 2080
Ontosein **4**: 1709
Opalmon **4**: 2060
OPC 1085 **1**: 154
OPC-31260 **1**: 300
Opcon **4**: 2029
Opdensit **2**: 699
Operational qualification **4**: 2143
Opertil **2**: 609
Ophthaine **2**: 545; **4**: 2003
Ophthalgan **4**: 2023
Ophthalmic administration **4**: 2084
Ophthalmic controlled-release system **4**: 2119
Ophthalmic solutions **4**: 2088
Ophthalmological agents **4**: 2001
Ophthetic **4**: 2003
Ophtocortin **4**: 2018
Opiates **2**: 386
Opioid agonists **2**: 389
Opioid antagonists **2**: 421
Opioid Peptides **3**: 1379
Opioid Receptors **3**: 1379
Opioid receptor types **2**: 387
Opioids **2**: 386
Opipramol **2**: 619
Opiran **2**: 602
Opisthorchis species **3**: 1268
Opportunistic infections **3**: 1109
Opsonins **4**: 1836
Opstin **4**: 2027
Optanox **2**: 508
Optazine **4**: 2029
Optef **4**: 2017
Opteron **1**: 41
Opthochlor **4**: 2011
Opthocort **4**: 2011
Opticrom **4**: 2019
Opti-Free **4**: 2038
Optigene 3 **4**: 2030
Optimil **2**: 514
Optinoxan **2**: 514
Optipranolol **4**: 2022
Optistin **4**: 2029
Optizyme **4**: 2040

2285

Opustan **4**: 1697
Orabolin **3**: 1592
Oracon **3**: 1598
Oradexon **3**: 1627
Oradrate **2**: 498
Oral administration **4**: 2083
Oral antidiabetic drugs **3**: 1653
Oralcon **4**: 2022
Oral contraceptives **3**: 1497
Oralep **2**: 602
Oral inhalation **4**: 2085
Oral vaccines **4**: 1762
Oranabol **3**: 1593
Oranyst **3**: 1238
Orap **2**: 602
Orasthin **3**: 1402
Ora Testryl **3**: 1590
Oratrol **4**: 2022
Orbinamon **2**: 597
Orciprenaline **2**: 869
Oretic **1**: 293
ORF 5513 **3**: 1514
Orfidal **2**: 522
Orfiril **2**: 477
Org-5222 **2**: 609
Orgallin **2**: 753
Orgametril **3**:
Orgasteron **3**: 1598
Orgatax **2**: 646
Orgatrax **2**: 915
Orgotein **4**: 1708
Orientomycin **3**: 1008
Orimeten **4**: 1924
Orix **1**: 210
Orlaam **2**: 406
Orlest **3**:
Ormaplatin **4**: 1921
Ormodon **2**: 518
Ornidazole **3**: 1156, 158
Ornipressin **3**: 1402
Ornoprostil **4**: 2060
Oromycosal **3**: 1208
Oronazol **3**: 1208
Orotic acid **2**: 769
Orotirelin **2**: 581
Orotyl-L-histidine-L-prolinamide TRH **2**: 581
Orphenadrine **2**: 567, 680
Orsanil **2**: 592
Orthoboric acid **4**: 2038
Orthoclon **4**: 1996
Orthocort **4**: 2017
Ortho Cyclen/Cilest **3**: 1500
Orthodox sleep **2**: 490
Ortho Drops **4**: 2016
Orthomyxoviridae **3**: 1174
Ortho-Novum **3**: , 500, 502, 502
Orthostatic hypotension **1**: 146

Ortho Tri-Cyclen **3**: 1502
Orudis **2**: 366; **4**: 1690
Osmoglyn **4**: 2023
Osmotic control
 of drug release **4**: 2115
Osnervan **2**: 564
Ospolot **2**: 486
Osteoarthritis **4**: 1674
Osteogenic Proteins **3**: 1361
Osyrol **3**: 1617
OT **3**: 1398
OTC products **1**: 5
Other fluoroquinolones **3**: 1126
Other hydantoins **2**: 475
Otic administration **4**: 2084
Otic solutions **4**: 2088
Otilonium bromide **2**: 695
Otrivin **4**: 2029
Ouabagenin **1**: 179
Ouabain **1**: 181, 90
Ouabain-Nativelle **1**: 190
Ovarid **3**: 1596
Ovestin **3**: 1583
Ovex **3**: 1583
Ovin **3**: 1598
Ovis Neu **3**: 1217
Ovocylin **3**: 1583
Ovosiston **3**: 1596
Ovsatol **3**: 1584
Ovulen **3**:
Oxabolone Cypionate **3**: 1592
1-Oxacephalothin **3**: 970
1-Oxacephems **3**: 969
Oxacillin **3**: 962
 staphylococcus aureus inhibition **3**: 1091
Oxacillin sodium **3**: 1021
Oxadethiapenam **3**: 970
Oxaflozane **2**: 629
Oxaliplatin **4**: 1921
Oxametacine **4**: 1681
Oxamniquine **3**: 1267, 283
Oxamorph **2**: 390
Oxandrolone **3**: 1594
oxantel **3**: 1274, 283
Oxapium iodide **2**: 696
Oxaprozine **2**: 381
Oxatomide **2**: 916
Oxazepam **2**: 515, 522, 638
 indications for use **2**: 519
 synthesis **2**: 520
Oxazinomycin **3**: 994
Oxcarbazepine **2**: 476
Oxcord **1**: 210
Oxepam **2**: 522
Oxetacaine **2**: 549
Oxford unit **3**: 958
Oxibuprokain Minims **4**: 2003

Oxiconazole   **3:** 1224; **4:** 1730
Oxipertine   **2:** 609
Oxiracetam   **2:** 651
Oxis   **2:** 883
Oxistat   **3:** 1225
Oxitropium bromide   **2:** 892
2-Oxo-1,3-benzoxazine-3(4H)-acetamide   **2:** 624
γ-oxo(1,1′biphenyl)-4-butanoic acid   **4:** 1686
4-(3-Oxobutyl)-1,2-diphenyl-pyrazolidine-3,5-dione   **2:** 372
Oxoformycin   **3:** 993
1-(5-Oxohexyl)theobromine   **1:** 48
Oxolinic acid   **3:** 1116, 118
5-Oxo-L-proline-L-leucine-L-prolinamide   **2:** 581
Oxomemazine   **2:** 921
2-Oxo-1-pyrrolidineacetamide   **2:** 650
22-Oxovincaleukoblastine   **4:** 1910
Oxpam   **2:** 522
Oxprenolol   **1:** 157, 226
Oxybuprocaine hydrochloride   **2:** 544; **4:** 2003
Oxybutynin   **2:** 704
Oxycodone   **2:** 396
OxyContine   **2:** 397
Oxydiazepam   **2:** 523
2′,4′-O-(Oxydistibylidyne)bis(D-gluconic acid)-2,4-Sb,Sb′-dioxide trisodium salt · nonahydrate   **3:** 1154
Oxygesic   **2:** 397
Oxylone   **3:** 1624
Oxymesterone   **3:** 1593
Oxymetazoline hydrochloride   **4:** 2029
Oxymetholone   **3:** 1590
Oxymorphone   **2:** 397
Oxymycin   **3:** 1008
Oxyntomodulin   **3:** 1428
Oxyphenbutazone   **2:** 373; **4:** 1694
Oxyphenonium bromide   **2:** 696
Oxyphenylbutazone   **2:** 373
Oxyphylline   **2:** 888
OxySept   **4:** 2039
Oxytetracycline   **3:** 954, 978, 1037; **4:** 2007, 2010, 2161
Oxytocin   **3:** 1398, 1402, 1542

P-1382   **2:** 797
Pabenol   **2:** 652
PACAP   **3:** 1425
Pacinol   **2:** 595
Pacinone   **2:** 643
Pacinox   **2:** 502
Pacisyn   **2:** 518
Paclitaxel   **4:** 1915
   annual sales   **4:** 1939
Pacteus   **1:** 153
Padeskin   **2:** 700
Padisal   **2:** 922
Padrin   **2:** 696

Pain
   classification   **2:** 343
   treatment   **2:** 341
Palatability   **4:** 2102
Palfium   **2:** 406
Palmitate hydrochloride   **4:** 2011
Paludrine   **3:** 1165
p-Aminosalicylic acid   **3:** 1141
Panaldine   **1:** 41
Pancopride   **2:** 796
Pancreastatin   **3:** 1449
Pancreatic Peptide   **3:** 1408
Pancreatin   **4:** 2040
Pancuronium bromide   **2:** 671
Panectyl   **2:** 591
Panergon   **2:** 699
Panfungol   **3:** 1208
Panlomyc   **3:** 1230
Panomifene   **4:** 1924
Pantocain   **2:** 545
Pantocaine   **2:** 545
Pantoprazole   **2:** 714, 717, 720
Pantozol   **2:** 721
Papain   **4:** 2040
Papaverine   **2:** 699
Papaverine-like spasmolytics   **2:** 698
Papilloma   **3:** 1174
Papovaviridae   **3:** 1174
Paquil   **3:** 1286
Paracetaldehyde   **2:** 497
Paracetamol   **2:** 348
Paracodin   **2:** 393
Paracrine effect   **3:** 1340
Paracrine secretion   **3:** 1536
Paradox sleep   **2:** 490
Paraffinum liquidum   **2:** 781
Paraflex   **2:** 677
Paragonimus species   **3:** 1267
Parainfluenza   **3:** 1174
Paral   **2:** 497
Paraldehyde   **2:** 497
Paraldehyde Thilo   **2:** 497
Paramethasone   **3:** 1629
   doses   **3:** 1605
Paramyxoviridae   **3:** 1174
Parapenzolate bromide   **2:** 696
Paraplatin   **4:** 1920
Parapulmonary anesthetics   **2:** 437
Paraquick   **2:** 645
Parasites vaccines   **4:** 1805
Parasitic diseases
   treatment with monoclonal antibodies   **4:** 1843
Parasiticides
   veterinary   **4:** 2173
Parasitic Worms   **3:** 1266
Parasympathetic system   **2:** 312
Parasympatholytics   **2:** 687

Parasympathomimetic agents
    antiglaumatons   **4**: 2023
Parathormone   **3**: 1546
Parathyroid Hormone   **3**: 1362
Parcil   **2**: 641
Paredrine   **4**: 2027
Parenteral administration   **4**: 2081
Parenteral solutions   **4**: 2088
Parenteral vaccines   **4**: 1762
Parest   **2**: 514
Parfenac   **2**: 354
Pargin   **3**: 1219
Pargitan   **2**: 563
Pargyline   **2**: 624
Pariprazole   **2**: 719
Paritane   **1**: 157
Parkemed   **2**: 352; **4**: 1697
Parkin   **2**: 566
Parkinane   **2**: 563
Parkinson's disease   **2**: 555
Parkinsonism Treatment   **2**: 555
Parlodel   **2**: 576
Parmilene   **2**: 514
Parnate   **2**: 625
Parnox   **2**: 514
Paromomycin   **3**: 984
Paromomycin I   **3**: 985
Paromomycin sulfate   **3**: 1045
Déparon   **2**: 618
Parotid hormone   **3**: 1421
Paroxetine   **2**: 629
Paroxysmal Tachycardia   **1**: 117
Parsidol   **2**: 566
Parsilid   **1**: 41
Partial seizures   **2**: 470
Partocon   **3**: 1402
Parvoviridae   **3**: 1174
Parvoviruses   **3**: 1174
Pasaden   **2**: 595, 640
PAS – Heyl   **3**: 1142
Paspertin   **2**: 745
Passive immunization   **4**: 1763
Passive targeting   **4**: 2121
Pasterior pituitary hormones   **3**: 1539
Pastes   **4**: 1715, 2090
Pasteurellaceae   **3**: 1110
Patch-clamp electrode   **2**: 321
Patent protection
    monoclonal antibodies   **4**: 1994
Pathilon   **2**: 697
Pathogens
    transmittable by blood   **1**: 280
Pavabid   **2**: 699
Pavakey   **2**: 699
Pavaspan   **2**: 699
Paveron   **2**: 699
Pavulon   **2**: 672

Paxel   **2**: 637
Paxipam   **2**: 643
Paxisyn   **2**: 518
Paxyl   **2**: 596
PCPIM   **3**: 1216
PD – 107779   **3**: 1125
PD-81565   **4**: 1881
PDGF   **1**: 33
Pecilocin   **3**: 1008, 1069
Peckle   **3**: 1217
Pectobloc   **1**: 158
PED   **3**: 1587
Pedesal   **3**: 1248
Pediderm   **3**: 1248
Pedimycose   **3**: 1248
Pedisafe   **3**: 1217
Pefloxacin   **3**: 1125
Peganone   **2**: 475
Pellegal   **2**: 700
Pelson   **2**: 518
Pemal   **2**: 478
Pembule   **2**: 509
Pemolin   **2**: 339
Pena-Vel II   **4**: 2037
Penbon   **2**: 509
Penem-3-carboxylic acid   **3**: 972
Penems   **3**: 971
Penfluridol   **2**: 601
Pengitoxin   **1**: 188
D-Penicillamine   **2**: 771; **4**: 1705
Penicillin   **3**: 953, 959
Penicillin-2-hydroxyprocaine
    biological activity   **3**: 958
Penicillin extraction   **3**: 1086
Penicillin F   **3**: 960
Penicillin formation   **3**: 1084
Penicillin G   **3**: 960
Penicillin G benzathine   **3**: 1017
Penicillin G potassium   **3**: 1017
Penicillin G procaine   **3**: 1017
Penicillin K   **3**: 960
Penicillin manufacture   **3**: 1080
Penicillin N   **3**: 960
Penicillin-*N*-ethylpiperidine   **3**: 958
Penicillin-*N,N'*-dibenzyl-ethylenediamine   **3**: 958
Penicillin O   **3**: 962, 958
Penicillin (P)
    staphylococcus aureus inhibition   **3**: 1091
Penicillins
    as ophthamological agents   **4**: 2006
    natural   **3**: 959
    semisynthetic   **3**: 960
    veterinary   **4**: 2160
Penicillin V   **3**: 1018
Penicillin V potassium   **3**: 1018
Penicillin X   **3**: 960

Pentaacetylgitoxin 1: 188
Pentadorm 2: 497
Pentaerythritchloral 2: 499
Pentaerythritol tetranitrate 1: 224
Pentaerythrityl tetranitrate 1: 224
Pentagastrin 3: 1419
Pentagestrone 3: 1596
Pentagit 1: 188
(2R,3S)-3,3′,4′,5,7-pentahydroxyflavan 2: 765
Pental 2: 509
4,4′-(Pentamethylenedioxy)dibenzamidine 3: 1149
Pentamethylmelamine 4: 1896
N,N,N′,N′,N″-Pentamethyl-1,3,5-triazine-2,4,6-triamine 4: 1896
Pentamidine 3: 1149
Pentamidine isethionate 3: 1149
2,2′-[1,5-pentanediylbis[oxy(3-oxo-3,1-propanediyl)]]-bis[1-[3,4-dimethoxyphenyl)methyl]-1,2,3,4-tetrahydro-6,7-dimethoxy-2-methylisoquinolinium]dibenzenesulfonate 2: 669
Pentazine 2: 594
Pentazocine 2: 419
Penthrane 2: 442
Pento AbZ 1: 49
Pentobarbital 2: 504, 509
Pentobarbitone 2: 509
Pentoflux 1: 49
Pentone 2: 509
PENTO-PUREN 1: 49
Pentorex 2: 463
Pentosol 2: 509
Pentostam 3: 1154
Pentostatin 4: 1881
Pentothal 2: 445
Pentothal-Natrium 2: 445
Pentoxifylline 1: 48
pentox von ct 1: 49
2-Pentylaminoacetamide 2: 624
Pepdul 2: 723
Pepinal 2: 506
Pepleomycin 3: 1002, 1056
Peplomycin 4: 1906
Peplomycin sulfate 3: 1056
Pepsin 2: 714
Pepstatin 3: 1063
Pepticinnamin E 4: 1937
Peptide abbreviations 3: 1345
Peptide and Protein Hormones 3: 1542
Peptide antibiotics 4: 2169
  economic aspects 3: 1094
Peptide Nomenclature 3: 1342
Peptide regulatory factors 3: 1340
Peptides and Protein Hormones 3: 1339
Peptides antibiotics 3: 1001
Peptidomimetics 4: 1935
Peptococcaceae 3: 1110

Perandren 3: 1590
Peraprin 2: 745
Perasthman Inhalat 2: 891
Perazine 2: 593
Percoten 3: 1615
Percutalgine 2: 348
Percutaneous absorption 4: 1716
Percutaneous–Transhepatic Lysis 2: 758
Perdipina 1: 210
Perdipine 1: 210
Perequil 2: 645
Perfane 1: 196
Perfudan 1: 50
Perfume ingredients 4: 2102
Pergolide 2: 577
Pergonal 3: 1357
Perhexiline 1: 229
Periactin 2: 924
Pericyazine 2: 593
Perifosine 4: 1938
Perimetazine 2: 496
Perimethazine 2: 496
Peripheral decarboxylase inhibitors 2: 569
Peripheral nervous system 2: 312
Periplogenin 1: 179
Periplum 1: 211
Peristil 2: 754
Peritol 2: 924
Peritrate 1: 224
Perkod 1: 39
Perlapine 2: 496
Permitil 2: 595
Pernocton 2: 509
Pernoston 2: 509
Pernovin 2: 925
Peroxinorm 4: 1709
Perphenazine 2: 594
Persantin 1: 39
Persantine 1: 39
Persedon 2: 512
Pertofran 2: 614
Pertussis prophylaxis 4: 1828
Pertussis vaccine 4: 1768
Perverme 3: 1285
Pervitin 2: 339, 461
Pethidine 2: 397
Petnidan 2: 478
Petrichloral 2: 499
Petrolatum 4: 1723, 727
Pevalip 3: 1219
Pevaryl 3: 1219, 1220
Pexaqualone 2: 514
PGA$_2$ 4: 2049
PGD$_2$ 4: 2049
PGE$_1$ 4: 2048, 2049

PGE$_2$ 4: 2048, 2049
PGE$_3$ 4: 2048
PGF$_1\alpha$ 4: 2049
PGF$_2\alpha$ 4: 2048, 2049
PGF$_2\alpha$ synthesis 4: 2058
PGI$_2$ 1: 37
PGI$_2$ sodium salt 4: 2049
PGI$_2$ synthesis 4: 2069
PH 3: 1421
PH-20 3: 1512
Phanodorm 2: 507
Phanotal 2: 507
Pharmaceutical containers 4: 2106
Pharmaceutical dosage forms 4: 2077
Pharmaceutical excipients 4: 2101
Pharmaceutical Industry 1: 3
Pharmaceutical inspection convention 4: 2151
Pharmaceutical inspection cooperation scheme 4: 2152
Pharmaceutical market 1: 5
Pharmaceutical necessities 4: 2101
Pharmacokinetic studies 4: 2129
Pharmacological testing 4: 2128
Pharmacopeias 4: 2086, 2152
Pharmacopoeias 3: 1088
Pharmacovigilance 4: 2136
Pharmorubicin 4: 1898
Phase II investigations 4: 2135
Phase I investigations 4: 2134
Phase IV investigations 4: 2136
Phases of clinical drug development 4: 2134
Phenacetin 2: 349
Phenamin 2: 912
Phenazine 2: 595
Phen Bar 2: 506
Phenbenzamine 2: 904, 907
Phenbutrazate 2: 465
Phencol 2: 645
Phendimetrazine 2: 465, 467
Phenelzine 2: 624
Phenemalum 2: 506
Phenergan 2: 921
Phenethicillin 3: 962
   biological activity 3: 958
Phenethicillin potassium 3: 1018
Phenethylhydrazine 2: 624
N-(1-Phenethylpiperidin-4-yl)-N-phenylpropionamide 2: 399
Pheneturide 2: 501
4-[2-[[6-[(2-phenetyl)amino]hexyl]amino]ethyl]-1,2-benzenediol 1: 194
Phenformin 3: 1660
Phenhydan 2: 475
Phenidine 2: 349
Phenindamine 2: 925
Pheniramine 2: 912
Phenmetrazine 2: 465, 467

Phenobal 2: 480
Phenobarbital 2: 473, 479, 504, 506
Phenobarbitone 2: 479, 506
Phenobarbyl 2: 506
Phenol 4: 1723
Phenolphthalein 2: 787
Phenoperidine 2: 453
Phenothiazine antiemetics 2: 744
Phenothiazine neuroleptics 2: 589
Phenothiazines antihistamines 2: 918
Phenoxene 2: 742
Phenoxybenzamine 2: 706
Phenoxymethylpenicillin 3: 962, 018, 958
Phenoxymethylpenicillin potassium 3: 1018
4-[3-(α-Phenoxy-p-tolyl)propylmorpholine hydrochloride 2: 551
Phenpentermine 2: 463
Phenprobamate 2: 645
Phenprocoumon 1: 15
Phensuximide 2: 478
Phentermine 2: 462, 467
Phentolamine 1: 87; 2: 706
Phenydan 1: 126
L-Phenylalanine mustard 4: 1887
Phenylalkylamines 1: 208
Phenylbutazone 2: 374; 4: 1694, 2177
2-phenylbutyrylurea 2: 501
Phenylephrine hydrochloride 4: 2026, 2029
N^1-{3-[(1-Phenylethyl)amino]propyl}bleomycinamide 4: 1906
Phenylethylbiguanide 3: 1660
D-Phenylglycine 3: 1012
β-Phenylisopropylamines 2: 460
Phenylmercury(II) acetate 4: 2037
Phenylmercury(II) nitrate 4: 2037
3-Phenyl-3-(4-methylphenyl)-N,N-dimethylpropanamine 2: 912
γ-Phenyl-N-(1-phenylethyl)benzenepropanamine 1: 208
1-Phenylpentanol 2: 750
2-Phenyl-1,3-propanediol dicarbamate 2: 485
1-Phenylpropanol 2: 750
Phenylpropanolamine 2: 463
17β-Phenylpropionyloxy-4-estren-3-one 3: 1592
3-Phenylpropyl carbamate 2: 645
1-Phenyl-2-propynyl carbamate 2: 500
6-Phenyl-2,4,7-pteridinetriamine 1: 75
{(+)-R-1-[(E)-3-(2-phenylpyrazolo[1,5-a]pyridin-3-yl) acrylol]-2-piperidine} 1: 301
5-Phenyl-2,4-thiazole diamine · HCl 2: 339
Phenyltoloxamine 2: 910
17-Phenyl-18,19,20-trinor-PGE$_2$ 4: 2054
Phenyltrope 4: 2028
Phenytoin 1: 125; 2: 473, 474
Phenytoin AWD 2: 475
Phenytoin sodium 2: 475
Pheochromocytoma 1: 148

PHI 3: 1423
Phlogont 2: 374; 4: 1695
PHM 3: 1423
Phosphalugel 2: 726
Phosphatidylcholine 2: 768
Phosphatidylinositol-4,5-diphosphate 2: 688
Phosphatidylserine 2: 658
1,1′,1″-Phosphinothioylidynetrisaziridine 4: 1889
Phosphodiesterase inhibitors 1: 194
    in asthma treatment 2: 897
Phospholine 4: 2025
Phospholine Iodide 4: 2025
Phospholinjodid 4: 2025
Phospholipase C 2: 688
Phosphonoacetic acid 3: 1183
Phosphonoformate 3: 1184
Phosphonoformate sodium 3: 1184
Phosphonoformic Acid 2: 763; 3: 1184
Photoallergic contact dermatitis 4: 1718
Photochemotherapy
    as psoriasis treatment 4: 1729
Phototherapy
    in psoriasis treatment 4: 1728
Phrenixol 2: 597
I Phrine 4: 2026
1-(2H)-phthalazinone hydrazone 1: 99
Phthalazinones antihistamines 2: 931
3-Phthalimido-2,6-dioxopiperidine 2: 511
Phthalylsulfathiazole 3: 1137
Phyllin 2: 888
Physcion 2: 790
Physeptone 2: 405
Physostigmine 2: 653; 4: 2025
α_1-PI 1: 261
PIC 4: 2151
PIC/S 4: 2152
I-Picamide 4: 2028
Picornaviridae 3: 1174
Picromycin 3: 997
Pidilat 1: 95, 210
Pidorubicin 4: 1897
Pigment colorants 4: 2102
Pigment stones 2: 754
Pilagan Liquifilm 4: 2025
Pills 4: 2093
Pilocar 4: 2024
I-Pilocarpine 4: 2024, 2024, 2032
Pilokair 4: 2024
Pilomann 4: 2038
Pilopine 4: 2024
Pilzcin 3: 1218
Pima-Biciron 3: 1236
Pimafucin 3: 1236
Pimagyn 3: 1236
Pimaricin 3: 1065, 1234, 1235, 1236
Pimethixene 2: 925
Pimilprost 4: 2062

Pimotid 2: 602
Pimozide 2: 601
Pinacidil 2: 706
Pinbetol 1: 158
Pindolol 1: 92, 57, 227
Pinicolic acid 3: 1326
Pinorubicin 4: 1898
Pinworm 3: 1273
Pioglitazone 3: 1666
PIP_2 2: 688
Pipamazine 2: 592
Pipamperone 2: 599
Pipecurium bromide 2: 672
Pipecuronium bromide 2: 672
Pipemidic acid 3: 1119
Piperacetazine 2: 593
Piperacillin 3: 962
Piperacillin sodium 3: 1023
Piperazine 3: 1272, 1273, 1284
Piperazines antihistamines 2: 913
Piperidinediones
    as sedatives 2: 511
3-(piperidine-1-yl)-propyl-4-amino-5-chloro-2-methoxy-
    benzoate hydrochloride 2: 807
Piperidinobutyrophenone neuroleptics 2: 597
5-[3-(4-Piperidino-4- carbamoylpiperidino)propyl]-
    10,11-dihydro-5H-dibenz[b, f]azepine 2: 603
2-Piperidinoethyl 1- cyclohexyl-1- cyclohexanecarboxy-
    late 2: 703
2-Piperidinoethyl 3-methyl-4-oxo-2-phenyl-4H-1-benzo-
    pyran-8-carboxylate 2: 704
2-(N-piperidinomethyl)-1,4-benzodioxane 2: 903
N-[3-[(α-piperidino-m-tolyl)oxy]-propyl]glycola-
    mide 2: 724
6-(1-Piperidinyl)-2,4-pyrimidinediamine-3-
    oxide 1: 101
2,2′,2″,2‴-[(4,8-Di-1-piperidinylpyrimido[5,4-d]pyrimi-
    dine-2,6-diyl)dinitrilo]tetraethanol 1: 39
(±)-N-(2-Piperidylmethyl)–2,5-bis-(2,2,2-trifluoroethox-
    y)benzamide 1: 128
Piperocaine hydrochloride 2: 543
Piperonil 2: 599
Pipnodine 2: 496
Piportil 2: 593
Piportil Depot 2: 593
Piportil Longum 2: 593
Piportyl 2: 593
Pipotiazine 2: 593
Pipoxizine 2: 917
Piracebral 2: 650
Piracetam 2: 650
Piralone 2: 522
Piranver 3: 1285
Piranver F 3: 1285
Pirarubicin 4: 1898
Pirbuterol 1: 194; 2: 876
Pirem 2: 882

Pirenzepine **2:** 715, 729
Piretanide **1:** 297
Piridolan **2:** 410
Piritramide **2:** 410
Piritrexim isothionate **4:** 1871
Pirium **2:** 602
Pirlindole **2:** 626
Piroheptine **2:** 566
Piromidic acid **3:** 1119
Piro-Phlogont **4:** 1697
Piroxicam **2:** 377; **4:** 1696
Pirprofen **4:** 1693
Pitocin **3:** 1402
Pitofenone **2:** 704
Pitressin **3:** 1402
Pitressin Tannat **3:** 1402
Pitrex **3:** 1248
Pituitary Adenylate Cyclase Activating Polypeptide **3:** 1425
Pituitary effector hormones **3:** 1539
Pituitary glandotropic hormones **3:** 1538
Pituitary hormone secretion **3:** 1540
Pivampicillin **3:** 963
Pivmecillinam **3:** 963, 1026
PKM **2:** 567
PK-Merz **2:** 574
PL **3:** 1434, 439
Placental Hormones **3:** 1546
Placental lactogen **3:** 1434, 1439, 1546
Placidel **2:** 507
Placidyl **2:** 497
Placinor **2:** 522
Plactamin **1:** 208
Planor **3:** 1598
Plantago seed **2:** 783
Planum **2:** 523
Plaquenil sulfate **3:** 1163
Plasil **2:** 745
Plasma
  frozen **1:** 272
Plasma Accelerator Globulin **1:** 255
Plasma donation **1:** 265
Plasma fractionation **1:** 267
Plasma in prophylaxis and therapy **4:** 1835
Plasmakinin system **4:** 1672
Plasmapheresis **1:** 266
Plasma proteinase inhibitors **1:** 259
Plasma proteins **1:** 251
Plasmids **3:** 1112
Plasminogen **1:** 252, 258
Plasminogen activators **1:** 276
Plasmodia **3:** 1159; **4:** 1813
Plasmodium **3:** 1159
Plasmodium falciparum **4:** 1813
Plastic containers **4:** 2106
Plastic dosage forms **4:** 2091
Platelet Affecting Drugs **1:** 30

Platelet Aggregation **1:** 30
Platelet-derived growth factor **1:** 33
platelets
  blood **1:** 249
  therapeutic use **1:** 271
Plavix **1:** 43
Plegine **2:** 465, 467
Plendil **1:** 211
Plicamycin **3:** 1066
Plimycon **3:** 1217
Pluripotent stem cells **1:** 240
PM 185184 **3:** 1157
PMSG **3:** 1545
Pneumadin **3:** 1399
Pneumococcus vaccine **4:** 1772
Pneumocystis carinii **3:** 1159, 168
PNS **2:** 312
PNU 140690 **3:** 1314
Podophyllotoxin **4:** 1912
Poen Caina **4:** 2003
Polagol **2:** 703
l-Polamidon **2:** 405
Polaramin **2:** 912
Polaronil **2:** 912
Polaxamers **4:** 2103
Pol gene **3:** 1292
Polibutin **2:** 702
Polio prophylaxis **4:** 1826
Polio vaccine **4:** 1792
Poliovirus **3:** 1174
Polistin T-Caps **2:** 908
Polocaine **2:** 547
Poloxamer 407 **4:** 2040
Poloxamine **4:** 2040
Polycain **3:** 1220
Polyclonal antibodies **4:** 1759
Poly[(dimethyliminio)-2-butene-1,4-diyl chloride] **4:** 2038
Polydocanol **2:** 552
Polyene antibiotics **3:** 954
Polyene antimycotics **3:** 1232
Polyethers **3:** 1094
Polyethylene glycol **4:** 2034, 2039
Poly(ethylene glycol) 4-isooctylphenyl ether **4:** 2039
Poly(ethylene glycol)monododecyl ether **2:** 552
Poly(ethylene glycol) monostearate **4:** 2039
Polyfungin A1 **3:** 1237
Polygris **3:** 1242
Poly(hexamethylenebiguanide) hydrochloride **4:** 2037
Polymers
  for viscosity adjustment **4:** 2103
Polymycin **4:** 2010
Polymyxin B **3:** 1057; **4:** 2011
Polymyxin B sulfate **4:** 2009
Polymyxins **3:** 1005
Polyoxin **3:** 1050
Polyoxins **3:** 991

Polyoxyethylated lauryl ether **4:** 2103
Polyoxyethylated octyl phenol **4:** 2103
Polyoxyethylene 50 stearate **4:** 2103
Polyoxyl 40 stearate **4:** 2039
Polypeptide antibiotics **4:** 2009
Polyporenic acid **3:** 1326
Poly-Pred **4:** 2016
Poly-Pred Liquifilm **4:** 2008, 2011
Polyquad **4:** 2038
Polyquarternium-1 **4:** 2038
Polyquat **4:** 2038
Polyquil **3:** 1286
Polysorbate 80 **4:** 2103
Polysporin **4:** 2009, 2010, 2011
Polythiazide **1:** 71, 294
Polytopic extrasystoles **1:** 117
Poly(vinyl alcohol) **4:** 2034, 2103
Polyvinylpyrrolidone **4:** 2103
Ponderal **2:** 462
Ponderax **2:** 462
Pondimin **2:** 467
Pondomin **2:** 462
Ponstan **2:** 352; **4:** 1697
Ponstel **2:** 352; **4:** 1697
Ponstyl **2:** 352
Pontal **4:** 1697
Pontocaine **4:** 2003
Porfanil **2:** 606
Porfiromycin **3:** 1010
Poriferastane **3:** 1323
Poriferasterol **3:** 1324
Pork Tapeworm **3:** 1270
Portal hypertension **1:** 148
Porter – Silber reaction **3:** 1606
Posdel **2:** 913
Positive inotropic agents **1:** 175
Positive inotropic drugs
  function **1:** 16
Positron-emission tomography **2:** 325
Postacton **3:** 1402
Postafen **2:** 916
Post-Coital Contraception **3:** 1508
Postganglionic sympathetic nerves **2:** 312
Postimmune serum **4:** 1836
Postnatal perinatal investigation **4:** 2132
Potassium aldadiene **3:** 1617
Potassium bromide **2:** 486
Potassium canrenoate **1:** 298
Potassium stearate **4:** 2102
Potent tinctures **4:** 2087
Povan 1959 **3:** 1286
Povan(yl) **3:** 1286, 286
Povidone **4:** 2034
Powders **4:** 2091
Poxviridae **3:** 1174
[5α,7α(R)]-4,5-Epoxy-3-hydroxy-6-methoxy-α,17-di-
  methylmorphinan-6-one **2:** 393

PP **3:** 1408
PPA **2:** 463
PPIs **2:** 714
Practolol **1:** 158, 227
Pragman Gelee **2:** 912
Pragmazone **2:** 632
Prajmalium bitartrate **1:** 127
Pramiel **2:** 745
Pramipexole **2:** 577
Pramiracetam **2:** 659
Pramocaine hydrochloride **2:** 550
Pranlukast **4:** 2064
Pranone **3:** 1598
Prantal **2:** 693
Pravachol **1:** 28
Pravasin **1:** 28
Pravastatin **1:** 28
Pravidel **2:** 576
Praxiten **2:** 522, 638
Prazene **2:** 639
Prazepam **2:** 639
Prazine **2:** 590
Prazinil **2:** 603
Praziquantel **3:** 1267, 1268, 1269, 1270, 1271, 1284
Prazosin **1:** 87; **2:** 706
Prealbumin **1:** 252
Preclinical drug testing **4:** 2126
Precose **3:** 1663
Predair **4:** 2017
Predair Forte **4:** 2017
Predalon **3:** 1357
Predef **3:** 1620
Predef 2 X **4:** 2178
Pred Forte **4:** 2017
Pred-G **4:** 2016
Pred-G Liquifilm **4:** 2008
Pred-G S.O.P. **4:** 2008
Pred Mild **4:** 2017
Prednicarbate **3:** 1636; **4:** 1722
Prednisolone **3:** 1617; **4:** 2017
  doses **3:** 1605
Prednison **3:** 1605
Prednisone **3:** 1617
Prednylidene **3:** 1628, 1605
Predsulfair **4:** 2005
Prefin **4:** 2027
Prefin Z Liquifilm **4:** 2037
Prefrin **4:** 2029
Preganglionic sympathetic nerves **2:** 312
9β,10α-Pregna-4,6-diene-3,20-dione **3:** 1597
Pregnane **3:** 1323
Pregnane gestagens **3:** 1596
Pregnane skeleton **3:** 1601
Pregnant mare's serum gonadotropin **3:** 1545
4-Pregnene-3,20-dione **3:** 1596
Pregnesin **3:** 1357

Prelis **1:** 156
Prelu-2 **2:** 465, 467
Preludin **2:** 465, 467
Premarin **3:** 1583
Prenalterol **1:** 193
Prenenolone **3:** 1565
Prenormine **1:** 152
Prent **1:** 93, 152
Prenylamine **1:** 208, 229
Prepro-EK A **3:** 1382
Prepro-EK B **3:** 1384
Preproenkephalin A **3:** 1382
Preproenkephalin B **3:** 1384
Presamine **2:** 614
Prescaina **4:** 2003
Prescal **1:** 212
Presdate **1:** 94, 55
Pre-Sert **4:** 2034
Presome **3:** 1583
Pressalolo **1:** 155
Pressure coating **4:** 2100
Presuren **2:** 452
Prevex **1:** 211
PRFs **3:** 1340
Priamide **2:** 694
Priatan **2:** 878
Pridinol **2:** 564
Prifinium bromide **2:** 696
Prilagin **3:** 1210
Prilocaine hydrochloride **2:** 546; **4:** 2004
Primacor **1:** 195
Primaquine **2:** 763; **3:** 1163
Primaquine phosphate **3:** 1163
Primary angina **1:** 220
Primary Biliary Liver Disease **2:** 765
Primatene **1:** 60
Primidone **2:** 480
Primobolam **3:** 1594
Primobolan **3:** 1594
Primogonyl **3:** 1357
Primolut Nor **3:**
Primperan **2:** 745
Prinzmetal's angina **1:** 220
Privin **4:** 2029
Privina **4:** 2029
Privine **4:** 2029
PRL **3:** 1434, 437, 545
Probarbital **2:** 504, 507
Probucol **1:** 26
Procainamide **1:** 14, 24
Procaine **2:** 540
Procaine hydrochloride **2:** 543; **4:** 2004
Procarbazine **4:** 1894
Procardia **1:** 95, 210
Procaterol **2:** 884
Processine **1:** 209
Process validation **4:** 2143

Procetofene **1:** 25
Prochlor-Iso **2:** 695
Prochlorperazine **2:** 594, 744
Proconvertin **1:** 256
Procorum **1:** 131, 208
Procrazine **2:** 591
Procyclidin **2:** 564
Procyclin **4:** 2061
Pro-Dafalgan **2:** 349
Pro-Diaban **3:** 1657
Prodipine **2:** 564
Prodorm **2:** 499, 514, 522, 638
Prodormol **2:** 509
Prodox **3:** 1596
Profact **4:** 1932
Profenamine **2:** 565
Proflax **1:** 160
Profénid **2:** 366
Profoliol **3:** 1583
Profundol **2:** 510
Progesic **4:** 1690
Progestasert **3:** 1507; **4:** 2085
Progesterone **3:** 1497, 1551, 1565, 1568, 1596
Progesterone Antagonists **3:** 1513, 1600
Progesterone synthesis **3:** 1571
Progesterone synthesis inhibitors **3:** 1600
Progestin **3:** 1596
Progestogen-Only Oral Contraceptives **3:** 1503
Proglumetacin maleate **4:** 1681
Progynon **3:** 1584
Progynon-Salbe **3:** 1583
Progynova **3:** 1584
Pro-Iso **2:** 695
Proketazine **2:** 595
Proklar **3:** 1136
Prolacam **2:** 576
Prolactin **3:** 1434, 437, 545
Proladone **2:** 397
Prolixan **4:** 1695
Prolixin **2:** 595
Prolonged-action dosage forms **4:** 2100
Prolonged release technology **4:** 2112
Proloprim **3:** 1133
Proluteasi **3:** 1596
Proluton **3:** 1596
Proluton-Depot **3:** 1596
L-Prolyl-L-leucylglycinamide **2:** 581
Promacil **2:** 681
Promactil **2:** 590
Promapar **2:** 590
Promazine **2:** 590
Promethazine **2:** 921
Promexin **2:** 590, 681
Pronestyl **1:** 125
Pronoctan **2:** 523
Prontosil **3:** 953

Proof of efficacy **4**: 2137
Proopiomelanocortin **3**: 1369
Proopiomelanocortin Hormones **3**: 1543
Propacetamol **2**: 349
Propaderm **3**: 1627
Propafenone **1**: 129
Propahexal **1**: 159
Propallyonal **2**: 504, 510
Propane-1,2,3-triol **4**: 2023
1,2-Propane-diol **4**: 2035
1,2,3-Propanetriol **2**: 782
1,2,3-Propanetriol trinitrate, glyceryl trinitrate **1**: 223
Propanidid **2**: 447
Propantan **2**: 447
Propantheline bromide **2**: 697
Propaphenin **2**: 591
Proparacaine hydrochloride **4**: 2003
Proparakain **2**: 545
Propavan **2**: 922
Propax **2**: 522, 638
Propellant Gases **4**: 2104
Propentofylline **2**: 656
17-(2-Propenyl)morphinan-3-ol **2**: 421
Properdin **1**: 252
Propicillin **3**: 962, 019
Propine **4**: 2020
Propiomazine **2**: 921
Propionibacteriaceae **3**: 1110
Propipocaine hydrochloride **2**: 550
Propitan **2**: 599
Propiverine **2**: 704
Propizepine **2**: 619
bis-(2-Propoxyethyl) 1,4-dihydro-2,6-dimethyl-4-(3-nitrophenyl)pyridine-3,5-dicarboxylate **1**: 211
1-(4-propoxyphenyl)-3-(1-piperidinyl)-1-propanone hydrochloride **2**: 550
Propranolol **1**: 15, 90, 29, 58, 226; **3**: 1528
Propulsin **2**: 728
Propyl 3-methoxy-4-(N,N-diethylcarbamoylmethoxy)phenylacetate **2**: 447
4-Propylajmalium hydrogen tartrate **1**: 127
(S)-(–)-1-Propyl-N-(2,6-dimethylphenyl)-2-piperidinecarboxamide hydrochloride monohydrate **2**: 547
Propylene glycol **4**: 2035
Propylhexedrine **2**: 462
2-Propylpentanoic acid **2**: 477
(–)-(S)-2-[N-Propyl-N-[2-(2-thienyl)ethyl]amino]-5-hydroxy-1,2,3,4-tetrahydronaphthalene **2**: 579
Propylthiouracil **3**: 1525, 529
 pharmacokinetic data **3**: 1527
1-(2-Propynyl)cyclohexanol carbamate **2**: 500
(R)-N-2-Propynyl-1-indanamine **2**: 571
Propyphenazone **2**: 371
Proquazone **4**: 1699
Proquinal **2**: 511
Proscillaridin **1**: 183, 91

Proscillaridin-4′-methyl ether **1**: 191
Prosedar **2**: 523
Proseryl **2**: 652
Prostacyclin **1**: 37
Prostaglandin F$_2${\tf="Pi1"\char97} **3**: 1509
Prostaglandin occurrence **4**: 2050
Prostaglandin ONO 802 **3**: 1509
Prostaglandins **2**: 727; **3**: 1554; **4**: 2045
 biosynthesis **4**: 2052
 physical properties **4**: 2049
Prostaglandins nomenclature **4**: 2046
Prostaglandin syntheses **4**: 2058
Prostaglandin Synthesis Inhibitors **1**: 34
Prostalene **4**: 2054, 2059
Prostandin **4**: 2060
Prostanes
 occurrence **4**: 2053
Prostanoic acid **4**: 2046
Prostanoid receptors **4**: 2054
Prostavasin **4**: 2060
Protaxil **4**: 1682
Protaxon **2**: 358; **4**: 1682
Protective colloids **4**: 2104
α_1 Proteinase inhibitor **1**: 252, 261
Proteinase inhibitor deficiencies **1**: 277
Proteinase inhibitors **1**: 259
Protein C **1**: 252, 257
Protein Hormones **3**: 1339
Protein S **1**: 258
 in blood plasma **1**: 251
Prothanon **2**: 919
Prothiaden **2**: 613
Prothil **3**: 1597
Prothipendyl **2**: 596
Prothrombin **1**: 252, 255
Prothyrid **3**: 1561
Protirelin **2**: 654; **3**: 1544
Protivar **3**: 1594
Protokylol **2**: 867
Proton Pump Inhibitors **2**: 716
Protosib **3**: 1154
Protoverine **3**: 1334
Protozoan infections
 chemotherapy **3**: 1146
Protozoan organisms **3**: 1147
Protriptyline **2**: 615
Provector **2**: 617
Provera **3**: 1596
Provest **3**: 1596
Proviron **3**: 1590
Provocholine **4**: 2032
Proxen **2**: 368; **4**: 1689
Proxil **4**: 1682
Proxymetacaine hydrochloride **2**: 544; **4**: 2003
Proxyphylline **2**: 888
Prozac **2**: 628

Prozin **2**: 591
Prozine **2**: 590
PS-5 **3**: 974
PS-6 **3**: 974
PS-7 **3**: 974
Pseudomonadaceae **3**: 1110
Pseudomonas aeruginosa vaccine **4**: 1777
Psicoben **2**: 599
Psicopax **2**: 522
Psoralens **4**: 1729
Psoriasis **4**: 1727
PST **3**: 1449
Psychopharmacological agents **2**: 587
Psychoses **2**: 588
Psychozine **2**: 591
Psyllii Semen **2**: 783
Psyllium **2**: 783
Psyquil **2**: 591, 744
PTH **3**: 1546
Ptimal **2**: 479
PTL **2**: 758
4-(5-p-tolyl-3-trifluoromethylpyrazol-1-yl)-benzenesulfonamide **2**: 381
PTU **3**: 1525, 529
Pulmadil **2**: 868
Pulmicort **2**: 894; **3**: 1636
Pulmonary anesthetics **2**: 436
Pumaprazole **2**: 733
Punaglandin **4**: 2051
Purification
  of interferon **4**: 1968
  of monoclonal antibodies **4**: 1991
Purification antibiotics **3**: 1087
Purine antagonists **4**: 1885
PUVA **4**: 1729
Pyladox **2**: 654
Pylapron **1**: 159
Pylorid **2**: 732
Pynastin **1**: 158
Pyramal **2**: 906
Pyramidal cells **2**: 310
Pyramidal tract **2**: 315
Pyrantel **3**: 1272, 273, 285
Pyrazidol **2**: 626
Pyrazinamide **3**: 1140
Pyrazinecarboxamide **3**: 1140
Pyribenzamine **2**: 907
Pyridine-3-carboxylic **1**: 23
Pyridine-3-carboxylic acid **1**: 23
4,4′-(2-Pyridinylmethylene)bisphenol,bis(hydrogensulfate) **2**: 788
4,4′-(2-Pyridinylmethylene) bisphenol diacetate, bis-(p-acetoxyphenyl)-(2-pyridyl)-methane **2**: 787
(E)-5-[[[3-pyridinyl[3-(trifluoromethyl)phenyl]methylene]amino]oxy]pentanoic acid **1**: 36
Pyridoglutethimide **4**: 1926

5-{[p-(2-pyridylsulfamoyl)phenyl]azo}salicylicacid **3**: 1136
Pyrilamine **4**: 1720, 2184
Pyrimethamine **3**: 1130, 164, 167
N^1-2-pyrimidinylsulfanilamide **3**: 1134
8-[4-[4-(2-Pyrimidinyl)-1-piperazinyl]butyl]-8-azaspiro[4.5]decane-7,9-dione **2**: 647
Pyrindamycin **4**: 1907
Pyrithioxin **4**: 1706
Pyrithyldione **2**: 511
Pyritinol **2**: 657; **4**: 1706
L-Pyro-2-aminoadipyl-L-leucyl-L-prolinamide **2**: 654
Pyrogenicity testing
  interferon **4**: 1975
Pyrogens
  removal **4**: 1971
4-(1-Pyrrolidinyl)-1-(2,4,6-trimethoxyphenyl)-1-butanone **1**: 50
Pyrrolnitrin **3**: 1011, 1070
Pyrvinium **3**: 1273, 1285
PYY **3**: 1411

Quality Assurance **4**: 2105
Quality control **4**: 2140
Quaname **2**: 645, 679
Quanil **2**: 679
Quantalan **1**: 22
Quartermaster suspension **4**: 2162
Quarzan **2**: 693
Quaser **1**: 208
Quazepam **2**: 515, 523
  synthesis **2**: 520
Quelicin chloride **2**: 674
Questran **1**: 22
Quide **2**: 593
Quik-Sept **4**: 2039
Quin-260 **2**: 676
Quinacrine **2**: 763
Quinacrine hydrochloride **3**: 1158
Quinaglute **1**: 124
Quinamm **2**: 676
Quinazolinones
  as sedatives **2**: 513
Quinbar **2**: 511
Quine **2**: 676
Quinestrol **3**: 1584
Quinethazone **1**: 295
Quingestanol acetate **3**: 1598
Quinidine **1**: 14, 24
Quinine sulfate **2**: 675, 676
Quinisocaine hydrochloride **2**: 551
Quinite **2**: 676
Quinolone antibacterial agents **3**: 1115
Quinpirole **2**: 579
Quinuprine **2**: 616
Quitaxon **2**: 613
Quotane **2**: 551

(R*,R*)-(±)-1,4-Dihydro-2,6-dimethyl-4-(3-nitrophenyl)-
  3,5-pyridinedicarboxylic acid   **1**: 212
(R*,S*)-(±)-α-(1-Aminoethyl)benzenemetha-
  nol   **2**: 463
R07-1051   **3**: 1152
(R)-(−)-1-[4,4-bis(3-Methyl-2-thienyl)-3-butenyl]-3-piper-
  idinecarboxylic acid hydrochloride   **2**: 484
R 14827   **3**: 1219
R 14889   **3**: 1209
R 15454   **3**: 1222
R 18134   **3**: 1209
(R)-α-([2-(3,4-Dimethoxyphenyl)ethylamino]methyl)-4-
  hydroxybenzenemethanol   **1**: 193
R25831   **3**: 1157
R28096   **3**: 1157
(R)-2-Amino-9-(3,4-dihydroxybutyl)-1,9-dihydro-6H-
  purin-6-one   **3**: 1183
(R)-3-(2-Deoxy-β-D-erythro-pentofuranosyl)-3,6,7,8-tet-
  rahydroimidazo[4,5-d][1,3]diazepin-8-ol
  4: 1881
R 41400   **3**: 1206
(R)-(−)-4-[1-hydroxy-2-(methylamino)ethyl]-1,2-benze-
  nediol, (−)-epinephrine   **1**: 60
R 42470   **3**: 1228
(R)-(−)-4-(2-Amino-1-hydroxyethyl)-1,2-benzenediol, (−)-
  norepinephrine   **1**: 60
R 51211   **3**: 1203
R 58735   **2**: 659
(R)-6-Dimethylamino-4,4-diphenylheptan-3-
  one   **2**: 404
(R − 802)   **3**: 1126
R83842   **4**: 1926
(R)-9-(3,4-Dihydroxybutyl)guanine   **3**: 1183
Rabeprazole   **2**: 719
Rabies   **3**: 1174
  treatment with monoclonal antibodies   **4**: 1843
Rabies prophylaxis   **4**: 1828
Rabies vaccine   **4**: 1796
Radanil   **3**: 1152
Radedorm   **2**: 518
Radenarcon   **2**: 449
Radeverm   **3**: 1282
Radioiodine   **3**: 1528
Radioligand Binding
  benzamides   **2**: 819
Radiopharmaceutical dosage forms   **4**: 2101
Ralofekt   **1**: 49
Raloxifene   **4**: 1922, 924
Raltitrexed   **4**: 1871
Ramatroban   **4**: 2063
Ramorelix   **4**: 1934
Randolectil   **2**: 594
Ranimustine   **4**: 1893
Ranitidine   **2**: 724
Ranitidine bismuth citrate   **2**: 732
Rantudil   **2**: 354; **4**: 1680
Ranvil   **1**: 210

Rapid eye movement sleep   **2**: 490
Rapifen   **2**: 403
Rasagiline   **2**: 571
Rastinon   **3**: 1658
1589–RB   **3**: 1125
(R)-DHBG   **3**: 1183
Reactine   **2**: 927
Reasec   **2**: 408
Reboxetine   **2**: 633
Rebriden   **1**: 212
Rec 15/1476   **3**: 1221
$M_1 - M_4$ receptor   **2**: 687
H_2-Receptor antagonists   **2**: 722
β-Receptor blocking agents   **1**: 121, 29
κ-Receptors   **2**: 387
α-Receptorsβ-Receptors   **1**: 138; **2**: 387
Recombinant coagulation   **1**: 270
Recombinant IFNα
  purification   **4**: 1969
Recombinant proteins   **1**: 270
Re-Covr injection   **4**: 2184
Rectal administration   **4**: 2084
Recycling isoelectric focusing   **4**: 1972
Red blood cells   **1**: 242
Redeptin   **2**: 601
Redul S   **3**: 1657
Reflex arc   **2**: 314
Refresh   **4**: 2034
Regelan   **1**: 25
Regenon   **2**: 464
Regenox   **2**: 512
Reglan   **2**: 745
Reglovis   **3**: 1598
Regulex   **2**: 782
Regulton   **1**: 65
Rela   **2**: 677
Relact   **2**: 518
Relafen   **2**: 361
Relasom   **2**: 677
Relaspium   **2**: 692
Relaxan   **2**: 671
Relaxants   **2**: 438
  muscle   **1**: 95
Relaxin   **3**: 1445
Relefact TRH   **3**: 1354
Relief   **4**: 2029
Relifex   **2**: 361
Reliveran   **2**: 745
Remdue   **2**: 522
Remestan   **2**: 523
Remicut   **2**: 932
Remifentanil   **2**: 403
Remivox   **1**: 128
Remnos   **2**: 518
Remoxipride   **2**: 607
REM sleep   **2**: 490
Renese   **1**: 294

Reneuron **2:** 628
Rengasil **4:** 1693
Renin **3:** 1547
Renin–angiotensin system **1:** 78
Renoquid **3:** 1135
Rentylin **1:** 49
Renzapride **2:** 796
   mode of action **2:** 798
   Receptor binding affinities **2:** 808
Reocorin **1:** 208
Reonin **1:** 208
Reoviridae **3:** 1174
Reoviruses **3:** 1174
Repaglinide **3:** 1658
Repeat-action oral tablets **4:** 2112
Repocal **2:** 509
Repoise **2:** 594
Reproduction studies **4:** 2132
Reproterol **2:** 872
Reptilase **1:** 48
Requip **2:** 578
RESA **4:** 1819
Rescriptor **3:** 1302
Rescula **4:** 2061
Reserpex **1:** 86
Reserpine **1:** 15, 85
Reservoir products **4:** 2114
Residual solvents **4:** 2142
Resimatil **2:** 480
Resistance
   of antimicrobics **3:** 1111
Resistance antibiotics **3:** 957
Resochin **3:** 1162
Resolve **4:** 2040
Resorcinol bronchodilators **2:** 869
Respacal **2:** 918
Respilac **2:** 872
Respiratory stimulants
   veterinary **4:** 2179
Restatin **3:** 1238
Restoril **2:** 523
Retcol **2:** 636
Reticulocytes **1:** 239, 242
Retinoic acid **4:** 1724, 728
Retinoids **4:** 1724
Retinol binding protein **1:** 252
Retrovir **3:** 1295
Retroviridae **3:** 1174
Retroviruses **3:** 1292
Reumacillin **4:** 1706
Reumadolor **4:** 1680
Reversed-phase high-performance liquid chromatography **4:** 1973
Reversed-phase HPLC **4:** 1973
Reverse T_3 **3:** 1558
Reverse transcriptase **3:** 1294
Reverse Transcriptase Inhibitors **3:** 1294

Rev gene **3:** 1292
Revia **2:** 423
Revivan **1:** 61
Revonal **2:** 514
Rewodina **4:** 1685
Rexitene **1:** 84
Rezulin **3:** 1666
R-factors **3:** 1112
RGH-2202 **2:** 581
Rhabdoviridae **3:** 1174
α-D-Rhamnose **1:** 180
Rheaform **3:** 1170
Rhein **2:** 790
Rheuma **4:** 1671
Rheumatoid arthritis **4:** 1674
Rheumon **2:** 350
Rheumox **4:** 1695
Rhinolast **2:** 931
Rhinovirus **3:** 1174
Rhodialothan **2:** 440
Rhodomycin **3:** 954
Rhodomycin A **3:** 981
Rhodomycin B **3:** 981
Rhoptry antigens **4:** 1818
Rhubarb **2:** 791
Riabal **2:** 696
Ribavirin **2:** 764; **3:** 1181
1-β-D-Ribofuranosyl-1H-1,2,4-triazole-3-carboxamide **2:** 764; **3:** 1181
Ribostamycin **3:** 985, 1044, 1986
Ricini oleum **2:** 789
Ricinoleic acid **2:** 789
Rickamicin **3:** 986
Rickettsiaceae **3:** 1110
Rickettsias **3:** 1110
Ridaura **4:** 1704
Ridauran **4:** 1704
Ridazin **2:** 592
Ridene **1:** 210
Ridogrel **1:** 36
Rifampicin **3:** 1000, 055
Rifampin **3:** 1055
Rifamycin **3:** 1000
Rifamycins **3:** 1000
Rift Valley fever **3:** 1174
Rift valley fever vaccine **4:** 1805
Rigidity
   in Parkinsonism **2:** 556
Rilmafazone **2:** 526
Rilutek **2:** 575
Riluzole **2:** 575
Rimantadine **3:** 1184
Rimiterol **2:** 867
Ring-infected erythrocyte surface antigen **4:** 1819
Rinlaxer **2:** 677
Riopan **2:** 727
Riporest **2:** 514

Risk factors **1:** 17
Risperidone **2:** 609
Risumic **1:** 65
Ritalin **2:** 339
Ritanserin **2:** 633
Ritmos Elle **1:** 133
Ritonavir **3:** 1310
Rivostatin **3:** 1238
Rivotril **2:** 481
Rize **2:** 641
Rizen **2:** 641
RLX **3:** 1445
Ro 09–0410 **3:** 1186
Ro 13-8996 **3:** 1224
Ro 14-4767 **3:** 1253
Ro 2-9915 **3:** 1238
Ro 318959 **3:** 1307
Robaxin **2:** 680
Robinul **2:** 694
Rochagan **3:** 1152
Rocornal **1:** 230
Rodavan **2:** 742
Rods **3:** 1110
Roeridorm **2:** 497
Rofecoxib **2:** 384; **4:** 1701, 2054
Rofenid **3:** 1208
Roferon
   clinical studies **4:** 1977
Rogletimide **4:** 1926
Rohpinol **2:** 520
Rohypnol **2:** 520
Roimal **4:** 1684
Roipnol **2:** 520
Rolipram **2:** 633
Rolitetracycline **3:** 978, 038
Romafen **4:** 1698
Romifidine **4:** 2181
Rompun **4:** 2181
Ronok **4:** 2060
Ropinirole **2:** 578
Ropivacaine monohydrate **2:** 547
Ropoxyl **2:** 652
Rosal **4:** 2060
Rosamicin **3:** 999
Rosaprostol **4:** 2060
EV Rose Bengal **4:** 2031, 2031
Rosiglitazone **3:** 1666
Rosoxacin **3:** 1120
Rotary die process **4:** 2093
Rotaviruses **3:** 1174
Roundworms **3:** 1272
Roxadyl **3:** 1121
Roxatidine **2:** 724
Roxiam **2:** 607
Roxicodone **2:** 397
Roxindole **2:** 609
Roxit **2:** 724

RP 14539 **3:** 1157
RP 17774 **3:** 1232
2168-RP **3:** 1154
RP-56976 **4:** 1915
(RRP) **1:** 120
RS 23597-190 **2:** 807
RS26306 **4:** 1934
RS-44872 **3:** 1226
RSV **3:** 1174
Ru–43–280 **3:** 1126
RU486 **3:** 1510
Rubella **3:** 1174
   treatment with monoclonal antibodies **4:** 1843
Rubella prophylaxis **4:** 1827
Rubella vaccine **4:** 1790
Rubeola **4:** 1787
Rufen **4:** 1688
Rumensin 60 **4:** 2170
Ruvamed **4:** 1697
RWJ-26251 **4:** 1880
Rycarden **1:** 210
Rydene **1:** 210
Rythmodan **1:** 125
Rythmodul **1:** 125
Rytmonorm **1:** 129

S-10036 **4:** 1893
S 145 **4:** 2054
254-S **4:** 1921
4S **1:** 27
S 710674 **3:** 1218
Sabeluzole **2:** 659
Sabril **2:** 482
Safety testing **4:** 2136
Safety testing of biotechnological products **4:** 2133
Saffan **2:** 453
Sagamicin **3:** 1043; **4:** 2008
Saizen **3:** 1437
Salbutamol **1:** 194; **2:** 874
Sales
   global **1:** 4
Salicylamide **2:** 348
Salicylates
   as antiinflammatory drugs **4:** 1677
Salicylic acid **4:** 1727
Salicylic acid acetate **1:** 35
Saligenin bronchodilators **2:** 874
ReNu Saline Solution **4:** 2037
Salinomycin **3:** 1064
Salipran **4:** 1678
Salmeterol **2:** 875
Saltatory conduction **2:** 311
Saltucin **1:** 294
Saluretics **1:** 292
Saluric **1:** 293
Salus-öl **2:** 781
Samandarin **3:** 1334

Sanasthmax 2: 894
Sanasthmyl 2: 894; 3: 1636
Sanato-Lax 2: 781
Sanato-Lax-forte 2: 781
Sandostatin 3: 1448
Sandosten 2: 918
Sandoz 1: 212
Sangivamycin 3: 991
Sanoma 2: 677
Sanorex 2: 466, 467
Santemycin 4: 2008
Sapogenins 3: 1332
Saquinavir 3: 1305
Saralasin 3: 1391
Sarenin 3: 1391
Sarisol 2: 509
Sarkomycin A 3: 1009
Saroten 2: 611
Sarpul 2: 651
Satric 3: 1145
Sawaxin 2: 657
SB 204070 2: 807
SB 207710 2: 819
SB 209247 4: 2066
SB-223030 4: 1924
SB-75 4: 1934
SB-88 4: 1934
SC-19220 4: 2054
SC-50410/53491 2: 796
SC-52246 2: 796
SC-53116
   Receptor binding affinities 2: 809
SC-53116/55822/49518 2: 796
SC-53228 4: 2066
Scale down 3: 1078
Scaling up
   in antibiotics production 3: 1075
Scandicain 2: 547
Scandine 1: 193
S.C.B.Tal 2: 511
Sch-13521 4: 1928
SCH 15 719 W 1: 155
Scherisolon 3: 1618
Scherofluron 3: 1620
Scheroson F 3: 1607
Schistosoma species 3: 1266
Schistosomiasis vaccines 4: 1806
Schwann cells 2: 311
Scillaren 1: 181
Scillarenin 1: 179
Scilliglaucosidin 1: 179
Scilliroside 1: 183
Scoline 2: 674
Scopoderm TTS 2: 562
Scopolamine 2: 562, 741
Scopolamine hydrobromide 4: 2028

Screening
   of antibiotics 3: 1073
Screening assays 4: 1985
Scymnol 3: 1328
Sédaland 2: 598
SDZ 205-557 2: 804
Sebar 2: 511
SEC 3: 1421
Secaps 2: 511
Secbutabarbital 2: 504, 508
Secbutobarbital 2: 508
Secbutobarbitone 2: 508
Secidin 1: 208
Secnidazole 3: 1156
Secobarbital 2: 504, 510
Secocaps 2: 511
Secogen 2: 511
Seconal 2: 511
Secondary angina 1: 220
Second messengers 2: 335
Secretin 3: 1421, 549
Secrodyl 3: 1598
Secrosteron 3: 1598
Sectral 1: 93, 52
Securit 2: 522
Securon 1: 131, 208
Sedalin 2: 591
Sedalone 2: 514
Sedanox 2: 509
Sedanxol 2: 597
Seda-Tablinen 2: 506
Sedatival 2: 522
Sedatives 2: 489, 495
Sedatromin 1: 209
Sedesterol 4: 2018
Sediston 2: 590
Sedokin 2: 522
Sedolatan 1: 208
Sedormid 2: 501
Sedotime 2: 641
Seflenyl 4: 1693
Seglor 1: 66
Segontin 1: 208
Seguril 1: 296
Seizures 2: 470
Sekretolin 3: 1423
Sekundal-D 2: 566
Selectol 1: 155
Selegiline 2: 570, 625
Selenium sulfide 4: 1732
Selepam 2: 523
Seles Beta 1: 152
Selezyme 2: 598
Selobloc 1: 152
Seloken 1: 156
Selopral 1: 156
Semikon 2: 906

Semisolid dosage forms **4:** 2089
Sempera **3:** 1204
Semprex **2:** 927
Senepax **2:** 522
Senna leaf **2:** 791
Sennoside A **2:** 791
Sensit **1:** 208
Sensival **2:** 615
Sensory neurons **2:** 309
Sentil **2:** 644
Seotal **2:** 511
Sepamit **1:** 210
Sepan **1:** 209
Separin **3:** 1248
Septa **3:** 1133
Ser-Ap-Es **1:** 99
Seratrodast **4:** 2062
Serax **2:** 522, 638
Serenace **2:** 598
Serendyl **2:** 636
Serenesil **2:** 497
Serenid **2:** 522, 638
Serenone **2:** 565
Serentil **2:** 592
Serepax **2:** 522, 638
Seresta **2:** 522, 638
Serevent **2:** 876
Sermion **2:** 655
Seromycin **3:** 1008
Seropam **2:** 627
Serosod **4:** 1709
Serotonin **2:** 458
  influence on sleep **2:** 491
Serotonin uptake inhibitors **2:** 611, 627
Seroxat **2:** 629
Serpasil **1:** 86
Serpasil-Apresoline **1:** 99
Sertaconazole **3:** 1225
Sertan **2:** 480
Sertofren **2:** 614
Sertraline **2:** 629
Serum cholinesterase **1:** 252
Servium **2:** 636
Setastine **2:** 926
Sex hormones **3:** 1564
Seychellogenin **3:** 1335
SF-733 **3:** 986
SF 83-627 **3:** 1250, 250
Sgd 301-76 **3:** 1224
Shigella vaccines **4:** 1774
Showdomycin **3:** 994
Sialic Acid Content of Mouse Vagina **3:** 1583
SIB1508Y **2:** 581
Sibelium **1:** 209; **2:** 915
Siccanin **3:** 1072
Sigacalm **2:** 522
Sigmart **1:** 231

Signal transduction inhibitors
  cancer therapy **4:** 1937
Silent ischemia **1:** 144
Silver nitrate **4:** 2012
silybin **2:** 765
Silychristin **2:** 765
silydianin **2:** 765
Silymarin **2:** 765
Simfibrate **1:** 25
Simplene **4:** 2020
Simplotan **3:** 1156
Simvastatin **1:** 28
Sincalide **3:** 1416
Sinequan **2:** 613
Sinetens **1:** 87
Singulair **2:** 897; **4:** 2064
Sinoatrial node (SA node) **1:** 115
Sinogan **2:** 591
Sintodian **2:** 600
Sinus bradycardia **1:** 117
Sinus tachycardia **1:** 117
Sipcar **2:** 619
Siplarol **2:** 597
Sipronolactone **1:** 298
Siptazin **1:** 209
Siqualine **2:** 595
Siquil **2:** 591
Sirdalud **2:** 682
Siros **3:** 1204
Sirtal **2:** 476
Sisomicin **3:** 986, 1047; **4:** 2008
Site-specific drug delivery **4:** 2121
Sito-Lande **1:** 22
Sitosterol **1:** 21, 22; **3:** 1324, 1567, 1569
Skelaxin **2:** 679
Skeletal Muscle Relaxants **2:** 667
SKF-S-104864-A **4:** 1914
Skiacol **4:** 2028
Skiatropine Blache **2:** 690
Skilar **3:** 1220
Skin
  functions and structure **4:** 1714
Skin diseases **4:** 1713
SL-75 212 **1:** 153
SLD 212 **1:** 153
Sleep
  physiology **2:** 489
Sleep-inducing anesthetics **2:** 437, 449
Sloprolol **1:** 159
Slow-Pren **1:** 157
Slow-Trasicor **1:** 157
SM-A **3:** 1439
Smallpox **3:** 1174
Smallpox prophylaxis **4:** 1828
Smallpox vaccine **4:** 1804
SM-C **3:** 1439
Smooth Muscle Relaxants **1:** 95

SN-11841   **4:** 1903
Soaclens   **4:** 2037
Sobril   **2:** 522, 638
Sobutyric Acids
   aryloxy   **1:** 24
Sodium 5-allyl-1-methyl-5-(1-methyl-2-pentynyl)barbiturate   **2:** 444
Sodium aurothiomalate   **4:** 1703
Sodium carboxymethyl cellulose   **4:** 2103
Sodium Channel Blockers   **1:** 120, 124
Sodium chloride   **4:** 2104
Sodium cromoglycate   **4:** 2018
Sodium dioctyl sulfosuccinate   **4:** 2102
Sodium 5,5-diphenylhydantoin   **2:** 475
Sodium dodecyl sulfate   **4:** 2040
Sodium dodecyl sulfate – polyacrylamide gel electrophoresis   **4:** 1973
Sodium γ-hydroxybutyrate   **2:** 449
Sodium ^{131}I iodohipparate injection   **4:** 2101
Sodium lauryl sulfate   **4:** 2040, 2102
Sodium Nitroprusside   **1:** 101
Sodium oleate   **4:** 2102
Sodium oxybate   **2:** 449
Sodium phosphonoformate   **2:** 763
Sodium picosulfate   **2:** 788
Sodium ^{32}P phosphate solution   **4:** 2101
Sodium stibogluconate   **3:** 1154
Sodium Sulamyd   **4:** 2005
Sodium sulfate   **2:** 785
Sodium thiosulfate   **4:** 1732
Sodium valproate   **2:** 477
Sofmin   **2:** 591
Sofradex   **4:** 2008
Soframycin   **4:** 2008
Soft Care   **4:** 2037
Soft Mate   **4:** 2037, 2037, 2039
Soft Mate ps   **4:** 2033
Solanax   **2:** 641
Solanidine   **3:** 1333
Solantyl   **2:** 475
Solaskil   **3:** 1281
Solasodine   **3:** 1333
Solaxin   **2:** 677
Solazine   **2:** 594
Solfa   **2:** 934
Solfoton   **2:** 506
Solganal   **4:** 1703
Solgol   **1:** 92, 56
Solian   **2:** 607
Solid dosage forms   **4:** 2091
Solosin   **2:** 888
Solu-Medrol   **3:** 1624
Solutions   **4:** 2086
Solvents   **4:** 2102
Soma   **2:** 677
Somatic cell fusion   **4:** 1983
Somatic nerves   **2:** 312

Somatoliberin   **3:** 1430, 432
Somatomedins   **3:** 1439, 549
Somatostatin   **3:** 1446
Somatotropin   **3:** 1435, 545
Somatotropin release inhibiting hormone   **3:** 1446
Somatotropin releasing hormone   **3:** 1432
Somaz   **2:** 523
Somberol   **2:** 514
Sombevrin   **2:** 447
Sombucaps   **2:** 511
Sombulex   **2:** 511
Sombutol   **2:** 509
Somelin   **2:** 524
Somilan   **2:** 498
Somio   **2:** 499
Somitran   **2:** 518
Somnafac   **2:** 514
Somnalchlor   **2:** 498
Somnased   **2:** 518
Somnatrol   **2:** 524
Somnesin   **2:** 497
Somnite   **2:** 518
Somnium   **2:** 514
Somnol   **2:** 522
Somnolin   **2:** 518
Somnopentyl   **2:** 509
Somnos   **2:** 498
Somnotropon   **2:** 514
Somonex   **2:** 566
Som-Pam   **2:** 522
Somsanit   **2:** 450
Sonabarb   **2:** 506
Sonbutal   **2:** 509
Soneryl   **2:** 506
Soothe   **4:** 2037
Soothe Eye Drops   **4:** 2030
Soprol   **1:** 153
OROS oral tablet system   **4:** 2115
Sorbic acid   **4:** 2039
Sorbitan trioleate   **4:** 2103
Sorbitol   **2:** 785
Sorbitolum   **2:** 785
Sorbitrate   **1:** 224
Sorenor   **2:** 525
Sorgoa   **3:** 1248
Soripal   **4:** 1684
Sortis   **1:** 29
Sotacor   **1:** 159
Sotalex   **1:** 130, 159
Sotalol   **1:** 130, 159, 226
Sotazide   **1:** 130
Sovelin   **2:** 514
Soventol   **2:** 917
Soverin   **2:** 514
Sovinal   **2:** 514
SK-Soxazole   **3:** 1134
SP   **3:** 1394

SP-10   **3:** 1512
Spacine   **2:** 696
Spaderizine   **1:** 209
Spadon   **2:** 700
Spaneph   **2:** 878
Spansule   **2:** 591
Spantin   **2:** 889
Spantol   **2:** 645
Sparteine   **1:** 126
Spasmex   **2:** 692
Spasmium   **2:** 700
Spasmo-Cibalgin-compositum S   **2:** 703
Spasmo-Cibalgine   **2:** 508
Spasmo-Cibalgin S   **2:** 703
Spasmoctyl   **2:** 695
Spasmodex   **2:** 703
Spasmolysin   **2:** 889
Spasmo-lyt   **2:** 692
Spasmolytics   **2:** 685
Spasmomen   **2:** 695
Spasmophen   **2:** 696
Spasmoril   **2:** 700
Spasmorin   **2:** 700
Spasmo-Urolong   **2:** 702
Spasms
   therapy   **2:** 689
Spasodil   **2:** 699
Spasuret   **2:** 704
Species origin prefixes   **3:** 1345
Spectam injectable   **4:** 2164
Spectam scour-halt   **4:** 2164
Spectazole   **3:** 1220
Spectinomycin   **3:** 984; **4:** 2163
Spectinomycin hydrochloride   **3:** 1048
Spectramedryn   **4:** 2018
Speda   **2:** 508
Sperm antigens   **3:** 1512
Spiclomazine   **2:** 593
Spierifex   **4:** 1678
Spinal reflex   **2:** 316
Spinasterol   **3:** 1324
Spiperone   **2:** 599
Spiramycin   **3:** 1053; **4:** 2166
Spiramycins   **3:** 998
Spirits   **4:** 2087
Spirochaetaceae   **3:** 1111
Spirochetes   **3:** 1111
Spirolair   **2:** 877
Spironolactone   **3:** 1616
Spiropent   **2:** 880
Spiropitan   **2:** 599
Spongesterol   **3:** 1324
Sporanox   **3:** 1204
Sporiderm   **3:** 1248
Sporiline   **3:** 1248
Sporozoans   **3:** 1159
Sporozoite malaria vaccines   **4:** 1815

SQ 26 445   **3:** 976
SQ 26655   **4:** 2054
SQ 26 776   **3:** 977
SQ 29548   **4:** 2054
Squalene epoxidase inhibitors   **3:** 1246, 1248
SRH   **3:** 1430, 432
Sériel   **2:** 644
SRIF   **3:** 1446
SRIH   **3:** 1446
Stability testing   **4:** 2144
Stabilizers   **4:** 2102
Stablon   **2:** 617
Stadadorm   **2:** 507
Stadium   **4:** 1698
Stadol   **2:** 416
Stalleril   **2:** 592
Stanaprol   **3:** 1594
Stanazolol   **3:** 1594
Standard immune serum globulin   **4:** 1822
Stangyl   **2:** 620
Staphylococcus aureus vaccine   **4:** 1778
Staphylomycin   **3:** 1060
Startonyl   **2:** 653
O-V Statin   **3:** 1238; **4:** 2013
Statrol   **4:** 2010
Staurodorm   **2:** 522
Stavudine   **3:** 1298
Stay Brite   **4:** 2037
Stediril   **3:** 1598
Stediril-d   **3:** 1598
Stelabid   **2:** 695
Stelazine   **2:** 594
Stellate cells   **2:** 310
Stem cell factor   **1:** 243
Stem cells
   therapeutic use   **1:** 272
Stemetil   **2:** 594
Stemex   **3:** 1632
Stendomycin   **3:** 1008
Steran   **3:** 1321
Steranabol   **3:** 1593
Steranabol ritardo   **3:** 1592
Sterandryl Retard   **3:** 1590
Stereomycin   **3:** 1238
Sterile bulk drug substances   **3:** 1087
Sterile Disinfecting Solution   **4:** 2037
Sterile Lens Lubricant   **4:** 2037
Sterilization   **3:** 1507
Steroid alkaloids   **3:** 1333
Steroid Hormones   **3:** 1330, 551
Steroid Lactones   **3:** 1335
Steroids   **3:** 1321
Steroid Sex Hormones   **3:** 1564
Sterols   **3:** 1322
Stesolid   **2:** 637
STH   **3:** 1435
Stickoxydul   **2:** 438

Stiemazol **3:** 1217
Stigmastane **3:** 1323
Stigmasterol **3:** 1324, 1567, 1569
Stilamin **3:** 1448
Stilbestrol derivatives **3:** 1586
Stimamizol **3:** 1281
Stimucortex **2:** 650
Stone Formation **2:** 755
Stoxil **3:** 1178; **4:** 2015
Stratum corneum **4:** 1714
Streptococcaceae **3:** 1110
Streptococcus pneumoniae vaccine **4:** 1772
Streptokinase **1:** 276
Streptomycin **3:** 953, 983; **4:** 2162
 staphylococcus aureus inhibition **3:** 1091
Streptomycin formation **3:** 1084
Streptomycin sulfate **3:** 1045; **4:** 2009
Streptothricin **3:** 953
Streptozotocin **4:** 1891
Stresam **2:** 646
Strodival **1:** 190
Stromba **3:** 1594
Strongyloides stercoralis **3:** 1274
Strophanthidin **1:** 179
Strophanthidol **1:** 179
k-Strophanthin-α **1:** 190
k-Strophanthoside **1:** 181, 82
Strophanthus glycosides **1:** 181
Stuart prower factor **1:** 257
Sturgeron **1:** 209
Stutgeron **1:** 209; **2:** 914
Stutgin **1:** 209
Styquin **4:** 2175
Subarachnoid hemorrhage **1:** 149
Subcutaneous implants **4:** 2119
Subcutaneous injection **4:** 2082
Sublimase **2:** 401
Sublimaze **2:** 454
Subranyl **2:** 696
Substance P **3:** 1394
Subtilisin A **4:** 2040
Succicuran **2:** 674
Succinyl-Asta **2:** 674
Succinylcholine chloride **2:** 674
Succinylsulfathiazole **3:** 1137
Succostrin chloride **2:** 674
Sucralfate **2:** 715, 730
Sucrose hydrogen sulfate basic aluminum salt **2:** 730
Sufenta **2:** 404
Sufentanil **2:** 403
Sugar coatings **4:** 2098
Suladrin **4:** 2005
Suladyne **3:** 1135, 136
Sulbactam sodium **3:** 1025
Sulbenicillin **3:** 962, 020
Sulconazole **3:** 1226; **4:** 1730
Sulcosyn **3:** 1227

Suldicyn **3:** 1227
Sulf-10 **4:** 2005
Sulfacetamide **4:** 1725
Sulfacetamide sodium **4:** 2005
Sulfacetimide **4:** 2005
Sulfacytine **3:** 1135
Sulfadiazine **3:** 1134, 1167; **4:** 2005, 2014, 2159
Sulfadimethoxine **4:** 2159
Sulfadimidine **4:** 2159
Sulfadoxine **3:** 1136; **4:** 2159
Sulfa drugs **3:** 1128
Sulfafurazole **3:** 1134
Sulfaguanidine **3:** 1137
Sulfair **4:** 2005
Sulfamerazine **4:** 2159
Sulfamerazine sodium **4:** 2005
Sulfamethazine **4:** 2159
Sulfamethizole **3:** 1135; **4:** 2005
Sulfamethoxazole **3:** 1132; **4:** 1725
Sulfamethoxazole – trimethoprim **3:** 1133
Sulfanilamide **1:** 71
N-Sulfanilyl-N'-n-butylurea **3:** 1655
Sulfapyridine **4:** 2159
Sulfaquinoxaline **4:** 2159
Sulfasalazine **3:** 1136
Sulfasuxidine **3:** 1137
Sulfathalidine **3:** 1137
Sulfazecin **3:** 976
Sulfhydryl antirheumatic drugs **4:** 1705
Sulfisoxazole **3:** 1134
Sulfisoxazole diolamine **4:** 2005
Sulfium **4:** 2005
Sulfonal **2:** 529
Sulfonalamide **3:** 1130
Sulfonamide antibacterial agents **3:** 1128
Sulfonamide Diuretics **1:** 71
Sulfonamides **2:** 486; **4:** 2004
 veterinary **4:** 2158
4,4′-Sulfonylbis(acetanilide) **3:** 1143
4,4′-Sulfonyldianiline **3:** 1142
Sulfonylureas antidiabetics **3:** 1654
Sulforidazine **2:** 592
Sulfoxone, sodium **3:** 1143
Sulindac **2:** 358; **4:** 1682
Sulphrin **4:** 2005
Sulpiride **2:** 605
Sulpred **4:** 2005
Sulpril **2:** 606
Sulprostone **3:** 1509; **4:** 2054, 2059
Sulqui **3:** 1282
Sultanol **2:** 875
Sulten-10 **4:** 2005
Sulthiam **2:** 486
Sultopride **2:** 606
Sulvina **3:** 1242
Supanate **2:** 700

Supatonin **3:** 1279
Suppojuvent Sedante **2:** 498
Suppoptanox **2:** 508
Suppositol **4:** 2104
Suppositories **4:** 2091
Suppository base excipients **4:** 2104
Suppressor cells **1:** 247
Suprahypothalamic centers **3:** 1539
Suprarenin **1:** 60; **2:** 865; **3:** 1562
Supratonin **1:** 65
Supraventricular extrasystoles **1:** 117
Supraventricular tachycardias **1:** 147
Suprecur **4:** 1932
Suramin **3:** 1148, 1188, 1277, 1286
Surfactants **4:** 2102
Surgam **2:** 369
Surgex **2:** 623
Suriclone **2:** 649
Surital **2:** 446
Surmontil **2:** 620
Surolan **3:** 1210
Sursum **2:** 623
Suspensions **4:** 2089
Sus-Phrine **1:** 60
Sustained-release technology **4:** 2112, 2114
Sustiva **3:** 1303
Suxamethonium chloride **2:** 674
Suxilep **2:** 478
Suxinutin **2:** 478
Sweeteners **4:** 2102
Sylvemid **2:** 611
Symmetrel **2:** 574; **3:** 1185
Sympathetic nervous system **2:** 312
  inhibitors **1:** 81
Sympathomimetic amines **1:** 138
Sympathomimetics **1:** 59, 91
Synacthen **3:** 1375
Synadrin **1:** 208
Synalar **3:** 1632
Synalgos **2:** 393
Synandrol F **3:** 1590
Synapse **2:** 310
Synaptic transmission **2:** 327
Synaptic vesicles **2:** 319
Synaptosomes **2:** 323
Synchrocept **4:** 2059
Synchrodyn **3:** 1375
Synchronized sleep **2:** 490
Syncurine **2:** 674
Synogil **3:** 1236
Synopen **2:** 905
Synthalin A **3:** 1660
Synthetic chemotherapeutic agents **3:** 1107
Synthroid **3:** 1561
Syntocinon **3:** 1402
Synval **2:** 513
Syrups **4:** 2087

Syscor **1:** 211
5-HT$_{1P}$
  receptor **2:** 815

Tabazone **4:** 1695
Tableting tools **4:** 2096
Tablets **4:** 2094
Tablet triturates **4:** 2093
Tabrium **2:** 636
Tacaryl **2:** 919, 921
Tachmalor **1:** 128
Tachyarrhythmia **1:** 120
Tachycardia **1:** 147, 173
Tachykinins **3:** 1394
Tacrine **2:** 653
Tactaran **2:** 596
Taeniarhynchus saginatus **3:** 1270
Taenia saginata **3:** 1270
Taenia solium **3:** 1270
Tafil **2:** 641
Tagamet **2:** 723
Tagathen **2:** 906
Taglutimide **2:** 511, 513
Talampicillin **3:** 963
Talampicillin hydrochloride **3:** 1025
Talbutal **2:** 504, 510
Tallysomycins **3:** 1002
Talofen **2:** 590
Taloxa **2:** 485
Talusin **1:** 191
Talwin **2:** 421
Tambocor **1:** 128
Tamoa **3:** 1285
Tamoxifen **3:** 1587; **4:** 1922
  annual sales **4:** 1939
Tanderil **2:** 374
Tapeworms **3:** 1269
Taractan **2:** 596
Tarasan **2:** 596
Taratan **2:** 596
Targeted drug delivery systems **4:** 2121
Tarivid **3:** 1126
Tarodyl **2:** 694
Tarodyn **2:** 694
Tarpan **2:** 924
Tasedan **2:** 524
Tasmolin **2:** 563
Tat-59 **4:** 1924
Tat gene **3:** 1292
Tauredon **4:** 1703
Taurine **2:** 329
Tavegil **2:** 909
Tavor **2:** 522, 638, 644
Taxagon **2:** 632
Taxilan **2:** 593
Taxoids **4:** 1915
Taxotere **4:** 1916

Tazarotene **4:** 1725, 1728
TBZ 6 **4:** 2174
T-cells **1:** 247; **4:** 1760
TCu-200 **3:** 1507
TCu-220C **3:** 1507
TCu-380A **3:** 1507
TearGard **4:** 2033
Tearisol **4:** 2034
Tears Naturale **4:** 2034
Tears Naturale R **4:** 2033
Tears Plus **4:** 2034, 2039
Teflurane **2:** 440
Tegretal **2:** 424
Tegretol **2:** 424, 476
Telazol **4:** 2180
Teldane **2:** 931
Teldrin **2:** 912
Telen **2:** 731
Telgin-G **2:** 909
Telipex-Retard **3:** 1590
Telopar **3:** 1284
Temaril **2:** 591, 919
Temaril-P **4:** 2185
Temazepam **2:** 515, 523
  indications for use **2:** 519
  synthesis **2:** 520
Tementil **2:** 594
Temesta **2:** 522, 638
Temgesic **2:** 415
Temserin **1:** 93
Tenalet **1:** 154
Tenalin **1:** 154
Teniamida **3:** 1282
Teniarene **3:** 1282
Teniposide **4:** 1912
Tennecetin **3:** 1236
Tenoblock **1:** 152
Tenormin **1:** 152
Tenoxicam **2:** 377
Tenserin **1:** 160
Tensilon **4:** 2032
Tenso **2:** 523
Tenuate **2:** 464, 467
Teolaxin **2:** 506
Teoptic **1:** 154
Teoremin **4:** 1681
Tepanil **2:** 464, 467
Tepilta suspension **2:** 549
Teralithe **2:** 631
Terazol **3:** 1230
Terbasmin **2:** 871
Terbinafine **3:** 1250; **4:** 1731
Terbutaline **2:** 871
Tercian **2:** 591
Terconal **3:** 1230
Terconazole **3:** 1228
Tercospor **3:** 1230

Terfenadine **2:** 930
Terfluzine **2:** 594
Terguride **2:** 576
Teriam **1:** 299
Terion **2:** 339
Terlipressin **3:** 1402
Ternelin **2:** 682
Terodiline **1:** 209
Terolut **3:** 1597
Teronac **2:** 466
Terra-Cortil **4:** 2016
Terramycin **3:** 954; **4:** 2007, 2010
Terramycin soluble powder **4:** 2161
Tersigat **2:** 892
2-(*tert*-butylamino)-1-(2-chlorophenyl)ethanol **2:** 881
2-(*Tert*-butylamino)-3′-chloropropiophenone **2:** 631
1-(5-[2-(*Tert*-butylamino)-1-hydroxyethyl]-2-hydroxyphenyl)urea **2:** 881
(±)5-[2-(*Tert*-butylamino)-1-hydroxyethyl]-*m*-phenylenebis(dimethylcarbamate) **2:** 885
2-(*Tert*-butylamino)-1-(4-hydroxy-3-hydroxymethylphenyl)ethanol **2:** 874
α^6-[(*Tert*-butylamino)methyl]-3-hydroxy-2,6-pyridinedimethanol **2:** 876
Tertroxin **3:** 1561
Terzolin **3:** 1208
Tesnol **1:** 90
Tesoprel **2:** 598
Testosid **3:** 1590
Testosteron "Berco" **3:** 1590
Testosterone **3:** 1551, 565, 588, 590
Testosterone antagonists **4:** 1928
Testosterone derivatives **3:** 1593
Testosterone enanthate **3:** 1590
Testosterone propionate **3:** 1590
Testosterone synthesis **3:** 1574
Testoviron Depot **3:** 1590
Testoviron T **3:** 1590
Testryl **3:** 1590
Tetanus
  treatment with monoclonal antibodies **4:** 1841
Tetanus prophylaxis **4:** 1828
Tetanus Vaccine **4:** 1766
(2,3,4,6-tetra-*O*-Acetyl-1-thio-β-D-glucopyranosato-*S*)(triethylphosphine)gold **4:** 1703
Tetracaine **4:** 2178
Tetracaine hydrochloride **2:** 545; **4:** 2003, 2004
Tetrachloroethylene **3:** 1273
Tetracosactide **3:** 1375
Tetracyclic Antidepressants **2:** 620
Tetracycline **3:** 978, 036; **4:** 1725, 2007, 2019, 2161
  staphylococcus aureus inhibition **3:** 1091
Tetracycline antibiotics
  as ophthamological agents **4:** 2007
Tetracyclines
  veterinary **4:** 2160

Tetracyclines antibiotics **3:** 977
  economic aspects **3:** 1094
(S)-N-(5,6,7,9-Tetrahydro-1,2,3,10-tetramethoxy-9-oxo-benzo[a]heptalen-7-yl)acetamide **4:** 1708
1,2,3,6-Tetrahydro-1,3-dimethyl-2,6-dioxo-7H-purine-7-acetic acid **2:** 566
1,2,3,6-Tetrahydro-1,3-dimethyl-2,6-dioxo-7-purineacetic acid **2:** 889
Tetrahydrofolates **4:** 1868
(±)-4-(5,6,7,8-Tetrahydroimidazo[1,5-a]pyridin-5-yl)-benzonitrile **4:** 1926
1,2,3,4-Tetrahydro-2-[(isopropylamino)methyl]–7-nitro-6-quinolinemethanol **3:** 1283
2,3,4,9-Tetrahydro-2-methyl-9-phenyl-1H-indeno[2,1-c]pyridine **2:** 925
2,3,4,5-Tetrahydro-2-methyl-5-(phenylmethyl)-1H-pyrido[4,3-p]indole **2:** 925
(E)–3-[2-(1,4,5,6-tetrahydro-1-methyl-2-pyrimidinyl)ethenyl]phenol 4,4′-methylenebis(3-hydroxy-2-naphthalenecarboxylate) **3:** 1283
(E)-1,4,5,6-tetrahydro-1-methyl-2-[2-(2-thienyl)vinyl]-pyrimidinium 4,4′-methylenebis(3-hydroxy-2-naphthalenecarboxylate) **3:** 1285
1,2,3,4-Tetrahydro-4-oxonaphthalene-1-spiro-3′-piperidine-2′,6′-dione **2:** 512
2-(1,2,3,4-Tetrahydro-1-naphthyl)-2-imidazoline hydrochloride **4:** 2030
L(–)–2,3,5,6-tetrahydro-6-phenylimidazo[2,1-b]thiazole hydrochloride **3:** 1280
4-(Tetrahydro-2H-1,2-thiazin-2-yl)benzenesulfonamide S,S-dioxide **2:** 486
Tetrahydrozoline hydrochloride **4:** 2030
L-Tetraiodothyronine **3:** 1555
3,5,3′,5′-Tetraiodothyronine **3:** 1551
L-3,5,3′,5′-Tetraiodothyronine **3:** 1523, 1558
4-(1,1,3,3-Tetramethylbutyl)phenol **4:** 2040
Tetramide **2:** 621
Tetronic **4:** 2040
tetroquinol **2:** 868
TFT-Ophtiole **4:** 2015
6-TG **4:** 1882
TGF-β **3:** 1358
Thalamonal **2:** 401, 454
Thalidomide **2:** 511
Theelin **3:** 1583
Theelol **3:** 1583
T-helper cells **1:** 247
Thenalidine **2:** 918
Thenfadil **2:** 907
Thenyldiamine **2:** 907
Thenylene **2:** 906
Theobroma oil **4:** 2104
Theofibrate **1:** 26
Theophylline **1:** 290, 301; **2:** 887
Theophylline ethylenediamine **2:** 888
Theophyllol **2:** 888
Thephorin **2:** 925

Theralax **2:** 788
Theralene **2:** 919
Theranabol **3:** 1593
Therapeutic index **4:** 2080
Therapeutic segments
  of the pharmaceutical market **1:** 7
Theraplix **2:** 529; **4:** 2022, 2023
ReNu Thermal Enzymatic Contact Lens Cleaner **4:** 2040
Theruhistin **2:** 920
Thevier **3:** 1561
Thiabendazole **3:** 1269, 1272, 1273, 1274, 1275, 1276, 1287; **4:** 2174
Thiacetazone **3:** 1142
Thiadipone **2:** 642
Thiadrine **2:** 878
Thialbarbital **2:** 446
Thiamazole **3:** 1530
  pharmacokinetic data **3:** 1527
Thiamphenicol **4:** 2167
Thiamylal **2:** 446
Thiazinamium **2:** 922
Thiazolidinediones **3:** 1665
2-(4-Thiazolyl)-1H-benzimidazole **3:** 1287
4′-(2-Thiazolylsulfamoyl)-phthalanilic acid **3:** 1137
4′-(2-Thiazolylsulfamoyl)-succinanilic acid monohydrate **3:** 1137
Thibenzazolin **3:** 1530
Thibenzazoline **3:** 1525
Thienamycin **3:** 974
Thiethylperazine **2:** 744
Thimerosal **3:** 1255; **4:** 2037
Thiobarbital **2:** 445
Thiobarbituric acids **2:** 445
Thio-Barbityral **2:** 445
2,2′-Thiobis(4,6-dichlorophenol) **3:** 1279
Thiobutabarbital sodium **2:** 446
Thiocarbamates antimycotics **3:** 1246
Thioctacid **2:** 771
Thioctic acid **2:** 770
(1-thio-D-Glucopyranosato)gold **4:** 1702
5-Thio-D-glucose **3:** 1514
6-Thioguanine **4:** 1728, 882
Thiomesterone **3:** 1593
Thionembutal **2:** 445
Thionylan **2:** 906
Thiopental "Lentia" **2:** 445
Thiopental sodium **2:** 445
Thiopropazate **2:** 595
Thioproperazine **2:** 594
Thiopurines **4:** 1882
Thioridazine **2:** 592
Thioril **2:** 592
Thioseconal **2:** 446
Thiosnifil-A **3:** 1136
Thio-TEPA **4:** 1889
Thiothixene **2:** 596

2307

Thioxanthene neuroleptics **2:** 596
Thiphenamil **2:** 705
Thiphenum **2:** 705
Thomasin **1:** 64
Thonzylamine **2:** 907
Thorazine **2:** 591, 681
N-(Ethoxycarbonyl)-3-(4-morpholinyl)sydnone
    imine **1:** 230
THP-ADM **4:** 1898
Théralène **2:** 591
Threadworm **3:** 1274
(–)-Threo-3-(3,4-dihydroxyphenyl)-L-serine **2:** 569
Thrombi **1:** 32
Thrombocyte count **1:** 239
Thrombocytes **1:** 249
    therapeutic use **1:** 271
Thromboembolic diseases **1:** 271
Thrombogenicity
    of blood components **1:** 282
Thromboplastin **1:** 255
Thrombopoietin **1:** 240, 243
Thromboxane A$_2$ **1:** 34; **4:** 2047
Thromboxanes **4:** 2047
    biosynthesis **4:** 2052
Thybon **3:** 1561
N-Ethyl-N-(4-pyridylmethyl)tropamide **4:** 2028
Thymidine phosphorylase **4:** 1873
Thymidylate synthase inhibitors **4:** 1871
Thyrex **3:** 1561
Thyroid Depressants **3:** 1524, 529
Thyroid Diseases **3:** 1557
Thyroid gland **3:** 1520
Thyroid hormones **3:** 1520, 1551, 1555
    pharmacology **3:** 1521
Thyroliberin **2:** 654; **3:** 1348, 353
Thyrotardin **3:** 1561
Thyrotherapeutic Agents **3:** 1519
Thyrotoxicosis **1:** 149
Thyrotropin **3:** 1348, 357, 544
Thyrotropin-releasing hormone **2:** 581, 654; **3:** 1544
Thyroxin binding glycoprotein **1:** 252
L-Thyroxine **3:** 1523, 551, 555
Thyroxine Analogs **1:** 22
L-Thyroxine Roche **3:** 1561
L-Thyroxin Henning **3:** 1561
Tiagabine **2:** 484
Tiamulin **3:** 1072; **4:** 2168
Tianeptine **2:** 617
Tiapamil **1:** 209
Tiapride **2:** 606
Tiaprizal **2:** 606
Tiaprofenic acid **2:** 368
Tibatin **3:** 1217
Tibricol **1:** 210
Ticarcillin **3:** 962
Ticarcillin disodium **3:** 1020
Ticinil **4:** 1694

Tick-borne encephalitis vaccine **4:** 1802
Ticlid **1:** 41
Ticlopidina **1:** 41
Ticlopidine **1:** 40
Ticolubant **4:** 2066
Ticrynafen **1:** 295
Tiemonium iodide **2:** 697
Tienor **2:** 641
Tifenamil **2:** 705
Tigan **2:** 911
Tigloidine **2:** 563
Tiglyssin **2:** 563
Tiklid **1:** 41
Tiklyd **1:** 41
Tilade **2:** 896, 935
Tilcotil **2:** 378
Tiletamine **4:** 2180
Tilidate **2:** 411
Tilidine **2:** 410
Timacor **1:** 93, 60
Timaxel **2:** 615
Timiperone **2:** 599
Timolate **1:** 93, 160
Timolol **1:** 93, 160, 226
Timolol maleate **4:** 2021
Timolol maleate Betim **1:** 160
Timonil **2:** 476
Timoptic **1:** 93, 160; **4:** 2021
Timoptol **1:** 93, 160; **4:** 2021
Timostenil **2:** 624
Timotre **1:** 160
Tinacidin **3:** 1248
Tinactin **3:** 1248
Tinatox **3:** 1248
Tinavet **3:** 1248
Tinctures **4:** 2087
Tindal **2:** 595
Tineafax **3:** 1248
Ting **3:** 1248
Tinidazole **3:** 1156, 1158
Tinset **2:** 916
TIO **3:** 1230
Tioconazole **3:** 1230; **4:** 1730
Tiotropium bromide **2:** 892
Tipranavir **3:** 1314
Tirashizin **2:** 564
Tiratricol **3:** 1523
Tiropramide **2:** 701
Tispol **2:** 700
Tissue factor **1:** 255
Tissue hormones **3:** 1549
Tissue-invading nematodes
    vaccines **4:** 1810
Tizanidine **2:** 681
T-Lymphocytes **1:** 247
TNF **4:** 1673
TNF inhibitors **4:** 1708

2308

Tobinal **2**: 511
TobraDex **4**: 2009, 2016
Tobra-Gobens **4**: 2009
Tobral **4**: 2009
Tobra-Laf **4**: 2009
TobraLex **4**: 2009
Tobramaxin **4**: 2009
Tobramycin **3**: 985, 1041; **4**: 2009
Tobrex **4**: 2009
Tocainide **1**: 126
Tofisopam **2**: 643
Togaviridae **3**: 1174
Tohpiride **2**: 606
Tolazamide **3**: 1657
Tolazoline **1**: 87
Tolbutamide **1**: 15; **3**: 1658
Tolcapone **2**: 572
Tolciclate **3**: 1246
Tolcyclamide **3**: 1658
Tolectin **2**: 360; **4**: 1684
Tolerability testing **4**: 2135
Toleran **2**: 445, 511
Tolerance **4**: 2129
Tolid **2**: 522
Toliman **1**: 209
Tolinase **3**: 1657
Tolindol **4**: 1682
Tolmetin **2**: 359; **4**: 1683
Tolmicen **3**: 1246
Tolmicol **3**: 1246
Tolnaftate **3**: 1247; **4**: 1731
Tolnate **2**: 596
Tolopelon **2**: 600
Toloxatone **2**: 626
Tolpropamine **2**: 912
Tolvin **2**: 621
Tolvon **2**: 621
Tomatidine **3**: 1333
Tombran **2**: 632
Tonamil **2**: 907
Tonaton **2**: 690
Tonibral **2**: 652
Tonicity adjusters **4**: 2103
Tonic seizures **2**: 470
Tonocard **1**: 126
Tonoftal **3**: 1248
Topagen **4**: 2177
Topamax **2**: 484
Topical anesthetics **2**: 539
Topical corticoid therapy **3**: 1604
Topical therapy **4**: 1715
Topicort **3**: 1634
Topiramate **2**: 483
Topisolon **3**: 1634
Topoisomerase **4**: 1915
Topoisomerase II **3**: 1117
Topotecin **4**: 1914

Topotecin **4**: 1914
Topral **2**: 606
Tora-Dol **2**: 380, 380
Toraflon **2**: 514
Toremifene **4**: 1922
Torental **1**: 49
Torinal **2**: 514
Tornalate **2**: 885
Tosmilen **4**: 2025
N-(Tosyl)-*N'*-(3-aza-bicyclo-[3.3.0.]-3-octyl)-
    urea **3**: 1656
N-(Tosyl)-*N'*-cyclohexylurea **3**: 1658
N-(Tosyl)-*N'*-(2-endo-hydroxy-3-endo-DL-bornyl)-
    urea **3**: 1655
N-(Tosyl)-*N'*-(hexahydro-1*H*-azepin-1-yl)-urea **3**: 1657
Total **4**: 2034
Toxicity on repeated administration **4**: 2131
Toxicity study requirements **4**: 2132
Toxicological testing **4**: 2130
Toxocara species **3**: 1275
Toxoplasma **3**: 1159
Toxoplasma gondii **3**: 1167
Toxoplasmosis **3**: 1167
Toyocamycin **3**: 991
T-PA **1**: 276
5-HTP-DP **2**: 815
Trachyl **2**: 391
Tracosal **1**: 157
Tracrium **2**: 670
Tradon **2**: 339
Tragacanth **2**: 784
Tragacantha **2**: 784
Tral **2**: 694
Tralanta **2**: 695
Tralen **3**: 1231
Tramadol **2**: 411
Tramal **2**: 412
Tramisol **4**: 2175
Trancalgyl **2**: 348
Trancin **2**: 595
Trancolon **2**: 695
Trandate **1**: 94, 155
Trankimazin **2**: 641
Tranquazine **2**: 590
Tranquilin **2**: 645
Tranquilizers **2**: 4951
    major **2**: 588
Trans-(±)-2-phenylcyclopropylamine **2**: 625
Trans-3,4bis(4-methoxyphenyl)-3-hexene **3**: 1586
Trans-3-Ethyl-2,5-dihydro-4-methyl-*N*-[2-[4-[[[[(4-
    methylcyclohexyl) amino]carbonyl]amino]sulfo-
    nyl]phenyl]ethyl] **3**: 1656
trans-3-Methyl-2-phenylmorpholine **2**: 465
Transapin **2**: 625
Transcortin **1**: 252
Transdermal drug delivery systems **4**: 2117
Transdermal patch **4**: 2118

2309

Transderm-nitro **1**: 223; **4**: 2118
Transderm scop **4**: 2118
Transderm-V **2**: 562
3,4-*Trans*-2,2-dimethyl-3-phenyl-4-[*p*-β-pyrrolidino-
  ethoxy)-phenyl]-7-methoxychromane **3**: 1509
Transene **2**: 639
Transferrin **1**: 252
(–)-*Trans*-4-(4-fluorophenyl)-3-{[3,4-(methylenedioxy)-
  phenoxy]-methyl}-piperidine **2**: 629
Transforming growth factor-β **3**: 1358, 359
Transithal **2**: 446
Transmission blocking immunity **4**: 1820
Transmitters
  neuro **2**: 328
Transplants
  in Parkinsonism treatment **2**: 581
[±(*trans*)]-Tetrachloro(1,2-cyclohexanediamine-*N,N'*]pla-
  tinum **4**: 1921
Trans,-trans,2,2'-[dimethylene-bis-(carbonyl-
  oxytrimethyl-ene)-bis-(1,2,3,4-tetrahydro-6,7,8-tri-
  methoxy-2-methyl **2**: 670
Tranxène **2**: 639
Tranxilène **2**: 639
Tranxilium **2**: 639
Tranylcypromine **2**: 625
Trapanal **2**: 445
Trapax **2**: 522
Trapidil **1**: 230
Trasacor **1**: 157
Trasicor **1**: 157
Traumacut **2**: 680
Travogen **3**: 1222
Travogyn **3**: 1222
Trazodil **2**: 632
Trazodone **2**: 632
Trédémine **3**: 1282
Trecalmo **2**: 641
Trecator – SC **3**: 1141
5-HT$_3$ receptor antagonists **2**: 800
5-HT$_4$ receptor antagonists **2**: 804
5-HT$_4$ Receptors
  electrophysiological actions **2**: 817
Trematodes **3**: 1266
Tremblex **2**: 565
Tremor
  in Parkinsonism **2**: 556
Trentadil **2**: 890
Trental **1**: 49
Trepidan **2**: 639
Tresaderm **4**: 2174
Tresortil **2**: 680
Trethylene **2**: 439
Tretinoin **4**: 1724, 1733
Trexan **2**: 423
TRH **2**: 654; **3**: 1544
Triacetin **4**: 1731
Triacetyloleandomycin **3**: 1052

Triaconazole **3**: 1228
Triamcinolone **3**: 1620; **4**: 2179
  doses **3**: 1605
Triamcinolone acetonide **3**: 1620
,4,7-Triamino-6-phenylpteridine **1**: 75
Triamterene **1**: 75, 299
Triasporin **3**: 1204
4,4'-(1-Triazene-1,3-diyl)bis(benzenecarboximidamide)-
  bis(*N*-acetylglycinate) **3**: 1150
Triazolam **2**: 515, 524
  indications for use **2**: 519
  metabolism **2**: 517
  synthesis **2**: 521
Triazole antimyotics **3**: 1201
4,4'-(1*H*-1,2,4-Triazol-1-ylmethylene)bis[benzoni-
  trile] **4**: 1926
2,2'-[5-(1*H*-1,2,4-Triazol-1-ylmethyl)-1,3-phenylene]-
  bis(2-methylpropionitrile **4**: 1926
Trichina **3**: 1275
Trichinella spiralis **3**: 1275
Trichloran **2**: 439
Trichlormethiazide **1**: 294
2,2,2-Trichloro-1,1-ethanediol **2**: 498
2,2,2-Trichloroethyl dihydrogen phosphate **2**: 498
Trichloroethylene **2**: 439
α-*O*-(2,2,2-Trichloroethylidene)-1,2-α-D-glucofuranose-
  (*R*) **2**: 499
N-(2,2,2-Trichloro-1-hydroxyethyl)-ethyl carba-
  mate **2**: 499
1,1,1-trichloro-2-methyl-2-propanol **4**: 2038
Trichomonas vaginalis **3**: 1155
Trichomoniasis **3**: 1155
Trichomycin **3**: 1066
Trichuris trichiura **3**: 1274
TriCiilest **3**: 1502
Triclofos **2**: 498
Tricloran **2**: 499
Triclos **2**: 499
Triclosyl **2**: 499
Tricocel **3**: 1285
Tricosten **3**: 1217
Tricuran **2**: 671
Tricyclic antidepressants **2**: 610
1-Tricyclo-[3.3.1.1^{3,7}]-decylamine, 1-adamantana-
  mine **2**: 573
[1(1*S*,2*R*),5(*S*)]-2,3,5-trideoxy-*N*-(2,3-dihydro-2-hydroxy-
  1*H*-inden-1 **3**: 1312
Tridihexethyl chloride **2**: 697
Tridione **2**: 479
Triethylenethiophosphoramide **4**: 1889
Triflucan **3**: 1202
Trifluoperazine **2**: 594
Trifluoper-Ez-Ets **2**: 594
Trifluopromazine **2**: 744
5'-Trifluoromethyl-2'-deoxyuridine **3**: 1178
2-Trifluoromethyl-10-[3-(4-methyl-1-piperazinyl)pro-
  pyl]-10*H*-phenothiazine **2**: 594

2-[3-(Trifluoromethyl)phenyl]aminobenzoic acid **2**: 351; **4**: 1697
2-(3-Trifluoromethylphenylamino)nicotinic acid, 2-[[3-(trifluoromethyl)phenyl]amino]-3-pyridinecarboxylic acid **2**: 352
2-{[3-(Trifluoromethyl)phenyl]amino}-3-pyridinecarboxylicacid **4**: 1698
Trifluorothymidine **3**: 1178
2,2,2-Trifluoro-1-vinyloxyethane **2**: 442
Trifluperidol **2**: 599
Triflupromazine **2**: 590, 591
Trifluridine **3**: 1178; **4**: 2015
Trifonazole **3**: 1212
Triglyceride level **1**: 20
Trigot **2**: 655
Trihexylphenidyl hydrochloride **2**: 563
Trihydrate **4**: 2006
1,3,5-Trihydro-2,6-dioxopyrimidine-4-carboxylic acid, uracil-6-carboxylic acid **2**: 769
N'-2,3,4-Trihydroxybenzyl-DL-serylhydrazide, DL-serine[2'-(2,3,4-trihydroxybenzyl)hydrazide] **2**: 569
3,7,12-Trihydroxycholanic acid **2**: 749
2',4',6'-Trihydroxypropiophenone **2**: 700
9α,11α,15α-Trihydroxy-5Z,13E-prostadienoic acid **4**: 2048
3,5,3'-Triiodothyroacetic acid **3**: 1523
Triiodothyronine **3**: 1555
3,5,3'-Triiodothyronine **3**: 1523, 1551
L-3,3',5'-Triiodothyronine **3**: 1558
L-3,5,3'-Triiodothyronine **3**: 1558
Trilafon **2**: 595
Trilene **2**: 439
Trileptal **2**: 477
Trilifan **2**: 595
Trilombrin **3**: 1285
Trimebutine **2**: 701
Trimedone **2**: 479
Trimeprazine **4**: 2185
Trimethadione **2**: 473, 479
Trimethobenzamide **2**: 910
Trimethoprim **3**: 1130, 1133; **4**: 1725
Trimethoquinol **2**: 868
(−)-1-(3,4,5-Trimethoxybenzyl)-1,2,3,4-tetrahydro-6,7-isoquinolinediol **2**: 868
Trimethyl-(1-methyl-2-phenothiazin-10-ylethyl)ammonium methyl sulfate **2**: 922
α,α,β-Trimethylbenzeneethanamine **2**: 463
Trimethylene **2**: 439
4,4'-(Trimethylenedioxy)bis(3-bromobenzamidine) **4**: 2014
3,5,5-Trimethyl-1,3-oxazolidine-2,4-dione **2**: 479
Trimethyltetradecylammonium bromide **4**: 2038
2,4,6-Trimethyl-1,3,5-trioxane **2**: 497
Trimetrexate **4**: 1871
Trimol **2**: 566
Trimpex **3**: 1133
Trimysten **3**: 1217
Tri-Norinyl **3**: 1502
Trional **2**: 529
Triovex **3**: 1583
3,7,12-Trioxocholanic acid **2**: 749
Tripamide **1**: 295
Tripelenamine **2**: 907
Tripelennamine **4**: 1720, 2184
Triperidol **2**: 599
Triphasil **3**: 1502
Triphenidone **2**: 563
Triphenylmethane dyes **3**: 1255
Triple-Gen **4**: 2016
Triprolidine **2**: 912
Triptil **2**: 616
Triptorelin **4**: 1932
Triquilar **3**: 1502
Trisorcin **4**: 1706
Tristep **3**: 1502
1,2,3-Tris(2-triethylammonium ethoxy)benzene triiodide **2**: 670
Triton X-100 **3**: 1495
Tritop **4**: 2178
Tritoqualine **2**: 933
Troches **4**: 2094
Trocinate **2**: 705
Troglitazone **3**: 1666
Trokonil **2**: 695
Trolamine laurate **4**: 2102
Troleandomycin **3**: 1052
Trolovol **4**: 1706
Tronothane **2**: 551
Tropane Alkaloid Parasympatholytics **2**: 690
Tropane alkaloids **2**: 561
Tropane Alkaloids antiemetics **2**: 741
1αH,5αH-tropan-3α-ol (±)-tropate **2**: 891
3α(1αH,5αH)-Tropanyl-DL-tropate **2**: 562
3α (1αH,5αH)-tropanyl-(±)-tropate **2**: 690
Tropergen **2**: 408
Tropicacyl **4**: 2028
Tropicamide **4**: 2028
Tropikamid Minims **4**: 2028
Tropimil **4**: 2028
Tropisetron **2**: 800
Trosid **3**: 1231
Trospium chloride **2**: 692
Trosyd **3**: 1231
Trosyl **3**: 1231
Trucallol **3**: 1326
Truxal **2**: 596
Truxaletten **2**: 596
Trypanosoma cruzi **3**: 1151
Trypanosoma gambiense **3**: 1147
Trypanosoma rhodesiense **3**: 1147, 1148
Tryparsam **3**: 1150
Tryparsamide **3**: 1150
Tryparsamidium **3**: 1150

Tryparsone **3:** 1150
Tryptizol **2:** 611
Tsetse fly **3:** 1148
TSH **3:** 1544
Tualone **2:** 514
Tuazol **2:** 514
Tuazolona **2:** 514
Tubadil **2:** 672
Tubarine **2:** 672
Tube dilution method **3:** 1090
Tuberactinomycin N **3:** 1061
Tubercidin **3:** 991
Tuberculosis
    chemotherapeutic agents **3:** 1139
Tuberculosis vaccine **4:** 1780
d-Tubocurarine chloride **2:** 672, 672
Tubulin binders **4:** 1909, 910
Tulip **4:** 1691
Tulobuterol **2:** 881
Tumor cells **3:** 956
Tumor necrosis factor **1:** 243; **4:** 1673, 707
Tumors **4:** 1866
Tumulosic acid **3:** 1326
5′-Tungsto-2-antimonate **3:** 1189
Turbocalcitonin **3:** 1367
Turimycin H_5 **3:** 998
Turinabol **3:** 1593
TXA_2 **1:** 34; **4:** 2047
TXA_2 synthesis **4:** 2069
Tybamate **2:** 645
Tybatran **2:** 645
Tylan 40 **4:** 2166
Tylan injection **4:** 2166
Tylciprine **2:** 625
Tylenol **2:** 349
Tylosin **3:** 999, 1054; **4:** 2165
Tylosin injection **4:** 2166
Tyloxapol **4:** 2040
Tymelyt **2:** 614
Typhoid fever vaccine **4:** 1770
Tyrimide **2:** 694
Tyrocidines **3:** 1005

U44069 **4:** 2054
U46619 **4:** 2054
U-77233 **4:** 1921
UK 49858 **3:** 1202
Ulbreval **2:** 446
Ulcogant **2:** 730
Ulgrax **2:** 646
Ultair **4:** 2064
Ultiva **2:** 403
Ultracain **2:** 548
Ultracorten **3:** 1618
Ultracorten-H **3:** 1618
Ultragris **3:** 1242
Ultralan **3:** 1634

Ultram **2:** 412
Ultranden **3:** 1590
Ultratears **4:** 2034
Ultraviolet radiation
    in psoriasis treatment **4:** 1728
Ultrazyme **4:** 2040
Unakalm **2:** 641
Undecylenic acid **3:** 1255; **4:** 1731
Uniblock **1:** 152
Uniloc **1:** 152
Unisom **2:** 496, 909
Unisomnia **2:** 518
Univer **1:** 208
Unoprostone isopropylester **4:** 2061
Unovis **3:** 1598
Urapidil **1:** 89
Urbadan **2:** 644
Urbanol **2:** 644
Urbanyl **2:** 644
Urbason **3:** 1624
Urea **4:** 1723, 1727, 2023
Ureaphil **4:** 2023
Urethane **2:** 499
Ureum **4:** 2023
Urispadol **2:** 704
Urispas **2:** 704
Urogenital flagellates **3:** 1155
Urogonadotropin **3:** 1357
Urokinase **1:** 276
Uromykol **3:** 1217
Uro-Ripirin **2:** 693
Ursodeoxycholic acid **2:** 756; **3:** 1329
Uskan **2:** 522
USP XIX **2:** 439
Ústrogen-Holzinger **3:** 1586
Ústrogynal sine **3:** 1584
Uterine Growth Test **3:** 1584
Utibid **3:** 1119
Util **2:** 497
UVB phototherapy **4:** 1728
Uvilon **3:** 1284
Uzarigenin **1:** 179
V2 receptor antagonists **1:** 300
Vaben **2:** 522
Vaccines **4:** 1753
    classification **4:** 1762
    contraceptive **3:** 1510
Vaccinia **3:** 1174; **4:** 1828
Vagantin **2:** 695
Vaginal Contraceptives **3:** 1495
Vaginal Opening Test **3:** 1583
Vaginal Rings **3:** 1506
Vaginal Smear Test **3:** 1582
Vagistat **3:** 1231
Vagotrope S **2:** 691
Vagran **2:** 620
Valamin **2:** 500

Valbazen 3: 1278
ValCaps 2: 678
Valcote 2: 477
Valdettamil 2: 502
Valdorm 2: 522
Valeans 2: 641
Valepotriate 2: 495
17β-Valeryloxy-1,3,5(10)-estratrien-3-ol 3: 1584
Validamycin 3: 1049
Validation 4: 2142
Valine-gramicidines 3: 1005
Valiquid 2: 637, 678
Valium 2: 481, 637, 678
Vallergan 2: 591
Valmane 2: 495
Valmid 2: 500
Valmidate 2: 500
Valoron N 2: 411
Valpin 2: 692
Valproic acid 2: 477
Valrelease 2: 637, 678; 4: 2115
Valtratum 2: 495
Valvular disease 1: 146
Vancil 3: 1283
Vancocin 4: 2012
Vancoled 4: 2012
Vancomycin 3: 1061; 4: 2012
Vancor 4: 2012
Vanpar 3: 1286
Vanquil 3: 1286
Vanquin 3: 1286
Vanticon 4: 2064
Varesal 2: 652
Varicella prophylaxis 4: 1829
Varicella vaccine 4: 1799
Varicosities 2: 310
Variola 4: 1804
Variolation 4: 1755
Variotin 3: 1008, 1069
Varson 2: 655
Vascor 1: 209
Vasectomy 3: 1507
Vasoactive agents
    ophthalmological 4: 2029
Vasoactive Drugs 1: 13
    function 1: 16
Vasoactive Intestinal Peptide 3: 1423
Vasocidin 4: 2005
Vasocon Regular 4: 2029
Vasoconstriction 1: 173
Vasodilators 1: 99
    Ca^{2+} channel antagonists 1: 95
    caronary artery 1: 230
    coronary arteriolar 1: 231
Vasodin 1: 210
Vasolan 1: 97, 208
Vasonase 1: 210

Vasopressin 3: 1398, 542
Vasosulf 4: 2027, 2029
Vasotocin 3: 1399
Vaspit 3: 1636
Vecuronium bromide 2: 672
VEE 3: 1174
Vehicles
    for dermatological drugs 4: 1717
Veillonellaceae 3: 1110
Velban 4: 1910
Velnacrine 2: 659
Veltrim dermatologic cream 4: 2171
Vems 2: 868
Venen 2: 913
Venesthene 2: 441
Venlafaxine 2: 633
Venom immunotherapy 2: 935
Ventaire 2: 867
Ventilat 2: 892
Ventolin 2: 875
Ventricular arrhythmias 1: 147
Ventricular extrasystoles 1: 117
Veracim 1: 208
Veractil 2: 591
Veralipride 2: 608
Veramex 1: 208
Verapamil 1: 15, 97, 131, 206, 208, 229
Verapril 2: 608
Veraptin 1: 208
Verdal 3: 1285
Verelan 1: 208
Veretran 2: 641
Verexamyl 1: 208
Veritab 2: 916
Verladyn 1: 66
Vermox 3: 1281
Veroletten 2: 506
Veronal 2: 506
Veroqual 2: 523
Vertebrate nervous system 2: 312
Verticine 3: 1334
Vertigon 2: 594
Vertimec 4: 2174
Very low density lipoproteins 1: 19
Vesprin 2: 591
Vestran 2: 639
Vetalar 2: 448; 4: 2180
Vetalog 4: 2179
Vetanarcol 2: 509
Veterinary drugs 4: 2157
Viadril 2: 452
Viadril G 2: 452
Viaductor 1: 133
Viaspera 2: 617
Vibrionaceae 3: 1110
Vicodin 2: 394
Vidarabin 2: 764

Vidarabine **2:** 764; **3:** 1179; **4:** 2015
Videx **3:** 1296
Vif gene **3:** 1292
Vigabatrin **2:** 481
Viloxazine **2:** 632
Viminol **2:** 412
Vinbarbital **2:** 509
Vinblastine **4:** 1910
Vinca alkaloids **4:** 1910
Vincaleukoblastine **4:** 1910
Vinconate **2:** 660
Vincristine **4:** 1910
Vindesine **4:** 1911
Vinether **2:** 441
Vinorelbine **4:** 1912
Vinpocetine **2:** 656
Vinydan **2:** 441
Vinylbarbital **2:** 508
Vinylbital **2:** 504, 508
Vinyl ether **2:** 441
Viodor **2:** 529
Vioform **3:** 1170
Viomycin **3:** 1008, 1061
Vioxx **2:** 385; **4:** 1701
VIP **3:** 1423
Viprinex **1:** 47
Vira-A **3:** 1179; **4:** 2015
Viracept **3:** 1308
Viral infections
  chemotherapy **3:** 1172
    treatment with monoclonal antibodies **4:** 1842
Viral Nucleic Acid Synthesis Inhibitors **2:** 763
Viral proteins **3:** 1172
Viral vaccines **4:** 1756, 787
Viramune **3:** 1301
Virazole **3:** 1181, 1182
Virginiamycin **3:** 1008, 1060; **4:** 2169
Virofral **2:** 574
Viroptic **3:** 1179; **4:** 2015
Virulence factors **3:** 1113
Viruses **3:** 1172
    transmission by plasma-derived products **1:** 280
Virus families **3:** 1173
Visalens Wetting Solution **4:** 2034
Visceralgina **2:** 697
Visceralgine **2:** 697
Visculose **4:** 2034
Visine A.C. **4:** 2030
Visine Eye Drops **4:** 2030
Visine L.R. Eye Drops **4:** 2029
Visine Plus **4:** 2030
Visken **1:** 158
Vistaril Parenteral **2:** 646
Vistosan **4:** 2027
Visudrisone **4:** 2018
Visumidriatic **4:** 2028
Vitamin A **4:** 1724

Vitamin D **3:** 1325
Vitamin D_2 **3:** 1637
Vivactil **2:** 616
Vivalan **2:** 632
Vivarint **2:** 632
Vividrin **4:** 2019
Vividyl **2:** 615
VLDL **1:** 19
VM-26 **4:** 1912
VML-295 **4:** 2065
Vodol **3:** 1210
Voglibose **3:** 1663, 1664
Volidan **3:** 1596
Volon **3:** 1622
Volon-A **3:** 1622
Voltarène **2:** 355
Voltaren **2:** 355; **4:** 1685
Voltarol **2:** 355
Voluntary movement **2:** 315
Vomex **2:** 743
Vomiting **2:** 741
Vontrol **2:** 743
von Willebrand disease **1:** 275
Vorozole **4:** 1926
VP **3:** 1398
VP-16-213 **4:** 1912
VSV **3:** 1174
VT **3:** 1399
Vuxolin **2:** 607
VZV **3:** 1174

Watsonius watsoni **3:** 1269
Wecobees **4:** 2104
WEE **3:** 1174
Wehedryl **2:** 566
Wellbatrin **2:** 631
Wellbutrin **2:** 631
Wet-N-Soak **4:** 2034
Wheat bran **2:** 784
Whey acid **2:** 769
Whipworm **3:** 1274
Whitfield's ointment **4:** 1732
Whooping cough vaccine **4:** 1768
Wildnil **4:** 2183
WIN 49375 **3:** 1126
Win 51711 **3:** 1187
Wincoram **1:** 195
Winstrol **3:** 1594
Wintomylon **3:** 1118
Witepsol **4:** 2104
Withaferin **3:** 1335
Wolff – Chaikoff effect **3:** 1526
Wolff – Parkinson – White Syndrome **1:** 147
WOS **1:** 27
WPW syndrome **1:** 147
Wuchereria bancrofti **3:** 1276
Wurster apparatus **4:** 2099

Wypax **2:** 522, 638
Wyseals **2:** 679
Wytensin **1:** 84

Xalatan **4:** 2060
Xamoterol **1:** 193
Xanax **2:** 641
Xani **4:** 1695
Xanor **2:** 641
Xanthine bronchodilators **2:** 886
Xantinol nicotinate **1:** 23
Xefo **2:** 376
Xerenal **2:** 613
Xipamide **1:** 295
X-Otag **2:** 567
X-ray computed tomography **2:** 325
Xylazine **4:** 2181
Xylocain **1:** 125; **2:** 546
Xylocaine **1:** 125; **2:** 546
Xylocard **1:** 125
Xyloctan **1:** 126
Xylonest **2:** 546
N-(2,6-xylyl)-diethylaminoacetamide **1:** 125

Y-8894 **2:** 659
5-(Hydroxymethyl)-3-m-tolyl-2-oxazolidinone **2:** 626
5-Hydroxytryptamine **2:** 329
5-Hydroxytryptophyl-5-hydroxytryptophan amide **2:** 815
Yellow fever **3:** 1174
Yellow fever vaccine **4:** 1801
YK-176 **4:** 1881
YL-704 A_3 **3:** 998
YL-704 B_3 **3:** 998
Ylestrol **3:** 1584
YM-14673 **2:** 581
Yobine **4:** 2182
Yodoxin **3:** 1170
Yohimbine **1:** 87; **4:** 2182
Yomesan **3:** 1282
Yurelax **2:** 678
(Z)-1-[2-[2-(4-Chlorophenoxy)ethoxy]-2-(2,4-dichlorophenyl)-1-methylethenyl]-1H-imidazole **3:** 1223
(Z)-1-(2,4-Dichlorophenyl)-2-(1H-imidazol-1-yl)ethanone O-[(2,4-dichlorophenyl)methyl]oxime **3:** 1224
(Z)-2-[p-(1,2-Diphenyl-1-butenyl)phenoxy]-N,N-dimethylethylamine **4:** 1922
(Z)-5-Fluoro-2-methyl-1-[4-(methylsulfinyl)phenyl]-methylene-1H-indene-3-acetic acid **2:** 358; **4:** 1682

Zacopride **2:** 609, 796
  Receptor binding affinities **2:** 808
Zadine **2:** 923

Zadipina **1:** 211
Zaditen **2:** 925
Zafirlukast **2:** 896; **4:** 1721, 2064
Zalain **3:** 1226
Zalcitabine **3:** 1297
Zantic **2:** 724
Zantrene **4:** 1903
Zapex **2:** 522
Zarontin **2:** 478
Zaroxolyn **1:** 295
Zaxopam **2:** 638
ZD-1033 **4:** 1926
ZD-182780 **4:** 1924
Zeasorb-AF **3:** 1248
Zeisin **2:** 636, 877
Zenas **1:** 29
Zentropil **1:** 126; **2:** 475
Zenusin **1:** 210
Zerit **3:** 1298
Zero-order release technologies **4:** 2112
Ziagen **3:** 1300
Zidovudine **3:** 1295
Zileuton **2:** 897; **4:** 2055
Zilueton **4:** 1721
Zincfrin **4:** 2027, 2029
Zinc oxide **4:** 1723
Zinc pyrithione **4:** 1732
Zinostatin **3:** 1062
Zirofalen **4:** 1683
ZK 110841 **4:** 2054
ZK 158252 **4:** 2066
ZK98 299 **3:** 1510
ZK98 734 **3:** 1510
Zn α_2 Glycoprotein **1:** 252
Zocor **1:** 28
Zoflam **4:** 1680
Zoladex **4:** 1932, 933
Zolazepam **4:** 2181
Zolpidem **2:** 529
Zondel **1:** 62
Zoniden **3:** 1231
Zonisamide **2:** 485
Zooecdysones **3:** 1331
Zopiclone **2:** 527
Zorprin **4:** 1678
Zovirax **2:** 763; **3:** 1178; **4:** 2014
Zuclopenthixol **2:** 597
Zydol **2:** 412
Zyflo **2:** 897
Zygomycin A **3:** 984
Zyklolat **4:** 2028
Zymosterol **3:** 1324

Ref
RS
156
.P47

37480

SOUTH UNIVERSITY
709 MALL BLVD.
SAVANNAH, GA 31406